BIOLOGY

the dynamic science

VOLUME 2

Peter J. Russell

Stephen L. Wolfe

Paul E. Hertz

Cecie Starr

Beverly McMillan

W9-CHX-488

THOMSON

BROOKS/COLE

™ Australia · Brazil · Canada · Mexico · Singapore · Spain · United Kingdom · United States

THOMSON
★
BROOKS/COLE

Biology: The Dynamic Science, Volume 2, First Edition
Peter J. Russell, Stephen L. Wolfe, Paul E. Hertz, Cecie Starr, Beverly McMillan

Vice President, Editor in Chief: Michelle Julet

Publisher: Yolanda Cossio

Managing Editor: Peggy Williams

Senior Development Editors: Mary Arbogast,
   Shelley Parlante

Development Editor: Christopher Delgado

Assistant Editor: Jessica Kuhn

Editorial Assistant: Rose Barlow

Technology Project Managers: Keli Amann, Kristina Razmara,
   Melinda Newfarmer

Marketing Manager: Kara Kindstrom

Development Project Manager: Terri Mynatt

Production Manager: Shelley Ryan

Creative Director: Rob Hugel

Art Directors: John Walker, Lee Friedman

Art Developers: Steve McEntee, Dragonfly Media Group

Print Buyer: Karen Hunt

Permissions Editor: Sarah D'Stair

Production Service: Graphic World Inc.

Production Service Manager: Suzanne Kastner

Text Designer: Jeanne Calabrese

Photo Researchers: Linda Sykes, Robin Samper

Copy Editor: Christy Goldfinch

Illustrators: Dragonfly Media Group, Steve McEntee,
   Precision Graphics, Dartmouth Publishing, Inc.

Cover Designer: Jeremy Mendes

Cover Image: © Leonardo Papini/Getty Images®

Cover Printer: Transcontinental Printing/Interglobe

Compositor: Graphic World Inc.

Printer: Transcontinental Printing/Interglobe

Printed in Canada
1 2 3 4 5 6 7 12 11 10 09 08 07

Library of Congress Control Number: 2007931665

Student Edition:
ISBN-13: 978-0-534-24966-3
ISBN-10: 0-534-24966-3
Volume 2:
ISBN-13: 978-0-495-01033-3
ISBN-10: 0-495-01033-2

For more information about our products, contact us at:
Thomson Learning Academic Resource Center
1-800-423-0563

For permission to use material from this text or product, submit a request online at http://www.thomsonrights.com.
Any additional questions about permissions can be submitted by e-mail to thomsonrights@thomson.com.

Thomson Higher Education
10 Davis Drive
Belmont, CA 94002-3098
USA

About the Authors

PETER J. RUSSELL received a B.Sc. in Biology from the University of Sussex, England, in 1968 and a Ph.D. in Genetics from Cornell University in 1972. He has been a member of the Biology faculty of Reed College since 1972; he is currently a Professor of Biology. He teaches a section of the introductory biology course, a genetics course, an advanced molecular genetics course, and a research literature course on molecular virology. In 1987 he received the Burlington Northern Faculty Achievement Award from Reed College in recognition of his excellence in teaching. Since 1986, he has been the author of a successful genetics textbook; current editions are *iGenetics: A Mendelian Approach, iGenetics: A Molecular Approach,* and *Essential iGenetics.* He wrote nine of the BioCoach Activities for The Biology Place. Peter Russell's research is in the area of molecular genetics, with a specific interest in characterizing the role of host genes in pathogenic RNA plant virus gene expression; yeast is used as the model host. His research has been funded by agencies including the National Institutes of Health, the National Science Foundation, and the American Cancer Society. He has published his research results in a variety of journals, including *Genetics, Journal of Bacteriology, Molecular and General Genetics, Nucleic Acids Research, Plasmid,* and *Molecular and Cellular Biology.* He has a long history of encouraging faculty research involving undergraduates, including cofounding the biology division of the Council on Undergraduate Research (CUR) in 1985. He was Principal Investigator/Program Director of an NSF Award for the Integration of Research and Education (AIRE) to Reed College, 1998–2002.

STEPHEN L. WOLFE received his Ph.D. from Johns Hopkins University and taught general biology and cell biology for many years at the University of California, Davis. He has a remarkable list of successful textbooks, including multiple editions of *Biology of the Cell, Biology: The Foundations, Cell Ultrastructure, Molecular and Cellular Biology,* and *Introduction to Cell and Molecular Biology.*

PAUL E. HERTZ was born and raised in New York City. He received a bachelor's degree in Biology at Stanford University in 1972, a master's degree in Biology at Harvard University in 1973, and a doctorate in Biology at Harvard University in 1977. While completing field research for the doctorate, he served on the Biology faculty of the University of Puerto Rico at Rio Piedras. After spending 2 years as an Isaac Walton Killam Postdoctoral Fellow at Dalhousie University, Hertz accepted a teaching position at Barnard College, where he has taught since 1979. He was named Ann Whit-ney Olin Professor of Biology in 2000, and he received The Barnard Award for Excellence in Teaching in 2007. In addition to his service on numerous college committees, Professor Hertz was Chair of Barnard's Biology Department for 8 years. He has also been the Program Director of the Hughes Science Pipeline Project at Barnard, an undergraduate curriculum and research program funded by the Howard Hughes Medical Institute, since its inception in 1992. The Pipleline Project includes the Intercollegiate Partnership, a program for local community college students that facilitates their transfer to 4-year colleges and universities. He teaches one semester of the introductory sequence for Biology majors and preprofessional students as well as lecture and laboratory courses in vertebrate zoology and ecology. Professor Hertz is an animal physiological ecologist with a specific research interest in the thermal biology of lizards. He has conducted fieldwork in the West Indies since the mid-1970s, most recently focusing on the lizards of Cuba. His work has been funded by the National Science Foundation, and he has published his research in such prestigious journals as *The American Naturalist, Ecology, Nature,* and *Oecologia.*

CECIE STARR is the author of best-selling biology textbooks. Her books include multiple editions of *Unity and Diversity of Life, Biology: Concepts and Applications,* and *Biology Today and Tomorrow.* Her original dream was to be an architect. She may not be building houses, but with the same care and attention to detail, she builds incredible books: *"I invite students into a chapter through an intriguing story. Once inside, they get the great windows that biologists construct on the world of life. Biology is not just another house. It is a conceptual mansion. I hope to do it justice."*

BEVERLY McMILLAN has been a science writer for more than 20 years and is coauthor of a college text in human biology, now in its seventh edition. She has worked extensively in educational and commercial publishing, including 8 years in editorial management positions in the college divisions of Random House and McGraw-Hill. In a multifaceted freelance career, Bev also has written or coauthored six trade books and numerous magazine and newspaper articles, as well as story panels for exhibitions at the Science Museum of Virginia and the San Francisco Exploratorium. She has worked as a radio producer and speechwriter for the University of California system and as a media relations advisor for the College of William and Mary. She holds undergraduate and graduate degrees from the University of California, Berkeley.

Preface

Welcome to *Biology: The Dynamic Science*. The title of our book reflects an explosive growth in the knowledge of living systems over the past few decades. Although this rapid pace of discovery makes biology the most exciting of all the natural sciences, it also makes it the most difficult to teach. How can college instructors—and, more important, college students—absorb the ever-growing body of ideas and information? The task is daunting, especially in introductory courses that provide a broad overview of the discipline.

Our primary goal in this text is to convey fundamental concepts while maintaining student interest in biology

In this entirely new textbook, we have applied our collective experience as college teachers, science writers, and researchers to create a readable and understandable introduction to our field. We provide students with straightforward explanations of fundamental concepts presented from the evolutionary viewpoint that binds all of the biological sciences together. Having watched our students struggle to navigate the many arcane details of college-level introductory biology, we have constantly reminded ourselves and each other to "include fewer facts, provide better explanations, and maintain the narrative flow," thereby enabling students to see the big picture. Clarity of presentation, a high level of organization, a seamless flow of topics within chapters, and spectacularly useful illustrations are central to our approach.

One of the main goals in this book is to sustain students' fascination with the living world instead of burying it under a mountain of disconnected facts. As teachers of biology, we encourage students to appreciate the dynamic nature of science by conveying our passion for biological research. We want to amaze students with *what* biologists know about the living world and *how* we know it. We also hope to excite them about the opportunities they will have to expand that knowledge. Inspired by our collective effort as teachers and authors, some of our students will take up the challenge and become biologists themselves, asking important new questions and answering them through their own innovative research. For students who pursue other career paths, we hope that they will leave their introductory—and perhaps only—biology courses armed with the knowledge and intellectual skills that allow them to evaluate future discoveries with a critical eye.

We emphasize that, through research, our understanding of biological systems is alive and constantly changing

In this book, we introduce students to a biologist's "ways of knowing." Scientists constantly integrate new observations, hypotheses, experiments, and insights with existing knowledge and ideas. To do this well, biology instructors must not simply introduce students to the current state of our knowledge. We must also foster an appreciation of the historical context within which that knowledge developed and identify the future directions that biological research is likely to take.

To achieve these goals, we explicitly base our presentation and explanations on the research that established the basic facts and principles of biology. Thus, a substantial proportion of each chapter focuses on studies that define the state of biological knowledge today. We describe recent research in straightforward terms, first identifying the question that inspired the work and relating it to the overall topic under discussion. Our research-oriented theme teaches students, through example, how to ask scientific questions and pose hypotheses, two key elements of the "scientific process."

Because advances in science occur against a background of past research, we also give students a feeling for how biologists of the past uncovered and formulated basic knowledge in the field. By fostering an appreciation of such discoveries, given the information and theories that were available to scientists in their own time, we can help students to better understand the successes and limitations of what we consider cutting edge today. This historical perspective also encourages students to view biology as a dynamic intellectual endeavor, and not just a list of facts and generalities to be memorized.

One of our greatest efforts has been to make the science of biology come alive by describing how biologists formulate hypotheses and evaluate them using hard-won data, how data sometimes tell only part of a story, and how studies often end up posing more questions than they answer. Although students often prefer to read about the "right" answer to a question, they must be encouraged to embrace "the unknown," those gaps in our knowledge that create opportunities for further research. An appreciation of what we *don't* know will draw more students into the field. And by defining *why* we don't understand interesting phenomena, we encourage students to follow paths dictated by their own curiosity. We hope that this approach will encourage students to make biology a part of their daily lives—to have informal discussions about new scientific discoveries, just as they do about politics, sports, or entertainment.

Special features establish a story line in every chapter and describe the process of science

In preparing this book, we developed several special features to help students broaden their understanding of the material presented and of the research process itself.

- The chapter openers, entitled *Why It Matters,* tell the story of how a researcher arrived at a key insight or how biological research solved a major societal problem or shed light on a fundamental process or phenomenon. These engaging, short vignettes are designed to capture students' imagination and whet their appetite for the topic that the chapter addresses.

- To complement this historical or practical perspective, each chapter closes with a brief essay, entitled *Unanswered Questions,* often prepared by an expert in the field. These essays identify important unresolved issues relating to the chapter topic and describe cutting-edge research that will advance our knowledge in the future.

- Each chapter also includes a short boxed essay, entitled *Insights from the Molecular Revolution,* which describes how molecular technologies allow scientists to answer questions that they could not have even posed 20 or 30 years ago. Each *Insight*

focuses on a single study and includes sufficient detail for its content to stand alone.

- Almost every chapter is further supplemented with one or more short boxed essays that *Focus on Research.* Some of these essays describe seminal studies that provided a new perspective on an important question. Others describe how basic research has solved everyday problems relating to health or the environment. Another set introduces model research organisms—such as *E. coli, Drosophila, Arabidopsis, Caenorhabditis,* and *Anolis*—and explains why they have been selected as subjects for in-depth analysis.

Spectacular illustrations enable students to visualize biological processes, relationships, and structures

Today's students are accustomed to receiving ideas and information visually, making the illustrations and photographs in a textbook more important than ever before. Our illustration program provides an exceptionally clear supplement to the narrative in a style that is consistent throughout the book. Graphs and anatomical drawings are annotated with interpretative explanations that lead students through the major points they convey.

Three types of specially designed Research Figures provide more detailed information about how biologists formulate and test specific hypotheses by gathering and interpreting data.

- *Research Method* figures provide examples of important techniques, such as gel electrophoresis, the use of radioisotopes, and cladistic analysis. Each *Research Method* figure leads a student through the technique's purpose and protocol and

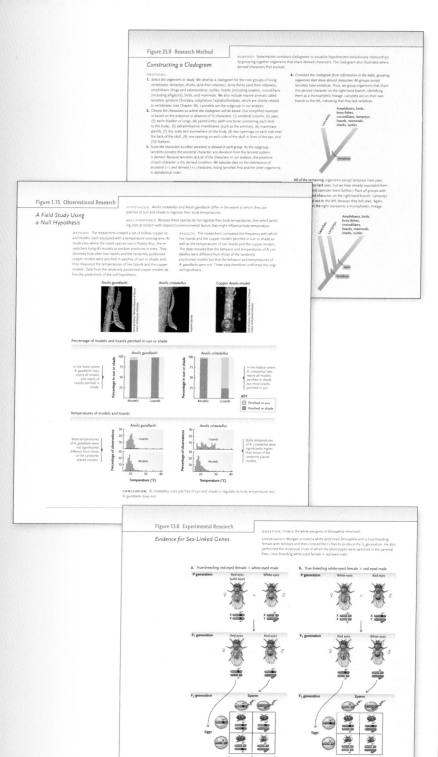

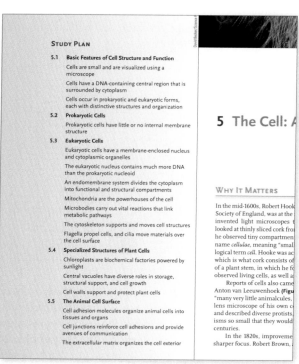

describes how scientists interpret the data it generates.

- *Observational Research* figures describe specific studies in which biologists have tested hypotheses by comparing systems under varying natural circumstances.

- *Experimental Research* figures describe specific studies in which researchers used both experimental and control treatments—either in the laboratory or in the field—to test hypotheses by manipulating the system they study.

Chapters are structured to emphasize the big picture and the most important concepts

As authors and college teachers, we know how easily students can get lost within a chapter that spans 15 pages or more. When students request advice about how to approach such a large task, we usually suggest that, after reading each section, they pause and quiz themselves on the material they have just encountered. After completing all of the sections in a chapter, they should quiz themselves again, even more rigorously, on the individual sections and, most important, on how the concepts developed in different sections fit together. To assist these efforts, we have adopted a structure for each chapter that will help students review concepts as they learn them.

- The organization within chapters presents material in digestible chunks, building on students' knowledge and understanding as they acquire it. Each major section covers one broad topic. Each subsection, titled with a declarative sentence that summarizes the main idea of its content, explores a narrower range of material.

- Whenever possible, we include the derivation of unfamiliar terms so that students will see connections between words that share etymological roots. Mastery of the technical language of biology will allow students to discuss ideas and processes precisely. At the same time, we have minimized the use of unnecessary jargon as much as possible.
- Sets of embedded *Study Break* questions follow every major section. These questions encourage students to pause at the end of a section and review what they have learned before going on to the next topic within the chapter. Short answers to these questions appear in an appendix.

End-of-chapter material encourages students to review concepts, test their knowledge, and think analytically

Supplementary materials at the end of each chapter help students review the material they have learned, assess their understanding, and think analytically as they apply the principles developed in the chapter to novel situations. Many of the end-of-chapter questions also serve as good starting points for class discussions or out-of-class assignments.

- A brief *Review* that references figures and tables in the chapter provides an outline summary of important ideas developed in the chapter. The *Reviews* are much too short to serve as a substitute for reading the chapter. Instead, students may use them as an outline of the material, filling in the details on their own.
- Each chapter also closes with a set of 10 multiple choice *Self-Test* questions that focus on factual material.

- Several open-ended *Questions for Discussion* emphasize concepts, the interpretation of data, and practical applications of the material.
- A question on *Experimental Analysis* asks students to consider how they would develop and test hypotheses about a situation that relates to the chapter's main topic.
- An *Evolution Link* question relates the subject of the chapter to evolutionary biology.
- The *How Would You Vote?* exercise allows students to weigh both sides of an issue by reading pro/con articles, and then making their opinion known through an online voting process.

We hope that, after reading parts of this textbook, you agree that we have developed a clear, fresh, and well-integrated introduction to biology as it is understood by researchers today. Just as important, we hope that our efforts will excite students about the research process and the new discoveries it generates.

Acknowledgments

We are grateful to the many people who have generously fostered the creation of this text

The creation of a new textbook is a colossal undertaking, and we could never have completed the task without the kind assistance of many people.

Jack Carey first conceived of this project and put together the author team. Michelle Julet and Yolanda Cossio have provided the support and encouragement necessary to move it forward to completion. Peggy Williams has served as the extraordinarily able coordinator of the authors, editors, reviewers, contributors, artists, and production team—we like to think of Peggy as the "cat herder."

Developmental Editors play nearly as large a role as the authors, interpreting and deconstructing reviewer comments and constantly making suggestions about how we could tighten the narrative and stay on course. Mary Arbogast has done banner service as a Developmental Editor, patiently working on the project since its inception. Shelley Parlante has provided very helpful guidance as the manuscript matured. Jody Larson and Catherine Murphy have offered useful comments on many of the chapters.

We are grateful to Christopher Delgado and Jessica Kuhn for coordinating the print supplements, and our Editorial Assistant Rose Barlow for managing all our reviewer information.

Many thanks to Keli Amann, Kristina Razmara, and Christopher Delgado, who were responsible for partnering with our technology authors and media advisory board in creating tools to support students in learning and instructors in teaching.

We appreciate the help of the production staff led by Shelley Ryan and Suzanne Kastner at Graphic

World. We thank our Creative Director Rob Hugel, Art Director John Walker.

The outstanding art program is the result of the collaborative talent, hard work, and dedication of a select group of people. The meticulous styling and planning of the program is credited to Steve McEntee and Dragonfly Media Group, led by Craig Durant and Mike Demaray. The DMG group created hundreds of complex, vibrant art pieces. Steve's role was crucial in overseeing the development and consistency of the art program; he was also the illustrator for the unique Research features.

We appreciate Kara Kindstrom, our Marketing Manager, and Terri Mynatt, our Development Project Manager, whose expertise ensured that you would know all about this new book.

Peter Russell thanks Stephen Arch of Reed College for valuable discussions and advice during the writing of the Unit Six chapters on Animal Structure and Function. Paul E. Hertz thanks Hilary Callahan, John Glendinning, and Brian Morton of Barnard College for their generous advice on many phases of this project, and John Alcock of Arizona State University and James Danoff-Burg of Columbia University for their contributions to the discussions of Animal Behavior and Conservation Biology, respectively. Paul would also like to thank Jamie Rauchman, for extraordinary patience and endless support as this book was written, and his thousands of past students, who have taught him at least as much as he has taught them.

We would also like to thank our advisors and contributors:

Media Advisory Board

Scott Bowling, Auburn University
Jennifer Jeffery, Wharton County Junior College
Shannon Lee, California State University, Northridge
Roderick M. Morgan, Grand Valley State University
Debra Pires, University of California, Los Angeles

Art Advisory Board

Lissa Leege, Georgia Southern University
Michael Meighan, University of California, Berkeley
Melissa Michael, University of Illinois at Urbana–Champaign
Craig Peebles, University of Pittsburgh
Laurel Roberts, University of Pittsburgh

Accuracy Checkers

Brent Ewers, University of Wyoming
Richard Falk, University of California, Davis
Michael Meighan, University of California, Berkeley
Michael Palladino, Monmouth University

End-of-Chapter Questions

Patricia Colberg, University of Wyoming
Elizabeth Godrick, Boston University

Student Study Guide

Carolyn Bunde, Idaho State University
William Kroll, Loyola University Chicago
Mark Sheridan, North Dakota State University
Jyoti Wagle, Houston Community College

Instructor's Resource Manual

Benjie Blair, Jacksonville State University
Nancy Boury, Idaho State University
Mark Meade, Jacksonville State University
Debra Pires, University of California, Los Angeles
James Rayburn, Jacksonville State University

Test Bank

Scott Bowling, Auburn University
Laurie Bradley, Hudson Valley Community College
Jose Egremy, Northwest Vista College
Darrel L. Murray, University of Illinois, Chicago
Jacalyn Newman, University of Pittsburgh
Mark Sugalski, Southern Polytechnic State University

Technology Authors

Catherine Black, Idaho State University
David Byres, Florida Community College, Jacksonville
Kevin Dixon, University of Illinois
Albia Dugger, Miami Dade College
Mary Durant, North Harris College
Brent Ewers, University of Wyoming
Debbie Folkerts, Auburn University
Stephen Kilpatrick, University of Pittsburgh
Laurel Roberts, University of Pittsburgh
Thomas Sasek, University of Louisiana, Monroe
Bruce Stallsmith, University of Alabama–Huntsville

Workshop and Focus Group Participants

Karl Aufderheide, *Texas A&M University*
Bob Bailey, *Central Michigan University*
John Bell, *Brigham Young University*
Catherine Black, *Idaho State University*
Hessel Bouma III, *Calvin College*

Scott Bowling, *Auburn University*
Bob Brick, *Blinn College, Bryan*
Randy Brooks, *Florida Atlantic University*
Nancy Burley, *University of California, Irvine*

Genevieve Chung, *Broward Community College*
Allison Cleveland, *University of South Florida*
Patricia Colberg, *University of Wyoming*

Jay Comeaux, *Louisiana State University*

Sehoya Cotner, *University of Minnesota*

Joe Cowles, *Virginia Tech*

Anita Davelos-Baines, *University of Texas, Pan American*

Donald Deters, *Bowling Green State University*

Kevin Dixon, *University of Illinois at Urbana–Champaign*

Jose Egremy, *Northwest Vista College*

Diana Elrod, *University of North Texas*

Zen Faulkes, *University of Texas–Pan American*

Elizabeth Godrick, *Boston University*

Barbara Haas, *Loyola University Chicago*

Julie Harless, *Montgomery College*

Jean Helgeson, *Collin County Community College*

Mark Hunter, *University of Michigan*

Andrew Jarosz, *Michigan State University, Montgomery College*

Jennifer Jeffery, *Wharton County Junior College*

John Jenkin, *Blinn College, Bryan*

Wendy Keenleyside, *University of Guelph*

Steve Kilpatrick, *University of Pittsburgh at Johnstown*

Gary Kuleck, *Loyola Marymount University*

Allen Kurta, *Eastern Michigan University*

Mark Lyford, *University of Wyoming*

Andrew McCubbin, *Washington State University*

Michael Meighan, *University of California, Berkeley*

John Merrill, *Michigan State University*

Richard Merritt, *Houston Community College, Northwest*

Melissa Michael, *University of Illinois at Urbana–Champaign*

James Mickle, *North Carolina State University*

Betsy Morgan, *Kingwood College*

Kenneth Mossman, *Arizona State University*

Darrel Murray, *University of Illinois, Chicago*

Jacalyn Newman, *University of Pittsburgh*

Dennis Nyberg, *University of Illinois–Chicago*

Bruce Ostrow, *Grand Valley State University–Allendale*

Craig Peebles, *University of Pittsburgh*

Nancy Pencoe, *University of West Georgia*

Mitch Price, *Pennsylvania State University*

Kelli Prior, *Finger Lakes Community College*

Laurel Roberts, *University of Pittsburgh*

Ann Rushing, *Baylor University*

Bruce Stallsmith, *University of Alabama–Huntsville*

David Tam, *University of North Texas*

Franklyn Te, *Miami Dade College*

Nanette Van Loon, *Borough of Manhattan Community College*

Alexander Wait, *Missouri State University*

Lisa Webb, *Christopher Newport University*

Larry Williams, *University of Houston*

Michelle Withers, *Louisiana State University*

Denise Woodward, *Pennsylvania State University*

Class Test Participants

Tamarah Adair, *Baylor University*

Idelissa Ayala, *Broward Community College–Central*

Tim Beagley, *Salt Lake Community College*

Catherine Black, *Idaho State University*

Laurie Bradley, *Hudson Valley Community College*

Mirjana Brockett, *Georgia Tech*

Carolyn Bunde, *Idaho State University*

John Cogan, *Ohio State University*

Anne M. Cusic, *University of Alabama–Birmingham*

Ingeborg Eley, *Hudson Valley Community College*

Brent Ewers, *University of Wyoming*

Miriam Ferzli, *North Carolina State University*

Debbie Folkerts, *Auburn University*

Mark Hens, *University of North Carolina, Greensboro*

Anna Hill, *University of Louisiana, Monroe*

Anne Hitt, *Oakland University*

Jennifer Jeffery, *Wharton County Junior College*

David Jones, *Dixie State College*

Wendy Keenleyside, *University of Guelph*

Brian Kinkle, *University of Cincinnati*

Brian Larkins, *University of Arizona*

Shannon Lee, *California State University, Northridge*

Harvey Liftin, *Broward Community College*

Jim Marinaccio, *Raritan Valley Community College*

Monica Marquez-Nelson, *Joliet Junior College*

Kelly Meckling, *University of Guelph*

Richard Merritt, *Houston Community College–Town and Country*

Russ Minton, *University of Louisiana, Monroe*

Necia Nichols, *Calhoun State Community College*

Nancy Rice, *Western Kentucky University*

Laurel Roberts, *University of Pittsburgh*

John Russell, *Calhoun State Community College*

Pramila Sen, *Houston Community College*

Reviewers

Heather Addy, *University of Calgary*

Adrienne Alaie-Petrillo, *Hunter College–CUNY*

Richard Allison, *Michigan State University*

Terry Allison, *University of Texas–Pan American*

Deborah Anderson, *Saint Norbert College*

Robert C. Anderson, *Idaho State University*

Andrew Andres, *University of Nevada–Las Vegas*

Steven M. Aquilani, *Delaware County Community College*

Jonathan W. Armbruster, *Auburn University*

Peter Armstrong, *University of California, Davis*

John N. Aronson, *University of Arizona*

Joe Arruda, *Pittsburg State University*

Karl Aufderheide, *Texas A&M University*

Charles Baer, *University of Florida*

Gary I. Baird, *Brigham Young University*

Aimee Bakken, *University of Washington*

Marica Bakovic, *University of Guelph*

Michael Baranski, *Catawba College*

Michael Barbour, *University of California, Davis*

Edward M. Barrows, *Georgetown University*

Anton Baudoin, *Virginia Tech*

Penelope H. Bauer, *Colorado State University*

Kevin Beach, *University of Tampa*

Mike Beach, *Southern Polytechnic State University*

Ruth Beattie, *University of Kentucky*

Robert Beckmann, *North Carolina State University*

Jane Beiswenger, *University of Wyoming*

Andrew Bendall, *University of Guelph*

Catherine Black, *Idaho State University*

Andrew Blaustein, *Oregon State University*

Anthony H. Bledsoe, *University of Pittsburgh*

Harriette Howard-Lee Block, *Prairie View A&M University*

Dennis Bogyo, *Valdosta State University*

David Bohr, *University of Michigan*

Emily Boone, *University of Richmond*

Hessel Bouma III, *Calvin College*

Nancy Boury, *Iowa State University*

Scott Bowling, *Auburn University*

Laurie Bradley, *Hudson Valley Community College*

William Bradshaw, *Brigham Young University*

J. D. Brammer, *North Dakota State University*

G. L. Brengelmann, *University of Washington*

Randy Brewton, *University of Tennessee–Knoxville*

Bob Brick, *Blinn College, Bryan*

Mirjana Brockett, *Georgia Tech*

William Bromer, *University of Saint Francis*

William Randy Brooks, *Florida Atlantic University–Boca Raton*

Mark Browning, *Purdue University*

Gary Brusca, *Humboldt State University*

Alan H. Brush, *University of Connecticut*

Arthur L. Buikema, Jr., *Virginia Tech*

Carolyn Bunde, *Idaho State University*

E. Robert Burns, *University of Arkansas for Medical Sciences*

Ruth Buskirk, *University of Texas–Austin*

David Byres, *Florida Community College, Jacksonville*

Christopher S. Campbell, *University of Maine*

Angelo Capparella, *Illinois State University*

Marcella D. Carabelli, *Broward Community College–North*

Jeffrey Carmichael, *University of North Dakota*

Bruce Carroll, *North Harris Montgomery Community College*

Robert Carroll, *East Carolina University*

Patrick Carter, *Washington State University*

Christine Case, *Skyline College*

Domenic Castignetti, *Loyola University Chicago–Lakeshore*

Jung H. Choi, *Georgia Tech*

Kent Christensen, *University Michigan School of Medicine*

John Cogan, *Ohio State University*

Linda T. Collins, *University of Tennessee–Chattanooga*

Lewis Coons, *University of Memphis*

Joe Cowles, *Virginia Tech*

George W. Cox, *San Diego State University*

David Crews, *University of Texas*

Paul V. Cupp, Jr., *Eastern Kentucky University*

Karen Curto, *University of Pittsburgh*

Anne M. Cusic, *University of Alabama–Birmingham*

David Dalton, *Reed College*

Frank Damiani, *Monmouth University*

Peter J. Davies, *Cornell University*

Fred Delcomyn, *University of Illinois at Urbana–Champaign*

Jerome Dempsey, *University of Wisconsin–Madison*

Philias Denette, *Delgado Community College–City Park*

Nancy G. Dengler, *University of Toronto*

Jonathan J. Dennis, *University of Alberta*

Daniel DerVartanian, *University of Georgia*

Donald Deters, *Bowling Green State University*

Kathryn Dickson, *CSU Fullerton*

Kevin Dixon, *University of Illinois at Urbana–Champaign*

Gordon Patrick Duffie, *Loyola University Chicago–Lakeshore*

Charles Duggins, *University of South Carolina*

Carolyn S. Dunn, *University North Carolina–Wilmington*

Roland R. Dute, *Auburn University*

Melinda Dwinell, *Medical College of Wisconsin*

Gerald Eck, *University of Washington*

Gordon Edlin, *University of Hawaii*

William Eickmeier, *Vanderbilt University*

Ingeborg Eley, *Hudson Valley Community College*

Paul R. Elliott, *Florida State University*

John A. Endler, *University of Exeter*

Brent Ewers, *University of Wyoming*

Daniel J. Fairbanks, *Brigham Young University*

Piotr G. Fajer, *Florida State University*

Richard H. Falk, *University of California, Davis*

Ibrahim Farah, *Jackson State University*

Jacqueline Fern, *Lane Community College*

Daniel P. Fitzsimons, *University of Wisconsin–Madison*

Daniel Flisser, *Camden County College*

R. G. Foster, *University of Virginia*

Dan Friderici, *Michigan State University*

J. W. Froehlich, *University of New Mexico*

Paul Garcia, *Houston Community College–SW*

Umadevi Garimella, *University of Central Arkansas*

Robert P. George, *University of Wyoming*

Stephen George, *Amherst College*

John Giannini, *St. Olaf College*

Joseph Glass, *Camden County College*

John Glendinning, *Barnard College*

Elizabeth Godrick, *Boston University*

Judith Goodenough, *University of Massachusetts Amherst*

H. Maurice Goodman, *University of Massachusetts Medical School*

Bruce Grant, *College of William and Mary*

Becky Green-Marroquin, *Los Angeles Valley College*

Christopher Gregg, *Louisiana State University*

Katharine B. Gregg, *West Virginia Wesleyan College*

John Griffin, *College of William and Mary*

Samuel Hammer, *Boston University*

Aslam Hassan, *University of Illinois at Urbana–Champaign, Veterinary Medicine*

Albert Herrera, *University of Southern California*

Wilford M. Hess, *Brigham Young University*

Martinez J. Hewlett, *University of Arizona*

Christopher Higgins, *Tarleton State University*

Phyllis C. Hirsch, *East Los Angeles College*

Carl Hoagstrom, *Ohio Northern University*

Stanton F. Hoegerman, *College of William and Mary*

Ronald W. Hoham, *Colgate University*

Margaret Hollyday, *Bryn Mawr College*

John E. Hoover, *Millersville University*

Howard Hosick, *Washington State University*

William Irby, *Georgia Southern*

John Ivy, *Texas A&M University*

Alice Jacklet, *SUNY Albany*

John D. Jackson, *North Hennepin Community College*

Jennifer Jeffery, *Wharton County Junior College*

John Jenkin, *Blinn College, Bryan*

Leonard R. Johnson, *University Tennessee College of Medicine*

Walter Judd, *University of Florida*

Prem S. Kahlon, *Tennessee State University*

Thomas C. Kane, *University of Cincinnati*

Peter Kareiva, *University of Washington*

Gordon I. Kaye, *Albany Medical College*

Greg Keller, *Eastern New Mexico University*

Stephen Kelso, *University of Illinois–Chicago*

Bryce Kendrick, *University of Waterloo*

Bretton Kent, *University of Maryland*

Jack L. Keyes, *Linfield College Portland Campus*

John Kimball, *Tufts University*

Hillar Klandorf, *West Virginia University*

Michael Klymkowsky, *University of Colorado–Boulder*

Loren Knapp, *University of South Carolina*

Ana Koshy, *Houston Community College–NW*

Kari Beth Krieger, *University of Wisconsin–Green Bay*

David T. Krohne, *Wabash College*

William Kroll, *Loyola University Chicago–Lakeshore*

Josepha Kurdziel, *University of Michigan*

Allen Kurta, *Eastern Michigan University*

Howard Kutchai, *University of Virginia*

Paul K. Lago, *University of Mississippi*

John Lammert, *Gustavus Adolphus College*

William L'Amoreaux, *College of Staten Island*

Brian Larkins, *University of Arizona*

William E. Lassiter, *University of North Carolina–Chapel Hill*

Shannon Lee, *California State University, Northridge*

Lissa Leege, *Georgia Southern University*

Matthew Levy, *Case Western Reserve University*

Harvey Liftin, *Broward Community College–Central*

Tom Lonergan, *University of New Orleans*

Lynn Mahaffy, *University of Delaware*

Alan Mann, *University of Pennsylvania*

Kathleen Marrs, *Indiana University Purdue University Indianapolis*

Robert Martinez, *Quinnipiac University*

Joyce B. Maxwell, *California State University, Northridge*

Jeffrey D. May, *Marshall University*

Geri Mayer, *Florida Atlantic University*

Jerry W. McClure, *Miami University*

Andrew G. McCubbin, *Washington State University*

Mark McGinley, *Texas Tech University*

F. M. Anne McNabb, *Virginia Tech*

Mark Meade, *Jacksonville State University*

Bradley Mehrtens, *University of Illinois at Urbana–Champaign*

Michael Meighan, *University of California, Berkeley*

Catherine Merovich, *West Virginia University*

Richard Merritt, *Houston Community College–Town and Country*

Ralph Meyer, *University of Cincinnati*

James E. "Jim" Mickle, *North Carolina State University*

Hector C. Miranda, Jr., *Texas Southern University*

Jasleen Mishra, *Houston Community College–SW*

David Mohrman, *University of Minnesota Medical School*

John M. Moore, *Taylor University*

David Morton, *Frostburg State University*

Alexander Motten, *Duke University*

Alan Muchlinski, *California State University, Los Angeles*

Michael Muller, *University of Illinois–Chicago*

Richard Murphy, *University of Virginia*

Darrel L. Murray, *University of Illinois–Chicago*

Allan Nelson, *Tarleton State University*

David H. Nelson, *University of South Alabama*

Jacalyn Newman, *University of Pittsburgh*

David O. Norris, *University of Colorado*

Bette Nybakken, *Hartnell College, California*

Tom Oeltmann, *Vanderbilt University*

Diana Oliveras, *University of Colorado–Boulder*

Alexander E. Olvido, *Virginia State University*

Karen Otto, *University of Tampa*

William W. Parson, *University of Washington School of Medicine*

James F. Payne, *University of Memphis*

Craig Peebles, *University of Pittsburgh*

Joe Pelliccia, *Bates College*

Susan Petro, *Rampao College of New Jersey*

Debra Pires, *University of California, Los Angeles*

Thomas Pitzer, *Florida International University*

Roberta Pollock, *Occidental College*

Jerry Purcell, *San Antonio College*

Kim Raun, *Wharton County Junior College*

Tara Reed, *University of Wisconsin–Green Bay*

Lynn Robbins, *Missouri State University*

Carolyn Roberson, *Roane State Community College*

Laurel Roberts, *University of Pittsburgh*

Kenneth Robinson, *Purdue University*

Frank A. Romano, *Jacksonville State University*

Michael R. Rose, *University of California, Irvine*

Michael S. Rosenzweig, *Virginia Tech*

Linda S. Ross, *Ohio University*

Ann Rushing, *Baylor University*

Linda Sabatino, *Suffolk Community College*

Tyson Sacco, *Cornell University*

Peter Sakaris, *Southern Polytechnic State University*

Frank B. Salisbury, *Utah State University*

Mark F. Sanders, *University of California, Davis*

Andrew Scala, *Dutchess Community College*

John Schiefelbein, *University of Michigan*

Deemah Schirf, *University of Texas–San Antonio*

Kathryn J. Schneider, *Hudson Valley Community College*

Jurgen Schnermann, *University Michigan School of Medicine*

Thomas W. Schoener, *University California, Davis*

Brian Shea, *Northwestern University*

Mark Sheridan, *North Dakota State University–Fargo*

Dennis Shevlin, *College of New Jersey*

Richard Showman, *University of South Carolina*

Bill Simcik, *Tomball College*

Robert Simons, *University of California, Los Angeles*

Roger Sloboda, *Dartmouth College*

Jerry W. Smith, *St. Petersburg College*

Nancy Solomon, *Miami University*

Bruce Stallsmith, *University of Alabama–Huntsville*

Karl Sternberg, *Western New England College*

Pat Steubing, *University of Nevada–Las Vegas*

Karen Steudel, *University of Wisconsin–Madison*

Richard D. Storey, *Colorado College*

Michael A. Sulzinski, *University of Scranton*

Marshall Sundberg, *Emporia State University*

David Tam, *University of North Texas*

David Tauck, *Santa Clara University*

Jeffrey Taylor, *Slippery Rock University*

Franklyn Te, *Miami Dade College*

Roger E. Thibault, *Bowling Green State University*

Megan Thomas, *University of Nevada–Las Vegas*

Patrick Thorpe, *Grand Valley State University–Allendale*

Ian Tizard, *Texas A&M University*

Robert Turner, *Western Oregon University*

Joe Vanable, *Purdue University*

Linda H. Vick, *North Park University*

J. Robert Waaland, *University of Washington*

Douglas Walker, *Wharton County Junior College*

James Bruce Walsh, *University of Arizona*

Fred Wasserman, *Boston University*

Edward Weiss, *Christopher Newport University*

Mark Weiss, *Wayne State University*

Adrian M. Wenner, *University of California, Santa Barbara*

Adrienne Williams, *University of California, Irvine*

Mary Wise, *Northern Virginia Community College*

Charles R. Wyttenbach, *University of Kansas*

Robert Yost, *Indiana University Purdue University Indianapolis*

Xinsheng Zhu, *University of Wisconsin–Madison*

Adrienne Zihlman, *University of California–Santa Cruz*

Brief Contents

Unit Three Evolutionary Biology

19 The Development of Evolutionary Thought 401
20 Microevolution: Genetic Changes within Populations 419
21 Speciation 443
22 Paleobiology and Macroevolution 463
23 Systematic Biology: Phylogeny and Classification 491

Unit Four Biodiversity

24 The Origin of Life 511
25 Prokaryotes and Viruses 525
26 Protists 549
27 Plants 575
28 Fungi 605

29 Animal Phylogeny, Acoelomates, and Protostomes 627
30 Deuterostomes: Vertebrates and Their Closest Relatives 667

Unit Seven Ecology and Behavior

49 Population Ecology 1125
50 Population Interactions and Community Ecology 1151
51 Ecosystems 1181
52 The Biosphere 1203
53 Biodiversity and Conservation Biology 1229
54 The Physiology and Genetics of Animal Behavior 1253
55 The Ecology and Evolution of Animal Behavior 1269

The chapters listed below are not included in Volume 2

1 Introduction to Biological Concepts and Research

Unit One Molecules and Cells

2 Life, Chemistry, and Water
3 Biological Molecules: The Carbon Compounds of Life
4 Energy, Enzymes, and Biological Reactions
5 The Cell: An Overview
6 Membranes and Transport
7 Cell Communication
8 Harvesting Chemical Energy: Cellular Respiration
9 Photosynthesis
10 Cell Division and Mitosis

Unit Two Genetics

11 Meiosis: The Cellular Basis of Sexual Reproduction
12 Mendel, Genes, and Inheritance
13 Genes, Chromosomes, and Human Genetics
14 DNA Structure, Replication, and Organization
15 From DNA to Protein
16 Control of Gene Expression
17 Bacterial and Viral Genetics
18 DNA Technologies and Genomics

Unit Five Plant Structure and Function

31 The Plant Body
32 Transport in Plants
33 Plant Nutrition
34 Reproduction and Development in Flowering Plants
35 Control of Plant Growth and Development

Unit Six Animal Structure and Function

36 Introduction to Animal Organization and Physiology
37 Information Flow and the Neuron
38 Nervous Systems
39 Sensory Systems
40 The Endocrine System
41 Muscles, Bones, and Body Movements
42 The Circulatory System
43 Defenses against Disease
44 Gas Exchange: The Respiratory System
45 Animal Nutrition
46 Regulating the Internal Environment
47 Animal Reproduction
48 Animal Development

Contents

UNIT THREE EVOLUTIONARY BIOLOGY **401**

19 Development of Evolutionary Thought 401

19.1 Recognition of Evolutionary Change **402**

19.2 Darwin's Journeys **405**

19.3 Evolutionary Biology since Darwin **411**

> FOCUS ON RESEARCH
> Basic Research: Charles Darwin's Life as a Scientist **408**
>
> INSIGHTS FROM THE MOLECULAR REVOLUTION
> Artificial Selection in the Test Tube **409**
>
> Figure 19.11 Experimental Research
> *How Exposure to Insecticide Fosters the Evolution of Insecticide Resistance* **412**
>
> Figure 19.14 Observational Research
> *How Differences in Amino Acid Sequences among Species Reflect Their Evolutionary Relationships* **415**

20 Microevolution: Genetic Changes within Populations 419

20.1 Variation in Natural Populations **420**

20.2 Population Genetics **423**

20.3 The Agents of Microevolution **425**

20.4 Maintaining Genetic and Phenotypic Variation **435**

20.5 Adaptation and Evolutionary Constraints **437**

> Figure 20.6 Experimental Research
> *Using Artificial Selection to Demonstrate That Activity Level in Mice Has a Genetic Basis* **423**
>
> FOCUS ON RESEARCH
> Basic Research: Using the Hardy-Weinberg Principle **426**
>
> INSIGHTS FROM THE MOLECULAR REVOLUTION
> Genetic Variation Preserved in Humpback Whales **429**
>
> Figure 20.10 Observational Research
> *Evidence for Stabilizing Selection in Humans* **431**
>
> Figure 20.11 Observational Research
> *How Opposing Forces of Directional Selection Produce Stabilizing Selection* **432**
>
> Figure 20.13 Experimental Research
> *Sexual Selection in Action* **434**

> Figure 20.15 Observational Research
> *Habitat Variation in Color and Striping Patterns of European Garden Snails* **437**
>
> Figure 20.16 Experimental Research
> *Demonstration of Frequency-Dependent Selection* **438**

21 Speciation 443

21.1 What Is a Species? **444**

21.2 Maintaining Reproductive Isolation **447**

21.3 The Geography of Speciation **449**

21.4 Genetic Mechanisms of Speciation **454**

> FOCUS ON RESEARCH
> Basic Research: Speciation in Hawaiian Fruit Flies **452**
>
> Figure 21.13 Observational Research
> *Evidence for Reproductive Isolation in Bent Grass* **454**
>
> INSIGHTS FROM THE MOLECULAR REVOLUTION
> Monkey-Flower Speciation **455**
>
> Figure 21.19 Observational Research
> *Chromosomal Similarities and Differences among the Great Apes* **458**

22 Paleobiology and Macroevolution 463

22.1 The Fossil Record **464**

22.2 Earth History, Biogeography, and Convergent Evolution **469**

22.3 Interpreting Evolutionary Lineages **473**

22.4 Macroevolutionary Trends in Morphology **477**

22.5 Macroevolutionary Trends in Biodiversity **480**

22.6 Evolutionary Developmental Biology **483**

> Figure 22.4 Research Method
> *Radiometric Dating* **468**
>
> FOCUS ON RESEARCH
> Basic Research: The Great American Interchange **472**
>
> Figure 22.12 Observational Research
> *Evidence Supporting the Punctuated Equilibrium Hypothesis* **476**

Figure 22.13 Observational Research
Evidence Supporting the Gradualist Hypothesis **478**

Figure 22.16 Observational Research
Paedomorphosis in Delphinium *Flowers* **481**

INSIGHTS FROM THE MOLECULAR REVOLUTION
Fancy Footwork from Fins to Fingers **486**

23 Systematic Biology: Phylogeny and Classification **491**

23.1 Systematic Biology: An Overview **492**

23.2 The Linnaean System of Taxonomy **493**

23.3 Organismal Traits as Systematic Characters **494**

23.4 Evaluating Systematic Characters **495**

23.5 Phylogenetic Inference and Classification **497**

23.6 Molecular Phylogenetics **501**

Figure 23.9 Research Method
Constructing a Cladogram **500**

INSIGHTS FROM THE MOLECULAR REVOLUTION
Whales with Cow Cousins? **502**

Figure 23.10 Observational Research
Using Amino Acid Sequences to Construct a Phylogenetic Tree **504**

Figure 23.11 Research Method
Aligning DNA Sequences **505**

UNIT FOUR BIODIVERSITY **511**

24 The Origin of Life **511**

24.1 The Formation of Molecules Necessary for Life **512**

24.2 The Origin of Cells **515**

24.3 The Origins of Eukaryotic Cells **519**

INSIGHTS FROM THE MOLECULAR REVOLUTION
Replicating the RNA World **518**

25 Prokaryotes and Viruses **525**

25.1 Prokaryotic Structure and Function **526**

25.2 The Domain Bacteria **534**

25.3 The Domain Archaea **537**

25.4 Viruses, Viroids, and Prions **540**

INSIGHTS FROM THE MOLECULAR REVOLUTION
Extreme but Still in Between **539**

26 Protists **549**

26.1 What Is a Protist? **550**

26.2 The Protist Groups **553**

FOCUS ON RESEARCH
Applied Research: Malaria and the *Plasmodium* Life Cycle **559**

INSIGHTS FROM THE MOLECULAR REVOLUTION
Getting the Slime Mold Act Together **566**

27 Plants **575**

27.1 The Transition to Life on Land **576**

27.2 Bryophytes: Nonvascular Land Plants **581**

27.3 Seedless Vascular Plants **585**

27.4 Gymnosperms: The First Seed Plants **590**

27.5 Angiosperms: Flowering Plants **595**

INSIGHTS FROM THE MOLECULAR REVOLUTION
The Powerful Genetic Toolkit for Studying Plant Evolution **597**

28 Fungi **605**

28.1 General Characteristics of Fungi **606**

28.2 Major Groups of Fungi **610**

28.3 Fungal Associations **620**

INSIGHTS FROM THE MOLECULAR REVOLUTION
There Was Probably a Fungus among Us **611**

FOCUS ON RESEARCH
Applied Research: Lichens as Monitors of Air Pollution's Biological Damage **621**

29 Animal Phylogeny, Acoelomates, and Protostomes **627**

29.1 What Is an Animal? **628**

29.2 Key Innovations in Animal Evolution **629**

29.3 An Overview of Animal Phylogeny and Classification **633**

29.4 Animals without Tissues: Parazoa **635**

29.5 Eumetazoans with Radial Symmetry **636**

29.6 Lophotrochozoan Protostomes **641**

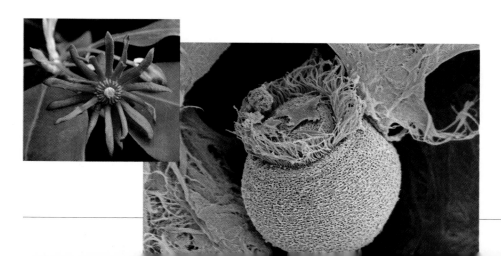

29.7 Ecdysozoan Protostomes **653**

> Focus on Research
> Applied Research: A Rogue's Gallery of Parasitic Worms **644**

> Focus on Research
> Model Research Organisms: *Caenorhabditis elegans* **654**

> Insights from the Molecular Revolution
> Unscrambling the Arthropods **656**

30 Deuterostomes: Vertebrates and Their Closest Relatives **667**

30.1 Invertebrate Deuterostomes **668**

30.2 Overview of the Phylum Chordata **671**

30.3 The Origin and Diversification of Vertebrates **674**

30.4 Agnathans: Hagfishes and Lampreys, Conodonts and Ostracoderms **677**

30.5 Jawed Fishes **678**

30.6 Early Tetrapods and Modern Amphibians **683**

30.7 The Origin and Mesozoic Radiations of Amniotes **686**

30.8 Testudines: Turtles **688**

30.9 Living Nonfeathered Diapsids: Sphenodontids, Squamates, and Crocodilians **689**

30.10 Aves: Birds **692**

30.11 Mammalia: Monotremes, Marsupials, and Placentals **695**

30.12 Nonhuman Primates **697**

30.13 The Evolution of Humans **702**

> Focus on Research
> Model Research Organisms: *Anolis* Lizards of the Caribbean **691**

> Insights from the Molecular Revolution
> The Guinea Pig Is Not a Rat **698**

Unit Seven Ecology and Behavior **1125**

49 Population Ecology **1125**

49.1 The Science of Ecology **1126**

49.2 Population Characteristics **1127**

49.3 Demography **1129**

49.4 The Evolution of Life Histories **1132**

49.5 Models of Population Growth **1133**

49.6 Population Regulation **1139**

49.7 Human Population Growth **1145**

> Figure 49.3 Research Method
> *Using Mark-Release-Recapture to Estimate Population Size* **1128**

> Insights from the Molecular Revolution
> Tracing Armadillo Paternity and Migration **1130**

> Focus on Research
> Basic Research: The Evolution of Life History Traits in Guppies **1134**

> Figure 49.16 Experimental Research
> *Evaluating Density-Dependent Interactions between Species* **1142**

50 Population Interactions and Community Ecology **1151**

50.1 Population Interactions **1152**

50.2 The Nature of Ecological Communities **1160**

50.3 Community Characteristics **1163**

50.4 Effects of Population Interactions on Community Characteristics **1166**

50.5 Effects of Disturbance on Community Characteristics **1167**

50.6 Ecological Succession: Responses to Disturbance **1170**

50.7 Variations in Species Richness among Communities **1174**

> Figure 50.8 Experimental Research
> *Gause's Experiments on Interspecific Competition in Paramecium* **1156**

> Figure 50.12 Experimental Research
> *Demonstration of Competition between Two Species of Barnacles* **1159**

> Insights from the Molecular Revolution
> Finding a Molecular Passport to Mutualism **1161**

> Figure 50.22 Experimental Research
> *Effect of a Predator on the Species Richness of Its Prey* **1168**

Figure 50.23 Experimental Research
The Complex Effects of an Herbivorous Snail on Algal Species Richness **1169**

FOCUS ON RESEARCH
Basic Research: Testing the Theory of Island Biogeography **1177**

51 Ecosystems **1181**

51.1 Energy Flow and Ecosystem Energetics **1182**

51.2 Nutrient Cycling in Ecosystems **1191**

51.3 Ecosystem Modeling **1199**

Figure 51.6 Observational Research
Energy Flow in the Silver Springs Ecosystem **1187**

INSIGHTS FROM THE MOLECULAR REVOLUTION
Fishing Fleets at Loggerheads with Sea Turtles **1190**

FOCUS ON RESEARCH
Basic Research: Studies of the Hubbard Brook Watershed **1193**

FOCUS ON RESEARCH
Applied Research: Disruption of the Carbon Cycle **1196**

52 The Biosphere **1203**

52.1 Environmental Diversity of the Biosphere **1205**

52.2 Organismal Responses to Environmental Variation **1209**

52.3 Terrestrial Biomes **1211**

52.4 Freshwater Biomes **1219**

52.5 Marine Biomes **1221**

INSIGHTS FROM THE MOLECULAR REVOLUTION
Fish Antifreeze Proteins **1210**

Figure 52.9 Observational Research
How Lizards Compensate for Altitudinal Variations in Environmental Temperature **1211**

FOCUS ON RESEARCH
Basic Research: Exploring the Rain Forest Canopy **1214**

Figure 52.24 Experimental Research
Artificial Eutrophication of a Lake **1222**

53 Biodiversity and Conservation Biology **1229**

53.1 The Benefits of Biodiversity **1230**

53.2 The Biodiversity Crisis **1232**

53.3 Biodiversity Hotspots **1239**

53.4 Conservation Biology: Principles and Theory **1241**

53.5 Conservation Biology: Practical Strategies and Economic Tools **1247**

Figure 53.4 Experimental Research
Predation on Songbird Nests in Forests and Forest Fragments **1232**

FOCUS ON RESEARCH
Applied Research: Biological Magnification **1236**

INSIGHTS FROM THE MOLECULAR REVOLUTION
Developing a DNA Barcode System **1242**

FOCUS ON RESEARCH
Applied Research: Preserving the Yellow-Bellied Glider **1244**

Figure 53.16 Observational Research
Metapopulation Structure of the Bay Checkerspot Butterfly **1245**

Figure 53.19 Experimental Research
Effect of Landscape Corridors on Plant Species Richness in Habitat Fragments **1247**

54 The Physiology and Genetics of Animal Behavior **1253**

54.1 Genetic and Environmental Contributions to Behavior **1254**

54.2 Instinctive Behaviors **1255**

54.3 Learned Behaviors **1257**

54.4 The Neurophysiological Control of Behavior **1259**

54.5 Hormones and Behavior **1260**

54.6 Nervous System Anatomy and Behavior **1263**

Figure 54.2 Experimental Research
The Role of Sign Stimuli in Parent-Offspring Interactions **1256**

INSIGHTS FROM THE MOLECULAR REVOLUTION
A Knockout by a Whisker **1258**

Figure 54.10 Experimental Research
Effects of the Social Environment on Brain Anatomy and Chemistry **1262**

Figure 54.12 Experimental Research
*Nervous System Structure and Appropriate
Behavioral Responses* **1265**

**55 The Ecology and Evolution of Animal
Behavior 1269**

55.1 Migration and Wayfinding **1270**

55.2 Habitat Selection and Territoriality **1274**

55.3 The Evolution of Communication **1276**

55.4 The Evolution of Reproductive Behavior
and Mating Systems **1279**

55.5 The Evolution of Social Behavior **1281**

55.6 An Evolutionary View of Human Social Behavior **1285**

Figure 55.4 Experimental Research
Using Landmarks to Find the Way Home **1272**

Figure 55.5 Experimental Research
*Experimental Analysis of the Indigo Bunting's
Star Compass* **1273**

Figure 55.18 Research Method
Calculating Degrees of Relatedness **1283**

INSIGHTS FROM THE MOLECULAR REVOLUTION
Unadorned Truths about Naked Mole-Rat Workers **1286**

Figure 55.21 Observational Research
An Evolutionary Analysis of Human Cruelty **1287**

Appendix A: Answers A-1

Appendix B: Classification A-11

Appendix C: Annotations to a Journal Article A-15

Glossary G-1

Credits C-1

Index I-1

A replica of H.M.S. *Beagle*, the ship that carried Charles Darwin on his round-the-world journey of discovery.

Christopher Ralling

19 Development of Evolutionary Thought

STUDY PLAN

19.1 Recognition of Evolutionary Change

Europeans integrated ideas from ancient Greek philosophy into Christian doctrine

Scientists slowly became aware of change in the natural world

Lamarck developed an early theory of biological evolution

Geologists recognized that Earth had changed over time

19.2 Darwin's Journeys

Darwin saw the world on the voyage of the *Beagle*

Darwin used common knowledge and several inferences to develop his theory

Darwin's theory revolutionized the way we think about the living world

19.3 Evolutionary Biology since Darwin

The modern synthesis created a unified theory of evolution

Research in many fields has provided evidence of evolutionary change

Some people misinterpret the theory of evolution

WHY IT MATTERS

On June 18, 1858, Charles Darwin received the shock of his life. Alfred Russel Wallace, a young naturalist working in the Asian tropics, had solicited Darwin's opinion of a short manuscript about how species change through time. Darwin quickly realized that Wallace had independently described a mechanism for biological evolution that was nearly identical to the one he had been studying for more than 20 years but had not yet described in print.

Like researchers today, scientists in the nineteenth century had to publish their work quickly to establish the "priority" on which scientific reputations are made. Darwin's friend and colleague, the geologist Charles Lyell, had encouraged him to publish a preliminary essay on evolution 2 years before Wallace's letter arrived. But Darwin procrastinated, and because Wallace was the first to prepare his work for publication, Darwin feared that history would credit the younger man with these new ideas. Despite his anxiety, Darwin forwarded Wallace's manuscript to Lyell, who passed it along to the botanist Joseph Hooker. Lyell and Hooker engineered a solution that gave credit to both men **(Figure 19.1)**. On July 1, 1858, papers by Darwin and Wallace were presented to the Linnaean Society of London, a prestigious scientific organization.

Charles Darwin   Alfred Russel Wallace

Courtesy George P. Darwin, Darwin Museum, Down House

Down House and The Royal College of Surgeons of England

Figure 19.1

Pioneers of evolutionary theory. Charles Darwin (1809–1882) and Alfred Russel Wallace (1823–1913) independently discovered the mechanism of natural selection.

Darwin worked feverishly after this harrowing experience, and his now-famous book, *On the Origin of Species by Means of Natural Selection,* was published on November 24, 1859. The first printing of 1250 copies sold out in one day. Today, we honor Darwin for developing the seminal idea about how biological evolution occurs and for the vast documentation that he accumulated over decades of study.

In *The Origin,* Darwin proposed that natural mechanisms produce and transform the diversity of life on Earth. His concept of evolution still forms the unifying intellectual paradigm within which all biological research is undertaken. Even when researchers do not address explicitly evolutionary questions, their observations, theories, hypotheses, and experiments are formulated with the implicit knowledge that all forms of life are related and have evolved from ancestral forms.

Biological evolution occurs in populations when specific *processes* cause the genomes of organisms to differ from those of their ancestors. These genetic changes, and the phenotypic modifications they cause, are the *products* of evolution. By studying the products of evolution, biologists strive to understand the processes that cause evolutionary change.

The theory of evolution is so widely accepted that most people cannot think about the biological world in any other way. But the biological changes implied by Darwin's ideas and by modern evolutionary theory had not been included in earlier worldviews.

19.1 Recognition of Evolutionary Change

The historical development of evolutionary theory is a fascinating tale of scientists struggling to reconcile evidence of change with a prevailing philosophy that change was impossible in a perfectly created universe.

Europeans Integrated Ideas from Ancient Greek Philosophy into Christian Doctrine

The Greek philosopher Aristotle (384–322 B.C.) was a keen observer of nature, and he is generally considered the first student of **natural history**, the branch of biology that examines the form and variety of organisms in their natural environments. Aristotle believed that both inanimate objects and living species had fixed characteristics. Careful study of their differences and similarities enabled him to create a ladder-like classification of nature from simplest to most complex forms: minerals ranked below plants, plants below animals, animals below humans, and humans below the gods of the spiritual realm.

By the fourteenth century, Europeans had merged Aristotle's classification with the biblical account of creation: all of the different kinds of organisms had been specially created by God, species could never change or become extinct, and new species could never arise. Biological research became dominated by **natural theology**, which sought to name and catalog all of God's creation. Careful study of each species would identify its position and purpose in the *Scala Naturae,* or Great Chain of Being, as Aristotle's ladder of life was called. In the eighteenth century, the Swedish botanist Carolus Linnaeus (1707–1778), who developed the science of **taxonomy**, the branch of biology that classifies organisms (see Chapter 23), undertook this important work *ad majorem Dei gloriam* ("for the greater glory of God").

Scholars also used a literal interpretation of scripture to date the time of creation precisely. By tabulating the human generations described in the Bible, they determined that the creation had occurred around 4000 B.C., making Earth a bit less than 6000 years old. Thus, Earth hardly seemed old enough for much change to have taken place.

Scientists Slowly Became Aware of Change in the Natural World

Modern science came of age in the fifteenth through eighteenth centuries. The English philosopher and statesman Sir Francis Bacon (1561–1626) established the importance of observation, experimentation, and inductive reasoning. Other scientists, notably Nicolaus Copernicus (1473–1543), Galileo Galilei (1564–1642), René Descartes (1596–1650), and Sir Isaac Newton (1643–1727), proposed mechanistic theories to explain physical events. In addition, three new disciplines—biogeography, comparative morphology, and geology—promoted a growing awareness of change.

Questions about Biogeography. As long as naturalists encountered organisms only from Europe and surrounding lands, the task of understanding the *Scala Naturae* was manageable. But global explorations in the fifteenth through seventeenth centuries provided

Ostrich *(Struthio camelus)* of Africa   Rhea *(Rhea americana)* of South America   Emu *(Dromaius novaehollandiae)* of Australia

Figure 19.2
Large, flightless birds. Three large bird species with greatly reduced wings occupy similar habitats in geographically separated regions.

naturalists with thousands of unknown plants and animals from Asia, sub-Saharan Africa, the Pacific Islands, and the Americas. Although some were similar to European species, others were new and very strange.

Studies of the world distribution of plants and animals, now called **biogeography,** raised puzzling questions. Was there no limit to the number of species created by God? Where did all these species fit in the *Scala Naturae*? If all species had been created in the Garden of Eden, why were the species found in Africa or Asia different from those found in Europe? Why was each species found only in certain places and not others **(Figure 19.2)?**

Questions about Comparative Morphology. When biologists began to compare the **morphology** (anatomical structure) of organisms, they discovered interesting similarities and differences. For example, the front legs of pigs, the flippers of dolphins, and the wings of bats differ markedly in size, shape, and function **(Figure 19.3).** But these appendages have similar locations in the animals' bodies; all are constructed of bones, muscles, and skin; and all develop similarly in the animals' embryos. If these limbs were specially created for different means of locomotion, why didn't the Creator use different materials and structures for walking, swimming, and flying?

Natural theologians answered that some general body plans were perfect, and there was no need to invent a new plan for every species. But a French scientist, George-Louis Leclerc (1707–1788), le Comte (Count) de Buffon, was still puzzled by the existence of body parts with no apparent function. For example, he noted that the feet of pigs and some other mammals have two toes that never touch the ground (see Figure 19.3). If each species is anatomically perfect for its particular way of life, why do useless structures exist?

Buffon proposed that some animals must have *changed* since their creation; he suggested that **vestigial structures,** the useless parts we observe today, must have functioned in ancestral organisms. Buffon offered no explanation of how functional structures became vestigial, but he clearly recognized

that some species were "conceived by Nature and produced by Time."

Questions about Fossils. By the mid-eighteenth century, geologists were mapping the **stratification,** or horizontal layering, of sedimentary rocks beneath the soil surface (see Figure 22.3). Different layers held different kinds of **fossils** (*fossilis* = dug up). Relatively small and simple fossils appeared in the deepest layers. Fossils in the layers above them were more complex. Those in the uppermost layers often resembled living organisms. Moreover, fossils found in any particular layer were often similar, even if they were collected from geographically separated sites. What were these fossils, and why did they vary more from one layer of

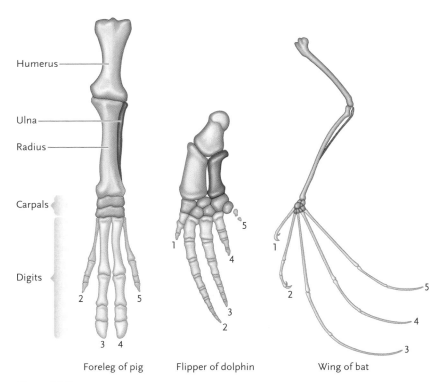

Humerus

Ulna

Radius

Carpals

Digits

Foreleg of pig   Flipper of dolphin   Wing of bat

Figure 19.3
Mammalian forelimbs and locomotion. Pigs use their legs to walk or run, dolphins use their flippers to swim, and bats use their wings to fly. Homologous bones are pictured in the same color, and digits (fingers) are numbered; pigs have lost the first digit over evolutionary time.

rock to another than from one geographical region to another?

Some scientists suggested that fossils were the remains of extinct organisms, but natural theology did not allow extinction. Thomas Jefferson, the third president of the United States and an amateur fossil hunter, thought that fossils were the remains of species that were now extremely rare; he believed that nature could not have "permitted any one race of her animals to become extinct" or "formed any link in her great works so weak [as] to be broken." He even asked Lewis and Clark to keep an eye out for giant ground sloths, now known to be extinct, during their exploration of the Pacific Northwest.

Georges Cuvier (1769–1832), a French zoologist and a founder of comparative morphology, as well as **paleobiology** (the study of ancient organisms), realized that the layers of fossils represented organisms that had lived at successive times in the past. He suggested that the abrupt changes between geological strata marked dramatic shifts in ancient environments. Cuvier and his followers developed the theory of **catastrophism**, reasoning that each layer of fossils represented the remains of organisms that had died in a local catastrophe, such as a flood. Somewhat different species then recolonized the area, and when another catastrophe struck, they formed a different set of fossils in the next higher layer.

Lamarck Developed an Early Theory of Biological Evolution

A contemporary of Cuvier and a student of Buffon, Jean Baptiste de Lamarck (1744–1829) proposed the first comprehensive theory of biological evolution based on specific mechanisms. He proposed that a metaphysical "perfecting principle" caused organisms to become better suited to their environments. Simple organisms evolved into more complex ones, moving up the ladder of life; microscopic organisms were replaced at the bottom by spontaneous generation.

Lamarck theorized that two mechanisms fostered evolutionary change. According to his *principle of use and disuse,* body parts grow in proportion to how much they are used, as anyone who "pumps iron" well knows. Conversely, structures that are not often used get weaker and shrink, such as the muscles of an arm immobilized in a cast. According to his second principle, the *inheritance of acquired characteristics,* changes that an animal acquires during its lifetime are inherited by its offspring. Thus, Lamarck argued

Figure 19.4
A great blue heron *(Ardea herodias).* Like many other wading birds, herons have long, stiltlike legs.

Rich Kirchner/Foto Natura/Photo Researchers, Inc.

that long-legged wading birds, such as herons **(Figure 19.4),** are descended from short-legged ancestors that stretched their legs to stay dry while feeding in shallow water. Their offspring inherited slightly longer legs, and after many generations, their legs became extremely long.

Today, we know that Lamarck's proposed mechanisms do not cause evolutionary change. Although muscles do grow larger through continued use, most structures do not respond in the way Lamarck predicted. Moreover, structural changes acquired during an organism's lifetime are not inherited by the next generation. Even in his own day, Lamarck's ideas were not widely accepted.

Despite the shortcomings of his theory, Lamarck made four tremendously important contributions to the development of an evolutionary worldview. First, he proposed that all species change through time. Second, he recognized that new characteristics are passed from one generation to the next. Third, he suggested that organisms change in response to their environments. And fourth, he hypothesized the existence of specific mechanisms that caused evolutionary change. The first three of these ideas became cornerstones of Darwin's evolutionary theory. Perhaps Lamarck's most important contribution was that he fostered discussion. By the mid-nineteenth century, most educated Europeans were talking about evolutionary change, whether they believed in it or not.

Geologists Recognized That Earth Had Changed over Time

In 1795, the Scottish geologist James Hutton (1726–1797) argued that slow and continuous physical processes, *acting over long periods of time,* produced Earth's major geological features; for example, the movement of water in a river slowly erodes the land and deposits sediments near the mouth of the river. Given enough time, erosion creates deep canyons, and sedimentation creates thick topsoil on flood plains. Hutton's **gradualism,** the view that Earth changed *slowly* over its history, contrasted sharply with Cuvier's catastrophism.

The English geologist Charles Lyell (1797–1875) championed and extended Hutton's ideas in an influential series of books, *Principles of Geology.* Lyell argued that the geological processes that sculpted Earth's surface over long periods of time—such as volcanic eruptions, earthquakes, erosion, and the formation and movement of glaciers—are exactly the same as the processes observed today. This concept, **uniformitarianism,** undermined any remaining notions of an unchanging Earth. Also, because geological processes proceed very slowly, it must have taken millions of years, not just a few thousand, to mold the landscape into its current configuration.

1. Why did the existence of vestigial structures make Buffon question the idea that living systems never changed?
2. What were Lamarck's contributions to an evolutionary worldview?
3. How do the concepts of gradualism and uniformitarianism in geology undermine the belief that Earth is only about 6000 years old?

Figure 19.5
Darwin's voyage. H.M.S. *Beagle* circumnavigated the globe between 1831 and 1836.

19.2 Darwin's Journeys

In 1831, in the midst of this intellectual ferment, young Charles Darwin wondered what to do with his life. Raised in a wealthy English household, he had always collected shells and studied the habits of insects and birds; he preferred hunting and fishing to classical studies. Despite lackluster performance as a student, Darwin was expected to continue the family tradition of practicing medicine. But he abandoned medical studies after 2 years. Instead, he followed his interest in natural history over the objections of his father, who reputedly told him, "You care for nothing but shooting, dogs, and rat-catching and you will be a disgrace to yourself and all of your family."

At the suggestion of his father, Darwin studied for a career as a clergyman, earning a degree at Cambridge University. There, he found a mentor in the Reverend John Henslow, a leading botanist, who arranged for Darwin to travel as the captain's dining companion aboard H.M.S. *Beagle,* a naval surveying ship. Darwin thus embarked on a sea voyage and an intellectual journey that altered the foundations of modern thought.

Darwin Saw the World on the Voyage of the *Beagle*

The *Beagle* sailed westward to map the coastline of South America and then circumnavigated the globe **(Figure 19.5).** When the ship's naturalist quit his post midjourney, Darwin replaced him in an unofficial capacity. For nearly 5 years Darwin toured the world, and because he suffered from seasickness, he seized every chance to go ashore. He collected plants and animals in Brazilian rain forests and fossils in Patagonia. He hiked the grasslands of the pampas and climbed the Andes in Chile. Armed with Henslow's parting gift, the first volume of Lyell's *Principles of Geology,* Darwin was primed to apply gradualism and uniformitarianism to the living world.

What Darwin Saw. When he began his travels, Darwin had no clue that biological evolution had produced the mind-boggling variety of species that he would en-

counter. Three broad sets of observations later helped him unravel the mystery of evolutionary change.

First, while exploring along the coast of Argentina, Darwin discovered fossils that often resembled organisms that inhabit the same region today. For example, despite an enormous size difference, living armadillos and fossilized glyptodonts had similar body armor, but they were unlike any other species known to science **(Figure 19.6).** If both species had been created at the same time and both were found in South America, why didn't glyptodonts live alongside armadillos? Darwin later wondered whether armadillos might be the living descendants of the now-extinct glyptodonts.

Second, Darwin observed that the animals he encountered in different South American habitats clearly resembled each other but differed from species that occupied similar habitats in Europe. For example, he

Charles R. Knight painting (negative CK211), Field Museum of Natural History, Chicago

Calvin Larsen/Photo Researchers, Inc.

Figure 19.6
Ancestors and descendants. An extinct glyptodont (top) probably weighed 300 to 400 times as much as its living descendant, a nine-banded armadillo *(Dasypus novemcinctus).*

a. South American nutria   **b.** European beaver

Figure 19.7
Morphologic differences in species from different continents. Darwin noted that **(a)** South American nutria *(Myocastor coypus)* and **(b)** European beavers *(Castor fiber)* differ in appearance, even though both species are aquatic rodents that feed on vegetation. Notice that nutria have long, round tails, whereas beavers have short, flat tails.

a. The Galápagos   **b.** Galápagos tortoise   **c.** Marine iguana

d. Blue-footed booby

Figure 19.8
The Galápagos. **(a)** Volcanic eruptions created the Galápagos archipelago (located 1000 km west of Ecuador) between 3 and 5 million years ago. **(b)** The islands were named for the giant tortoises found there (in Spanish, *galápa* means tortoise); this tortoise *(Geochelone elephantopus)* is native to Isla Santa Cruz. **(c)** Marine iguanas *(Amblyrhynchus cristatus)* dive into the Pacific Ocean to feed on algae. **(d)** A male blue-footed booby *(Sula nebouxii)* engages in a courtship display.

a. *Certhidea olivacea*   **b.** *Geospiza scandens*   **c.** *Geospiza magnirostris*   **d.** *Camarhynchus pallidus*

Figure 19.9

Bill shape and food habits. The 13 finch species that inhabit the Galápagos are descended from a common ancestor, a seed-eating ground finch that migrated to the islands from South America. **(a)** *Certhidea olivacea* uses its slender bill to probe for insects in vegetation. **(b)** *Geospiza scandens* has a medium-sized bill suitable for eating cactus flowers and fruit. **(c)** *Geospiza magnirostris* uses its thick, strong bill to crush cactus seeds. **(d)** *Camarhynchus pallidus* uses its bill to hammer at bark and to hold cactus spines, with which it probes for wood-boring insects, such as termites.

noted that nutria *(Myocastor coypus),* a semiaquatic rodent in South America, bore a closer resemblance to rodent species from the mountains or grasslands of that continent than it did to the European beaver *(Castor fiber),* another semiaquatic rodent that had once been common in England **(Figure 19.7).** Why did animals from markedly different South American environments resemble each other, and why were animals that lived in similar environments on separate continents different? Darwin later understood that animals in South America resembled each other because they had inherited their similarities from a common ancestor.

Third, Darwin observed fascinating patterns in the distributions of species on the Galápagos **(Figure 19.8).** There he found strange and wonderful creatures, including giant tortoises and lizards that dove into the sea to feed on algae. Darwin quickly noted that the animals on different islands varied slightly in form. Indeed, experienced sailors could easily identify a tortoise's island of origin by the shape of its shell. Moreover, many species resembled those on the distant South American mainland. Why did so many different organisms occupy one small island cluster, and why did these species resemble others from the nearest continent? Darwin later hypothesized that the plants and animals of the Galápagos were descended from South American ancestors, and that each species had changed after being isolated on a particular island.

Darwin's Reflections after His Voyage. The *Beagle* returned to England in 1836, and Darwin began his first notebook on the *Transmutation of Species* the fol-

lowing year. He realized that changes in species over time provided the only plausible explanation for his observations.

A diverse group of finches from the Galápagos **(Figure 19.9)** provided the single greatest spark for Darwin's work. He had noticed great variability in the shapes of their bills, but he had incorrectly assumed that birds on different islands belonged to the same species. Thus, he had not recorded the island where he had captured each specimen. Luckily, the *Beagle*'s captain, Robert Fitzroy, had more thoroughly documented his own collection, allowing Darwin to study the relationships and geographical distributions of a dozen species. As Darwin reviewed his data, he began to focus on two aspects of a general problem. Why were the finches on a particular island slightly different from those on nearby islands, and how did all these different species arise?

Darwin Used Common Knowledge and Several Inferences to Develop His Theory

With a substantial inheritance and burdened by chronic illness, Darwin led a reclusive life as he embarked on an intellectual journey every bit as exciting as his voyage on the *Beagle* (see *Focus on Research*). His lifetime goal was to accumulate evidence of evolutionary change and identify the mechanism that caused it.

Selective Breeding and Heredity. Having grown up in the country, Darwin was well aware that "like begets like"; that is, offspring frequently resemble their parents. Plant and animal breeders had applied this basic truth of inheritance for thousands of years. By

Basic Research: Charles Darwin's Life as a Scientist

Darwin's observations during the voyage of H.M.S. *Beagle* convinced him that species change through time, and that natural processes produced Earth's biodiversity. He spent the rest of his life gathering data to support his ideas and unravel the workings of natural selection.

Shortly after the *Beagle* returned to England in 1836, Darwin began his first notebook on the "transmutation of species." But he put his study of evolution aside while he wrote up the geological and biological research that he had undertaken during the voyage. This task took him 10 years to complete—twice as long as the journey itself. The results of these efforts were numerous articles and several books, including the now famous *Journal of the Voyage of the Beagle*, published in 1839.

After preparing a sketch of his ideas about evolution in 1844, Darwin continued to write up his observations from the voyage. But he had trouble classifying one species of barnacle, a small marine invertebrate, which he had collected in Chile. For the next 8 years he studied barnacles, examining more than 10,000 specimens and revising the entire classification of these animals. His colleagues saw this study as a strange diversion from his work on evolution, but Darwin's detailed examination of barnacle anatomy sharpened his observational skills and provided a test case in which he could apply his ideas about descent with modification to a large and diverse group of organisms. He published four volumes about barnacles in 1854.

While studying barnacles, Darwin continued to think about "the species question." He kept notebooks about variation in plants and animals, focusing on variation that was amplified by selective breeding. He was a tireless collector of facts, which he sought from every possible source. He badgered dog breeders, horse farmers, and horticulturists with long lists of questions about their work. His enthusiasm was infectious, and workers throughout the world supplied him with data and specimens. Darwin was also an eager and skilled experimentalist, and he took up pigeon breeding, marveling at the huge variety of morphological traits that he and other breeders could produce. In the late 1850s, a communication from another naturalist, Alfred Russel Wallace, forced him to finally complete *The Origin*, which revolutionized the study of biology.

Even after *The Origin* was published, Darwin continued to gather facts and write about evolution, working almost up to the day he died in 1882 at age 74. He published a detailed analysis of how earthworms improve the soil *(The Formation of Vegetable Mould through the Action of Worms)* and wrote books on several botanical topics, among them plants that eat animals *(Insectivorous Plants)*, pollination and fertilization systems *(Fertilisation in Orchids* and *The Effects of Self- and Cross-Fertilisation)*, and the tendency of plants to grow toward sunlight *(The Power of Movement in Plants)*. Darwin's work always had an evolutionary focus, however, and he produced several revisions of *The Origin*, as well as books on artificial selection *(Variation of Animals and Plants under Domestication)*, human ancestry *(The Descent of Man)*, and animal behavior *(The Expression of the Emotions in Men and Animals)*.

Darwin's study. Darwin undertook most of his life's work in this room at Down House. He hesitated to discard old papers and specimens, believing that he would find a use for them as soon as they were carried away in the trash.

William Perlman/Star Ledger/Corbis

selectively breeding individuals with favorable characteristics, they enhanced those traits in future generations.

Farmers use selective breeding to improve domesticated plants and animals. If one cow produces more milk than any other, the farmer selectively breeds her (rather than others), hoping that her offspring will also be good milk producers. Although the mechanism of heredity was not yet understood, this principle had been applied countless times to produce bigger beets, plumper pigs, and fancier pigeons (see Figure 1.10). Darwin was well aware of this process, which he called **artificial selection**, but he puzzled over how it could operate in nature. (*Insights from the Molecular Revolu-*

Artificial Selection in the Test Tube

From Darwin's time until very recently, artificial selection was the province of plant and animal breeders, who chose individuals with desired traits to be the parents of the next generation. Now the laborious and time-consuming techniques of the breeders have been bypassed by rapid artificial selection experiments on DNA and protein molecules in the test tube.

One example of artificial selection in the test tube was provided by John J. Toole and his colleagues at Gilead Sciences in Foster City, California. They were interested in developing DNA molecules that could interfere with blood clotting by binding to thrombin, a blood protein that forms a major part of blood clots. The DNA could be used to treat people who are in danger of developing blood clots that might clog arteries in the heart, brain, or other critical organs. Nucleic acid molecules would be particularly useful as anticlotting agents because, unlike the proteins now used for this purpose, they rarely induce an immune reaction in the person being treated.

The investigators began their experiments by using a commercially available apparatus to make short, artificial DNA molecules of random sequence. They ran the apparatus long enough to produce more than 10^{13} (10 trillion!) different DNA sequences, and then made multiple copies of the sequences using the polymerase chain reaction (PCR; see Section 18.1). To select for DNA molecules that could bind to thrombin, they poured the entire DNA preparation through a column that contained thrombin molecules attached to glass beads. Only a few sequences among the trillions, about 0.01% of the total DNA sample, were able to bind strongly to thrombin. The researchers used PCR to multiply the sequences they had captured, generating 10 trillion "progeny" molecules. These progeny DNA molecules were poured through another column that contained thrombin molecules attached to glass beads. This time, a larger percentage of the molecules bound strongly to the thrombin molecules. These strongly binding DNA molecules were then used as the "parents" to generate another 10 trillion progeny. After five repetitions of the total process, producing five generations of DNA molecules, 40% of the DNA molecules in the preparation could recognize and bind strongly to thrombin.

The final products of the artificial selection were tested for their ability to interfere with the activity of thrombin in the blood clotting reaction. These experiments were successful; the anti-thrombin DNA molecules are being tested in monkeys and baboons, in which they appear to work effectively as anticlotting agents.

Toole and his team thus mimicked the evolutionary process on the molecular scale. Their experimental process selected DNA molecules that could bind to thrombin from the many random nucleotide sequences available in the test tube. The sequences that survived the selection test produced the greatest number of progeny molecules in the next generation. The same selection pressure, exerted over five generations of progeny molecules, greatly increased the percentage that could bind strongly to the protein. As a result, the DNA population evolved in the test tube from one with little or no ability to bind thrombin to one with high ability.

This approach is being used in many laboratories to develop DNA and RNA molecules with desired functions. By starting with DNA molecules that encode enzymes, researchers hope to select biological catalysts that can speed chemical reactions with scientific, medical, or industrial purposes.

---

tion describes how modern researchers apply artificial selection to molecules in a test tube.)

The Struggle for Existence. Darwin had a revelation about how selective breeding could occur naturally when he read the famous publication by Thomas Malthus, *Essay on the Principles of Population*. Malthus, an English clergyman and economist, was worried about the fate of the nation's poor. England's population was growing much faster than its agricultural capacity, and with individuals competing for limited food resources, some would inevitably starve.

Darwin applied Malthus's argument to organisms in nature. Species typically produce many more offspring than are needed to replace the parent generation, yet the world is not overrun with sunflowers, tortoises, or bears. Darwin even calculated that, if its reproduction went unchecked, a single pair of elephants, the slowest breeding animal known, would leave roughly 19 million descendants after only 750 years. Happily for us (and all other species that might get underfoot), the world is not so crowded with elephants. Instead, some members of every population survive and reproduce, whereas others die without reproducing.

Darwin's Inferences. Darwin's discovery of a mechanism for evolutionary change required him to infer the nature of a process that no one had envisioned, much less documented **(Table 19.1)**. First, individuals within populations vary in size, form, color, behavior, and other characteristics. Second, many of these variations are hereditary. What if variations in hereditary traits enabled some individuals to survive and reproduce more than others? Organisms with favorable traits would leave many offspring, whereas those that lacked favorable traits would die leaving

Table 19.1 Darwin's Observations and Inferences about Evolution by Means of Natural Selection

Observations	Inferences	
Most organisms produce more than one or two offspring.		
Populations do not increase in size indefinitely.	Individuals within a population compete for limited resources.	
Food and other resources are limited for most populations.		A population's characteristics will change over the generations as advantageous, heritable characteristics become more common.
Individuals within populations exhibit variability in many characteristics.	Hereditary characteristics may allow some individuals to survive longer and reproduce more than others.	
Many variations have a genetic basis that is inherited by subsequent generations.		

few, if any, descendants. Thus, favorable hereditary traits would become more common in the next generation. If the next generation was subjected to the same process of selection, the traits would be even more common in the third generation. Because this process is analogous to artificial selection, Darwin called it **natural selection.**

As an evolutionary mechanism, natural selection favors **adaptive traits,** genetically based characteristics that make organisms more likely to survive and reproduce. And by favoring individuals that are well adapted to the environments in which they live, natural selection causes species to change through time. As shown in Figure 19.9, each species of Galápagos finch has a distinctive bill. Variations in bill size and shape make some birds better adapted for crushing seeds and others for capturing insects. Imagine an island where large seeds were the only food available; individuals with a stout bill would be more likely to survive and reproduce than would birds with slender bills. These favored individuals would pass the genes that produce stout bills to their descendants, and after many generations, their bills might resemble those of *Geospiza magnirostris* (see Figure 19.9c). Natural selection also changes nonmorphologic characteristics of populations; for example, insect populations that are exposed to insecticides develop resistance to these toxic chemicals over time (see Figure 19.11).

Darwin realized that natural selection could also account for striking differences between populations and, given enough time, for the production of new species. For example, suppose that small insects were the only food available to finches on a different island. Birds with long thin bills might be favored by natural selection, and the population of finches might eventually possess a bill shaped like that of *Certhidea olivacea* (see Figure 19.9a). If we apply parallel reasoning to the many characteristics that affect survival and reproduction, natural selection would cause the populations to become more different over time, a process called **evolutionary divergence.**

Darwin's Theory Revolutionized the Way We Think about the Living World

It would be hard to overestimate the impact of Darwin's theory on Western thought. In *The Origin,* Darwin proposed a logical mechanism for evolutionary change and provided enough supporting evidence to convince the educated public.

Darwin argued that all the organisms that have ever lived arose through **descent with modification,** the evolutionary alteration and diversification of ancestral species. He envisioned this pattern of descent as a tree growing through time **(Figure 19.10).** The base of the

Present

Time

Origin of life

Figure 19.10
The tree of life. Darwin envisioned the history of life as a tree. Branching points represent the origins of new lineages; branches that do not reach the top represent extinct groups.

trunk represents the ancestor of all organisms. Branching points above it represent the evolutionary divergence of ancestors into their descendants. Each limb represents a body plan suitable for a particular way of life; smaller branches represent more narrowly defined groups of organisms; and the uppermost twigs represent living species.

Darwin proposed natural selection as the mechanism that drives evolutionary change. In fact, most of *The Origin* was an explanation of how natural selection acted on the variability within groups of organisms, preserving favorable traits and eliminating unfavorable ones.

Four characteristics distinguish Darwin's theory from earlier explanations of biological diversity and adaptive traits:

1. Darwin provided purely physical, rather than spiritual, explanations about the origins of biological diversity.
2. Darwin recognized that evolutionary change occurs in groups of organisms, rather than in individuals: some members of a group survive and reproduce more successfully than others.
3. Darwin described evolution as a multistage process: variations arise within groups, natural selection eliminates unsuccessful variations, and the next generation inherits successful variations.
4. Like Lamarck, Darwin understood that evolution occurs because some organisms function better than others *in a particular environment.*

What is most amazing about Darwin's intellectual achievement is that he knew nothing about Mendelian genetics (see Chapter 12). Thus, he had no clear idea of how variation arose or how it was passed from one generation to the next.

Evolution was a popular topic in Victorian England, and Darwin's theory was both praised and ridiculed. Although he had not speculated about the evolution of humans in *The Origin,* many readers were quick to extrapolate Darwin's ideas to our own species. Needless to say, certain influential Victorians were not amused by the suggestion that humans and apes share a common ancestry.

Nevertheless, Darwin's painstaking logic and careful documentation convinced most readers that evolution really does take place. Thomas Huxley, so staunch an advocate that he was known as "Darwin's bulldog," summed up the reaction of many when he quipped that the theory was so obvious, once articulated, that he was surprised he had not thought of it himself. Darwin's vision of common ancestry quickly became the intellectual framework for nearly all biological research. Many readers, however, did not readily accept the mechanism of natural selection. The major stumbling block was that Darwin had not provided any plausible theory of heredity.

STUDY BREAK

1. What observations that Darwin made on his round-the-world voyage influenced his later thoughts about evolution?
2. How did Darwin's understanding of artificial selection enable him to envision the process of natural selection?
3. What were the four great intellectual triumphs of Darwin's theory?

19.3 Evolutionary Biology since Darwin

Although Gregor Mendel published his work on genetics in 1866, it was not well known in England until 1900. At that time, scientists perceived a fundamental conflict between Darwin's and Mendel's theories. One problem was that Darwin had used complex characteristics, such as the structure of bird bills, to illustrate how natural selection worked. We now know that at least several genes often control such traits. By contrast, Mendel had studied simpler characteristics, such as the height of pea plants (see Chapter 12). A single gene often controls simple traits, which is one reason Mendel could interpret his experimental results so clearly. Biologists had a hard time applying Mendel's straightforward experimental results to Darwin's complex examples.

A second problem arose because Darwin believed that biological evolution occurred gradually over many generations. However, early twentieth-century geneticists, focusing on simple traits such as those Mendel had studied, sometimes observed very rapid and dramatic changes in certain characteristics. A widely accepted theory, *mutationism* suggested that evolution occurred in spurts, induced by the chance appearance of "hopeful monsters," rather than by gradual change.

The Modern Synthesis Created a Unified Theory of Evolution

In the 1910s and 1920s, geneticists and mathematicians forged a critical link between Darwinism and Mendelism. The new discipline, **population genetics**, recognized the importance of genetic variation as the raw material of evolution. Population geneticists constructed mathematical models, which applied equally well to simple and complex traits, to predict how natural selection and other processes influence a population's genetics.

In the 1930s and 1940s, a unified theory of evolution, the **modern synthesis,** interpreted data from biogeography, comparative morphology, comparative

Figure 19.11 Experimental Research

How Exposure to Insecticide Fosters the Evolution of Insecticide Resistance

QUESTION: Does exposure to insecticide foster the evolution of insecticide resistance in insect populations?

EXPERIMENT: Researchers studied samples of wild mosquitoes *(Anopheles culicifacies)* captured at a small village in India, where public health officials frequently sprayed the insecticide dichloro-diphenyl-trichloroethane (DDT) to control these pests. For each test, the researchers exposed samples of mosquitoes to a 4% concentration of DDT for 1 hour and then measured the percentage that died during the next 24 hours. Tests were repeated 12 months and 16 months after the first experiment.

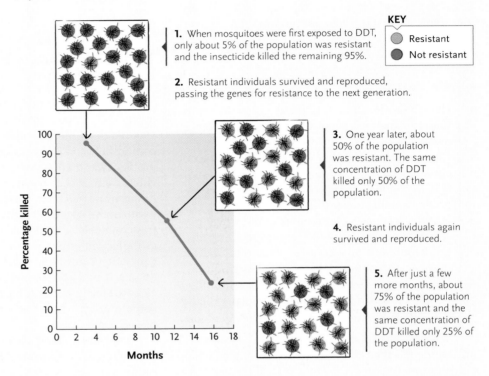

KEY
- Resistant
- Not resistant

1. When mosquitoes were first exposed to DDT, only about 5% of the population was resistant and the insecticide killed the remaining 95%.

2. Resistant individuals survived and reproduced, passing the genes for resistance to the next generation.

3. One year later, about 50% of the population was resistant. The same concentration of DDT killed only 50% of the population.

4. Resistant individuals again survived and reproduced.

5. After just a few more months, about 75% of the population was resistant and the same concentration of DDT killed only 25% of the population.

RESULTS: Over the course of the experiment, smaller and smaller percentages of the mosquitoes died after their exposure to the test concentration of the insecticide.

CONCLUSION: The indiscriminate use of DDT established natural selection that favored DDT-resistant individuals. Exposure to DDT therefore fostered the evolution of an adaptive resistance to DDT in the mosquito population.

embryology, paleontology, and taxonomy within an evolutionary framework. The authors of the modern synthesis focused on evolutionary change within populations, and although they considered natural selection the primary mechanism of evolution, they acknowledged the importance of other processes (see Chapter 20). Proponents of the modern synthesis also embraced Darwin's idea of gradualism and deemphasized the significance of mutations that changed traits suddenly and dramatically.

The modern synthesis also tried to link the two levels of evolutionary change that Darwin had identified: microevolution and macroevolution. **Microevolution** describes the small-scale genetic changes that populations undergo, often in response to shifting environmental circumstances; a small evolutionary shift in the size of the bill of a finch species is an example of microevolution. **Macroevolution** describes large-scale patterns in the history of life, such as the appearance and then relatively sudden disappearance of gigantic dinosaurs. According to the modern synthesis, macroevolution results from the gradual accumulation of microevolutionary changes, but researchers are just beginning to unravel the genetic mechanisms that establish a relationship between these two levels of evolutionary change (see Chapter 22).

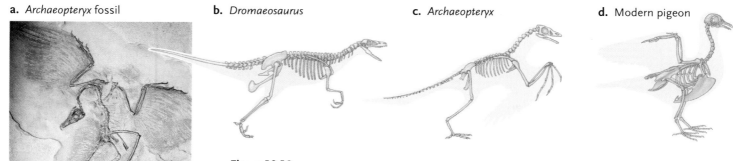

a. *Archaeopteryx* fossil **b.** *Dromaeosaurus* **c.** *Archaeopteryx* **d.** Modern pigeon

P. Morris/Ardea, London

Figure 19.12

Bird ancestry. **(a)** One of the few known fossils of *Archaeopteryx lithographica*, from limestone deposits more than 140 million years old. **(b)** *Dromaeosaurus* was a small, bipedal dinosaur that had teeth, long limbs with toes and fingers, and a long, bony tail. **(c)** *Archaeopteryx* shared those three traits with *Dromaeosaurus*, but it also had feathers and hollow bones, characteristics that it shares with modern birds. **(d)** Modern birds, such as the pigeon, have long limbs similar to those of *Dromaeosaurus* and *Archaeopteryx*, but their fingers and bony tails are greatly reduced; like *Archaeopteryx*, their bodies are covered with feathers, but a horny bill has replaced their teeth.

Research in Many Fields Has Provided Evidence of Evolutionary Change

During the past 100 years, scientists have assembled a huge and compelling body of evidence from many biological disciplines indicating that biological evolution is a fact of life on Earth.

Adaptation by Natural Selection. Biologists interpret the products of natural selection as evolutionary adaptations. For example, the wings of birds, which have been modified by evolutionary processes over millions of years, have an obvious function that helps these animals survive and reproduce. Sometimes, however, natural selection operates on a short time scale, as illustrated by the development of pesticide resistance in insects. When we first use a new pesticide, a low concentration often kills a large percentage of the pests. However, just by chance, a few insects may have genetic characteristics that confer resistance to the poison. The surviving individuals produce offspring, many of which inherit the resistance. As a result, a given concentration of the poison kills a smaller percentage of insects in the next generation; therefore, over time, the entire population may become highly resistant **(Figure 19.11).**

The Fossil Record. Because evolution results from the modification of existing species, Darwin's theory proposes that all species that have ever lived are genetically related. The fossil record documents such continuity, providing clear evidence of ongoing change in many **biological lineages**, evolutionary sequences of ancestral organisms and their descendants (see Chapter 22). For example, the evolution of modern birds can be traced from a dinosaur ancestor through fossils such as *Archaeopteryx lithographica* **(Figure 19.12).** This species, discovered only 2 years after *The Origin* was published, resembled both dinosaurs and birds. Like small carnivorous dinosaurs, *Archaeopteryx* walked on its hind legs and had teeth, claws on its forelimbs, and a long, bony tail. Like modern birds, it had hollow bones, an enlarged sternum, and feathers that covered its body.

Historical Biogeography. Analyses of **historical biogeography**, the study of the geographical distributions of plants and animals in relation to their evolutionary history, are generally consistent with Darwin's theory of evolution. Species on oceanic islands often closely resemble species on the nearest mainland, suggesting that the island and mainland species share a common ancestry. Moreover, species on a continental land mass are clearly related to one another and are often distinct from those on other continents. For example, monkeys in South America have long, prehensile tails and broad noses, traits that they inherited from a shared South American ancestor. By contrast, monkeys in Africa and Asia evolved from a different common ancestor in the Old World, and their shorter tails and narrower noses distinguish them from their American cousins.

Comparative Morphology. Other evidence of evolution comes from **comparative morphology**, analyses of the structure of living and extinct organisms. Such analyses are based on the comparison of **homologous traits**, characteristics that are similar in two species because they inherited the genetic basis of the trait from their common ancestor. For example, the forelimbs of all four-legged vertebrates are homologous because they evolved from a common ancestor with a forelimb composed of the same component parts (see Figure 19.3, which shows homologous bones in the same color). Even though the shapes of the bones are different in pigs, dolphins, and bats, similarities in the three limbs are apparent. The differences in structural details arose over evolutionary time, allowing pigs to walk, dolphins to swim, and bats to fly. The arms of humans and the wings of birds are also constructed of comparable elements, suggesting that they, too, share a common ancestor with the three species illustrated.

Comparative Embryology. The early embryos of different species within a major group of organisms are often strikingly similar. For example, certain components of the circulatory system emerge in all vertebrate embryos at corresponding stages of development **(Figure 19.13)**. In addition, the early embryos of humans and other four-limbed vertebrates possess gill pouches (similar to those in adult fishes) and a tiny tail. These embryonic similarities indicate that fishes, amphibians, reptiles, birds, and mammals all evolved from a common ancestor. Additional genetic instructions have also evolved, causing their adult morphology to diverge.

Comparative Molecular Biology. The genes and proteins of different species also contain information about evolutionary relationships. The very existence of a common genetic code is powerful evidence for the relatedness of all forms of life. Moreover, some genes and their protein products are present in most living organisms, an observation that is most easily explained by the hypothesis of common ancestry. For example, cytochrome *c*, a protein involved in cellular respiration (see Section 8.4), is found within the mitochondria of all eukaryotic organisms. Evolutionary processes have modified the gene that codes for this protein, establishing variations in its amino acid sequence among different groups of organisms. Closely related species—for example, humans and their fellow primates, chimpanzees and rhesus monkeys—exhibit few differences in the amino acid sequence; more distantly related organisms, such as humans and yeast, exhibit many differences **(Figure 19.14)**.

Some People Misinterpret the Theory of Evolution

The theory of evolution has always been a contentious subject because it challenges deeply held traditional views of how living organisms originated. Many of Darwin's contemporaries were dismayed by the suggestion that all organisms share a common ancestry. Some people even misinterpreted this assertion as "humans evolved from chimpanzees or gorillas." But the theory of evolution makes no such claims. Instead, it suggests that humans and apes are descended from an apelike common ancestor (see Section 30.13). In other words, an ancient population of organisms left descendants, which now include the living species of apes, as well as our own species. Moreover, the theory recognizes that evolution is an ongoing process: humans and apes have been evolving up until this very moment and will continue to evolve for as long as their descendants persist.

Early in the twentieth century, some scientists embraced the notion of **orthogenesis,** or progressive, goal-oriented evolution. This idea, derived from the *Scala Naturae,* suggests that evolution produces new species with the goal of improvement "in mind." We now know that evolution proceeds as an ongoing process of dynamic adjustment, not toward any fixed goal. Natural selection preserves the genes of organisms that function well in particular environments, but it cannot predict future environmental change. Imagine a population of plants with genes that affect how well they function under wet versus dry conditions. After a 5-year drought, the population would include mostly dry-adapted plants. If a series of wet years follows the drought, these plants will be poorly adapted to the altered conditions. The process that favored drought-adapted plants operated under the prevailing dry conditions, not in anticipation of how conditions might change in the future.

Evolution is the core theory of modern biology because its explanatory power touches on every aspect of the living world. And the application of molecular techniques to the study of evolutionary biology has greatly enhanced our knowledge. Despite some common misunderstandings about what the theory predicts, the study of evolution is alive and

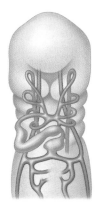
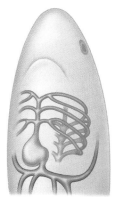

Human embryo   Adult shark

Figure 19.13
Embryologic clues to evolutionary history. Related species often show similar patterns of embryonic development. The aortic arches (red), a two-chambered heart (orange), and a set of veins (blue) in an early human embryo are also present in the embryos of other vertebrates. These structures persist into adulthood in some fishes, such as sharks.

Figure 19.14 Observational Research

How Differences in Amino Acid Sequences among Species Reflect Their Evolutionary Relationships

HYPOTHESIS: The genetic instructions coding for proteins are more similar in closely related species than they are in more distantly related species.

PREDICTION: The amino acid sequences for a particular protein will be more similar in closely related species than in more distantly related species.

OBSERVATIONAL METHODS: Researchers gathered the amino acid sequences for the protein cytochrome *c* from a variety of organisms and compared them with the 104 amino acid sequence of this protein in humans.

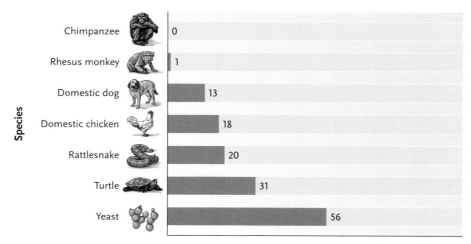

Number of amino acids that differ from the human sequence

RESULTS: Species that are closely related to humans, such as chimpanzees and rhesus monkeys, have amino acid sequences that are identical or nearly identical to the sequence in humans. More distantly related species, such as turtles and yeasts, exhibit sequences that are quite different from the sequence in humans.

CONCLUSION: Closely related species have very similar amino acid sequences in their proteins, reflecting similarities in their genetic makeup. More distantly related species exhibit substantial differences in amino acid sequences, reflecting the genetic divergence among them.

well. In fact, in late 2005, *Science* magazine, a prestigious scientific journal devoted to all of the natural sciences, declared "Evolution in Action" as the breakthrough of the year. The editorial staff cited exciting recent discoveries about genetic differences among organisms ranging from bacteria to humans, mechanisms that promote species formation, and the regulatory genes that may bridge the gap between microevolution and macroevolution.

In the remaining chapters of this unit you will discover how contemporary evolutionary theory explains changes at every level of biological organization from adaptive modifications within populations (see Chapter 20), to the development of new species (see Chapter 21), to the history of life (see Chapter

22), and the classification of all organisms on Earth (see Chapter 23).

STUDY BREAK

1. What two problems slowed the acceptance of Darwin's theory among scientists?
2. What is the difference between microevolution and macroevolution?
3. What types of data provide evidence that evolution has adapted organisms to their environments and promoted the diversification of species?

What determines whether a species adapts to a changing environment or becomes extinct?

Natural selection has produced marvelous adaptations in every species on Earth, and we know that evolutionary adaptation to certain environmental changes has allowed many species to persist. But we also know that more than 99% of the species that have ever lived became extinct, evidently because they failed to adapt to changes in climate, natural competitors or enemies, or other environmental factors. But what kinds of genetic variation are required for adaptation, and what kinds of characteristics must evolve to allow survival? This is a critical question today, because human activities are changing environments so rapidly and drastically that many species face the threat of extinction. Can aquatic species adapt to various kinds of water pollution? Can animals and plants that lived in prairies adapt to different habitats, now that most prairies have been destroyed? Can Arctic species adapt to changes in climate as human production of carbon dioxide increases Earth's average temperature faster than ever before?

Is adaptation by natural selection responsible for most of the genetic differences between species?

New genetic variations sometimes become more common within populations or species because the proteins for which they code are advantageous and preserved by natural selection. But biologists who study molecular evolution have discovered that a large part of the genome in most organisms (about 98% of the human genome, for example) does not code for proteins and therefore appears to have no function. If this observation is generally correct, why do the noncoding parts of genomes exist? Are evolutionary changes in noncoding regions and the differences in noncoding sequences among species adaptive? For example, only about 1% of the DNA base pairs differ between hu-

man and chimpanzee genomes—but this amounts to about 34 million base-pair differences altogether, at least 60,000 of which alter the amino acid sequences of proteins. How can we determine which of these differences are adaptive and which differences underlie the unique characteristics of humans?

How do pathways of embryonic development evolve?

The characteristics of adult organisms are the product of developmental events, starting with the fertilized egg, that include growth in size, changes in the shape of various body parts, and the differentiation of cell types. These processes are largely controlled by genes, with input from the environment. Although biologists are beginning to learn how the genetic foundations of developmental processes evolve, many questions remain. For example, how do genetic changes induce differences in the branching patterns of antlers among species of deer, or differences in the length of the tails of monkeys and apes (including humans), or differences in the number and size of scales among species of lizards? We know that the proteins forming the lens of the eye are actually enzymes that play different roles in other cells, and that they have been "recruited" to form the lens, but what mechanisms induce them to assume this new role? And why do different enzymes form the lens in eyes of birds and mammals? Evolutionary developmental biology, which is discussed in Chapter 22, is one of the most active, exciting fields in biology at this time.

Douglas J. Futuyma is Distinguished Professor in the Department of Evolution and Ecology at Stony Brook University. His research interests focus on speciation and the evolution of ecological interactions among species, and in particular on insect–plant interactions. Learn more about his work at http://life.bio.sunysb.edu/ee/people/futuyindex.html.

Review

Go to **Thomson**NOW™ at www.thomsonedu.com/login to access quizzing, animations, exercises, articles, and personalized homework help.

19.1 Recognition of Evolutionary Change

- Ancient Greek philosophers classified the natural world, ranking inanimate objects and living organisms from simple to complex.

- Natural theologians, who merged Greek philosophy with the biblical account of creation, believed that all species were specially created and perfectly adapted. Existing species could not change or become extinct, and new species could not arise. Studies in biogeography, comparative morphology, and paleontology led scientists to wonder whether species might change through time (Figures 19.2 and 19.3).

- Lamarck developed the first comprehensive theory of biological evolution; he proposed that species evolved into more complex forms that functioned better in their environments. He hypothesized that structures in an organism changed when they were used, and that those changes were inherited by the organism's offspring. Experiments have refuted Lamarck's proposed mechanisms.

- Two geologists, Hutton and Lyell, recognized that major features on Earth were created by the long-term action of the very

slow geological processes that scientists observe today. Their insights suggested that Earth was much older than natural theologians had supposed.

19.2 Darwin's Journeys

- Darwin's observations during his voyage on the *Beagle* provided much of the data and inspiration for the development of his theory of evolution (Figures 19.5–19.8).

- Darwin based the theory of evolution by means of natural selection on three inferences: (1) individuals within a population compete for limited resources, (2) hereditary characteristics allow some individuals to survive longer and reproduce more than others, and (3) a population's characteristics change over time as advantageous heritable characteristics become more common (Table 19.1).

- Darwin also proposed that the accumulation of differences fostered by natural selection could cause populations to diverge over time. Such evolutionary divergence can lead to the production of new species, which can, in turn, give rise to new evolutionary lineages (Figures 19.9 and 19.10).

Animation: The Galápagos
Animation: Finches of the Galpágos

19.3 Evolutionary Biology since Darwin

- Scientists working in population genetics developed theories of evolutionary change by integrating Darwin's ideas with Mendel's research on genetics.

- In the 1930s and 1940s, the modern synthesis provided a unified view of evolution that drew on studies from many biological disciplines. It emphasized evolution within populations, the central role of variation in the evolutionary process, and the gradualism of evolutionary change.

- Studies of adaptation, the fossil record, historical biogeography, comparative morphology, comparative embryology, and comparative molecular biology provide compelling evidence of evolutionary change (Figures 19.11–19.14).

- Evolutionary biology is an active field of study, and the application of molecular techniques is yielding new answers to old questions.

Questions

Self-Test Questions

1. Which of the following statements about evolutionary studies is *not* true?
   a. Biologists study the products of evolution to understand the processes causing it.
   b. Biologists design molecular experiments to examine evolutionary processes operating over short time periods.
   c. Biologists study the inheritance of characteristics that a parent acquired during its lifetime.
   d. Biologists study variation in homologous structures among related organisms.
   e. Biologists examine why a huge variety of species may inhabit a small island cluster.

2. Which of the following ideas is *not* included in Darwin's theory?
   a. All organisms that have ever existed arose through evolutionary modifications of ancestral species.
   b. The great variety of species alive today resulted from the diversification of ancestral species.
   c. Natural selection drives some evolutionary change.
   d. Natural selection preserves favorable traits.
   e. Natural selection eliminates adaptive traits.

3. The father of taxonomy is:
   a. Charles Darwin.
   b. Charles Lyell.
   c. Alfred Wallace.
   d. Carolus Linnaeus.
   e. Jean Baptiste de Lamarck.

4. The wings of birds, the legs of pigs, and the flippers of whales provide an example of:
   a. vestigial structures.
   b. homologous structures.
   c. acquired characteristics.
   d. artificial selection.
   e. uniformitarianism.

5. Which of the following statements is *not* compatible with Darwin's theory?
   a. All organisms have arisen by descent with modification.
   b. Evolution has altered and diversified ancestral species.
   c. Evolution occurs in individuals rather than in groups.
   d. Natural selection eliminates unsuccessful variations.
   e. Evolution occurs because some individuals function better than others in a particular environment.

6. Which of the following does *not* contribute to the study of evolution?
   a. population genetics
   b. inheritance of acquired characteristics
   c. the fossil record
   d. DNA sequencing
   e. comparative morphology

7. Which of the following could be an example of microevolution?
   a. a slight change in a bird population's color due to a small genetic change in the population
   b. large differences between fossils found near the ground surface and those found in deep rock layers
   c. the sudden disappearance of an entire genus
   d. the direct evolutionary link between living primates and humans
   e. a flood that drowns all members of a population

8. Which of the following ideas proposed by Lamarck was *not* included in Darwin's theory?
   a. Organisms change in response to their environments.
   b. Changes that an organism acquires during its lifetime are passed to its offspring.
   c. All species change with time.
   d. Genetic changes may be passed from one generation to the next.
   e. Specific mechanisms cause evolutionary change.

9. Medical advances now allow many people who suffer from genetic diseases to survive and reproduce. These advances:
   a. refute Darwin's theory.
   b. support Lamarck's theory.
   c. disprove descent with modification.
   d. reduce the effects of natural selection.
   e. eliminate adaptive traits.

10. The belief that evolution is progressive or goal-oriented is called:
    a. gradualism.
    b. uniformitarianism.
    c. taxonomy.
    d. orthogenesis.
    e. the modern synthesis.

Questions for Discussion

1. Explain why the characteristics we see in living organisms adapt them to the environments in which their ancestors lived rather than to the environments in which they live today.

2. Imagine a population of mice that includes both brown and black individuals. They live in a habitat with brown soil, where predatory hawks can see black mice more easily than they can see brown ones. Design a study that would allow you to determine whether the brown mice are better adapted to this environment than black mice.

3. Find examples from popular publications or advertisements for consumer products that misrepresent the theory of biological evolution. Explain how the theory is misrepresented.

Experimental Analysis

Design an experiment to test Lamarck's hypothesis that characteristics acquired during an organism's lifetime are inherited by their offspring. (You may wish to review the components of a well-designed experiment in Chapter 1 before formulating your answer.) Can you think of examples of acquired characteristics that are *not* inherited by offspring?

Evolution Link

Identify three discoveries or inventions that have changed how humans are affected by natural selection. Describe in detail how each discovery influences survival or reproduction in our species.

How Would You Vote?

A large asteroid could obliterate civilization and much of Earth's biodiversity. Should nations around the world contribute to locating and tracking asteroids? Go to www.thomsonedu.com/login to investigate both sides of the issue and then vote.

Phenotypic variation. The frog *Dendrobates pumilio* exhibits dramatic color variation in populations that inhabit the Bocas del Toro Islands, Panama.

© Mark Moffett/Foto Natura/Minden Pictures

STUDY PLAN

20.1 Variation in Natural Populations

Evolutionary biologists describe and quantify phenotypic variation

Phenotypic variation can have genetic and environmental causes

Several processes generate genetic variation

Populations often contain substantial genetic variation

20.2 Population Genetics

All populations have a genetic structure

The Hardy-Weinberg principle is a null model that defines how evolution does not occur

20.3 The Agents of Microevolution

Mutations create new genetic variations

Gene flow introduces novel genetic variants into populations

Genetic drift reduces genetic variability within populations

Natural selection shapes genetic variability by favoring some traits over others

Sexual selection often exaggerates showy structures in males

Nonrandom mating can influence genotype frequencies

20.4 Maintaining Genetic and Phenotypic Variation

Diploidy can hide recessive alleles from the action of natural selection

Natural selection can maintain balanced polymorphisms

Some genetic variations may be selectively neutral

20.5 Adaptation and Evolutionary Constraints

Scientists construct hypotheses about the evolution of adaptive traits

Several factors constrain adaptive evolution

20 Microevolution: Genetic Changes within Populations

WHY IT MATTERS

On November 28, 1942, at the height of American involvement in World War II, a disastrous fire killed more than 400 people in Boston's Cocoanut Grove nightclub. Many more would have died later but for a new experimental drug, penicillin. A product of *Penicillium* mold, penicillin fought the usually fatal infections of *Staphylococcus aureus,* a bacterium that enters the body through damaged skin. Penicillin was the first antibiotic drug based on a naturally occurring substance that kills bacteria.

Until the disaster at the Cocoanut Grove, the production and use of penicillin had been a closely guarded military secret. But after its public debut, the pharmaceutical industry hailed penicillin as a wonder drug, promoting its use for the treatment of the many diseases caused by infectious microorganisms. Penicillin became widely available as an over-the-counter remedy, and Americans dosed themselves with it, hoping to cure all sorts of ills **(Figure 20.1).** But in 1945, Alexander Fleming, the scientist who discovered penicillin, predicted that some bacteria could survive low doses, and that the offspring of those germs would be more resistant to its effects. In 1946—just 4 years after penicillin's use in Boston—14% of the *Staphylococcus* strains

Figure 20.1
Selling penicillin. This ad, from a 1944 issue of *Life* magazine, credits penicillin with saving the lives of wounded soldiers.

microorganisms—along with the genes that confer antibiotic resistance—become more common in later generations. In other words, bacterial strains adapt to antibiotics through the evolutionary process of selection. Our use of antibiotics is comparable to artificial selection by plant and animal breeders (see Chapter 19), but when we use antibiotics, we inadvertently select for the success of organisms that we are trying to eradicate.

The evolution of antibiotic resistance in bacteria is an example of **microevolution**, which is a heritable change in the genetics of a population. A **population** of organisms includes all the individuals of a single species that live together in the same place and time. Today, when scientists study microevolution, they analyze variation—the differences between individuals—in natural populations and determine how and why these variations are inherited. Darwin recognized the importance of heritable variation within populations; he also realized that natural selection can change the pattern of variation in a population from one generation to the next. Scientists have since learned that microevolutionary change results from several processes, not just natural selection, and that sometimes these processes counteract each other.

In this chapter, we first examine the extensive variation that exists within natural populations. We then take a detailed look at the most important processes that alter genetic variation within populations, causing microevolutionary change. Finally, we consider how microevolution can fine-tune the functioning of populations within their environments.

isolated from patients in a London hospital were resistant. By 1950, more than half the strains were resistant.

Scientists and physicians have discovered numerous antibiotics since the 1940s, and many strains of bacteria have developed resistance to these drugs. In fact, according to the Centers for Disease Control and Prevention, between 30,000 and 40,000 Americans die each year from infection by antibiotic-resistant bacteria.

How do bacteria become resistant to antibiotics? The genomes of bacteria—like those of all other organisms—vary among individuals, and some bacteria have genetic traits that allow them to withstand attack by antibiotics. When we administer antibiotics to an infected patient, we create an environment favoring bacteria that are even slightly resistant to the drug. The surviving bacteria reproduce, and resistant

20.1 Variation in Natural Populations

In some species, individuals vary dramatically in appearance; but in most species, the members of a population look pretty much alike **(Figure 20.2)**. Even those that look alike, such as the *Cerion* snails on the right in Figure 20.2, are not identical, however. With a scale and ruler, you could detect differences in their weight as well as in

a. European garden snails

b. Bahaman land snails

Figure 20.2
Phenotypic variation. **(a)** Shells of the European garden snail *(Cepaea nemoralis)* from a population in Scotland vary considerably in appearance. **(b)** By contrast, shells of *Cerion christophei* from a population in the Bahamas look very similar.

the length and diameter of their shells. With suitable techniques, you could also document variations in their individual biochemistry, physiology, internal anatomy, and behavior. All of these are examples of **phenotypic variation**, differences in appearance or function that are passed from generation to generation.

Evolutionary Biologists Describe and Quantify Phenotypic Variation

Darwin's theory recognized the importance of heritable phenotypic variation, and today, microevolutionary studies often begin by assessing phenotypic variation within populations. Most characters exhibit **quantitative variation**: individuals differ in small, incremental ways. If you weighed everyone in your biology class, for example, you would see that weight varies almost continuously from your lightest to your heaviest classmate. Humans also exhibit quantitative variation in the length of their toes, the number of hairs on their heads, and their height, as discussed in Chapter 12.

We usually display data on quantitative variation in a bar graph or, if the sample is large enough, as a curve **(Figure 20.3).** The width of the curve is proportional to the variability—the amount of variation—among individuals, and the *mean* describes the average value of the character. As you will see shortly, natural selection often changes the mean value of a character or its variability within populations.

Other characters, like those Mendel studied (see Section 12.1), exhibit **qualitative variation:** they exist in two or more discrete states, and intermediate forms are often absent. Snow geese, for example, have *either* blue *or* white feathers **(Figure 20.4).** The existence of discrete variants of a character is called a **polymorphism** (*poly* = many; *morphos* = form); we describe such traits as *polymorphic.* The *Cepaea nemoralis* snail shells in Figure 20.2a are polymorphic in background color, number of stripes, and color of stripes. Biochemical polymorphisms, like the human A, B, AB, and O blood groups (described in Section 12.2), are also common.

We describe phenotypic polymorphisms quantitatively by calculating the percentage or *frequency* of each trait. For example, if you counted 123 blue snow geese and 369 white ones in a population of 492 geese, the frequency of the blue phenotype would be 123/492 or 0.25, and the frequency of the white phenotype would be 369/492 or 0.75.

Phenotypic Variation Can Have Genetic and Environmental Causes

Phenotypic variation within populations may be caused by genetic differences between individuals, by differences in the environmental factors that individuals experience, or by an interaction between genetics and the environment. As a result, genetic and pheno-

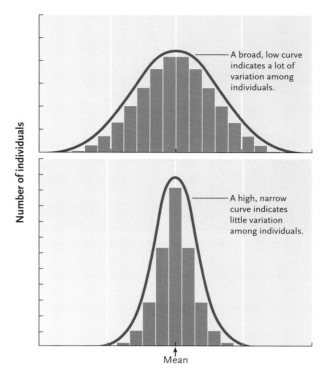

Figure 20.3

Quantitative variation. Many traits vary continuously among members of a population, and a bar graph of the data often approximates a bell-shaped curve. The mean defines the average value of the trait in the population, and the width of the curve is proportional to the variability among individuals.

typic variations may not be perfectly correlated. Under some circumstances, organisms with different genotypes exhibit the same phenotype. For example, the black coloration of some rock pocket mice from Arizona is caused by certain mutations in the *Mc1r* gene (see Section 1.2); but black mice from New Mexico do not share those mutations—that is, they have different genotypes—even though they exhibit the same phenotype. On the other hand, organisms with the same genotype sometimes exhibit different phenotypes. For

Figure 20.4

Qualitative variation. Individual snow geese *(Chen caerulescens)* are either blue or white. Although both colors are present in many populations, geese tend to associate with others of the same color.

Figure 20.5

Environmental effects on phenotype. Soil acidity affects the expression of the gene controlling flower color in the common garden plant *Hydrangea macrophylla*. When grown in acid soil, it produces deep blue flowers. In neutral or alkaline soil, its flowers are bright pink.

example, the acidity of soil influences flower color in some plants **(Figure 20.5).**

Knowing whether phenotypic variation is caused by genetic differences, environmental factors, or an interaction of the two is important because *only genetically based variation is subject to evolutionary change.* Moreover, knowing the causes of phenotypic variation has important practical applications. Suppose, for example, that one field of wheat produced more grain than another. If a difference in the availability of nutrients or water caused the difference in yield, a farmer might choose to fertilize or irrigate the less productive field. But if the difference in productivity resulted from genetic differences between plants in the two fields, a farmer might plant only the more productive genotype. Because environmental factors can influence the expression of genes, an organism's phenotype is frequently the product of an interaction between its genotype and its environment. In our hypothetical example, the farmer may maximize yield by fertilizing and irrigating the better genotype of wheat.

How can we determine whether phenotypic variation is caused by environmental factors or by genetic differences? We can test for an environmental cause experimentally by changing one environmental variable and measuring the effects on genetically similar subjects. You can try this yourself by growing some cuttings from an ivy plant in shade and other cuttings from the same plant in full sun. Although they all have the same genotype, the cuttings grown in sun will produce smaller leaves and shorter stems.

Breeding experiments can demonstrate the genetic basis of phenotypic variation. For example, Mendel inferred the genetic basis of qualitative traits, such as flower color in peas, by crossing plants with different phenotypes. Moreover, traits that vary quantitatively will respond to artificial selection only if the variation has some genetic basis. For example, re-

searchers observed that individual house mice (*Mus domesticus*) differ in activity levels, as measured by how much they use an exercise wheel and how fast they run. John G. Swallow, Patrick A. Carter, and Theodore Garland, Jr., then at the University of Wisconsin at Madison, used artificial selection to produce lines of mice that exhibit increased wheel-running behavior, demonstrating that the observed differences in these two aspects of activity level have a genetic basis **(Figure 20.6).**

Breeding experiments are not always practical, however, particularly for organisms with long generation times. Ethical concerns also render these techniques unthinkable for humans. Instead, researchers sometimes study the inheritance of particular traits by analyzing genealogical pedigrees, as discussed in Section 13.2, but this approach often provides poor results for analyses of complex traits.

Several Processes Generate Genetic Variation

Genetic variation, the raw material molded by microevolutionary processes, has two potential sources: the production of new alleles and the rearrangement of existing alleles. Most new alleles probably arise from small scale mutations in DNA (described later in this chapter). The rearrangement of existing alleles into new combinations can result from larger scale changes in chromosome structure or number and from several forms of genetic recombination, including crossing over between homologous chromosomes during meiosis, the independent assortment of nonhomologous chromosomes during meiosis, and random fertilizations between genetically different sperm and eggs.

The shuffling of *existing* alleles into new combinations can produce an extraordinary number of novel genotypes and phenotypes in the next generation. By one estimate, more than 10^{600} combinations of alleles are possible in human gametes, yet there are fewer than 10^{10} humans alive today. So unless you have an identical twin, it is extremely unlikely that another person with your genotype has ever lived or ever will.

Populations Often Contain Substantial Genetic Variation

How much genetic variation actually exists within populations? In the 1960s, evolutionary biologists began to use gel electrophoresis (see Figure 18.7) to identify biochemical polymorphisms in diverse organisms. This technique separates two or more forms of a given protein if they differ significantly in shape, mass, or net electrical charge. The identification of a protein polymorphism allows researchers to infer genetic variation at the locus coding for that protein.

Figure 20.6　Experimental Research

Using Artificial Selection to Demonstrate That Activity Level in Mice Has a Genetic Basis

QUESTION: Do observed differences in activity level among house mice have a genetic basis?

EXPERIMENT: Swallow, Carter, and Garland knew that a phenotypic character responds to artificial selection only if it has a genetic, rather than an environmental, basis. In an experiment with house mice *(Mus domesticus)*, they selected for the phenotypic character of increased wheel-running activity. In four experimental lines, they bred those mice that ran the most. Four other lines, in which breeders were selected at random with respect to activity level, served as controls.

RESULTS: After 10 generations of artificial selection, mice in the experimental lines ran longer distances and ran faster than mice in the control lines. Thus, artificial selection on wheel-running activity in house mice increased **(a)** the distance that mice run per day and **(b)** their average speed. The data illustrate responses of females in four experimental lines and four control lines. Males showed similar responses.

a. Distance run

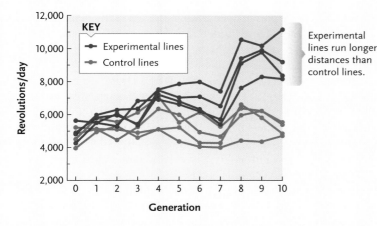

b. Average speed

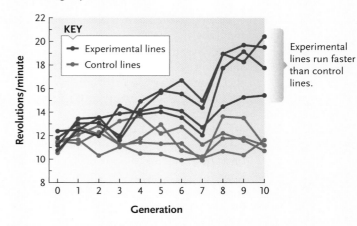

CONCLUSION: Because two measures of activity level responded to artificial selection, researchers concluded that variation in this behavioral character has a genetic basis.

Researchers discovered much more genetic variation than anyone had imagined. For example, nearly half the loci surveyed in many populations of plants and invertebrates are polymorphic. Moreover, gel electrophoresis actually underestimates genetic variation because it doesn't detect different amino acid substitutions if the proteins for which they code migrate at the same rate.

Advances in molecular biology now allow scientists to survey genetic variation directly, and researchers have accumulated an astounding knowledge of the structure of DNA and its nucleotide sequences. In general, studies of chromosomal and mitochondrial DNA suggest that every locus exhibits some variability in its nucleotide sequence. The variability is apparent in comparisons of individuals from a single population, populations of one species, and related species. However, some variations detected in the protein-coding regions of DNA may not affect phenotypes because, as explained on page 426, they do not change the amino acid sequences of the proteins for which the genes code.

STUDY BREAK

1. If a population of skunks includes some individuals with stripes and others with spots, would you describe the variation as quantitative or qualitative?
2. In the experiment on house mice described in Figure 20.6, how did researchers demonstrate that variations in activity level had a genetic basis?
3. What factors contribute to phenotypic variation in a population?

20.2　Population Genetics

To predict how certain factors may influence genetic variation, population geneticists first describe the genetic structure of a population. They then create hypotheses, which they formalize in mathematical models, to describe how evolutionary processes may change the genetic structure under specified conditions. Finally, researchers test the predictions of these models to evaluate the ideas about evolution that are embodied within them.

All Populations Have a Genetic Structure

Populations are made up of individuals, each with its own genotype. In diploid organisms, which have pairs of homologous chromosomes, an individual's genotype includes two alleles at every gene locus. The sum of all alleles at all gene loci in all individuals is called the population's **gene pool.**

To describe the structure of a gene pool, scientists first identify the genotypes in a representative sample and calculate **genotype frequencies**, the percentages of individuals possessing each genotype. Knowing that each diploid organism has two alleles (either two copies of the same allele or two different alleles) at each gene locus, a scientist can then calculate **allele frequencies**, the relative abundances of the different alleles. For a locus with two alleles, scientists use the symbol p to identify the frequency of one allele, and q the frequency of the other.

The calculation of genotype and allele frequencies for the two alleles at the gene locus governing flower color in snapdragons (genus *Antirrhinum*) is straightforward **(Table 20.1)**. This locus is easy to study because it exhibits incomplete dominance (see Section 12.2). Individuals that are homozygous for the C^R allele ($C^R C^R$) have red flowers; those homozygous for the C^W allele ($C^W C^W$) have white flowers; and heterozygotes ($C^R C^W$) have pink flowers. Genotype frequencies represent how the C^R and C^W alleles are distributed among individuals. In this example, examination of the plants reveals that 45% of individuals have the $C^R C^R$ genotype, 50% have the heterozygous $C^R C^W$

genotype, and the remaining 5% have the $C^W C^W$ genotype. Allele frequencies represent the commonness or rarity of each allele in the gene pool. As calculated in the table, 70% of the alleles in the population are C^R and 30% are C^W. Remember that for a gene locus with two alleles, there are three genotype frequencies, but only two allele frequencies (p and q). The sum of the three genotype frequencies must equal 1; so must the sum of the two allele frequencies.

The Hardy-Weinberg Principle Is a Null Model That Defines How Evolution Does Not Occur

When designing experiments, scientists often use control treatments to evaluate the effect of a particular factor: the control tells us what we would see if the experimental treatment had no effect. As you may recall from the hypothetical example presented in Chapter 1 (see Figure 1.14), to determine whether fertilizer has an effect on plant growth, you must compare the growth of fertilized plants (the experimental treatment) to the growth of plants that received no fertilizer (the control treatment). However, in studies that use observational rather than experimental data, there is often no suitable control. In such cases, investigators develop conceptual models, called **null models**, which predict what they would see if a particular factor had no effect. Null models serve as theoretical reference points against which observations can be evaluated.

Early in the twentieth century, geneticists were puzzled by the persistence of recessive traits because they assumed that natural selection replaced recessive or rare alleles with dominant or common ones. An English mathematician, G. H. Hardy, and a German physician, Wilhelm Weinberg, tackled this problem independently in 1908. Their analysis, now known as the

| Table 20.1 | Calculation of Genotype Frequencies and Allele Frequencies for the Snapdragon Flower Color Locus |

Because each diploid individual has two alleles at each gene locus, the entire sample of 1000 individuals has a total of 2000 alleles at the C locus.

Flower Color Phenotype	Genotype	Number of Individuals	Genotype Frequency[1]	Total Number of C^R Alleles[2]	Total Number of C^W Alleles[2]
Red	$C^R C^R$	450	450/1000 = 0.45	2 × 450 = 900	0 × 450 = 0
Pink	$C^R C^W$	500	500/1000 = 0.50	1 × 500 = 500	1 × 500 = 500
White	$C^W C^W$	50	50/1000 = 0.05	0 × 50 = 0	2 × 50 = 100
	Total	1000	0.45 + 0.50 + 0.05 = 1.0	1400	600

Calculate allele frequencies using the total of 1400 + 600 = 2000 alleles in the sample:

$$p = \text{frequency of } C^R \text{ allele} = 1400/2000 = 0.7$$
$$q = \text{frequency of } C^W \text{ allele} + 600/2000 = 0.3$$
$$p + q = 0.7 + 0.3 = 1.0$$

[1]Genotype frequency = the number of individuals possessing a particular genotype divided by the total number of individuals in the sample.
[2]Total number of C^R or C^W alleles = the number of C^R or C^W alleles present in one individual with a particular genotype multiplied by the number of individuals with that genotype.

Hardy-Weinberg principle, specifies the conditions under which a population of diploid organisms achieves **genetic equilibrium**, the point at which neither allele frequencies nor genotype frequencies change in succeeding generations. Their work also showed that dominant alleles need not replace recessive ones, and that the shuffling of genes in sexual reproduction does not in itself cause the gene pool to change.

The Hardy-Weinberg principle is a mathematical model that describes how genotype frequencies are established in sexually reproducing organisms. According to this model, genetic equilibrium is possible only if *all* of the following conditions are met:

1. No mutations are occurring.
2. The population is closed to migration from other populations.
3. The population is infinite in size.
4. All genotypes in the population survive and reproduce equally well.
5. Individuals in the population mate randomly with respect to genotypes.

If the conditions of the model are met, the allele frequencies of the population will never change, and the genotype frequencies will stop changing after one generation. In short, under these restrictive conditions, microevolution will *not* occur (see *Focus on Research*). The Hardy-Weinberg principle is thus a null model that serves as a reference point for evaluating the circumstances under which evolution *may* occur.

If a population's genotype frequencies do not match the predictions of this model or if its allele frequencies change over time, microevolution may be occurring. Determining which of the model's conditions are not met is a first step in understanding how and why the gene pool is changing. Natural populations never fully meet all five requirements simultaneously, but they often come pretty close.

STUDY BREAK

1. What is the difference between the genotype frequencies and the allele frequencies in a population?
2. Why is the Hardy-Weinberg principle considered a null model of evolution?
3. If the conditions of the Hardy-Weinberg principle are met, when will genotype frequencies stop changing?

20.3 The Agents of Microevolution

A population's allele frequencies will change over time if conditions of the Hardy-Weinberg model are violated. The processes that foster microevolutionary

Table 20.2 **Agents of Microevolutionary Change**

Agent	Definition	Effect on Genetic Variation
Mutation	A heritable change in DNA	Introduces new genetic variation into population
Gene flow	Change in allele frequencies as individuals join a population and reproduce	May introduce genetic variation from another population
Genetic drift	Random changes in allele frequencies caused by chance events	Reduces genetic variation, especially in small populations; can eliminate alleles
Natural selection	Differential survivorship or reproduction of individuals with different genotypes	One allele can replace another or allelic variation can be preserved
Nonrandom mating	Choice of mates based on their phenotypes and genotypes	Does not directly affect allele frequencies, but usually prevents genetic equilibrium

change—which include mutation, gene flow, genetic drift, natural selection, and nonrandom mating—are summarized in **Table 20.2.**

Mutations Create New Genetic Variations

A **mutation** is a spontaneous and heritable change in DNA. Mutations are rare events; during any particular breeding season, between one gamete in 100,000 and one in 1 million will include a new mutation at a particular gene locus. New mutations are so infrequent, in fact, that they exert little or no immediate effect on allele frequencies in most populations. But over evolutionary time scales, their numbers are significant—mutations have been accumulating in biological lineages for billions of years. And because it is a mechanism through which entirely new genetic variations arise, *mutation is a major source of heritable variation.*

For most animals, only mutations in the germ line (the cell lineage that produces gametes) are heritable; mutations in other cell lineages have no direct effect on the next generation. In plants, however, mutations may occur in meristem cells, which eventually produce flowers as well as nonreproductive structures (see Chapter 31); in such cases, a mutation may be passed to the next generation and ultimately influence the gene pool.

Deleterious mutations alter an individual's structure, function, or behavior in harmful ways. In mammals, for example, a protein called collagen is an essential component of most extracellular structures. Several simple mutations in humans cause forms of Ehlers-Danlos syndrome, a disruption of collagen synthesis that may result in loose skin, weak joints, or sudden death from the rupture of major blood vessels, the colon, or the uterus.

By definition, *lethal mutations* cause the death of organisms carrying them. If a lethal allele is dominant, both homozygous and heterozygous carriers suffer

FOCUS ON RESEARCH

Basic Research: Using the Hardy-Weinberg Principle

To see how the Hardy-Weinberg principle can be applied, we will analyze the snapdragon flower color locus, using the hypothetical population of 1000 plants described in Table 20.1. This locus includes two alleles—C^R (with its frequency designated as p) and C^W (with its frequency designated as q)—and three genotypes—homozygous C^RC^R, heterozygous C^RC^W, and homozygous C^WC^W. Table 20.1 lists the number of plants with each genotype: 450 have red flowers (C^RC^R), 500 have pink flowers (C^RC^W), and 50 have white flowers (C^WC^W). It also shows the calculation of both the genotype frequencies ($C^RC^R = 0.45$, $C^RC^W = 0.50$, and $C^WC^W = 0.05$) and the allele frequencies ($p = 0.7$ and $q = 0.3$) for the population.

Let's assume for simplicity that each individual produces only two gametes and that both gametes contribute to the production of offspring. This assumption is unrealistic, of course, but it meets the Hardy-Weinberg requirement that all individuals in the population contribute equally to the next generation. In each parent, the two alleles segregate and end up in different gametes:

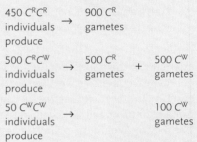

450 C^RC^R individuals produce → 900 C^R gametes

500 C^RC^W individuals produce → 500 C^R gametes + 500 C^W gametes

50 C^WC^W individuals produce → 100 C^W gametes

You can readily see that 1400 of the 2000 total gametes carry the C^R allele and 600 carry the C^W allele. The frequency of C^R gametes is 1400/2000 or 0.7, which is equal to p; the frequency of C^W gametes is 600/2000 or 0.3, which is equal to q. Thus, the allele frequencies in the gametes are exactly the same as the allele frequencies in the parent generation—it could not be

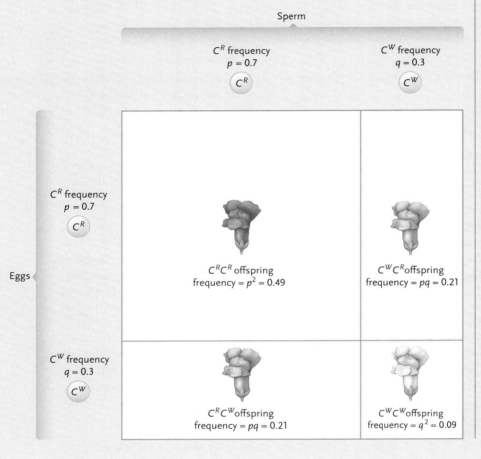

Sperm

C^R frequency $p = 0.7$ — C^R

C^W frequency $q = 0.3$ — C^W

Eggs

C^R frequency $p = 0.7$ — C^R

C^W frequency $q = 0.3$ — C^W

C^RC^R offspring frequency = $p^2 = 0.49$

C^WC^R offspring frequency = $pq = 0.21$

C^RC^W offspring frequency = $pq = 0.21$

C^WC^W offspring frequency = $q^2 = 0.09$

from its effects; if recessive, it affects only homozygous recessive individuals. A lethal mutation that causes death before the individual reproduces is eliminated from the population.

Neutral mutations are neither harmful nor helpful. Recall from Section 15.1 that in the construction of a polypeptide chain, a particular amino acid can be specified by several different codons. As a result, some DNA sequence changes—especially certain changes at the third nucleotide of the codon—do not alter the amino acid sequence. Not surprisingly, mutations at the third position appear to persist longer in populations than those at the first two positions. Other mutations may change an organism's phenotype without influencing its survival and reproduction. A neutral mutation might even be beneficial later if the environment changes.

Sometimes a change in DNA produces an *advantageous mutation,* which confers some benefit on an individual that carries it. However slight the advantage, natural selection may preserve the new allele and even increase its frequency over time. Once the mutation has been passed to a new generation, other agents of microevolution determine its long-term fate.

Gene Flow Introduces Novel Genetic Variants into Populations

Organisms or their gametes (for example, pollen) sometimes move from one population to another. If the immigrants reproduce, they may introduce novel alleles into the population they have joined. This phenomenon, called **gene flow**, violates the Hardy-Weinberg requirement that populations must be closed to migration.

otherwise because each gamete carries one allele at each locus.

Now assume that these gametes, both sperm and eggs, encounter each other at random. In other words, individuals reproduce without regard to the genotype of a potential mate. We can visualize the process of random mating in the mating table on the left.

We can also describe the consequences of random mating—$(p + q)$ sperm fertilizing $(p + q)$ eggs—with an equation that predicts the genotype frequencies in the offspring generation:

$$(p + q) \times (p + q) = p^2 + 2pq + q^2$$

If the population is at genetic equilibrium for this locus, p^2 is the predicted frequency of the $C^R C^R$ genotype, $2pq$ the predicted frequency of the $C^R C^W$ genotype, and q^2 the predicted frequency of the $C^W C^W$ genotype. Using the gamete frequencies determined above, we can calculate the predicted genotype frequencies in the next generation:

frequency of $C^R C^R =$
$$p^2 = (0.7 \times 0.7) = 0.49$$

frequency of $C^R C^W =$
$$2pq = 2(0.7 \times 0.3) = 0.42$$

frequency of $C^W C^W =$
$$q^2 = (0.3 \times 0.3) = 0.09$$

Notice that the predicted genotype frequencies in the offspring generation have changed from those in the parent generation: the frequency of heterozygous individuals has decreased, and the frequencies of both types of homozygous individuals have increased. This result occurred because the starting population was *not already* in equilibrium at this gene locus. In other words, the distribution of parent genotypes did not conform to the predicted $p^2 + 2pq + q^2$ distribution.

The 2000 gametes in our hypothetical population produced 1000 offspring. Using the genotype frequencies we just calculated, we can predict how many offspring will carry each genotype:

490 red ($C^R C^R$)
420 pink ($C^R C^W$)
90 white ($C^W C^W$)

In a real study, we would examine the offspring to see how well their numbers match these predictions.

What about the allele frequencies in the offspring? The Hardy-Weinberg principle predicts that they did not change. Let's calculate them and see. Using the method shown in Table 20.1 and the prime symbol (') to indicate offspring allele frequencies:

$$p' = ([2 \times 490] + 420)/2000 =$$
$$1400/2000 = 0.7$$

$$q' = ([2 \times 90] + 420)/2000 =$$
$$600/2000 = 0.3$$

You can see from this calculation that the allele frequencies did not change from one generation to the next, even though the alleles were rearranged to produce different proportions of the three genotypes. Thus, the population is now at genetic equilibrium for the flower color locus; neither the genotype frequencies nor the allele frequencies will change in succeeding generations as long as the population meets the conditions specified in the Hardy-Weinberg model.

To verify this, you can calculate the allele frequencies of the gametes for this offspring generation and predict the genotype frequencies and allele frequencies for a third generation. You could continue calculating until you ran out of either paper or patience, but these frequencies will not change.

Researchers use calculations like these to determine whether an actual population is near its predicted genetic equilibrium for one or more gene loci. When they discover that a population is not at equilibrium, they infer that microevolution is occurring and can investigate the factors that might be responsible.

Gene flow is common in some animal species. For example, young male baboons typically move from one local population to another after experiencing aggressive behavior by older males. And many marine invertebrates disperse long distances as larvae carried by ocean currents.

Dispersal agents, such as pollen-carrying wind or seed-carrying animals, are responsible for gene flow in most plant populations. For example, blue jays foster gene flow among populations of oaks by carrying acorns from nut-bearing trees to their winter caches, which may be as much as a mile away **(Figure 20.7)**. Transported acorns that go uneaten may germinate and contribute to the gene pool of a neighboring oak population.

Documenting gene flow among populations is not always easy, particularly if it occurs infrequently. Researchers can use phenotypic or genetic markers to

Figure 20.7

Gene flow. Blue jays *(Cyanocitta cristata)* serve as agents of gene flow for oaks (genus *Quercus*) when they carry acorns from one oak population to another. An uneaten acorn may germinate and contribute to the gene pool of the population into which it was carried.

identify immigrants in a population, but they must also demonstrate that immigrants reproduced, thereby contributing to the gene pool of their adopted population. In the San Francisco Bay area, for example, Bay checkerspot butterflies *(Euphydryas editha bayensis)* rarely move from one population to another because they are poor fliers (see Figure 53.16). When adult females do change populations, it is often late in the breeding season, and their offspring have virtually no chance of finding enough food to mature. Thus, many immigrant females do not foster gene flow because they do not contribute to the gene pool of the population they join.

The evolutionary importance of gene flow depends upon the degree of genetic differentiation between populations and the rate of gene flow between them. If two gene pools are very different, a little gene flow may increase genetic variability within the population that receives immigrants, and it will make the two populations more similar. But if populations are already genetically similar, even lots of gene flow will have little effect.

Genetic Drift Reduces Genetic Variability within Populations

Chance events sometimes cause the allele frequencies in a population to change unpredictably. This phenomenon, known as **genetic drift**, has especially dramatic effects on small populations, which clearly violate the Hardy-Weinberg assumption of infinite population size.

A simple analogy clarifies why genetic drift is more pronounced in small populations than in large ones. When individuals reproduce, male and female gametes often pair up randomly, as though the allele in any particular sperm or ovum was determined by a coin toss. Imagine that "heads" specifies the *R* allele and "tails" specifies the *r* allele. If the two alleles are equally common (that is, their frequencies, *p* and *q*, are both equal to 0.5), heads should be as likely an outcome as tails. But if you toss the coin 20 or 30 times to simulate random mating in a small population, you won't often see a 50-50 ratio of heads and tails. Sometimes heads will predominate and sometimes tails will—just by chance. Tossing the coin 500 times to simulate random mating in a somewhat larger population is more likely to produce a 50-50 ratio of heads and tails. And if you tossed the coin 5000 times, you would get even closer to a 50-50 ratio.

Chance deviations from expected results—which cause genetic drift—occur whenever organisms engage in sexual reproduction, simply because their population sizes are not infinitely large. But genetic drift is particularly common in small populations because only a few individuals contribute to the gene pool and because any given allele is present in very few individuals.

Genetic drift generally leads to the loss of alleles and reduced genetic variability. Two general circum-stances, population bottlenecks and founder effects, often foster genetic drift.

Population Bottlenecks. On occasion, a stressful factor such as disease, starvation, or drought kills a great many individuals and eliminates some alleles from a population, producing a **population bottleneck.** This cause of genetic drift greatly reduces genetic variation even if the population numbers later rebound.

In the late nineteenth century, for example, hunters nearly wiped out northern elephant seals *(Mirounga angustirostris)* along the Pacific coast of North America **(Figure 20.8).** Since the 1880s, when the species received protected status, the population has increased to more than 30,000, all descended from a group of about 20 survivors. Today the population exhibits no variation in 24 proteins studied by gel electrophoresis. This low level of genetic variation, which is unique among seal species, is consistent with the hypothesis that genetic drift eliminated many alleles when the population experienced the bottleneck.

Founder Effect. When a few individuals colonize a distant locality and start a new population, they carry only a small sample of the parent population's genetic variation. By chance, some alleles may be totally missing from the new population, whereas other alleles that were rare "back home" might occur at relatively high frequencies. This change in the gene pool is called the **founder effect.**

The human medical literature provides some of the best-documented examples of the founder effect. The Old Order Amish, an essentially closed religious community in Lancaster County, Pennsylvania, have an exceptionally high incidence of Ellis–van Creveld syndrome, a genetic disorder caused by a recessive allele. In the homozygous state, the allele produces dwarfism, shortened limbs, and polydactyly (extra fin-

Figure 20.8

Population bottleneck. Northern elephant seals *(Mirounga angustirostris)* at the Año Nuevo State Reserve in California are descended from a population that was decimated by hunting late in the nineteenth century. In this photo, two large bulls fight to control a harem of females.

Frans Lanting/Minden Pictures

Genetic Variation Preserved in Humpback Whales

For centuries, hunters slaughtered humpback whales *(Megaptera novaeangliae)* for their meat and oil. By 1966, when an international agreement limited whale hunting, the worldwide population of humpbacks had been reduced to fewer than 5000 individuals. These survivors were distributed among three distinct populations in the North Atlantic, North Pacific, and Southern oceans. Since the hunting agreement was imposed, the populations have recovered to include more than 20,000 individuals.

The derivation of present-day humpback populations from the relatively small number surviving in 1966 is of concern because the population bottleneck may have reduced genetic variability. Such a loss could have adverse effects on the surviving population's reproductive capacity, resistance to disease, and ability to survive unfavorable environmental changes.

How serious was the bottleneck for the surviving humpback whales? A large group of researchers working in Hawaii, the continental United States, Australia, South Africa, Canada, Mexico, and the Dominican Republic set out to answer this question, using molecular techniques to measure the amount of genetic variability in the surviving whale populations.

The researchers chose mitochondrial DNA (mtDNA) for their measurements because it is small, it is easily extracted and identified, and almost all of its variability comes from chance mutations that occur at a steady rate rather than from genetic recombination (see Section 13.5). Except for the few changes produced by mutations since the population bottleneck (which can be estimated from the mutation rate and subtracted from the total), the variability of mtDNA should be the amount remaining from the population that existed before the bottleneck.

Using biopsy darts, the researchers obtained small skin samples from 90 humpback whales distributed among the three oceanic populations. They extracted the mtDNA from the skin samples and amplified it using the polymerase chain reaction (see Figure 18.6). They then isolated a 463-base-pair segment containing the promoters and replication origin for mtDNA, along with spacer sequences. The DNA base sequence was determined for each sample.

The researchers were surprised to find that the mtDNA sequence variation was relatively high in most of their sample, between 76% and 82% of the average variation found in all animal species studied to date. However, a subpopulation of the north Pacific population living near Hawaii showed low genetic variability; in fact, no variability at all was detected in the mtDNA segment of this subpopulation. Why the Hawaiian humpbacks have no variability in the mtDNA segment examined is unclear. One possibility is that this subpopulation originated recently, perhaps during the twentieth century. Information supporting this idea comes from whaling records, which list no sightings or catches of humpbacks in the Hawaiian region during the nineteenth century. Furthermore, the native Hawaiian people have no legends or words describing whales of the humpback type (baleen whales). Perhaps the subpopulation was started by a few whales with the same genetic make-up in the mtDNA region, providing an example of the founder effect.

With the exception of this Hawaiian subpopulation, humpback whales appear to have retained genetic variability comparable to other animals. This retention of variability in the face of near extinction may result from the whales' relatively long generation time. Because they have a potential life span of about 50 years, some individuals that survived the period of commercial hunting are still alive today. The researchers suggest that enough of these long-lived individuals survived to provide a reservoir of variability from the old populations.

These results indicate that the hunting ban came in time to prevent a significant loss of genetic variability in humpback whales. Hopefully, the same is true of other whale species that were hunted nearly to extinction.

gers). Genetic analysis suggests that, although this syndrome affects less than 1% of the Amish in Lancaster County, as many as 13% may be heterozygous carriers of the allele. All of the individuals exhibiting the syndrome are descended from one couple who helped found the community in the mid-1700s.

Conservation Implications. Genetic drift has important implications for conservation biology. By definition, endangered species experience severe population bottlenecks, which result in the loss of genetic variability. Moreover, the small number of individuals available for captive breeding programs may not fully represent a species' genetic diversity. Without such variation, no matter how large a population may become in the future, it will be less resistant to diseases or less able to cope with environmental change.

For example, scientists believe that an environmental catastrophe produced a population bottleneck in the African cheetah *(Acinonyx jubatus)* 10,000 years ago. Cheetahs today are remarkably uniform in genetic make-up. Their populations are highly susceptible to diseases; they also have a high proportion of sperm cell abnormalities and a reduced reproductive capacity. Thus, limited genetic variation, as well as small numbers, threatens populations of endangered species. *Insights from the Molecular Revolution* describes techniques used to determine whether hunting has had the same effect on humpback whales.

Natural Selection Shapes Genetic Variability by Favoring Some Traits over Others

The Hardy-Weinberg model requires all genotypes in a population to survive and reproduce equally well. But as you know from Section 19.2, heritable traits enable some individuals to survive better and reproduce more than others. **Natural selection** is the process by which such traits become more common in subsequent generations. Thus, natural selection violates a requirement of the Hardy-Weinberg equilibrium.

Although natural selection can change allele frequencies, *it is the phenotype of an individual organism,* *rather than any particular allele, that is successful or not.* When individuals survive and reproduce, their alleles—both favorable and unfavorable—are passed to the next generation. Of course, an organism with harmful or lethal dominant alleles will probably die before reproducing, and all the alleles it carries will share that unhappy fate, even those that are advantageous.

To evaluate reproductive success, evolutionary biologists consider **relative fitness**, the number of surviving offspring that an individual produces compared with the number left by others in the population. Thus, a particular allele will increase in frequency in the next generation if individuals carrying that allele leave *more*

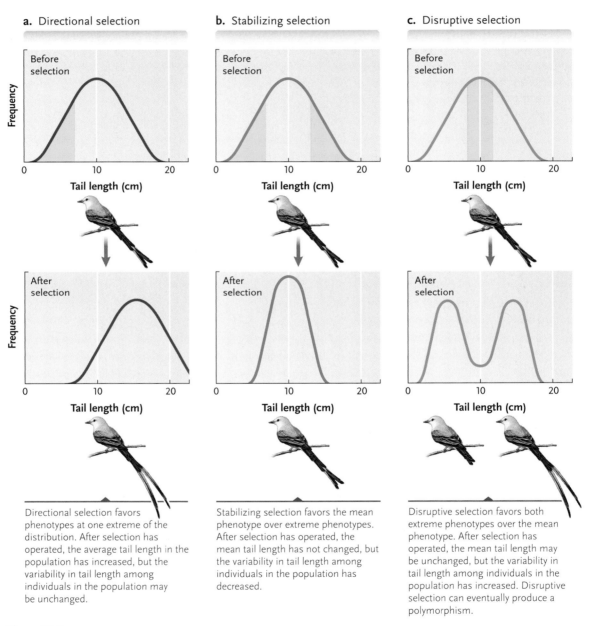

a. Directional selection

Directional selection favors phenotypes at one extreme of the distribution. After selection has operated, the average tail length in the population has increased, but the variability in tail length among individuals in the population may be unchanged.

b. Stabilizing selection

Stabilizing selection favors the mean phenotype over extreme phenotypes. After selection has operated, the mean tail length has not changed, but the variability in tail length among individuals in the population has decreased.

c. Disruptive selection

Disruptive selection favors both extreme phenotypes over the mean phenotype. After selection has operated, the mean tail length may be unchanged, but the variability in tail length among individuals in the population has increased. Disruptive selection can eventually produce a polymorphism.

Figure 20.9
Three modes of natural selection. This hypothetical example uses tail length of birds as the quantitative trait subject to selection. The yellow shading in the top graphs indicates phenotypes that natural selection does *not* favor. Notice that the area under each curve is constant because each curve presents the frequencies of all phenotypes in the population. When stabilizing selection **(b)** reduces variability in the trait, the curve becomes higher and narrower.

offspring than individuals carrying other alleles. Differences in the *relative* success of individuals are the essence of natural selection.

Natural selection tests fitness differences at nearly every stage of the life cycle. One plant may be fitter than others in the population because its seeds survive colder conditions, because the arrangement of its leaves captures sunlight more efficiently, or because its flowers are more attractive to pollinators. However, natural selection exerts little or no effect on traits that appear during an individual's postreproductive life. For example, Huntington disease, a dominant-allele disorder that first strikes humans after the age of 40, is not subject to strong selection. Carriers of the disease-causing allele reproduce before the onset of the condition, passing it to the next generation.

Biologists measure the effects of natural selection on phenotypic variation by recording changes in the mean and variability of characters over time (see Figure 20.3). Three modes of natural selection have been identified: directional selection, stabilizing selection, and disruptive selection **(Figure 20.9)**.

Directional Selection. Traits undergo **directional selection** when individuals near one end of the phenotypic spectrum have the highest relative fitness. Directional selection shifts a trait away from the existing mean and toward the favored extreme (see Figure 20.9a). After selection, the trait's mean value is higher or lower than before.

Directional selection is extremely common. For example, predatory fish promote directional selection for larger body size in guppies when they selectively feed on the smallest individuals in a guppy population (see *Focus on Research* in Chapter 49). And most cases of artificial selection, including the experiment on the activity levels of house mice, are directional, aimed at increasing or decreasing specific phenotypic traits. Humans routinely use directional selection to produce domestic animals and crops with desired characteristics, such as the small size of chihuahuas and the intense "bite" of chili peppers.

Stabilizing Selection. Traits undergo **stabilizing selection** when individuals expressing intermediate phenotypes have the highest relative fitness (see Figure 20.9b). By eliminating phenotypic extremes, stabilizing selection reduces genetic and phenotypic variation and increases the frequency of intermediate phenotypes. Stabilizing selection is probably the most common mode of natural selection, affecting many familiar traits. For example, very small and very large human newborns are less likely to survive than those born at an intermediate weight **(Figure 20.10)**.

Warren G. Abrahamson and Arthur E. Weis of Bucknell University have shown that opposing forces of directional selection can sometimes produce an overall pattern of stabilizing selection **(Figure 20.11)**.

Figure 20.10 Observational Research

Evidence for Stabilizing Selection in Humans

HYPOTHESIS: Human birth weight has been adjusted by natural selection.

NULL HYPOTHESIS: Natural selection has not affected human birth weight.

METHOD: Two noted human geneticists, Luigi Cavalli-Sforza and Sir Walter Bodmer of Stanford University, collected data on the variability in human birth weight, a character exhibiting quantitative variation, and on the mortality rates of babies born at different weights. The researchers then searched for a relationship between birth weight and mortality rate by plotting both data sets on the same graph. A lack of correlation between birth weight and mortality rate would support the null hypothesis.

RESULTS: When plotted together on the same graph, the bar graph (birth weight) and the curve (mortality rate) illustrate that the mean birth weight is very close to the optimum birth weight (the weight at which mortality is lowest). The two data sets also show that few babies are born at the very low and very high weights associated with high mortality.

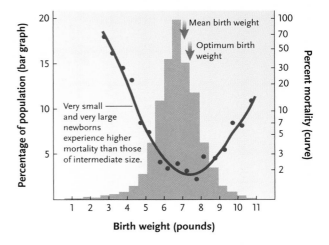

CONCLUSION: The shapes and positions of the birth weight bar graph and the mortality rate curve suggest that stabilizing selection has adjusted human birth weight to an average of 7 to 8 pounds.

The gallmaking fly *(Eurosta solidaginis)* is a small insect that feeds on the tall goldenrod plant *(Solidago altissima)*. When a fly larva hatches from its egg, it bores into a goldenrod stem, and the plant responds by producing a spherical growth deformity called a gall. The larva feeds on plant tissues inside the gall. Galls vary dramatically in size; genetic experiments indicate that gall size is a heritable trait of the fly, although plant genotype also has an effect.

Fly larvae inside galls are subjected to two opposing patterns of directional selection. On one hand, a tiny wasp *(Eurytoma gigantea)* parasitizes gallmaking flies by laying eggs in fly larvae inside their galls. After hatching, the young wasps feed on the fly larvae, killing them in the process. However, adult wasps are

Figure 20.11 Observational Research

How Opposing Forces of Directional Selection Produce Stabilizing Selection

HYPOTHESIS: The size of galls made by larvae of the gallmaking fly *(Eurosta solidaginis)* is governed by conflicting selection pressures established by parasitic wasps and predatory birds.

PREDICTION: Gallmaking flies that produce galls of intermediate size will be more likely to survive than those that make either small galls or large galls.

METHOD: Abrahamson and his colleagues surveyed galls made by the larvae of the gallmaking fly in Pennsylvania. They measured the diameters of the galls they encountered, and, for those galls in which the larvae had died, they determined whether they had been killed by **(a)** a parasitic wasp *(Eurytoma gigantea)* or **(b)** a predatory bird, such as the downy woodpecker *(Dendrocopus pubescens)*.

a. *Eurytoma gigantea,* a parasitic wasp

Forrest W. Buchanan/Visuals Unlimited

b. *Dendrocopus pubescens,* a predatory bird

Gregory K. Scott/Photo Researchers, Inc.

RESULTS: Tiny wasps are more likely to parasitize gallmaking fly larvae inside small galls **(c),** fostering directional selection in favor of large galls. By contrast, birds usually feed on fly larvae inside large galls **(d),** fostering directional selection in favor of small galls. These opposing patterns of directional selection create stabilizing selection for the size of galls that the fly larvae make **(e).**

c.

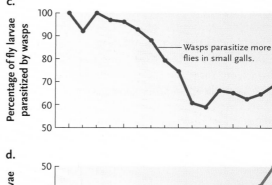

Wasps parasitize more flies in small galls.

d.

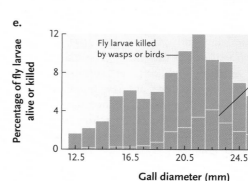

Birds consume more flies in large galls.

e.

Fly larvae killed by wasps or birds

Fly larvae alive in galls

Gall diameter (mm)

CONCLUSION: Because wasps preferentially parasitize fly larvae in small galls and birds preferentially eat fly larvae in large galls, the opposing forces of directional selection establish an overall pattern of stabilizing selection in favor of medium-sized galls.

so small that they cannot easily penetrate the thick walls of a large gall; they generally lay eggs in fly larvae occupying small galls. Thus, wasps establish directional selection favoring flies that produce large galls, which are less likely to be parasitized. On the other hand, several bird species open galls to feed on mature fly larvae; these predators preferentially open large galls, fostering directional selection in favor of small galls.

In about one-third of the populations surveyed in central Pennsylvania, wasps and birds attacked galls with equal frequency, and flies producing galls of intermediate size had the highest survival rate. The smallest and largest galls—as well as the genetic pre-

Geospiza conirostris

Heather Angel

Birds with long bills open cactus fruits to feed on the fleshy pulp.

Birds with intermediate bills may be favored during nondrought years when many types of food are available.

Birds with deep bills strip bark from trees to locate insects.

Figure 20.12
Disruptive selection. Cactus finches *(Geospiza conirostris)* on Genovesa exhibit extreme variability in the size and shape of their bills.

disposition to make very small or very large galls—were eliminated from the population.

Disruptive Selection. Traits undergo **disruptive selection** when extreme phenotypes have higher relative fitness than intermediate phenotypes (see Figure 20.9c). Thus, alleles producing extreme phenotypes become more common, promoting polymorphism. Under natural conditions, disruptive selection is much less common than directional selection and stabilizing selection.

Peter Grant of Princeton University, the world's expert on the ecology and evolution of the Galápagos finches, has analyzed a likely case of disruptive selection on the size and shape of the bill in a population of cactus finches *(Geospiza conirostris)* on the island of Genovesa. During normal weather cycles the finches feed on ripe cactus fruits, seeds, and exposed insects. During drought years, when food is scarce, they also search for insects by stripping bark from the branches of bushes and trees.

During the long drought of 1977, about 70% of the cactus finches on Genovesa died; the survivors exhibited unusually high variability in their bills **(Figure 20.12)**. Grant suggested that this morphological variability allowed birds to specialize on particular foods. Birds that stripped bark from branches to look for insects had particularly deep bills, and birds that opened cactus fruits to feed on the fleshy interior had especially long bills. Thus, birds with extreme bill phenotypes appeared to feed efficiently on specific resources, establishing disruptive selection on the size and shape of their bills. The selection may be particularly strong when drought limits the variety and overall availability of food. However, intermediate bill morphologies may be favored during nondrought years when insects and small seeds are abundant.

Sexual Selection Often Exaggerates Showy Structures in Males

Darwin hypothesized that a special process, which he called **sexual selection**, has fostered the evolution of showy structures—such as brightly colored feathers, long tails, or impressive antlers—as well as elaborate courtship behavior in the males of many animal spe-

cies. Sexual selection encompasses two related processes. As the result of *intersexual selection* (that is, selection based on the interactions between males and females), males produce these otherwise useless structures simply because females find them irresistibly attractive. Under *intrasexual selection* (that is, selection based on the interactions between members of the same sex), males use their large body size, antlers, or tusks to intimidate, injure, or kill rival males. In many species, sexual selection is the most probable cause of **sexual dimorphism**, differences in the size or appearance of males and females.

Like directional selection, sexual selection pushes phenotypes toward one extreme. But the products of sexual selection are sometimes bizarre—such as the ridiculously long tail feathers of male African widowbirds. How could evolutionary processes favor the production of such costly structures? Malte Andersson of the University of Gothenburg, Sweden, conducted a field experiment to determine whether the long tail feathers were the product of either intersexual selection or intrasexual selection **(Figure 20.13)**. Male widowbirds compete vigorously for favored patches of habitat in which they court females. After surveying the behavior of birds under natural conditions, Andersson lengthened the tails of some males, shortened those of others, and left some males essentially unaltered to serve as controls. His results suggest that females are more strongly attracted to males with long tails than to males with short tails, but that tail length had no effect on a male's ability to compete with other males for space in the habitat. Thus, the long tail of the African widowbird is a product of intersexual selection, not intrasexual selection. Behavioral aspects of sexual selection are described further in Chapter 55.

Nonrandom Mating Can Influence Genotype Frequencies

The Hardy-Weinberg model requires individuals to select mates randomly with respect to their genotypes. This requirement is, in fact, often met; humans, for example, generally marry one another in total ignorance of their genotypes for digestive enzymes or blood types.

Nevertheless, many organisms mate nonrandomly, selecting a mate with a particular phenotype

Figure 20.13 Experimental Research

Sexual Selection in Action

QUESTION: Is the long tail of the male long-tailed widowbird *(Euplectes progne)* the product of intrasexual selection, intersexual selection, or both?

EXPERIMENT: Andersson counted the number of females that associated with individual male widowbirds in the grasslands of Kenya. He then shortened the tails of some individuals by cutting the feathers, lengthened the tails of others by gluing feather extensions to their tails, and left a third group essentially unaltered as a control. One month later, he again counted the number of females associating with each male and compared the results from the three groups.

RESULTS: Males with experimentally lengthened tails attracted more than twice as many mates as males in the control group, and males with experimentally shortened tails attracted fewer. Andersson observed no differences in the ability of altered males and control group males to maintain their display areas.

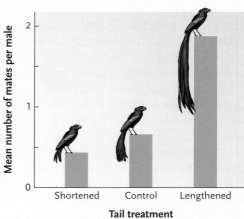

© 2008 Josef Hlasak

CONCLUSION: Female widowbirds clearly prefer males with experimentally lengthened tails to those with normal tails or experimentally shortened tails. Tail length had no obvious effect on the interactions between males. Thus, the long tail of male widowbirds is the product of intersexual selection.

and underlying genotype. Snow geese, for example, usually select mates of their own color, and a tall woman is more likely to marry a tall man than a short man. If no one phenotype is preferred by all potential mates, nonrandom mating does not establish selection for one phenotype over another. But because individuals with similar genetically based phenotypes mate with each other, the next generation will contain fewer heterozygous offspring than the Hardy-Weinberg model predicts.

Inbreeding is a special form of nonrandom mating in which individuals that are genetically related mate with each other. Self-fertilization in plants (see Chapter 34) and a few animals (see Chapter 47) is an extreme example of inbreeding because offspring are produced from the gametes of a single parent. However, other organisms that live in small, relatively closed populations often mate with related individuals. Because relatives often carry the same alleles, inbreeding generally increases the frequency of homozygous genotypes and decreases the frequency of heterozygotes. Thus, recessive phenotypes are often expressed.

For example, the high incidence of Ellis–van Creveld syndrome among the Old Order Amish population, mentioned earlier, is caused by inbreeding. Although the founder effect originally established the disease-causing allele in this population, inbreeding increases the likelihood that it will be expressed. Most human societies discourage matings between genetically close relatives, thereby reducing inbreeding and the production of recessive homozygotes.

STUDY BREAK

1. Which agents of microevolution tend to increase genetic variation within populations, and which ones tend to decrease it?
2. Which mode of natural selection increases the representation of the average phenotype in a population?
3. In what way is sexual selection like directional selection?

20.4 Maintaining Genetic and Phenotypic Variation

Evolutionary biologists continue to discover extraordinary amounts of genetic and phenotypic variation in most natural populations. How can so much variation persist in the face of stabilizing selection and genetic drift?

Diploidy Can Hide Recessive Alleles from the Action of Natural Selection

The diploid condition reduces the effectiveness of natural selection in eliminating harmful recessive alleles from a population. Although such alleles are disadvantageous in the homozygous state, they may have little or no effect on heterozygotes. Thus, recessive alleles can be protected from natural selection by the phenotypic expression of the dominant allele.

In most cases, the masking of recessive alleles in heterozygotes makes it almost impossible to eliminate them completely through selective breeding. Experimentally, we can prevent homozygous recessive organisms from mating. But, as the frequency of a recessive allele decreases, an increasing proportion of its remaining copies is "hidden" in heterozygotes **(Table 20.3).** Thus, the diploid state preserves recessive alleles at low frequencies, at least in large populations. In small populations, a combination of natural selection and genetic drift can eliminate harmful recessive alleles.

Natural Selection Can Maintain Balanced Polymorphisms

A **balanced polymorphism** is one in which two or more phenotypes are maintained in fairly stable proportions over many generations. Natural selection preserves balanced polymorphisms when heterozygotes have higher relative fitness, when different alleles are favored in different environments, and when the rarity of a phenotype provides an advantage.

Heterozygote Advantage. A balanced polymorphism can be maintained by **heterozygote advantage**, when heterozygotes for a particular locus have higher relative fitness than either homozygote. The best-documented example of heterozygote advantage is the maintenance of the *HbS* (sickle) allele, which codes for a defective form of hemoglobin in humans. As you learned in Chapter 12, hemoglobin is an oxygen-transporting molecule in red blood cells. The hemoglobin produced by the *HbS* allele differs from normal hemoglobin (coded by the *HbA* allele) by just one amino acid. In *HbS*/*HbS* homozygotes, the faulty hemoglobin forms long fibrous chains under low oxygen conditions, causing red blood cells to assume a sickle shape (as shown

Table 20.3 **Masking of Recessive Alleles in Diploid Organisms**

When a recessive allele is common in a population (top), most copies of the allele are present in homozygotes. But when the allele is rare (bottom), most copies of it exist in heterozygotes. Thus, rare alleles that are completely recessive are protected from the action of natural selection because they are masked by dominant alleles in heterozygous individuals.

Frequency of Allele *a*	Genotype Frequencies*			% of Allele *a* Copies in	
	AA	Aa	aa	Aa	aa
0.99	0.0001	0.0198	0.9801	1	99
0.90	0.0100	0.1800	0.8100	10	90
0.75	0.0625	0.3750	0.5625	25	75
0.50	0.2500	0.5000	0.2500	50	50
0.25	0.5625	0.3750	0.0625	75	25
0.10	0.8100	0.1800	0.0100	90	10
0.01	0.9801	0.0198	0.0001	99	1

*Population is assumed to be in genetic equilibrium.

in Figure 12.1). Homozygous *HbS*/*HbS* individuals often die of sickle-cell disease before reproducing, yet in tropical and subtropical Africa, *HbS*/*HbA* heterozygotes make up nearly 25% of many populations.

Why is the harmful allele maintained at such high frequency? It turns out that sickle-cell disease is most common in regions where malarial parasites infect red blood cells in humans **(Figure 20.14).** When heterozygous *HbA*/*HbS* individuals contract malaria, their infected red blood cells assume the same sickle shape as those of homozygous *HbS*/*HbS* individuals. The sickled cells lose potassium, killing the parasites, which limits their spread within the infected individual. Heterozygous individuals often survive malaria because the parasites do not multiply quickly inside them; their immune systems can effectively fight the infection; and they retain a large population of uninfected red blood cells. Homozygous *HbA*/*HbA* individuals are also subject to malarial infection, but because their infected cells do not sickle, the parasites multiply rapidly, causing a severe infection with a high mortality rate.

Therefore, *HbA*/*HbS* heterozygotes have greater resistance to malaria and are more likely to survive severe infections in areas where malaria is prevalent. Natural selection preserves the *HbS* allele in these populations because heterozygotes in malaria-prone areas have higher relative fitness than homozygotes for the normal *HbA* allele.

Selection in Varying Environments. Genetic variability can also be maintained within a population when different alleles are favored in different places or at different times. For example, the shells of European garden

a. Distribution of *HbS* allele

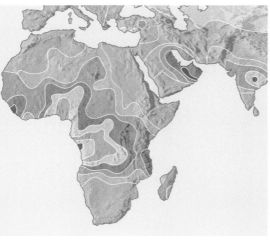

b. Distribution of malarial parasite

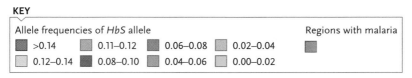

KEY

Allele frequencies of *HbS* allele				Regions with malaria
■ >0.14	■ 0.11–0.12	■ 0.06–0.08	□ 0.02–0.04	■
□ 0.12–0.14	■ 0.08–0.10	■ 0.04–0.06	□ 0.00–0.02	

Figure 20.14

Heterozygote advantage. The distribution of the *HbS* allele **(a)**, which causes sickle-cell disease in homozygotes, roughly matches the distribution of the malarial parasite *Plasmodium falciparum* **(b)** in southern Europe, Africa, the Middle East, and India. Gene flow among human populations has carried the *HbS* allele to some malaria-free regions.

snails range in color from nearly white to pink, yellow, or brown, and may be patterned by one to five stripes of varying color (see Figure 20.2a). This polymorphism, which is relatively stable through time, is controlled by several gene loci. The variability in color and in striping pattern can be partially explained by selection for camouflage in different habitats.

Predation by song thrushes *(Turdus ericetorum)* is a major agent of selection on the color and pattern of these snails in England. When a thrush finds a snail, it smacks it against a rock to break the shell. The bird eats the snail, but leaves the shell near its "anvil." Researchers used the broken shells near an anvil to compare the phenotypes of captured snails to a random sample of the entire snail population. Their analyses indicated that thrushes are visual predators, usually capturing snails that are easy to find. Thus, well-camouflaged snails survive, and the alleles that specify their phenotypes increase in frequency.

The success of camouflage varies with habitat, however; local subpopulations of the snail, which occupy different habitats, often differ markedly in shell color and pattern. The predators eliminate the most conspicuous individuals in each habitat; thus, natural selection differs from place to place **(Figure 20.15)**. In woods where the ground is covered with dead leaves, snails with unstriped pink or brown shells predominate. In hedges and fields, where the vegetation in-cludes thin stems and grass, snails with striped yellow shells are the most common. In populations that span several habitats, selection preserves different alleles in different places, thus maintaining variability in the population as a whole.

Frequency-Dependent Selection. Sometimes genetic variability is maintained in a population simply because rare phenotypes—whatever they happen to be—have higher relative fitness than more common phenotypes. The rare phenotype will increase in frequency until it becomes so common that it loses its advantage. Such phenomena are examples of **frequency-dependent selection** because the selective advantage enjoyed by a particular phenotype depends on its frequency in the population.

Predator-prey interactions can establish frequency-dependent selection because predators often focus their attention on the most common types of prey (see Chapter 50). For example, the aquatic insects called water boatmen occur in three different shades of brown. When all three shades are available at moderate frequencies, fish preferentially feed on the darkest individuals, which are the least camouflaged. But if any one phenotype is very common, fish will learn to focus their attention on that phenotype (see Chapter 54), consuming it in disproportionately large numbers **(Figure 20.16)**.

Some Genetic Variations May Be Selectively Neutral

Many biologists believe that some genetic variations are neither preserved nor eliminated by natural selection. According to the **neutral variation hypothesis**, some of the genetic variation at loci coding for enzymes and other soluble proteins is **selectively neutral**. Even if various alleles code for slightly different amino acid sequences in proteins, the different forms of the proteins may function equally well. In those cases, natural selection would not favor some alleles over others.

Biologists who support the neutral variation hypothesis do not question the role of natural selection in producing complex anatomical structures or useful biochemical traits. They also recognize that selection reduces the frequency of harmful alleles. But they argue that we should not simply assume that every genetic variant that persists in a population has been preserved by natural selection. In practice, it is often very difficult to test the natural variation hypothesis because the fitness effects of different alleles are often subtle and vary with small changes in the environment.

The neutral variation hypothesis helps to explain why we see different levels of genetic variation in different populations. It proposes that genetic variation is directly proportional to a population's size and the length of time over which variations have accumulated. Small populations experience fewer mutations than large populations simply because they include fewer replicating genomes. Small populations also lose rare alleles more readily through genetic drift. Thus, small populations should exhibit less genetic variation than large ones, and a population, like the northern elephant seals, that has experienced a recent population bottleneck should exhibit an exceptionally low level of genetic variation. These predictions of the neutral variation hypothesis are generally supported by empirical data.

STUDY BREAK

1. How does the diploid condition protect harmful recessive alleles from natural selection?
2. What is a balanced polymorphism?
3. Why is the allele that causes sickle-cell disease very rare in human populations that are native to northern Europe?

20.5 Adaptation and Evolutionary Constraints

Although natural selection preserves alleles that confer high relative fitness on the individuals that carry them, researchers are cautious about interpreting the benefits that particular traits may provide.

Figure 20.15 Observational Research

Habitat Variation in Color and Striping Patterns of European Garden Snails

HYPOTHESIS: Genetically based variations in the shell color and striping patterns of the European garden snail *(Cepaea nemoralis)* differ substantially from one type of vegetation to another because birds and other visual predators establish strong selection for camouflage in local populations.

PREDICTION: Snails with plain, dark-colored shells will be most abundant in woodland habitats, but snails with striped, light-colored shells will be most abundant in hedges and fields.

METHOD: Two British researchers, A. J. Cain and P. M. Shepard, surveyed the distribution of color and striping patterns of snails in many local populations. They plotted the data on a graph showing the percentage of snails with yellow shells versus the percentage of snails with striped shells, noting the vegetation type where each local population lived.

RESULTS: The shell color and striping patterns of snails living in a particular vegetation type tend to be clustered on the graph, reflecting phenotypic differences that enable the snails to be camouflaged in different habitats. Thus, the alleles that control these characters vary from one local population to another.

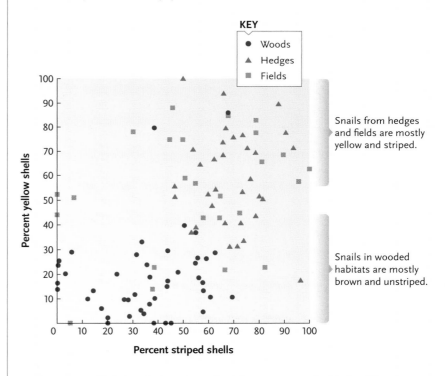

CONCLUSION: Variations in the color and striping patterns on the shells of European garden snails allow most snails to be camouflaged in whatever habitat they occupy. Because these traits are genetically based, the frequencies of the alleles that control them also differ among snails living in different vegetation types. Natural selection therefore favors different alleles in different local populations, maintaining genetic variability in populations that span several vegetation types.

Figure 20.16 Experimental Research

Demonstration of Frequency-Dependent Selection

QUESTION: How does the frequency of a prey type influence the likelihood that it will be captured by predators?

EXPERIMENT: Water boatmen *(Sigara distincta)* occur in three color forms, which vary in the effectiveness of their camouflage. Researchers offered different proportions of the three color forms to predatory fishes in the laboratory and recorded how many of each form were eaten.

RESULTS: When all three phenotypes were available, predatory fishes consumed a disproportionately large number of the most common form, thereby reducing its frequency in the population.

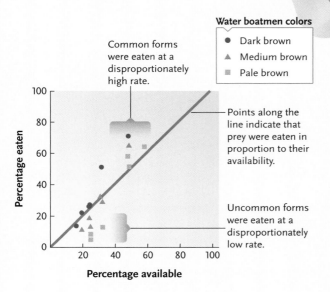

CONCLUSION: Predators tend to feed disproportionately on whatever form of their prey is most abundant, thereby reducing its frequency in the prey population.

Scientists Construct Hypotheses about the Evolution of Adaptive Traits

An **adaptive trait** is any product of natural selection that increases the relative fitness of an organism in its environment. **Adaptation** is the accumulation of adaptive traits over time, and this book describes many examples. The change in the oxygen-binding capacity of hemoglobin in response to carbon dioxide concentration, the water-retaining structures and special photosynthetic pathways of desert plants, and the warning coloration of poisonous animals can all be interpreted as adaptive traits.

In fact, we can concoct an adaptive explanation for almost any characteristic we observe in nature. But such explanations are just fanciful stories unless they are framed as testable hypotheses about the relative fitness of different phenotypes and genotypes. Unfor-

tunately, evolutionary biologists cannot always conduct straightforward experiments because they sometimes study traits that do not vary much within a population or species. In such cases, they may compare variations of a trait in closely related species living in different environments. For example, one can test how the traits of desert plants are adaptive by comparing them to traits in related species from moister habitats.

When biologists try to unravel how and why a particular characteristic evolved, they must also remember that a trait they observe today may have had a different function in the past. For example, the structure of the shoulder joint in birds allows them to move their wings first upward and backward and then downward and forward during flapping flight. But analyses of the fossil record reveal that this adaptation, which is essential for flight, did not originate in birds: some predatory nonflying dinosaurs, including the ancestors of birds, had a similarly constructed shoulder joint. Researchers hypothesize that these fast-running predators may have struck at prey with a flapping motion similar to that used by modern birds. Thus, the structure of the shoulder may have first evolved as an adaptation for capturing prey, and only later proved useful for flapping flight. This hypothesis—however plausible it may be—cannot be tested by direct experimentation because the nonflying ancestors of bird have been extinct for millions of years. Instead, evolutionary biologists must use anatomical studies of birds and their ancestors as well as theoretical models about the mechanics of movement to challenge and refine the hypothesis.

Finally, although evolution has produced all the characteristics of organisms, not all are necessarily adaptive. Some traits may be the products of chance events and genetic drift. Others are produced by alleles that were selected for unrelated reasons (see Section 12.2). And still other characteristics result from the action of basic physical laws. For example, the seeds of many plants fall to the ground when they mature, reflecting the inevitable effect of gravity.

Several Factors Constrain Adaptive Evolution

When we analyze the structure and function of an organism, we often marvel at how well adapted it is to its environment and mode of life. However, the adaptive traits of most organisms are compromises produced by competing selection pressures. Sea turtles, for example, must lay their eggs on beaches because their embryos cannot acquire oxygen under water. Although flippers allow females to crawl to nesting sites on beaches, they are not ideally suited for terrestrial locomotion. Their structure reflects their primary function in underwater locomotion.

Moreover, no organism can be perfectly adapted to its environment because environments change over

What are the evolutionary forces affecting molecular variation within populations?

This question may sound like a simple restatement of the entire chapter you have just read, but it is one of the *fundamental* questions in population genetics today—and we have only begun to scratch its surface. The Hardy-Weinberg principle provides a useful null hypothesis, but since we know that evolution happens routinely, that null hypothesis is very frequently rejected. Recent studies have attempted to address this question using theoretical models, extensive DNA sequence data, and detailed measures of recombination rate.

Recombination generates new variation, and, most importantly, it causes the evolutionary forces acting on some genes to become independent of forces acting on other genes. Let's imagine that genes A and B are on the same chromosome, as shown in this depiction of chromosomes sampled from different individuals within a population:

A	B
A	b
A	B
A	b
A	b
A	B
a	b

Gene B has two alleles (B and b), but they have no phenotypic effect, and natural selection does not act on them. Suppose that a new advantageous allele at gene A (designated a) arises in one chromosome. If there is no recombination between genes A and B, then as allele a spreads in the population by selection, so too will allele b, even though there was no selection directly favoring the b allele. This effect of selection on nearby genes is called a *selective sweep*. By contrast, if genes A and B frequently recombine, then allele a may not remain associated with allele b. Under frequent recombination, the spread of allele a may have little or no effect on gene B: sometimes a will be associated with b, but at other times a will be associated with B.

In the 1990s, evolutionary geneticists were greatly excited by several studies that identified a strong and positive relationship between the recombination rate between particular genes and the amount of genetic variation within those genes. In other words, genes that experienced a lot of recombination also exhibited a great deal of variability. This relationship is consistent with the hypothesis that natural selection often occurs throughout the genome—new advantageous alleles arise frequently, and the impact of their "sweeps" is proportional to their recombination rates. This relationship between recombination and genetic variation was first documented in *Drosophila* (fruit flies) by Chip

Aquadro and his team at Cornell University, but it has since been demonstrated in humans and various plants. Hence, this pattern appears to be very general.

However, our initial interpretation may be too simplistic. Brian Charlesworth, then at the University of Chicago, suggested that the observed pattern may result from the frequent appearance of detrimental mutations that eliminate variation in regions of low recombination—called *background selection*—rather than from sweeps associated with the spread of advantageous alleles. Given that detrimental mutations arise far more frequently than advantageous ones, background selection surely explains some of this general pattern, and perhaps much of it.

An alternative hypothesis that may explain the relationship between recombination rate and genetic variation suggests that recombination rate and the level of genetic variation may be mechanistically connected. A direct connection may operate if recombination itself induces mutations, resulting in higher mutation rates in regions of high recombination. Alternatively, the connection may be indirect: recombination rate is known to be related to the base composition in specific regions of the genome, and base composition is known to influence mutation rates. In 2006, Chris Spencer and his colleagues at Oxford University examined the impact of recombination rates on patterns of nucleotide variation at a very fine scale across the human genome. They found that recombination rates had very local effects on variation, an observation that is consistent with the alternative hypothesis of a mechanistic connection between recombination and mutation rate; their results are not consistent with explanations involving natural selection.

Although biologists first thought that the observed relationship between recombination rate and genetic variation had solved questions about the evolutionary forces that affect molecular variation, this observation has become a puzzle in and of itself. Many of us continue to address this question, now using whole-genome sequences and theoretical and empirical tools for estimating recombination rates. We know that the "final answer" will be that all of the processes described above contribute to this relationship, but knowing their specific contributions will help us understand how, how much, and what kinds of natural selection shape variation within genomes.

 Mohamed Noor is an associate professor of biology at Duke University. His research interests include speciation and evolutionary genetics, and recombination. To learn more about his research go to http://www.biology.duke.edu/noorlab/Noorlab.html.

Dr. Noor was a PhD student with Dr. Jerry Coyne, who contributed the Unanswered Questions for Chapter 21.

time. When selection occurs in a population, it preserves alleles that are successful under the prevailing environmental conditions. Thus, each generation is adapted to the environmental conditions under which its parents lived. If the environment changes from one generation to the next, adaptation will always lag behind.

Another constraint on the evolution of adaptive traits is historical. Natural selection is not an engineer that designs new organisms from scratch. Instead, it acts on new mutations and existing genetic variation. Because new mutations are fairly rare, natural selection works primarily with alleles that have been present for many generations. Thus, adaptive changes in the morphology of an organism are almost inevitably based on small modifications of existing structures. The bipedal (two-footed) posture of humans, for example, evolved from the quadrupedal (four-footed) posture of our ancestors. Natural selection did not produce an entirely new skeletal design to accompany this radical behavioral shift. Instead, existing characteristics of the spinal column and the musculature of the legs and back were modified, albeit imperfectly, for an upright stance.

The agents of evolution cause microevolutionary changes in the gene pools of populations. In the next chapter, we examine how microevolution in different populations can cause their gene pools to diverge. The extent of genetic divergence is sometimes sufficient to cause the populations to evolve into different species.

STUDY BREAK

1. How can a biologist test whether a trait is adaptive?
2. Why are most organisms adapted to the environments in which their parents lived?

Review

Go to ThomsonNOW at www.thomsonedu.com/login to access quizzing, animations, exercises, articles, and personalized homework help.

20.1 Variation in Natural Populations

- Phenotypic traits exhibit either quantitative or qualitative variation within populations of all organisms (Figures 20.2 and 20.3).
- Genetic variation, environmental factors, or an interaction between the two cause phenotypic variation within populations. Only genetically based phenotypic variation is heritable and subject to evolutionary change.
- Genetic variation arises within populations largely through mutation and genetic recombination. Artificial selection experiments and analyses of protein and DNA sequences reveal that most populations include significant genetic variation (Figure 20.6).

20.2 Population Genetics

- All the alleles in a population comprise its gene pool, which can be described in terms of allele frequencies and genotype frequencies.
- The Hardy-Weinberg principle of genetic equilibrium is a null model that describes the conditions under which microevolution will not occur: mutations do not occur; populations are closed to migration; populations are infinitely large; natural selection does not operate; and individuals select mates at random. Microevolution, a change in allele frequencies through time, occurs in populations when the restrictive requirements of the model are not met.

Animation: How to find out if a population is evolving

20.3 The Agents of Microevolution

- Several processes cause microevolution in populations. Mutation introduces completely new genetic variation. Gene flow carries novel genetic variation into a population through the arrival and reproduction of immigrants. Genetic drift causes random changes in allele frequencies, especially in small populations. Natural selection occurs when the genotypes of some individuals enable them to survive and reproduce more than others. Nonrandom mating within a population can cause its genotype frequencies to depart from the predictions of the Hardy-Weinberg equilibrium.

- Natural selection alters phenotypic variation in one of three ways (Figure 20.9). Directional selection increases or decreases the mean value of a trait, shifting it toward a phenotypic extreme. Stabilizing selection increases the frequency of the mean phenotype and reduces variability in the trait (Figure 20.10). Disruptive selection increases the frequencies of extreme phenotypes and decreases the frequency of intermediate phenotypes (Figure 20.12).
- Sexual selection promotes the evolution of exaggerated structures and behaviors (Figure 20.13).
- Although nonrandom mating does not change allele frequencies, it can affect genotype frequencies, producing more homozygotes and fewer heterozygotes than the Hardy-Weinberg model predicts.

Animation: Directional selection

Animation: Change in moth population

Animation: Stabilizing selection

Animation: Disruptive selection

Animation: Disruptive selection among African finches

Animation: Simulation of genetic drift

20.4 Maintaining Genetic and Phenotypic Variation

- Diploidy can maintain genetic variation in a population if alleles coding for recessive traits are not expressed in heterozygotes and are thus hidden from natural selection.
- Polymorphisms are maintained in populations when heterozygotes have higher relative fitness than both homozygotes (Figure 20.14), when natural selection occurs in variable environments (Figure 20.15), or when the relative fitness of a phenotype varies with its frequency in the population (Figure 20.16).
- Some biologists believe that many genetic variations are selectively neutral, conferring neither advantages nor disadvantages on the individuals that carry them. The neutral variation hypothesis explains why large populations and those that have not experienced a recent population bottleneck exhibit the highest levels of genetic variation.

Animation: Distribution of sickle-cell trait

Animation: Life cycle of *Plasmodium*

20.5 Adaptation and Evolutionary Constraints

- Adaptive traits increase the relative fitness of individuals carrying them. Adaptive explanations of traits must be framed as testable hypotheses.

- Natural selection cannot result in perfectly adapted organisms because most adaptive traits represent compromises among conflicting needs; because most environments are constantly changing; and because natural selection can affect only existing genetic variation.

    Animation: Adaptation to what?

Questions

Self-Test Questions

1. Which of the following represents an example of qualitative phenotypic variation?
   a. the lengths of people's toes
   b. the body sizes of pigeons
   c. human ABO blood groups
   d. the birth weights of humans
   e. the number of leaves on oak trees

2. A population of mice is at Hardy-Weinberg equilibrium at a gene locus that controls fur color. The locus has two alleles, M and m. A genetic analysis of one population reveals that 60% of its gametes carry the M allele. What percentage of mice contains both the M and m alleles?
   a. 60%   d. 36%
   b. 48%   e. 16%
   c. 40%

3. If the genotype frequencies in a population are 0.60 AA, 0.20 Aa, and 0.20 aa, and if the requirements of the Hardy-Weinberg principle apply, the genotype frequencies in the offspring generation will be:
   a. 0.60 AA, 0.20 Aa, 0.20 aa.
   b. 0.36 AA, 0.60 Aa, 0.04 aa.
   c. 0.49 AA, 0.42 Aa, 0.09 aa.
   d. 0.70 AA, 0.00 Aa, 0.30 aa.
   e. 0.64 AA, 0.32 Aa, 0.04 aa.

4. The reason spontaneous mutations do not have an immediate effect on allele frequencies in a large population is that:
   a. mutations are random events, and mutations may be either beneficial or harmful.
   b. mutations usually occur in males and have little effect on eggs.
   c. many mutations exert their effects after an organism has stopped reproducing.
   d. mutations are so rare that mutated alleles are greatly outnumbered by nonmutated alleles.
   e. most mutations do not change the amino acid sequence of a protein.

5. The phenomenon in which chance events cause unpredictable changes in allele frequencies is called:
   a. gene flow.
   b. genetic drift.
   c. inbreeding.
   d. balanced polymorphism.
   e. stabilizing selection.

6. An Eastern European immigrant carrying the allele for Tay Sachs disease settled in a small village on the St. Lawrence River. Many generations later, the frequency of the allele in that village is statistically higher than it is in the immigrant's homeland. The high frequency of the allele in the village probably provides an example of:
   a. natural selection.
   b. the concept of relative fitness.

   c. the Hardy-Weinberg genetic equilibrium.
   d. phenotypic variation.
   e. the founder effect.

7. If a storm kills many small sparrows in a population, but only a few medium-sized and large ones, which type of selection is probably operating?
   a. directional selection
   b. stabilizing selection
   c. disruptive selection
   d. intersexual selection
   e. intrasexual selection

8. Which of the following phenomena explains why the allele for sickle-cell hemoglobin is common in some tropical and subtropical areas where the malaria parasite is prevalent?
   a. balanced polymorphism
   b. heterozygote advantage
   c. sexual dimorphism
   d. neutral selection
   e. stabilizing selection

9. The neutral variation hypothesis proposes that:
   a. complex structures in most organisms have not been fostered by natural selection.
   b. most mutations have a strongly harmful effect.
   c. some mutations are not affected by natural selection.
   d. natural selection cannot counteract the action of gene flow.
   e. large populations are subject to stronger natural selection than small populations.

10. Phenotypic characteristics that increase the fitness of individuals are called:
    a. mutations.
    b. founder effects.
    c. heterozygote advantages.
    d. adaptive traits.
    e. polymorphisms.

Questions for Discussion

1. Most large commercial farms routinely administer antibiotics to farm animals to prevent the rapid spread of diseases through a flock or herd. Explain why you think that this practice is either wise or unwise.

2. Many human diseases are caused by recessive alleles that are not expressed in heterozygotes. Explain why it is almost impossible to eliminate such genetic traits from human populations.

3. Using two types of beans to represent two alleles at the same gene locus, design an exercise to illustrate how population size affects genetic drift.

4. In what ways are the effects of sexual selection, disruptive selection, and nonrandom mating different? How are they similar?

Experimental Analysis

Design an experiment to test the hypothesis that the differences in size among adult guppies are determined by the amount of food they eat rather than by genetic factors.

Evolution Link

Captive breeding programs for endangered species often have access to a limited supply of animals for a breeding stock. As a result, their offspring are at risk of being highly inbred. Why and how might zoological gardens and conservation organizations avoid or minimize inbreeding?

How Would You Vote?

The symptoms of Huntington disease and some other genetically based diseases in humans appear only after the carriers of the disease-causing allele have already reproduced. As a result, they pass the alleles to their offspring and the disease persists in the population. Do you think that all people should be screened for disease-causing alleles and that carriers of such alleles should be discouraged or even prevented from having children? Go to www.thomsonedu.com/login to investigate both sides of the issue and then vote.

Two closely related species of parrot, the scarlet macaw *(Ara chloroptera)* and the blue and yellow macaw *(Ara arauna)*, perching together in the Amazon jungle of Peru.

© Mickey Gibson/Animals, Animals—Earth Scenes

21 Speciation

STUDY PLAN

21.1 What Is a Species?

The morphological species concept is a practical way to identify species

The biological and phylogenetic species concepts derive from evolutionary theory

Many species exhibit substantial geographical variation

21.2 Maintaining Reproductive Isolation

Prezygotic isolating mechanisms prevent the production of hybrid individuals

Postzygotic isolating mechanisms reduce the success of hybrid individuals

21.3 The Geography of Speciation

Allopatric speciation occurs between geographically separated populations

Parapatric speciation may occur between adjacent populations

Sympatric speciation occurs within one continuously distributed population

21.4 Genetic Mechanisms of Speciation

Genetic divergence in allopatric populations can lead to speciation

Polyploidy is a common mechanism of sympatric speciation in plants

Chromosome alterations can foster speciation

WHY IT MATTERS

In 1927, nearly 100 years after Darwin boarded the *Beagle,* a young German naturalist named Ernst Mayr embarked on his own journey, to the highlands of New Guinea. He was searching for rare "birds of paradise," no trace of which had been seen in Europe since plume hunters had returned years before with ornate and colorful feathers that were used to decorate ladies' hats **(Figure 21.1).** On his trek through the remote Arfak Mountains, Mayr identified 137 bird species (including many birds of paradise) based on differences in their size, plumage, color, and other external characteristics.

To Mayr's surprise, the native Papuans—who were untrained in the ways of Western science, but who hunted these birds for food and feathers—had their own names for 136 of the 137 species he had identified. The close match between the two lists confirmed Mayr's belief that the *species* is a fundamental level of organization in nature. Each species has a unique combination of genes underlying its distinctive appearance and habits. Thus, people who observe them closely—whether indigenous hunters or Western scientists—can often distinguish one species from another.

Figure 21.1

Birds of paradise. A male Count Raggi's bird of paradise *(Paradisaea raggiana)* has clearly attracted the attention of a female (the smaller, less colorful bird) with his showy plumage and conspicuous display. There are 43 known bird of paradise species, 35 of them found only on the island of New Guinea.

Mayr also discovered some remarkable patterns in the geographical distributions of the bird species in New Guinea. For example, each mountain range he explored was home to some species that lived nowhere else. Closely related species often lived on different mountaintops, separated by deep valleys of unsuitable habitat. In 1942, Mayr published the book *Systematics and the Origin of Species,* in which he described the role of geography in the evolution of new species; the book quickly became a cornerstone of the modern synthesis (which was outlined in Section 19.3).

What mechanisms produce distinct species? As you discovered in Chapter 20, microevolutionary processes alter the pattern and extent of genetic and phenotypic variation within populations. When these processes differ between populations, the populations will diverge, and they may eventually become so different that we recognize them as distinct species. Although Darwin's famous book was titled *On the Origin of Species,* he didn't dwell on the question of *how* new species arise. But the concept of **speciation**—the process of species formation—was implicit in his insight that similar species often share inherited characteristics and a common ancestry. Darwin also recognized that "descent with modification" had generated the amazing diversity of organisms on Earth.

Today evolutionary biologists view speciation as a *process,* a series of events that occur through time. However, they usually study the *products* of speciation, species that are alive today. Because they can rarely witness the process of speciation from start to finish,

scientists make inferences about it by studying organisms in various stages of species formation. In this chapter, we consider four major topics: how biologists define and recognize species; how species maintain their genetic identity; how the geographical distributions of organisms influence speciation; and how different genetic mechanisms produce new species.

21.1 What Is a Species?

Like the hunters of the Arfak Mountains, most of us recognize the different species that we encounter every day. We can distinguish a cat from a dog and sunflowers from roses. The concept of species is based on our perception that Earth's biological diversity is packaged in discrete, recognizable units, and not as a continuum of forms grading into one another. As evolutionary scientists learn more about the causes of microevolution, they refine our understanding of what a species really is.

The Morphological Species Concept Is a Practical Way to Identify Species

Biologists often describe new species on the basis of visible anatomical characteristics, a process that dates back to Linnaeus' classification of organisms in the eighteenth century (described in Chapter 23). This approach is based on the **morphological species concept,** the idea that all individuals of a species share measurable traits that distinguish them from individuals of other species.

The morphological species concept has many practical applications. For example, paleobiologists use morphological criteria to identify the species of fossilized organisms (see Chapter 22). And because we can observe the external traits of organisms in nature, field guides to plants and animals list diagnostic (that is, distinguishing) physical characters that allow us to recognize them **(Figure 21.2).**

Nevertheless, relying exclusively on a morphological approach can present problems. Consider the variation in the shells of *Cepaea nemoralis* (shown earlier in Figure 20.2). How could anyone imagine that so variable a collection of shells represents just one species

Yellow-throated warbler　　　Myrtle warbler

Figure 21.2

Diagnostic characters. Yellow-throated warblers *(Dendroica dominica)* and myrtle warblers *(Dendroica coronata)* can be distinguished by the color of feathers on the throat and rump.

of snail? Moreover, morphology does not help us distinguish some closely related species that are nearly identical in appearance. Finally, morphological species definitions tell us little about the evolutionary processes that produce new species.

The Biological and Phylogenetic Species Concepts Derive from Evolutionary Theory

The **biological species concept** emphasizes the dynamic nature of species. Ernst Mayr defined biological species as "groups of . . . interbreeding natural populations that are reproductively isolated from [do not produce fertile offspring with] other such groups." The concept is based on reproductive criteria and is easy to apply, at least in principle: if the members of two populations interbreed and produce fertile offspring *under natural conditions,* they belong to the same species; their fertile offspring will, in turn, produce the next generation of that species. If two populations do not interbreed in nature, or fail to produce fertile offspring when they do, they belong to different species.

The biological species concept defines species in terms of population genetics and evolutionary theory. The first half of Mayr's definition notes the genetic *cohesiveness* of species: populations of the same species experience gene flow, which mixes their genetic material. Thus, we can think of a species as one large gene pool, which may be subdivided into local populations.

The second part of the biological species concept emphasizes the genetic *distinctness* of each species. Because populations of different species are reproductively isolated, they cannot exchange genetic information. In fact, the process of speciation is frequently defined as the evolution of reproductive isolation between populations.

The biological species concept also explains why individuals of a species generally look alike: members of the same gene pool share genetic traits that determine their appearance. Individuals of different species generally do not resemble one another as closely because they share fewer genetic characteristics. In practice, biologists often use similarities or differences in morphological traits as convenient markers of genetic similarity or reproductive isolation.

However, the biological species concept does not apply to the many forms of life that reproduce asexually, including most bacteria; some protists, fungi, and plants; and a few animals. In these species, individuals don't interbreed, so it is pointless to ask whether different populations do. Similarly, we cannot use the biological species concept to study extinct organisms, because we have little or no data on their reproductive habits. These species must all be defined using morphological or biochemical criteria. Yet, despite its limitations, the biological species concept currently provides the best evolutionary definition of a sexually reproducing species.

Recognizing the limitations of the biological species concept, some researchers have proposed a **phylogenetic species concept.** Using both morphological and genetic sequence data, scientists first reconstruct the evolutionary tree for the populations of interest. They then define a phylogenetic species as a cluster of populations—the tiniest twigs on the tree—that emerge from the same small branch. Thus, a phylogenetic species comprises populations that share a recent evolutionary history. We will consider this approach for defining species as well as more inclusive evolutionary groups in Chapter 23.

Many Species Exhibit Substantial Geographical Variation

Populations change in response to shifting environments, and separate populations of a species frequently differ both genetically and phenotypically. Neighboring populations often have shared characteristics because they live in similar environments, exchange individuals, and experience comparable patterns of natural selection. Widely separated populations, by contrast, may live under different conditions and experience different patterns of selection; because gene flow is less likely to occur between distant populations, their gene pools and phenotypes often differ.

When geographically separated populations of a species exhibit dramatic, easily recognized phenotypic variation, biologists may identify them as different **subspecies (Figure 21.3),** which are local variants of a species. Individuals from different subspecies usually interbreed where their geographical distributions

Figure 21.3
Subspecies. Five subspecies of rat snake *(Elaphe obsoleta)* in eastern North America differ in color and in the presence or absence of stripes or blotches.

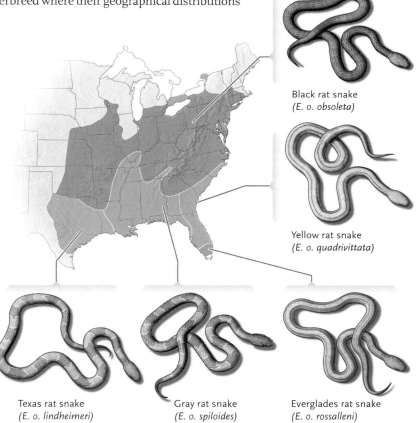

Black rat snake
(E. o. obsoleta)

Yellow rat snake
(E. o. quadrivittata)

Texas rat snake
(E. o. lindheimeri)

Gray rat snake
(E. o. spiloides)

Everglades rat snake
(E. o. rossalleni)

Figure 21.4

Ring species. Six of the seven subspecies of the salamander *Ensatina eschscholtzii* are distributed in a ring around California's Central Valley. Subspecies often interbreed where their geographical distributions overlap. However, the two subspecies that nearly close the ring in the south (marked with an arrow), the Monterey salamander and the yellow-blotched salamander, rarely interbreed.

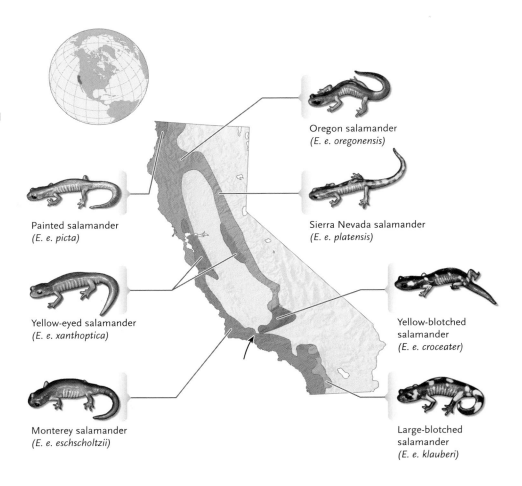

Oregon salamander
(*E. e. oregonensis*)

Painted salamander
(*E. e. picta*)

Sierra Nevada salamander
(*E. e. platensis*)

Yellow-eyed salamander
(*E. e. xanthoptica*)

Yellow-blotched salamander
(*E. e. croceater*)

Monterey salamander
(*E. e. eschscholtzii*)

Large-blotched salamander
(*E. e. klauberi*)

meet, and their offspring often exhibit intermediate phenotypes. Biologists sometimes use the word "race" as shorthand for the term "subspecies."

Various patterns of geographical variation have provided great insight into the speciation process. Two of the best-studied patterns are *ring species* and *clinal variation*.

Ring Species. Some plant and animal species have a ring-shaped geographical distribution that surrounds uninhabitable terrain. Adjacent populations of these so-called **ring species** can exchange genetic material directly, but gene flow between distant populations occurs only through the intermediary populations.

The lungless salamander *Ensatina eschscholtzii*, an example of a ring species, is widely distributed in the coastal mountains and the Sierra Nevada of California, but it cannot survive in the hot, dry Central Valley **(Figure 21.4)**. Seven subspecies differ in biochemical traits, color, size, and ecology. Individuals from adjacent subspecies often interbreed where their geographical distributions overlap, and intermediate phenotypes are fairly common. But at the southern end of the Central Valley, adjacent subspecies rarely interbreed. Apparently, they have differentiated to such an extent that they can no longer exchange genetic material directly.

Are the southernmost populations of this salamander subspecies or different species? A biologist who saw *only* the southern populations, which coexist without interbreeding, might define them as separate species. However, they still have the potential to exchange genetic material through the intervening populations that form the ring. Hence, biologists recognize these populations as belonging to the same species. Most likely, the southern subspecies are in an intermediate stage of species formation.

Clinal Variation. When a species is distributed over a large, environmentally diverse area, some traits may exhibit a **cline**, a pattern of smooth variation along a geographical gradient. Clinal variation usually results from gene flow between adjacent populations that are each adapting to slightly different conditions. For example, many birds and mammals in the northern hemisphere show clinal variation in body size **(Figure 21.5)** and the relative length of their appendages: in general, populations living in colder environments have larger bodies and shorter appendages, a pattern that is usually interpreted as a mechanism to conserve heat (see Chapter 46). If a cline extends over a large geographical gradient, populations at the opposite ends may be very different.

Despite the geographical variation that many species exhibit, most closely related species are genetically and morphologically different from each other. In the next section, we consider the mechanisms that maintain the genetic distinctness of closely related species by preventing their gene pools from mixing.

21.2 Maintaining Reproductive Isolation

Reproductive isolation is central to the biological species concept. A **reproductive isolating mechanism** is a biological characteristic that prevents the gene pools of two species from mixing. Biologists classify reproductive isolating mechanisms into two categories (summarized in **Table 21.1**): **prezygotic isolating mechanisms** exert their effects before the production of a zygote, or fertilized egg, and **postzygotic isolating mechanisms** operate after zygote formation. These isolating mechanisms are not mutually exclusive; two or more of them may operate simultaneously.

Prezygotic Isolating Mechanisms Prevent the Production of Hybrid Individuals

Biologists have identified five mechanisms that can prevent interspecific (between species) matings or fertilizations, and thus prevent the production of hybrid (mixed species) offspring. These five prezygotic mechanisms are *ecological, temporal, behavioral, mechanical,* and *gametic isolation.*

Species living in the same geographical region may experience **ecological isolation** if they live in different habitats. For example, lions and tigers were both common in India until the mid-nineteenth century, when hunters virtually exterminated the Asian lions. However, because lions live in open grasslands and tigers in dense forests, the two species did not encounter one another and did not interbreed. Lion-tiger hybrids are sometimes born in captivity, but do not occur under natural conditions.

Species living in the same habitat can experience **temporal isolation** if they mate at different times of day or different times of year. For example, the fruit flies *Drosophila persimilis* and *Drosophila pseudo-obscura* overlap extensively in their geographical distributions, but they do not interbreed, in part because *D. persimilis* mates in the morning and *D. pseudo-obscura* in the

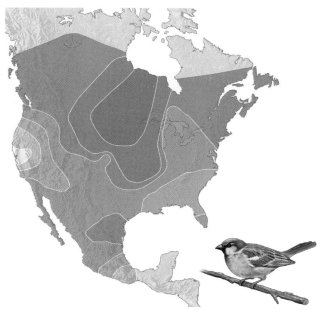

Figure 21.5

Clinal variation. House sparrows *(Passer domesticus)* exhibit clinal variation in overall body size, which was summarized from measurements of 16 skeletal features. Darker shading indicates larger size.

Table 21.1	Reproductive Isolating Mechanisms	
Timing Relative to Fertilization	**Mechanism**	**Mode of Action**
Prezygotic ("premating") mechanisms	Ecological isolation	Species live in different habitats
	Temporal isolation	Species breed at different times
	Behavioral isolation	Species cannot communicate
	Mechanical isolation	Species cannot physically mate
	Gametic isolation	Species have nonmatching receptors on gametes
Postzygotic ("postmating") mechanisms	Hybrid inviability	Hybrid offspring do not complete development
	Hybrid sterility	Hybrid offspring cannot produce gametes
	Hybrid breakdown	Hybrid offspring have reduced survival or fertility

KEY

1	P. consimilis	**4**	P. collustrans	**7**	P. ignitus
2	P. brimleyi	**5**	P. marginellus	**8**	P. pyralis
3	P. carolinus	**6**	P. consanguineus	**9**	P. granulatus

Figure 21.6

Behavioral reproductive isolation. Male fireflies (*Photinus* species) use bioluminescent signals to attract potential mates. The different flight paths and flashing patterns of males in nine North American species are represented here. Females respond only to the display given by males of their own species.

(Courtesy of James E. Lloyd. Miscellaneous Publications of the Museum of Zoology of the University of Michigan, 130:1–195, 1966.)

afternoon. Two species of pine in California are reproductively isolated where their geographical distributions overlap: even though both rely on the wind to carry male gametes (pollen grains) to female gametes (ova) in other cones, *Pinus radiata* releases pollen in February and *Pinus muricata* releases pollen in April.

Many animals rely on specific signals, which often differ dramatically between species, to identify the species of a potential mate. **Behavioral isolation** results when the signals used by one species are not recognized by another. For example, female birds rely on the song, color, and displays of males to identify members of their own species. Similarly, female fireflies identify males by their flashing patterns **(Figure 21.6)**. These behaviors (collectively called *courtship displays*) are often so complicated that signals sent by one species are like a foreign language that another species simply does not understand.

Mate choice by females and sexual selection (discussed in Section 20.3) generally drive the evolution of mate recognition signals. Females often spend substantial energy in reproduction, and choosing an appropriate mate—that is, a male of her own species—is critically important for the production of successful young. By contrast, a female that mates with a male from a different species is unlikely to leave any surviving offspring at all. Over time, the number of males with recognizable traits, as well as the number of females able to recognize the traits, increases in the population.

Differences in the structure of reproductive organs or other body parts—**mechanical isolation**—may prevent individuals of different species from interbreeding. In particular, many plants have anatomical features that allow only certain pollinators, usually particular bird or insect species, to collect and distribute pollen (see Chapter 27). For example, the flowers and nectar of two native California plants, the monkey-flowers *Mimulus lewisii* and *Mimulus cardinalis,* attract different animal pollinators **(Figure 21.7)**. *Mimulus lewisii* is pollinated by bumblebees. It has shallow pink flowers with broad petals that provide a landing platform for the bees. Bright yellow streaks on the petals serve as "nectar guides," directing bumblebees to the short nectar tube and reproductive parts, which are located among the petals. Bees enter the flowers to drink their concentrated nectar, and they pick up and deliver pollen as they brush against the reproductive parts of the flowers. *Mimulus cardinalis,* by contrast, is pollinated by hummingbirds. It has long red flowers with no yellow streaks, and the reproductive parts extend above the petals. The red color attracts hummingbirds but lies outside the color range detected by bumblebees. The nectar of *M. cardinalis* is more dilute than that of *M. lewisii* but is produced in much greater quantity, making it easier for hummingbirds to

Mimulus lewisii   *Mimulus cardinalis*

Figure 21.7

Mechanical reproductive isolation. Because of differences in floral structure, two species of monkey-flower attract different animal pollinators. *Mimulus lewisii* attracts bumblebees and *Mimulus cardinalis* attracts hummingbirds.

ingest. When a hummingbird visits *M. cardinalis* flowers, it pushes its long bill down the nectar tube, and its forehead touches the reproductive parts, picking up and delivering pollen. Recent research has demonstrated that where the two monkey-flower species grow side-by-side, animal pollinators restrict their visits to either one species or the other 98% of the time, providing nearly complete reproductive isolation.

Even when individuals of different species mate, **gametic isolation,** an incompatibility between the sperm of one species and the eggs of another, may prevent fertilization. Many marine invertebrates release gametes into the environment for external fertilization. The sperm and eggs of each species recognize one another's complementary surface proteins (see Chapter 47), but the surface proteins on the gametes of different species don't match. In animals with internal fertilization, sperm of one species may not survive and function within the reproductive tract of another. Interspecific matings between some *Drosophila* species, for example, induce a reaction in the female's reproductive tract that blocks "foreign" sperm from reaching eggs. Parallel physiological incompatibilities between a pollen tube and a stigma prevent interspecific fertilization in some plants.

Postzygotic Isolating Mechanisms Reduce the Success of Hybrid Individuals

If prezygotic isolating mechanisms between two closely related species are incomplete or ineffective, sperm from one species sometimes fertilizes an egg of the other species. In such cases the two species will be reproductively isolated if their offspring, called interspecific (between species) hybrids, have lower fitness than those produced by intraspecific (within species) matings. Three postzygotic isolating mechanisms—*hybrid inviability, hybrid sterility,* and *hybrid breakdown*—can reduce the fitness of hybrid individuals.

Many genes govern the complex processes that transform a zygote into a mature organism. Hybrid individuals have two sets of developmental instructions, one from each parent species, which may not interact properly for the successful completion of embryonic development. As a result, hybrid organisms frequently die as embryos or at an early age, a phenomenon called **hybrid inviability.** For example, domestic sheep and goats can mate and fertilize one another's ova, but the hybrid embryos always die before coming to term, presumably because the developmental programs of the two parent species are incompatible.

Although some hybrids between closely related species develop into healthy and vigorous adults, they may not produce functional gametes. This **hybrid sterility** often results when the parent species differ in the number or structure of their chromosomes, which cannot pair properly during meiosis. Such hybrids have zero fitness because they leave no descendants. The most familiar example is a mule, the product of mating be-

Figure 21.8
Interspecific hybrids. Horses and zebroids (hybrid offspring of horses and zebras) run in a mixed herd. Zebroids are usually sterile.

tween a female horse ($2n = 64$) and a male donkey ($2n = 62$). Zebroids, the offspring of matings between horses and zebras, are also usually sterile **(Figure 21.8)**.

Some first-generation hybrids (F_1; see Section 12.1) are healthy and fully fertile. They can breed with other hybrids and with both parental species. However, the second generation (F_2), produced by matings between F_1 hybrids, or between F_1 hybrids and either parental species, may exhibit reduced survival or fertility, a phenomenon known as **hybrid breakdown.** For example, experimental crosses between *Drosophila* species may produce functional hybrids, but their offspring experience a high rate of chromosomal abnormalities and harmful types of genetic recombination. Thus, reproductive isolation is maintained between the species because there is little long-term mixing of their gene pools.

STUDY BREAK

1. What is the difference between prezygotic and postzygotic isolating mechanisms?
2. When a male duck of one species performed a courtship display to a female of another species, she interpreted his behavior as aggressive rather than amorous. What type of reproductive isolating mechanism does this scenario illustrate?

21.3 The Geography of Speciation

As Ernst Mayr recognized, geography has a huge impact on whether gene pools have the opportunity to mix. Biologists define three modes of speciation, based on the geographical relationship of populations as they become

1 At first, a population is distributed over a large geographical area.

2 A geographical change, such as the advance of a narrow glacier, separates the original population, creating a barrier to gene flow.

3 In the absence of gene flow, the separated populations evolve independently and diverge into different species.

4 When the glacier later melts, allowing individuals of the two species to come into secondary contact, they do not interbreed.

Figure 21.9
The model of allopatric speciation and secondary contact.

reproductively isolated: *allopatric speciation* (*allo* = different; *patria* = homeland), *parapatric speciation* (*para* = beside), and *sympatric speciation* (*sym* = together).

Allopatric Speciation Occurs between Geographically Separated Populations

Allopatric speciation may take place when a physical barrier subdivides a large population or when a small population becomes separated from a species' main geographical distribution. Probably the most common mode of speciation in large animals, allopatric speciation occurs in two stages. First, two populations become *geographically* separated, preventing gene flow between them. Then, as the populations experience distinct mutations as well as different patterns of natural selection and genetic drift, they may accumulate genetic differences that isolate them *reproductively*.

Geographical separation sometimes occurs when a barrier divides a large population into two or more units **(Figure 21.9)**. For example, hurricanes may create new channels that divide low coastal islands and the populations inhabiting them. Uplifting mountains or landmasses as well as advancing glaciers can also pro-

duce barriers that subdivide populations. The uplift of the Isthmus of Panama, caused by movements of Earth's crust about five million years ago (see the *Focus on Research* in Chapter 22), separated a once-continuous shallow sea into the eastern tropical Pacific Ocean and the western tropical Atlantic Ocean. Populations of marine organisms were subdivided by this event, and pairs of closely related species now live on either side of this divide **(Figure 21.10)**.

In other cases, small populations may become isolated at the edge of a species' geographical distribution. Such peripheral populations often differ genetically from the central population because they are adapted to somewhat different environments. Once a small population is isolated, genetic drift and natural selection as well as limited gene flow from the parent population foster further genetic differentiation. In time, the accumulated genetic differences may lead to reproductive isolation.

Populations on oceanic islands represent extreme examples of this phenomenon. Founder effects, an example of genetic drift (see Section 20.3), make the populations genetically distinct. And on oceanic archipelagos, such as the Galápagos and Hawaiian islands, individuals from one island may colonize nearby islands, found-

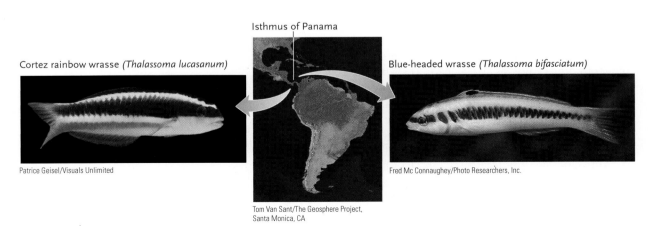

Cortez rainbow wrasse *(Thalassoma lucasanum)*

Isthmus of Panama

Blue-headed wrasse *(Thalassoma bifasciatum)*

Patrice Geisel/Visuals Unlimited

Tom Van Sant/The Geosphere Project, Santa Monica, CA

Fred Mc Connaughey/Photo Researchers, Inc.

Figure 21.10
Geographical separation. The uplift of the Isthmus of Panama divided an ancestral wrasse population. The Cortez rainbow wrasse now occupies the eastern Pacific Ocean, and the blue-headed wrasse now occupies the western Atlantic Ocean.

ing populations that differentiate into distinct species. Each island may experience multiple invasions, and the process may be repeated many times within the archipelago, leading to the evolution of a **species cluster**, a group of closely related species recently descended from a common ancestor **(Figure 21.11)**. The nearly 800 species of fruit flies on the Hawaiian Islands, described in *Focus on Research*, form several species clusters.

Sometimes, allopatric populations reestablish contact when a geographical barrier is eliminated or breached (see Figure 21.9, step 4). This *secondary contact* provides a test of whether or not the populations have diverged into separate species. If their gene pools did not differentiate much during geographical separation, the populations will interbreed and merge. But if the populations have differentiated enough to be reproductively isolated, they have become separate species.

During the early stages of secondary contact, prezygotic reproductive isolation may be incomplete. Some members of each population may mate with individuals from the other, producing viable, fertile offspring, in areas called **hybrid zones**. Although some hybrid zones have persisted for hundreds or thousands of years **(Figure 21.12)**, they are generally narrow, and ecological or geographical factors maintain the separation of the gene pools for the majority of individuals in both species.

If hybrid offspring have lower fitness than those produced within each population, natural selection will favor individuals that mate only with members of their own population. Recent studies of *Drosophila* suggest that this phenomenon, called **reinforcement**, enhances reproductive isolation that had begun to develop while the populations were geographically separated. Thus, natural selection may promote the evolution of prezygotic isolating mechanisms.

Parapatric Speciation May Occur between Adjacent Populations

Sometimes a single species is distributed across a discontinuity in environmental conditions, such as a major change in soil type. Although organisms on one side of the discontinuity may interbreed freely with those on the other side, natural selection may favor different alleles on either side, limiting gene flow. In such cases, **parapatric speciation**—speciation arising between adjacent populations—may occur if hybrid offspring have low relative fitness.

Some strains of bent grass *(Agrostis tenuis)*, a common pasture plant in Great Britain, have the physiological ability to grow on mine tailings where the soil is heavily polluted by copper or other metals. Plants of the copper-tolerant strains grow well on polluted soils, but plants of the pasture strain do not. Conversely, copper-tolerant plants don't survive as well as pasture plants on unpolluted soils. These strains often grow within a few meters of each other where polluted and unpolluted soils form an intricate mosaic. Because

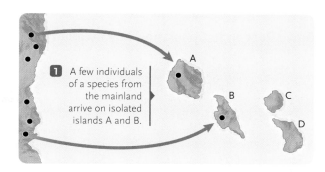

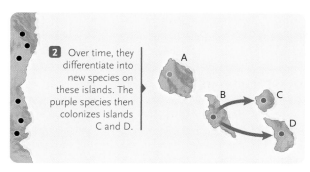

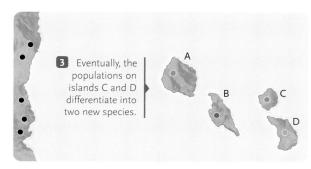

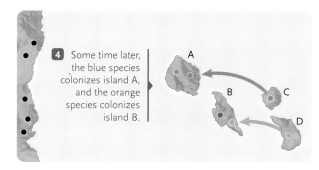

Figure 21.11

Evolution of a species cluster on an archipelago. Letters identify four islands in a hypothetical archipelago, and colored dots represent different species. The ancestor of all the species is represented by black dots on the mainland. At the end of the process, islands A and B are each occupied by two species, and islands C and D are each occupied by one species, all of which evolved on the islands.

bent grass is wind-pollinated, pollen is readily transferred from one strain to another.

Thomas McNeilly and Janis Antonovics of University College of North Wales crossed these strains in the laboratory and determined that they are still fully interfertile. However, copper-tolerant plants flower about one week earlier than nearby pasture plants, which promotes prezygotic (temporal) isola-

Basic Research: Speciation in Hawaiian Fruit Flies

After Darwin published his analyses of island species, evolutionary biologists realized that oceanic archipelagos provide "natural laboratories" for studies of speciation. The islands of the Hawaiian archipelago have been geographically isolated throughout their history, lying at least 3200 km (1900 miles) from the nearest continents or other islands **(Figure a).** They were built by undersea volcanic eruptions over hundreds of thousands of years and emerged from the sea from northwest to southeast: Kauai is at least 5 million years old, and Hawaii, the "Big Island," is less than 1 million years old. Individual islands differ in maximum elevation and include a wide range of habitats, from dry zones of sparse vegetation to wet tracts of lush forest.

Resident species must have arrived from distant mainland localities or evolved on the islands from colonizing ancestors. The islands' isolation, differ-

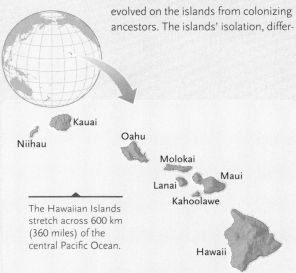

Kauai

Niihau

Oahu

Molokai

Maui

Lanai

Kahoolawe

Hawaii

The Hawaiian Islands stretch across 600 km (360 miles) of the central Pacific Ocean.

Figure a
The Hawaiian Islands

Drosophila heteroneura   *Drosophila silvestris*

Kenneth Y. Kaneshiro, University of Hawaii

Kenneth Y. Kaneshiro, University of Hawaii

Figure b
Two *Drosophila* species in which the males' head shapes differ.

ent ages, and geographical and ecological complexity provide environmental conditions that foster repeated interisland colonizations followed by allopatric speciation events. Thus, it is not surprising that species clusters have evolved in several groups of organisms (including flowering plants, insects, and birds).

Nearly 800 species of fruit flies have been identified on the archipelago, and most species live on only one island. Biologists used many characters to identify the different fruit fly species, including external and internal anatomy, cell structure, chromosome structure, ecology, and mating behavior. Their data suggest that the vast majority of native Hawaiian species arose from one ancestral species that colonized the archipelago long ago, probably from eastern Asia. After repeated speciation events, the fruit flies of the Hawaiian Islands represent more than 25% of all known fruit fly species.

Hampton Carson, now of the University of Hawaii, has spearheaded studies on the evolutionary relationships of Hawaiian fruit flies. He and his colleagues have gathered data on hundreds of fly species—a daunting task. Most species are sexually dimorphic. Although the females of different species may be similar in appearance, the males of even closely related species differ in virtually every aspect of their external anatomy: body size, head shape, and the structure of their eyes, antennae, mouthparts, bristles, legs, and wings. Their mating behavior and choice of mating sites also vary dramatically.

Nevertheless, closely related species on different islands occupy comparable habitats and associate with related plant species. Carson suggests that speciation in these flies resulted from the evolution of different genetically determined *mating systems,* the behaviors and sexual characteristics that males display when seeking a mate. The mating systems serve as prezygotic isolating mechanisms.

The 100 or more species of "picture-wing" *Drosophila,* relatively large flies with patterns on their wings, illustrate the evolution of a species cluster. Carson and his colleagues used similarities and differences in the banding patterns on

the flies' giant salivary chromosomes (described in the *Focus on Research* in Chapter 13), to trace the evolutionary origin of species on the younger islands by identifying their closest relatives on the older islands. Their analysis of 26 species on Hawaii, the youngest island, suggests that flies from the older islands colonized Hawaii at least 19 different times, and each founder population evolved into a new species there. Additional species apparently evolved when lava flows on Hawaii subdivided existing populations.

Among the picture-wing fruit flies, some interspecies matings result in hybrid sterility or hybrid breakdown. But for the majority of species, prezygotic reproductive isolation is maintained by differences in their mating systems. For example, *Drosophila silvestris* and *Drosophila heteroneura,* which produce healthy and fertile hybrids in the laboratory, have similar geographical distributions; however, differences in courtship behavior and in the shape of the males' heads, a characteristic that females use to recognize males of their own species **(Figure b),** keep these two species reproductively isolated. In nature, they hybridize only in one small geographical area.

The work of Carson and his colleagues suggests that most speciation in Hawaiian *Drosophila* has resulted from founder effects. When a fertile female—or a small group of males and females—moves to a new island, this founding population responds to novel selection pressures in its new environment. Sexual selection then exaggerates distinctive morphological and behavioral characteristics, maintaining the population's reproductive isolation from its new neighbors. The tremendous variety of Hawaiian fruit flies has undoubtedly been produced by repeated colonizations of newer islands by flies from older islands and by the back-colonization of older islands by newly evolved species. Thus, they represent what evolutionary biologists describe as an *adaptive radiation,* a cluster of closely related species that are ecologically different (as described further in Chapter 22).

Bullock's oriole

Baltimore oriole

KEY
- ■ Bullock's oriole
- ▨ Hybrid zone
- □ Baltimore oriole

Figure 21.12

Hybrid zones. Males of the Baltimore oriole *(Icterus galbula)* and Bullock's oriole *(Icterus bullockii)* differ in color and courtship song. The populations have maintained a hybrid zone for hundreds of years, and once were considered subspecies of the same species. The American Ornithologists' Union recognized them as separate species in 1997. They now hybridize less frequently than they once did, leading some researchers to suggest that their reproductive isolation evolved recently.

tion of the two strains **(Figure 21.13)**. If the flowering times become further separated, the two strains may attain complete reproductive isolation and become separate species.

Some biologists argue that the places where parapatric populations of bent grass interbreed are really hybrid zones where allopatric populations have established secondary contact. Unfortunately, there is no way to determine whether the hybridizing populations were parapatric or allopatric in the past. Thus, a thorough evaluation of the parapatric speciation hypothesis must await the development of techniques that enable biologists to distinguish clearly between the products of allopatric and parapatric speciation.

Sympatric Speciation Occurs within One Continuously Distributed Population

In **sympatric speciation**, reproductive isolation evolves between distinct subgroups that arise within one population. Models of sympatric speciation do not require that the populations be either geographically or environmentally separated as their gene pools diverge. We examine below general models of sympatric speciation in animals and plants; the genetic basis of sympatric speciation is one of the topics we consider in the next section.

Insects that feed on just one or two plant species are among the animals most likely to evolve by sympatric speciation. These insects generally carry out most important life cycle activities on or near their "host" plants. Adults mate on the host plant; females lay their eggs on it; and larvae feed on the host plant's tissues, eventually developing into adults, which initiate another round of the life cycle. Host plant choice is

genetically determined in many insect species. In others, individuals associate with the host plant species they ate as larvae.

Theoretically, a genetic mutation could suddenly change some insects' choice of host plant. Mutant individuals would shift their life cycle activities to the new host, and then interact primarily with others preferring the same new host, an example of ecological isolation. These individuals would collectively form a separate subpopulation, called a **host race.** Reproductive isolation could evolve between different host races if the individuals of each host race are more likely to mate with members of their own host race than with members of another. Some biologists criticize this model, however, because it assumes that the genes controlling two traits, the insects' host plant choice and their mating preferences, change simultaneously. Moreover, host plant choice is controlled by multiple gene loci in some insect species, and it is clearly influenced by prior experience in others.

The apple maggot *(Rhagoletis pomonella)* is the most thoroughly studied example of possible sympatric speciation in animals **(Figure 21.14).** This fly's natural host plant in eastern North America is the hawthorn *(Crataegus* species), but at least two host races have appeared in little more than 100 years. The larvae of a new host race were first discovered feeding on apples in New York state in the 1860s. In the 1960s, a cherry-feeding host race appeared in Wisconsin.

Recent research has shown that variations at just a few gene loci underlie differences in the feeding preferences of *Rhagoletis* host races; other genetic differences cause the host races to develop at different rates. Moreover, adults of the three races mate during different summer months. Nevertheless, individuals

Figure 21.13 Observational Research

Evidence for Reproductive Isolation in Bent Grass

QUESTION: Do adjacent populations of bent grass *(Agrostis tenuis)* living on different soil types exhibit any signs of reproductive isolation?

HYPOTHESIS: McNeilly and Antonovics hypothesized that adjacent populations of bent grass flowered at slightly different times, which could foster prezygotic reproductive isolation between them.

METHODS: On a late summer day in 1965, the researchers compared the flowers of bent grass growing on polluted soil at a copper mine with those of plants growing on unpolluted soil in a nearby pasture. A meter-wide stretch of polluted pasture (indicated by cross-hatching) formed a boundary between the two populations. Researchers assigned a score to every flower, with immature flowers scored as 3 and mature flowers as 4.

RESULTS: On the day that they were surveyed, flowers of the copper-tolerant plants had higher scores, indicating that they were more mature—and thus would complete pollination earlier—than the flowers of the pasture plants.

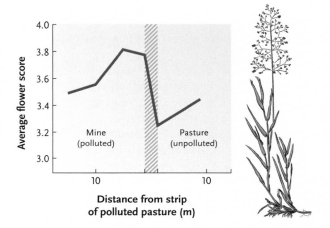

CONCLUSION: Because adjacent populations of bent grass flower at slightly different times, temporal reproductive isolation may be developing between them.

Figure 21.14
Sympatric speciation in animals. Male and female apple maggots *(Rhagoletis pomonella)* court on a hawthorn leaf. The female will later lay her eggs on the fruit, and the offspring will feed, mate, and lay their eggs on hawthorns as well.

ing with individuals of the parent species. Nearly half of all flowering plant species are polyploid, including many important crops and ornamental species. The genetic mechanisms that produce polyploid individuals in plant populations are well understood; we describe them in detail as part of a larger discussion of the genetics of speciation.

STUDY BREAK

1. What are the two stages required for allopatric speciation?
2. What factor appears to promote parapatric speciation in bent grass?
3. Why might insects from different host races be unlikely to mate with each other?

21.4 Genetic Mechanisms of Speciation

What genetic changes lead to reproductive isolation between populations, and how do these changes arise? In this section we examine three genetic mechanisms that can lead to reproductive isolation: *genetic divergence* between allopatric populations, *polyploidy* in sympatric populations, and *chromosome alterations,* which occur independently of the geographical distributions of populations.

Genetic Divergence in Allopatric Populations Can Lead to Speciation

In the absence of gene flow, geographically separated populations inevitably accumulate genetic differences. Most postzygotic isolating mechanisms probably develop as accidental by-products of mutation, genetic

show no particular preference for mates of their own host race, at least under simplified laboratory conditions. Thus, although behavioral isolation has not developed between races, ecological and temporal isolation may separate adults in nature. Researchers are still not certain that the different host races are reproductively isolated under natural conditions.

Sympatric speciation often occurs in plants through a genetic phenomenon, **polyploidy,** in which an individual receives one or more *extra* copies of the entire haploid complement of chromosomes (see Section 13.3). As we explain in the next section, polyploidy can lead to speciation because these large-scale genetic changes may prevent polyploid individuals from breed-

Monkey-Flower Speciation

Reproductive isolation is the primary criterion that biologists use to distinguish species. A molecular study by H. D. Bradshaw and his coworkers at the University of Washington indicates that the amount of genetic change required to establish reproductive isolation, and thus new species, may be surprisingly small in some cases.

These scientists studied two monkey-flower species, *Mimulus lewisii* and *Mimulus cardinalis,* that experience mechanical reproductive isolation because differences in flower structure keep bumblebees or hummingbirds from carrying pollen from one species to the other (see Figure 21.7). Although these species do not hybridize in nature, they are easily crossed in the laboratory and produce fertile hybrids. The F_2 offspring of the laboratory crosses have flowers with various forms intermediate between the parental *lewisii* and *cardinalis* types, indicating that several gene loci control the traits separating the species. But how many?

Relatively little is known about the genetics of the two monkey-flower species, so a direct genetic analysis of their hereditary differences was impractical. Instead, the investigators studied 153 randomly chosen DNA sequences distributed throughout the haploid number of eight chromosomes in the two species. They correlated the distribution of these sequences with the distribution of flower traits in 93 plants of the F_2 generation. Some of the DNA sequences segregated so closely with a particular trait, such as yellow pigment, that they are almost certainly located near that trait in the chromosomes. Because the sequences can pair with complementary DNA in the chromosomes, the investigators used them as "probes" to find the sites in the chromosomes from which they originated. From the close linkage of the sequences to the traits, the investigators could estimate the positions and approximate number of genes that establish reproductive isolation.

Their results indicate that reproductive isolation of *M. lewisii* and *M. cardinalis* results from differences in eight floral traits—the amount of (1) anthocyanin pigments and (2) carotenoid pigments in petals; (3) flower width; (4) petal width; (5) nectar volume; (6) nectar concentration; and the lengths of the stalks supporting the (7) male and (8) female reproductive parts. Although the investigators could not directly determine the number of genes controlling each trait, the characteristics of the traits, their locations at eight sites on six of the chromosomes, and their pattern of inheritance make it most likely that each trait is controlled by a single gene, giving a likely minimum of eight genes. Thus mutations in as few as eight genes may have established reproductive isolation and speciation in the monkey-flowers.

This research was the first in which random differences in DNA sequences were used to answer the fundamental evolutionary question of how much genetic change is needed to produce a new species.

---

drift, and natural selection. Note, however, that natural selection cannot promote the evolution of reproductive isolating mechanisms between *allopatric* populations directly: individuals in such populations do not encounter one another and therefore have no opportunity to produce hybrid offspring. And if there are no hybrid offspring, natural selection cannot select against the matings that would have produced them. Nevertheless, natural selection may sometimes foster adaptive changes that create postzygotic reproductive isolation between populations when they later reestablish contact. And, if postzygotic isolating mechanisms reduce the fitness of hybrid offspring, natural selection can reinforce the evolution of prezygotic isolating mechanisms.

How much genetic divergence is necessary for speciation to occur? To understand the genetic basis of speciation in closely related species, researchers first identify the specific causes of reproductive isolation. They then use standard techniques of genetic analysis along with new molecular approaches such as gene mapping and sequencing to analyze the genetic mechanisms that establish reproductive isolation. As explained in *Insights from the Molecular Revolution,* these techniques now allow researchers to determine the minimum number of genes responsible for reproductive isolation in particular pairs of species.

In cases of postzygotic reproductive isolation, mutations in at least a few gene loci establish reproductive isolation. For example, if two common aquarium fishes, swordtails *(Xiphophorus helleri)* and platys *(Xiphophorus maculatus),* mate, two genes induce the development of lethal tumors in their hybrid offspring. When hybrid sterility is the primary cause of reproductive isolation between *Drosophila* species, at least 5 gene loci are responsible. About 55 gene loci contribute to postzygotic reproductive isolation between the toads *Bombina bombina* and *Bombina variegata.*

In cases of prezygotic reproductive isolation, some mechanisms have a surprisingly simple genetic basis. For example, a single mutation reverses the direction of coiling in the shells of some snail species. Snails with shells that coil in opposite directions cannot approach each other closely enough to mate, making reproduction between them mechanically impossible.

Many traits that now function as prezygotic isolating mechanisms may originally have evolved in response to sexual selection (described in Section 20.3).

Figure 21.15

Sexual selection and prezygotic isolation. In closely related species, such as mallard ducks *(Anas platyrhynchos)* and pintails *(Anas acuta)*, males have much more distinctive coloration than females, a sure sign of sexual selection at work.

Mallards

Pintails

This evolutionary process exaggerates showy structures and courtship behaviors in males, the traits that females use to identify appropriate mates. When two species encounter one another on secondary contact, these traits may also prevent interspecific mating. For example, many closely related duck species exhibit dramatic variation in the appearance of males, but not females **(Figure 21.15)**, an almost certain sign of sexual selection. Yet these species hybridize readily in captivity, producing offspring that are both viable and fertile. Speciation in these birds probably resulted from geographical isolation and sexual selection without significant genetic divergence: only a few morphological and behavioral characters are responsible for their reproductive isolation. Thus, sometimes the evolution of reproductive isolation may not require much genetic change at all.

Polyploidy Is a Common Mechanism of Sympatric Speciation in Plants

Polyploidy is common among plants, and it may be an important factor in the evolution of some fish, amphibian, and reptile species. Polyploid individuals can arise from chromosome duplications within a single species (autopolyploidy) or through hybridization of different species (allopolyploidy).

Autopolyploidy. In **autopolyploidy (Figure 21.16),** a diploid ($2n$) individual may produce, for example, tetraploid ($4n$) offspring, each of which has four complete chromosome sets. Autopolyploidy often results when gametes, through an error in either mitosis or meiosis, spontaneously receive the same number of chromosomes as a somatic cell. Such gametes are called **unreduced gametes** because their chromosome number has not been reduced compared with that of somatic cells.

Diploid pollen can fertilize the diploid ovules of a self-fertilizing individual, or it may fertilize diploid ovules on another plant with unreduced gametes. The resulting tetraploid offspring can reproduce either by self-pollination or by breeding with other tetraploid individuals. However, a tetraploid plant cannot produce fertile offspring by hybridizing with its diploid parents. The fusion of a diploid gamete with a normal haploid gamete produces a triploid ($3n$) offspring, which is usually sterile because its odd number of chromosomes cannot segregate properly during meiosis. Thus, the tetraploid is reproductively isolated from the original diploid population. Many species of grasses, shrubs, and ornamental plants, including violets, chrysanthemums, and nasturtiums, are autopolyploids, having anywhere from four to 20 complete chromosome sets.

Allopolyploidy. In **allopolyploidy (Figure 21.17),** two closely related species hybridize and subsequently form polyploid offspring. Hybrid offspring are sterile if the two parent species have diverged enough that their chromosomes do not pair properly during meiosis. However, if the hybrid's chromosome number is doubled, the chromosome complement of the gametes is also doubled, producing homologous chromosomes that *can* pair during meiosis. The hybrid can then produce polyploid gametes and, through self-fertilization or fertilization with other doubled hybrids, establish a population of a new polyploid species. Compared with speciation by genetic divergence, speciation by allopolyploidy is extremely rapid, causing a new species to arise in one generation without geographical isolation.

Meiosis

Self-fertilization

$2n = 6$

$4n = 12$

Diploid parent karyotype

Through an error in meiosis, a spontaneous doubling of chromosomes produces diploid gametes.

Fertilization of one diploid gamete by another produces a tetraploid zygote (offspring).

Figure 21.16

Speciation by autopolyploidy in plants. A spontaneous doubling of chromosomes during meiosis produces diploid gametes. If the plant fertilizes itself, a tetraploid zygote will be produced.

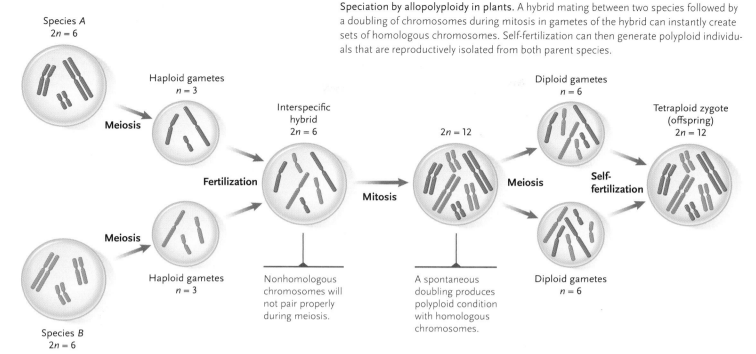

Figure 21.17

Speciation by allopolyploidy in plants. A hybrid mating between two species followed by a doubling of chromosomes during mitosis in gametes of the hybrid can instantly create sets of homologous chromosomes. Self-fertilization can then generate polyploid individuals that are reproductively isolated from both parent species.

Species *A*
2*n* = 6

Haploid gametes
n = 3

Meiosis

Interspecific hybrid
2*n* = 6

Fertilization

Mitosis

2*n* = 12

Diploid gametes
n = 6

Meiosis

Self-fertilization

Tetraploid zygote (offspring)
2*n* = 12

Meiosis

Haploid gametes
n = 3

Species *B*
2*n* = 6

Nonhomologous chromosomes will not pair properly during meiosis.

A spontaneous doubling produces polyploid condition with homologous chromosomes.

Diploid gametes
n = 6

Even when sterile, polyploids are often robust, growing larger than either parent species. For that reason, both autopolyploids and allopolyploids have been important to agriculture. For example, the wheat used to make flour *(Triticum aestivum)* has six sets of chromosomes **(Figure 21.18)**. Other polyploid crop plants include plantains (cooking bananas), coffee, cotton, potatoes, sugarcane, and tobacco.

Plant breeders often try to increase the probability of forming an allopolyploid by using chemicals that

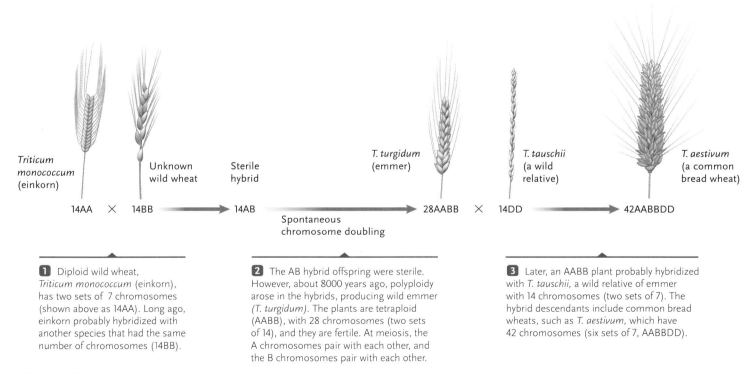

Triticum monococcum (einkorn)

Unknown wild wheat

Sterile hybrid

T. turgidum (emmer)

T. tauschii (a wild relative)

T. aestivum (a common bread wheat)

14AA ✕ 14BB ⟶ 14AB ⟶ 28AABB ✕ 14DD ⟶ 42AABBDD

Spontaneous chromosome doubling

1 Diploid wild wheat, *Triticum monococcum* (einkorn), has two sets of 7 chromosomes (shown above as 14AA). Long ago, einkorn probably hybridized with another species that had the same number of chromosomes (14BB).

2 The AB hybrid offspring were sterile. However, about 8000 years ago, polyploidy arose in the hybrids, producing wild emmer *(T. turgidum)*. The plants are tetraploid (AABB), with 28 chromosomes (two sets of 14), and they are fertile. At meiosis, the A chromosomes pair with each other, and the B chromosomes pair with each other.

3 Later, an AABB plant probably hybridized with *T. tauschii*, a wild relative of emmer with 14 chromosomes (two sets of 7). The hybrid descendants include common bread wheats, such as *T. aestivum*, which have 42 chromosomes (six sets of 7, AABBDD).

Figure 21.18

The evolution of wheat *(Triticum)*. Cultivated wheat grains more than 11,000 years old have been found in the Eastern Mediterranean region. Researchers believe that speciation in wheat occurred through hybridization and polyploidy.

Figure 21.19 Observational Research

Chromosomal Similarities and Differences among the Great Apes

QUESTION: Does chromosome structure differ between humans and their closest relatives among the great apes?

HYPOTHESIS: Yunis and Prakash hypothesized that chromosome structure would differ markedly between humans and their close relatives among the apes: chimpanzees, gorillas, and orangutans.

METHODS: The researchers used Giemsa stain to visualize the banding patterns on metaphase chromosome preparations from humans, chimpanzees, gorillas, and orangutans. By matching the banding patterns on the chromosomes, the researchers verified that they were comparing the same segments of the genomes in the four species. They then searched for similarities and differences in the structure of the chromosomes.

RESULTS: The analysis of human chromosome 2 reveals that it was produced by the fusion of two smaller chromosomes that are still present in the other three species. Although the position of the centromere in human chromosome 2 matches that of the centromere in one of the chimpanzee chromosomes, in gorillas and orangutans it falls within an inverted segment of the chromosome.

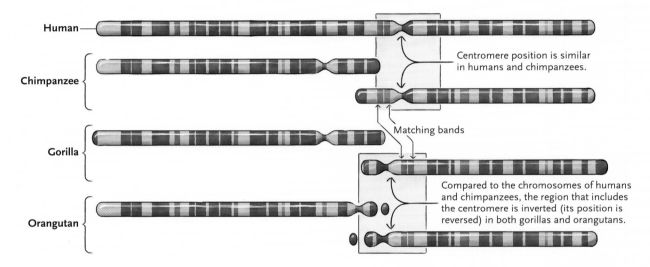

Centromere position is similar in humans and chimpanzees.

Matching bands

Compared to the chromosomes of humans and chimpanzees, the region that includes the centromere is inverted (its position is reversed) in both gorillas and orangutans.

CONCLUSION: Differences in chromosome structure between humans and both gorillas and orangutans are more pronounced than they are between humans and chimpanzees. Structural differences in the chromosomes of these four species may contribute to their reproductive isolation.

---

foster nondisjunction of chromosomes during mitosis. In the first such experiment, undertaken in the 1920s, scientists crossed a radish and a cabbage, hoping to develop a plant with both edible roots and leaves. Instead, the new species, *Raphanobrassica,* combined the least desirable characteristics of each parent, growing a cabbagelike root and radishlike leaves. Recent experiments have been more successful. For example, plant scientists have produced an allopolyploid grain, triticale, that has the disease-resistance of its rye parent and the high productivity of its wheat parent.

Chromosome Alterations Can Foster Speciation

Other changes in chromosome structure or number may also foster speciation. Closely related species often have a substantial number of chromosome differences between them, including inversions, translocations, deletions, and duplications (described in Section 13.3). These differences may foster postzygotic isolation.

In 1982, Jorge J. Yunis and Om Prakash of the University of Minnesota Medical School compared the chromosome structures of humans and their closest relatives among the apes—chimpanzees, gorillas, and orangutans—by examining the *banding patterns* in stained chromosome preparations. In all species, banding patterns vary from one chromosome segment to another. When researchers find identical banding patterns in chromosome segments from two or more related species, they know that they are examining comparable portions of the species' genomes. Thus, the banding patterns allow scientists to identify specific chromosome segments and compare their positions in the chromosomes of different species.

Nearly all of the 1000 bands that Yunis and Prakash identified are present in humans and in the three ape species. However, the banding patterns revealed that whole sections of chromosomes have been rearranged over evolutionary time **(Figure 21.19)**. For example, humans have a diploid chromosome complement of 46 chromosomes, whereas chimpanzees, gorillas, and orangutans have 48. The difference can be traced to the fusion (that is, the joining together) of two ancestral

chromosomes into chromosome 2 of humans; the ancestral chromosomes are separate in the other three species.

Moreover, banding patterns suggest that the position of the centromere in human chromosome 2 closely matches that of a centromere in one of the chimpanzee chromosomes, reflecting their close evolutionary relationship. But this centromere falls within an inverted region of the chromosome in gorillas and orangutans, reflecting their evolutionary divergence from chimpanzees and humans. (Recall from Section 13.3 that an inverted chromosome segment has a reversed orientation, so the order of genes on it is reversed relative to the order in a segment that is not inverted.) Nevertheless, humans and chimps differ from each other in centromeric inversions in six other chromosomes.

How might such chromosome rearrangements promote speciation? In a paper published in 2003, Arcadi Navarro of the Universitat Pompeu Fabra in Spain and Nick H. Barton of the University of Edin-

UNANSWERED QUESTIONS

Do asexual organisms form species?

As you learned in this chapter, the biological species concept applies only to sexually reproducing organisms because only those organisms can evolve barriers to gene flow (asexual organisms reproduce more or less clonally). Nevertheless, research is starting to show that organisms whose reproduction is almost entirely asexual, such as bacteria, seem to form distinct and discrete clusters in nature. (These clusters could be considered "species.") That is, bacteria and other asexual forms may be as distinct as the species of birds described by Ernst Mayr in New Guinea. Workers are now studying the many species of bacteria in nature (only a small number of which have been discovered) to see if they indeed fall into distinct groups. If they do, then scientists will need a special theory, independent of reproductive isolation, to explain this distinctness. Scientists are now working on theories of whether the existence of discrete ecological niches in nature might explain the possible discreteness of asexual "species."

How often does speciation occur allopatrically versus sympatrically or parapatrically?

Scientists do not know how often speciation occurs between populations that are completely isolated geographically (allopatric speciation) compared with how often it occurs in populations that exchange genes (parapatric or sympatric speciation). The relative frequency of these modes of speciation in nature is an active area of research. The ongoing work includes studies on small isolated islands: if an invading species divides into two or more species in this situation, it probably did so sympatrically or parapatrically, since geographical isolation of populations in small islands is unlikely. In addition, biologists are reconstructing the evolutionary history of speciation using molecular tools and correlating this history with the species' geographical distributions. If

this line of research were to show, for example, that the most closely related pairs of species always had geographically isolated distributions, it would imply that speciation was usually allopatric. These lines of research should eventually answer the controversial question of the relative frequency of various forms of speciation.

What are the genetic changes underlying speciation?

Biologists know a great deal about the types of reproductive isolation that prevent gene flow between species, but almost nothing about their genetic bases. Which genes control the difference between flower shape in monkey-flower species? Which genes lead to inviability and sterility of *Drosophila* hybrids? Which genes cause species of ducks to preferentially mate with members of their own species over members of other species? Do the genetic changes that lead to reproductive isolation tend to occur repeatedly at the same genes in a group of organisms, or at different genes? Do the changes occur mostly in protein-coding regions of genes, or in the noncoding regions that control the production of proteins? Were the changes produced by natural selection or by genetic drift? Biologists are now isolating "speciation genes" and sequencing their DNA. With only a handful of such genes known, and all of these causing hybrid sterility or inviability, there will undoubtedly be a lot to learn about the genetics of speciation in the next decade.

Jerry Coyne conducts research on speciation and teaches at the University of Chicago. To learn more about his research go to http://pondside.uchicago.edu/ecol-evol/faculty/coyne_j.html.

burgh in Scotland compared the rates of evolution in protein-coding genes that lie within rearranged chromosome segments of humans and chimpanzees to those in genes outside the rearranged segments. They discovered that proteins evolved more than twice as quickly in the rearranged chromosome segments. Navarro and Barton reasoned that because chromosome rearrangements inhibit chromosome pairing and recombination during meiosis, new genetic variations favored by natural selection would be conserved within the rearranged segments. These variations accumulate over time, contributing to genetic divergence between populations with the rearrangement and those without it. Thus, chromosome rearrangements can be a trigger for speciation: once a chromosome rearrangement becomes established within a population, that population will diverge more rapidly from populations lacking the rearrangement. The genetic divergence eventually causes reproductive isolation.

In the next chapter we consider the effects of speciation over vast spans of time as we examine paleobiology and patterns of macroevolution.

STUDY BREAK

1. How can natural selection promote reproductive isolation in allopatric populations?
2. What group of organisms has frequently undergone speciation by polyploidy?

Review

Go to **Thomson**NOW™ at www.thomsonedu.com/login to access quizzing, animations, exercises, articles, and personalized homework help.

21.1 What Is a Species?

- In practice, most biologists describe, identify, and recognize species on the basis of morphological characteristics that serve as indicators of their genetic similarity to or divergence from other species (Figure 21.2).

- The biological species concept defines species as groups of interbreeding populations that are reproductively isolated from populations of other species in nature. A biological species thus represents a gene pool within which genetic material is potentially shared among populations. The biological species concept cannot be applied to organisms that reproduce only asexually, to those that are extinct, or to geographically separated populations. The phylogenetic species concept defines a species as a group of populations with a recently shared evolutionary history.

- Most species exhibit geographical variation of phenotypic and genetic traits. When marked geographical variation in phenotypes is discontinuous, biologists sometimes name subspecies (Figure 21.3). In ring species, populations are distributed in a ring around unsuitable habitat (Figure 21.4). Many species exhibit clinal variation of characteristics, which change smoothly over a geographical gradient (Figure 21.5).

  Animation: Morphological differences within a species

21.2 Maintaining Reproductive Isolation

- Reproductive isolating mechanisms are characteristics that prevent two species from interbreeding.

- Prezygotic isolating mechanisms either prevent individuals of different species from mating or prevent fertilization between their gametes. Prezygotic isolation occurs because species live in different habitats, breed at different times, use different courtship behavior (Figure 21.6), or differ anatomically (Figure 21.7). Prezygotic isolation can also result from genetic and physiological incompatibilities between male and female gametes.

- Postzygotic isolating mechanisms reduce the fitness of interspecific hybrids through hybrid inviability, hybrid sterility (Figure 21.8), or hybrid breakdown.

  Animation: Reproductive isolating mechanisms

  Animation: Temporal isolation among cicadas

21.3 The Geography of Speciation

- The model of allopatric speciation proposes that speciation results from divergent evolution in geographically separated populations (Figures 21.9–21.11). If allopatric populations accumulate enough genetic differences, they will be reproductively isolated upon secondary contact. Nevertheless, some species hybridize over small areas of secondary contact (Figure 21.12).

- The model of parapatric speciation suggests that reproductive isolation can evolve between parts of a population that occupy opposite sides of an environmental discontinuity (Figure 21.13).

- A model of sympatric speciation in insects suggests that reproductive isolation may evolve between host races that rarely contact one another under natural conditions (Figure 21.14). Sympatric speciation commonly occurs in flowering plants by allopolyploidy.

  Animation: Models of speciation

  Animation: Allopatric speciation on an archipelago

  Animation: Sympatric speciation in wheat

21.4 Genetic Mechanisms of Speciation

- Allopatric populations inevitably accumulate genetic differences, some of which contribute to their reproductive isolation. Reproductive isolating mechanisms evolve as by-products of genetic changes that occur during divergence. Prezygotic isolating mechanisms may evolve in populations experiencing secondary contact (Figure 21.15).

- We cannot yet generalize about how many gene loci participate in the process of speciation, but at least several gene loci are usually involved.

- Speciation by polyploidy in flowering plants involves the duplication of an entire chromosome complement through nondisjunction of chromosomes during meiosis or mitosis. Polyploids can arise among the offspring of a single species (autopolyploidy; Figure 21.16) or, more commonly, after hybridization between closely related species (allopolyploidy; Figures 21.17 and 21.18).

- Chromosome alterations can promote speciation by fostering the genetic divergence of, and reproductive isolation between, populations with different numbers of chromosomes or different chromosome structure (Figure 21.19).

Questions

Self-Test Questions

1. The biological species concept defines species on the basis of:
   a. reproductive characteristics.
   b. biochemical characteristics.
   c. morphological characteristics.
   d. behavioral characteristics.
   e. all of the above

2. Biologists can apply the biological species concept *only* to species that:
   a. reproduce asexually.
   b. lived in the past.
   c. are allopatric to each other.
   d. hybridize in captivity.
   e. reproduce sexually.

3. A characteristic that exhibits smooth changes in populations distributed along a geographical gradient is called a:
   a. ring species.
   b. subspecies.
   c. cline.
   d. hybrid breakdown.
   e. subspecies.

4. If two species of holly (genus *Ilex*) flower during different months, their gene pools may be kept separate by:
   a. mechanical isolation.
   b. ecological isolation.
   c. gametic isolation.
   d. temporal isolation.
   e. behavioral isolation.

5. Prezygotic isolating mechanisms:
   a. reduce the fitness of hybrid offspring.
   b. generally prevent individuals of different species from producing zygotes.
   c. are found only in animals.
   d. are found only in plants.
   e. are observed only in organisms that reproduce asexually.

6. In the model of allopatric speciation, the geographical separation of two populations:
   a. is sufficient for speciation to occur.
   b. occurs only after speciation is complete.
   c. allows gene flow between them.
   d. reduces the relative fitness of hybrid offspring.
   e. inhibits gene flow between them.

7. Adjacent populations that produce hybrid offspring with low relative fitness may be undergoing:
   a. clinal isolation.
   b. parapatric speciation.
   c. allopatric speciation.
   d. sympatric speciation.
   e. geographical isolation.

8. An animal breeder, attempting to cross a llama with an alpaca for finer wool, found that the hybrid offspring rarely lived more than a few weeks. This outcome probably resulted from:
   a. genetic drift.
   b. prezygotic reproductive isolation.
   c. postzygotic reproductive isolation.
   d. sympatric speciation.
   e. polyploidy.

9. Which of the following could be an example of allopolyploidy?
   a. One parent has 32 chromosomes, the other has 10, and their offspring have 42.
   b. Gametes and somatic cells have the same number of chromosomes.
   c. Chromosome number increases by one in a gamete and in the offspring it produces.
   d. Chromosome number decreases by one in a gamete and in the offspring it produces.
   e. Chromosome number in the offspring is exactly half of what it is in the parents.

10. Which of the following genetic characteristics is shared by humans and chimpanzees?
    a. They have the same number of chromosomes.
    b. The position of the centromere on human chromosome 2 matches the position of a centromere on a chimpanzee chromosome.
    c. A fusion of ancestral chromosomes formed chromosome 2.
    d. Centromeres on all of their chromosomes fall within inverted chromosome segments.
    e. all of the above

Questions for Discussion

1. All domestic dogs are classified as members of the species *Canis familiaris*. But it is hard to imagine how a tiny Chihuahua could breed with a gigantic Great Dane. Do you think that artificial selection for different breeds of dogs will eventually create different dog species?

2. Human populations often differ dramatically in external morphological characteristics. On what basis are all human populations classified as a single species?

3. If intermediate populations in a ring species go extinct, eliminating the possibility of gene flow between populations at the two ends of the ring, would you now identify those remaining populations as full species? Explain your answer.

Experimental Analysis

Design an experiment to test whether populations of birds on different islands belong to the same species.

Evolution Link

How do human activities (such as destruction of natural habitats, diversion of rivers, and the construction of buildings) influence the chances that new species of plants and animals will evolve in the future? Frame your answer in terms of the geographical and genetic factors that foster speciation.

How Would You Vote?

Often, when a species is at the brink of extinction, some individuals are captured and brought to zoos for captive breeding programs. Some people object to this practice. They say that keeping a species alive in a zoo is a distraction from more meaningful conservation efforts, and captive animals seldom are successfully restored to the wild. Do you support captive breeding of highly endangered species? Go to www.thomsonedu.com/login to investigate both sides of the issue and then vote.

Fossil of a dragonfly *(Cordulagomphus tuberculatus)* from the Cretaceous period, discovered in Ceara Province, Brazil.

Courtesy Lowcountry Geologic

STUDY PLAN

22.1 The Fossil Record

Fossils form when organisms are buried by sediments or preserved in oxygen-poor environments

The fossil record provides an incomplete portrait of life in the past

Scientists assign relative and absolute dates to geological strata and the fossils they contain

Fossils provide abundant information about life in the past

22.2 Earth History, Biogeography, and Convergent Evolution

Geological processes have often changed Earth's physical environment

Historical biogeography explains the broad geographical distributions of organisms

Convergent evolution produces similar adaptations in distantly related organisms

22.3 Interpreting Evolutionary Lineages

Modern horses are living representatives of a once-diverse lineage

Evolutionary biologists debate the mode and tempo of macroevolution

22.4 Macroevolutionary Trends in Morphology

The body size of organisms has generally increased over time

Morphological complexity has also generally increased over time

Several phenomena trigger the evolution of morphological novelties

22.5 Macroevolutionary Trends in Biodiversity

Adaptive radiations are clusters of related species with diverse ecological adaptations

Extinctions have been common in the history of life

Biodiversity has increased repeatedly over evolutionary history

22.6 Evolutionary Developmental Biology

Most animals share the same genetic tool kit that regulates their development

Evolutionary changes in developmental switches may account for much evolutionary change

22 Paleobiology and Macroevolution

WHY IT MATTERS

In January 1796, Georges Cuvier surprised his audience at the National Institute of Sciences and Arts in Paris by suggesting that fossils were the remains of species that no longer lived on Earth. Natural historians had long recognized the organic origin of fossils, but they did not believe that any creature could become extinct. They thought that the species preserved as fossils still lived in remote and inaccessible places.

Cuvier realized that he could not use the abundant fossils of small marine animals to demonstrate the reality of extinction: these species might still live in the deep sea or other unexplored regions. However, he reasoned that the world was already so well explored that scientists were unlikely to discover any new large terrestrial mammals. Thus, if he could show that fossilized mammals were different from living mammals, he could logically conclude that the fossilized species were truly extinct.

Now credited as the founder of comparative morphology, Cuvier thought that animals were essentially like machines. Each anatomical structure was a crucial part of a perfectly integrated whole. For example, a carnivore requires limbs to pursue prey, claws to catch it, teeth

463

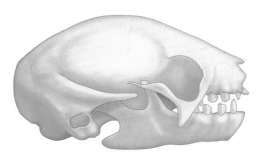

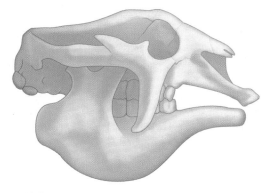

Figure 22.1

Comparing living organisms to fossils. Georges Cuvier compared the skull of a living sloth (top) to a fossilized skull from Paraguay (bottom). The fossilized skull has been reduced in size to facilitate the comparison.

to tear its flesh, and internal organs to digest meat. Thus, from the study of a few critical parts, a knowledgeable anatomist could make reasonable inferences about an animal's overall structure.

Cuvier is also recognized as the founder of paleobiology because he used the anatomy of living species to analyze fossils, which are rarely complete. Paleobiologists often use their knowledge of comparative morphology to make inferences about missing parts. Thus, when asked to analyze a large fossilized skull from Paraguay, Cuvier compared it to specimens in the museum and declared it to be a sloth **(Figure 22.1).** But living sloths are small, whereas this specimen was gigantic, so Cuvier concluded that it was extinct. If such a large species were still living, naturalists would surely have discovered it while exploring South America.

Cuvier studied fossils of other large mammals, especially elephants and rhinoceroses. In every case, he demonstrated that fossilized species were anatomically different from living species. And because no one had seen living examples of the fossilized species, Cuvier concluded that they must be extinct. In 1812, he produced a multivolume treatise in which he acknowledged Earth's great age and documented the appearance and disappearance of species over time. He even noted that fossils lying near the ground surface more closely resembled living species than did

those that were deeply buried. Despite these extraordinary insights, Cuvier never embraced the concept of evolution. If all anatomical features of an animal's body were perfectly integrated, as he believed, how could any part change without upsetting that delicate functional balance?

Cuvier was an early student of macroevolution, the large-scale changes in morphology and diversity that characterize the 3.8-billion-year history of life. Macroevolution has occurred over so vast a span of time and space that the evidence for it is fundamentally different from that for microevolution and speciation. In this chapter we consider what paleobiology and the new field of evolutionary developmental biology tell us about macroevolutionary patterns.

22.1 The Fossil Record

Paleobiologists discover, describe, and name new fossil species and analyze the morphology and ecology of extinct organisms. Because fossils provide physical evidence of life in the past, they are our primary sources of data about the evolutionary history of many organisms.

Fossils Form When Organisms Are Buried by Sediments or Preserved in Oxygen-Poor Environments

Most fossils form in sedimentary rocks. Rain and runoff constantly erode the land, carrying fine particles of rock and soil downstream to a swamp, a lake, or the sea. Particles settle to the bottom as sediments, forming successive layers over millions of years. The weight of newer sediments compresses the older layers beneath them into a solid-matter matrix: sand into sandstone and silt or mud into shale. Fossils form within the layers when the remains of organisms are buried in the accumulating sediments.

The process of fossilization is a race against time because the soft remains of organisms are quickly consumed by scavengers or decomposed by microorganisms. Thus, fossils usually preserve the details of hard structures, such as the bones, teeth, and shells of animals and the wood, leaves, and pollen of plants. During fossilization, dissolved minerals replace some parts molecule by molecule, leaving a fossil made of stone **(Figure 22.2a);** other fossils form as molds, casts, or impressions in material that is later transformed into solid rock **(Figure 22.2b).**

In some environments, the near absence of oxygen prevents decomposition, and even soft-bodied organisms are preserved. Some insects, plants, and tiny lizards and frogs are embedded in amber, the fossilized resin of coniferous trees **(Figure 22.2c).** Other organisms are preserved in glacial ice, coal, tar pits, or the highly acidic water of peat bogs **(Figure**

22.2d). Sometimes organisms are so well preserved that researchers can examine their internal anatomy, cell structure, and food in their digestive tracts. Biologists have even analyzed samples of DNA from a 40-million-year-old magnolia leaf.

a. Petrified wood

George H. H. Huey/Corbis

b. An invertebrate

Neville Pledge/South Australian Museum

c. Insects in amber

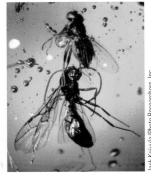

Jack Koivula/Photo Researchers, Inc.

d. Mammoth in permafrost

Novosti/Photo Researchers, Inc.

Figure 22.2
Fossils. **(a)** Petrified wood, from the Petrified Forest National Park in Arizona, formed when minerals replaced the wood of dead trees molecule by molecule. **(b)** The soft tissues of an invertebrate (genus *Dickinsonia*) from the Proterozoic era were preserved as an impression in very fine sediments. **(c)** This 30-million-year-old fly (above) and wasp were trapped in the oozing resin of a coniferous tree and are now encased in amber. **(d)** A frozen baby mammoth (genus *Mammonteus*) that lived about 40,000 years ago was discovered embedded in Siberian permafrost in 1989.

The Fossil Record Provides an Incomplete Portrait of Life in the Past

The 300,000 described fossil species represent less than 1% of all the species that have ever lived. Several factors make the fossil record incomplete. First, soft-bodied organisms do not fossilize as easily as species with hard body parts. Moreover, we are unlikely to find the fossilized remains of species that were rare and locally distributed. Finally, fossils rarely form in habitats where sediments do not accumulate, such as mountain forests. The most common fossils are those of hard-bodied, widespread, and abundant organisms that lived in swamps or shallow seas, where sedimentation is ongoing.

Most fossils are composed of stone, but they don't last forever. Many are deformed by pressure from overlying rocks or destroyed by geological disturbances like volcanic eruptions and earthquakes. Once they are exposed on Earth's surface, where scientists are most likely to find them, rain and wind cause them to erode. Because the effects of these destructive processes are additive, old fossils are much less common than those formed more recently.

Scientists Assign Relative and Absolute Dates to Geological Strata and the Fossils They Contain

The sediments found in any one place form distinctive strata (layers) that differ in color, mineral composition, particle size, and thickness **(Figure 22.3)**. If they have not been disturbed, the strata are arranged in the order in which they formed, with the youngest layers on top. However, strata are sometimes uplifted, warped, or even inverted by geological processes.

Geologists of the early nineteenth century deduced that the fossils discovered in a particular sedimentary stratum, no matter where it is found, represent organisms that lived and died at roughly the same time in the past. Because each stratum formed at a specific time, the sequence of fossils in the lowest (oldest) to the highest (newest) strata reveals their *relative ages*. Geologists used the sequence of strata and their distinctive fossil assemblages to establish the geological time scale **(Table 22.1)**.

Although the geological time scale provides a relative dating system for sedimentary

David Noble/FPG/Getty Images

Figure 22.3
Geological strata in the Grand Canyon. Millions of years of sedimentation in an old ocean basin produced layers of rock that differ in color and particle size. Tectonic forces later lifted the land above sea level, and the flow of the Colorado River carved this natural wonder.

Table 22.1 The Geological Time Scale and Major Evolutionary Events

Eons (Duration drawn to scale)
Cenozoic
Mesozoic
Paleozoic
Phanerozoic
Proterozoic

Eon	Era	Period	Epoch	Millions of Years Ago	Major Evolutionary Events
Phanerozoic	Cenozoic	Quaternary	Holocene	0.01	
			Pleistocene	1.7	Origin of humans; major glaciations
		Tertiary	Pliocene	5.2	Origin of ape-like human ancestors
			Miocene	23	Angiosperms and mammals further diversify and dominate terrestrial habitats
			Oligocene	33.4	Divergence of primates; origin of apes
			Eocene	55	Angiosperms and insects diversify; modern orders of mammals differentiate
			Paleocene	65	Grasslands and deciduous woodlands spread; modern birds and mammals diversify; continents approach current positions
	Mesozoic	Cretaceous		144	Many lineages diversify: angiosperms, insects, marine invertebrates, fishes, dinosaurs; asteroid impact causes mass extinction at end of period, eliminating dinosaurs and many other groups
		Jurassic		206	Gymnosperms abundant in terrestrial habitats; first angiosperms; modern fishes diversify; dinosaurs diversify and dominate terrestrial habitats; frogs, salamanders, lizards, and birds appear; continents continue to separate
		Triassic		251	Predatory fishes and reptiles dominate oceans; gymnosperms dominate terrestrial habitats; radiation of dinosaurs; origin of mammals; Pangaea starts to break up; mass extinction at end of period

Era	Period	Millions of years ago	Events
Phanerozoic (continued)	Permian	290	Insects, amphibians, and reptiles abundant and diverse in swamp forests; some reptiles colonize oceans; fishes colonize freshwater habitats; continents coalesce into Pangaea, causing glaciation and decline in sea level; mass extinction at end of period eliminates 85% of species
	Carboniferous	354	Vascular plants form large swamp forests; first seed plants and flying insects; amphibians diversify; first reptiles appear
Paleozoic	Devonian	417	Terrestrial vascular plants diversify; fungi and invertebrates colonize land; first insects appear; first amphibians colonize land; major glaciation at end of period causes mass extinction, mostly of marine life
	Silurian	443	Jawless fishes diversify; first jawed fishes; first vascular plants on land
	Ordovician	490	Major radiations of marine invertebrates and fishes; major glaciation at end of period causes mass extinction of marine life
	Cambrian	543	Diverse radiation of modern animal phyla (Cambrian explosion); simple marine communities
Proterozoic		2500	High concentration of oxygen in atmosphere; origin of aerobic metabolism; origin of eukaryotic cells; evolution and diversification of protists, fungi, soft-bodied animals
Archaean		3800	Evolution of prokaryotes, including anaerobic bacteria and photosynthetic bacteria; oxygen starts to accumulate in atmosphere
		4600	Formation of Earth at start of era; Earth's crust, atmosphere, and oceans form; origin of life at end of era

Figure 22.4 Research Method

PURPOSE: Radiometric dating allows researchers to estimate the absolute age of a rock sample or fossil.

Radiometric Dating

PROTOCOL:

1. Knowing the approximate age of a rock or fossil, select a radioisotope that has an appropriate half-life. Because different radioisotopes have half-lives ranging from seconds to billions of years, it is usually possible to choose one that brackets the estimated age of the sample under study. For example, if you think that your fossil is more than 10 million years old, you might use uranium-235. The half-life of ^{235}U, which decays into the lead isotope ^{207}Pb, is about 700 million years. Or if you think that your fossil is less than 70,000 years old, you might select carbon-14. The half-life of ^{14}C, which decays into the nitrogen isotope ^{14}N, is 5730 years.

Radioisotopes Commonly Used in Radiometric Dating

Radioisotope (Unstable)	More Stable Breakdown Product	Half-Life (Years)	Useful Range (Years)
Samarium-147 ⟶	Neodymium-143	106 billion	>100 million
Rubidium-87 ⟶	Strontium-87	48 billion	>10 million
Thorium-232 ⟶	Lead-208	14 billion	>10 million
Uranium-238 ⟶	Lead-206	4.5 billion	>10 million
Uranium-235 ⟶	Lead-207	700 million	>10 million
Potassium-40 ⟶	Argon-40	1.25 billion	>100,000
Carbon-14 ⟶	Nitrogen-14	5730	<70,000

2. Prepare a sample of the material and measure the quantities of the parent radioisotope and its more stable breakdown product.

INTERPRETING THE RESULTS: Compare the relative quantities of the parent radioisotope and its breakdown product (or some other stable isotope) to determine what percentage of the original parent radioisotope remains in the sample. Then use a graph of radioactive decay for that isotope to determine how many half-lives have passed since the sample formed.

In newly formed rock, 100% of the parent isotope is present.

After one half-life, 50% remains.

After two half-lives, 25% remains.

Percentage of parent isotope remaining vs *Number of half-lives*

Knowing the number of half-lives that have passed allows you to estimate the age of the sample.

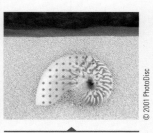

A living mollusk absorbed trace amounts of ^{14}C, a rare radioisotope of carbon, and large amounts of ^{12}C, which is the more stable and common isotope of carbon.

When the mollusk died, it was buried in sand and fossilized. From the moment of its death, the ratio of ^{14}C to ^{12}C began to decline through radioactive decay. Because the half-life of ^{14}C is 5730 years, half of the original ^{14}C was eliminated from the fossil in 5730 years and half of what remained was eliminated in another 5730 years.

After the fossil was discovered, a scientist determined that its ^{14}C to ^{12}C ratio was one-eighth of the ^{14}C to ^{12}C ratio in living organisms. Thus, radioactive decay had proceeded for three half-lives—or about 17,000 years—since the mollusk's death.

strata, it does not tell us how old the rocks and fossils actually are. But many rocks contain radioisotopes, which, from the moment they form, begin to break down into other, more stable elements. The breakdown proceeds at a steady rate that is unaffected by chemical reactions or environmental conditions such as temperature or pressure. Using a technique called **radiometric dating**, scientists can estimate the age of a rock by noting how much of an unstable "parent" isotope has decayed to another form. By measuring the relative amounts of the parent radioisotope and its breakdown products and comparing this ratio with the isotope's **half-life**—the time it takes for half of a given

amount of radioisotope to decay—researchers can estimate the *absolute age* of the rock **(Figure 22.4).** Table 22.1 presents these age estimates along with the major geological and evolutionary events of each period.

Radiometric dating works best with volcanic rocks, which form when lava cools and solidifies. But most fossils are found in sedimentary rocks. To date sedimentary fossils, scientists determine the age of volcanic rocks from the same strata. Using this method, investigators have linked fossils to deposits that are hundreds of millions of years old.

Fossils that still contain organic matter, such as the remains of bones or wood, can be dated directly by measuring their content of the radioactive carbon isotope ^{14}C, which decays to ^{14}N. Living organisms absorb traces of ^{14}C and large quantities of ^{12}C, a stable carbon isotope, from the environment and incorporate them into biological molecules. As long as an organism is still alive, its ^{14}C content remains constant because any ^{14}C that decays is replaced by the uptake of other ^{14}C atoms. But as soon as the organism dies, no further replacement occurs and ^{14}C begins its steady radioactive decay. Scientists use the ratio of ^{14}C to ^{12}C present in a fossil to determine its age, as explained in Figure 22.4.

To develop a feeling for geological time, imagine the 4.5-billion-year history of Earth scaled onto an annual calendar; each day represents a little over 12 million years. The planet was formed on January 1. Animal life originated in mid-November, dinosaurs lived between December 14 and December 26, and the primate ancestors of modern humans appeared during the last 4 hours of December 31.

Fossils Provide Abundant Information about Life in the Past

Imperfect as it is, the fossil record provides our only direct information about life in the past. Fossilized skeletons, shells, stems, leaves, and flowers tell us about the size and appearance of ancient animals and plants. The fossil record also allows scientists to see how structures were modified as they became adapted for specialized uses (see Figure 19.3). Moreover, fossils chronicle the proliferation and extinction of evolutionary lineages and provide data on their past geographical distributions.

Fossils can also provide indirect data about behavior, physiology, and ecology. For example, the fossilized footprints of some dinosaurs suggest that adults surrounded their young when the group moved, perhaps to protect them from predators. Complex scrolls of bone in the nasal passages of early mammals suggest that they had a well-developed sense of smell, and fossilized teeth and dung provide data about the diets of extinct animals. The study of fossilized pollen allows paleobiologists to reconstruct the vegetation and climate of ancient sites. The changing arrays of fossils that document biological evolution partly reflect large-scale shifts in Earth's physical environments, a topic that we explore in the next section.

22.2 Earth History, Biogeography, and Convergent Evolution

Organisms interact constantly with their environments. Some of these interactions have caused fundamental changes in Earth's physical environment, such as the development of an oxidizing atmosphere (see Chapter 24). In this section we consider other aspects of Earth's history and their profound effects on living systems.

Geological Processes Have Often Changed Earth's Physical Environment

Long-term shifts in geography and climate—as well as brief but catastrophic events—have significantly altered environments on Earth. Major geological and climatic shifts occur because the planet's crust is in motion.

According to the theory of **plate tectonics**, Earth's crust is broken into irregularly shaped plates of rock that float on its semisolid mantle **(Figure 22.5)**. Currents in the mantle cause the plates—and the continents embedded in them—to move, a phenomenon called **continental drift.** About 250 million years ago, Earth's landmasses coalesced into a single supercontinent called Pangaea; continental drift later separated Pangaea into a northern continent, Laurasia, and a southern continent, Gondwana. Laurasia and Gondwana subsequently broke into the continents we know today **(Figure 22.6).**

The drifting continents induced global changes in Earth's climate. For example, the movement of continents toward the poles encouraged the formation of glaciers, which caused temperature and rainfall to decrease worldwide. As a result of complex continental movements, Earth's average temperature has fluctuated widely. During one geologically recent cold spell (about 20,000 years ago), the northern polar ice cap extended into southern Indiana and Ohio.

Unpredictable events have also changed physical environments on Earth. Massive volcanic eruptions and asteroid impacts have occasionally altered the planet's atmosphere and climate drastically. These cataclysmic events have sometimes caused many forms of life to disappear over relatively short periods of geological time.

Historical Biogeography Explains the Broad Geographical Distributions of Organisms

More than a century after Darwin published his observations, the theory of plate tectonics refocused attention on biogeography. Historical biogeographers try to explain how organisms acquired their geographical distributions over evolutionary time.

Continuous and Disjunct Distributions. Many species have a **continuous distribution**: they live in suitable habitats throughout a geographical area. For example, herring gulls *(Larus argentatus)* live along the coastlines of all northern continents. Continuous distributions usually require no special historical explanation.

Other groups exhibit **disjunct distributions**, in which closely related species live in widely separated locations. For example, magnolia trees *(Magnolia* species)

a. Earth's crustal plates

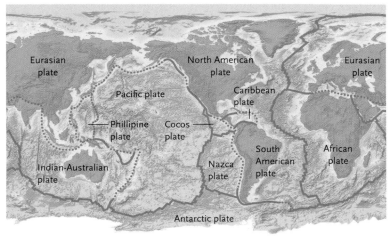

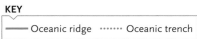

b. Model of plate tectonics

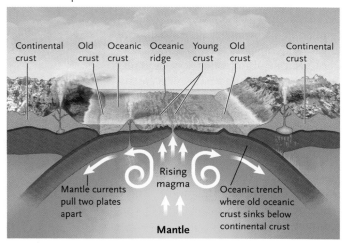

Figure 22.5
Plate tectonics. **(a)** Earth's crust is broken into large, rigid plates. New crust is added at oceanic ridges, and old crust is recycled into the mantle at oceanic trenches. **(b)** Oceanic ridges form where pressure in the mantle forces magma (molten rock) through fissures in the sea floor. Mantle currents pull the plates apart on either side of the ridge, forcing the sea floor to move laterally away from the ridge. This phenomenon, seafloor spreading, is widening the Atlantic Ocean about 3 cm per year. Oceanic trenches form where plates collide. The heavier oceanic crust sinks below the lighter continental crust, and it is recycled into the mantle, a process called subduction. The highest mountain ranges (including the Rockies, Himalayas, Alps, and Andes) formed where subduction uplifted continental crust. Earthquakes and volcanoes are also common near trenches.

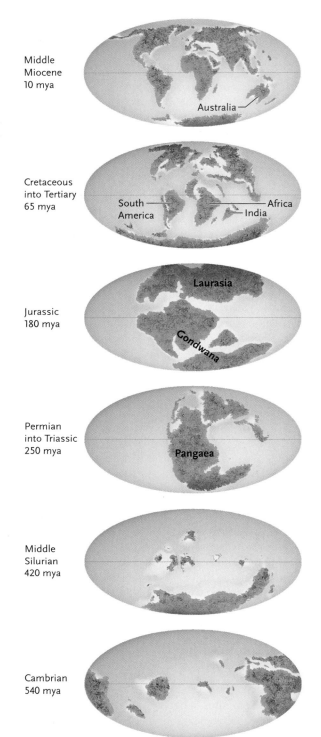

Figure 22.6
History of long-term changes in the positions of continents. Earth's many landmasses coalesced during the Permian period, forming the supercontinent Pangaea. About 180 million years ago (mya), Pangaea separated into Gondwana and Laurasia. Then Gondwana began to break apart. Africa and India pulled away first, opening the South Atlantic and Indian Oceans. Australia separated from Antarctica about 55 million years ago and slowly drifted northward. South America separated from Antarctica shortly thereafter. Laurasia remained nearly intact until 43 million years ago when North America and Greenland together separated from Europe and Asia. Movement of the continents also changed the shapes and the sizes of the oceans.

occur in parts of North, Central, and South America as well as in China and Southeast Asia, but nowhere in between.

Two phenomena—dispersal and vicariance—create disjunct distributions. **Dispersal** is the movement of organisms away from their place of origin; it can produce a disjunct distribution if a new population becomes established on the far side of a geographical barrier. **Vicariance** is the fragmentation of a continuous geographical distribution by external factors. Over the course of evolutionary history, dispersal and vicariance have together influenced the geographical distributions of organisms on a very grand scale (see *Focus on Research*).

Biogeographical Realms. For species that were widespread in the Mesozoic era, Pangaea's breakup was a powerful vicariant experience. The subsequent geographical isolation of continents fostered the evolution of distinctive regional **biotas** (all organisms living in a region). Alfred Russel Wallace used the biotas to define six **biogeographical realms**, which we still recognize today **(Figure 22.7).**

The Australian and Neotropical realms, which have been geographically isolated since the Mesozoic, contain many **endemic species** (those that occur nowhere else on Earth). The Australian realm, in particular, has had no complete land connection to any other continent for approximately 55 million years. As a result, Australia's mammalian fauna (all the mammals living in the region) is unique, made up almost entirely of endemic marsupials. Few native placental mammals occur in Australia because the placental lineage arose elsewhere after Australia had become isolated.

The biotas of the Nearctic and Palearctic realms are, by contrast, fairly similar. North America and Eurasia were frequently connected by land bridges; eastern North America was attached to Western Europe until the breakup of Laurasia 43 million years ago, and northwestern North America had periodic contact with northeastern Asia over the Bering land bridge during much of the past 60 million years.

Convergent Evolution Produces Similar Adaptations in Distantly Related Organisms

Distantly related species living in different biogeographical realms are sometimes very similar in appearance. For example, the overall form of cactuses in the Americas is almost identical to that of spurges in Africa **(Figure 22.8).** But these lineages arose independently long after those continents had separated; thus, cactuses and spurges did not inherit their similarities from a shared ancestor. Their overall resemblance is the product of **convergent evolution,** the evolution of similar adaptations in distantly related organisms that occupy similar environments.

Figure 22.7
Wallace's biogeographical realms. Each realm contains a distinctive biota.

Convergent evolution also creates similarities in distantly related animals that use the same mode of locomotion. Some marine fishes, birds, and mammals have torpedo-shaped bodies and appendages modified for swimming in strong ocean currents **(Figure 22.9).** Even entire faunas can develop convergent morphologies. For example, the marsupial mammals of Australia and placental mammals of North America—groups that arose long after the breakup of Pangaea—include many pairs of morphologically convergent species that also occupy similar habitats and feed on similar foods. To understand convergent evolution as well as most other macroevolutionary patterns, biologists must analyze the evolutionary history of individual lineages, a sometimes-controversial activity that we consider next.

STUDY BREAK

1. How did the process of continental drift affect the geographical distributions of organisms?
2. Why do some distantly related species that live in different biogeographical realms sometimes resemble each other?

a. Cactus   **b.** Spurge

Figure 22.8
Convergent evolution in plants. **(a)** *Echinocereus* and other North American cactuses (family Cactaceae) are strikingly similar to **(b)** *Euphorbia* and other African spurges (family Euphorbiaceae). Convergent evolution adapted both groups to desert environments with thick, water-storing stems, spiny structures that discourage animals from feeding on them, CAM photosynthesis (see Section 9.4), and stomata that open only at night.

Basic Research: The Great American Interchange

a. Jurassic Period

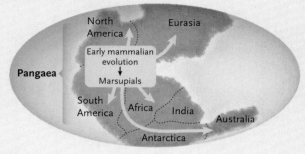

b. Cretaceous Period

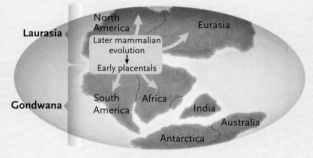

c. Miocene Epoch

d. Pliocene Epoch

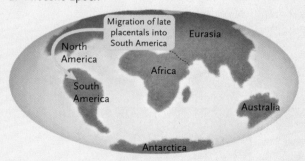

Figures a–d
Dispersal and vicariance changed the geographical distributions of marsupial and placental mammals.

Paleobiologists reconstruct the biogeographical history of a lineage by dating its fossils and mapping their geographical distributions at specific times in the past. The complex evolutionary history of mammals, especially in North and South America, illustrates the effects of dispersal and vicariance.

Mammals first arose in western Pangaea (now part of North America), where they diverged into several evolutionary lineages. The earliest marsupials (whose young complete development in a pouch on the mother's belly) dispersed to the future Eurasia, Africa, South America, Antarctica, and Australia during the Jurassic period **(Figure a).** Somewhat later, but before the continents completely separated during the Cretaceous period, the earliest placentals (whose young complete development within the mother's uterus) also dispersed from North America into Eurasia, Africa, and South America **(Figure b).**

The breakup of Pangaea did not destroy all of these dispersal paths immediately. Persistent land connections allowed many organisms to migrate freely between North America and Eurasia throughout the Cretaceous period. Modern placentals further diversified in Eurasia during the Miocene epoch, and these new forms quickly dispersed back into North America **(Figure c).** As a result, the mammalian faunas of these continents have always been very similar.

By contrast, Australia and South America experienced substantial geographical isolation, particularly after the breakup of Gondwana. In South America, a distinctive mammalian fauna evolved from the marsupials and early placentals that had arrived during the Mesozoic. Small marsupials fed primarily on insects and other invertebrates, but larger species, including a marsupial saber-toothed cat, ate other vertebrates. The early South

American placentals gave rise to many large ungulates (hoofed herbivorous mammals) as well as to armadillos, sloths, and anteaters, some of which still live in South America today.

Periodic dispersal events slowly added to South America's mammal fauna. For example, rodents and primates first arrived about 25 million years ago, during the Oligocene epoch. They probably came from Africa by island-hopping across the slowly widening South Atlantic. These rodents eventually gave rise to guinea pigs and their relatives, and these primates to all the living New World monkeys. South America then began to drift northwest toward North America. By the late Miocene epoch, 6 million to 8 million years ago, North American rodents and raccoons were able to disperse into South America across the narrow water gap. By about 3 million years ago, in the Pliocene epoch, the Panamanian land bridge was established between North and South America, allowing mammals to migrate in both directions **(Figure d).**

A group of paleobiologists led by Larry G. Marshall of the Field Museum of Natural History in Chicago and S. David Webb of the Florida State Museum and the University of Florida have determined that about 10% of the mammal species on each continent dispersed across the land bridge to the other side. But North America—with its greater size and long-standing connections to Eurasia—had a greater variety of mammals than South America. Thus, more different types of mammals moved from north to south than in the opposite direction. Dispersal was so extensive during the Pliocene that paleobiologists describe these movements as the Great American Interchange.

The dispersal of so many northern mammals into South America fundamentally changed its ecological communities. Carnivorous cats, dogs, and weasels and herbivorous camels, deer, elephants, horses, rabbits, and tapirs swept into the continent. Many new arrivals were wildly successful, apparently because they ate resources that native South American mammals were not using. Moreover, the northern immigrants had high rates of speciation, producing numerous descendant species.

As climates periodically cooled during the Pleistocene epoch, many mammals became extinct on both continents. Descendants of the northern immigrants have fared well over the long term: about half the mammals in South America today—including all the cats, llamas, tapirs, and many rodents—are the descendants of northern ancestors. Most South American species that moved north were not as successful. Perhaps they could not adapt to physical conditions in the north, especially during the Pleistocene glaciations; or perhaps they could not prevail in competition with mammals that were already there. Today, relatively few mammals of southern origin persist in North America. Armadillos, monkeys, and anteaters are restricted to the southernmost parts. Only one opossum species (*Didelphis virginiana*, **Figure e**) and one porcupine species (*Erethizon dorsatum*) have moved further north.

Figure e
Opossums *(Didelphis virginiana)* are among the few mammals of South American origin that are successful in North America.

a. Shark

b. Penguin

c. Porpoise

Figure 22.9
Convergent evolution in marine vertebrates. Convergent evolution produced similar body forms and appendages in distantly related marine predators: **(a)** sharks, which are cartilaginous fishes; **(b)** penguins, which are birds; and **(c)** porpoises, which are mammals. The resemblances are superficial, however. The tails of sharks are vertical, whereas those of penguins and porpoises are horizontal. Penguins also lack fins along their backs.

22.3 Interpreting Evolutionary Lineages

As newly discovered fossils demand the reinterpretation of old hypotheses, biologists constantly refine their ideas about the history of life. The evolution of horses is a case in point.

Modern Horses Are Living Representatives of a Once-Diverse Lineage

The earliest known ancestors of modern horses were first identified by Othniel C. Marsh of Yale University just a year after Darwin published *On the Origin of Species*. These early horses, *Hyracotherium*, stood 25 to 50 cm high and weighed no more than 20 kg. Their toes (four on the front feet and three on the hind) were each capped with a tiny hoof, but the animals walked on soft pads as dogs do today. Their faces were short, their teeth were small, and they browsed on soft leaves in woodland habitats.

In 1879, Marsh published his analysis of 60 million years of horse family history. He described the evolution of this group of mammals as a sequence of stages from the tiny *Hyracotherium* through intermediates represented by *Mesohippus, Merychippus,* and *Pliohippus* to the modern *Equus* **(Figure 22.10a).** (Each of these names refers to a genus, a group of closely related species.) Marsh inferred a pattern of descent characterized by gradual, directional evolution in several skeletal features. Changes in the legs and feet allowed horses to run more quickly, and changes in the face and teeth accompanied a switch in diet from soft leaves to tough grasses.

a. Marsh's reconstruction of horse evolution

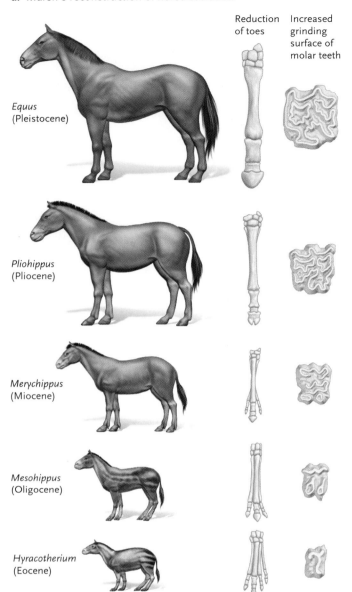

Reduction of toes

Increased grinding surface of molar teeth

Equus (Pleistocene)

Pliohippus (Pliocene)

Merychippus (Miocene)

Mesohippus (Oligocene)

Hyracotherium (Eocene)

b. Modern reconstruction of horse evolution

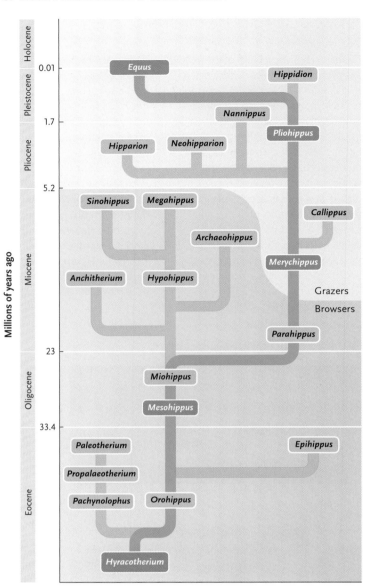

Millions of years ago

Holocene

Pleistocene — 0.01

1.7

Pliocene — 5.2

Miocene

23

Oligocene — 33.4

Eocene

Equus

Hippidion

Nannippus

Pliohippus

Hipparion   Neohipparion

Sinohippus   Megahippus

Callippus

Archaeohippus

Merychippus

Anchitherium   Hypohippus

Grazers

Browsers

Parahippus

Miohippus

Mesohippus

Paleotherium

Epihippus

Propalaeotherium

Pachynolophus   Orohippus

Hyracotherium

c. Changes in body size of horse species over time

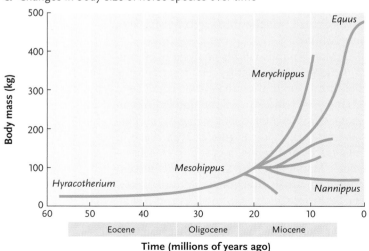

Body mass (kg)

500

400

300

200

100

Equus

Merychippus

Hyracotherium

Mesohippus

Nannippus

60   50   40   30   20   10   0

Eocene   Oligocene   Miocene

Time (millions of years ago)

Figure 22.10

Evolution of horses. **(a)** Marsh depicted the evolution of horses as a linear pattern of descent characterized by an increase in body size, a reduction in the number of toes, increased fusion of the bones in the lower leg, elongation of the face, and an increase in the size of grinding teeth at the back of the mouth. **(b)** Recent studies revealed that the horse family includes numerous evolutionary branches with variable morphology. The horses in Marsh's analysis are highlighted in green. **(c)** Although many branches of the lineage evolved a larger body size, some remained as small as the earliest horses.

The fossil record for horses is superb, and we now have fossils of more than 100 extinct species from five continents. These data reveal a macroevolutionary history very different from Marsh's interpretation. *Hyracotherium* was not gradually transformed into *Equus* along a linear track. Instead, the evolutionary tree for horses was highly branched **(Figure 22.10b),** and *Hyracotherium*'s descendants differed in size, number of toes, tooth structure, and other traits. Although many branches of this lineage lived in the Miocene and Pliocene epochs, all but one are now extinct. The species of the genus *Equus* living today (horses, donkeys, and zebras) are the surviving tips of that one branch.

When we study extinct organisms, we tend to focus on traits that characterize modern species. Marsh, for example, assumed that the differences between *Hyracotherium* and *Equus* were typical of the changes that characterized the group's evolutionary history. But not all fossil horses were larger **(Figure 22.10c),** had fewer toes, or were better adapted to feed on grass than their ancestors. And if a branch other than *Equus* had survived, Marsh's description of trends in horse evolution would have been very different. All evolutionary lineages have extinct branches, and any attempt to trace a linear evolutionary path—as Marsh did for horses and many people do for humans—imposes artificial order on an inherently disorderly history.

Evolutionary Biologists Debate the Mode and Tempo of Macroevolution

What evolutionary processes produce the numerous branches of a lineage such as the horse lineage, and over what time scale does a lineage evolve?

Modes of Evolutionary Change. The species that paleobiologists identify may arise by one of two modes, or processes of change, called anagenesis and cladogenesis. **Anagenesis** refers to the accumulation of changes in a lineage as it adapts to changing environments. If morphological changes are large, we may give the organisms different names at different times in their history. One might say, for example, that Species A from the late Mesozoic era had evolved into Species B from the middle Cenozoic era **(Figure 22.11a).** Anagenesis does not increase the number of species—it is the evolutionary transformation of an existing species rather than the production of new ones.

Cladogenesis refers to the evolution of two or more descendant species from a common ancestor. If the fossilized remains of the descendants are distinct, paleobiologists will recognize them as different species **(Figure 22.11b).** Cladogenesis does increase the number of species on Earth.

Tempo of Morphological Change. Macroevolutionists have developed two alternative hypotheses to describe the tempo, or timing, of morphological change.

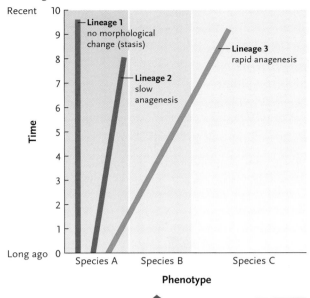

a. Anagenesis

Three lineages begin with the same phenotype, identified as fossil Species A, at time 0. The rate of evolutionary change is shown by the angle of the line for each lineage: lineage 1 undergoes no change over time, lineage 2 changes slowly, and lineage 3 changes so rapidly that its phenotype shifts far to the right in the graph. Paleobiologists might assign different names to the fossils of lineage 3 at different times in its history—Species A at time 1, Species B at times 2 through 6, and Species C at times 7 through 9—even though no additional species evolved. By contrast, fossils of lineages 1 and 2 change so little over time that they would be identified as Species A throughout their evolutionary history.

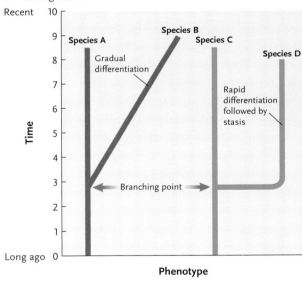

b. Cladogenesis

Each branching point represents a new line of descent. The branching may show either gradual (left) or rapid (right) morphological differentiation from the parent species.

Figure 22.11
Patterns of evolution. In these hypothetical examples, the vertical axis represents geological time and the horizontal axis represents variation in phenotypic traits.

Figure 22.12 Observational Research

Evidence Supporting the Punctuated Equilibrium Hypothesis

HYPOTHESIS: The punctuated equilibrium hypothesis states that most morphological change within evolutionary lineages appears during periods of rapid speciation.

PREDICTION: The fossil record will reveal that most species experienced relatively little morphological change for long periods of time, but that new, morphologically distinctive species arose suddenly.

METHOD: Cheetham examined numerous fossilized samples of populations of a small marine invertebrate, the ectoproct *Metrarabdotos*, from the Dominican Republic. He measured 46 morphological characters in populations representing 18 species and then used a complex statistical analysis to summarize how morphologically different the species are.

RESULTS: The morphology of most *Metrarabdotos* species changed very little over millions of years, but new species, morphologically very different from their ancestors, often appeared suddenly in the fossil record. Each dot in the graph represents a sample of a population from the fossil record. The horizontal axis reflects the overall morphological difference between samples of one species over time or between samples of ancestral species and descendant species. The dashed lines represent gaps in the fossil record for this genus.

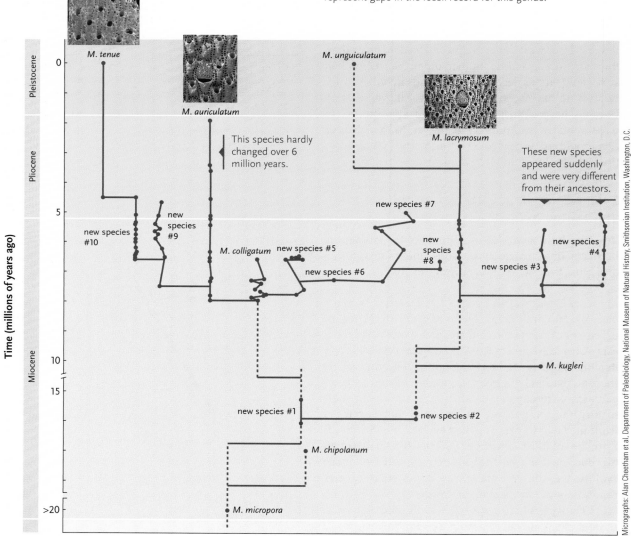

This species hardly changed over 6 million years.

These new species appeared suddenly and were very different from their ancestors.

Micrographs: Alan Cheetham et al. Department of Paleobiology, National Museum of Natural History, Smithsonian Institution, Washington, D.C.

Overall morphological difference

CONCLUSION: The *Metrarabdotos* lineage exhibits a pattern of morphological evolution that is consistent with the predictions of the punctuated equilibrium hypothesis.

According to the **gradualist hypothesis**, large changes result from the slow, continuous accumulation of small changes over time. If this hypothesis is correct, we might expect to find a series of transitional fossils that document gradual evolution. In fact, we rarely find evidence of perfectly gradual change in any lineage. Most species appear suddenly in a particular stratum, persist for some time with little change, and then disappear from the fossil record. Then another species with different traits suddenly appears in the next higher stratum.

In the early 1970s, Niles Eldredge of the American Museum of Natural History and Stephen Jay Gould of Harvard University published an explanation for the absence of transitional forms, or "missing links." Their **punctuated equilibrium hypothesis** suggested that speciation usually occurs in isolated populations at the edge of a species' geographical distribution. Such populations experience substantial genetic drift and distinctive patterns of natural selection (as described in Section 21.3). According to this hypothesis, morphological variations arise rapidly during cladogenesis. Thus, most species exhibit long periods of morphological equilibrium or stasis (little change in form), punctuated by brief periods of cladogenesis and rapid morphological evolution. If this hypothesis is correct, transitional forms live only for short periods of geological time in small, localized populations—the very conditions that discourage broad representation in the fossil record. Darwin himself used this line of reasoning to explain puzzling gaps in the fossil record: new species appear as fossils only after they become abundant and widespread and begin a period of morphological stasis.

Some evolutionists point to flaws in the punctuated equilibrium hypothesis. First, rapid morphological evolution frequently occurs without cladogenesis. For example, in North America, variations in the body size of house sparrows evolved within 100 years without the appearance of new sparrow species (see Figure 21.5). Furthermore, geographical variation in most widespread species (see Section 21.1) provides compelling evidence of morphological evolution without speciation.

Second, critics challenge the hypothesis' definition of rapid morphological change, particularly given our inability to resolve time precisely in the fossil record. To a paleobiologist with a geological perspective, "instantaneous" events occur over tens or hundreds of thousands of years. But to a population geneticist, those time scales may encompass thousands of generations, ample time for gradual microevolutionary change.

Third, examples of evolutionary stasis may not be as static as they appear. Alternating periods of directional selection that favor opposite patterns of change could produce the appearance of stasis. For example, if natural selection favored slight increases in body size

for 2000 years and then favored slight decreases for the next 2000 years, paleobiologists would probably detect no change in body size at all.

The fossil record provides some support for both hypotheses. A punctuated pattern is evident in the evolutionary history of *Metrarabdotos*, a genus of ectoprocts from the Caribbean Sea. Ectoprocts are small colonial animals that build hard skeletons (see Figure 29.15a), the details of which are well preserved in fossils. Alan Cheetham of the Smithsonian Institution measured 46 morphological traits in fossils of 18 *Metrarabdotos* species. He then used a summary statistic to describe the morphological difference between populations of a single species over time and between ancestral species and their descendants. His results indicate that most species did not change much over millions of years, but new species, which were morphologically different from their ancestors, often appeared quite suddenly **(Figure 22.12)**.

By contrast, a study of Ordovician trilobites supports the gradualist hypothesis of evolution. The number of "ribs" in their tail region changed continuously over 3 million years. The change was so gradual that a sample from any given stratum was almost always intermediate between samples from the strata just above and below it. The changes in rib number probably evolved without cladogenesis **(Figure 22.13)**.

The punctuationalist and gradualist hypotheses represent extremes on a continuum of possible macroevolutionary patterns. The mode and tempo of evolution vary among lineages, and both viewpoints are validated by data on some organisms but not others. Although some biologists still question the punctuated equilibrium hypothesis, its publication rekindled interest in paleobiology and macroevolution, inspiring much new research. Some of the most interesting results have focused on morphological changes within lineages and on long-term changes in the number of living species.

STUDY BREAK

1. Did the horse lineage undergo a steady increase in body size over its evolutionary history?
2. How do the predictions of the gradualist and the punctuationalist hypotheses differ?

22.4 Macroevolutionary Trends in Morphology

Some evolutionary lineages exhibit trends toward larger size and greater morphological complexity, and others are marked by the development of novel structures.

Figure 22.13 Observational Research

Evidence Supporting the Gradualist Hypothesis

HYPOTHESIS: The gradualist hypothesis states that most morphological change within evolutionary lineages results from the accumulation of small, incremental changes over long periods of time.

PREDICTION: The fossil record will reveal that the morphology of fossils from a given stratum will be intermediate between those of fossils from the strata immediately below and above it.

METHOD: Peter R. Sheldon of Trinity College, Dublin, Ireland, counted the number of "ribs" in the tail region of the exoskeletons of approximately 15,000 trilobite fossils from central Wales, United Kingdom. The fossils had formed over a span of about 3 million years during the Ordovician period. Sheldon plotted the mean number of ribs found in successive samples of each lineage.

RESULTS: Sheldon's data reveal gradual changes in the mean number of "ribs" in these animals with no evidence of speciation.

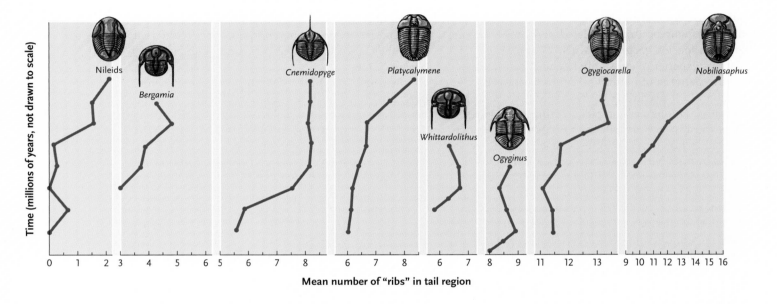

CONCLUSION: Morphological changes in Ordovician trilobites from central Wales are consistent with the predictions of the gradualist hypothesis.

The Body Size of Organisms Has Generally Increased over Time

Body size affects most aspects of an organism's physiology and ecology. When we look at the entire history of life, organisms have generally become larger over time. The earliest organisms were tiny, as most still are today. But the change from replicating molecules to acellular, unicellular, and finally multicellular organization must have demanded an increase in body size.

Within evolutionary lineages, increases in body size are not universal, but they are common. The nineteenth-century paleobiologist Edward Drinker Cope first noted this trend toward larger body size, now known as *Cope's Rule,* in vertebrates. Although Cope's Rule also applies to some invertebrate and plant lineages, no one has conducted a truly broad survey to test the generality of the hypothesis. Insects, for example,

are a major exception to Cope's Rule. Most insects have remained small since their appearance in the Devonian, probably because of the constraints imposed by an external skeleton (see Section 29.7).

We can readily imagine why natural selection may sometimes favor larger size. Large organisms maintain a more constant internal environment than small ones. They may also have access to a wider range of resources, harvest resources more efficiently, and be less likely to be captured by predators. Moreover, larger females may produce more young, and larger males may have greater access to mates. Unfortunately, we cannot test such hypotheses about extinct life forms directly. We can only analyze past events with an understanding of how natural selection affects organisms living today.

In the 1970s, Steven Stanley of Johns Hopkins University proposed an explanation for how macroevolutionary trends may develop. He suggested that certain traits might make some species more likely to undergo speciation than others. This mechanism, called **species selection**, is analogous to natural selection. In natural selection, the evolutionary success of an individual is measured by the number of its surviving offspring. In species selection, the evolutionary success of a species is measured by the number of its descendant species. Thus, the traits of species that frequently undergo cladogenesis become more common, establishing a trend within a lineage. For example, if large species leave more descendant species than small ones do, the number of large species will increase faster than the number of small species. As a result, the average size of species in the lineage will increase over time. Stanley's hypothesis has not been widely tested.

Morphological Complexity Has Also Generally Increased over Time

In general, the evolutionary increase in size has been accompanied by an increase in morphological complexity. Among contemporary organisms, for example, species with large body size have a greater variety of cell types than do species with small body size. We can probably assume that new cell types arose when larger organisms first evolved.

However, under some circumstances, natural selection has simplified traits. The single toe and fused leg bones of modern horses are stronger, but mechanically less complex, than the ancestral structures in *Hyracotherium*. Similarly, snakes, which evolved from lizards with well-developed legs, have lost their limbs entirely. These changes, which increase the efficiency of locomotion, represent decreases in morphological complexity.

Several Phenomena Trigger the Evolution of Morphological Novelties

Sometimes a trait that is adaptive in one context turns out to be advantageous in another. Natural selection may then modify the trait to enhance its new function. Such **preadaptations** are just lucky accidents; they never evolve *in anticipation* of future evolutionary needs.

John Ostrom of Yale University described how some carnivorous dinosaurs, the immediate ancestors of *Archaeopteryx* and modern birds, were preadapted for flight (see Figure 19.12). These small, agile creatures were bipedal with lightweight hollow bones and long forelimbs to capture prey; some even had rudimentary feathers that may have retained body heat. But all these traits evolved because they were useful adaptations in highly active and mobile predators, not because they would someday allow flight.

The morphology of individuals sometimes changes over time because of **allometric growth** (*allo* = different;

metro = measure), the differential growth of body parts. In humans, for example, the relative sizes of different body parts change because human heads, torsos, and limbs grow at different rates **(Figure 22.14a)**.

Allometric growth can also create morphological differences in closely related species. For example, the skulls of chimpanzees and humans are similar in newborns, but markedly different in adults **(Figure 22.14b)**. Some regions of the chimp skull grow much faster than others, while the proportions of the human skull

Figure 22.14
Examples of allometric growth.

a. Allometric growth in humans

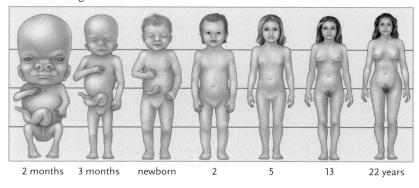

2 months   3 months   newborn   2   5   13   22 years

Humans exhibit allometric growth from prenatal development until adulthood. Our heads grow more slowly than other body parts; our legs grow faster.

b. Differential growth in the skulls of chimpanzees and humans

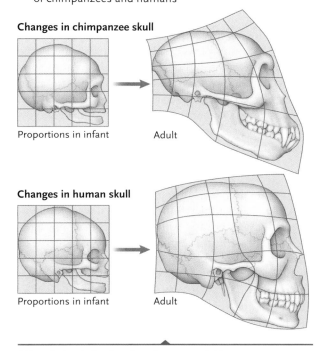

Changes in chimpanzee skull

Proportions in infant   Adult

Changes in human skull

Proportions in infant   Adult

Although the skulls of newborn humans and chimpanzees are remarkably similar, differential patterns of growth make them diverge during development. Imagine that the skulls are painted on a blue rubber sheet marked with a grid. Stretching the sheet deforms the grid in particular ways, mimicking the differential growth of various parts of the skull.

Figure 22.15

Paedomorphosis in salamanders. Some small-mouthed salamanders *(Ambystoma tal-poideum)* undergo metamorphosis, losing their gills and developing lungs (left). Others are paedomorphic: they retain juvenile morphological characteristics, such as gills, after attaining sexual maturity (right).

David Scott/SERL

change much less. Differences in the adult skulls may simply reflect changes in one or a few genes that regulate the pattern of growth.

Changes in the timing of developmental events, called **heterochrony** (*hetero* = different; *chronos* = time), also cause the morphology of closely related species to differ. **Paedomorphosis** (*paedo* = child; *morpho* = form), the development of reproductive capability in an organism with juvenile characteristics, is a common form of heterochrony.

Many salamanders, for example, undergo metamorphosis from an aquatic juvenile into a morphologically distinct terrestrial adult. However, populations of several species are paedomorphic—they grow to adult size and become reproductively mature without changing to the adult form **(Figure 22.15).** The evolutionary change causing these differences may be surprisingly simple. In amphibians, including salamanders, the hormone thyroxine induces metamorphosis (see Chapter 40). Paedomorphosis could result from a mutation that either reduces thyroxine production or limits the responsiveness of some developmental processes to thyroxine concentration.

Changes in developmental rates also influence the morphology of plants **(Figure 22.16).** The flower of a larkspur species, *Delphinium decorum,* includes a ring of petals that guide bees to its nectar tube and structures on which bees can perch. By contrast, *Delphinium nudicaule,* a more recently evolved species, has tight flowers that attract hummingbird pollinators, which can hover in front of the flowers. Slower development in *D. nudicaule* flowers causes the structural difference: a mature flower in the descendant species resembles an unopened (juvenile) flower of the ancestral species.

Novel morphological structures, such as the wings of birds, often appear suddenly in the fossil record. How do novel features evolve? Scientists have identified several mechanisms including preadaptation, allometric growth, and heterochrony. We describe new research about the genetic basis of some morphological innovations in the last section of this chapter.

22.5 Macroevolutionary Trends in Biodiversity

The number of species living on Earth—its overall **biodiversity**—changes over time as a result of both adaptive radiation and extinction.

Adaptive Radiations Are Clusters of Related Species with Diverse Ecological Adaptations

In some lineages, rapid speciation produces a cluster of closely related species that occupy different habitats or consume different foods; we describe such a lineage as an **adaptive radiation.** The Galápagos finches **(Figure 22.17)** and the Hawaiian fruit flies described in Chapter 21 are examples of adaptive radiations.

Adaptive radiation usually occurs after an ancestral species moves into an unfilled **adaptive zone,** a general way of life. Browsing on soft leaves in the forest is the adaptive zone that early horses occupied, and grazing on grass in open habitats is the adaptive zone that horses occupy today. Feeding on plastic in landfills might become an adaptive zone in the future if some organism develops the ability to digest that now-abundant resource.

An organism may move into a new adaptive zone after the chance evolution of a key morphological innovation that allows it to use the environment in a unique way. For example, the dehydration-resistant eggs of early reptiles enabled them to complete their life cycle on land, opening terrestrial habitats to them. Similarly, the evolution of flowers that attract insect pollinators was a key innovation in the history of flowering plants.

An adaptive zone may also open up after the demise of a successful group. Mammals, for example, were relatively inconspicuous during their first 150 million years on Earth, presumably because dinosaurs dominated terrestrial habitats. But after dinosaurs declined in the late Mesozoic era, mammals underwent an explosive adaptive radiation. Today they are the dominant vertebrates in many terrestrial habitats.

Extinctions Have Been Common in the History of Life

Increased biodiversity is counteracted by **extinction,** the death of the last individual in a species or the last species in a lineage. Paleobiologists recognize two distinct patterns of extinction in the fossil record, background extinction and mass extinction.

Species and lineages have been going extinct since life first appeared. We should expect species to disappear at some low rate, the **background extinction rate;** as environments change, poorly adapted organisms will not survive and reproduce. In all likelihood, more than 99.9% of the species that have ever lived are now extinct. David Raup of the University of Chicago has suggested that, on average, as many as 10% of species go extinct every million years and that more than 50% go extinct every 100 million years. Thus, the history of life has been characterized by an ongoing turnover of species.

On at least five occasions, extinction rates rose well above the background rate. During these **mass extinctions**, large numbers of species and lineages died out over relatively short periods of geological time **(Figure 22.18).** The Permian extinction was the most severe: more than 85% of the species alive at that time—including all trilobites, many amphibians, and the trees of the coal swamp forests—disappeared forever. During the last mass extinction, at the end of the Cretaceous, half the species on Earth, including most dinosaurs, became extinct. A sixth mass extinction, potentially the largest of all, may be occurring now as a result of human degradation of the environment (see Chapter 53).

Different factors were responsible for the five mass extinctions. Some were probably caused by tectonic activity and associated changes in climate. For example, the Ordovician extinction occurred after Gondwana moved toward the South Pole, triggering a glaciation that cooled the world's climate and lowered sea levels. The Permian extinction coincided with a major glaciation and a decline in sea level induced by the formation of Pangaea.

Many researchers believe that an asteroid impact caused the Cretaceous mass extinction. The resulting dust cloud may have blocked the sunlight necessary for photosynthesis, setting up a chain reaction of extinctions that began with microscopic marine organisms. Geological evidence supports this hypothesis. Rocks dating to the end of the Cretaceous period (65 million years ago) contain a highly concentrated layer of iridium, a metal that is rare on Earth but common in asteroids. The impact from an iridium-laden asteroid only 10 km in diameter could have caused an explosion equivalent to that of a billion tons of TNT, scattering iridium dust around the world. Geologists have identified the Chicxulub crater, 180 km in diameter, on the edge of Mexico's Yucatán peninsula as the likely site of the impact.

Although scientists agree that an asteroid struck Earth at that time, many question its precise relationship to the mass extinction. Dinosaurs had begun their decline at least 8 million years earlier, but many persisted for at least 40,000 years after the impact. More-

Figure 22.16 Observational Research

Paedomorphosis in Delphinium *Flowers*

HYPOTHESIS: The narrow tubular shape of the flowers of *Delphinium nudicaule*, which are pollinated by hummingbirds, is the product of paedomorphosis, the retention of juvenile characteristics in a reproductive adult.

PREDICTION: The flowers of *D. nudicaule* grow more slowly and mature at an earlier stage of development than those of *Delphinium decorum*, a species with broad, open flowers that are pollinated by bees.

D. decorum

D. nudicaule

METHOD: Edward O. Guerrant of the University of California at Berkeley measured 42 bud and flower characteristics in *D. nudicaule* and *D. decorum* as their flowers developed and used the number of days since the completion of meiosis in pollen grains as a measure of flower maturity. He then used a complex statistical analysis to compare the characteristics of the buds and flowers of both species.

RESULTS: The mature flowers of *D. nudicaule* resemble the buds of both species more closely than they resemble the flowers of *D. decorum*. Although the time required for maturation of the reproductive structures is similar in the two species, the rate of petal growth (measured as petal blade length) is slower in *D. nudicaule*. As a result, the mature flowers of *D. nudicaule* do not open as widely as those of *D. decorum*. Because of these morphological differences, bees can pollinate flowers of *D. decorum*, but they can't land on the flowers of *D. nudicaule*, which are instead pollinated by hummingbirds.

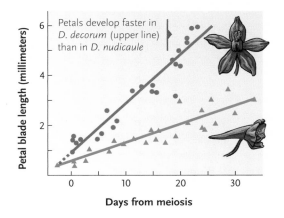

CONCLUSION: The narrower and more tubular shape of *D. nudicaule* flowers, which mature at an earlier stage of development than *D. decorum* flowers, is the product of paedomorphosis.

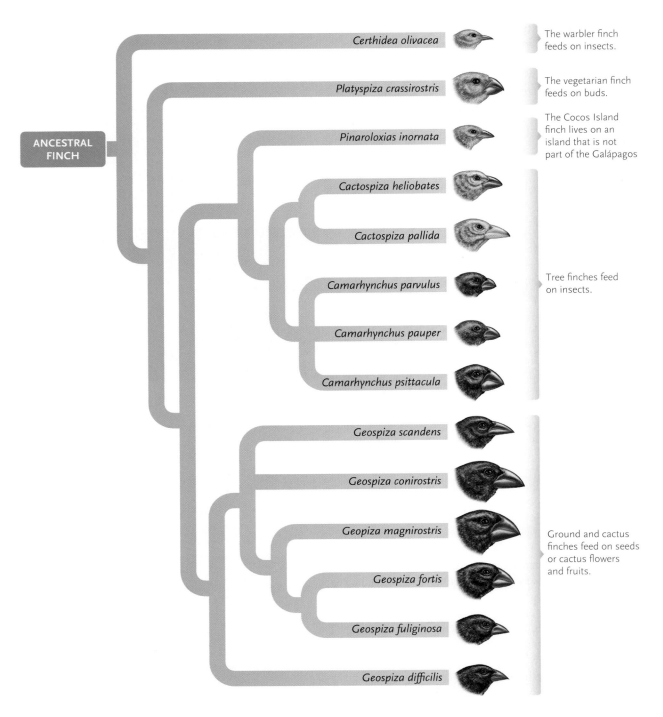

The warbler finch feeds on insects.

The vegetarian finch feeds on buds.

The Cocos Island finch lives on an island that is not part of the Galápagos

Tree finches feed on insects.

Ground and cactus finches feed on seeds or cactus flowers and fruits.

Certhidea olivacea

Platyspiza crassirostris

Pinaroloxias inornata

Cactospiza heliobates

Cactospiza pallida

Camarhynchus parvulus

Camarhynchus pauper

Camarhynchus psittacula

Geospiza scandens

Geospiza conirostris

Geopiza magnirostris

Geospiza fortis

Geospiza fuliginosa

Geospiza difficilis

ANCESTRAL FINCH

Figure 22.17
Adaptive radiation. The 14 species of Galápagos finches are descended from one ancestral species.

over, other groups of organisms did not suddenly disappear, as one would expect after a global calamity. Instead, the Cretaceous extinction took place over tens of thousands of years.

Biodiversity Has Increased Repeatedly over Evolutionary History

Although mass extinctions temporarily reduce biodiversity, they also create evolutionary opportunities. Some species survive because they have highly adaptive traits, large population sizes, or widespread distributions. And some surviving species undergo adaptive

radiation, filling adaptive zones that mass extinctions made available.

Sometimes the success of one lineage comes at the expense of another. Although the diversity of terrestrial vascular plants has increased almost continuously since the Devonian period, this trend includes booms and busts in several lineages **(Figure 22.19).** Ferns and conifers recovered rapidly after the Permian extinction, maintaining their diversity until the end of the Mesozoic era. However, angiosperms, which arose and diversified in the late Jurassic and early Cretaceous periods, may have hastened the decline of these groups by replacing them in many environments.

The superb fossil record left by certain marine animals reveals three major periods of adaptive radiation (Figure 22.20). The first occurred during the Cambrian, more than 500 million years ago, when all animal phyla, the major categories of animal life, arose. Most of these phyla became extinct, and a second wave of radiations established the dominant Paleozoic fauna during the Ordovician period. A third evolutionary fauna emerged in the Triassic period, right after the great Permian extinction; it produced the immediate ancestors of modern marine animals. The diversity of marine animals has increased consistently since the early Triassic, in large measure because of continental drift. As continents and shallow seas became increasingly isolated, regional biotas diversified independently of one another, increasing worldwide biodiversity.

Historical increases in biodiversity can also be attributed to the evolution of ecological interactions. For example, the number of plant species found *within* fossil assemblages has increased over time, suggesting the evolution of mechanisms that allow more species to coexist. In addition, insects diversified dramatically in the Cretaceous period, possibly because the angiosperms created a new adaptive zone for them. New insect species then provided a novel set of pollinators that may have stimulated the radiation of angiosperms. Such long-term evolutionary interactions between ecologically intertwined lineages have played an important role in structuring ecological communities, which are described more fully in Chapter 50.

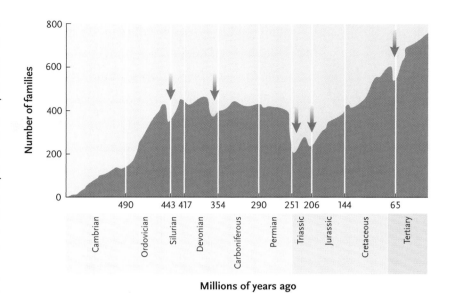

Figure 22.18

Mass extinctions. Biodiversity, indicated by the height of the dark blue area in the graph, was temporarily reduced by at least five mass extinctions (arrows) during the history of life. The data presented in this graph record the family-level diversity of marine animals. A family is a group of genera descended from a common ancestor.

STUDY BREAK

1. What factors might allow a population of organisms to occupy a new adaptive zone?
2. Did the mass extinction at the end of the Cretaceous period occur quickly or over a long period of time?
3. When did the first major adaptive radiation of animals occur?

22.6 Evolutionary Developmental Biology

Historically, evolutionary biologists compared the embryos of different species to study their evolutionary history (see Figure 19.13), but they often worked independently from scientists studying the embryonic development of organisms. As a result, evolutionary biologists were unable to construct a coherent picture of the specific developmental mechanisms that contributed to morphological innovations. Since the late 1980s, however, advances in molecular genetics have allowed scientists to explore the genomes of organisms in great detail, fostering a new approach to these studies. **Evolutionary developmental biology**—evo-devo, for short—asks how evolutionary changes in the genes regulating embryonic development can lead to changes in body shape and form.

The study of the genetics of embryonic development helps us understand macroevolutionary trends

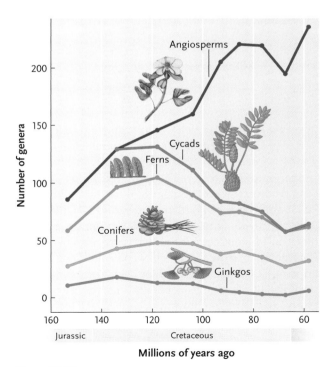

Figure 22.19

History of vascular plant diversity. The diversity of angiosperms increased during the Mesozoic era as the diversity of other groups declined.

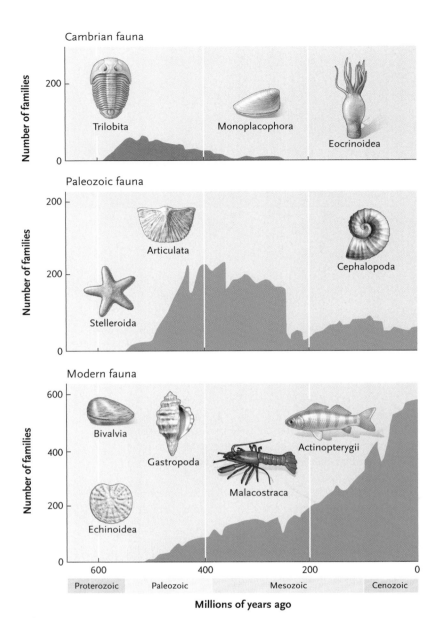

Cambrian fauna

Trilobita
Monoplacophora
Eocrinoidea

Paleozoic fauna

Articulata
Cephalopoda
Stelleroida

Modern fauna

Bivalvia
Gastropoda
Actinopterygii
Malacostraca
Echinoidea

Proterozoic | Paleozoic | Mesozoic | Cenozoic

Millions of years ago

Figure 22.20

History of marine animal diversity. Marine animals have undergone three major radiations. Few remnants of the Cambrian fauna remain alive today.

because changes in genes that regulate development often promote the evolution of morphological innovations. Moreover, the resulting changes in body plan have sometimes fostered adaptive radiations, increasing biodiversity over geological time. In the life cycle of a multicellular organism (see Figure 1.7), the many different body parts of the adult develop in a highly controlled sequence of steps that is specified by genetic instructions in the single cell of a fertilized egg. Developmental biologists study how regulatory genes control the development of phenotypes and their variations. (Gene regulation was described in Chapter 16.) When these genes code for transcription factors that bind regulatory sites on DNA, either activating or repressing the expression of other genes that contribute to an organism's form, they are called **homeotic genes** (described further in Chapters 34 and 48). In this section we describe a few intriguing discoveries about the

genetic mechanisms that underlie some macroevolutionary trends in animals.

Most Animals Share the Same Genetic Tool Kit That Regulates Their Development

Comparisons of genome sequence data reveal that most animals, regardless of their complexity or position in the tree of life, share a set of several hundred homeotic genes that control their development. Collectively, these genes have been dubbed the "genetic tool kit," because they govern the basic design of the body plan by controlling the activity of thousands of other genes. Some of the tool-kit genes must be at least 500 million years old, because all living animals inherited them from a common ancestor alive at that time. Some of the same tool-kit genes are also present in plants, fungi, and prokaryotes, suggesting that those genes may date back to the earliest forms of life.

Structurally, tool-kit genes do not differ much among the animals that possess them, and they generally play the same role in development for all species. For example, genes in the *Hox* family control the overall body plan of animals. All *Hox* genes include a 180-nucleotide sequence called a **homeobox**, which codes for a **homeodomain**, part of a protein that functions as a transcription factor. When bound to a regulatory site on a strand of DNA, the homeodomain either activates or represses a downstream gene involved in development.

Among other functions, *Hox* genes specify where appendages—wings in flies and legs in mice—will develop on the animal's body. They do so by producing transcription factors that activate the genes specifying wings or legs in the body regions where these appendages typically grow. The different *Hox* genes, which are expressed at different positions along the head-to-tail axis of a developing embryo, are arranged on a chromosome in the same sequence in which they are expressed in the body. Remarkably, the *Hox* genes and their relative positions on chromosomes have been conserved by evolution; nearly identical genes are found in animals as different as fruit flies and mice **(Figure 22.21)**. Genes with comparable functions control aspects of development in plants (see Chapter 34).

Another example of a highly conserved and widely distributed tool-kit gene, the *Pax-6* gene, triggers the formation of light-sensing organs as diverse as the eye spots in flatworms, the compound eyes of insects and other arthropods, and the camera eyes of vertebrates (see Chapter 39). Like the *Hox* genes, *Pax-6* also contains a homeobox, indicating that the protein for which it codes either activates or represses gene transcription. The proteins coded by *Pax-6* in different animals are so similar that when researchers genetically engineered fruit fly larvae to express the *Pax-6* gene taken from a squid or a mouse, the flies responded by developing eyes. The induced eyes were, however, fruit fly eyes—not squid eyes or mouse eyes. Thus, *Pax-6* triggers activity in the genes that carry the specific instructions for making an eye typical of the species. Apparently, the ancient genetic sequence for *Pax-6*, the master regulatory gene for eye development, has been conserved over the hundreds of millions of years since the common ancestor of squids, fruit flies, and mice lived.

Evolutionary Changes in Developmental Switches May Account for Much Evolutionary Change

If most animals share the same tool-kit genes, how has evolution produced different body plans among species? What makes a squid, a fruit fly, and a mouse different? Researchers in evo-devo have proposed that morphological differences among species arise when mutations alter the effects of developmental regulatory genes. As you will discover in Chapter 48, the developmental programs of animals involve complex networks of many interacting genes. Varying combinations of tool-kit genes may be expressed at different times and in different body regions. According to this hypothesis, the several hundred tool-kit genes encode proteins that work either as activators or repressors in a multitude of possible combinations. Thus, they can generate an unimaginably large number of different gene expression patterns, each with the potential to alter morphology.

Sean Carroll of the Howard Hughes Medical Institute and the University of Wisconsin at Madison has described the regulatory sites that transcription factors can bind as *switches,* like those we use to turn lights on or off. When a combination of transcription factors turns on a regulatory switch, a gene further downstream is activated. When transcription factors turn off a regulatory switch, a downstream gene is inactivated.

Although all the cells in an animal contain exactly the same set of genes, the differential expression of genes in different body regions and at different times during embryonic development causes different structures to be made. Allometric growth can result from evolutionary changes in developmental switches that cause certain body parts to grow larger or faster than others. Similarly, heterochrony can be explained as an evolutionary change in the switches that either delays the development of adult characteristics or speeds up the development of reproductive maturity.

If Carroll's hypothesis is correct, morphological novelties arise when evolutionary changes in developmental switches alter the expression patterns of *existing* genes. This view contrasts markedly with the explanation proposed in the modern synthesis (the unified theory of evolution described in Chapter 19), that most morphological novelties arise as mutations slowly accumulate in genes that carry the blueprints for building particular structures. According to the modern synthe-

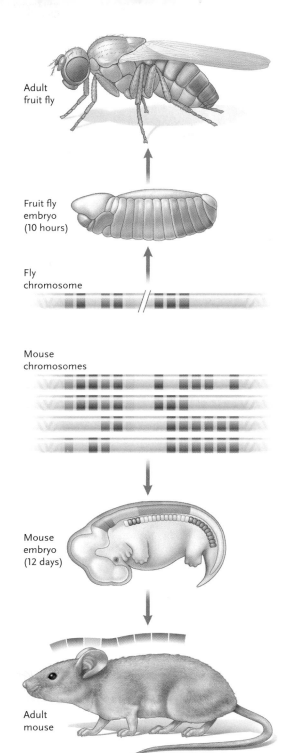

Figure 22.21

Hox genes. The linear sequence of *Hox* genes on chromosomes and the expression of *Hox* genes in different body regions have been conserved by evolution. Each color-coded band on the chromosomes in the illustration represents a different gene in the *Hox* family of genes. Fruit flies have one set of *Hox* genes, which are arranged on a single chromosome in the same order that they are expressed in the fruit fly embryo. Like all mammals, mice have four sets of *Hox* genes, arranged on four chromosomes that are expressed in mouse embryos in the same order as the *Hox* genes in fruit flies. The illustrations of the adult fruit fly and mouse show the adult body regions that are influenced by the expression of *Hox* genes in their embryos.

Labels in figure: Adult fruit fly; Fruit fly embryo (10 hours); Fly chromosome; Mouse chromosomes; Mouse embryo (12 days); Adult mouse

Fancy Footwork from Fins to Fingers

Both fishes and tetrapods (four-footed animals) have two pairs of appendages, one anterior and one posterior.

a. Fishes

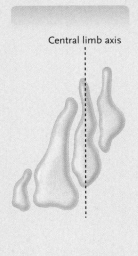

Central limb axis

Bones in the fin of a fish develop from centers of cartilage formation along a central axis (dashed line).

b. Tetrapods

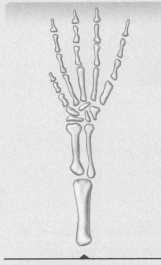

Bones in the limb and digits of a tetrapod also develop from centers of cartilage formation in the central axis.

c. Fishes

d. Tetrapods

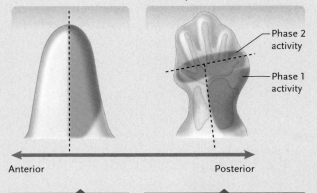

Phase 2 activity

Phase 1 activity

Anterior   Posterior

During development of the fin in fishes, *HoxD* genes become active in cells posterior to the central axis of the fin (shown in blue).

During development of the limb and digits in tetrapods, *HoxD* genes first become active in cells posterior to the central axis of the limb (blue). Later, these genes are active in a band of cells perpendicular to the central axis of the limb (green).

They develop similarly in both groups during early embryonic stages. They start out as buds of mesoderm—the middle of the three primary embryonic tissue layers—and thicken by increased cell division. As the buds elongate, cartilage is deposited at localized centers, and bones of the appendages later form in these centers.

In fishes, the bones develop along a central axis from the base to the tip of the limb **(Figure a)**. In tetrapods, the centers of cartilage formation generate the long bones of the limb and the five digits of the foot **(Figure b)**. In humans, the digits are the thumb and fingers of the hand or the toes of the foot.

For some time, evolutionists wondered whether the digits of tetrapods were modifications of the bones radiating from the central limb axis in fishes or novel evolutionary structures. Molecular research by Paolo Sordino, Frank van der Hoeven, and Denis Duboule at the University of Geneva in Switzerland indicates that animal digits are a morphological novelty.

In all animals with paired anterior and posterior appendages, groups of homeobox genes control their development. Sordino and his colleagues compared the activity of a group called the *HoxD* genes in the zebrafish *Danio rerio* (a common aquarium fish and a model research organism) with the previously known *HoxD* patterns in birds and rodents.

To begin their work, the researchers used the DNA of a rodent *HoxD* gene as a probe to search for similar genes in fragmented zebrafish DNA. The probe paired with fragments of zebrafish DNA that, when cloned and sequenced, proved to include three *HoxD* genes—*HoxD-11*, *HoxD-12*, and *HoxD-13*—arranged in the same order

as in rodents. They tested the activity of the *HoxD* genes in developing zebrafish limbs by using a nucleic acid probe that could pair with mRNA products of the genes. The probe was linked to a blue dye molecule, so that cells in which a particular *HoxD* gene was active would appear blue in the light microscope. The investigators found that in zebrafish, the *HoxD* genes became active in cells along the posterior side of the central axis **(Figure c)**. As fin development neared completion, activity of the *HoxD* genes dropped off.

Using the same techniques to study tetrapods, the investigators found that the *HoxD* genes were activated in two distinct phases **(Figure d)**. In phase 1, gene activity was restricted to the posterior half of the limb, as it is in zebrafish; this period of activity corresponds to development of the long limb bones. Later, in phase 2, the *HoxD* genes became active in a band of cells perpendicular to the central axis; the cartilage centers that form the bones of the digits develop in this anterior-posterior band. Sordino and his colleagues found no equivalent band of activity in zebrafish; the *HoxD* gene activity remained restricted to a single phase along the posterior half of the fin.

Thus, the phase of *HoxD* gene activity corresponding to development of the digits is a separate pattern unique to tetrapods; it therefore appears to be a morphological novelty. If this is the case, fishes probably have no bones homologous to the five digits of the hand and foot, which were added as new structures during the evolutionary events that split the ancestors of fishes from those of four-footed animals.

sis, the accumulated mutations eventually create *new* genes that specify the creation of new structures.

Although Carroll's hypothesis argues that changes in genes regulating development cause most morphological change, proponents of evo-devo recognize that mutations in developmental regulatory genes and their effects on morphology are subject to the action of the same microevolutionary processes—natural selection, genetic drift, and gene flow—that influence the frequencies of genotypes and phenotypes in populations. Thus, every morphological change induced by a mutation in a homeotic gene or in a developmental switch is tested by the success or failure of the individual that carries it.

Numerous studies have shown that changes in the expression of homeotic genes can have dramatic effects on morphology. *Insights from the Molecular Revolution* explains how a change in the number and expression of *Hox* genes produced a striking alteration in the structure of vertebrate appendages.

In another example, researchers have determined how an adaptive morphological change in a small fish, the three-spined stickleback (*Gasterosteus aculeatus*), results from the deactivation of a homeotic gene. The freshwater stickleback populations in North American lakes are the descendants of marine ancestors that colonized the lakes after the retreat of glaciers between 10,000 and 20,000 years ago. Marine sticklebacks have bony armor along their sides and prominent spines; lake-dwelling sticklebacks have greatly reduced armor and, in many populations, lack spines on their pelvic fins **(Figure 22.22)**.

Natural selection has apparently fostered these morphological differences in response to the dominant predators in each habitat. In marine environments, long spines prevent some predatory fishes from

Unanswered Questions

Does morphological evolution always proceed gradually or can it occur in great leaps and bounds?

As you read in this chapter, biologists disagree about whether evolutionary changes in morphology can occur very rapidly. Although biologists have proposed various hypotheses to explain the abrupt changes that we sometimes find in the fossil record, evo-devo studies provide insight into one mechanism for how dramatic changes can arise: the spatial redeployment of homeotic genes. *Homeosis* is defined as the complete replacement of one type of organ with another. In one famous example, a *Hox* gene mutation in *Drosophila* replaces the antennae with legs. If such a mutation were to occur in nature, the organism would probably not have a selective advantage. But what if it did? These kinds of mutant phenotypes first inspired Richard Goldschmidt to develop his idea of the "hopeful monster" early in the twentieth century. Stephen Jay Gould later revised and updated this idea in the context of his punctuated equilibrium hypothesis about the tempo and mode of evolution. The hypothesis suggests that if, on very rare occasions, truly dramatic morphological changes provide a selective advantage, they may lead to the rapid formation of a new species based on only a few genetic differences. What types of organisms are the most likely to exhibit homeotic change in an evolutionary context? The best candidates are those with highly modular bodies made up of repeating units—like the segments of an insect or the bones in the spine of a vertebrate. Such animals often express different organ identities in different modular units—such as the antennae, claws, and legs of a lobster—and these identities may be redeployed to different positions along the body axis. Plants are among the most modular organisms on Earth. They produce serially repeated structures—a leaf, a bud at the base of the leaf, and a stem—to generate their bodies. Many exciting and promising questions in plant evo-devo relate to how evolutionary homeosis may have generated rapid change in plant morphology.

Has homeosis contributed to the appearance and diversification of the angiosperms?

The sudden appearance of flowering plants, the angiosperms, in the fossil record so puzzled Charles Darwin that he dubbed their evolution an "abominable mystery." How did the gymnosperms, which always bear their male and female reproductive structures separately, give rise to the hermaphroditic (that is, bearing both male and female structures) flower? Our current understanding of the genetics of floral developmental provides a simple solution to this puzzle: the genetic program controlling floral organ identity is homeotic. Thus, it is possible for very simple genetic changes to transform an entirely male set of reproductive organs into a combination of male and female parts. Such models have been outlined by Günter Theissen at the Friedrich-Schiller-Universität in Germany as well as David Baum and Lena Hileman at the University of Wisconsin and University of Kansas, respectively. In addition to fostering the origin of the angiosperms, homeosis may have played a role in the group's diversification. Commonly observed shifts in the morphology of sepals and petals or in the number of stamens (male reproductive structures) are suggestive of homeotic changes. These examples are more suitable for experimental verification than the question on the origin of the angiosperms is, because they are much more recent occurrences. Although these hypotheses are very attractive, they remain to be confirmed through molecular genetic analyses.

How have new floral organ identity programs evolved?

The homeotic scenarios described here involve spatial shifts in the expression of preexisting identity programs, such as a stamen developing where there was previously a petal. But what about cases where a whole new type of floral organ appears? Across the angiosperms there are many examples of flowers that have more than the four most common types of organs—sepals, petals, stamens, and carpels, described in Chapter 34—which suggests that new organ identity programs must have evolved. In such instances, do the new organs evolve through modification of preexisting identity programs, or are entirely new gene pathways recruited? Studies using a new model plant for genetics, *Aquilegia*—commonly known as columbine—suggest that a fifth type of floral organ has evolved through modification of the stamen identity program. Many questions about how this process actually occurred remain unanswered. Was the derivation of the new program achieved through just a few genetic changes, or many? Did it involve changes in regulatory gene function, or only shifts in gene expression? There is much more to learn about how completely new types of floral organs have evolved.

Elena M. Kramer is the John L. Loeb Associate Professor of Biology at Harvard University, where she studies the evolution of the genetic mechanisms controlling floral development. To learn more about Dr. Kramer's research go to http://www.oeb.harvard.edu/faculty/kramer/index.htm.

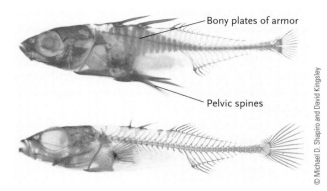

Bony plates of armor

Pelvic spines

© Michael D. Shapiro and David Kingsley

Figure 22.22

Sticklebacks. Marine populations (top) of three-spined sticklebacks *(Gasterosteus aculeatus)* have prominent bony plates along their sides and large spines on their dorsal and pelvic fins. Many freshwater populations of the same species (bottom) lack the bony plates and spines. Pelvic spines do not develop in the freshwater fishes because they do not express the *Pitx1* gene in their fin buds during embryonic development. The skeletons of these specimens, each about 8 cm long, were dyed bright red.

swallowing sticklebacks. But long spines are a liability in lakes, where voracious dragonfly larvae grab hold of sticklebacks by their spines and then devour them; freshwater sticklebacks that lack spines are more likely to escape from their clutches.

The presence or absence of spines on the pelvic fins of these fishes is governed by the expression of the gene *Pitx1*. Pelvic spines are part of the pelvic fin skeleton, the fishes' equivalent of a hind limb. In fact, *Pitx1* also contributes to the development of hind limbs in four-legged vertebrates as well as certain glands and sensory organs in the head. In long-spined marine sticklebacks, *Pitx1* is expressed in the embryonic buds from which pelvic fins develop, promoting the development of spines. But *Pitx1* is not expressed in the fin buds of the freshwater sticklebacks; hence, pelvic spines do not develop. However, freshwater sticklebacks have not *lost* the *Pitx1* gene; it is still expressed elsewhere in the fishes' bodies. Apparently, a mutation somehow blocks its expression in the developing pelvic fin, thereby blocking the production of pelvic spines.

In the next chapter, we examine how biologists explore the evolutionary relationships among the many species they encounter and how they organize that information into a useful framework for researchers in every biological discipline. In the following unit, we will revisit evo-devo and examine some recent discoveries about how changes in homeotic genes have diversified body plans in the major evolutionary groups of organisms.

STUDY BREAK

1. What evidence suggests that many developmental control genes have been conserved by evolution?
2. What genetic factor is apparently responsible for the presence or absence of spines in stickleback fish?

Review

Go to ThomsonNOW™ at www.thomsonedu.com/login to access quizzing, animations, exercises, articles, and personalized homework help.

22.1 The Fossil Record

- Fossils are the parts of organisms preserved in sedimentary rocks or in oxygen-poor environments (Figure 22.2).
- The fossil record is incomplete because few organisms fossilize completely, because some organisms are more likely to fossilize than others, and because natural processes destroy many fossils.
- Fossils provide a relative dating system, the geological time scale, for the strata in which they occur (Figure 22.3). Radiometric dating techniques establish the absolute age of rocks and fossils (Figure 22.4 and Table 22.1).
- The fossil record provides data on changes in morphology, biogeography, ecology, and behavior of organisms; some fossils also contain biological molecules.

Animation: Radioisotope decay

Animation: Radiometric dating

Animation: Geologic time scale

22.2 Earth History, Biogeography, and Convergent Evolution

- Earth's crust is composed of plates of solid rock that float on a semisolid mantle (Figure 22.5). New crust is constantly generated and old crust is recycled, and currents in the mantle cause the continents to move over geological time (Figure 22.6). Continental movements cause variations in patterns of glaciation, sea level, and climate. Asteroid impacts and volcanic eruptions have also influenced the environment.
- Disjunct distributions of species are produced by dispersal and vicariance. Dispersal results in a disjunct distribution when a new population is established on the far side of a barrier. Vicariance results in a disjunct distribution when external factors such as continental drift fragment the landscape.
- Continental drift has created six major biogeographical realms, each with a characteristic biota (Figure 22.7).
- Convergent evolution produces similar adaptations in distantly related species that live in similar environments (Figures 22.8 and 22.9).

Animation: Plate margins

Animation: Geologic forces

22.3 Interpreting Evolutionary Lineages

- The horse lineage is complex and highly branched. It includes species of various sizes and diverse morphological adaptations (Figure 22.10).

- Anagenesis and cladogenesis produce morphological change (Figure 22.11). Anagenesis is the accumulation of many small changes in a species over long periods of time. Cladogenesis is the evolutionary division of an ancestral species into multiple descendant species. Only cladogenesis increases the number of species living at a particular time.

- The gradualist hypothesis of evolution suggests that major morphological changes result from the accumulation of small changes over long periods of time. The punctuated equilibrium hypothesis suggests that most morphological evolution occurs during short periods of cladogenesis. Both patterns occur in nature (Figures 22.12 and 22.13).

Animation: Evolutionary tree diagrams

22.4 Macroevolutionary Trends in Morphology

- In many lineages, body size has increased over evolutionary history. Evolutionary trends may be produced by species selection if certain traits are associated with high rates of speciation.

- In a general way, morphological complexity has also increased over evolutionary time, although certain morphological features have become simplified in some lineages.

- A preadaptation is a trait that turns out to be useful in a new environmental context even before natural selection refines its form. Morphological novelties can arise from evolutionary changes in the relative growth rates of body parts (Figure 22.14), the timing of developmental events (Figures 22.15 and 22.16),

or changes in homeobox genes that control developmental processes.

Animation: Morphological divergence

22.5 Macroevolutionary Trends in Biodiversity

- Adaptive radiation produces morphological diversity within lineages (Figure 22.17).

- Extinction decreases species diversity. Mass extinctions have occurred at least five times in the history of life (Figure 22.18). Tectonic activity, climatic change, and asteroid strikes are probable causes of mass extinctions.

- Biodiversity has increased since life first evolved, partly in response to increased geographical separation of the continents and partly because complex interactions evolve among existing species (Figures 22.19 and 22.20).

22.6 Evolutionary Developmental Biology

- Evolutionary developmental biology—evo-devo—examines how evolutionary changes in genes that regulate embryonic development can foster changes in body shape and form.

- Most organisms share an ancient tool kit of several hundred genes that regulate the expression of thousands of genes involved in development (Figure 22.21). The tool-kit genes produce transcription factors that collectively activate or repress genes in a complex developmental network. *Hox* genes control aspects of the overall body plan of animals, and the *Pax-6* gene triggers the development of light-sensing organs.

- Evolutionary changes in developmental switches may account for many morphological changes. The differential expression of the *Pitx1* gene in sticklebacks determines whether or not a fish grows pelvic spines (Figure 22.22).

Questions

Self-Test Questions

1. The fossil record:
   a. provides direct evidence about life in the past.
   b. supports the punctuated equilibrium hypothesis, but not the gradualist hypothesis.
   c. provides abundant data about rare species with local distributions.
   d. is equally good for all organisms that ever lived.
   e. provides no evidence about the physiology or behavior of ancient organisms.

2. The absolute age of a geological stratum is determined by:
   a. the thickness of its rocks.
   b. the particle size in its rocks.
   c. the types of fossils found within it.
   d. anagenetic analysis.
   e. radiometric dating techniques.

3. The observation that fossils of *Premedosaurus* are found only in Argentina and Northern Europe provides an example of:
   a. a continuous distribution.
   b. a disjunct distribution.
   c. species selection.
   d. allometry.
   e. gradualism.

4. The evolutionary history of horses demonstrates that:
   a. modern horses are the direct, lineal descendants of the earliest horses.
   b. the leg bones of modern horses are more complex than those of the earliest horses.

   c. horses have always had specialized teeth that allow them to feed on tough grasses.
   d. horses diversified greatly, but only a few types survived to the present.
   e. the first horses lived in open, grassy habitats.

5. The punctuated equilibrium hypothesis:
   a. recognizes that morphological evolution may occur slowly or quickly.
   b. suggests that major morphological novelties can arise by anagenesis.
   c. may help explain why there are so many "missing links" in the fossil record.
   d. suggests that the fossil record is usually complete.
   e. links mass extinctions to the impact of asteroids striking Earth.

6. Macroevolutionary trends in body size could be caused by:
   a. plate tectonics.
   b. paedomorphosis.
   c. species selection.
   d. heterochrony.
   e. convergent evolution.

7. The differential growth of body parts is called:
   a. allometry.
   b. paedomorphosis.
   c. heterochrony.
   d. cladogenesis.
   e. preadaptation.

8. Preadaptations are traits that:
   a. prepare some organisms for future environmental changes.
   b. appear in lineages as a result of an adaptive radiation.
   c. evolve in anticipation of a species' future needs.

        d. are useful in new situations before natural selection changes them.
        e. occur in animals, but not in plants.
    9. Adaptive radiations often follow mass extinctions because:
        a. mass extinctions limit the impact of species selection.
        b. mass extinctions foster allometry and heterochrony.
        c. mass extinctions decimate all forms of life on Earth.
        d. species that undergo frequent cladogenesis survive mass extinctions.
        e. extinctions open adaptive zones that had been previously occupied.
    10. Homeotic genes are defined as genes that:
        a. bind directly to a regulatory site on DNA.
        b. code for transcription factors activating or repressing genes that influence an organism's form.
        c. determine whether or not a morphological innovation leads to an adaptive radiation.
        d. have been inherited from an ancient ancestor by nearly all forms of life.
        e. help biologists differentiate between plants and animals.

Questions for Discussion

1. Many millions of years from now, continental drift may obliterate the Pacific Ocean, pushing North America into physical contact with Asia. What effects might these events have on the organisms living at that time?

2. The species selection hypothesis measures evolutionary success in terms of the number of descendant species that a given species produces. Should our species, *Homo sapiens,* be considered successful under this definition?

3. Extinctions are common in the history of life. Why are biologists alarmed by the current wave of extinctions caused by human activity?

Experimental Analysis

Design a study to determine whether the wings of birds, bats, and insects and their ability to fly are the products of convergent evolution.

Evolution Link

The geological evolution of Earth has had an obvious effect on biological evolution. Consider the reverse: How has the evolution of different organisms, such as photosynthetic microorganisms or humans, changed the physical environment on Earth?

How Would You Vote?

Scientifically important fossils are sometimes found on privately owned land, creating disputes about who owns the fossils and how they should be used. For example, ownership of "Sue," the largest *Tyrannosaurus* fossil ever discovered, had to be settled in a court of law. Although the fossil was unearthed on a privately owned ranch on a Sioux Indian reservation, the land was held in trust by the U.S. government. The government argued that the fossil was public property, but the court eventually decided that the rancher owned the fossil. He could keep it or dispose of it however he chose. He sold it at auction for more than $8 million, the highest price ever paid for a fossil. A group of corporate sponsors raised the funds to buy the fossil on behalf of the Field Museum in Chicago, where it is now on public display. Do you think that scientifically important specimens should be the property of any one individual, or should they belong to the government, a museum, or some other research institution? Go to www.thomsonedu.com/login to investigate both sides of the issue and then vote.

A new plant species from Idaho. Sacajawea's bitterroot *(Lewisia sacajaweana)* was formally described in 2006. It is named in honor of Sacajawea, the Native American woman who guided Lewis and Clark in their exploration of the Pacific Northwest in the early 1800s.

Courtesy of U.S. Forest Service, Boise National Forest (Kathryn M. Beall photo)

STUDY PLAN

23.1   Systematic Biology: An Overview

The twin goals of systematics are reconstruction of evolutionary history and classification of species

Systematics provides essential information for all of the biological sciences

23.2   The Linnaean System of Taxonomy

Linnaeus developed the system of binomial nomenclature

The taxonomic hierarchy organizes huge amounts of systematic data

23.3   Organismal Traits as Systematic Characters

Morphological characters provide abundant clues to evolutionary relationships

Behavioral characters offer additional data when species are not morphologically distinct

23.4   Evaluating Systematic Characters

Characters must be independent markers of underlying genetic similarity and differentiation

Only homologous characters provide data about evolutionary relationships

Systematists focus attention on derived versions of characters

23.5   Phylogenetic Inference and Classification

Many systematic studies rely on the principles of monophyly and parsimony

Traditional evolutionary systematics was based on Linnaeus' methods

Cladistics uses shared derived characters to trace evolutionary history

23.6   Molecular Phylogenetics

Molecular characters have both advantages and disadvantages over organismal characters

Variations in the rates at which molecules evolve govern the molecules chosen for phylogenetic analyses

The analysis of molecular characters requires specialized approaches

Molecular phylogenetics has clarified many evolutionary relationships

23   Systematic Biology: Phylogeny and Classification

WHY IT MATTERS

Mention the word "malaria," and people envision the tropics: explorers wander through the jungle in pith helmets and sleep under mosquito netting; clouds of insects hover nearby, ready to infect them with *Plasmodium*, the parasite that causes this disease. You may be surprised to learn, however, that less than 100 years ago, malaria was also a serious threat in the southeastern United States and much of western Europe.

Scientists puzzled over the cause of malaria for thousands of years. Hippocrates, a Greek physician who worked in the fifth century B.C., knew that people who lived near malodorous marshes often suffered from fevers and swollen spleens. Indeed, the name malaria is derived from the Latin for "bad air." By 1900, scientists had established that mosquitoes, *Plasmodium*'s intermediate hosts, transmit the parasite to humans. Mosquitoes breed in standing water, and anyone living nearby is likely to suffer their bites.

Until the 1920s, scientists thought that the mosquito species *Anopheles maculipennis* carried malaria in Europe. But some areas with huge mosquito populations had little human malaria, whereas other areas had relatively few mosquitoes and a high incidence of the disease.

Then, a French researcher reported variation in the mosquitoes, and Dutch scientists identified two forms of the "species," only one of which seemed to carry malaria. The breakthrough came in 1924, when a retired public health inspector in Italy discovered that individual mosquitoes—all thought to be the same species—produced eggs with one of six distinctive surface patterns **(Figure 23.1)**.

Further research revealed that the name *Anopheles maculipennis* had been applied to six separate mosquito species. Although the adults of these species are very similar, their eggs are clearly different. The species are reproductively isolated from each other, and they differ ecologically: some breed in brackish coastal marshes, others in freshwater inland marshes, and still others in slow-moving streams. Only some of these species have a preference for human blood, and researchers eventually determined that only three of them routinely transmit malaria to humans.

These discoveries explained why the geographical distributions of mosquitoes and malaria did not always match. And government agencies could finally fight malaria by eradicating the disease-carrying species. Health workers drained marshes to prevent mosquitoes from breeding. They applied insecticides to kill mosquito larvae or introduced *Gambusia,* the mosquito

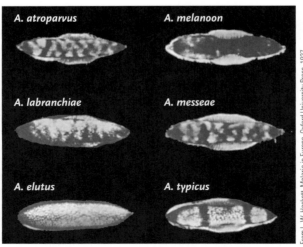

A. atroparvus

A. melanoon

A. labranchiae

A. messeae

A. elutus

A. typicus

From L. W. Hackett, Malaria in Europe, Oxford University Press, 1937

Figure 23.1

Eggs of European mosquitoes. Differences in surface patterns on the eggs of *Anopheles* mosquitoes in Europe helped researchers identify six separate species. The adults of all six species look remarkably alike. An adult *Anopheles atroparvus* is illustrated.

fish, which eats them. These targeted control programs were very successful.

The eradication of malaria in Europe owes a debt to **systematics**, the branch of biology that studies the diversity of life and its evolutionary relationships. Systematic biologists—systematists for short—identify, describe, name, and classify organisms, organizing their observations within a framework that reflects evolutionary relationships. In this chapter we first describe the goals of systematics and the traditional classification scheme that has been used for more than 200 years. Next we consider some of the evidence that systematists use and how that evidence must be interpreted to infer evolutionary relationships. Finally, we consider the analytical methods that contemporary systematists embrace.

23.1 Systematic Biology: An Overview

By organizing information about the biological world, systematics facilitates research in all fields of biology.

The Twin Goals of Systematics Are Reconstruction of Evolutionary History and Classification of Species

The science of systematics has two major goals. One is to reconstruct the **phylogeny**, the evolutionary history, of a group of organisms. Phylogenies are illustrated in **phylogenetic trees**, formal hypotheses that identify likely relationships among species. Like all hypotheses, they are revised as scientists gather new data.

The second goal of systematics is **taxonomy**, the identification and naming of species and their placement in a classification. A **classification** is an arrangement of organisms into hierarchical groups that reflect their relatedness. Most systematists want classifications to mirror phylogenetic history and, thus, the pattern of branching evolution.

Systematics Provides Essential Information for All of the Biological Sciences

Systematics is sometimes maligned as "stamp collecting" by those who think that systematists just collect, describe, and maintain specimens. In fact, systematists study the patterns of phenotypic and genetic variation discussed in Chapters 20 and 21. Thus, their work enhances our understanding of microevolution, speciation, adaptive radiation, and extinction. While studying these phenomena, systematists also prepare guidebooks to biodiversity.

The ability to identify species is also crucial for controlling agricultural pests and agents of disease, such as malaria-carrying mosquitoes. Systematics also helps us to identify endangered species, manage wildlife effectively, and choose wild plants and animals for selective breeding and genetic engineering projects.

Data collected and organized by systematists also allows biologists to select appropriate organisms for their work. Most biological experiments are first conducted with individuals of a single species, because each species is a closed genetic system that may respond uniquely to experimental conditions. If a researcher inadvertently used two species, and these species responded differently, the mixed results probably wouldn't make much sense.

Finally, accurate phylogenetic trees are essential components of the comparative method, which biologists use to analyze evolutionary processes. Without a good phylogenetic hypothesis, we could not distinguish similarities inherited from a common ancestor from those that evolved independently in response to similar environments. For example, if biologists did not know the ancestry of sharks, penguins, and porpoises, they could not determine that their similarities were produced by convergent evolution (see Figure 22.9).

STUDY BREAK

1. What is the difference between a phylogenetic tree and a classification?
2. How does work in systematics allow biologists to select appropriate organisms for research?

23.2 The Linnaean System of Taxonomy

The practice of naming and classifying organisms originated with the Swedish naturalist Carl von Linné (1707–1778), better known by his Latinized name, Carolus Linnaeus. A professor at the University of Uppsala, Linnaeus sent ill-prepared students around the world to gather specimens, losing perhaps a third of his followers to the rigors of their expeditions. Although not a commendable student adviser, Linnaeus developed the basic system of naming and classifying organisms still in use today.

Linnaeus Developed the System of Binomial Nomenclature

Linnaeus invented the system of **binomial nomenclature**, in which species are assigned a Latinized two-part name, or **binomial.** The first part identifies a group of species with similar morphology, called a **genus** (plural, *genera*). The second part is the **specific epithet,** or species name.

A combination of the generic name and the specific epithet provides a unique name for every species. For example, *Ursus maritimus* is the polar bear and

Ursus arctos is the brown bear. By convention, the first letter of a generic name is always capitalized; the specific epithet is never capitalized; and the entire binomial is italicized. In addition, the specific epithet is never used without the full or abbreviated generic name preceding it because the same specific epithet is often given to species in different genera. For instance, *Ursus americanus* is the American black bear, *Homarus americanus* is the Atlantic lobster, and *Bufo americanus* is the American toad. If you were to order just "*americanus*" for dinner, you might be dismayed when your plate arrived—unless you have an adventurous palate!

Nonscientists often use different common names to identify a species. For example, *Bothrops asper,* a poisonous snake native to Central and South America, is called *barba amarilla* (meaning "yellow beard") in some places and *cola blanca* (meaning "white tail") in others; biologists have recorded about 50 local names for this species. Adding to the confusion, the same common name is sometimes used for several different species. Binomials, however, allow people everywhere to discuss organisms unambiguously.

Many binomials are descriptive of the organism or its habitat. *Asparagus horridus,* for example, is a spiny plant. Other species, such as the South American bird *Rhea darwinii,* are named for notable biologists. The naming of newly discovered species follows a formal process of publishing a description of the species in a scientific journal. International commissions meet periodically to settle disputes about scientific names.

The Taxonomic Hierarchy Organizes Huge Amounts of Systematic Data

Linnaeus described and named thousands of species on the basis of their similarities and differences. Keeping track of so many species was no easy task, so he devised a **taxonomic hierarchy** for arranging organisms into ever more inclusive categories **(Figure 23.2)**. A **family** is a group of genera that closely resemble one another. Similar families are grouped into **orders,** similar orders into **classes,** similar classes into **phyla** (singular, *phylum*), and similar phyla into **kingdoms.** Finally, all life on Earth is classified into three **domains,** described in Section 1.3. The organisms included within any category of the taxonomic hierarchy compose a **taxon** (plural, *taxa*). Woodpeckers, for example, are a taxon (Picidae) at the family level, and pine trees are a taxon *(Pinus)* at the genus level.

Linnaeus did not believe in evolution. His goals were to illuminate the details of God's creation and to devise a practical way for naturalists to keep track of their discoveries. Nevertheless, the taxonomic hierarchy he defined was easily applied to Darwin's concept of branching evolution, which is itself a hierarchical phenomenon. As we discussed in the preceding two chapters, ancestral species give rise to descendant species through repeated branching of a lineage. Organ-

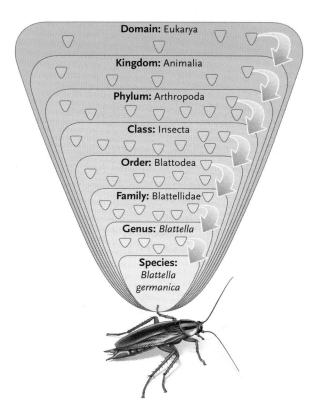

Figure 23.2

The Linnaean hierarchy of classification. The classification of a common household pest, the German cockroach (*Blattella germanica*), illustrates the nested hierarchy that Linnaeus developed. The German cockroach is one of many closely related species classified together in the genus *Blattella*, which is in turn one of nine genera in the family Blattellidae. Six distinctive cockroach families compose the order Blattodea, one of about 30 orders grouped into the class Insecta. The phylum Arthropoda contains about a dozen classes of animals, including insects, horseshoe crabs, spiders, crabs, and centipedes. Arthropoda is one of approximately 30 phyla, each representing a major lineage and body plan, within the kingdom Animalia. The classification of animal diversity is described in detail in Chapters 29 and 30.

isms in the same genus generally share a fairly recent common ancestor, whereas those assigned only to the same class or phylum share a common ancestor from the more distant past.

STUDY BREAK

1. How does the system of binomial nomenclature minimize ambiguity in the naming and identification of species?
2. Which taxonomic category is immediately above family? Which is immediately below it?

23.3 Organismal Traits as Systematic Characters

Systematists compare organisms and then group species that share certain characteristics. Linnaeus focused on external anatomy. For example, he defined

birds as a class of oviparous (egg-laying) animals with feathered bodies, two wings, two feet, and a bony beak. No other animals possess all these characteristics, which distinguish birds from "quadrupeds" (his term for mammals), "amphibians" (among which he included reptiles), fishes, insects, and "worms."

For roughly 200 years, systematists building on Linnaeus' work relied on a variety of organismal traits to analyze evolutionary relationships and classify organisms: chromosomal anatomy; details of physiological functioning; the morphology of subcellular structures, cells, organ systems, and whole organisms; and patterns of behavior. Today, systematists often focus on the molecular sequences of nucleic acids and proteins (see Section 23.6). Here we consider two commonly studied organismal characteristics: *morphological traits* and *behavioral traits*.

Morphological Characters Provide Abundant Clues to Evolutionary Relationships

Morphological differences often reflect genetic differences between organisms (see Section 20.1), and they are easy to measure in preserved or living specimens. Moreover, morphological characteristics are often clearly preserved in the fossil record, allowing the comparison of living species with their extinct relatives.

Useful morphological traits vary from group to group. In flowering plants, the details of flower anatomy often reveal common ancestry. Among vertebrates, the presence or absence of scales, feathers, and fur as well as the structure of the skull help scientists to reconstruct the evolutionary history of major groups. Sometimes systematists use obscure characters of unknown function. But differences in the number of scales on the back of a lizard or in the curvature of a vein in the wing of a bee may be good indicators of the genetic differentiation that accompanied or followed speciation—even if we do not know *why* these differences evolved.

Sometimes we rely on characteristics found only in the earliest stages of an organism's life cycle to provide evidence of evolutionary relationships. As described in Chapter 30, analyses of the embryos of vertebrates reveal that they are rather closely related to sea cucumbers, sea stars, and sea urchins and even more closely related to a group of nearly shapeless marine invertebrates called sea squirts or tunicates.

Behavioral Characters Offer Additional Data When Species Are Not Morphologically Distinct

Sometimes external morphology cannot be used to differentiate species. For example, two species of treefrog (*Hyla versicolor* and *Hyla chrysoscelis*) commonly occur together in forests of the central and eastern United

States. Both species have bumpy skin and adhesive pads on their toes that enable them to climb vegetation. They also have gray backs, white bellies, yellowish-orange coloration on their thighs, and large white spots below their eyes. The frogs are so similar that even experts cannot easily tell them apart.

How do we know that these frogs represent two species? During the breeding season, males of each species use a distinctive mating call to attract females **(Figure 23.3)**. The difference in calls is a prezygotic reproductive isolating mechanism that prevents females from mating with males of a different species (see Section 21.2). Prezygotic isolating mechanisms are excellent systematic characters because they are often the traits that animals themselves use to recognize members of their own species. The two frog species also differ in chromosome number—*Hyla chrysoscelis* is diploid and *Hyla versicolor* is tetraploid—which is a postzygotic isolating mechanism.

Hyla versicolor   *Hyla chrysoscelis*

Figure 23.3

Look-alike frog species. The frogs *Hyla versicolor* and *Hyla chrysoscelis* are so similar in appearance that one photo can depict both species. Male mating calls, visualized in sound spectrograms for the two species, are very different. The spectrograms, which depict call frequency on the vertical axis and time on the horizontal axis, show that *H. chrysoscelis* has a faster trill rate.

(Sound spectrograms from The Amphibians and Reptiles of Missouri, by T. R. Johnson © 1987 by the Conservation Commission of the State of Missouri. Reprinted by permission.)

STUDY BREAK

1. Why are morphological traits often helpful in tracing the evolutionary relationships within a group of organisms?
2. Why are prezygotic isolating mechanisms useful characters for systematic studies of animals?

23.4 Evaluating Systematic Characters

With a wealth of traits available for analysis, systematists use several guidelines to select characters for study. In this section we examine the most important of these principles.

Characters Must Be Independent Markers of Underlying Genetic Similarity and Differentiation

Ideally, systematists would create phylogenetic hypotheses and classifications by analyzing the genetic changes that caused speciation and differentiation. But in many cases they have had to rely on phenotypic traits as indicators of genetic similarity or divergence. Thus, systematists study traits in which phenotypic variation reflects genetic differences; they exclude differences caused by environmental variation (see Section 20.1).

Characters must also be genetically *independent,* reflecting different parts of the organisms' genomes. This precaution is necessary because different organismal characters can have the same genetic basis—and we want to use each genetic variation only once in an analysis. For example, tropical *Anolis* lizards climb

trees using small adhesive pads on the underside of their toes. The number of pads varies from species to species, and researchers have used the number of pads on the fourth toe of the left hind foot as a systematic character. They do not also use the number of pads on the fourth toe of the *right* hind foot as a separate character, because the same genes almost certainly control the number of pads on the toes of both feet.

Only Homologous Characters Provide Data about Evolutionary Relationships

A basic premise of systematic analyses is that phenotypic similarities between organisms reflect their underlying genetic similarities. As you may recall from Figure 19.3, species that are morphologically similar have often inherited the genetic basis of their resemblance from a common ancestor. Similarities that result from shared ancestry, such as the four limbs of all tetrapod vertebrates, are called **homologies** (or homologous characters). *Systematic analyses rely on the comparison of homologous characters as indicators of common ancestry and genetic relatedness.*

Even though homologous structures were inherited from a common ancestor, they may differ greatly among species, especially if their function has changed. For example, the stapes, a bone in the middle ear of tetrapod vertebrates, evolved from—and is therefore homologous to—the hyomandibula, a bone that supported the jaw joint of early fishes. The ancestral function of the bone is retained in some modern fishes, but its structure, position, and function are different in tetrapods **(Figure 23.4)**.

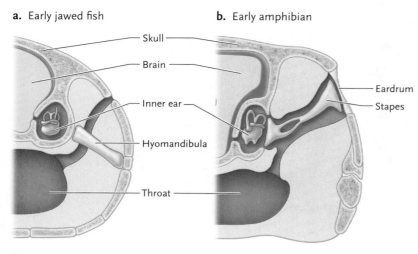

a. Early jawed fish **b.** Early amphibian

Skull

Brain

Inner ear

Hyomandibula

Throat

Eardrum

Stapes

Figure 23.4

Homologous bones, different structure and function. The hyomandibula, which braced the jaw joint against the skull in early jawed fishes **(a),** is homologous to the stapes, which transmits sound to the inner ear in the four-legged vertebrates, exemplified here by an early amphibian **(b).** Both diagrams show a cross section through the head just behind the jaw joint.

and in their relationship to surrounding structures. For example, the bones within the wings of birds and bats are homologous **(Figure 23.5).** Both wings include the same basic structural elements with similar spatial relationships to each other and to the bones that attach the wing to the rest of the skeleton. However, the large flat surfaces of their wings are homoplasious, the products of convergent evolution. The bird's wing is made of feathers, whereas the bat's wing is formed of skin.

Second, homologous characters emerge from comparable embryonic structures and grow in similar ways during development. Systematists have put great stock in embryological indications of homology on the assumption that evolution has conserved the pattern of embryonic development in related organisms. Indeed, recent discoveries in evolutionary development biology (described in Section 22.6 and explored further in Chapters 29 and 30) have revealed that the genetic controls of developmental pathways are very similar across a wide variety of organisms.

Systematists Focus Attention on Derived Versions of Characters

In all evolutionary lineages, some characteristics evolve slowly and others evolve rapidly, a phenomenon called **mosaic evolution.** Because mosaic evolution is pervasive, every species displays a mixture of **ancestral characters** (old forms of traits) and **derived characters** (new forms of traits). Derived characters provide the most useful information about evolutionary relationships because once a derived character becomes established, it is usually present in all of that species' descendants. Thus, unless they are lost or replaced by newer characters over evolutionary time, derived characters serve as markers for entire evolutionary lineages.

As you know from the discussion of convergent evolution in Section 22.2, organisms that are not closely related sometimes bear a striking resemblance to one another, especially when they live in similar environments. Phenotypic similarities that evolved independently in different lineages are called **homoplasies** (or homoplasious characters). Some biologists use the terms *analogies* or *analogous characters* for homoplasious characters that serve a similar function in different species. *Systematists exclude homoplasies from their analyses, because homoplasies provide no information about shared ancestry or genetic relatedness.*

If homoplasies are similar and homologies are sometimes different, how can we tell them apart? First, homologous structures are similar in anatomical detail

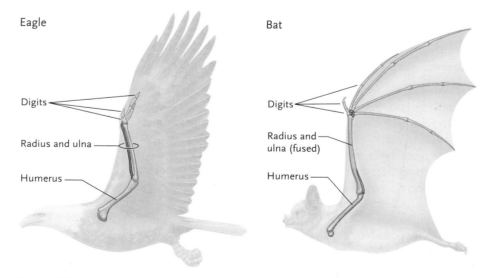

Eagle

Bat

Digits

Radius and ulna

Humerus

Digits

Radius and ulna (fused)

Humerus

Figure 23.5

Assessing homology. The wing skeletons of birds and bats are homologous structures with the same basic elements. However, the flat wing surfaces are homoplasious structures.

a. Caddis fly

b. Orange palm dart butterfly

c. Monarch butterfly

Figure 23.6

Outgroup comparison. Most adult insects, like the **(a)** caddis fly (family Limnephilidae) and the **(b)** orange palm dart butterfly *(Cephrenes auglades)*, have six walking legs. This comparison of butterflies with other insects suggests that the four walking legs of the **(c)** monarch butterfly *(Danaus plexippus)* represents the derived character state.

Systematists score characters as either ancestral or derived only when comparing them among organisms. Thus, any particular character is derived *only in relation to* what occurs in other organisms—either an older version of the same character or, in the case of an entirely new trait, the absence of it altogether. For example, most species of animals lack a vertebral column and the other components of an internal skeleton. However, one animal lineage—the vertebrates, including fishes, amphibians, reptiles, birds, and mammals—has those structures. Thus, when systematists compare vertebrates to all of the animals that lack a vertebral column, they score the absence of a vertebral column as the ancestral condition and the presence of a vertebral column as derived.

How can systematists distinguish between ancestral and derived characters? In other words, how can they determine the direction in which a character has evolved? The fossil record, if it is detailed enough, can provide unambiguous information. For example, biologists are confident that the presence of a vertebral column is a derived character because fossils of the earliest animals lack that structure.

For some traits, researchers use embryological evidence. Derived characters often appear later during embryonic development as modifications of an ancestral developmental plan. Recall, for example, that the early embryos of mammals first develop fishlike features in their circulatory systems (as shown in Figure 19.13) and only later develop the characteristic adult morphology. This developmental sequence suggests that the two-chambered linear hearts of fishes are ancestral, and that the four-chambered, double-loop hearts of mammals are derived.

Systematists frequently use a technique called **outgroup comparison** to identify ancestral and derived characters by comparing the group under study to more distantly related species that are not otherwise included in the analysis. Most modern butterflies, for example, have six walking legs, but species in two families have four walking legs and two small, nonwalking

legs. Which is the ancestral character state, and which is derived? Outgroup comparison with other insects, most of which have six walking legs as adults, suggests that six walking legs is ancestral and four is derived **(Figure 23.6).**

STUDY BREAK

1. Why do systematists use homologous characters in their phylogenetic analyses?
2. What is outgroup comparison?

23.5 Phylogenetic Inference and Classification

After exploring two guiding principles of research in systematics, we describe how systematists use their analyses of organismal characters to reconstruct phylogenetic histories and create classifications.

Many Systematic Studies Rely on the Principles of Monophyly and Parsimony

Phylogenetic trees portray the evolutionary diversification of lineages as a hierarchy that reflects the branching pattern of evolution. Each branch represents the descendants of a single ancestral species. When converting the phylogenetic tree into a classification, systematists use the **principle of monophyly**; that is, they try to define **monophyletic taxa**, each of which contains a single ancestral species and all of its descendants **(Figure 23.7).** By contrast, **polyphyletic taxa**—which systematists never intentionally define—would include species from separate evolutionary lineages. A taxon that included convergent species, such as sharks, penguins, and dolphins, would be polyphyletic. **Paraphyletic taxa** each contain an ancestor and some, but not all, of

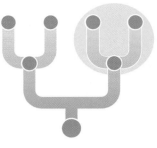

Monophyletic taxon

A monophyletic taxon includes
an ancestral species and all of
its descendants.

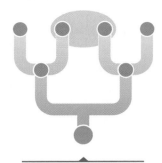

Polyphyletic taxon

A polyphyletic taxon includes
species from different
evolutionary lineages.

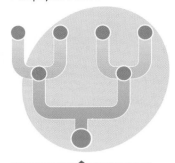

Paraphyletic taxon

A paraphyletic taxon includes an
ancestral species and only some
of its descendants.

Figure 23.7
Defining taxa in a classification. Systematists can create different classifications from the same phylogenetic tree by identifying different groups of species as a single taxon (shaded).

its descendants. For example, the traditional taxon Reptilia is paraphyletic, as described in the next section. These distinctions are crucial when making classifications.

Many systematists also strive to create *parsimonious* phylogenetic hypotheses, which means that they include the fewest possible evolutionary changes to account for the diversity within a lineage. According to the **principle of parsimony**, any particular evolutionary change is an unlikely event; therefore it is extremely unlikely that the same change evolved twice in one lineage. For example, phylogenetic trees place all birds on a single branch, implying that feathered wings evolved once in their common ancestor. This hypothesis is more parsimonious than one proposing that feathered wings evolved independently in two or more vertebrate lineages.

Traditional Evolutionary Systematics Was Based on Linnaeus' Methods

For a century after Darwin published *On the Origin of Species*, most systematists followed Linnaeus' practice of using phenotypic similarities and differences to infer evolutionary relationships. This approach, called **traditional evolutionary systematics**, groups together species that share both ancestral and derived characters. For example, mammals are defined by their internal skeleton, vertebral column, and four limbs—all ancestral characters among the tetrapod vertebrates—as well as hair, mammary glands, and a four-chambered heart—all of which are derived characters.

The classifications produced by traditional systematics reflect both evolutionary branching and morphological divergence **(Figure 23.8a).** For example, among the tetrapod vertebrates, the amphibian and mammalian lineages each diverged early, followed shortly thereafter by the turtle lineage. The remaining organisms then diverged into two groups: lepidosaurs gave rise to lizards and snakes, and archosaurs gave rise to crocodilians, dinosaurs, and birds. Thus, although

crocodilians outwardly resemble lizards, they share a more recent common ancestor with birds. Birds differ from crocodilians because birds experienced substantial morphological change when they emerged as a distinct group.

Even though the phylogenetic tree shows six living groups, the traditional classification recognizes only four classes of tetrapod vertebrates: Amphibia, Mammalia, Reptilia, and Aves (birds). These groups are given equal ranking because each represents a distinctive body plan and way of life. The class Reptilia, however, is clearly a paraphyletic taxon: it includes *some* of the descendants of the common ancestor labeled A in Figure 23.8a, namely turtles, lizards, snakes, and crocodilians; but it omits birds, and thus does not include *all* descendants.

Traditional evolutionary systematists justify this definition of the Reptilia because it includes morphologically similar animals with close evolutionary relationships. Crocodilians are classified with lizards, snakes, and turtles because they share a distant common ancestry and are covered with dry, scaly skin. Traditional systematists also argue that the key innovations initiating the adaptive radiation of birds—wings, feathers, high metabolic rates, and flight—represent such extreme divergence from the ancestral morphology that birds merit recognition as a separate class.

Cladistics Uses Shared Derived Characters to Trace Evolutionary History

In the 1950s and 1960s, some researchers criticized classifications that were based on two distinct phenomena, branching evolution and morphological divergence, as inherently unclear. After all, how can we tell *why* two groups are classified in the same higher taxon? They may have shared a recent common ancestor, as did lizards and snakes. Alternatively, they may have retained similar ancestral characteristics after being separated on different branches of a phylogenetic tree, as is the case for lizards and crocodilians.

a. Traditional phylogenetic tree with classification

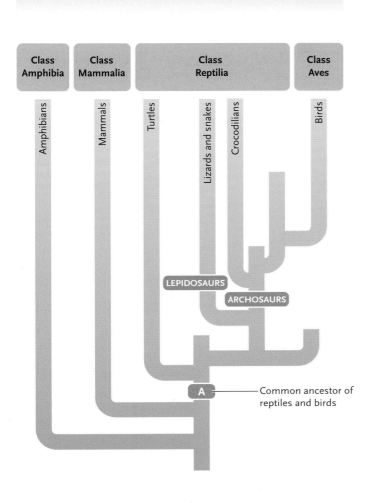

b. Cladogram with classification

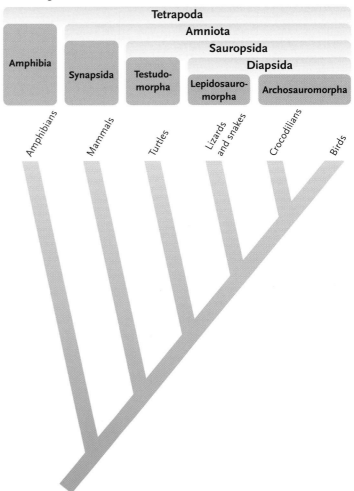

Figure 23.8
Phylogenetic trees and classifications for tetrapod verte- brates. **(a)** Tradi- tional and **(b)** cla- distic approaches produce different phylogenetic trees and classifications. Classifications are presented above the trees.

To avoid such confusion, many systematists quickly followed the philosophical and analytical lead of Willi Hennig, a German entomologist, who pub- lished an influential book, *Phylogenetic Systematics,* in 1966. Hennig and his followers argued that classifica- tions should be based solely on evolutionary relation- ships. **Cladistics,** as this approach is known, produces phylogenetic hypotheses and classifications that reflect only the branching pattern of evolution; it ignores mor- phological divergence altogether.

Cladists group together only species that *share de- rived characters.* For example, cladists argue that mam- mals form a monophyletic lineage—a **clade**—because they possess a unique set of derived characters: hair, mammary glands, reduction of bones in the lower jaw, and a four-chambered heart. The ancestral characters found in mammals—internal skeleton, vertebral col- umn, and four legs—do not distinguish them from other tetrapod vertebrates, so these traits are excluded from the analysis.

The phylogenetic trees produced by cladists, called **cladograms,** thus illustrate the hypothesized sequence of evolutionary branchings, with a hypothetical ancestor at each branching point **(Figure 23.8b).** They portray strictly monophyletic groups and are usually constructed using the principle of parsimony. Once a researcher identifies derived, homologous characters, constructing a cladogram is straightforward **(Figure 23.9).**

The classifications produced by cladistic analysis often differ radically from those of traditional evolu- tionary systematics (compare the two parts of Figure 23.8). Pairs of higher taxa are defined directly from the two-way branching pattern of the cladogram. Thus, the clade Tetrapoda (the traditional amphibians, reptiles, birds, and mammals) is divided into two taxa, the Am- phibia (tetrapods that do not have an amnion, as dis- cussed in Section 30.3) and the Amniota (tetrapods that have an amnion). The Amniota is subdivided into two taxa on the basis of skull morphology and other char- acteristics: Synapsida (mammals) and Sauropsida (turtles, lizards, snakes, crocodilians, and birds). The Sauropsida is further divided into the Testudomorpha (turtles) and the Diapsida (lizards and snakes, crocodil- ians, and birds). Finally, the Diapsida is subdivided

Figure 23.9 Research Method

Constructing a Cladogram

PURPOSE: Systematists construct cladograms to visualize hypothesized evolutionary relationships by grouping together organisms that share derived characters. The cladogram also illustrates where derived characters first evolved.

PROTOCOL:

1. *Select the organisms to study.* To demonstrate the method, we develop a cladogram for the nine groups of living vertebrates: lampreys, sharks (and their relatives), bony fishes (and their relatives), amphibians (frogs and salamanders), turtles, lizards (including snakes), crocodilians (including alligators), birds, and mammals. We also include marine animals called lancelets (phylum Chordata, subphylum Cephalochordata), which are closely related to vertebrates (see Chapter 30). Lancelets are the outgroup in our analysis.

2. *Choose the characters on which the cladogram will be based.* Our simplified example is based on the presence or absence of 10 characters: (1) vertebral column, (2) jaws, (3) swim bladder or lungs, (4) paired limbs (with one bone connecting each limb to the body), (5) extraembryonic membranes (such as the amnion), (6) mammary glands, (7) dry, scaly skin somewhere on the body, (8) two openings on each side near the back of the skull, (9) one opening on each side of the skull in front of the eye, and (10) feathers.

3. *Score the characters as either ancestral or derived in each group.* As the outgroup, lancelets possess the ancestral character; any deviation from the lancelet pattern is derived. Because lancelets lack all of the characters in our analysis, the presence of each character is the derived condition. We tabulate data on the distribution of ancestral (−) and derived (+) characters, listing lancelets first and the other organisms in alphabetical order.

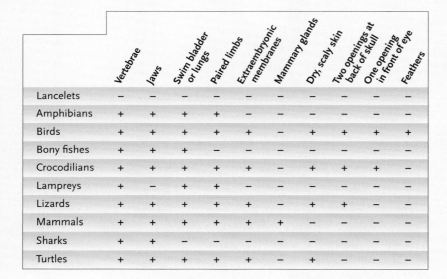

	Vertebrae	Jaws	Swim bladder or lungs	Paired limbs	Extraembryonic membranes	Mammary glands	Dry, scaly skin	Two openings at back of skull	One opening in front of eye	Feathers
Lancelets	−	−	−	−	−	−	−	−	−	−
Amphibians	+	+	+	+	−	−	−	−	−	−
Birds	+	+	+	+	+	−	+	+	+	+
Bony fishes	+	+	+	−	−	−	−	−	−	−
Crocodilians	+	+	+	+	+	−	+	+	+	−
Lampreys	+	−	+	+	−	−	−	−	−	−
Lizards	+	+	+	+	+	−	+	+	−	−
Mammals	+	+	+	+	+	+	−	−	−	−
Sharks	+	+	−	−	−	−	−	−	−	−
Turtles	+	+	+	+	+	−	+	−	−	−

4. *Construct the cladogram from information in the table, grouping organisms that share derived characters.* All groups except lancelets have vertebrae. Thus, we group organisms that share this derived character on the right-hand branch, identifying them as a monophyletic lineage. Lancelets are on their own branch to the left, indicating that they lack vertebrae.

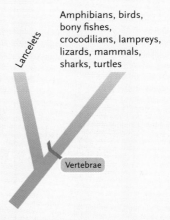

All of the remaining organisms except lampreys have jaws. (Lancelets also lack jaws, but we have already separated them out, and do not consider them further.) Place all groups with jaws, a derived character, on the right-hand branch. Lampreys are separated out to the left, because they lack jaws. Again, the branch on the right represents a monophyletic lineage.

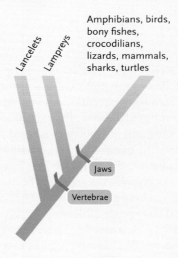

into two more recently evolved taxa—the Lepidosauromorpha (lizards and snakes) and the Archosauromorpha (crocodilians and birds). Thus, a strictly cladistic classification exactly parallels the pattern of branching evolution that produced the organisms included in the classification. These parallels are the essence and strength of the cladistic method.

Most biologists now use the cladistic approach because of its evolutionary focus, clear goals, and precise methods. In fact, some systematists advocate abandoning the Linnaean hierarchy for classifying and naming organisms. They propose using a strictly cladistic system, called **PhyloCode**, that identifies and names clades instead of pigeonholing or-

5. *Construct the rest of the cladogram using the same step-by-step procedure to separate the remaining groups.* In our completed cladogram, seven groups share a swim bladder or lungs; six share paired limbs; and five have extraembryonic membranes during development. Some groups are distinguished by the unique presence of a derived character, such as feathers in birds.

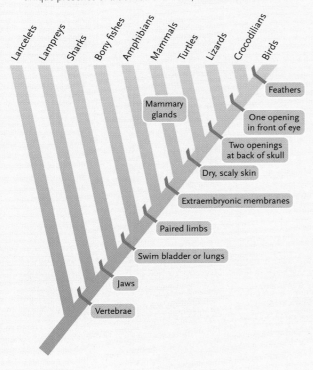

INTERPRETING THE RESULTS: Although cladograms provide information about evolutionary relationships, the common ancestors represented by the branch points are often hypothetical. You can tell from the cladogram, however, that birds are more closely related to lizards than they are to turtles. Follow the branches of the cladogram from birds and lizards back to their intersection, or node. Next, trace the branches of birds and turtles to their node. You can see that the bird–turtle node is closer to the bottom of the cladogram than the bird–lizard node. Nodes that are closer to the bottom of the cladogram indicate a more distant common ancestry than those closer to the top.

ganisms into the familiar taxonomic levels embraced by more traditional systematists. Although traditional evolutionary systematics has guided many people's understanding of biological diversity, we use a cladistic approach to describe evolutionary lineages and taxa in the Biodiversity unit that follows this chapter.

STUDY BREAK

1. How does a monophyletic taxon differ from a polyphyletic taxon?
2. Why is the traditionally defined group Reptilia a paraphyletic taxon?
3. What characteristics are used to group organisms in a cladistic analysis?

23.6 Molecular Phylogenetics

Most systematists now conduct phylogenetic analyses using molecular characters, such as the nucleotide base sequences of DNA and RNA or the amino acid sequences of the proteins for which they code. Because DNA is inherited, shared changes in molecular sequences—insertions, deletions, or substitutions—provide clues to the evolutionary relationships of organisms. Technological advances have automated many of the necessary laboratory techniques, and analytical software makes it easy to compare new data to information filed in data banks accessible over the Internet.

Molecular Characters Have Both Advantages and Disadvantages over Organismal Characters

Molecular sequences have certain practical advantages over organismal characters. First, they provide abundant data: every amino acid in a protein and every base in a nucleic acid can serve as a separate, independent character for analysis. Moreover, because many genes have been conserved by evolution, molecular sequences can be compared between distantly related organisms that share no organismal characteristics. Molecular characters can also be used to study closely related species with only minor morphological differences. Finally, many proteins and nucleic acids are not directly affected by the developmental or environmental factors that cause nongenetic morphological variations such as those described in Section 20.1.

Molecular characters have certain drawbacks, however. For example, only four alternative character states (the four nucleotide bases) exist at each position in a DNA or RNA sequence and only 20 alternative character states (the 20 amino acids) at each position in a protein. (You may want to review Sections 14.2 and 15.1 on the structure of these molecules.) And if two species have the same nucleotide base substitution at a given position in a DNA segment, their similarity may well have evolved independently. As a result, systematists often find it difficult to verify that molecular similarities were inherited from a common ancestor.

For organismal characters, biologists can establish that similarities are homologous by analyzing the characters' embryonic development or details of their func-

Whales with Cow Cousins?

More than 50 million years ago, whales evolved from terrestrial mammals into streamlined creatures, spectacularly adapted to life in the sea. But which mammals were their ancestors? Using morphological comparisons of living and fossil species, evolutionists had hypothesized that modern cetaceans—whales, dolphins, and porpoises—evolved from wolflike mammals called mesonychians. However, recent work by molecular biologists suggests that cetaceans are part of a lineage that includes an ungulate ancestor of cows and hippopotamuses.

Several molecular studies support this surprising conclusion. Mitsuru Shimamura and his colleagues at the Tokyo Institute of Technology and other Japanese institutions examined the distribution of transposable elements (TEs) in whales and ungulates. TEs are sequences that move to new locations in DNA (see Section 17.3). The TEs that the researchers studied in whales move by making RNA copies of themselves; the RNA copies then act as templates for making DNA copies, which are inserted in new locations. The mechanism leaves the original copy still in place in the DNA.

The TEs studied by the Shimamura team are called SINEs (for *Short INter-*spersed *Elements*). These elements, which occur only in mammals, are particularly useful for evolutionary studies because the pattern by which they duplicate and move to new locations is unique in each evolutionary lineage. If SINEs occur at the same sites in the nuclear DNA of several species, those species are likely to be members of the same lineage.

To begin their work, Shimamura and his coworkers isolated two types of SINEs from whales, which they designated CHR-1 and CHR-2. They found that the DNA of these SINEs could pair with sequences in the nuclear DNA of hippos, cows, and other ruminants, but not with sequences of pigs and camels. This result showed that the CHR-1 and CHR-2 SINEs are present in whales, cows, and hippos but not in pigs and camels.

The researchers then used similar techniques to work out the locations of the SINEs in the DNA, with particular focus on SINEs that may have inserted into known protein-encoding genes. SINEs can insert into genes without serious damage if they do so in introns, the surplus segments that are transcribed but spliced out of the messenger RNA copy of the gene (see Section 15.3). To find genes containing the SINEs, the researchers added probes—labeled DNA sequences that could pair with CHR-1 and CHR-2—to DNA preparations containing all the genes of the species under study. They also searched through electronic databanks of known gene sequences of the species, looking for genes with introns containing either of the two SINEs.

The probes and computer searches produced seven "hits" among protein-encoding genes. Three CHR-1 insertions were found at the same locations in genes of cetaceans, ruminants, and hippos, but were absent from these locations in camels and pigs. The results indicate that the SINEs inserted at these locations in a common ancestor of cetaceans, ruminants, and hippos after camels and pigs had split off as a separate group (see **figure**). Additionally, some other SINEs evidently inserted later, after an evolutionary split had separated the ruminants and cetaceans. Two CHR-1 insertions were found in ruminants but not in cetaceans, hippos, camels, or pigs; two CHR-2 insertions were found only in cetaceans. These data enabled the investigators to construct the phylogenetic tree shown in the figure; the gene loci within which they found CHR-1 and CHR-2 insertions are labeled on the branches of the tree. Molecular studies testing the distribution of other DNA sequences, including mitochondrial DNA, support the close relationships between whales and cows suggested by the Shimamura experiments.

Some evolutionists contested the conclusions from molecular studies because they considered the database too limited and because morphological studies supported other hypotheses. Pigs, ruminants, camels, and hippos share a mobile heel joint that is different from the nonmobile joint in all other mammals. With their greatly reduced hind limbs, modern whales have no heel joint; but a land-living fossil believed to be an ancestor of whales has a nonmobile heel joint. Further, the teeth of pigs, ruminants, camels, and hippos are different from those of cetaceans. These morphological characters support a traditional classification in which ruminants, pigs, camels, and hippos form one lineage, and cetaceans a separate one. However, in 2001, Philip D. Gingerich of the University of Michigan and his colleagues in Pakistan reported the discovery of two ancient whale fossils, both of which had mobile heel joints. These new findings provide strong evidence in support of the conclusion that whales are closely related to ruminants and hippos.

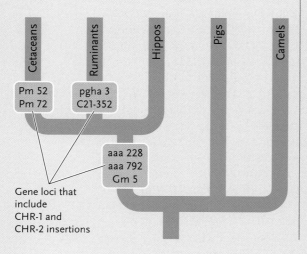

Gene loci that include CHR-1 and CHR-2 insertions

tion. But molecular characters have no embryonic development, and biologists still do not understand the functional significance of most molecular differences. Despite these disadvantages, molecular characters represent the genome directly, and researchers use them with great success in phylogenetic analyses. *Insights from the Molecular Revolution* describes an example using sequences called transposable elements.

Variations in the Rates at Which Molecules Evolve Govern the Molecules Chosen for Phylogenetic Analyses

Although molecular phylogenetics is based on the observation that many molecules have been conserved by evolution, different adaptive changes and neutral mutations accumulate in separate lineages from the moment they first diverge. Mutations in some types of DNA appear to arise at a relatively constant rate. Thus, differences in the DNA sequences of two species can serve as a **molecular clock**, indexing their time of divergence. Large differences imply divergence in the distant past, whereas small differences suggest a more recent common ancestor.

Because mosaic evolution exists at the molecular level, different molecules exhibit individual rates of change, and every molecule is an independent clock, ticking at its own rate. Researchers study different molecules to track evolutionary divergences that occurred over different time scales. For example, mitochondrial DNA (mtDNA) evolves relatively quickly; it is useful for dating evolutionary divergences that occurred within the last few million years. Studies of mtDNA have illuminated aspects of the evolutionary history of humans, as described in Section 30.13. By contrast, chloroplast DNA (cpDNA) and genes that encode ribosomal RNA evolve much more slowly, providing information about divergences that date back hundreds of millions of years.

To synchronize molecular clocks, some researchers study DNA sequences that are not parts of protein-encoding genes. Because they don't affect protein structure, mutations in these sequences are probably not often eliminated by natural selection. Thus, the sequence differences between species in noncoding regions probably result from mutation alone and therefore reflect the ticking of the molecular clock more directly. Some researchers also calibrate molecular clocks to the fossil record, so that actual times of divergence can be predicted from molecular data with a fair degree of certainty.

The Analysis of Molecular Characters Requires Specialized Approaches

Molecular phylogenetics relies on the same basic logic that underlies analyses based on organismal characters: species that diverged recently from a common ancestor should share many similarities in their molecular sequences, whereas more distantly related species should exhibit fewer similarities. Nevertheless, the practice of molecular phylogenetics is based on a set of distinctive methods.

Determining the Molecular Sequence. After selecting a protein molecule or appropriate segment of a nucleic acid for analysis, systematists determine the exact sequence of amino acids (in the case of proteins) or nucleotide bases (in the case of DNA or RNA) that compose the molecule.

Amino acid sequencing allows systematists to compare the primary structure of protein molecules directly. As you may recall from Chapter 15, the amino acid sequence of a protein is determined by the sequence of nucleotide bases in the gene encoding that protein. When two species exhibit similar amino acid sequences for the same protein, systematists infer their genetic similarity and evolutionary relationship. For example, researchers have used sequence data from the protein cytochrome *c* to construct a phylogenetic tree for organisms as different as slime molds, vascular plants, and humans **(Figure 23.10)**.

Most systematic studies are now based, at least in part, on DNA sequencing data, which provide a detailed view of the genetic material that evolutionary processes change. The polymerase chain reaction (PCR) makes it easy for researchers to produce numerous copies of specific segments of DNA for comparison (see Section 18.1). This technique allows scientists to sequence minute quantities of DNA taken from dried or preserved specimens in museums and even from some fossils.

Aligning Molecular Sequences. Before comparing molecular sequences from different organisms, systematists must ensure that the homologous sequences being compared are properly "aligned." In other words, they must be certain that they are comparing nucleotide bases or amino acids at exactly the same positions in the nucleic acid or protein molecule. This crucial step is necessary because mutations often change the length of a DNA sequence and the relative locations of specific positions through the insertion or deletion of base pairs (see Section 15.4). Such mutations make sequence comparisons more difficult; but, by determining where such insertions or deletions have occurred, systematists can match up the positions of—in other words, *align*—the nucleotides for comparison. Although alignments can be done "by eye" in many cases, most systematists use computer programs to accomplish this task. **Figure 23.11** provides a simplified example of this step in the process.

Constructing Phylogenetic Trees. Once the molecules are aligned, a systematist can compare the nucleotide base or amino acid sequences to determine whether

Figure 23.10 Observational Research

Using Amino Acid Sequences to Construct a Phylogenetic Tree

HYPOTHESIS: Because the amino acid sequences of proteins change over evolutionary time, sequence differences between organisms should reflect their evolutionary relationships.

PREDICTION: Closely related species will exhibit similar amino acid sequences, whereas more distantly related species will exhibit greater differences in their amino acid sequences.

METHOD: Researchers determined the amino acid sequence of cytochrome *c*, a protein in the electron transport system that has been conserved by evolution, using samples from a wide variety of eukaryotic species classified in four kingdoms. They compared the data derived from the different species and used the sequences to construct a phylogenetic tree.

RESULTS: The amino acid sequence of cytochrome *c* is surprisingly similar in distantly related organisms that diverged from a common ancestor hundreds of millions of years ago. Gold shading marks the amino acids that are identical in the sequences for yeast (top row), wheat (middle row), and human (bottom row). Abbreviations for the amino acids listed below are derived from those in Figure 3.15.

Yeast

$^+$NH$_3$- GDVEKGKKIFIMKCSQCHTVEKGGKHKTGPNLHGLFGRKTGQAPGYSYTAANKNKGIIWGEDTLMEYLENPKKYIPGTKMIFVGIKKKEERADLIAYLKKATNE -COO$^-$

Wheat

$^+$NH$_3$- ASFSEAPPGNPDAGAKIFKTKCAQCHTVDAGAGHKQGPNLHGLFGRQSGTTAGYSYSAANKNKAVEWEENTLYDYLLNPKKYIPGTKMVFPGLKKPQDRADLIAYLKKATSS -COO$^-$

Human

$^+$NH$_3$- TEFKAGSAKKGATLFKTRCLQCHTVEKGGPHKVGPNLHGIFGRHSGQAEGYSYTDANIKKNVLWDENNMSEYLTNPKKYIPGTKMAFGGLKKKEKDRNDLITYLKKACE -COO$^-$

The phylogenetic tree based on similarities and differences in cytochrome *c* sequences is remarkably consistent with trees based on organismal characters. The vertical axis gives the approximate time of each evolutionary branching, estimated from the amino acid sequence data.

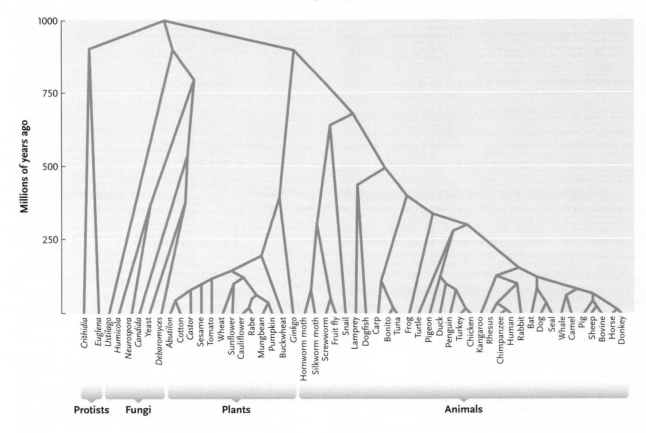

CONCLUSION: Amino acid sequence data can be used to construct phylogenetic trees for species that share essentially no organismal characteristics.

mutations or other processes have produced evolutionary changes in the sequences. The similarities and differences can then be used to reconstruct the phylogenetic tree. Every phylogenetic tree is a hypothesis about evolutionary relationships, and different assumptions can yield multiple alternative trees for any data set. Indeed, systematists have developed several approaches for comparing molecular sequences and constructing trees.

For DNA sequences, the simplest approach is to count the number of similarities and differences between every pair of organisms being compared. Systematists use such data to estimate the *genetic distances* between species and to construct a phylogenetic tree by grouping together those organisms that exhibit the smallest genetic distances. However, this approach reconstructs phylogenies with both ancestral and derived characters, the same criticism that was leveled against traditional evolutionary systematics.

An alternative approach for converting molecular sequence data into a phylogenetic tree follows a cladistic method, using the principle of parsimony, which requires the identification of ancestral and derived character states. In other words, systematists must determine, for each position in the sequence, which nucleotide base is ancestral and which is derived. As is the case for organismal characters, the analysis of homologous sequences in a designated outgroup can provide that information. Under the parsimony approach, a computer program then tests all possible phylogenetic trees and identifies the one that accounts for the diversity of organisms in the group with the fewest evolutionary changes in molecular sequences.

In recent years, researchers have faulted the parsimony approach because identical changes in nucleotides often arise independently. To avoid this problem, systematists have begun using a series of sophisticated statistical techniques collectively called *maximum likelihood methods*. This approach reconstructs phylogenetic history from molecular sequence data by making assumptions about variations in the rate at which different segments of DNA evolve. These statistical models can take into account variations in the rates of evolution between genes or between species as well as changes in evolutionary rates over time. Maximum likelihood programs construct numerous alternative phylogenetic trees and estimate how likely it is that each tree represents the true evolutionary history. Systematists then accept the phylogenetic tree that is most likely to be true—until more data are available.

Molecular Phylogenetics Has Clarified Many Evolutionary Relationships

As you will see in the next unit, molecular phylogenetics has enabled systematists to resolve some longstanding disputes about evolutionary relationships.

Figure 23.11 Research Method

Aligning DNA Sequences

PURPOSE: The insertion or deletion of base pairs often changes the length of a DNA sequence and the relative locations of specific positions along its length. Systematists must therefore "align" the sequences that they are comparing. This procedure ensures that the nucleotide bases being compared are at exactly the same positions in the nucleic acid molecules. By determining where insertions or deletions have occurred, systematists can match up the positions of—in other words, *align*—the nucleotides for comparison. In this hypothetical example, imagine that the DNA segments were obtained from three different species. A comparable procedure is used to align the amino acid sequences of proteins.

PROTOCOL:

1. Before alignment, three DNA segments differ in length and exhibit nucleotide differences in many positions.

Segment A `AATTGACCTTCTAAGTGTAAT`
Segment B `AATTGAGCCTTCTAAGTCTAAT`
Segment C `AATTGATTCTAAGTGTAAT`

2. The computer program detects similar sequences in parts of the three segments.

Segment A `AATTGACCTTCTAAGTGTAAT`
Segment B `AATTGAGCCTTCTAAGTCTAAT`
Segment C `AATTGATTCTAAGTGTAAT`

3. The three segments are aligned under the hypotheses that segment B included a one-nucleotide insertion and segment C had experienced a two-nucleotide deletion.

One-nucleotide insertion

Segment A `AATTGA` `CCTTCTAAGTGTAAT`
Segment B `AATTGA` `G` `CCTTCTAAGTCTAAT`
Segment C `AATTGA` `TTCTAAGTGTAAT`

Two-nucleotide deletion

INTERPRETING THE RESULTS: After alignment, the sequences can be compared at every position. In addition to the one-nucleotide insertion in segment B and the two-nucleotide deletion in segment C, the comparison reveals one nucleotide substitution in segment B.

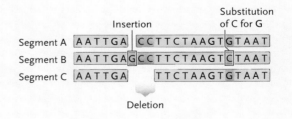

As one example, analyses of morphological data had produced conflicting hypotheses about the origin and relationships of flowering plants. In 1999, four teams of researchers, analyzing different parts of flowering plant genomes, independently identified

Figure 23.12

The ancestral flowering plant. DNA sequencing studies identified *Amborella trichopoda* as a living representative of the earliest group of flowering plants.

Amborella branch

Thomas J. Lemieux, University of Colorado

Amborella flower

Sandra Floyd, University of Colorado

Amborella trichopoda, a bush native to the South Pacific island of New Caledonia, as a living representative of the most ancient group of flowering plants yet discovered **(Figure 23.12)**. The first team to publish their results, Sarah Mathews and Michael Donoghue of Harvard University, studied phytochrome genes (*PHYA* and *PHYC*) that had duplicated early in the evolutionary history of this group. Other researchers, who studied chloroplast, mitochondrial, and ribosomal sequences, obtained similar results, providing strong support for this phylogenetic hypothesis.

On a very grand scale, molecular phylogenetics has revolutionized our view of the entire tree of life. The first efforts to create a phylogenetic tree for all forms of life were based on morphological analyses. However, these analyses did not resolve branches of the tree containing prokaryotes, which lack significant structural variability, or the relationships of those branches to eukaryotes.

In the 1960s and early 1970s, biologists organized living systems into five kingdoms. All prokaryotes were grouped into the kingdom Monera. The eukaryotic organisms were grouped into four kingdoms: Fungi, Plantae, Animalia, and Protista. The Protista was always recognized as a polyphyletic "grab bag" of unicellular or acellular eukaryotic organisms. Unfortunately, phylogenetic analyses based on morphology were unable to sort these organisms into distinct evolutionary lineages.

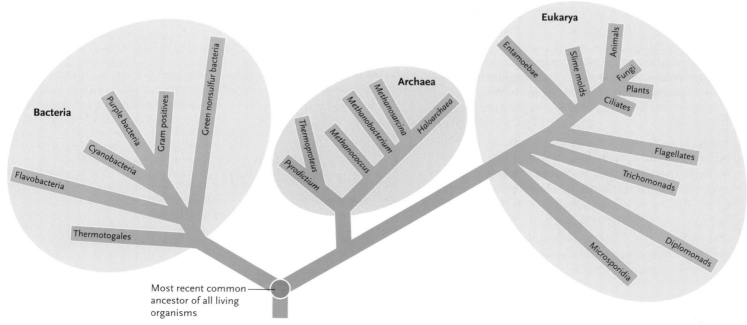

Figure 23.13

Three domains: the tree of life. Carl R. Woese's analysis of rRNA sequences suggests that all living organisms can be classified into one of three domains: Bacteria, Archaea, and Eukarya.

In the 1970s, biologists realized that molecular phylogenetics provides an alternative approach. They simply needed to identify and analyze molecules that have been conserved by evolution over billions of years. Carl R. Woese, a microbiologist at the University of Illinois at Urbana-Champaign, identified the small subunit of ribosomal RNA as a suitable molecule for analysis. Ribosomes, the structures that translate messenger RNA molecules into proteins (see Section 15.1), are remarkably similar in all forms of life. They are apparently so essential to cellular

processes that the genes specifying ribosomal structure exhibit similarities in their nucleotide sequences in organisms from bacteria to humans. Thus, it is possible to sequence these genes and align them for analysis.

The phylogenetic tree based on rRNA sequences divides living organisms into three primary lineages called domains: Bacteria, Archaea, and Eukarya **(Figure 23.13).** According to this hypothesis, two domains, Bacteria and Archaea, consist of prokaryotic organisms, and one, Eukarya, consists of eukaryotes. Bacteria in-

UNANSWERED QUESTIONS

Should we abandon the traditional Linnaean hierarchy in favor of a more evolutionary classification?

Diligent Kindly Professors Cannot Often Fail Good Students—or some equally silly mnemonic device for remembering the Linnaean taxonomic hierarchy—is all that many students recall about systematics. Even if they remember the underlying rank names—is G for "group" or "genus"?—they often forget that Linnaeus conceived his system of classification more than a century before Darwin articulated his theory of evolution, which revolutionized our understanding of biological diversity. In the roughly 150 years since Darwin published *On the Origin of Species*, systematists have sought not only to categorize life's diversity but, more importantly, to understand its origins. The broad relevance of studies in systematics has become increasingly clear as biologists have discovered that systematic principles are as important to tracing the emergence and spread of avian flu as they are to distinguishing a duck from a dove.

As we approach the sesquicentennial of Darwin's theory, its impact becomes increasingly revolutionary. Perhaps the most striking recent example is a call for the complete abandonment of the Linnaean taxonomic hierarchy. Although biologists thought they had reconciled the perspectives of Darwin and Linnaeus, a growing minority of systematists now believe that any effort to catalog and categorize life's diversity must be explicitly phylogenetic and free of the arbitrary ranks that Linnaeus invented. This movement, which has been codified in the PhyloCode initiative, is fueled largely by newly available molecular data, vastly improved phylogenetic methodologies, and increasingly fast computers. These advances offer the potential to reconstruct accurate and fully resolved phylogenetic trees at a scale never before possible. For the first time, biologists see real progress in accurately reconstructing the entire tree of life. Although we are still far from achieving this goal, every day millions of new, phylogenetically informative DNA fragments are being sequenced and analyzed by thousands of computers running around the clock.

Although PhyloCode's synthesis of taxonomy and evolutionary systematics may be long overdue, this attempted coup is not without controversy. For example, some systematists contend that such a radical revision of our taxonomic system will introduce confusion and in-

stability in the naming of species. Even the revolution's adherents recognize that we still face many challenging limitations to the synthesis between taxonomic practice and Darwinian principles. Nowhere is this more evident than in the definition of species.

During Linnaeus' time, species were viewed as immutable natural types created by God. Darwin, however, formulated his theory on the principle that species change over time. Although the truth of this basic hypothesis is no longer a subject of debate, its practical implications for delimiting species boundaries and understanding how new species form are among the most exciting areas of study in modern systematics. Most practicing systematists view species as real (that is, biologically meaningful) categories, but the criteria for recognizing species vary dramatically among systematists working on different types of organisms (plants versus animals, or organisms that reproduce asexually versus those that reproduce sexually). Biologists are now beginning to use new molecular tools to address the challenge of understanding the origin of new species. Using these tools and sophisticated genetic experiments, evolutionary biologists are beginning to probe the precise genetic basis of species. Over the past decade a small number of "speciation genes" have been identified; more such discoveries are sure to follow in the coming years. Although many of these studies have been restricted to model research organisms, such as fruit flies, the new tools offered by the fields of genomics and bioinformatics offer the potential to address similar questions in an increasingly broad array of organisms.

Simply put, the systematics of today is not that of your grandparents. Given the enormous challenge involved in categorizing and understanding the origin and evolutionary relationships of millions of species, many additional changes are on the horizon. For the next generation of systematists, however, a better mnemonic to remember may be "Keep Probing Charles' *Origin* For Good Systematics."

Rich Glor conducts research on the evolution of *Anolis* lizards at the University of Rochester. To learn more about his research, go to http://www.lacertilia.com.

cludes well-known microorganisms, and Archaea includes microorganisms that live in physiologically harsh environments, such as hot springs or very salty habitats. Eukarya includes the familiar animals, plants, and fungi, as well as the many lineages formerly included among the Protista. The next unit of this book is devoted to detailed analyses of the biology and evolutionary relationships between and within these three domains.

STUDY BREAK

1. What are three advantages of using molecular characters in phylogenetic analyses?
2. How can molecular sequence data be used as a molecular clock?
3. Why was a phylogenetic analysis of prokaryotes based on molecular sequence data more successful than the analysis based on morphological data?

Review

Go to **Thomson**NOW™ at www.thomsonedu.com/login to access quizzing, animations, exercises, articles, and personalized homework help.

23.1 Systematic Biology: An Overview

- Systematic biology has two goals: the reconstruction of evolutionary history and the naming and classification of organisms. Phylogenetic trees and classifications are hypotheses about the relationships of organisms.

- By providing a guide to biological diversity, systematics allows biologists to identify species for research, for the control of harmful organisms, and for conservation (Figure 23.1).

23.2 The Linnaean System of Taxonomy

- Linnaeus invented a system of binomial nomenclature in which each species receives a unique two-part name.

- Species are organized into a taxonomic hierarchy (Figure 23.2), which reflects the pattern of branching evolution. Species classified in the same genus or family have a more recent common ancestor than species classified only in the same class or phylum.

Animation: Classification systems

23.3 Organismal Traits as Systematic Characters

- Systematists have always studied organismal characters, such as morphology, chromosome structure and number, physiology, and behavior.

- Morphological traits often allow the reconstruction of a group's phylogeny, that is, its evolutionary history.

- Behavioral characters are useful for understanding the relationships of animals that are not morphologically different (Figure 23.3).

Animation: Evolutionary tree for plants

23.4 Evaluating Systematic Characters

- Systematists study characters that are genetically independent, reflecting different parts of the organisms' genomes.

- Most systematists use homologous characters that reflect genetic similarities and differences among species (Figures 23.4 and 23.5).

- Because characters evolve at different rates, systematists select traits that evolved at a rate consistent with the timing of branching evolution.

- Systematists base their analyses on derived versions of homologous traits (Figure 23.6).

23.5 Phylogenetic Inference and Classification

- Phylogenetic trees and classifications include only monophyletic taxa, each of which contains a single ancestral species and all of its descendants (Figure 23.7). Many systematists create parsimonious phylogenies, which include the fewest possible evolutionary changes to account for the diversity within a lineage.

- Traditional evolutionary systematics emphasizes branching evolution and morphological divergence. Using both ancestral and derived characters, this approach sometimes creates classifications with paraphyletic taxa, which include an ancestor and some, but not all, of its descendants (Figure 23.8a).

- Cladistics emphasizes only evolutionary branching to define monophyletic taxa (Figure 23.8b). Cladists create phylogenetic hypotheses and classifications by grouping organisms that share derived characters (Figure 23.9).

Animation: Constructing a cladogram

Animation: Interpreting a cladogram

Animation: Current evolutionary tree

23.6 Molecular Phylogenetics

- Contemporary systematists use the structure of proteins and nucleic acids in their analyses. Molecular characters provide abundant data and can be compared among many morphologically distinct forms of life, but because molecular similarities in different species may have evolved independently, systematists cannot always verify that they were inherited from a common ancestor.

- Molecular characters may act as molecular clocks, providing data that allows researchers to determine the times when lineages first diverged (Figure 23.10).

- The use of molecular characters in phylogenetic studies requires the sequencing and alignment of molecules (Figure 23.11). Several methods, including genetic distances, parsimony, and maximum likelihood, have been proposed for the construction of phylogenetic trees.

- Molecular phylogenetics has clarified relationships among the flowering plants (Figure 23.12) and provided insights into the evolutionary relationships of all organisms (Figure 23.13).

Animation: Cytochrome c comparison

Questions

Self-Test Questions

1. The evolutionary history of a group of organisms is called its:
   a. classification.
   b. taxonomy.
   c. phylogeny.
   d. domain.
   e. outgroup.

2. In the Linnaean hierarchy, the organisms classified within the same taxonomic category are called:
   a. a phylum.
   b. a taxon.
   c. a genus.
   d. a binomial.
   e. an epithet.

3. When systematists study morphological or behavioral traits to reconstruct the evolutionary history of a group of animals, they assume that:
   a. similarities and differences in phenotypic characters reflect underlying genetic similarities and differences.
   b. the animals use exactly the same traits to identify appropriate mates.
   c. differences in these traits caused speciation in the past.
   d. the adaptive value of these traits can be explained.
   e. variations in these traits are produced by environmental effects during development.

4. Which statement best describes the concept of mosaic evolution?
   a. Some phenotypic variation is caused by environmental factors.
   b. Homologous characters are those that different organisms inherit from a common ancestor.
   c. Different organismal traits may reflect the same part of an organism's genome.
   d. Some characters evolve more quickly than others.
   e. The fossil record provides clues about the ancestral versions of characters.

5. Which of the following pairs of structures are homoplasious?
   a. the wing skeleton of a bird and the wing skeleton of a bat
   b. the wing of a bird and the wing of a fly
   c. the eye of a fish and the eye of a human
   d. the bones in the foot of a duck and the bones in the foot of a chicken
   e. the adhesive toe pads on the right hind foot of an *Anolis* lizard and those on the left hind foot

6. Which of the following does *not* help systematists determine which version of a morphological character is ancestral and which is derived?
   a. outgroup comparison
   b. patterns of embryonic development
   c. studies of the fossil record
   d. studies of the character in more related species
   e. dating of the character by molecular clocks

7. In a cladistic analysis, a systematist groups together organisms that share:
   a. derived homologous traits.
   b. derived homoplasious traits.
   c. ancestral homologous traits.
   d. ancestral homoplasious traits.
   e. all of the above.

8. A monophyletic taxon is one that contains:
   a. an ancestor and all of its descendants.
   b. an ancestor and some of its descendants.
   c. organisms from different evolutionary lineages.
   d. an ancestor and those descendants that still resemble it.
   e. organisms that resemble each other because they live in similar environments.

9. Which of the following is *not* an advantage of using molecular characters in a systematic analysis?
   a. Molecular characters provide abundant data.
   b. Systematists can compare molecules among species that are morphologically very similar.
   c. Systematists can compare molecules among species that share few morphological characters.
   d. Amino acid sequences in proteins are generally not influenced by environmental factors.
   e. Systematists can easily determine whether base substitutions in the DNA of two species are homologous.

10. To construct a cladogram by applying the principles of parsimony to molecular sequence data, one would:
    a. start by making assumptions about variations in the rates at which different DNA segments evolve.
    b. group together organisms that share the largest number of ancestral sequences.
    c. group together organisms that share derived sequences, matching the groups to those defined by morphological characters.
    d. group together organisms that share derived sequences, minimizing the number of hypothesized evolutionary changes.
    e. identify derived sequences by studying the embryology of the organisms.

Questions for Discussion

1. Systematists use both amino acid sequences and DNA sequences to determine evolutionary relationships. Think about the genetic code (Section 15.1), and explain why phylogenetic hypotheses based on DNA sequences may be more accurate than those based on amino acid sequences.

2. Traditional evolutionary systematists identify the Reptilia as one class of vertebrates, even though we know that this taxon is paraphyletic. Describe the advantages and disadvantages of defining paraphyletic taxa in a classification.

3. The following table provides information about the distribution of ancestral and derived states for six systematic characters (labeled 1 through 6) in five species (labeled A through E). A "d" means that the species has the derived form of the character, and an "a" means that it has the ancestral form. Construct a cladogram for the five species using the principle of parsimony; in other words, assume that each derived character evolved only once in this group of organisms. Mark the branches of the cladogram to show where each character changed from the ancestral to the derived state.

Species	Character					
	1	2	3	4	5	6
A	a	a	a	a	a	a
B	d	a	a	a	a	d
C	d	d	d	a	a	a
D	d	d	d	a	d	a
E	d	d	a	d	a	a

4. Imagine that you are a systematist studying a group of little-known flowering plants. You discover that the phylogenetic tree based on flower morphology differs dramatically from the phylogenetic tree based on DNA sequences. How would you try to resolve the discrepancy? Which tree would you believe is more accurate?

5. Create an imaginary phylogenetic tree for an ancestral species and its 10 descendants. Circle a monophyletic group, a polyphyletic group, and a paraphyletic group on the tree. Explain why the groups you identify match the definitions of the three types of groups.

Experimental Analysis

Imagine that you are trying to determine the evolutionary relationships among six groups of animals that look very much alike because they have few measurable morphological characters. What data would you collect to reconstruct their phylogenetic history?

Evolution Link

How do the two models of macroevolution (gradualist versus punctuated equilibrium) relate to the philosophies of phylogenetic inference espoused by traditional evolutionary systematists and cladists? You may want to review material in Section 22.3 before answering this question.

Black smoker hydrothermal vents on the ocean floor. Many scientists support the theory that life developed near hydrothermal vents, where superheated, mineral-rich water is found.

Dr. Ken Macdonald/SPL/Photo Researchers, Inc.

STUDY PLAN

24.1 The Formation of Molecules Necessary for Life

Conditions on primordial Earth led to the formation of organic molecules

The Oparin-Haldane hypothesis initiated scientific investigations into the origin of life

Chemistry simulation experiments support the Oparin-Haldane hypothesis

Scientists have new theories about the sites for the origin of life

24.2 The Origin of Cells

Protocells formed with some of the properties of life

Living cells may have developed from protocells

Prokaryotic cells were the first living cells

Subsequent events increased the oxidizing nature of the atmosphere

24.3 The Origins of Eukaryotic Cells

The endosymbiont hypothesis proposes that mitochondria and chloroplasts evolved from ingested prokaryotes

Several lines of evidence support the endosymbiont hypothesis

Eukaryotic cells may have evolved from a common ancestral line shared with archaeans

Multicellular eukaryotes probably evolved in colonies of cells

Life may have been the inevitable consequence of the physical conditions of the primitive Earth

24 The Origin of Life

WHY IT MATTERS

In 1927, Belgian priest and astronomer George Lemaître proposed the Big Bang Theory, which is now the dominant scientific theory about the origin of the universe. According to this theory, an incomprehensibly vast explosion about 14 billion years ago produced the matter and energy of our universe. Most of the matter was initially distributed in clouds of gas and dust; some of these clouds still exist today **(Figure 24.1)**. As the universe expanded, gravitational attraction caused the dust clouds to condense in some regions into more concentrated collections of matter. In our small corner of the early universe, the dust clouds condensed into the sun and its surrounding planets, including Earth.

Earth is estimated to have formed approximately 4.6 billion years ago, when it condensed out of cosmic dust and began its long transition into the environment we know today. There is no record of the time when life first formed, but microscopic deposits resembling bacteria have been found in Australia, in rocks laid down as sediments about 3.5 billion years ago during the Archaean era (inset to Figure 24.1). If these deposits are actually fossil prokaryotes, then life may have appeared during the first billion years or so of Earth's existence.

511

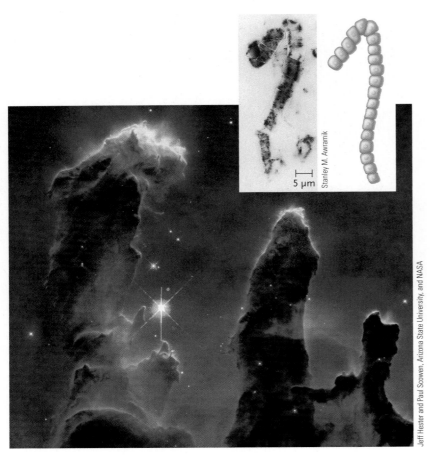

Figure 24.1

The Eagle nebula, a cloud of gas and dust particles some 7000 light years from Earth. Gas is condensing and forming stars, and perhaps planets, in this nebula. The inset shows structures that are believed to be a strand of fossil prokaryote cells in a rock sample 3.5 billion years old.

Figure 24.2 outlines the key events in the early evolution of life, which we will examine in this chapter. The earliest events are uncertain, but probably include the formation of organic molecules and the development of **protocells**, primitive cell-like structures that have some of the properties of life and that might have been the precursors of cells. Prokaryotic cells arose during the first billion years or so after the formation of Earth, and about 500 million years later some of them developed the capacity to perform photosynthesis, which released oxygen into the atmosphere. The oxygen-enriched environment was probably essential to the development of the first eukaryotic cells, which may have occurred as long as 2.2 billion years ago.

24.1 The Formation of Molecules Necessary for Life

All present-day living cells are complex; they have (1) a boundary membrane separating the cell interior from the exterior; (2) one or more nucleic acid coding molecules located in a nuclear region (a nucleus in eukaryotes and a nucleoid region in prokaryotes); (3) a

system using the coded information to make proteins and, through them, other biological molecules; and (4) a metabolic system providing energy for these activities. Because these systems are so complex, it is highly unlikely that living cells appeared suddenly from nonliving matter. Rather, there must have been a transition from nonliving to living matter.

No fossils or other records exist to inform us about this transition, but much evidence supports the idea that life did emerge from the nonliving world. Living organisms are composed entirely of elements common in the nonliving, physical world on Earth and throughout the universe. Moreover, all of the reactions that sustain life are elaborations of those in the physical world. Most scientists study the origin of life by assuming that it originated from nonliving matter on Earth, through chemical and physical processes no different from those operating today. Hypotheses made under these assumptions are testable to the extent that the chemical and physical processes can be duplicated in the laboratory.

But some scientists have not ruled out an extraterrestrial origin of life. Analysis of meteorites has shown that they contain some organic molecules characteristic of living organisms. Could a living cell or organism have arrived in such a way? Most scientists believe it is unlikely that a cell or an organism could have survived a long journey in space, even if protected from radiation, or that it could have survived intense heating while traveling through Earth's atmosphere and the actual impact with Earth. However, other scientists argue that conditions inside some meteorites might have been less extreme and allowed "life" to continue. At this point the hypothesis that life arrived on Earth by interplanetary transport cannot be ruled out. Nonetheless, even if a living organism arrived from space and

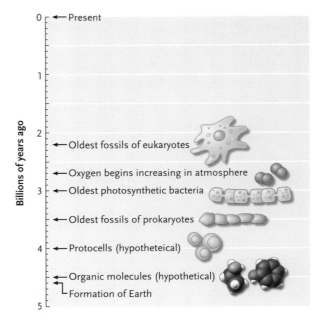

Figure 24.2

A timeline for the evolution of cellular life.

spawned a population on this planet, life would still have had to arise from nonliving matter in a similar way on the organism's home planet.

Conditions on Primordial Earth Led to the Formation of Organic Molecules

As we noted in the introduction, astronomers estimate that our solar system condensed from an interstellar dust cloud some 4.6 billion years ago. Intense heat and pressure generated in the central region of the cloud by the condensation set off a thermonuclear reaction that established the star of our solar system, the sun. The remainder of the spiraling dust and gas condensed into the planets and other bodies orbiting the sun.

Gravitational compression caused internal temperatures in our planet to rise to 1000° to 3000°C, causing its matter to melt and stratify into layers. Metallic elements sank to the core and lighter substances, such as silicates, carbides, and sulfides of the metallic elements, floated to the surface **(Figure 24.3)**. As the planet radiated away some of its heat, the surface layers cooled and solidified into the rocks of the crust. Earth's gravitational pull was strong enough to hold an atmosphere around the planet, derived partly from the original dust cloud and partly from gases released from the planet's interior as it cooled.

Primordial Earth met several basic conditions necessary for life to begin. Although its gravitational pull was strong enough to retain an atmosphere, it was not strong enough to compress the atmospheric gases into liquid form. Earth's distance from the sun was such that, on average, sunlight warmed the surface enough to keep much of the liquid water (much of which may have come from icy objects from the main asteroid belt colliding with Earth) from freezing, but not enough to boil the water. This allowed liquid water to accumulate in rivers, lakes, and seas. *Liquid water is essential for the chemistry of biological systems* (see Chapter 2).

Evaporation of water at the surface would have contributed water vapor to the atmosphere. Besides water vapor, the primordial atmosphere probably contained hydrogen and nitrogen molecules. Erupting volcanoes probably released large quantities of hydrogen sulfide, carbon dioxide, and carbon monoxide. Any molecular oxygen would have reacted with elements of the crust and atmosphere to form oxides. Spontaneous reactions of hydrogen, nitrogen, and carbon would have produced ammonia (NH_3) and methane (CH_4).

As Earth's surface cooled, natural sources of energy caused chemical bonds to break and reform, leading to the formation of organic molecules. In addition to sunlight and electrical discharges during storms, radioactivity from atomic decay and heat from volcanoes, geysers, and hydrothermal (hot water) vents in the sea floor all acted on the primordial atmosphere and crust—as they still do today. As many as a half-billion years may have passed before the concentrations of organic mol-

Figure 24.3
An artist's depiction of Earth during its early cooling stage, still too hot to support life.

ecules reached levels where their interactions formed more complex organic substances. We now consider the current thinking about how simple molecules were converted into the key molecules of life.

The Oparin-Haldane Hypothesis Initiated Scientific Investigations into the Origin of Life

Scientific efforts to explain the origin of life began with a major hypothesis proposed independently in the 1920s by two investigators, Aleksandr I. Oparin, a Russian plant biochemist at Moscow State University in Russia, and J. B. S. Haldane, a Scottish geneticist and evolutionary biologist at Cambridge University in England. Their hypothesis rested on the critical assumption that Earth's primordial atmosphere was radically different from today's atmosphere. They proposed that, rather than being an oxygen-rich (oxidizing) atmosphere as it is now, the early atmosphere was composed of substances such as hydrogen (H_2), methane (CH_4), ammonia (NH_3), and water, which are *fully reduced*—they contain the maximum possible number of electrons and hydrogens (see Section 8.1). These substances, they concluded, would have given the primordial atmosphere a *reducing* character; it contained an abundance of electrons and hydrogens available for reduction reactions, which could create organic molecules from inorganic elements and compounds. Energy to drive the reductions, according to the hypothesis, came from solar energy and other natural sources such as the electrical energy of lightning in atmospheric storms.

The absence of oxygen in the primitive atmosphere is essential to the Oparin-Haldane hypothesis. Oxygen can reverse reductions by removing electrons and hydrogens from organic molecules (see Section 8.1). In other words, if oxygen was present, the newly formed molecules would have been broken down quickly by oxidation.

Oparin and Haldane proposed that reductions occurring on the primordial Earth produced great quantities of organic molecules. The molecules accumulated because the two main routes by which such substances break down today, chemical attack by oxygen and decay by microorganisms, could not take place. According to Oparin and Haldane's hypothesis, the organic substances would have became so concentrated that the oceans and other bodies of water resembled a "prebiotic soup."

Oparin and Haldane assumed that these highly concentrated organic molecules would tend to aggregate in random combinations and that, by chance, some of the combinations were able to carry out one or more primitive reactions characteristic of life, such as increasing in mass by adding new materials. Later, scientists reasoned that these combinations were able to compete successfully against less efficient combinations for space and materials in the organic soup. As a result, they persisted and became more numerous.

Chemistry Simulation Experiments Support the Oparin-Haldane Hypothesis

In the 1950s, new discoveries in chemistry provided direct support for the most basic proposals of Oparin and Haldane's hypothesis. In 1953, Stanley L. Miller, a graduate student in Harold Urey's laboratory at the University of Chicago, tested the hypothesis by creating a laboratory simulation of conditions Oparin and Haldane believed existed on early Earth. Miller placed components of a reducing atmosphere—hydrogen, methane, ammonia, and water vapor—in a closed apparatus and exposed the gases to an energy source in the form of continuously sparking electrodes **(Figure 24.4).** Water vapor was added to the "atmosphere" by boiling water in one part of the apparatus, and it was

removed by cooling and condensation in another part. After running the apparatus for only a week, Miller found a large assortment of organic compounds in the water, including urea, amino acids, and lactic, formic, and acetic acids. In fact, as much as 15% of the carbon was now in the form of organic compounds. Two percent of the carbon was in the form of amino acids, which form easily under sufficiently reducing conditions. The significance of the finding at the time was enormous: amino acids, which are essential to cellular life, could be made under the conditions scientists believed existed on early Earth.

Other chemicals have been tested in the Miller-Urey apparatus. For example, hydrogen cyanide (HCN) and formaldehyde (CH_2O) were considered likely to have been among the earliest substances formed in the primitive atmosphere. When HCN and CH_2O molecules were added to the simulated primitive atmosphere in Miller's apparatus, all the building blocks of complex biological molecules were produced. Among the products were amino acids; fatty acids; the purine and pyrimidine building blocks of nucleic acids; sugars such as glyceraldehyde, ribose, deoxyribose, glucose, fructose, mannose, and xylose; and phospholipids, which form the lipid bilayers of biological membranes.

The synthesis of complex biological molecules in a reducing atmosphere in the Miller-Urey experiment supported the Oparin-Haldane hypothesis. However, it is only a conjecture that a reducing atmosphere was present at the time key organic molecules were formed on early Earth. Indeed, current thinking is that early Earth's atmosphere was not reducing but that it contained large amounts of oxidants such as CO_2 and N_2. In such an oxidizing atmosphere, any organic molecules generated spontaneously in the environment would be oxidized quickly back to inorganic forms by combination with the oxygen in the atmosphere. This is supported experimentally: running the Miller-Urey experiment in the presence of oxygen results in essentially no organic molecules. Moreover, amino acids cannot be produced in such an atmosphere, making the origin of life impossible.

In addition, the Miller-Urey experiment required the input of a large amount of energy. In the experiment, energy was provided continuously, but in the atmosphere of early Earth it would have been delivered, at best, intermittently from lightning storms. Scientists think that amino acids and other organic compounds may well have formed under these conditions, but not in the amounts seen in the laboratory experiment.

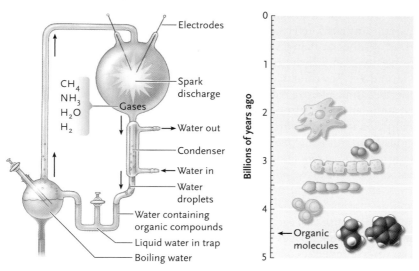

Figure 24.4

The Miller-Urey apparatus demonstrating that organic molecules can be synthesized spontaneously under conditions simulating the primordial Earth. Operation for 1 week converted 15% of the carbon in the "atmosphere" inside the apparatus into a surprising variety of organic compounds.

(Redrawn from an original courtesy of S. L. Miller. Copyright 1955 by the American Chemical Society.)

Scientists Have New Theories about the Sites for the Origin of Life

If organic compounds were not generated in a reducing atmosphere, how else could they have arisen? Scientists have developed a number of theories. All of them assume the presence of liquid water, which is a reasonable assumption. Remember that water is essential for the chemistry of biological systems (see Chapter 2). Two of the more reasonable theories are described here.

One current theory for the origin of life, which has significant support among scientists, is that life developed near hydrothermal vents in the sea floor. Many such vents exist in today's oceans, emitting bursts of mineral-rich water superheated to up to 400°C by submarine volcanoes. Scientists exploring hydrothermal vents find complex ecosystems associated with them.

Life might have originated near oceanic hydrothermal vents because reducing conditions existed there along with an abundance of the chemicals essential for life. Even now, there are high levels of hydrogen gas, methane, and ammonia around the vents. Indeed, based on simulation experiments, scientists believe that hydrothermal vents could have produced a lot more organic material than that generated in the Miller-Urey experiment. However, if life did evolve near hydrothermal vents, we would expect many present hydrothermal-vent life forms to be ancient. This is not the case: in most cases these organisms are closely related to modern non-vent organisms. Critics of the hydrothermal-vent origin of life theory also argue that the temperature at the vents is too high to permit the origin of life. The critics argue that, at the high temperature found at vents, the organic molecules are too unstable and would be destroyed as soon as they form. Supporters of the theory counter that the necessary organic molecules for life are formed not at the vent itself, but somewhere in the gradient between the hot water at the vent and the near-freezing water surrounding the vent.

Scientists debate whether organic molecules could be produced in the temperature gradient in the amounts needed. Recently, Koichiro Matsuno and his colleagues at Nagaoka University of Technology in Japan assembled an artificial system simulating the environment of ocean bottom hydrothermal vents, and added the feature of cycling materials between heat and cold. This feature accommodated the possibility that chemical products made near the vents were quenched in the surrounding colder water and then reentered the vent area where they could undergo further reactions. Their experiments demonstrated that amino acids are formed and that they can polymerize into short polypeptides under these conditions. They argue that the amounts are sufficient to form complicated molecules.

Another theory is that some organic compounds had an extraterrestrial origin. Interestingly, many of the compounds made in the Miller-Urey experiment exist in outer space. For example, a meteorite that fell on Murchison, Australia, in 1969 contained more than 90 amino acids, only 19 of which are found on Earth. Since amino acids appear to be able to survive in outer space, they could potentially have been present when Earth was formed. And perhaps other organic compounds arrived by meteor or comet impact.

24.2 The Origin of Cells

Whether organic molecules originated in the atmosphere, in hydrothermal vents, or in outer space, they still do not qualify as life. In this section, we discuss the key stage in the origin of life, the formation of the first cells.

Protocells Formed with Some of the Properties of Life

How did organic building blocks such as amino acids assemble into macromolecules such as proteins and nucleic acids? To answer this question, researchers have proposed and tested several processes. One process is the concentration of subunits by the evaporation of water. Another is *dehydration synthesis (condensations)*, in which subunits assemble into larger molecules through removal of the elements of a molecule of water (see Section 3.1). Experiments with these processes under simulated conditions showed that both evaporation and condensation reactions can produce polypeptide chains from amino acids, polysaccharides from glucose and other monosaccharides, and nucleotides and nucleic acids from nitrogenous bases, ribose, and phosphates.

Scientists reason that spontaneous condensations and other reactions produced significant quantities of all the major biological molecules over the hundreds of millions of years following the initial formation of Earth. They hypothesize that the accumulation of organic matter set up the conditions necessary for the next stage, the chance assembly of molecules into aggregations that became membrane-bound to form primitive protocells. Protocells are key to the origin of life, because life depends upon reactions occurring in a controlled and sequestered environment, the cell. Researchers have proposed several mechanisms for the assembly of organic molecules into aggregates, each of which has been successfully duplicated in laboratory experiments simulating primordial conditions. Two of those mechanisms are absorption into clays and lipid bilayer assembly.

Absorption into Clays. Could clays have provided an ideal environment for molecular aggregation and interaction on the primitive Earth? Clays consist of very thin layers of minerals separated by layers of water only a few nanometers thick. The layered structure readily absorbs ions and organic molecules and promotes their interactions, including condensations and other assembly reactions. Clays can also store potential energy, and therefore could have channeled some of the energy into reactions taking place inside them.

Several experiments have supported these proposals. For example, Noam Lahav at the Hebrew University in Israel and Sherwood Chang of NASA's Ames Research Center added amino acids to clays and exposed the mixtures to water-content changes and fluctuating temperatures, as they might be in a tidal flat. After several cycles of the fluctuating conditions, polypeptides were detected in the clays. Other researchers found that RNA nucleotides linked to phosphate chains could combine into RNA-like molecules in clays. Accumulation of these and other macromolecules in the clays could have provided an environment in which they could react to carry out the first reactions of life.

However, even if molecules became organized in clay and some of the reactions of life commenced, it is not clear how a lipid bilayer membrane could have formed around them. Such a membrane is necessary to organize the molecules into protocells, the presumed precursors of cells. (The biological importance of lipid bilayers and membranes are discussed in Sections 2.4, 3.4, and 6.1.)

Lipid Bilayer Assembly. In the 1950s, R. J. Goldacre at Chester Beatty Research Institute, London, hypothesized that protocells could have formed starting with lipid bilayers that had assembled spontaneously. In the 1970s, David W. Deamer at the University of California at Davis and other investigators tested this hypothesis, finding that phospholipids and some other types of lipid molecules could form under simulated conditions. The phospholipids self-assembled readily into bilayers when suspended in water (see Section 6.1). Often, the bilayers rounded up into stable, closed vesicles consisting of a continuous-boundary "membrane" surrounding an inner space **(Figure 24.5)**.

Further tests showed that the bilayers formed in these experiments have many properties of living membranes. For example, they can incorporate proteins onto their surfaces or into the hydrophobic membrane interior, and they form vesicles that can trap other substances in the fluid enclosed by the membrane. Potentially, on early Earth, the concentration of organic molecules in such vesicles could have stimulated their growth and eventual fragmentation into smaller vesicles, providing a primitive form of reproduction. These mechanisms of aggregation, as well as others, may have worked separately or together to form protocells.

Living Cells May Have Developed from Protocells

Eventually the chemical reactions taking place in the primitive protocells became organized enough to make the transition to living cells. Of the several critical events necessary for this transition, we will look closely at two: the development of pathways that captured and harnessed the energy required to drive molecular synthesis and the development of a system for the storage, replication, and translation of information for protein synthesis. Remember that proteins are the catalysts for most cellular reactions.

Development of Energy-Harnessing Reaction Pathways. Oxidation-reduction reactions (see Section 8.1) were probably among the initial energy-releasing reactions of the primitive protocells. In an oxidation, electrons are removed from a substance; the removal releases free energy that can be used to drive synthesis and other reactions. In a reduction, electrons are added to a substance; the added electrons provide energy that can contribute to the formation of complex molecules from simpler building blocks.

At first the electrons removed in an oxidation would have been transferred directly to the substances being reduced, in a one-step process. However, the greater efficiency of stepwise energy release would have favored development of intermediate carriers and opened the way for primitive electron transfer systems. Evolved from those primitive systems are the present-day electron transfer systems of mitochondria and chloroplasts (see Sections 8.4 and 9.2).

As part of the energy-harnessing reactions, ATP became established as the coupling agent that links energy-releasing reactions to those requiring energy. ATP may first have entered protocells as one of many organic molecules absorbed from the primitive environment. Initially, it was probably simply hydrolyzed into ADP and phosphate as an energy source. Later, as protocells developed, some of the free energy released

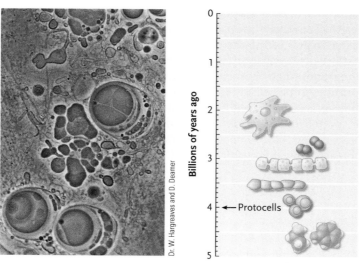

Dr. W. Hargreaves and D. Deamer

Billions of years ago

4 ← Protocells

Figure 24.5
An electron micrograph of vesicles of various sizes and shapes assembled from phospholipids synthesized under simulated primordial conditions. When the vesicles are more highly magnified than in this micrograph, their walls can be seen to consist of a lipid bilayer.

during electron transfer was probably used to synthesize ATP directly from ADP and inorganic phosphate. Because of the efficiency and versatility of energy transfer by ATP, it gradually became the primary substance connecting energy-releasing and energy-requiring reactions in early cells.

Origin of the Information System. A system that could store, reproduce, and translate the information required for protein synthesis was a second critical event for the transition from protocells to living cells. How the information system developed is crucial to the understanding of the origin of life.

In contemporary organisms, information flows from DNA to RNA to protein. This nucleic acid–based information system depends mostly on enzymatic proteins for replication, transcription, and translation of the nucleic acids. However, the specificity of enzymatic proteins depends on their amino acid sequences, which are determined by the sequences of nucleotides in nucleic acids. Thus, proteins depend on nucleic acids for their structure, and nucleic acids depend on proteins to catalyze their activities. How could one have appeared before the other? Scientists believe the information system developed in stages, although the order of the steps is a subject of debate. There are two main hypotheses: the RNA-first hypothesis and the protein-first hypothesis.

The *RNA-first hypothesis* states that the first genes and enzymes were RNA molecules. That is, *ribozymes*—RNA molecules capable of catalyzing biochemical reactions—may have functioned both as informational molecules and as catalysts in protocells, without requiring protein enzymes for catalytic reactions (ribozymes are discussed in Section 4.6). Thus, a self-catalyzed "RNA world" may have been the first step in the development of an information system.

Ribozymes may have originally developed by the chance assembly of RNA nucleotides taking part in oxidative and other metabolic reactions in protocells (RNA nucleotides such as ATP, NAD, and coenzyme A form important parts of many metabolic pathways, including glycolysis, respiration, and photosynthesis; see Chapters 8 and 9). The RNA molecules then developed the capacity to replicate themselves and other RNA molecules. That is, these RNA molecules acted both as templates—like mRNA—and as catalysts—like ribosomal RNA (see Section 15.4). Then ribozymes could replicate ribozymes, with no need for protein enzymes. Such self-replicating systems may have provided the basis of an RNA-based informational system, and founded the RNA world. *Insights from the Molecular Revolution* describes an experiment in which ribozymes that can replicate RNA were generated in a test tube.

In the RNA world, DNA would have developed as a subsequent step. At first, DNA nucleotides may have been produced by random removal of an oxygen atom from the ribose subunits of the RNA nucleotides. At some point, the DNA nucleotides paired with the RNA informational molecules, and were assembled into complementary copies of the RNA sequences. Some modern day viruses carry out this RNA-to-DNA reaction using the enzyme reverse transcriptase (see Section 18.1). Once the DNA copies were made, selection may have favored DNA as the informational storage molecule because it has greater chemical stability and can be assembled into much longer coding sequences than RNA. RNA was left to function at intermediate steps between the stored information in DNA and protein synthesis, as it still does today.

As the RNA-based information system evolved, some RNAs may have acted as tRNA-like molecules, linking to amino acids and pairing with the RNA informational molecules. These associations could have led to the assembly of polypeptides of ordered sequence—the development of an RNA genetic code. When DNA took over information storage from RNA, the code would have been transferred to DNA.

Modern analysis of the ribosome, the organelle responsible for translation of mRNA (see Section 15.4) has shown that the enzyme that catalyzes the formation of a peptide bond between amino acids is a property of one of the RNA molecules of the ribosome. This finding supports the proposal that, in addition to replicating themselves, RNA molecules also generated the first proteins.

The second hypothesis, the *protein-first hypothesis,* states that proteins were the first informational molecules to arise. Then, once complex enzymes developed within protocells, nucleic acids—both DNA and RNA—were assembled enzymatically from small molecules, and replication and transcription processes developed.

Of course, we have no way of knowing exactly how life originated. Sifting through the various models and theories we can perhaps agree that there were some basic steps: (1) the abiotic (nonliving) synthesis of organic molecules such as amino acids; (2) the assembly of complex organic molecules from simple molecules, including protein or RNA or both; and (3) the aggregation of complex organic molecules inside membrane-bound protocells. Once the information system had developed in the protocells, and the protocells could divide, they had become true living cells. The advent of living cells marked the beginning of biological evolution, which depends on cells that can reproduce and pass on information to their descendants.

Prokaryotic Cells Were the First Living Cells

The change to biological evolution set the stage for the appearance of all the features of cellular life. One of these features was a nuclear region that contained the DNA of the coding system and the mechanisms replicating the DNA and transcribing it into RNA. Another feature was a cytoplasmic region containing ribosomes and the enzymes required to translate RNA informa-

Replicating the RNA World

The discovery of ribozymes led to the proposal that an RNA world was the first step in the evolution of a molecular information system that could store, reproduce, and translate the information required for protein synthesis. In an RNA world, RNA molecules would have to act both as templates for their own replication and as catalysts to carry out the replication. The catalytic ability of RNA molecules has been amply demonstrated, but could they carry out RNA replication?

Wendy K. Johnston and her coworkers at the Massachusetts Institute of Technology decided to answer this question by using a ribozyme (a catalytic RNA) as the starting point for developing an RNA molecule that could replicate itself, as might have happened during the evolution of cellular life. The ribozyme they chose is an RNA ligase, which can catalyze one of the most fundamental reactions of replication, linking together short chains of nucleotides.

To achieve their feat, Johnston and her coworkers used a technique that accomplishes molecular evolution in a

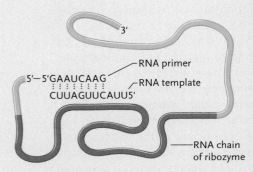

The general arrangement of sequences in the ribozymes used by Johnston and her coworkers. The blue and green portions are sequences added to provide raw materials for test-tube evolution. The template shown is three nucleotides longer than the primer and would require a ribozyme to add two nucleotides to the primer to be selected as successful (the last nucleotide in the template cannot be copied).

test tube (the RNA ligase used to start their experiments was the product of an earlier test-tube experiment). They assembled a reaction mixture containing the RNA ligase with an added a 76-nucleotide RNA strand of random sequence to serve as a template for self-replication. They then generated 1×10^{15} versions (a quintillion!) of the ligase with different sequences concentrated in the added strand. To the mutated versions in a test tube they added RNA nucleoside triphosphates (NTPs), an RNA template chain, and an RNA primer, with the RNA primer linked covalently to the ribozyme (see **figure**).

In the initial run, the template was only two bases longer than the primer, so to be successful a ribozyme had only to add two nucleotides to the primer. To detect the successful ribozymes, the investigators used RNA nucleoside triphosphates that were tagged with a chemical label. Any ribozymes that added the nucleotides to the primer would become labeled and thus be identifiable among the unsuccessful ribozymes in the test tube.

After the first round of selection, the investigators selected the labeled ribozyme variants, which had added nucleotides to the primer, and multiplied them using PCR (the polymerase chain reaction; see Section 18.1). They then added all the elements to the test tube for another round of replication and selection. This cycle of replication and selection was repeated through 18 successive rounds. As part of the process, additional mutations were induced in the ribozymes after round 10, and the selection pressure was increased by several methods. One was to make the template longer in successive rounds, so that the ribozymes had to add more nucleotides to the primer to be successful. Another was to alter the se-

quence of the template chain, so that the ribozymes had to be able to copy a template of any sequence to be successful. Also, the investigators shortened the time allowed for replication in successive runs, so that ribozymes had to work faster to be successful.

By the 18th round of replication, the selection process had produced a ribozyme that could replicate an RNA template 14 nucleotides longer than the primer. The template could be of any sequence. In addition, the template did not have to be covalently linked to the ribozyme for replication to occur.

To check on the accuracy of replication, the investigators gave the 18th-round ribozyme a template chain that was 11 nucleotides longer than the primer and then sequenced 100 of the complementary chains produced by the ribozyme. Of the replication products, 89 of 100 were precise complementary copies, all matched exactly to the template. In the remaining 11 products, only 12 base mismatches were found, slightly more than one base mismatch per copy.

Thus, the selected ribozyme was able to work as an RNA polymerase, faithfully replicating an RNA template into a complementary copy and thereby meeting a major requirement for an RNA world. The research continues, with further test-tube selection experiments designed to increase the accuracy of replication, the length of the template, and the rate of replication. These are small steps compared to the enormous task involved in the evolution of a full-fledged information system, but it is likely that life evolved in the same pattern, through the accumulation of small changes over hundreds of millions of years of molecular trial and error.

tion into sequences of amino acids in proteins. The cytoplasm also contained an oxidative system supplying chemical energy for protein synthesis and assembly of other required molecules. A mechanism of cell division also evolved, allowing replicated DNA to be distributed equally between daughter cells. All these

systems were enclosed by a membrane controlling the flow of molecules and ions in and out of the cell. The stages leading to this level may have taken more than a billion years, occupying the period from Earth's formation 4.6 billion years ago to the earliest known prokaryotic fossils, dated as 3.5 billion years old.

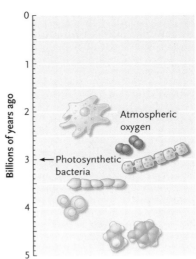

Bill Bachmann/Photo Researchers, Inc.

Figure 24.6
Stromatolites exposed at low tide in Western Australia's Shark Bay. These mounds, which consist of mineral deposits made by photosynthetic cyanobacteria, are about 2000 years old; they are highly similar in structure to fossil stromatolites that formed more than 3.0 billion years ago. As a result of photosynthesis by cyanobacteria, oxygen began to accumulate in the atmosphere.

Subsequent Events Increased the Oxidizing Nature of the Atmosphere

According to Richard E. Dickerson of UCLA and others, the earliest form of photosynthesis evolved about 3.5 billion years ago in the early prokaryotes. This form of photosynthesis probably used electron donors such as hydrogen sulfide (H_2S) that do not release oxygen. However, at some point, an enzymatic system evolved that could use the most abundant molecule of the environment, water (H_2O), as the electron donor for photosynthesis. This reaction split water into protons, electrons, and oxygen, which was released into the atmosphere.

The oxygen released by the water-splitting reaction accumulated in the atmosphere and set the stage for the development of electron transfer systems using oxygen as the final electron acceptor. These transfer systems arose when some cells developed cytochromes that could deliver low-energy electrons to oxygen (see Section 8.4). These cells were able to tap the greatest possible amount of energy from the electrons before releasing them from electron transfer, making the cells highly successful in their environment.

When might water-splitting photosynthesizers have appeared? A possible answer to this question has been found in rock formations laid down at least 3 billion years ago. These rocks contain **stromatolites**, fossils of ancient prokaryotes (cyanobacteria) that carried out photosynthesis by the water-splitting reaction **(Figure 24.6)**. Thus, oxygen-producing bacteria were present at least 3 billion years ago and perhaps evolved soon after the first prokaryotes appeared. Scientists believe that it may have taken another billion years for oxygen to accumulate to significant quantities in the atmosphere.

These major events established the preconditions for the evolution of eukaryotic cells. The next section traces this evolution, which was pivotal to the later evolution of large-scale multicellularity and the plants, animals, and the other organisms of the domain Eukarya.

STUDY BREAK

Several mechanisms have been proposed for the assembly of organic molecules into protocells. Why is the model involving a lipid bilayer membrane a particularly attractive one?

24.3 The Origins of Eukaryotic Cells

Present-day eukaryotic cells have several interrelated characteristics that distinguish them from prokaryotes: (1) the separation of DNA and cytoplasm by a nuclear envelope; (2) the presence in the cytoplasm of membrane-bound compartments with specialized metabolic and synthetic functions—mitochondria, chloroplasts, the endoplasmic reticulum (ER), and the Golgi complex, among others; and (3) highly specialized motor (contractile) proteins that move cells and internal cell parts. In this section we discuss how eukaryotes most probably evolved from associations of prokaryotes.

The Endosymbiont Hypothesis Proposes that Mitochondria and Chloroplasts Evolved from Ingested Prokaryotes

The **endosymbiont hypothesis**, put forward by Lynn Margulis at the University of Massachusetts, Amherst, proposes that the membranous organelles of eukaryotic cells, the mitochondria and chloroplasts, may each have originated from symbiotic (mutually advantageous) relationships between two prokaryotic cells **(Figure 24.7)**.

Mitochondria began to develop when photosynthetic and nonphotosynthetic prokaryotes coexisted in an oxygen-rich atmosphere. The nonphotosynthetic prokaryotes fed themselves by ingesting organic molecules from their environment. These prokaryotes included both anaerobes, unable to use oxygen as the final acceptor for electron transfer, and aerobes, fully

Figure 24.7
The endosymbiont hypothesis. Mitochondria and chloroplasts of eukaryotic cells are thought to have originated from various bacteria that lived as endosymbionts within other cells.

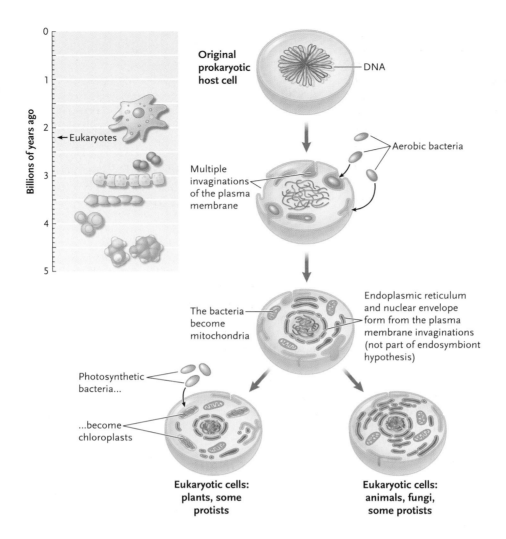

Original prokaryotic host cell

DNA

Aerobic bacteria

Multiple invaginations of the plasma membrane

The bacteria become mitochondria

Endoplasmic reticulum and nuclear envelope form from the plasma membrane invaginations (not part of endosymbiont hypothesis)

Photosynthetic bacteria...

...become chloroplasts

Eukaryotic cells: plants, some protists

Eukaryotic cells: animals, fungi, some protists

Billions of years ago

Eukaryotes

capable of using oxygen. Only the aerobes could fully exploit the energy stored in organic molecules, but predatory anaerobes could capture that energy by eating aerobic cells. These anaerobic prokaryotes had become efficient predators, and lived by ingesting other cells. Among the ingested cells were some aerobic prokaryotes; instead of being digested, some of them persisted in the cytoplasm of the predators and continued to respire aerobically in their new location. They had become *endosymbionts,* organisms that live symbiotically within a host cell. The cytoplasm of the host anaerobe, formerly limited to the use of organic molecules as final electron acceptors, was now home to an aerobe capable of carrying out the much more efficient transfer of electrons to oxygen.

As a part of the transition to a true eukaryotic cell, the cell also evolved to acquire other membranous structures, the major ones being the nuclear envelope, the ER, and the Golgi complex. Endocytosis, the process of infolding of the plasma membrane (see Figure 5.14), is believed to be responsible for the evolution of these structures. (These events are not part of the endosymbiont hypothesis.) Researchers believe that, in cell lines leading from prokaryotes to eukaryotes, pockets of the plasma membrane formed during endocyto-

sis may have extended inward and surrounded the nuclear region. Some of these membranes fused around the DNA, forming the nuclear envelope and, hence, the nucleus. The remaining membranes formed vesicles in the cytoplasm that gave rise to the ER and the Golgi complex **(Figure 24.8).**

Next, according to the endosymbiont hypothesis, many functions duplicated in the aerobic endosymbiont were taken over by the host cell. As part of this transfer of function, most of the genes of the aerobe moved to the cell nucleus and became integrated into the host cell's DNA. At the same time, the host anaerobe became dependent for its survival on the respiratory capacity of the symbiotic aerobe. The ingested aerobe presumably benefited as well, because the host cell brought in large quantities of food molecules to be oxidized. This gradual process of mutual adaptation culminated in transformation of the cytoplasmic aerobes into mitochondria. The first eukaryotic cells had appeared, the ancestors of all modern-day eukaryotes.

The endosymbiont hypothesis proposes that a similar mechanism led to the appearance of the membrane-bound plastids (the general term for chloroplasts and related organelles, both photosynthetic and nonphotosynthetic) some time after mitochondria evolved. Plastids originated when aerobic cells that had mitochondria, but were unable to carry out photosynthesis, ingested photosynthetic prokaryotes resembling present-day cyanobacteria (see Figure 24.7). These photosynthetic prokaryotes gradually changed into plastids by evolutionary processes similar to those that produced mitochondria. The cells with both plastids and mitochondria founded the cell lines that gave rise to the modern eukaryotic algae and plants.

Several Lines of Evidence Support the Endosymbiont Hypothesis

Researchers reasoned that if the endosymbiont hypothesis is correct, then both mitochondria and plastids would have structures and biochemical reactions more like those of prokaryotes than those of eukaryotes. This has been shown to be the case. For example, both organelles typically contain circular DNA molecules that closely resemble prokaryotic DNA, and code for rRNAs and ribosomes that resemble prokaryotic forms.

Another line of evidence supports a key assumption of the endosymbiont hypothesis by showing that engulfed cells or organelles can survive in the cytoplasm of the ingesting cell. Among animals, no less than 150 living genera, distributed among 11 phyla, include species that contain eukaryotic algae or cyanobacteria as residents in the cytoplasm of their cells. For example, larvae of the marine snail *Elysia* initially contain no chloroplasts, but after they begin feeding on algae, chloroplasts from the algal cells are taken up into the cells lining the gut. When the larvae develop into adult snails, the chloroplasts continue to carry out photosynthesis in their new location and produce carbohydrates that are used by the snails. The uptake of functional chloroplasts has also been observed among the Protoctista (the protists; see Chapter 26); **Figure 24.9** shows a protist with chloroplasts that closely resemble cyanobacteria.

How long did it take for evolutionary mechanisms to produce fully eukaryotic cells? The oldest known fossil eukaryotes are 2.2 billion years old. If prokaryotic cells first evolved some 3.5 billion years ago, it took up to 1.3 billion years for eukaryotic cells to evolve from prokaryotes (see Figure 24.2). If so, this long interval probably reflects the complexity of the adaptations leading from prokaryotic to eukaryotic cells. Of course, it is possible that eukaryotic cells evolved more quickly, and we have yet to find the evidence.

Eukaryotic Cells May Have Evolved from a Common Ancestral Line Shared with Archaeans

The system of classification that has gained acceptance among biologists, and the one used in this book, groups all living organisms into three domains. One domain, the Eukarya, contains the eukaryotes. The sec-

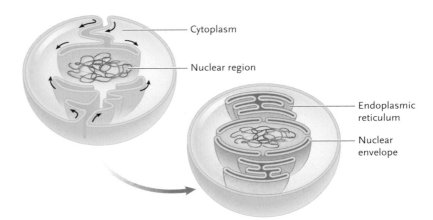

Figure 24.8
A hypothetical route for formation of the nuclear envelope and endoplasmic reticulum, through segments of the plasma membrane that were brought into the cytoplasm by endocytosis.

ond domain, the Bacteria, includes one of two groups of prokaryotes, the bacteria, which consists of both photosynthesizing and nonphotosynthesizing species. The third domain, the Archaea, contains the other group of prokaryotes, many of which inhabit extreme environments, including highly saline environments and hot springs.

There is little question that the three domains originated from a common ancestral cell line, because all share common fundamental characteristics—they all use the same genetic code, for example, and DNA and RNA molecules carry out the same basic functions in transcription and translation. However, the events leading from this common ancestry to the three domains of life remain unclear. The most difficult questions surround the role of the archaeans in both bacterial and eukaryotic evolution.

Archaeans have some features that are typical of bacteria, including a genome organized into a single, circular DNA molecule that is suspended in a nuclear region of the cytoplasm with no surrounding nuclear envelope. There are no membrane-bound organelles in the cytoplasm equivalent to mitochondria, chloroplasts, the ER, or the Golgi complex. However, the archaeans also have some features that are typically eukaryotic. One is the presence of interrupting, noncoding sequences called introns (see Section 15.3) in their genes;

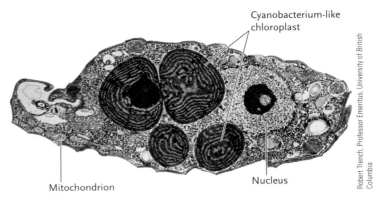

Cyanobacterium-like chloroplast

Mitochondrion

Nucleus

Robert Trench, Professor Emeritus, University of British Columbia

Figure 24.9
Cyanophora paradoxa, a protist with chloroplasts that closely resemble cyanobacteria without cell walls.

in some archaean genes that have counterparts in eukaryotes, the introns occur in exactly the same positions. By contrast, introns are rare or nonexistent in bacteria. The archaeans also have some characteristics that are unique to their domain, including features of gene and rRNA sequences, and features of cell wall and plasma membrane structure that are found nowhere else among living organisms. The characteristics shared by archaeans and eukaryotes suggest that their roots may lie in a common ancestral line that split off from the line leading to bacteria. At some point, this ancestral line split into the lines leading to archaeans and eukaryotes.

Multicellular Eukaryotes Probably Evolved in Colonies of Cells

The first eukaryotes were unicellular. They are the ancestors of the present-day diversity of unicellular eu-

karyotes. Multicellular eukaryotes evolved from unicellular eukaryotes and then diverged to produce the present-day multicellular eukaryotes. Molecular clock analysis indicates the first multicellular eukaryote likely arose between 800 and 1000 million years ago, while the first fossil records (of small algae) are from 600 to 800 million years ago.

According to the prevalent theory, multicellular eukaryotes arose by the congregation of cells of the same species into a colony. The ability to act in a coordinated way, probably increased the capacity of colonies to adapt to changes in the environment. Subsequently, differentiation of cells into various specialized cell types with distinct functions produced organisms with a wider range of capabilities and adaptability. Cell differentiation in a colony would have required cell signals that affected gene expression. That is, because each cell in the colony has the same genome, the development of specific func-

UNANSWERED QUESTIONS

What was the first polymer of life?

As discussed in this chapter, many researchers hypothesize that RNA was the first polymer of life because it both self-replicates and can catalyze chemical reactions. In addition, it is neatly connected with contemporary life, which is based on nucleic acids and proteins. There are several problems with this hypothesis, however. One of them is that the synthesis of RNA building blocks, nucleotides, and in particular their ribose fragment, is quite difficult under primordial conditions. To circumvent this difficulty, several researchers have proposed that other genetic polymers, whose monomers are simpler to synthesize, might have preceded RNA. For example, Albert Eschenmoser of the Swiss Federal Institute of Technology in Zurich, Switzerland, replaced ribose with the sugar pyranose, and Peter Nielsen from the University of Copenhagen, Denmark, synthesized a polymer with a peptide-like backbone. These polymers are stable and capable of self-replication. Another popular proposal is that the initial complement of nucleic acid bases was different from A, U, G, and C. One reason for this proposal is poor stability of cytosine in water. Although we have no evidence that transitional polymers were present on early Earth, it is important to realize that alternatives to nucleic acids exist and might be used by life elsewhere.

The protein-first hypothesis is currently not in favor with scientists, even though these polymers are excellent catalysts of chemical reactions and their building blocks, amino acids, existed on prebiotic Earth. This is because there is no known mechanism for proteins to self-replicate. Some researchers speculate that a limited replication of proteins is possible. An alternative hypothesis, supported by computer simulations, is that replication of individual polymers was not necessary at the origin of life and, instead, the reproduction of protein functions in a population was initially sufficient. Currently, neither view has much experimental support, but as we learn more about the structure and functions of small proteins major surprises might be in store.

Can we recreate protocells in a laboratory?

As you read this chapter, you must have noticed that our knowledge about the origin of life is still incomplete. But do we know enough to

test our understanding by building in a laboratory a simple life capable of self-reproduction and Darwinian evolution? Several groups of scientists are attempting to do just that. Conceptually, the simplest design is "the minimal RNA cell" proposed by Jack Szostak from Harvard Medical School. It consists of only two ribozymes encapsulated in a membrane-bound structure. One of them is capable of copying both ribozymes; the other catalyzes the synthesis of the membrane-forming molecules from their precursors. In principle, such a system could self-reproduce and undergo evolution through mutations of the ribozymes. However, the apparent simplicity of this construct is somewhat deceiving—no actual ribozymes that function together in this way are currently known. An international team of scientists is attempting to build a simple cell using a set of already existing components, as originally proposed by Steen Rasmussen from Los Alamos National Laboratory and Liaohai Chen from Argonne National Laboratory. This cell would differ from everything we know, however, and would therefore represent an example of "alien life." Craig Venter and several other researchers have taken yet another approach. Starting with a simple, contemporary microorganism as a template, they are trying to delete nonessential genes or substitute natural or synthetic genes that are smaller in size. So far, each of these strategies has encountered a surprising number of conceptual and technical difficulties, and none has been successful. This shows that synthesizing life is more complex that one would expect. If any of these efforts eventually succeeds, it will open the doors not only to many new investigations on the origin of life on Earth but also to the exploration of alternative forms of life and to applications of artificial cells in biotechnology and medicine.

Andrew Pohorille heads the NASA Center for Computational Astrobiology and Fundamental Biology at NASA's Ames Research Center. He is also professor of Chemistry and Pharmaceutical Chemistry at the University of California, San Francisco. For his work on the origin of life he was awarded the 2002 NASA Exceptional Scientific Achievement Medal.

tions (phenotypes) would require intracellular signals that would change the program of gene regulation. Over time, as genomes evolved, the division of function among cells led to the evolution of the tissues and organ systems of complex eukaryotes.

Multicellularity evolved several times in early eukaryotes, producing a number of lineages of algae as well as the ancestors of present day fungi, plants, and animals.

Life May Have Been the Inevitable Consequence of the Physical Conditions of the Primitive Earth

The events outlined in this chapter, leading from Earth's origin to the appearance of eukaryotic cells, may seem improbable. But, as scientist and author George Wald of Harvard University put it, given the total time span of these events, more than 3.5 billion years, "the impossible becomes possible, the possible probable, and the probable virtually certain. One has only to wait; time itself performs the miracles."

Some researchers go a step further and maintain that the evolution of life on our planet was an inevitable outcome of the initial physical and chemical conditions established by Earth's origin, among them a reducing atmosphere (at least in some locations), a size that generates moderate gravitational forces, and a distance from the Sun that results in average surface temperatures between the freezing and boiling points of water. Given the same conditions and sufficient time, according to these scientists, it is inevitable that life has evolved or is evolving now on other planets in the universe.

The chapters to follow in this unit trace the course of evolution and its products after eukaryotic cells were added to the prokaryotes already on Earth. Among prokaryotes, evolution established two major groups, the Bacteria and Archaea; among eukaryotes, further evolution established the protists, fungi, plants, and animals. The survey begins in the next chapter with a description of present-day Bacteria and Archaea, and of the viruses that infect prokaryotes and eukaryotes.

STUDY BREAK

Summarize the key points of the theory of endosymbiont origins for mitochondria and chloroplasts.

Review

Go to **Thomson**NOW™ at www.thomsonedu.com/login to access quizzing, animations, exercises, articles, and personalized homework help.

24.1 The Formation of Molecules Necessary for Life

- Living cells are characterized by a boundary membrane, one or more nucleic acid coding molecules in a nuclear region, a system for using the coded information to make proteins, and a metabolic system providing energy for those activities.

- Oparin and Haldane independently hypothesized that life arose de novo under the conditions they thought prevailed on the primitive Earth, including a reducing atmosphere that lacked oxygen. Reduction reactions, fueled by natural energy acting on the primitive atmosphere, produced organic molecules. Random aggregations of these molecules were able to carry out primitive reactions characteristic of life that gradually became more complex until life appeared. Chemistry simulation experiments support the hypothesis that organic molecules would form under these conditions (Figure 24.4).

- Present thinking is that early Earth's atmosphere was not reducing, but in fact contained significant amounts of oxidants. This has caused skepticism about Oparin and Haldane's hypothesis. One new theory proposes that life developed near hydrothermal vents in the sea floor.

Animation: Miller's reaction chamber experiment

Animation: Milestones in the history of life

24.2 The Origin of Cells

- Organic molecules produced in early Earth's environment by chance formed aggregates that became membrane-bound in protocells, primitive cell-like structures with some of the properties of life. Protocells may have been the precursors of cells (Figure 24.5).

- Next, living cells may have developed from protocells by the development of several critical components, notably energy-harnessing pathways, and a system based on nucleic acids that could store and pass on the information required to make proteins.

- Subsequently, fully cellular life evolved, with a nuclear region containing DNA and the mechanisms for copying its information into RNA messages; a cytoplasmic region containing systems for utilizing energy and systems for translating RNA messages into proteins; a membrane separating the cell from its surroundings; and a reproductive system duplicating the informational molecules and dividing them among daughter cells.

- The first living cells were prokaryotes. Eventually, some early cells developed the capacity to carry out photosynthesis using water as an electron donor; the oxygen produced as a byproduct accumulated and the oxidizing character of Earth's atmosphere increased. From this time on organic molecules produced in the environment were quickly broken down by oxidation, and life could arise only from preexisting life, as in today's world.

24.3 The Origins of Eukaryotic Cells

- According to the endosymbiont hypothesis, mitochondria developed from ingested prokaryotes that were capable of using oxygen as final electron acceptor; chloroplasts developed from ingested cyanobacteria (Figure 24.7).

- Eukaryotic structures such as the ER, Golgi complex, and nuclear envelope appeared through infoldings of the plasma membrane as a part of endocytosis (Figure 24.8).

- Multicellular eukaryotes probably evolved by differentiation of cells of the same species that had congregated into colonies. Multicellularity evolved several times, producing lineages of several algae and ancestors of fungi, plants, and animals.

Animation: Eukaryotic evolution

Questions

Self-Test Questions

1. Earth was formed _____ years ago, whereas the oldest known living cell formed about _____ years ago.
   a. 400×10^3; 3.6×10^6
   b. 4.6×10^9; 1.0×10^9
   c. 3.8×10^9; 4.6×10^7
   d. 4.6×10^9; 3.5×10^9
   e. 2.0×10^9; 600×10^6

2. Which of the following is *not* a characteristic of all living organisms?
   a. They replicate genetic information and convert the information into proteins.
   b. They pass genetic information between generations.
   c. They get energy from molecules in a controlled fashion.
   d. They use external energy to drive internal reactions requiring energy.
   e. They use mitochondria to transform energy for their cells' needs.

3. The greatest leap in evolution is from:
   a. nonlife to prokaryotes.
   b. prokaryotes to one-celled eukaryotes.
   c. ancient archaeans to modern archaeans.
   d. one-celled eukaryotes to fungi.
   e. one-celled eukaryotes to insects.

4. According to the Oparin-Haldane hypothesis, the atmosphere when life began was believed to be composed primarily of:
   a. H_2O, N_2, and CO_2.
   b. H_2, H_2O, NH_3, and CH_4.
   c. H_2O, N_2, O_2, and CO_2.
   d. O_2 and no H_2.
   e. H_2 only.

5. The Miller-Urey experiment:
   a. was based on the belief the atmosphere was oxidizing.
   b. was able to synthesize amino acids and macromolecules from reduced gases.
   c. did not require much energy or a continuous energy source to keep synthesizing.
   d. did not require water to produce organic molecules.
   e. used free oxygen as a reactant.

6. An unknown organism was found in a park. It was one-celled, had no nuclear membrane around its DNA, and contained no mitochondria and no chloroplasts. It belongs to the group:
   a. eukaryotes.
   b. vertebrates or plants.
   c. bacteria or archaea.
   d. plants or fungi.
   e. fungi.

7. Hydrothermal vents are theorized as sources for the origin of life because:
   a. the temperature of the water around them supports most life.
   b. most organic molecules undergo dehydration synthesis at high temperatures.
   c. the amino acids degrade in the colder water soon after synthesis.
   d. water is needed by living things.
   e. reducing conditions with needed molecules surround them.

8. The proposed first macromolecule for the beginning of life is:
   a. DNA to code the cell's activities.
   b. protein to be used in cell functions.
   c. ribozymes to act as information and catalytic molecules.
   d. H_2O as needed by all living things.
   e. chlorophyll for photosynthesis.

9. As part of the evolution of eukaryotic cell, endocytosis, the process of infolding of the plasma membrane, led to the formation of:
   a. chromosomes.
   b. the cell wall.
   c. ribosomes.
   d. the nuclear envelope.
   e. microtubules.

10. Which of the following is *not* part of the evidence supporting the theory of endosymbiosis: Both mitochondria and plastids:
    a. are each the size of many bacterial cells.
    b. have structures and biochemical reactions more like prokaryotes than eukaryotes.
    c. code mRNA, rRNA, and tRNA similar to prokaryotes.
    d. contain circular DNA.
    e. have DNA similar to nuclear DNA.

Questions for Discussion

1. What evidence supports the idea that life originated through inanimate chemical processes?

2. Explain, in terms of hydrophilic and hydrophobic interactions, how protocells might have formed in water from aggregations of lipids, proteins, and nucleic acids.

3. What conditions would likely be necessary for a planet located elsewhere in the universe to evolve life similar to that on Earth?

4. Most scientists agree that life on Earth can arise only from pre-existing life, but also that life could have originated spontaneously on the primordial Earth. Can you reconcile these seemingly contradictory statements?

Experimental Analysis

Suppose you discover a hot springs-fed pool on a remote mountain never before explored by humans. In the pool you find a cellular life form that appears to be prokaryotic. What experiments would you do to distinguish between the alternative hypotheses that this organism evolved on Earth from ancestral prokaryotes or is descended from a life form that arrived at that location in a meteorite?

Evolution Link

In the evolution unit, you learned how changes in the environment can foster evolutionary changes in biological systems. How have changing biological systems influenced the evolution of changes in Earth's physical environment?

How Would You Vote?

Private companies make millions of dollars selling an enzyme first isolated from cells in Yellowstone National Park. Should the federal government let private companies bioprospect within the boundaries of national parks, as long as it shares in the profits from any discoveries? Go to www.thomsonedu.com/login to investigate both sides of the issue and then vote.

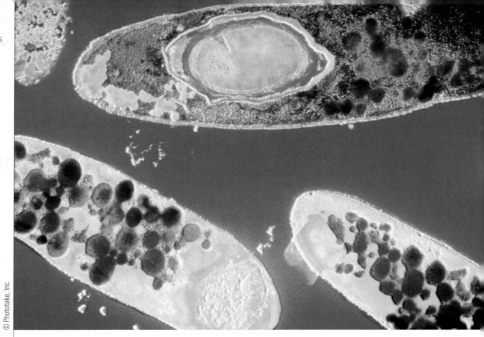

The bacterium *Clostridium butyricum*, one of the *Clostridium* species that produces the toxin botulin (colorized TEM).

© Phototake, Inc.

STUDY PLAN

25.1 Prokaryotic Structure and Function

Prokaryotes are simple in structure compared with eukaryotic cells

Prokaryotes have the greatest metabolic diversity of all living organisms

Prokaryotes differ in whether oxygen can be used in their metabolism

Prokaryotes fix and metabolize nitrogen

Prokaryotes reproduce asexually or, rarely, by a form of sexual reproduction

In nature, bacteria may live in communities attached to a surface

25.2 The Domain Bacteria

Molecular studies reveal more than a dozen evolutionary branches in the Bacteria

Bacteria cause diseases by several mechanisms

Pathogenic bacteria commonly develop resistance to antibiotics

25.3 The Domain Archaea

Archaea have some unique characteristics

Molecular studies reveal three evolutionary branches in the Archaea

25.4 Viruses, Viroids, and Prions

Viral structure is reduced to the minimum necessary to transmit nucleic acid molecules from one host cell to another

Viruses infect bacterial, animal, and plant cells by similar pathways

Viral infections are typically difficult to treat

Viruses may have evolved from fragments of cellular DNA or RNA

Viroids and prions are infective agents even simpler in structure than viruses

25 Prokaryotes and Viruses

WHY IT MATTERS

You wait in line with anticipation at a fast-food restaurant, biding your time until you reach the counter and get your hamburger. Somewhere in the back of your mind may be the worry that the hamburger will contain bacteria that could make you sick or even cost you your life. The hamburger you receive will be well done, almost to the crispy stage, because of that fear. Not too many years ago, people were sickened, and a few even died, because their fast-food hamburgers were contaminated by a pathogenic strain of the bacterium *Escherichia coli,* the normally harmless bacteria that inhabit our intestinal tract. Since then, fast-food restaurants have cooked their hamburgers well beyond the point required to kill any lurking *E. coli* or other pathogenic bacteria.

The bacterium *E. coli* is a prokaryote, an organism lacking a true nucleus. Prokaryotes, the main topic of this chapter, are the smallest organisms of the world **(Figure 25.1).** Few species are more than 1 to 2 μm long; from 500 to 1000 of them would fit side by side across the dot above this letter "i."

Prokaryotes are small, but their total collective mass (their *biomass*) on Earth may be greater than that of all plant life. They colonize

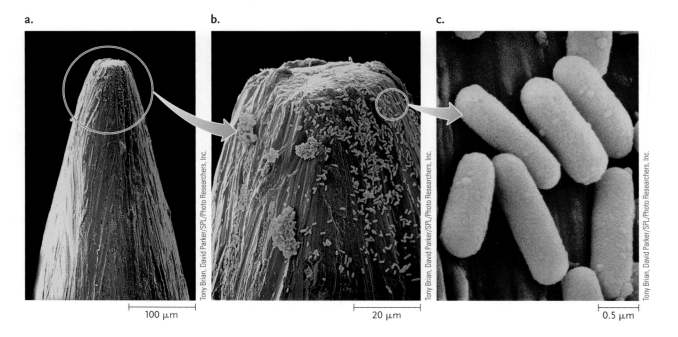

100 μm   20 μm   0.5 μm

Tony Brian, David Parker/SPL/Photo Researchers, Inc.

Figure 25.1

Bacillus bacteria on the point of a pin. Cells magnified **(a)** 70 times, **(b)** 350 times, and **(c)** 14,000 times.

every niche on Earth that supports life, meaning that they are found essentially everywhere. Huge numbers of bacteria inhabit surfaces and cavities of the human body, including the skin, the mouth and nasal passages, the large intestine, and the vagina. Collectively, the bacteria in and on the human body outnumber all the cells in the body.

Biologists classify prokaryotes into two of the three domains of life, the **Archaea** and the **Bacteria** (the third domain, the **Eukarya**, includes all eukaryotes). Bacteria are the prokaryotic organisms most familiar to us, including many types responsible for diseases of humans and other animals and many other types found in a wide variety of ecosystems. Many of the Archaea (*archaios* = ancient) live under conditions so extreme, including high salinity, acidity, or temperature, that their environments cannot be tolerated by other organisms, including bacteria.

As a group, prokaryotes have a wide range of metabolic capabilities. Their metabolic activities are crucial for maintenance of the biosphere. In particular, prokaryotes are the key players in the life-sustaining recycling of the elements carbon, nitrogen, and oxygen, and this recycling is necessary to sustain life. For example, prokaryotes are involved in breaking down organic material in dead plants and animals, releasing carbon dioxide that is used for plant growth. Prokaryotes are also the only living source of nitrogen, an element essential for all life. And a significant amount of the oxygen in the atmosphere originates from bacterial photosynthesis. An illustration of prokaryotes' importance is Biosphere 2, an attempt by scientists to build a completely closed ecosystem in Arizona. The attempt failed, in part because the researchers did not have a complete enough understanding of the activities of the microorganisms in the soil. Through respiration by soil microorganisms, the oxygen level in the Biosphere

structure decreased to lower-than-expected levels and the ecosystem ceased to be self-sustaining. This small-scale example illustrates the essential role of prokaryotes in enabling life of all forms to exist.

Prokaryotes also have a great impact on the lives of humans. Among other things, they are important for the production of certain foods, they carry out chemical reactions that are of importance in industry, they are used for the production of pharmaceutical products, they cause diseases, and they are used for bioremediation of polluted sites.

Viruses, the other subject of this chapter, are also extremely important in the biosphere. Smaller still than prokaryotes, viruses are present in the environment in even greater numbers than bacteria. In some aquatic habitats, viruses that infect bacteria alone exist at concentrations approaching 100 million per milliliter! Viruses are classified separately from the three domains of life because they are considered to be nonliving. However, viruses of one kind or another can infect the cells of just about every kind of living organism.

25.1 Prokaryotic Structure and Function

Prokaryotes show great diversity in their ability to colonize areas that can sustain life. Their cells are small, but relatively complex in organization. For instance, although they do not have a membrane-bound nucleus or organelles, their DNA and some proteins are localized in particular places. They vary in how their cell membrane is protected, and some species have specialized surface structures that protect them from their environment or that enable them to move actively. Prokaryotes also show great diversity in the

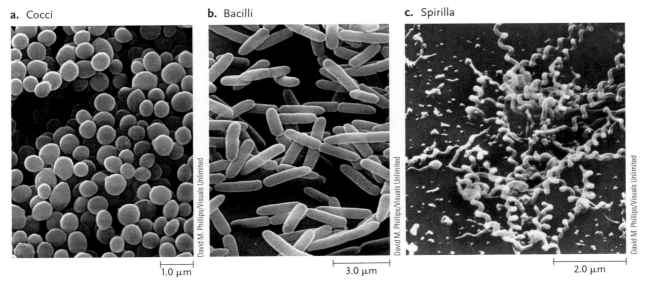

a. Cocci   **b.** Bacilli   **c.** Spirilla

1.0 μm   3.0 μm   2.0 μm

Figure 25.2

Common shapes among prokaryotes. **(a)** Scanning electron microscope (SEM) image of *Micrococcus*, a coccus bacterium; **(b)** SEM of *Salmonella*, a bacillus bacterium; **(c)** SEM of *Spiroplasma*, a spiral prokaryote of the spirillum type.

ways they obtain energy and in their metabolic activities.

The diversity of prokaryotes has arisen through rapid adaptation to their environments as a result of evolution by natural selection. Genetic variability in prokaryotic populations, the basis for this rapid adaptation, derives largely from mutation, and to a lesser degree from transfer of genes between organisms by transformation, transduction, and conjugation (see Chapter 17). Since prokaryotes have much shorter generation times than eukaryotes, and small genomes (roughly 1000 times smaller than an average eukaryote), prokaryotes have roughly 1000 times more mutations per gene, per unit time, per individual than is the case for eukaryotes. Further, prokaryotes typically have much larger population sizes than eukaryotes, contributing to their greater genetic variability. In short, prokaryotes have an enormous capacity to adapt and this has been key to their evolutionary success.

Prokaryotes Are Simple in Structure Compared with Eukaryotic Cells

Prokaryote cells examined under an electron microscope typically reveal little more than a cell wall and plasma membrane surrounding a cytoplasm with DNA concentrated in one region and ribosomes scattered throughout. They have no cytoplasmic organelles equivalent to the mitochondria, chloroplasts, endoplasmic reticulum, or Golgi complex of eukaryotic cells. With few exceptions, the reactions carried out by these organelles in eukaryotes are distributed between the cytoplasmic solution and the plasma membrane in prokaryotes.

Three shapes are common among prokaryotes: spherical, rodlike, and spiral **(Figure 25.2).** The spherical prokaryotes are **cocci** (singular, *coccus* = berry). Cylindrical or rod-shaped prokaryotes are **bacilli** (singular, *bacillus* = small staff or rod). The spiral prokaryotes are the **vibrios** (*vibrare* = to vibrate), which are curved and commalike, and the **spirilla** (singular, spirillum), which are twisted helically like a corkscrew. Among the prokaryotes of all structural types are some that live singly and others that link into chains or aggregates of cells.

Internal Structures. The genome of most prokaryotes consists of a single, circular DNA molecule called the *prokaryotic chromosome*. There are exceptions: a few bacterial species, for example the causative agent of Lyme disease *(Borrelia borgdorfri)*, have a linear chromosome. Genome sequencing projects have shown that the range of genome sizes among bacteria and archaeans is about 20-fold, with the smallest genome, that of *Mycoplasma genitalium,* being about 580,000 bp. In all prokaryotes, the chromosome is packed into an area of the cell called the **nucleoid.** There is no nucleolus in the nucleoid, and it has no boundary membranes equivalent to the nuclear envelope of eukaryotes **(Figure 25.3).**

Besides the DNA of the nucleoid, many prokaryotes also contain small circles of DNA called **plasmids,** distributed in the cytoplasm. The plasmids, which often contain genes with functions that supplement those in the nucleoid, contain a replication origin that allows them to replicate along with the nucleoid DNA and be passed on during cell division (see Section 14.5).

Prokaryotic ribosomes are smaller than eukaryotic ribosomes and contain fewer proteins and RNA molecules. Archaeal ribosomes resemble those of bacteria

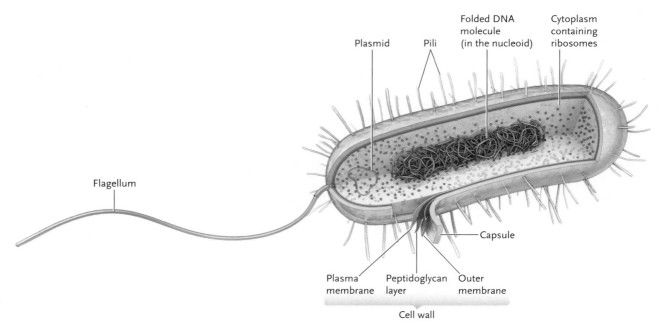

Plasmid   Pili   Folded DNA molecule (in the nucleoid)   Cytoplasm containing ribosomes

Flagellum

Capsule

Plasma membrane   Peptidoglycan layer   Outer membrane

Cell wall

Figure 25.3
The structures of a bacterial cell.

in size, but differ in structure. Scientists have demonstrated that, with some differences in detail, bacterial ribosomes carry out protein synthesis by the same mechanisms as those of eukaryotes (see Section 15.4). Interestingly, protein synthesis in archaeans is a combination of bacterial and eukaryotic processes, with some unique archaeal features. As a result, antibiotics that stop bacterial infections by targeting ribosome activity do not stop protein synthesis of archaeans.

Some prokaryotes are capable of photosynthesis. These microorganisms have membranous structures corresponding to those that carry out photosynthesis in plants, but they are organized differently.

The cytoplasm of many prokaryotes also contains storage granules holding glycogen, lipids, phosphates, or other materials. The stored material is used as an energy reserve or a source of building blocks for synthetic reactions.

Prokaryotic Cell Walls. All prokaryotic cells are bounded by a plasma membrane. This membrane must withstand both high intracellular osmotic pressures and the action of natural chemicals in the environment that have detergent properties. Most prokaryotes have one or more layers of materials coating the plasma membrane that provide the necessary protection.

Bacteria typically are surrounded by a cell wall that lies outside the plasma membrane. The primary structural molecules of bacterial cell walls are **peptidoglycans,** polymeric substances formed from a polysaccharide backbone tied together by short polypeptides. The peptidoglycans vary in chemical structure among different bacterial species.

Differences in bacterial cell wall composition are important clinically. In 1882, Hans Christian Gram, a Danish physician, developed a staining method to dis-

tinguish in bodily fluids two types of bacteria, each of which could cause pneumonia. In this **Gram stain technique,** an investigator treats bacteria with the dye crystal violet and then with iodine, which fixes the dye to the cell wall. Next the bacteria are washed with alcohol, and then treated with a second strain, either fuchsin or safranin. Bacteria that appear purple after these steps have retained the crystal violet stain; they are **Gram-positive.** Bacteria that appear pink after these steps have lost the crystal violet stain in the alcohol wash and are stained pink with the second dye; they are **Gram-negative.** (Gram-positive cells also react with the second dye, but the stain does not affect the color imparted by the crystal violet.)

The staining difference reflects differences in the cell walls of the bacteria **(Figure 25.4).** The cell wall of Gram-positive bacteria consists of a thick peptidoglycan layer (see Figure 25.4a). In contrast, the cell wall of Gram-negative bacteria consists of a thin layer of peptidoglycans (see Figure 25.4b). Outside of the thin cell wall is an additional boundary membrane, called the **outer membrane,** which covers the peptidoglycan layer. The outer membrane contains **lipopolysaccharides,** assembled from lipid and polysaccharide subunits found nowhere else in nature. The outer membrane protects Gram-negative bacteria from potentially damaging substances in their environment. For example, the outer membrane of *E. coli* protects it from the detergent effects of bile released into the intestinal tract, which otherwise would lyse (break open) the bacterium and kill it.

Rapidly distinguishing between Gram-positive and Gram-negative bacteria is important for determining the first line of treatment for bacterial-caused human diseases. Most pathogenic bacteria are Gram-negative species; their outer membrane protects them against the body's defense systems and blocks the en-

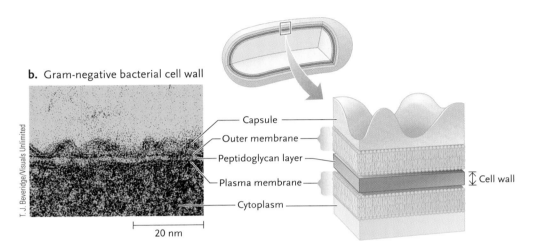

a. Gram-positive bacterial cell wall

Peptidoglycan layer

Plasma membrane

Cytoplasm

20 nm

Capsule may be present

Cell wall

T. J. Beveridge/Visuals Unlimited

b. Gram-negative bacterial cell wall

Capsule

Outer membrane

Peptidoglycan layer

Plasma membrane

Cytoplasm

20 nm

Cell wall

T. J. Beveridge/Visuals Unlimited

Figure 25.4
Cell wall structure in Gram-positive and Gram-negative bacteria. **(a)** The thick cell wall in Gram-positive bacteria. **(b)** The thin cell wall of Gram-negative bacteria.

try of drugs such as antibiotics. For example, the antibiotic penicillin blocks new bacterial cell wall formation by inhibiting peptidoglycan crosslinking. The weakened cell wall soon leads to the death of the bacterium. Penicillin is effective against Gram-positive pathogens, but it is less effective against Gram-negative pathogens because their outer membrane inhibits entry of the antibiotic.

Many Gram-positive and Gram-negative bacteria are surrounded by a slime coat typically composed of polysaccharides. When the slime is attached to the cells, it is a **capsule (Figure 25.5),** and when it is loosely associated with the cells, it is a **slime layer,** although there is no sharp distinction between the two. Depending on the species, the capsule ranges from a layer that is thinner than the cell wall to many times thicker than the entire cell. Slime typically is essential for survival of the bacteria in natural environments. For example, the slime helps protect the cells from desiccation and antibiotics.

In many bacteria, the capsule prevents bacterial viruses and molecules such as enzymes, antibiotics, and antibodies from reaching the cell surface. In many pathogenic bacteria, the presence or absence of the protective capsule differentiates infective from noninfective forms. For example, normal *Streptococcus pneumoniae* bacteria are capsulated and are virulent, caus-

ing severe pneumonia in humans and other mammals. Mutant *S. pneumoniae* without capsules are nonvirulent and can easily be eliminated by the body's immune system if they are injected into mice or other animals (see Section 14.1).

Flagella and Pili. Many bacteria and archaeans can move actively through liquids and across wet surfaces. The most common mechanism for movement involves

Figure 25.5
The capsule surrounding the cell wall of *Rhizobium*, a Gram-negative soil bacterium.

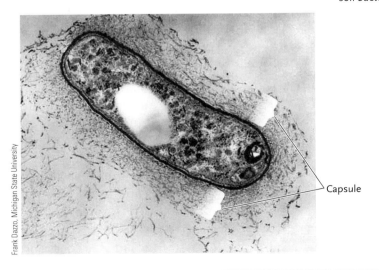

Capsule

Frank Dazzo, Michigan State University

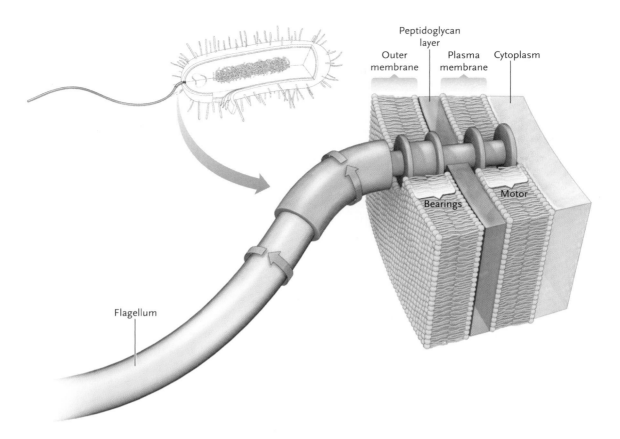

Figure 25.6
A flagellum of a Gram-negative bacterium. A proton (H⁺) gradient drives the motor, which rotates the flagellum in a counterclockwise direction.

the action of **flagella** (singular, flagellum, meaning whip) extending from the cell wall **(Figure 25.6).** These flagella are much smaller and simpler than the flagella of eukaryotic cells and contain no microtubules (eukaryotic flagella are discussed in Section 5.3).

Bacterial flagella consist of a helical fiber of protein that rotates in a socket in the cell wall and plasma membrane, much like the propeller of a boat. The rotation, produced by what is essentially a tiny electric motor, pushes the cell through liquid. The motor is powered by a gradient of hydrogen or sodium ions, which flow through it as positive charges, creating an electrical repulsion that makes the flagellum rotate.

Archaeal flagella are analogous, not homologous, to bacterial flagella. That is, they carry out the same function, but the genes for the two types of flagellar systems are different.

Some bacteria and archaeans have rigid shafts of protein called **pili** (singular, pilus) extending from their cell walls **(Figure 25.7).** Among bacteria, pili are characteristic primarily of Gram-negative bacteria; relatively few Gram-positive bacteria produce these structures. A recognition protein at the tip of a pilus allows bacterial cells to adhere to other cells. One type, called *sex pili,* allows bacterial cells to adhere to each other as a prelude to conjugation, a primitive form of sexual reproduction (see Section 17.1). Other types help bacteria to bind to animal cells. For example, *Neisseria gonorrhoeae,* the Gram-negative bacterium

that causes gonorrhea, has pili that allow it to attach to cells of the throat, eye, urogenital tract, or rectum in humans.

In sum, prokaryotes are simpler and less structurally diverse than eukaryotic cells. However, bacteria are much more diverse metabolically, as we will now explore.

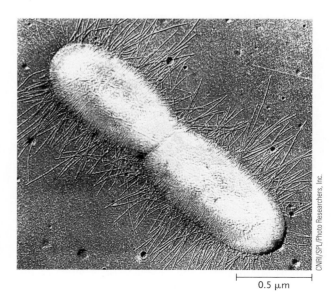

Figure 25.7
Pili extending from the surface of a dividing *E. coli* bacterium.

Prokaryotes Have the Greatest Metabolic Diversity of All Living Organisms

All organisms take in carbon and energy in some form, but prokaryotes show the greatest diversity in their modes of securing these resources (**Figure 25.8**). Some prokaryotes are **autotrophs** (*auto* = self; *troph* = nourishment), meaning that they, like plants, obtain carbon from an inorganic molecule, CO_2. (Note that, while CO_2 contains a carbon atom, oxides containing carbon are considered inorganic molecules.) Others are **heterotrophs**, meaning that they, like humans and other animals, obtain carbon from organic molecules. Bacterial heterotrophs obtain carbon from the organic molecules of living hosts, or from organic molecules in the products, wastes, or remains of dead organisms.

Prokaryotes are also divided according to the source of the energy they use to drive biological activities. **Chemotrophs** (*chemo* = chemical; *troph* = nourishment) obtain energy by oxidizing inorganic or organic substances, while **phototrophs** obtain energy from light. Combining carbon source and energy gives us the following four types (see Figure 25.8):

1. **Chemoautotrophs:** Prokaryotic chemoautotrophs obtain energy by oxidizing inorganic substances such as hydrogen, iron, sulfur, ammonia, nitrites, and nitrates and use CO_2 as their carbon source. They use the electrons they remove in the oxidations to make organic molecules by reducing CO_2 or to provide energy for ATP synthesis (using an electron transfer system embedded in the plasma membrane). Chemoautotrophs occur widely among the prokaryotes, including many bacteria and most archaeans, but are not found among eukaryotes.
2. **Chemoheterotrophs:** Prokaryotic chemoheterotrophs oxidize organic molecules as their energy source and obtain carbon in organic form. They include most of the bacteria that cause disease in humans, domestic animals, and plants and many bacteria responsible for decomposing matter. They are the largest prokaryotic group in terms of numbers of species.
3. **Photoautotrophs:** Photoautotrophs are photosynthetic organisms that use light as their energy source and CO_2 as their carbon source. They include several groups of bacteria, for example, the *cyanobacteria,* the *green sulfur bacteria,* and the *purple sulfur bacteria,* as well as plants and many protists. The cyanobacteria use water as their source of electrons for reducing CO_2, while the two types of sulfur bacteria use sulfur or sulfur compounds.
4. **Photoheterotrophs:** Photoheterotrophs use light as their ultimate energy source but obtain carbon in organic form rather than as CO_2. Photoheterotrophs are limited to two groups of bacteria, the *green* and *purple nonsulfur bacteria.* "Nonsulfur"

		Energy source	
		Oxidation of molecules*	**Light**
Carbon source	CO_2	**CHEMOAUTOTROPH** Found in some bacteria and archaeans; not found in eukaryotes	**PHOTOAUTOTROPH** Found in some photosynthetic bacteria, in some protists, and in plants
	Organic molecules	**CHEMOHETEROTROPH** Include some bacteria and archaeans, and also in protists, fungi, animals, and plants	**PHOTOHETEROTROPH** Found in some photosynthetic bacteria

*Inorganic molecules for chemoautotrophs and organic molecules for chemoheterotrophs.

Figure 25.8

Modes of nutrition among Bacteria and Archaea. All four modes of nutrition occur in the Bacteria with chemoheterotrophs as the most common type; among the Archaea, chemoautotrophs are most common, while others are chemoheterotrophs.

indicates they are unable to oxidize sulfur or other inorganic substances as an ultimate source of electrons for reductions; instead, they use a variety of substrates, including H_2, alcohols, or organic acids.

Prokaryotes Differ in Whether Oxygen Can Be Used in Their Metabolism

Prokaryotes also differ in how their metabolic systems function with respect to oxygen (see Chapter 8). **Aerobes** require oxygen for cellular respiration (in other words, oxygen is the final electron acceptor for that process); **obligate aerobes** cannot grow without oxygen. **Anaerobes** do not require oxygen to live. **Obligate anaerobes** are poisoned by oxygen, and survive either by fermentation, in which organic molecules are the final electron acceptors (see Section 8.5), or by a form of respiration in which inorganic molecules such as nitrate ions (NO_3^-) or sulfate ions (SO_4^{2-}) are used as final electron acceptors. **Facultative anaerobes** use O_2 when it is present, but under anaerobic conditions, they live by fermentation.

Prokaryotes Fix and Metabolize Nitrogen

Nitrogen is a component of amino acids and nucleotides and, hence, is of vital importance for the cell. Prokaryotes are able to metabolize nitrogen in many forms. For example, a number of bacteria and archaeans are able to reduce atmospheric nitrogen (N_2, the major component of Earth's atmosphere) to ammonia (NH_3), a process called **nitrogen fixation.** The ammonia is quickly ionized to ammonium (NH_4^+), which the cell then uses in biosynthetic pathways to produce nitrogen-containing molecules such as amino acids and nucleic acids. Nitrogen fixation is an exclusively prokaryotic process and is the only means of replenishing the nitrogen sources used by most microorganisms and by all plants and animals. In other words, all organisms use nitrogen fixed by bacteria. Examples of nitrogen-fixing bacteria are some of the cyanobacteria and *Azotobacter* among free-living bacteria and *Rhizobium* among bacteria that are symbiotic with plants (see Chapter 33).

Not all bacteria convert fixed nitrogen directly into organic molecules. Some bacteria carry out **nitrification**, the conversion of ammonium (NH_4^+) to nitrate (NO_3^-). This is carried out in two steps by two types of *nitrifying bacteria*. One type of nitrifying bacteria converts ammonium to nitrite (NO_2^-) (for example, *Nitrosomonas*), while the other converts nitrite to nitrate (for example, *Nitrobacter*). Because of this specialization, both types of nitrifying bacteria are usually present in soils and water, with some converting ammonium to nitrite and others using that nitrite to produce nitrate. The nitrate can be used by plants and fungi to incorporate nitrogen into organic molecules. Animals obtain nitrogen in organic form by eating other organisms.

In sum, nitrification makes nitrogen available to many other organisms, including plants and animals and bacteria that cannot metabolize ammonia. You will learn more about nitrogen metabolism in connection with the nitrogen cycle (see Chapter 51). The metabolic versatility of the prokaryotes is one factor that accounts for their abundance and persistence on the planet; another factor is their impressive reproductive capacity.

Prokaryotes Reproduce Asexually or, Rarely, by a Form of Sexual Reproduction

In prokaryotes, asexual reproduction is the normal mode of reproduction. In this process, a parent cell divides by binary fission into two daughter cells that are exact genetic copies of the parent (see Figure 10.18).

Conjugation, in which two parent cells join or "mate," occurs in some bacterial and archaeal species. Conjugation depends upon genes carried by a plasmid that replicates separately from the prokaryotic chromosome. Usually only the plasmid is passed on during conjugation, but in some bacteria, the plasmid integrates into the chromosome of the host so that host genes transfer from one parent (donor) to the other (recipient). Genetic recombination then occurs, thereby achieving a prokaryotic form of sexual reproduction. The recombinant cell divides to produce daughter cells that differ in genetic information from either parent. (Conjugation and the transfer of host genes between bacterial cells is described in Section 17.1.)

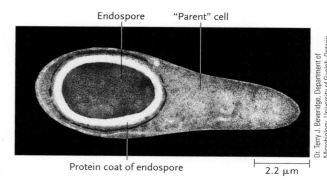

Endospore   "Parent" cell

Protein coat of endospore   2.2 μm

Dr. Terry J. Beveridge, Department of Microbiology, University of Guelph, Ontario, Canada/Biological Photo Service

Figure 25.9

A developing endospore of the bacterium *Clostridium tetani*, a dangerous pathogen that causes tetanus.

A small number of bacteria can produce an **endospore**, so-called because it develops *within* the cell **(Figure 25.9)**. The endospore, which typically develops when environmental conditions become unfavorable, is metabolically inactive and highly resistant to heat, desiccation, and attack by enzymes or other chemical agents. When an endospore forms, binary fission cuts the parent cell into parts of unequal size. The larger cell then envelops the smaller one and surrounds it with a tough, chemically resistant protein coat; the smaller cell develops into the endospore. Rupture of the larger cell releases the endospore to the environment. If environmental conditions become favorable for growth, the spore germinates: it becomes permeable, water enters the cell, its surface coat breaks, and the cell is released in a metabolically active form.

No one is certain how long endospores can survive. There are claims that endospores survive for thousands or millions of years, but the data are controversial.

In Nature, Bacteria May Live in Communities Attached to a Surface

Researchers grow prokaryotes as individuals in liquid cultures or as isolated colonies on solid media. The results from studies using pure cultures have been crucial in developing an understanding of, among many other things, the nature of the genetic material, DNA replication, gene expression, and gene regulation. But, since pure cultures are extremely rare in nature, some of the information learned from them may not apply to populations of prokaryotes in nature.

Researchers have discovered that, in nature, prokaryotes may live in communities where they interact in a variety of ways. The communities may consist of one or more species of bacteria, or archaeans, or both bacteria and archaeans. Eukaryotic microorganisms may also be in the communities. One important type of prokaryotic community is known as a **biofilm**, which consists of a complex aggregation of microorganisms attached to a surface. Benefits of biofilm formation to prokaryotes include adherence of the organisms to hospitable surfaces, the transfer of genes between species, and living off the products of other organisms in the biofilm. Biofilms form on any surface with sufficient water and nutrients for prokaryotes to grow. For instance, they may be found on lake surfaces, on rocks in freshwater or marine environments (making them slippery), surrounding plant roots and root hairs, and on animal tissues such as intestinal mucosa and teeth (human dental plaque is a biofilm).

Biofilms have practical consequences for humans, both beneficial and detrimental. On the beneficial side, for example, biofilms on solid supports are used in sewage treatment plants for processing organic matter before the water is discharged, and they can be effective in bioremediation (biological clean-up) of toxic organic molecules contaminating the groundwater. On the detrimental side, however, biofilms can be harmful to human health. For example, biofilms adhere to many kinds of surgical equipment and supplies, including catheters and synthetic implants such as pacemakers and artificial joints. When pathogenic bacteria are involved, infections occur. Those infections are difficult to treat, because pathogenic bacteria in a biofilm are up to 1000 times more resistant to antibiotics than are the same bacteria in liquid cultures. Other examples of medical conditions resulting from activities of biofilms include middle-ear infections, bacterial endocarditis (an infection of the heart's inner lining or the heart valves), and Legionnaire's disease (an acute respiratory infection caused by breathing in pieces of biofilms containing the pathogenic bacterium *Legionnella*).

How does a biofilm form? Imagine a surface, living or environmental, over which water containing nutrients is flowing **(Figure 25.10)**. The surface rapidly becomes coated with polymeric organic molecules from the liquid, such as polysaccharides or glycoproteins. Once the surface is conditioned with organic molecules, free bacteria attach in a reversible manner in a matter of seconds (see Figure 25.10, step 1). If the bacteria remain attached, the association may become irreversible (step 2), at which point the bacteria grow and divide on the surface (step 3). Next, the physiology of the bacteria changes and the cells begin to secrete *extracellular polymer substances* (EPS), a slimy, gluelike substance similar to the molecules found in bacterial slime layers. EPS extends between cells in the mixture, forming a matrix that binds cells to each other and anchors the complex to the surface, thereby establishing the biofilm (step 4). The slime layer entraps a variety of materials, such as dead cells and insoluble minerals. Over time, other organisms are attracted to and join the biofilm; depending on the environment, these may include other bacterial species, algae, fungi, or protozoa, producing diverse microbial communities (step 5).

Genomic and proteomic studies have shown that the changes in prokaryote physiology accompanying the formation of a biofilm result from marked changes in the prokaryote's gene expression pattern. In effect, the prokaryote becomes a significantly different organism. This change has large implications when pathogenic bacteria are involved, for example, because most research on the control of those bacteria is done with liquid cultures. The challenge now is to devise new treatment strategies for biofilm-caused diseases. If we can gain a better understanding of the genetic changes involved in the transition from free-floating to biofilm state, then perhaps we can devise treatments that will switch the bacteria back to the free-living state, where they are more susceptible to antibiotics.

In sum, we must recognize that rather than living as individuals as once was thought, prokaryotes typically live in communities in nature. Much remains to be learned about how bacteria form a biofilm, how the change in gene expression during the transition is regulated, and how they interact.

In the next two sections, we describe the major groups of prokaryotes.

Figure 25.10
Steps in the formation of a biofilm.

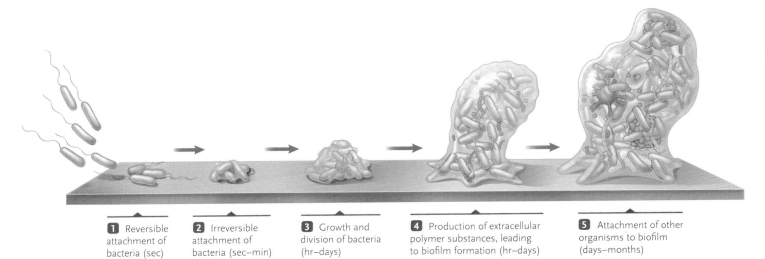

1 Reversible attachment of bacteria (sec)

2 Irreversible attachment of bacteria (sec–min)

3 Growth and division of bacteria (hr–days)

4 Production of extracellular polymer substances, leading to biofilm formation (hr–days)

5 Attachment of other organisms to biofilm (days–months)

25.2 The Domain Bacteria

Prokaryote classification has been revolutionized by molecular techniques that allow researchers to obtain and compare bacterial DNA, RNA, and protein sequences as tests of relatedness and evolutionary origin. Ribosomal RNA (rRNA) sequences, which are present in all organisms, have been most widely used in the evolutionary studies of prokaryotes. Under the assumption that mutations causing sequence changes occur at constant rates, researchers use the degree of sequence divergence to estimate how much time has passed since any two species shared the same ancestor (see Section 23.6). The sequencing studies thus provide a means to trace the evolutionary origins of prokaryotes and to place them in taxonomic groups. In this way, prokaryotes have been classified into the domains Bacteria and Archaea (the Eukarya is the third domain of life) **(Figure 25.11).** Researchers have identified several evolutionary branches within each prokaryote domain. In the future, full genomic sequences will

Figure 25.11
An abbreviated phylogenetic tree of prokaryotes.

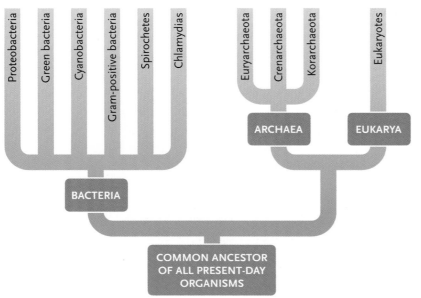

likely be compared to refine this taxonomic classification. We discuss the major groups of the domain Bacteria in this section and of the domain Archaea in the next section.

Molecular Studies Reveal More Than a Dozen Evolutionary Branches in the Bacteria

Sequencing studies reveal that bacteria have more than 12 distinct and separate evolutionary branches, variously called kingdoms, subkingdoms, phyla, or divisions. Although all these groups are of significance to science, medicine, and the human economy, we restrict our discussion to six that are particularly important—the proteobacteria, the green bacteria, the cyanobacteria, the Gram-positive bacteria, the spirochetes, and the chlamydias (see Figure 25.11).

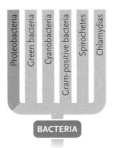

Proteobacteria: The Purple Bacteria and Their Relatives. The proteobacteria are a highly diverse group of Gram-negative bacteria that scientists hypothesize derive from a purple, photosynthesizing evolutionary ancestor. Many present-day species retain those characteristics, carrying out photosynthesis as either photoautotrophs (the purple sulfur bacteria) or photoheterotrophs (the purple nonsulfur bacteria). "Purple" refers to the color given to the cells by their photosynthetic pigment, a type of chlorophyll distinct from that of plants. Proteobacteria carry out a type of photosynthesis that does not use water as an electron donor and does not release oxygen as a by-product of photosynthesis.

Other present-day proteobacteria are chemoheterotrophs that are thought to have evolved as an evolutionary branch following the loss of photosynthetic capabilities in an early proteobacterium. The evolutionary ancestors of mitochondria are considered likely to have been ancient nonphotosynthetic proteobacteria.

Among the chemoheterotrophs classified with the proteobacteria are bacteria that cause human diseases such as bubonic plague, Legionnaire's disease, gonorrhea, and various forms of gastroenteritis and dysentery; bacterial plant pathogens that cause rots, scabs, and wilts; and the colon-inhabiting *E. coli* (shown dividing in Figure 25.7). The proteobacteria also include both free-living and symbiotic nitrogen-fixing bacteria.

Among the more unusual nonphotosynthetic proteobacteria are the myxobacteria, which form colonies held together by the slime they produce. Enzymes secreted by the colonies digest "prey"—other bacteria, primarily—that become stuck in the slime. When environmental conditions become unfavorable, as when soil nutrients or water are depleted, myxobacteria form a *fruiting body* **(Figure 25.12),** which contains clusters of

Figure 25.12
The fruiting body of *Chondromyces crocatus*, a myxobacterium. Cells of this species collect together to form the fruiting body.

Hans Reichenbach, Gesellschaft for Biotechnologische Forschung, Braunschweig, Germany

spores. When the fruiting body bursts, the spores disperse and form new colonies.

Green Bacteria. The green bacteria are a diverse group of Gram-negative photosynthesizers with photosynthetic pigments that give the cells a green color. The pigments are a form of chlorophyll distinct from the chlorophyll of plants. Like the purple bacteria, they do not release oxygen as a byproduct of photosynthesis. Green bacteria occur in two subgroups: green sulfur bacteria, which are photoautotrophs, and green nonsulfur bacteria, which are photoheterotrophs. The green sulfur bacteria are fairly closely related to the Archaea and are usually found in hot springs. The green nonsulfur bacteria are found typically in marine and high-salt environments.

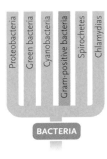

Cyanobacteria. The cyanobacteria **(Figure 25.13)** are Gram-negative photoautotrophs that have a blue-green color and carry out photosynthesis by the same pathways as eukaryotic algae and plants, using the same chlorophyll as in plants as their primary photosynthetic pigment. They release oxygen as a by-product of photosyn-

thesis. The first appearance of oxygen in quantity in Earth's atmosphere depended on the activities of ancient cyanobacteria.

The direct ancestors of present-day cyanobacteria were the first organisms to use the water-splitting reactions of photosynthesis. As such, they were critical to the appearance of oxygen in the atmosphere, which allowed the evolutionary development of aerobic organisms. Chloroplasts probably evolved from early cyanobacteria that were incorporated into the cytoplasm of primitive eukaryotes, which eventually gave rise to the algae and higher plants (see Section 24.3). Besides releasing oxygen, present-day cyanobacteria help fix nitrogen into organic compounds in aquatic habitats and in lichens, which are symbiotic organisms consisting of a cyanobacterium with a filamentous fungus (see Chapter 28).

Gram-Positive Bacteria. The large group of Gram-positive bacteria contains many species that live primarily as chemoheterotrophs. One species, *Bacillus subtilis,* is studied by biochemists and geneticists almost as extensively as is *E. coli.* A number of Gram-positive bacteria cause human diseases, including *Bacillus an-*

a.

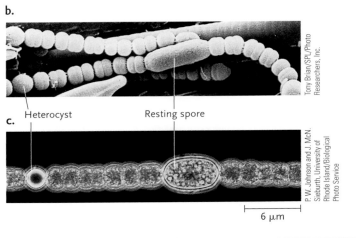

Dr. Jeremy Burgess/SPL/Photo Researchers, Inc.

b.

Tony Brian/SPL/Photo Researchers, Inc.

c.

Heterocyst   Resting spore

P. W. Johnson and J. McN. Sieburth, University of Rhode Island/Biological Photo Service

6 μm

Figure 25.13
Cyanobacteria.
(a) A population of cyanobacteria covering the surface of a pond. **(b)** and **(c)** Chains of cyanobacterial cells. Some cells in the chains form spores. The heterocyst is a specialized cell that fixes nitrogen.

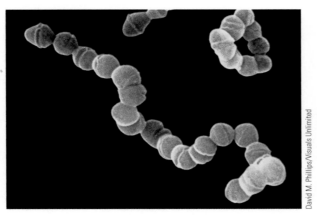

Figure 25.14
Streptococcus bacteria forming the long chains of cells typical of many species in this genus.

thracis, which causes anthrax and has been much in the news as a possible terrorist weapon; *Staphylococcus,* which causes some forms of food poisoning, skin infections such as pimples and boils, toxic shock syndrome, pneumonia, and meningitis; and *Streptococcus* **(Figure 25.14),** which causes strep throat, some forms of pneumonia, scarlet fever, and kidney infections. Nevertheless, some Gram-positive bacteria are beneficial; *Lactobacillus,* for example, carries out the lactic acid fermentation used in the production of pickles, sauerkraut, and yogurt. One unusual group of bacteria, the mycoplasmas, is placed among the Gram-positive bacteria by molecular studies even though they are Gram-negative. Their staining reaction reflects that they are naked cells that secondarily lost their cell walls in evolution. Some mycoplasmas, with diameters from 0.1 to 0.2 μm, are the smallest known cells.

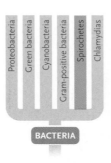

Spirochetes. The spirochetes are Gram-negative bacteria with helically spiraled bodies and an unusual form of movement in which bacterial flagella, embedded in the cytoplasm, cause the entire cell to twist in a corkscrew pattern. Their corkscrew movements enable them to move in viscous environments such as mud and sewage, where they are common. Two spirochetes, *Treponema denticola* and *Treponema vincentii,* are more or less

Figure 25.15
Cells of *Chlamydia trachomatis* inside a human cell. This bacterium is a major infectious cause of human eye and genital disease.

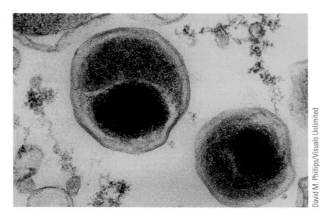

harmless inhabitants of the human mouth; another species, *Treponema pallidum,* is the cause of syphilis. Other pathogenic spirochetes cause relapsing fever and Lyme disease. Beneficial spirochetes in termite intestines aid in the digestion of plant fiber.

Chlamydias. The chlamydias are structurally unusual among bacteria because, although they are Gram-negative and have cell walls with a membrane outside of them, they lack peptidoglycans. All the known chlamydias are intracellular parasites that cause various diseases in animals. One bacterium of this group, *Chlamydia trachomatis* **(Figure 25.15),** is responsible for one of the most common sexually transmitted infections of the urinary and reproductive tracts of humans. The same bacterium causes trachoma, an infection of the cornea that is the leading cause of preventable blindness in humans.

Bacteria Cause Diseases by Several Mechanisms

As you have just learned, some bacteria cause diseases while others are beneficial. Here we focus on pathogenic bacteria.

Bacteria vary in the pathways by which they cause diseases. A number of bacterial lineages produce **exotoxins,** toxic proteins that leak from or are secreted from the bacterium and interfere with the biochemical processes of body cells in various ways. For example, the exotoxin of the Gram-positive bacterium *Clostridium botulinum* is found as a contaminant of poorly preserved foods, and causes botulism. The botulism exotoxin is one of the most poisonous substances known: a few nanograms can cause illness, and a few hundred grams could kill every human on Earth. It acts by interfering with the transmission of nerve impulses. The muscle paralysis produced by the exotoxin can be fatal if the muscles that control breathing are affected. Interestingly, the botulism exotoxin, with the brand name Botox, is used in low doses for the cosmetic removal of wrinkles, and in the treatment of migraine headaches, involuntary contraction of the eye muscles, and some other medical conditions.

Some other bacteria cause disease through **endotoxins.** Endotoxins are not released by living cells as exotoxins are; instead, they are lipopolysaccharides released from the outer membrane that surrounds cell walls when bacteria die and lyse. Endotoxins are natural components of the outer membrane of all Gram-negative bacteria, which include *E. coli, Salmonella,* and *Shigella.* These lipopolysaccharides cause disease by overstimulating the host's immune system, often triggering inflammation. Endotoxin release has different effects, depending on the bacterial species and the site of infection, that include typhoid or other fevers, diarrhea, and, in severe cases, organ failure and death. For example, *Salmonella typhi,* the cause of typhoid, enter the human intestines and penetrate the intestinal wall, eventually ending up in the lymph nodes. There they multiply, and some of the cells die and lyse, releasing endotoxins into the bloodstream. This both triggers the host's immune response and causes blood poisoning, a serious medical condition in which the circulatory system becomes dysfunctional. If the infection is not successfully treated, the condition can progress to multiple organ system failure and, eventually, death.

Some bacteria release **exoenzymes,** enzymatic proteins that digest plasma membranes and cause cells of the infected host to rupture and die. Exoenzymes may also digest extracellular materials such as collagen, causing connective tissue diseases. Some exoenzymes attack red or white blood cells, leading to anemias, impairment of the immune response, or interference with blood clotting. Among the bacteria that release exoenzymes are *Streptococcus, Staphylococcus,* and *Clostridium.* Necrotizing fasciitis (flesh-eating disease), the spectacularly destructive and rapid degeneration of subcutaneous tissues in the skin, is caused by an exoenzyme released by *Streptococcus* and some other bacteria.

Some of the ill effects of bacteria have little to do with exotoxins, endotoxins, or exoenzymes, but are caused purely by the body's responses to infection. The severe pneumonia caused by *Streptococcus pneumoniae,* for example, results from massive accumulation of fluid and white blood cells in the lungs in response to the infection. The white blood cells have little effect on the bacteria, however, because of the bacterial cell's protective capsule. As the fluid, white blood cells, and bacteria continue to accumulate, they block air passages in the lungs and severely impair breathing.

Pathogenic Bacteria Commonly Develop Resistance to Antibiotics

Antibiotics are routinely used to treat bacterial infections. These substances, produced as defensive molecules by some bacteria and fungi, or by chemical synthesis, kill or inhibit the growth of other microbial species. For example, streptomycins, produced by soil bacteria, block protein synthesis in their targets. Penicillins, produced by fungi, prevent formation of covalent bonds that hold bacterial cell walls together, weakening the wall and causing the cells to rupture.

Many pathogenic bacteria develop resistance to antibiotics through mutations that allow them to break down the drugs or otherwise counteract their effects (see *Why It Matters,* Chapter 20). Resistance is also acquired through genes carried on plasmids, picked up by conjugation or on DNA brought into pathogens by other pathways such as transformation and transduction (see Section 17.1). Taking antibiotics routinely in mild doses, or failing to complete a prescribed dosage, contribute to the development of resistance by selecting strains that can survive in the presence of the drug. Overprescription of antibiotics for colds and other virus-caused diseases can also promote bacterial resistance. That is, viruses are unaffected by antibiotics, but the presence of antibiotics in the system can lead to resistance as just described. Antibacterial agents that may promote resistance are also commonly included in such commercial products as soaps, detergents, and deodorants. Resistance is a form of evolutionary adaptation; antibiotics alter the bacterium's environment, conferring a reproductive advantage on those strains best adapted to the altered conditions.

The development of resistant strains has made tuberculosis, cholera, typhoid fever, gonorrhea, "staph," and other diseases caused by bacteria difficult to treat with antibiotics. For example, as recently as 1988, drug-resistant strains of *Streptococcus pneumoniae,* which causes pneumonia, meningitis, and middle-ear infections, were practically unheard of in the United States. Now, resistant strains of *S. pneumoniae* are common and increasingly difficult to treat.

In this section, you have seen that bacteria thrive in nearly every habitat on Earth, including the human body. However, some members of the second prokaryotic domain, the Archaea, the subject of the next section, live in habitats that are too forbidding even for the bacteria.

STUDY BREAK

1. What methodologies have been used to classify prokaryotes?
2. What were the likely characteristics of the evolutionary ancestor of present-day proteobacteria?
3. What are the differences between the way photosynthesis is carried out by photosynthetic Proteobacteria and by cyanobacteria?
4. What is an exotoxin, an endotoxin, and an exoenzyme, and how do they differ with respect to how they cause disease?

25.3 The Domain Archaea

Archaea were first discovered in 1977, and scientists believed they were bacteria. However, research showed that they have some eukaryotic features, some bacterial features, and some features that are unique to the group (also discussed in Section 24.3; **Table 25.1** compares the characteristics of Bacteria, Archaea, and Eukarya). Based on research by Carl Woese and his colleagues that compared their DNA and rRNA sequences with those of other organisms, Archaea were

Table 25.1 Characteristics of the Bacteria, Archaea, and Eukarya

Characteristic	Bacteria	Archaea	Eukarya
DNA arrangement	Single, circular in most, but some linear and/or multiple	Single, circular	Multiple linear molecules
Chromosomal proteins	Prokaryotic histone-like proteins	Five eukaryotic histones	Five eukaryotic histones
Genes arranged in operons	Yes	Yes	No
Nuclear envelope	No	No	Yes
Mitochondria	No	No	Yes
Chloroplasts	No	No	Yes
Peptidoglycans in cell wall	Present	Present but modified, or absent	Absent
Membrane lipids	Unbranched; linked by ester linkages	Branched; linked by ether linkages	Unbranched; linked by ester linkages
RNA polymerase	One type	Multiple types	Multiple types
Ribosomal proteins	Prokaryotic	Some prokaryotic, some eukaryotic	Eukaryotic
First amino acid placed in proteins	Formylmethionine	Methionine	Methionine
Aminoacyl-tRNA synthetases	Prokaryotic	Eukaryotic	Eukaryotic
Cell division proteins	Prokaryotic	Prokaryotic	Eukaryotic
Proteins of energy metabolism	Prokaryotic	Prokaryotic	Eukaryotic
Sensitivity to chloramphenicol and streptomycin	Yes	No	No

subsequently classified as a separate domain of life. (*Insights from the Molecular Revolution* describes the research that first revealed the complete DNA sequence of an archaean.) Scientists use sequencing studies and the archeans' unique characteristics to identify the organisms in this group.

Archaea Have Some Unique Characteristics

The first-studied Archaea were found in extreme environments, such as hot springs, hydrothermal vents on the ocean floor, and salt lakes **(Figure 25.16)**. For that reason, these prokaryotes were called *extremophiles*

("extreme lovers"). Subsequently archaeans have also been found living in normal environments; like bacteria, these are *mesophiles*.

Many Archaea are chemoautotrophs that obtain energy by oxidizing inorganic substances, while others are chemoheterotrophs that oxidize organic molecules. No known member of the Archaea has been shown to be pathogenic.

The cell structure of archaeans is basically prokaryotic. Among their unique characteristics are certain features of the plasma membrane and cell wall. The lipid molecules in archaean plasma membranes have a chemical bond between the hydrocarbon chains and

Figure 25.16
Typically extreme archaean habitats.
(a) Highly saline water in Great Salt Lake, Utah, colored red-purple by Archaea.
(b) Hot, sulfur-rich water in Emerald Pool, Yellowstone National Park, colored brightly by the oxidative activity of archaeans, which converts H_2S to elemental sulfur.

a.

Barry Rokeach

b.

© Alan L. Detrick/Science Source/Photo Researchers, Inc.

Extreme but Still in Between

In 1996 Carol J. Bult, Carl R. Woese, J. Craig Venter, and 37 other scientists at the Institute for Genomic Research obtained the complete DNA sequence of the archaean *Methanococcus jannaschii*. It was the first archaean genome to be sequenced. The results were obtained by sequencing randomly chosen overlapping DNA fragments from a DNA library until the entire genome was completed (the whole-genome shotgun approach, described in Section 18.3). Comparisons of the *Methanococcus* sequence with bacterial sequences and that of a eukaryote, the brewer's yeast *Saccharomyces cerevisiae*, give strong support to the proposal that the Archaea are a separate domain of living organisms.

Many archaeans have a lifestyle clearly different from those of the bacteria and eukaryotes, and *Methanococcus* is no exception. It was first discovered by the deep-sea submarine *Alvin* in a hot-water vent at a depth of more than 2600 m. It can live at temperatures as high as 94°C, only a few degrees less than the temperature of boiling water, and can tolerate pressure as high as 200 times the pressure of air at sea level.

The *Methanococcus* main genome, which includes 1,664,976 base pairs, was found to contain 1682 protein-encoding sequences. Two plasmids also contain protein-encoding genes, one plasmid with 44 genes and the other with 12. Of the total of 1738 protein-encoding sequences, only 38%—less than half—could be given probable identities based on sequence similarities with those of genes coding for known proteins in other organisms. Some of the sequences were similar to proteins of bacteria, and some to those of eukaryotes. Among the eukaryote-like genes are those encoding all five of the histone chromosomal proteins typical of eukaryotes and eukaryotic forms of the enzymes carrying out DNA replication and RNA transcription.

Other identified genes encode proteins unique to the Archaea, such as those encoding some enzymes and other proteins of the pathway reducing CO_2 to methane. Many other proteins with no known counterparts in the Bacteria or Eukarya are among the unidentified 62%, demonstrating the unique character of the Archaea and providing a rich lode of new proteins for mining by molecular biologists and other scientists.

---

glycerol unlike that in the plasma membranes of all other organisms. The difference is significant because the exceptional linkage is more resistant to disruption, making the plasma membranes of the Archaea more tolerant of the extreme environmental conditions under which many of these organisms live.

The cell walls of some archaeans are assembled from molecules related to the peptidoglycans, but with different molecular components and bonding structure. Others have walls assembled from proteins or polysaccharides instead of peptidoglycans. The cell walls of archaeans are as resistant to physical disruption as the plasma membrane is; some archaeans can be boiled in strong detergents without disruption. Different archaeans stain as either Gram-positive or Gram-negative.

Molecular Studies Reveal Three Evolutionary Branches in the Archaea

Based on differences between the rRNA coding sequences in their genomes, the domain Archaea is divided into three groups (see Figure 25.11). Two major groups, the **Euryarchaeota** and the **Crenarchaeota**, contain archaeans that have been cultured and examined in the laboratory. The third group, the **Korarchaeota**, has been recognized solely on the basis of rRNA coding sequences in DNA taken from environmental samples. A fourth group, the **Nanoarchaeota**, was proposed based on rRNA sequence analysis of a thermophilic archaean found in a symbiotic relationship with an-

other thermophilic archaean. Genome sequence comparisons have now shown that the Nanoarchaeota are most probably a subgroup of the Euryarchaeota.

Euryarchaeota. The Euryarchaeota are found in different extreme environments. They include methogens, which produce methane; extreme halophiles, which live in high concentrations of salt; and some extreme thermophiles, which live under high-temperature conditions.

Methanogens (methane generators) live in reducing environments **(Figure 25.17)**, and represent about one half of all known species of archaeans. All known methanogens belong to the Euryarchaeota. Examples are *Methanococcus* and *Methanobacterium*. Methanogens are obligate anaerobes,

Figure 25.17
A colony of the methanogenic archaean *Methanosarcina*, which lives in the sulfurous, waterlogged soils of marshes and swamps.

meaning they are killed by oxygen. They are found in the anoxic (oxygen-lacking) sediments of swamps, lakes, marshes, and sewage works, as well as in more moderate environments, such as the rumen of cattle, sheep, and camels; the large intestine of dogs and humans; and the hindguts of insects such as termites and cockroaches. Methanogens generate energy by converting at least ten different substrates such as carbon dioxide and hydrogen gas, methanol, or acetate into methane gas (CH_4), which is released into the atmosphere. A single species may use two or three substrates, for example converting carbon dioxide and hydrogen into methane and water.

Halophiles are salt-loving organisms. Extreme halophilic Archaea live in highly saline (salty) environments such as the Great Salt Lake or the Dead Sea, and on foods preserved by salting. Moreover, they require a high concentration of salt to live: they need a minimum NaCl concentration of about 1.5 M (about 9% solution), and can live in a fully saturated solution (5.5 M, or 32%). All known extreme halophilic Archaea belong to the Euryarchaeota. Most are aerobic chemoheterotrophs; they obtain energy from sugars, alcohols, and amino acids using pathways similar to those of bacteria. Examples are *Halobacterium* and *Natrosobacterium. Halobacterium,* like a number of extreme halophiles, uses light as a secondary energy source supplementing the oxidations that are its primary source of energy.

Extreme thermophiles live in extremely hot environments. Extreme thermophilic Archaea live in thermal areas such as ocean floor hydrothermal vents and hot springs such as those in Yellowstone National Park. Their optimal temperature range for growth is 70° to 95°C, approaching the boiling point of water. By comparison, no eukaryotic organism is known to live at a temperature higher than 60°C. Some extreme thermophiles are members of the Euryarchaeota. Some of them, such as *Pyrophilus,* are obligate anaerobes, while others, such as *Thermoplasma,* are facultative anaerobes that grow on a variety of organic compounds.

Crenarchaeota. The group Crenarchaeota contains most of the extreme thermophiles. Their optimal temperature range of 75° to 105°C is higher than that of the Euryarchaeota. Most are unable to grow at temperatures below 70°C. The most thermophilic member of the group, *Pyrobolus,* grows optimally at 106°C, but dies below 90°C. It can also grow at 113°C and survive an hour of autoclaving at 121°C! *Pyrobolus* lives in ocean floor hydrothermal vents where the pressure makes it possible to have temperatures above 100°C, the boiling point of water on Earth's surface.

Also within this group are **psychrophiles** ("cold loving"), organisms that grow optimally in cold temperatures in the range −10 to 20°C. These organisms are found mostly in the Antarctic and Arctic oceans, which are frozen most of the year, and in the intense cold at ocean depths.

Mesophilic members of the Crenarchaeota comprise a large part of plankton in cool, marine waters where they are food sources for other marine organisms. As yet, no individual species of these archaeans has been isolated and characterized.

Crenarchaeota archaeans exhibit a wide range of metabolism with regard to oxygen, including obligate anaerobes, facultative anaerobes, and aerobes.

Korarchaeota. The group Korarchaeota has been recognized solely on the basis of analyzing rRNA sequences in DNA obtained from marine and terrestrial hydrothermal environments, such as the Obsidian Pool at Yellowstone National Park. To date, no members of this group have been isolated and cultivated in the lab, and nothing is known about their physiology. They are the oldest lineage in the domain Archaea according to molecular data.

Thermophilic archaeans are important commercially. For example, enzymes from some species are used in basic and applied research, such as the thermostable DNA polymerase used in the polymerase chain reaction (PCR; see Chapter 18). Other enzymes from thermophilic archaeans are being tested for addition to detergents, where it is hoped that they will be active under high temperatures and acidic pH.

From the highly varied prokaryotes, we now turn to the viruses, which occur in most environments in even greater numbers than bacteria and archaeans. The next section also discusses prions and viroids, infective agents that are even simpler and smaller than viruses.

STUDY BREAK

1. What distinguishes members of the Archaea from members of the Bacteria and Eukarya?
2. How does a methanogen obtain its energy? In which group or groups of Archaea are methanogens found?
3. Where do extreme halophilic archaeans live? How do they obtain energy? In which group or groups of Archaea are the extreme halophiles found?
4. What are extreme thermophiles and psychrophiles?

25.4 Viruses, Viroids, and Prions

A **virus** (Latin for poison) is a biological particle that can infect the cells of a living organism. Viral infections usually have detrimental effects on their hosts.

The study of viruses is called *virology,* and researchers studying viruses are known as *virologists.*

All viruses contain a nucleic acid molecule (the genome), surrounded by a layer of protein called the **coat** or **capsid.** The complete virus particle is also called a **virion.** Viruses are considered nonliving primarily because they have no metabolic system of their own to provide energy for their life cycles; instead, they are dependent upon the host cells they infect for that function. That is, expression of virus genes in an infected host cell directs that cell to use its own machinery to duplicate the virus. However, their genome contains all the information required to convert host cells to the duplication of viruses of the same type. Although they are considered to be nonliving material, viruses are classified by the International Committee on Taxonomy of Viruses into orders, families, genera, and species using several criteria, including size and structure, type and number of nucleic acid molecules, method of replication of the nucleic acid molecules inside host cells, host range, and infective cycle. More than 4000 species of viruses have been classified into more than 80 families according to these criteria.

One or more kinds of viruses probably infect all living organisms. Usually a virus infects only a single species or a few closely related species. (A virus may even infect only one organ system, or a single tissue or cell type in its host.) However, some viruses are able to infect unrelated species, either naturally or after mutating. For example, some humans have contracted bird flu from being infected with the natural bird flu virus as a result of contact with virus-infected birds. At least 65 deaths of people in Asia have been attributed to bird flu. The bird flu virus has the potential to mutate to give efficient human-to-human transmission, raising significant concern about the possibility of a worldwide epidemic of bird flu virus infections of humans, with the possibility of millions of deaths. Of the viral families, 21 include viruses that cause human diseases. Viruses also cause diseases of wild and domestic animals; plant viruses cause annual losses of millions of tons of crops, especially cereals, potatoes, sugar beets, and sugar cane. (**Table 25.2** lists some important families that infect animals.)

The effects of viruses on the organisms they infect range from undetectable, through merely bothersome, to seriously debilitating or lethal. For instance, some viral infections of humans, such as those causing cold sores, chicken pox, and the common cold, are usually little more than a nuisance to healthy adults. Others, including AIDS, encephalitis, yellow fever, and smallpox, are among the most severe and deadly human diseases. While most viruses have detrimental effects, some may be considered beneficial. One of the primary reasons why bacteria do not completely overrun the planet is that they are destroyed in incredibly huge numbers by viruses known as **bacteriophages,** or **phages** for short (*phagein* = to eat; see Chapters 14 and 17 for the use of phages in important discoveries in

Table 25.2	Major Animal Viruses		
Viral Family	Envelope	Nucleic Acid	Diseases
Adenoviruses	No	ds DNA	Respiratory infections, tumors
Flaviviruses	Yes	ss RNA	Yellow fever, dengue, hepatitis C
Hepadnaviruses	Yes	ds DNA	Hepatitis B
Herpesviruses	Yes	ds DNA	
H. simplex I			Oral herpes, cold sores
H. simplex II			Genital herpes
Varicella-zoster			Chicken pox, shingles
Orthomyxovirus	Yes	ss RNA	Influenza
Papovaviruses	No	ds DNA	Benign and malignant warts
Paramyxoviruses	Yes	ss RNA	Measles, mumps, pneumonia
Picornaviruses	No	ss RNA	
Enteroviruses			Polio, hemorrhagic eye disease, gastroenteritis
Rhinoviruses			Common cold
Hepatitis A virus			Hepatitis A
Apthovirus			Foot-and-mouth disease in livestock
Poxviruses	Yes	ds DNA	Smallpox, cowpox
Retroviruses	Yes	ss RNA	
HTLV I, II			T-cell leukemia
HIV			AIDS
Rhabdoviruses	Yes	ss RNA	Rabies, other animal diseases

ss = single-stranded; ds = double-stranded.

molecular biology and bacterial genetics). Viruses also provide a natural means to control some insect pests.

Viral Structure Is Reduced to the Minimum Necessary to Transmit Nucleic Acid Molecules from One Host Cell to Another

The nucleic acid genome of a virus, depending on the viral type, may be either DNA or RNA, in either double- or single-stranded form. The nucleic acid molecule contains genes encoding proteins of the viral coat, and often also enzymes required to duplicate the genome. The simplest viral nucleic acid molecules contain only a few genes, but those of the most complex viruses may contain a hundred or more.

Some viruses have coats assembled from protein molecules of a single type; more complex viruses have coats made up of several different proteins—in some, 50 or more, including the recognition proteins that bind to host cells. The particles of some viruses also contain the DNA or RNA polymerase enzymes required for viral nucleic acid replication and an enzyme that attacks cell walls or membranes.

Most viruses take one of two basic structural forms, helical or polyhedral. In **helical viruses** the protein subunits assemble in a rodlike spiral around the genome **(Figure 25.18a)**. A number of viruses that infect plant cells are helical. In **polyhedral viruses** the coat proteins form triangular units that fit together like the parts of a geodesic sphere **(Figure 25.18b)**. The polyhedral viruses include forms that infect animals, plants, and bacteria. In some polyhedral viruses, protein spikes that provide host cell recognition extend from the corners where the facets fit together. Some viruses, the **enveloped viruses**, are covered by a surface membrane derived from their host cells; both enveloped helical and enveloped polyhedral viruses are known **(Figure 25.18c)**. For example, HIV (for *human immunodeficiency virus*), the virus that causes AIDS, is an enveloped polyhedral virus. Protein spikes extend through the membrane, giving the particle its recognition and adhesion functions.

A number of bacteriophages with DNA genomes, such as T2 (see Section 14.1), have a **tail** attached at one side of a polyhedral **head**, forming what is known as a **complex virus (Figure 25.18d)**. The genome is packed into the head; the tail is made up of proteins forming a collar, sheath, baseplate, and tail fibers. The tail has recognition proteins at its tip and, once attached to a host cell, functions as a sort of syringe that injects the DNA genome into the cell.

Viruses Infect Bacterial, Animal, and Plant Cells by Similar Pathways

Free viral particles move by random molecular motions until they contact the surface of a host cell. For infection to occur, the virus or the viral genome must then enter the cell. Inside the cell, typically the viral genes are expressed, leading to replication of the viral genome and assembly of progeny viruses. The viruses are then released from the host cell, a process that often ruptures the host cell, killing it.

Infection of Bacterial Cells. Bacteriophages vary as to whether they have a DNA or an RNA genome, and whether that nucleic acid is double-stranded or single-stranded. A DNA bacteriophage such as the virulent phage T2 (see Figure 25.18d) infects a bacterial cell and goes through the *lytic cycle,* in which the host cell is killed in each cycle of infection (described in Section 17.2 and Figure 17.9). In brief review, the lytic cycle of phage T2 is as follows: After the phage attaches to a host cell, an enzyme present in the baseplate of the viral coat, *lysozyme,* digests a hole in the cell wall through which the DNA of the phage enters the bacterium while the proteins of the viral coat remain outside. Once inside the bacterium, expression of phage genes directs the replication of the phage DNA, synthesis of phage proteins, and assembly of progeny phage particles. Next, the phage directs synthesis of a phage-encoded lysozyme enzyme that lyses the bacterial cell wall, causing the cell to rupture and releasing the progeny phages to the surroundings where they can infect other bacteria.

Some bacteriophages alternate between a lytic cycle and a *lysogenic cycle,* in which the viral DNA inserts into the host cell DNA and production of new viral particles is delayed (see Section 17.2). During the lysogenic cycle, the integrated viral DNA, known as the *prophage,* remains partially or completely inactive, but is replicated and passed on with the host DNA to all descendants of the infected cell. In response to certain environmental signals, the prophage loops out of the chromosome and the lytic cycle of the phage proceeds.

Infection of Animal Cells. Viruses infecting animal cells follow a similar pattern except that both the viral coat and genome, which is DNA or RNA depending

Figure 25.18
Viral structure. The tobacco mosaic virus in **(a)** assembles from more than 2000 identical protein subunits.

a. Helical virus (tobacco mosaic virus)

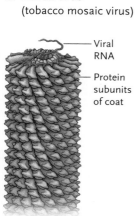
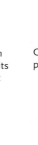

Viral RNA

Protein subunits of coat

b. Polyhedral virus (adenovirus)

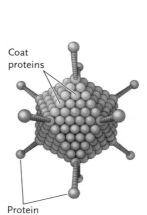

Coat proteins

Protein spikes

c. Enveloped virus (HIV)

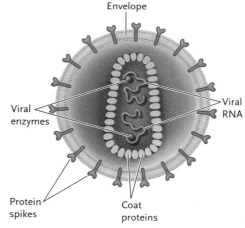

Envelope

Viral enzymes

Viral RNA

Protein spikes

Coat proteins

d. Complex polyhedral virus (T-even bacteriophage)

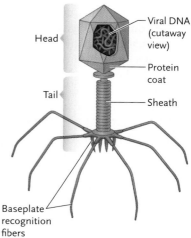

Head

Tail

Viral DNA (cutaway view)

Protein coat

Sheath

Baseplate recognition fibers

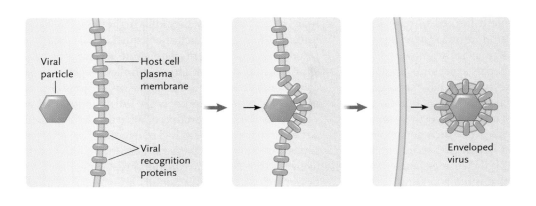

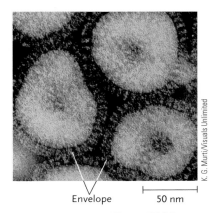

Envelope   50 nm

Figure 25.19
How enveloped viruses acquire their envelope. The micrograph shows the influenza virus with its envelope. Note the recognition proteins studding the envelope.

on the viral type, enter a host cell. Depending on the virus, removal of the coat to release the genome occurs during or after cell entry; the envelope does not enter the cell.

Viruses without an envelope, such as adenovirus (DNA genome) and poliovirus (RNA genome), bind by their recognition proteins to the plasma membrane and are then taken into the host cell by endocytosis. The virus coat and genome of enveloped viruses, such as herpesviruses and pox viruses (DNA genome), and HIV and influenza virus (RNA genome), enter the host cell by fusion of their envelope with the host cell plasma membrane.

Once inside the host cell, the genome directs the synthesis of additional viral particles by basically the same pathways as bacterial viruses. Newly completed viruses that do not acquire an envelope are released by rupture of the cell's plasma membrane, typically killing the cell. In contrast, most enveloped viruses receive their envelope as they pass through the plasma membrane, usually without breaking the membrane **(Figure 25.19).** This pattern of viral release typically does not injure the host cell.

The vast majority of animal virus infections are asymptomatic; pathogenesis—the causation of disease—is of no value to the virus. However, there are a number of pathogenic viruses, and they cause diseases in a variety of ways. Some viruses, for instance, cause cell death when progeny viruses are released from the cell. This can lead to massive cell death, destroying vital tissues such as nervous tissue or white or red blood cells, or causing lesions such as ulcers in skin and mucous membranes. Some other viruses release cellular molecules when infected cells break down, which can induce fever and inflammation. Yet other viruses alter gene function when they insert into the host cell DNA, leading to cancer and other abnormalities.

Some animal viruses enter a **latent phase** in which the virus remains in the cell in an inactive form: the viral nucleic acid is present in the cytoplasm or nuclear DNA, but no complete viral particles or viral release can be detected. (The latent phase is similar to the lysogenic cycle that is part of the life cycle of some bacteriophages.) At some point, the latent phase may end

as the viral DNA is replicated in quantity, coat proteins are made, and completed viral particles are released from the cell. The herpesviruses causing oral and genital ulcers in humans remains in a latent phase of this type in the cytoplasm of some body cells for the life of the individual. At times, particularly during periods of metabolic stress, the virus becomes active in some cells, directing viral replication and causing ulcers to form as cells break down during viral release.

Infection of Plant Cells. Plant viruses may be rodlike or polyhedral; although most include RNA as their nucleic acid, some contain DNA. None of the known plant viruses have envelopes. Plant viruses enter cells through mechanical injuries to leaves and stems or through transmission from plant to plant by biting and feeding insects such as leaf hoppers and aphids, by nematode worms, and by pollen during fertilization. Plant viruses can also be transmitted from generation to generation in seeds. Once inside a cell, plant viruses replicate in the same patterns as animal viruses. Within plants, virus particles pass from infected to healthy cells through plasmodesmata, the openings in cell walls that directly connect plant cells, and through the vascular system.

Plant viruses are generally named and classified by the type of plant they infect and their most visible effects. *Tomato bushy stunt virus,* for example, causes dwarfing and overgrowth of leaves and stems of tomato plants, and *tobacco mosaic virus* causes a mosaic-like pattern of spots on leaves of tobacco plants. Most species of crop plants can be infected by at least one destructive virus.

The tobacco mosaic virus was the first virus to be isolated, crystallized, disassembled, and reassembled in the test tube, and the first viral structure to be established in full molecular detail (see Figure 25.18a).

Viral Infections Are Typically Difficult to Treat

Viral infections are unaffected by the antibiotics and other treatment methods used for bacterial infections. As a result, many viral infections are allowed to run their course, with treatment limited to relieving the symptoms while the natural immune defenses of the patient attack the virus. Some viruses, however, cause serious and sometimes deadly symptoms upon infection and, consequently, researchers have spent considerable effort to develop antiviral drugs to treat them. Many of these drugs fight the virus directly by targeting a stage of the viral life cycle; they include amantidine (inhibits hepatitis B and hepatitis C virus entry into cells), acyclovir (analog of nucleosides [*analog* means it is chemically similar] that inhibits replication of the genomes of herpesviruses), and zanamivir (inhibits release of influenza virus particles from cells).

The influenza virus illustrates the difficulties inherent in treating viral diseases. The influenza type A virus (see Figure 25.19) causes flu epidemics that sweep over the world each year. It has many unusual features that tend to keep it a step ahead of efforts to counteract its infections. One is the genome of the virus, which consists of eight separate pieces of RNA. When two different influenza viruses infect the same individual, the pieces can assemble in random combinations derived from either parent virus. The new combinations can change the protein coat of the virus, making it unrecognizable to antibodies developed against either parent virus. Antibodies are highly specific protein molecules produced by the immune system that recognize and bind to foreign proteins originating from a pathogen (see Chapter 43). The invisibility to antibodies means that new virus strains can infect people who have already had the flu or who have had flu shots that stimulate the formation of antibodies effective only against the earlier strains of the virus. Random mutations in the RNA genome of the virus add to the variations in the coat proteins that make previously formed antibodies ineffective.

Luckily, most flu infections, although debilitating, are not dangerous, except for individuals who are very young or very old or who have compromised immune systems. However, some flu epidemics have been devastatingly lethal. The worst recorded example is the epidemic of 1918. A strain of influenza virus known as the Spanish flu infected approximately 20% of the world's 1.8 billion people, killing about 50 million.

The exact type of virus responsible for this deadly epidemic was finally determined in 2005, when researchers led by Jeffrey Taubenberger at the U.S. Armed Forces Institute of Pathology reconstructed the genome of the virus and produced infectious, pathogenic viruses in the laboratory. The team worked mainly with tissue from a 1918 flu victim found in permafrost in Alaska. Using modern DNA technology (see Chapter 18), they pieced together the sequences of the virus's eight genes and characterized their protein products. They also transformed clones of the genes into animal cells and were able to produce complete virus particles. These newly reconstructed 1918 viruses are about 50 times more virulent than modern-day hu-

man influenza viruses; they kill a higher percentage of mice and kill them much more quickly, for instance. (All of these experiments were done with appropriate approval and under highly controlled experimental conditions.) By studying the 1918 virus genome and its pathogenicity, the researchers are learning how highly virulent viruses can be produced. What they have learned so far is that the 1918 virus had mutations in polymerase genes for replicating the viral genome in host cells, likely making this strain capable of replicating more efficiently. Some of the mutations are similar to those found in bird flu viruses, including the one causing human deaths in Asia. The scientists believe that the 1918 flu virus likely arose directly from a bird flu virus and not from an assembly of RNA genome segments from a bird flu virus and a human flu virus in the same cell. If this is true, the concern about a devastating bird flu epidemic in the near future is well founded.

Other human viruses are also considered to have evolved from a virus that previously infected other animals. HIV is one of these; until the second half of the twentieth century, infections of this virus were apparently restricted almost entirely to African monkeys. Now the virus infects nearly 40 million people worldwide, with the greatest concentration of infected individuals in sub-Saharan Africa.

Viruses May Have Evolved from Fragments of Cellular DNA or RNA

Where did viruses come from? Because viruses can duplicate only by infecting a host cell, they probably evolved after cells appeared. They may represent fragments of DNA molecules that once formed part of the genetic material of living cells, or an RNA copy of such a fragment. In some way, the fragments became surrounded by a protective layer of protein with recognition functions and escaped from their parent cells. As viruses evolved, the information encoded in the core of the virus became reduced to a set of directions for producing more viral particles of the same kind.

Viroids and Prions Are Infective Agents Even Simpler in Structure than Viruses

Viroids, first discovered in 1971, are plant pathogens that consist of strands or circles of RNA, smaller than any viral DNA or RNA molecule, that have no protein coat. Some of the infective RNAs acting as viroids contain fewer than 300 nucleotides. Infection by viroids can rapidly destroy entire fields of citrus, potatoes, tomatoes, coconut palms, and other crop plants.

The manner in which viroids cause disease remains ill defined. In fact, researchers believe that there is more than one mechanism. Some recent research has defined one pathway to disease in which viroid RNA activates a protein kinase (an enzyme that adds phosphate groups to proteins) in plants. This process leads to a reduction in protein synthesis and protein activity, and disease symptoms result.

Prions, named by Stanley Prusiner of the University of California San Francisco in 1982 as a loose acronym for *pro*teinaceous *in*fectious particles, are the only known infectious agents that do not include a nucleic acid molecule.

Prions have been identified as the causal agents of certain diseases that degenerate the nervous system in mammals. One of these diseases is *scrapie,* a brain disease that causes sheep to rub against fences, rocks, or trees until they scrape off most of their wool. Another prion-based disease is bovine spongiform encephalopathy (BSE), also called *mad cow disease* **(Figure 25.20)**. The disease produces spongy holes and deposits of proteinaceous material in brain tissue. In 1996, 150,000 cattle in Great Britain died from an outbreak of BSE, which was traced to cattle feed containing ground-up tissues of sheep that had died of scrapie. Humans are subject to a fatal prion infection called *Creutzfeldt-Jakob disease (CJD)*. The symptoms of CJD include rapid mental deterioration, loss of vision and speech, and paralysis; autopsies show spongy holes and deposits in brain tissue similar to those of cattle with BSE. Classic CJD occurs as a result of the spontaneous transformation of normal proteins into prion proteins. Fewer than 300 cases a year occur in the United States. Variant CJD

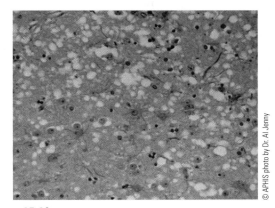

Figure 25.20
Bovine spongiform encephalopathy (BSE). The light-colored patches in this section from a brain damaged by BSE are areas where tissue has been destroyed.

Unanswered Questions

Do viruses infect archaeans?

Viruses of bacteria, and of the many types of eukaryotes, are well defined morphologically and molecularly. Do viruses infect members of the Archaea? If so, do these viruses resemble known viruses?

Mark Young's research group at Montana State University has focused on characterizing viruses from extreme thermophilic archaeans belonging to Crenoarchaeota. The researchers have discovered a number of viruses in archaeans from Yellowstone National Park acidic thermal areas. The morphology and molecular features of some of these viruses are novel and unrelated to those of any other known viruses. Young's group has sequenced the genomes of several of these new viruses, and their results indicate that the genes they carry have little or no similarity to known genes. Another archaean virus from the same area has a morphology also found in viruses of Bacteria and Eukarya. This result is of evolutionary significance because it suggests that the structure of the virus particle existed before the separation of each domain. The long-term goal of the research is to determine the mechanisms by which the viruses replicate in their extremely hot environment and to use them as tools for characterizing the special mechanisms the organisms use to survive at high temperatures. The research will also contribute to our understanding of the role viruses played in evolution.

How can West Nile virus be controlled?

West Nile virus is typically spread by mosquitoes. Usually, a mosquito becomes a carrier after biting a bird infected with the virus, and it then transmits the virus to other birds. Infected mosquitoes can also transmit the virus to humans and a number of other hosts, such as horses.

West Nile virus first entered the United States in 1999, and a number of humans have been infected. Humans infected with West Nile virus usually have mild symptoms such as fever, headache, body aches, rash, and swollen lymph glands. In some infected individuals, though, the virus enters the brain, where it can cause meningitis (inflammation of the lining of the brain and spinal cord) or encephalitis (inflammation of the brain), both of which can be fatal.

Researchers are trying to understand the infection cycle of the virus and how the virus causes disease. Specific research questions include how the virus replicates in the host and how the virus spreads through the body. Answers to these questions should aid efforts to develop effective vaccines and drugs to prevent and treat this disease. (At present, there is no vaccine for humans; one is available for horses.)

How do prion proteins move within the brain?

The brain-wasting diseases caused by prions are not well understood, despite much research. We know that prion proteins invade nerve cells and ultimately lead to fatal degeneration of the nervous system. To understand disease progression, scientists have investigated how prion proteins move through the nervous system. Using labeled-protein techniques, researchers have tracked infectious prion proteins from sites of infection up to the brain. Recently, the research groups of Bryon Caughey at the Rocky Mountain Laboratories in Montana, and Marco Prado at the University of Minas Gerais, Brazil, followed prion proteins as they invaded mouse brain cells growing in tissue culture. One exciting observation in these experiments was that prion proteins moved through the wirelike projections of the nerve cells to points of contact with other cells. Perhaps in a living organism, the prion proteins would be able to cross into the adjacent cell. The results are heralded as a significant step toward developing therapies to stop the spread of brain-wasting diseases by blocking the pathways by which prion proteins invade cells, replicate, move within the cell, and invade adjacent cells.

Peter J. Russell

is a form of the disease caused by eating nervous tissue containing meat or meat products from cattle with BSE. Another prion-based disease of humans, *kuru,* originally spread among cannibals in New Guinea, who became infected by eating raw human brain during ritual feasts following the death of an individual.

For several decades, scientists had hypothesized that a slow virus—a disease-causing virus with a long incubation period and gradual onset of pathogenicity—was responsible for these diseases. Prusiner was the first to hypothesize that infectious proteins were responsible. The research community mostly rejected this hypothesis out of hand because they held to the dogma that infectious agents required genes in the form of DNA or RNA to cause disease. Prusiner obtained experimental data supporting his hypothesis, and showed that prions are proteins normally made in the cell that misfold and cause other proteins of the same type to misfold, thereby "replicating" structural information from one molecule to the next. Typically, the misfolded prion proteins aggregate, whereas the normal proteins do not. If a misfolded prion protein is transferred from one animal to another, infection occurs; the transferred prions cause the recipient's proteins to misfold and eventually symptoms of the neurodegenerative disease characteristic of the prion will develop. Proteins with prion behavior are also found

naturally in yeast and other fungi; no diseases are associated with these prions. Prusiner received a Nobel Prize in 1997 for his discovery of prions.

The diseases caused by prions share symptoms that include loss of motor control, dementia, and eventually death. Progression of the disease is slow but there is no present cure. Under the microscope, aggregates of misfolded proteins called amyloid fibers are seen in brain tissues; the accumulation of these proteins in the brain is the likely cause of the brain damage in animals with prion diseases. The normal forms of the prion proteins are found on the surface of many types of cells, including brain cells. However, scientists do not know the function of the protein's normal form.

We began this chapter with prokaryotes, the simplest living organisms, and we end with still simpler entities, viruses, viroids, and prions, which are derived from living organisms and retain only some of the properties of life. In the next six chapters we turn to life at its most complex: the eukaryotic kingdoms of protists, plants, fungi, and animals.

STUDY BREAK

Distinguish between a virus, a viroid, and a prion.

Review

Go to Thomson**NOW** at www.thomsonedu.com/login to access quizzing, animations, exercises, articles, and personalized homework help.

25.1 Prokaryotic Structure and Function

- Three shapes are common in prokaryotes: spherical, rodlike, and spiral (Figure 25.2).
- Prokaryotic genomes typically consist of a single, circular DNA molecule packaged into the nucleoid. Many prokaryotic species also contain plasmids, which replicate independently of the main DNA (Figure 25.3).
- Gram-positive bacteria have a cell wall consisting of a thick peptidoglycan layer. Gram-negative bacteria have a thin peptidoglycan layer. The thin cell wall is surrounded by an outer membrane (Figure 25.4).
- A polysaccharide capsule or slime layer surrounds many bacteria. This sticky, slimy layer both protects the bacteria and helps them adhere to surfaces (Figure 25.5).
- Some prokaryotes have flagella, corkscrew-shaped protein fibers that rotate like propellers, and pili, protein shafts that help bacterial cells adhere to each other or to eukaryotic cells (Figure 25.7).
- Prokaryotes show great diversity in their modes of obtaining energy and carbon. Chemoautotrophs obtain energy by oxidizing inorganic substrates and use carbon dioxide as their carbon source. Chemoheterotrophs obtain both energy and carbon from organic molecules. Photoautotrophs are photosynthetic organisms that use light as a source of energy and carbon dioxide as their carbon source. Photoheterotrophs use light as a source of energy and obtain their carbon from organic molecules (Figure 25.8).

- Some prokaryotes are capable of nitrogen fixation, the conversion of atmospheric nitrogen to ammonia; others are responsible for nitrification, the two-step conversion of ammonium to nitrate.
- Prokaryotes normally reproduce asexually by binary fission. Some prokaryotes are capable of conjugation, in which part of the DNA of one cell is transferred to another cell.
- In nature, prokaryotes may live in an interacting community, such as a biofilm. Biofilms have both detrimental and beneficial consequences; they can harm human health, but they also can be effective in, for example, bioremediation (Figure 25.10).

Animation: Prokaryotic body plan

Animation: Gram staining

Animation: Prokaryotic fission

Animation: Prokaryotic conjunction

25.2 The Domain Bacteria

- Bacteria are divided into more than a dozen evolutionary branches (Figure 25.11).
- The proteobacteria are Gram-negative bacteria that include purple sulfur (photoautotrophic) and nonsulfur (photoheterotrophic) photosynthetic species, and nonphotosynthetic species. Free-living proteobacteria include the spore-forming myxobacteria and species that fix nitrogen (Figure 25.12).
- The green bacteria are Gram-negative and include sulfur (photoautotrophic) and nonsulfur (photoheterotrophic) photosynthetic bacteria.

- The cyanobacteria are Gram-negative photoautotrophs that carry out photosynthesis and release oxygen as a by-product (Figure 25.13).
- The Gram-positive bacteria are primarily chemoheterotrophs that include many pathogenic species (Figure 25.14).
- The spirochetes are spiral-shaped bacteria that are propelled by twisting movements produced by the rotation of flagella.
- Chlamydias are Gram-negative intracellular parasites that cause various diseases in animals. They have cell walls with an outer membrane, but they lack peptidoglycans (Figure 25.15).
- Bacteria cause disease through exotoxins, endotoxins, and exoenzymes.
- Pathogenic bacteria may develop resistance to antibiotics through mutation of their own genes, or by acquiring resistance genes from other bacteria or plasmids.

Animation: Examples of Eubacteria

25.3 The Domain Archaea

- The Archaea have some features that are like those of bacteria, others that are eukaryotic, and some that are uniquely archaean (Table 25.1).
- The archaean plasma membrane contains unusual lipid molecules. The cell walls of archaeans consist of distinct molecules similar to peptidoglycans, or of protein or polysaccharide molecules.

- The Archaea are classified into three groups. The Euryarchaeota include the methanogens, the extreme halophiles, and some extreme thermophiles. The Crenarchaeota contain most of the archaean extreme thermophiles, as well as psychrophiles and mesophiles. Obligate anaerobes, facultative anaerobes, and aerobes are found among the Crenarchaeota. The Korarchaeota are recognized only on the basis of sequences in DNA samples.

25.4 Viruses, Viroids, and Prions

- Viruses are nonliving infective agents. A free virus particle consists of a nucleic acid genome enclosed in a protein coat. Recognition proteins enabling the virus to attach to host cells extend from the surface of infectious viruses (Figure 25.18).
- Viruses reproduce by entering a host cell and directing the cellular machinery to make new particles of the same kind.
- Viruses are unaffected by antibiotics and most other treatment methods; hence, infections caused by them are difficult to treat.
- Viroids, which infect crop plants, consist only of a very small, single-stranded RNA molecule. Prions, which cause brain diseases in some animals, are infectious proteins with no associated nucleic acid. Prions are misfolded versions of normal cellular proteins that can induce other normal proteins to misfold.

Animation: Body plans of viruses

Animation: Bacteriophage multiplication cycles

Questions

Self-Test Questions

1. A urologist identifies cells in a man's urethra as bacterial. Which of the following descriptions applies to the cells?
   a. They have sex pili, which give them motility.
   b. They have flagella, which allow them to remain in one position in the urethral tube.
   c. They are covered by a capsule, which enables them to multiply quickly.
   d. They are covered by pili, which keep them attached to the urethral walls.
   e. They contain a peptidoglycan cell wall, which gives them buoyancy to float in the fluids of the urethra.

2. A bacterium that uses nitrites as its only energy source was found in a deep salt mine. It is a:
   a. chemoautotroph.
   b. parasite.
   c. photoautotroph.
   d. heterotroph.
   e. photoheterotroph.

3. The _____ are all oxygen-producing photoautotrophs.
   a. spirochetes
   b. chlamydias
   c. cyanobacteria
   d. Gram-positive bacteria
   e. proteobacteria

4. At the health center, a fecal sample was taken from a feverish student. Organisms with corkscrew-like flagella and no endomembranes but with cell walls were isolated as the cause for the illness. These organisms belong to the group:
   a. protists with nuclei.
   b. bacteria with ribosomes.
   c. fungi with endoplasmic reticulum.
   d. plants with chloroplasts.
   e. Archaea with Golgi bodies.

5. Which of the following is *not* a property of an endospore?
   a. resistant to boiling—must be autoclaved to be killed
   b. metabolically inactive
   c. can survive millions of years
   d. provides a method to preserve bacterial DNA under harsh conditions
   e. is a means that bacterial cells use to multiply

6. Each bacterial cell is traditionally thought to act independently of others. An exception to this is:
   a. biofilm aggregates.
   b. photosynthesis.
   c. peptidoglycan layering.
   d. toxin release.
   e. facultative anaerobic metabolism.

7. Penicillin, an antibiotic, inhibits the formation of cross-links between sugar groups in peptidoglycan. Bacteria treated with penicillin should be:
   a. aerobic.
   b. anaerobic.
   c. Gram-negative.
   d. Gram-positive.
   e. flagellated.

8. The best choice when using/prescribing antibiotics is to:
   a. increase the dosage when the original amount does not work.
   b. determine the kind of bacterium causing the problem.
   c. stop taking the antibiotic when you feel better but the prescription has not run out.
   d. ask the doctor to prescribe a drug as a precaution for an infection you do not have.
   e. choose soaps that are labeled "antibacterial."

9. When a virus enters the lysogenic stage:
   a. the viral DNA is replicated outside the host cell.
   b. it enters the host cell and kills it immediately.
   c. it enters the host cell, picks up host DNA, and leaves the cell unharmed.
   d. it sits on the host cell plasma membrane with which it covers itself and then leaves the cell.
   e. The viral DNA integrates into the host genome.

10. An infectious material is isolated from a nerve cell. It contains protein with amino acid sequences identical to the host protein but no nucleic acids. It belongs to the group:
   a. prions.
   b. Archaea.
   c. toxin producers.
   d. viroids.
   e. sporeformers.

Questions for Discussion

1. The digestive tract of newborn chicks is free of bacteria until they eat food that has been exposed to the feces of adult chickens. The ingested bacteria establishes a population in the digestive tract that is beneficial for the digestion of food. However, if *Salmonella* are present in the adult feces, this bacterium, which can be pathogenic for humans who ingest it, may become established in the digestive tracts of the chicks. To eliminate the possibility that *Salmonella* might become established, should farmers feed newborn chicks a mixture of harmless known bacteria from a lab culture, or a mixture of unknown fecal bacteria from healthy adult chickens? Design an experiment to answer this question.

2. Investigators in Australia found that mats of pond scum formed by the bacterium *Botyrococcus braunii* decayed into a substance resembling crude oil when the ponds dried up. Formulate a hypothesis explaining how this process may have contributed to Earth's oil deposits.

3. What rules would you suggest to prevent the spread of mad cow disease (BSE)?

Experimental Analysis

Suppose you isolate a previously unknown virus that has caused infection in humans. Describe how you would show experimentally to what virus genus and species this new virus is most closely related.

Evolution Link

Prion diseases cause similar fatal brain degeneration in a large number of animals, including human, baboon, chimpanzee, mule deer, cow, sheep, pig, golden hamster, rat, mouse, and rabbit. Can you make any evolutionary hypotheses based on this observation? How might you determine the evolutionary relationships of prion proteins?

How Would You Vote?

Eliminating mosquitoes is the best defense against West Nile virus. Many local agencies are spraying pesticides wherever mosquitoes are likely to breed. Some people fear ecological disruptions and bad effects on health and say spraying will never eliminate all mosquitoes anyway. Would you support a spraying program in your community? Go to www.thomsonedu.com/login to investigate both sides of the issue and then vote.

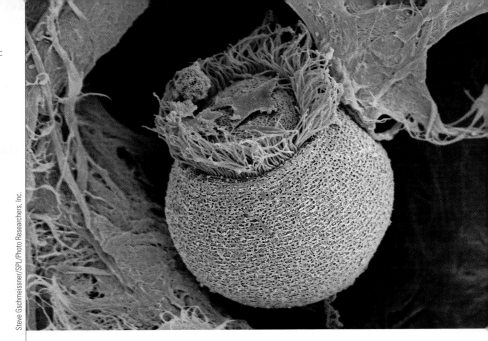

A ciliated protozoan, a type of protist (colorized SEM). This protozoan lives in water, feeding on bacteria and decaying organic matter.

Steve Gschmeissner/SPL/Photo Researchers, Inc.

26 Protists

STUDY PLAN

26.1 What Is a Protist?

Protists are most easily classified by what they are not

Protist diversity is reflected in their metabolism, reproduction, structure, and habitat

26.2 The Protist Groups

The Excavates lack mitochondria

The Discicristates include the euglenoids and kinetoplastids, which are motile protists

The Alveolates have complex cytoplasmic structures and use flagella or cilia to move

The Heterokonts include the largest protists, the brown algae

The Cercozoa are amoebas with filamentous pseudopods

The Amoebozoa includes most amoebas and two types of slime molds

The Archaeplastida include the red and green algae, and land plants

The Opisthokonts include the choanoflagellates, which may be the ancestors of animals

In several protist groups, plastids evolved from endosymbionts

WHY IT MATTERS

Go for a swim just about anywhere in the natural world and you will share the water with multitudes of diverse organisms called protists. Like their most ancient ancestors, almost all of these eukaryotic species are aquatic. Structurally, single-celled protists are the simplest of all eukaryotes. Although most are microscopic in size, many have had or have a significant impact on the world. For example, the protist *Phytophthora infestans,* a water mold also referred to as a downy mildew, infects valuable crop plants such as potatoes. Pototoes were the main food staple in Europe in the nineteenth century, and *P. infestans* destroyed potato crops, causing potato famines that spread across Europe; millions died in these famines. In Ireland, for instance, the growing seasons between 1845 and 1860 were cool and damp. Year after year, *P. infestans* spores spread along thin films of water on the plants. Late blight, a rotting of plant parts, became epidemic. One-third of the Irish population starved to death, died of typhoid fever (a secondary effect), or fled to other countries.

Today, related species threaten forests in the United States, Europe, and Australia. For example, when conditions favored its growth, *Phytophthora ramorum,* started an epidemic of sudden oak

549

death in California, during which tens of thousands of oak trees have died. As the name suggests, infected trees die rapidly. The first sign of infection is a dark red-to-black sap oozing from the bark surface. The pathogen has now jumped to madrones, redwoods, and certain other trees and shrubs. Cascading ecological changes resulting from tree death caused by this pathogen will reduce sources of food and shelter for forest species.

Protists are the subject of this chapter. Also known as *protoctists*, they are members of the kingdom **Protoctista.** Protists are the results of the varied early branching of eukaryotic evolution. They are abundant on Earth and play key ecological, economic, and medical roles in the world's biological communities.

We begin this chapter with a discussion of the identity of the protists. As our discussion will show, the members of the kingdom Protoctista are so diverse that they are best defined as what they are not—that is, by contrasting them with other kingdoms.

26.1 What Is a Protist?

Protists are easily the most varied of all Earth's creatures. **Figure 26.1** shows a number of protists, illustrating their great diversity. Protists include both microscopic single-celled and large multicellular organisms. They may inhabit aquatic environments, moist soils, or the bodies of animals and other organisms, and they may live as predators, photosynthesizers, parasites, or decomposers. The extreme diversity of the group has made the protists so difficult to classify that their status as a kingdom remains highly unsettled.

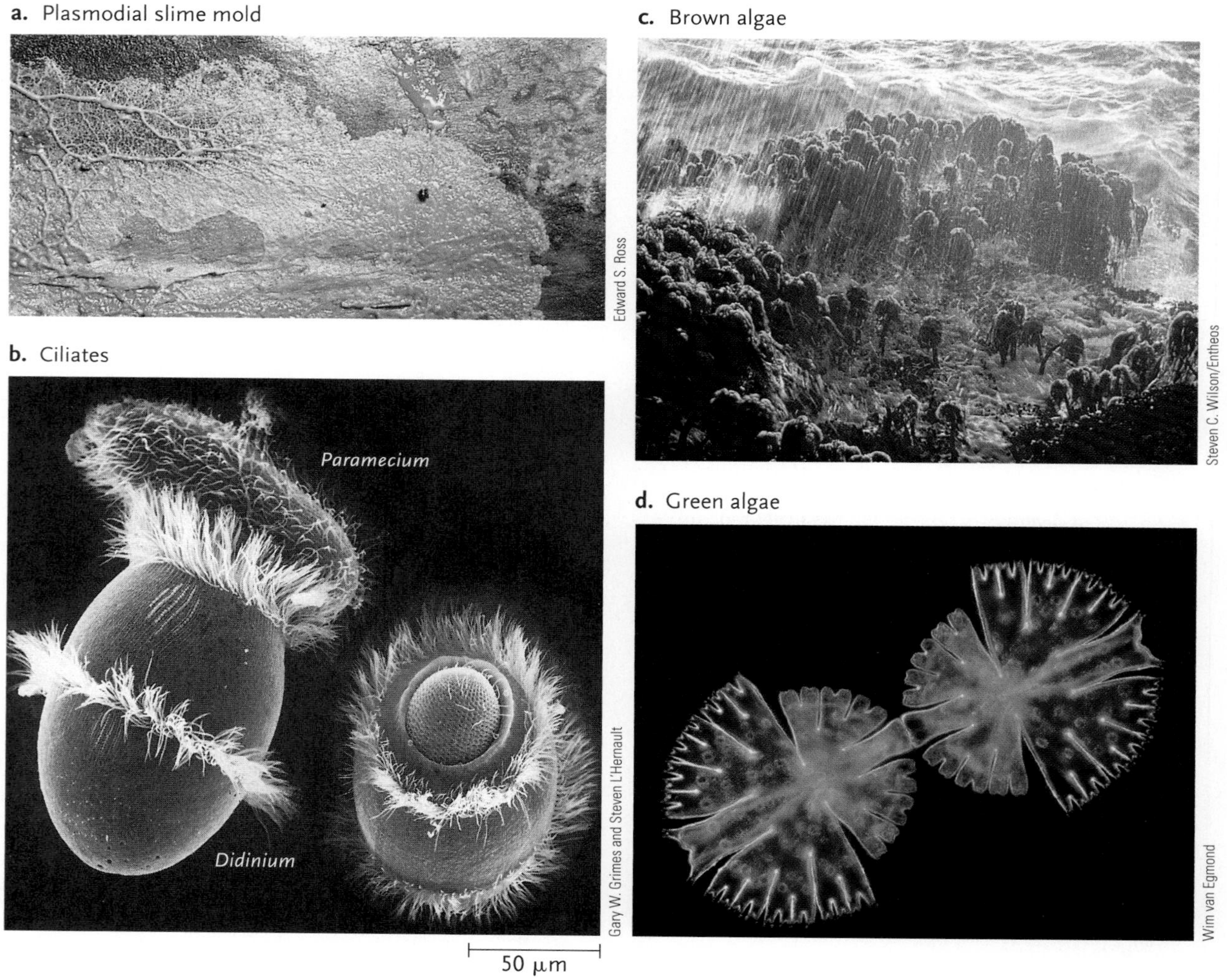

a. Plasmodial slime mold

c. Brown algae

Edward S. Ross

Steven C. Wilson/Entheos

b. Ciliates

Paramecium

Didinium

Gary W. Grimes and Steven L'Hernault

50 μm

d. Green algae

Wim van Egmond

Figure 26.1

A sampling of protist diversity. **(a)** *Physarum*, a plasmodial slime mold (yellow shape, lower part of figure) migrating over a rotting log. **(b)** *Didinium*, a ciliate, consuming another ciliate, *Paramecium*. **(c)** *Postelsia palmaeformis* (the sea palm), a brown alga, thriving in the surf pounding a California coast. **(d)** *Micrasterias*, a single-celled green alga, here shown dividing in two.

Protists Are Most Easily Classified by What They Are Not

The one reasonable certainty about protist classification is that the organisms lumped together in the kingdom Protoctista are not prokaryotes, fungi, plants, or animals.

Because protists are eukaryotes, the boundary between them and prokaryotes is clear and obvious. Unlike prokaryotes, protists have a true nucleus, with multiple, linear chromosomes. In addition to cytoplasmic organelles—including mitochondria (in most but not all species), endoplasmic reticulum, Golgi complex, and chloroplasts (in some species)—protists and other eukaryotes have microtubules and microfilaments, which provide motility and cytoskeletal support. They reproduce asexually by mitosis or sexually by meiosis and union of sperm and egg cells, rather than by binary fission as do prokaryotes.

The phylogenetic relationship between protists and other eukaryotes is more complex **(Figure 26.2)**. From its beginning, the eukaryotic family tree branched out in many directions. All of the organisms in the eukaryotic lineages consist of protists except for three groups, the animals, land plants, and fungi, which arose from protist ancestors. Although some protists have features that resemble those of the fungi, plants, or animals, several characteristics are distinctive. For instance, cell wall components in protists differ from those of the fungi (molds and yeasts, for example). In contrast to land plants, protists lack highly differentiated structures equivalent to true roots, stems, and leaves; they also lack the protective structures that encase developing embryos in plants. Protists are distinguished from animals by their lack of highly differentiated structures such as limbs and a heart, and by the absence of features such as nerve cells, complex developmental stages, and an internal digestive tract. Pro-

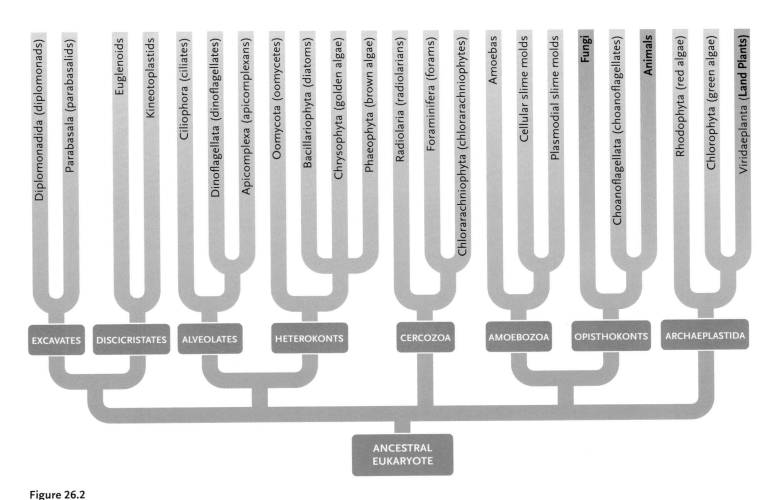

Figure 26.2

The phylogenetic relationship between the evolutionary groups within the kingdom Protoctista and the other eukaryotes. The Archaeplastida include the land plants of the kingdom Plantae (boxed), and the Opisthokonts include the animals of the kingdom Animalia and the fungi of the kingdom Fungi (boxed). The tree was constructed based on a consensus of molecular and ultrastructural data.

tists also lack collagen, the characteristic extracellular support protein of animals. Thus, the kingdom Protoctista is a catchall group that includes all the eukaryotes that are not fungi, plants, or animals.

Until recently, the protists were classified into phyla according to criteria such as body form, modes of nutrition and movement, and forms of meiosis and mitosis. However, comparisons of nucleic acid and amino acid sequences, now considered the most informative method for determining the evolutionary relationships of protists, show that most of the organisms previously grouped together do not share a common lineage. Further, many organisms within the phyla are no more closely related to each other than they are to the fungi, plants, or animals.

Given the extreme diversity of the protists, some evolutionists maintain that the kingdom Protoctista is actually a collection of many kingdoms—as many as 30, depending on differing evaluations of the lineages indicated by the sequence comparisons. Evolutionary lineages within the kingdoms are variously described as clades, subkingdoms, or phyla, and the existing schemes are constantly revised as new information is obtained.

For simplicity, we retain the Protoctista as a single kingdom in this book, with the understanding that it is a collection of largely unrelated organisms placed together for convenience. We will refer to the major evolutionary clusterings indicated by molecular and structural comparisons as groups (see Figure 26.2).

Figure 26.3
A ciliate, *Paramecium*, showing the cytoplasmic structures typical of many protists.
(Top: Frieder Sauer/ Bruce Coleman Ltd.; bottom: Redrawn from V. & J. Pearse and M. & R. Buchsbaum, Living Invertebrates, The Boxwood Press, 1987.)

Protist Diversity Is Reflected in Their Metabolism, Reproduction, Structure, and Habitat

As you might expect from the broad range of organisms included in the kingdom, protists are highly diverse in metabolism, reproduction, structure, and habitat.

Metabolism. Almost all protists are aerobic organisms that live either as heterotrophs—by obtaining their organic molecules from other organisms—or as autotrophs—by producing organic molecules for themselves by photosynthesis. Among the heterotrophs, some protists obtain organic molecules by directly ingesting part or all of other organisms and digesting them internally. Others absorb organic molecules from their environment. A few protists can live as either heterotrophs or autotrophs.

Reproduction. Reproduction may be asexual by mitosis or sexual by meiotic cell division and formation of gametes. In protists that reproduce by both mitosis and meiosis, the two modes of cell division are combined into a **life cycle** that is highly distinctive among the different protist groups.

Structure. Many protists live as single cells or as **colonies** in which individual cells show little or no differentiation and are potentially independent. Within colonies, individuals use cell signaling to cooperate on tasks such as feeding or movement. Some protists are large multicellular organisms, in which cells are differentiated and completely interdependent. For example, seaweeds are multicellular marine protists that include the largest and most differentiated organisms of the group; their structures include a hodlfast to secure the organism to the rocks, leaflike fronds, and, in some cases, an air bladder for flotation. The giant kelp of coastal waters rival forest trees in size.

Some single-celled and colonial protists have complex intracellular structures, some found nowhere else among living organisms **(Figure 26.3).** For example, many freshwater protists have a mechanism to maintain water balance in and out of the cell to prevent lysis. Excess water entering cells by osmosis (see Section 6.3) is handled using a specialized cytoplasmic organelle, the **contractile vacuole.** The contractile vacuole gradually fills with water; when it reaches maximum size it moves to the plasma membrane and forcibly contracts, expelling the water to the outside through a pore in the membrane. Many protists also have **food vacuoles** that digest prey or other organic material engulfed by the cells. Enzymes secreted into the food vacuoles digest the organic molecules; any remaining undigested matter is expelled to the outside by a mechanism similar to the expulsion of water by contractile vacuoles.

The cells of some protists are supported by an external cell wall, or by an internal or external shell built up

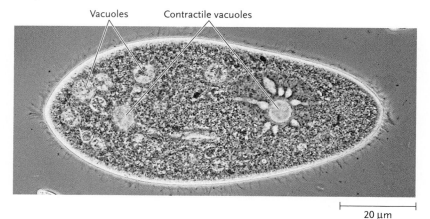

Vacuoles   Contractile vacuoles

20 μm

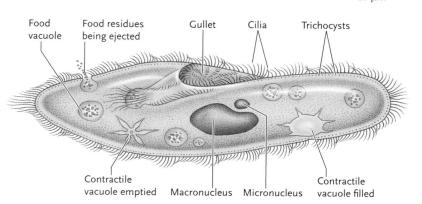

Food vacuole   Food residues being ejected   Gullet   Cilia   Trichocysts

Contractile vacuole emptied   Macronucleus   Micronucleus   Contractile vacuole filled

from organic or mineral matter; in some, the shell takes on highly elaborate forms. Other protists have a **pellicle**, a layer of supportive protein fibers located inside the cell, just under the plasma membrane, providing strength and flexibility instead of a cell wall (see Figure 26.5).

Almost all protists have structures providing motility at some time during their life cycle. Some move by amoeboid motion, in which the cell extends one or more lobes of cytoplasm called **pseudopodia** ("false feet"; see Figure 26.15). The rest of the cytoplasm and the nucleus then flow into a pseudopodium, completing the movement. Other protists move by the beating of flagella or cilia (see Section 5.3; cilia are essentially the same as flagella, except that cilia are often shorter and occur in greater numbers on a cell). In some protists, cilia are arranged in complex patterns, with an equally complex network of microtubules and other cytoskeletal fibers supporting the cilia under the plasma membrane. Among the protists are the most complex single cells known because of the wide variety of cytoplasmic structures they have.

Habitat. Protists live in aqueous habitats, including aquatic or moist terrestrial locations such as oceans, freshwater lakes, ponds, streams, and moist soils, and within host organisms. In bodies of water, small photosynthetic protists collectively make up the **phytoplankton** (*phytos* = plant; *planktos* = drifting), the abundant organisms that capture the energy of sunlight in nearly all aquatic habitats. These photosynthetic protists provide organic substances and oxygen for heterotrophic bacteria and protists and for the small crustaceans and animal larvae that are the primary constituents of **zooplankton** (*zoe* = life, usually meaning animal life); although protists are not animals, biologists often include them among the zooplankton. The phytoplankton and the larger multicellular protists forming seaweeds collectively account for about half of the total organic matter produced by photosynthesis.

In the moist soils of terrestrial environments, protists play important roles among the detritus feeders that recycle matter from organic back to inorganic form. In their roles in phytoplankton, zooplankton, and as detritus feeders, protists are enormously important in the world ecosystem.

Protists that live in host organisms are parasites, obtaining nutrients from the host. Indeed, many of the parasites that have significant effects on human health are protists, causing diseases such as malaria, sleeping sickness, and giardiasis.

STUDY BREAK

What distinguishes protists from prokaryotes? What distinguishes them from fungi, plants, and animals?

26.2 The Protist Groups

This section considers the biological features of each of the groups of protists included in Figure 26.2. This taxonomic tree represents a current consensus, based both on molecular data, such as comparative genomics, and on fine structures that have a distinctive form in a particular group.

The Excavates Lack Mitochondria

All members of the Excavates are single-celled animal parasites that lack mitochondria and move by means of flagella; most have a hollow (excavated) ventral feeding groove. Because they lack mitochondria, they are limited to glycolysis as an ATP source. However, the nuclei of Excavates contain genes derived from mitochondria, meaning that the ancestors of these protists probably had mitochondria. They may have lost their mitochondria as an adaptation to the parasitic way of life, in which oxygen is in short supply. We consider two groups here, the Diplomonadida and the Parabasala.

Diplomonadida. Diplomonad cells have two nuclei and move by means of multiple freely beating flagella. In addition to lacking mitochondria, they also lack a clearly defined endoplasmic reticulum and Golgi complex. The best-known representative of the group, *Giardia lamblia* **(Figure 26.4a)**, infects the mammalian intestinal tract, inducing severe diarrhea and abdominal cramps. *Giardia* is spread by contamination of water with feces, in which resistant cysts of the protist can be present in large numbers. So many streams and lakes in wilderness areas of the United States have become contaminated with *Giardia* cysts that hikers must boil water from these sources before drinking it, or pass it through filters able to remove particles as small as 1 μm. Treating water with chemicals such as chlorine or iodine does not kill the cysts.

Parabasala. In addition to freely beating flagella, species among the Parabasala have a sort of fin called an **undulating membrane**, formed by a flagellum buried in a fold of the cytoplasm. The buried flagellum allows parabasalans to move through thick and viscous fluids. Among the Parabasala are the trichomonads, including *Trichomonas vaginalis* **(Figure 26.4b)**, a worldwide nuisance responsible for infections of the urinary and reproductive tracts in both men and women. The infective trichomonad is passed from person to person primarily, but not exclusively, by sexual intercourse. It lives in the vagina in women and in the urethra of both sexes. The

a. *Giardia lamblia*

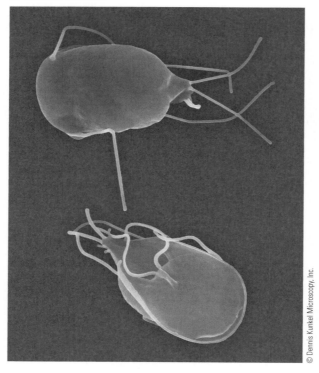

© Dennis Kunkel Microscopy, Inc.

b. *Trichomonas vaginalis*

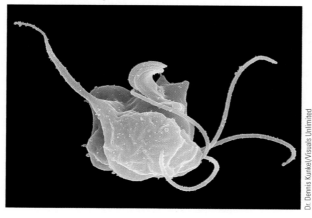

Dr. Dennis Kunkel/Visuals Unlimited

Figure 26.4
The Diplomonad-ida and Parabasala of the Excavates. **(a)** A diplomonad, *Giardia lamblia,* that causes intestinal disturbances. **(b)** A parabasalid, *Trichomonas vaginalis,* that causes a sexually transmitted disease, trichomoniasis.

infection is usually symptomless in men, but in women *T. vaginalis* can cause severe inflammation and irritation of the vagina and vulva. It is easily cured by drugs.

The Discicristates Include the Euglenoids and Kinetoplastids, Which Are Motile Protists

The **Discicristates** are named for their disc-shaped mitochondrial cristae (inner mitochondrial membranes). The group includes about 1800 species, almost all single-celled, highly motile cells that swim by means of flagella. While most are photosynthetic, some are facultative heterotrophs; some can even alternate between photosynthesis and life as a heterotroph. The best-known members of this group are the euglenoids, with *Euglena gracilis* **(Figure 26.5)** as the best-known species. Another group, the parasitic kinetoplastids, includes some organisms responsible for human diseases. A commonly used nontaxonomic name for protists of the Discicristates is *protozoa* ("first animal"), referring to their similarity to animals with respect to ingesting food and moving by themselves.

Euglenoids. With the exception of a few marine species, the euglenoids inhabit freshwater ponds, streams, and lakes. Most are autotrophs that carry out photosynthesis by the same mechanisms as plants, using the same photosynthetic pigments, including chlorophylls *a* and *b* and β-carotene. Many of the photosynthetic euglenoids, including *E. gracilis,* can also live as heterotrophs by absorbing organic molecules through the plasma membrane. Some euglenoids lack chloroplasts and live entirely as heterotrophs.

Euglena gracilis and other euglenoids have a profusion of cytoplasmic organelles, including a contractile vacuole and, in photosynthetic species, chloroplasts (see Figure 26.5). Rather than an external cell wall, the euglenoids have a spirally grooved pellicle formed from transparent, protein-rich material. Most of the photosynthetic euglenoids, including *E. gracilis,* have an *eyespot* containing carotenoid pigment granules in association with a light-sensitive structure. The eyespot is part of a sensory mechanism that stimulates cells to swim toward moderately bright light or away from intensely bright light so that the organism is in light conditions for optimal photosynthetic activity. The cells swim by whiplike movements of flagella that extend from a pocketlike depression at one end of the cell. Most have two flagella, one rudimentary and short, the other long.

Kinetoplastids. The kinetoplastids are a group of nonphotosynthetic, heterotrophic cells that live as animal parasites **(Figure 26.6)**. Their name reflects the structure of the single mitochondrion in a cell of this group, which contains a large DNA-protein deposit called a *kinetoplast*. Most kinetoplastids have a leading and a trailing flagellum, which are used for movement. In some cases, the trailing flagellum is attached to the side of the cell, forming an undulating membrane that is often used to enable the organism to glide along or attach to surfaces.

The kinetoplastids include the trypanosomes, responsible for several diseases afflicting millions of humans in tropical regions. *Trypanosoma brucei* (see Figure 26.6) causes African sleeping sickness, transmitted from one host to another by bites of the tsetse fly. Early symptoms include fever, headaches, rashes, and anemia. Untreated, the disease damages the cen-

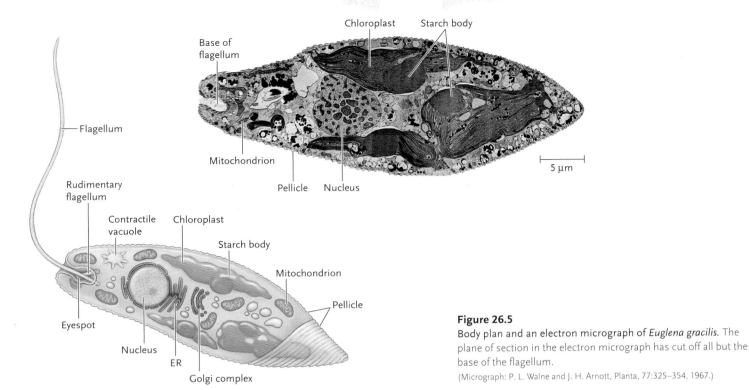

Chloroplast   Starch body

Base of
flagellum

Flagellum

Mitochondrion

Pellicle   Nucleus

5 μm

Rudimentary
flagellum

Contractile
vacuole   Chloroplast

Starch body

Mitochondrion

Pellicle

Eyespot

Nucleus

ER

Golgi complex

Figure 26.5

Body plan and an electron micrograph of *Euglena gracilis*. The plane of section in the electron micrograph has cut off all but the base of the flagellum.

(Micrograph: P. L. Walne and J. H. Arnott, *Planta*, 77:325–354, 1967.)

tral nervous system, leading to a sleeplike coma and eventual death. The disease has proved difficult to control because the same trypanosome infects wild mammals, providing an inexhaustible reservoir for the parasite. Other trypanosomes, also transmitted by insects, cause Chagas disease in the southwestern United States and Central and South America, and leishmaniasis in the tropics. Humans with Chagas disease have an enlarged liver and spleen and may experience severe brain and heart damage; people with leishmaniasis have skin sores and ulcers that may become very deep and disfiguring, particularly to the face.

The Alveolates Have Complex Cytoplasmic Structures and Use Flagella or Cilia to Move

The Alveolates are so called because they have small, membrane-bound vesicles called *alveoli* (*alvus* = belly) in a layer just under the plasma membrane. The Alveolates include two motile, primarily free-living groups, the Ciliophora and Dinoflagellata, and a nonmotile, parasitic group, the Apicomplexa.

Ciliophora: The Ciliates. The Ciliophora—the ciliates—includes nearly 10,000 known species of primarily single-celled but highly complex heterotrophic organisms that swim by means of cilia (see Figures 26.1b and 26.3). Ciliates were among the first organisms observed

in the seventeenth century by the pioneering microscopist Anton van Leeuwenhoek. Essentially any sample of pond water or bottom mud contains a wealth of these creatures.

The organisms in the Ciliophora have many highly developed organelles, including a mouthlike gullet lined with cilia; structures that exude mucins, toxins, or other defensive and offensive materials from the cell surface; contractile vacuoles; and complex systems of

Figure 26.6

Trypanosoma brucei, the parasitic kinetoplastid that causes African sleeping sickness.

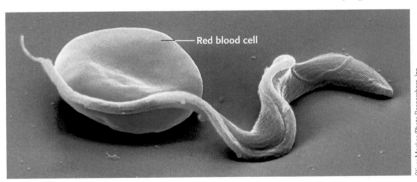

Red blood cell

Oliver Meckes/Photo Researchers, Inc.

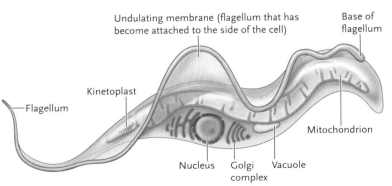

Undulating membrane (flagellum that has become attached to the side of the cell)

Base of flagellum

Flagellum

Kinetoplast

Mitochondrion

Nucleus   Golgi complex   Vacuole

food vacuoles. A pellicle reinforces cell shape. A complex cytoskeletal network of microtubules and other fibers anchors the cilia just below the pellicle and coordinates the ciliary beating. The cilia can stop and reverse their beating in synchrony, allowing ciliates to stop, back up, and turn if they encounter negative stimuli. Evidence that the cytoskeletal network organizes ciliary beating comes from microsurgical experiments in which a segment of the body surface was cut out and reinserted in the opposite direction. The cilia in the reversed segment beat in the opposite direction to those on the rest of the organism.

The ciliates are the only eukaryotes that have two types of nuclei in each cell: one or more small nuclei called *micronuclei,* and a single larger *macronucleus* (see Figure 26.3b). A **micronucleus** is a diploid nucleus that contains a complete complement of genes. It functions primarily in cellular reproduction, which may be asexual or sexual. The number of micronuclei present depends on the species. The **macronucleus** develops from a micronucleus, but loses all genes except those required for basic "housekeeping" functions of the cell and for ribosomal RNAs. These DNA sequences are duplicated many times, greatly increasing its capacity to transcribe the mRNAs needed for these functions, and the rRNAs needed to make ribosomes.

In asexual reproduction by mitosis, both types of nuclei replicate their DNA, divide, and are passed on to daughter cells. In sexual reproduction of *Paramecium,* for example, two cells **conjugate** by first forming a cytoplasmic bridge **(Figure 26.7).** Next, the micronucleus in each cell undergoes meiosis, producing four haploid micronuclei. In a series of steps, three of the four micronuclei in each cell degenerate, and the macronucleus also begins degenerating. The remaining micronucleus divides by mitosis, and one of the two micronuclei in each cell then passes through the cytoplasmic bridge into the other cell (step 5). The two haploid micronuclei in each cell now fuse to form a diploid micronucleus, with pairs of homologous chromosomes, one from each of the original parents. The two cells then separate. Through a further series of divisions, the micronucleus in each cell gives rise to two micronuclei and two macronuclei. Finally, each cell divides to produce two daughter cells, completing sexual reproduction.

Ciliates abound in freshwater and marine habitats, where they feed voraciously on bacteria, algae, and each other. *Paramecium* is a typical member of the group (see Figure 26.3). Its rows of cilia drive it through its watery habitat, rotating the cell on its long axis while it moves forward, or backs and turns. The cilia also sweep water laden with prey and food particles into the gullet, where food vacuoles form. The ciliate digests food in the vacuoles and eliminates indigestible material through an anal pore. Contractile vacuoles with elaborate, raylike extensions remove excess water from the cytoplasm and expel it to the outside. When under attack or otherwise stressed, *Paramecium* discharges many dartlike protein threads from surface organelles called **trichocysts.**

Some ciliates live individually while others are colonial. Certain ciliates are animal parasites; others live and reproduce in their hosts as mutually beneficial symbionts. (*Symbiosis* is the interaction between two organisms living together in close association, sometimes one inside another.) A compartment of the stomach of cattle and other grazing animals contains large numbers of symbiotic ciliates that digest the cellulose in their host's plant diet. The animals then digest the excess ciliates.

One ciliate, *Balantidium coli,* is a human intestinal parasite that causes diarrhea, with stools typically containing blood and pus. It is passed on when humans eat food contaminated by the feces of animals infected by *Balantidium,* particularly pigs. Less than 1% of the human population is infected worldwide.

Dinoflagellata: The Dinoflagellates. Of over 4000 known dinoflagellate species, most are single-celled organisms in marine phytoplankton. They live as heterotrophs or autotrophs; many can carry out both modes of nutrition. Some contain algae as symbionts. Typically, they have a shell formed from cellulose plates **(Figure 26.8).** The beating of flagella, which fit into grooves in the plates, makes dinoflagellates spin like a top (*dinos* = spinning).

The cytoplasmic structures of dinoflagellates include mitochondria, chloroplasts in photosynthetic species, and other internal membrane systems characteristic of eukaryotes. The photosynthetic dinoflagellates contain chlorophylls *a* and *c* along with accessory pigments that make them golden-brown or brown; algal symbionts give some a green, blue, or red color.

Their abundance in phytoplankton makes dinoflagellates a major primary producer of ocean ecosystems. Some species live as symbionts in the tissues of other marine organisms such as jellyfish, sea anemones, corals, and mollusks. For example, dinoflagellates in coral use the coral's carbon dioxide and nitrogenous waste, while supplying 90% of the coral's nutrition. The vast numbers of dinoflagellates living as photosynthetic symbionts in tropical coral reefs allow the reefs to reach massive size; without the dinoflagellates many coral species would die.

Some dinoflagellates are **bioluminescent**—they glow or release a flash of light, particularly when disturbed. The production of light depends on the enzyme *luciferase* and its substrate *luciferin,* in forms similar to the system that produces light in fireflies. Dinoflagellate fluorescence can make the sea glow in the wake of a boat at night and coat nocturnal surfers and swimmers with a ghostly light.

At times dinoflagellate populations grow to such large numbers that they color the seas red, orange, or brown. The resulting **red tides** are common in spring and summer months along the warmer coasts of the world, including all the U.S. coasts. Some red-tide dinoflagellates produce a toxin that interferes with nerve

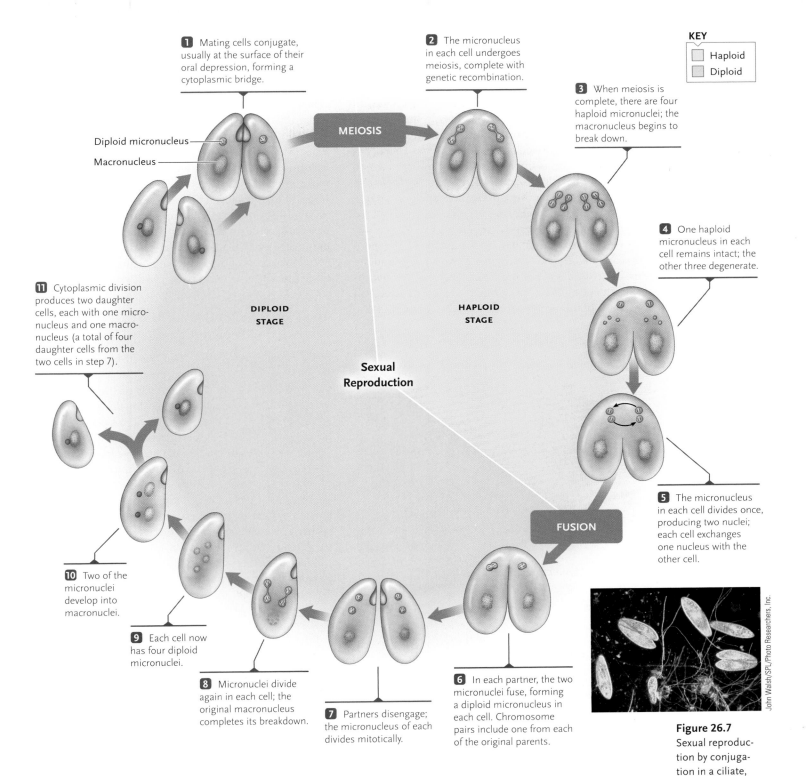

1 Mating cells conjugate, usually at the surface of their oral depression, forming a cytoplasmic bridge.

2 The micronucleus in each cell undergoes meiosis, complete with genetic recombination.

3 When meiosis is complete, there are four haploid micronuclei; the macronucleus begins to break down.

4 One haploid micronucleus in each cell remains intact; the other three degenerate.

5 The micronucleus in each cell divides once, producing two nuclei; each cell exchanges one nucleus with the other cell.

6 In each partner, the two micronuclei fuse, forming a diploid micronucleus in each cell. Chromosome pairs include one from each of the original parents.

7 Partners disengage; the micronucleus of each divides mitotically.

8 Micronuclei divide again in each cell; the original macronucleus completes its breakdown.

9 Each cell now has four diploid micronuclei.

10 Two of the micronuclei develop into macronuclei.

11 Cytoplasmic division produces two daughter cells, each with one micronucleus and one macronucleus (a total of four daughter cells from the two cells in step 7).

Diploid micronucleus

Macronucleus

KEY
☐ Haploid
☐ Diploid

MEIOSIS

FUSION

DIPLOID STAGE

HAPLOID STAGE

Sexual Reproduction

John Walsh/SPL/Photo Researchers, Inc.

Figure 26.7
Sexual reproduction by conjugation in a ciliate, *Paramecium.*

function in animals that ingest these protists. Fish that feed on plankton, and birds that feed on the fish, may be killed in huge numbers by the toxin. Dinoflagellate toxin does not noticeably affect clams, oysters, and other mollusks, but it becomes concentrated in their tissues. Eating the tainted mollusks can cause respiratory failure and death for humans and other animals. The toxin is especially deadly for mammals because it paralyzes the diaphragm and other muscles required for breathing.

Apicomplexa. The apicomplexans are all nonmotile parasites of animals. They absorb nutrients through their plasma membranes rather than by engulfing food particles, and they lack food vacuoles. They get their name from the *apical complex,* a group of organelles at one end of the cell that functions in attachment and invasion of host cells.

Typically, apicomplexan life cycles involve both asexual and sexual reproduction. All the apicomplexans, which includes almost 4000 known species, pro-

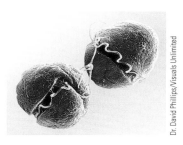

Figure 26.8
Karenia brevis, a toxin-producing dinoflagellate.

duce infective sporelike stages called *sporozoites.* The sporozoites reproduce asexually in cells they infect, eventually bursting them, which releases the progeny to infect new cells. At some point they generate specialized cells that form gametes; fusion of gametes produces resistant cells known as *cysts.* Usually, a host is infected by ingesting cysts, which divide to produce sporozoites. This basic life cycle pattern varies considerably among the apicomplexans, and many of these organisms use more than one host species for different stages of their life cycle.

One apicomplexan genus, *Plasmodium,* is responsible for malaria, one of the most widespread and debilitating diseases of humans. The disease is transmitted by the bite of 60 different species of mosquitoes, all members of the genus *Anopheles.* Although the disease is now rare in the United States, *Anopheles* mosquitoes are common enough to spread malaria if *Plasmodium* is introduced by travelers from other countries. The infective cycle of *Plasmodium,* described in *Focus on Research,* is representative of the complex life cycles of apicomplexans.

Another organism in this group, *Toxoplasma,* has a sexual phase of its life cycle in cats and asexual phases in humans, cattle, pigs, and other animals. Cysts of the parasite in the feces of infected cats are spread in household and garden dust. Humans ingesting or inhaling the cysts develop toxoplasmosis, a disease that is usually mild in adults but can cause severe brain damage or even death to a fetus. Because of the danger of toxoplasmosis, pregnant women should avoid emptying litter boxes or otherwise cleaning up after a cat.

The Heterokonts Include the Largest Protists, the Brown Algae

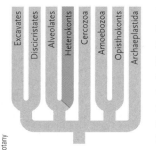

The Heterokonts (*hetero* = different; *kontos* = pole, referring to the flagellum) are named for their two different flagella: one with hollow tripartite hairs that give the flagellum a "hairy" appearance and a second one that is plain. The flagella occur only on reproductive cells such as eggs and sperm, except in the golden algae, in which cells are flagellated in all stages. The heterokonts include the Oomycota (water molds, white rusts, and mildews—formerly classified as fungi), Bacillariophyta (diatoms), Chrysophyta (golden algae), and Phaeophyta (brown algae).

Oomycota: Water Molds, White Rusts, and Downy Mildews. The Oomycota **(Figure 26.9)** are funguslike heterokonts that lack chloroplasts and live as heterotrophs. Like fungi, they secrete enzymes that digest the complex molecules of surrounding dead or alive organic matter into simpler substances small enough to be absorbed into their cells. The water molds live almost exclusively in freshwater lakes and streams or moist terrestrial habitats; the white rusts and downy mildews are parasites of plants. Oomycota may reproduce asexually or sexually.

Like fungi, many Oomycota grow as microscopic, nonmotile filaments called **hyphae** (singular, hypha), which form a network called a **mycelium (Figure 26.10).** Other features, however, set the Oomycota apart from the fungi; chief among them are differences in nucleotide sequence, which clearly indicate close evolutionary relationships to the heterokonts rather than to the fungi. Further, nuclei in hyphae are diploid in the Oomycota, rather than haploid as in the fungi, and repro-

a. Water mold

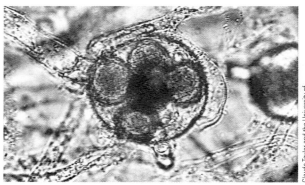

b. Water mold infecting fish

c. Downy mildew

Figure 26.9
Oomycota. **(a)** The water mold *Saprolegnia parasitica.* **(b)** *S. parasitica* growing as cottony white fibers on the tail of an aquarium fish. **(c)** A downy mildew, *Plasmopara viticola,* growing on grapes. At times it has nearly destroyed vineyards in Europe and North America.

Applied Research: Malaria and the *Plasmodium* Life Cycle

Although malaria is uncommon in the United States, it is a major epidemic in many other parts of the world. From 300 million to 500 million people become infected with malaria each year in tropical regions, including Africa, India, southeast Asia, the Middle East, Oceania, and Central and South America. Of these, about 2 million die each year, twice as many as from AIDS worldwide. It is particularly deadly for children younger than 6. In many countries where malaria is common, people are often infected repeatedly, with new infections occurring alongside preexisting infections.

Four different species of the apicomplexan genus *Plasmodium* cause malaria. In the life cycle of the parasites (see **figure**), sporozoites develop in the female *Anopheles* mosquito, which transmits them by its bite to human or bird hosts. The infecting parasites divide repeatedly in their hosts, initially in liver cells and then in red blood cells. Their growth causes red blood cells to rupture in regular cycles every 48 or 72 hours, depending on the *Plasmodium* species. The ruptured red blood cells clog vessels and release the parasite's metabolic wastes, causing cycles of chills and fever.

The victim's immune system is ineffective because, during most of the infective cycle, the parasite is inside body cells and thus "hidden" from antibodies. Further, *Plasmodium* regularly changes its surface molecules, continually producing new forms that are not recognized by antibodies developed against a previous form. In this way, the parasite keeps one step ahead of the immune system, often making malarial infections essentially permanent.

Travelers in countries with high rates of malaria are advised to use antimalarial drugs such as chloroquine, quinine, or quinidine as a preventative. However, many *Plasmodium* strains in Africa, India, and southeast Asia have developed resistance to the drugs. Vaccines have proved difficult to develop; because vaccines work by inducing the production of antibodies that recognize surface groups on the parasites, they are defeated by the same mechanisms the parasite uses inside the body to keep one step ahead of the immune reaction.

While in a malarial region, travelers should avoid exposure to mosquitoes by remaining indoors from dusk until dawn and sleeping inside mosquito nets treated with insect repellent. When out of doors, travelers should wear clothes that expose as little skin as possible and are thick enough to prevent mosquitoes from biting through the cloth. An insect repellent containing DEET should be spread on any skin that is exposed.

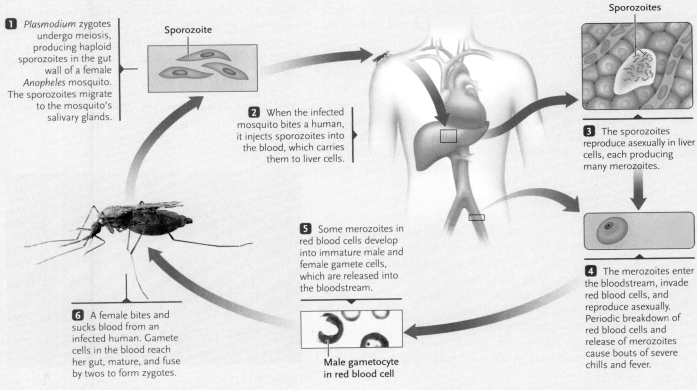

1 *Plasmodium* zygotes undergo meiosis, producing haploid sporozoites in the gut wall of a female *Anopheles* mosquito. The sporozoites migrate to the mosquito's salivary glands.

Sporozoite

2 When the infected mosquito bites a human, it injects sporozoites into the blood, which carries them to liver cells.

Sporozoites

3 The sporozoites reproduce asexually in liver cells, each producing many merozoites.

4 The merozoites enter the bloodstream, invade red blood cells, and reproduce asexually. Periodic breakdown of red blood cells and release of merozoites cause bouts of severe chills and fever.

5 Some merozoites in red blood cells develop into immature male and female gamete cells, which are released into the bloodstream.

Male gametocyte in red blood cell

6 A female bites and sucks blood from an infected human. Gamete cells in the blood reach her gut, mature, and fuse by twos to form zygotes.

Life cycle of a *Plasmodium* species that causes malaria.

(Photo: Sinclair Stammers/Photo Researchers, Inc.; micrograph: Steven L'Hernault.)

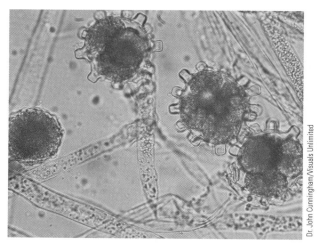

Figure 26.10
The funguslike body form of the Oomycota, consisting of filaments called hyphae, which grow into a network called a mycelium.

ductive cells are flagellated and motile; fungi have no motile stages. Finally, the cell walls of most Oomycota contain cellulose (see Figure 3.7c); fungal cell walls instead contain a different polysaccharide, chitin (see Figure 3.7d).

Most water molds are key decomposers of both aquatic and moist terrestrial habitats. Dead animal or plant material immersed in water commonly becomes coated with cottony water molds. Other water molds parasitize living aquatic animals, such as the mold growing on the fish shown in Figure 26.9a. The white rusts and downy mildews are parasites of land plants (see Figure 26.9c).

Some water molds have had drastic effects on human history. *P. infestans,* a water mold that causes rotting of potato and tomato plants, was responsible for the Irish potato famine of 1845 to 1860. In this famine more than a million people, a third of Ireland's population, starved to death. Many of the survivors migrated in large numbers to other countries, including the United States and Canada.

Bacillariophyta: Diatoms. The Bacillariophyta, or diatoms, are single-celled organisms that are covered by a glassy silica shell, which is intricately formed and beautiful in many species. The two halves of the shell fit together like the top and bottom of a candy box **(Figure 26.11).** Substances move to and from the plasma membrane through elaborately patterned perforations in the shell. Although flagella are present only in gametes, many diatoms move by an unusual mechanism in which a secretion released through grooves in the shell propels them in a gliding motion.

Diatoms are autotrophs that carry out photosynthesis by pathways similar to those of plants. The primary photosynthetic organisms of marine plankton, they fix more carbon into organic material than any other planktonic organism. They are also abundant in freshwater habitats as both phytoplankton and bottom-dwelling species. Although most diatoms are free living, some are symbionts inside other marine protists. One diatom, *Pseudonitzschia,* produces a toxic amino acid that can accumulate in shellfish. The amino acid, which acts as a nerve poison, causes amnesic shellfish poisoning when ingested by humans; the poisoning can be fatal.

Asexual reproduction in diatoms occurs by mitosis followed by a form of cytoplasmic division in which each daughter cell receives either the top or bottom half of the parent shell. The daughter cell then secretes the missing half, which becomes the smaller, inside shell of the box. The daughter cell receiving the larger top half grows to the same size as the parent shell, but the cell receiving the smaller bottom half is limited to the size of this shell. As asexual divisions continue, the cells receiving bottom halves become progressively smaller. Very small diatoms may switch to a sexual mode of reproduction; they enter meiosis and produce flagellated gametes, which lose their shells and fuse in pairs to form a zygote. The zygote grows to normal size before secreting a completely new shell with full-size top and bottom halves.

The shells of diatoms are common in fossil deposits. In fact, more diatoms are known as fossils than as living species—some 35,000 extinct species have been described as compared with 7000 living species. For about 180 million years the shells of diatoms have been accumulating into thick layers of sediment at the bottom of lakes and seas. Since diatoms store food as oil, fossil diatoms may be a source of oil in many oil deposits.

Grinding the fossilized shells into a fine powder produces *diatomaceous earth,* which is used in abra-

Figure 26.11
Diatom shells. Depending on the species, the shells are either radially or bilaterally symmetrical, as seen in this sample.

a. Golden alga

b. Brown alga, *Macrocystis*

c. Brown alga, *Postelsia palmaeformis*

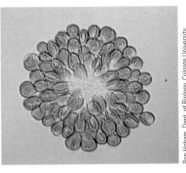

Figure 26.12

Golden and brown algae. **(a)** A microscopic, swimming colony of *Synura,* a golden alga. Each cell bears two flagellae, which are not visible in this light micrograph. **(b)** A brown alga, *Macrocystis.* Note the whitish gas bladders that keep the blades floating. **(c)** The holdfast, stemlike stipes, and leaflike blades, as seen in another brown alga, the sea palm *Postelsia palmaeformis.*

sives and filters, as an insulating material, and as a pesticide. Diatomaceous earth kills crawling insects by abrading their exoskeleton, causing them to dehydrate and die. Insect larvae are killed in the same way. Insects also die when they eat the powder but larger animals, including humans, are unaffected by it.

Chrysophyta: Golden Algae. Most golden algae **(Figure 26.12a)** are colonial forms in which each cell of the colony bears a pair of flagella. The golden algae have glassy shells, but in the form of plates or scales rather than in the candy-box form of the diatoms.

Nearly all chrysophytes are autotrophs and carry out photosynthesis using pathways similar to those of plants. Their color is due to a brownish carotenoid pigment, fucoxanthin, which masks the green color of the chlorophylls. Golden algae are important in freshwater habitats and in "nanoplankton," a community of marine phytoplankton composed of huge numbers of extremely small cells. During the spring and fall, "blooms" of golden algae can give a fishy taste and brownish color to the water.

Phaeophyta: Brown Algae. The brown algae (*phaios* = brown) are photosynthetic autotrophs that range from microscopic forms to giant kelps reaching lengths of 50 m or more (Figure 26.1c and **Figure 26.12b** and **c**). Their color is also due to fucoxanthin. Their cell walls contain cellulose and a mucilaginous polysaccharide, alginic acid.

Nearly all of the 1500 known phaeophyte species inhabit temperate or cool coastal marine waters. The kelps form vast underwater forests; fragments of these algae litter the beaches in coastal regions where they grow. Great masses of another brown alga, *Sargassum,* float in an area of the mid-Atlantic Ocean called the Sargasso Sea, which covers millions of

square kilometers between the Azores and the Bahamas.

Kelps are the largest and most complex of all protists. Their tissues are differentiated into leaflike *blades,* stalklike *stipes,* and rootlike *holdfasts* that anchor them to the bottom. Hollow, gas-filled bladders give buoyancy to the stipes and blades and help keep them upright. The stalks of some kelps contain tube-like vessels, similar to the vascular elements of plants, which rapidly distribute dissolved sugars and other products of photosynthesis throughout the body of the alga.

Life cycles among the brown algae are typically complex and in many species consist of alternating haploid and diploid generations **(Figure 26.13).** The large structures that we recognize as kelps and other brown seaweeds are diploid **sporophytes,** so called because they give rise to haploid spores by meiosis. The spores, which are flagellated swimming cells, germinate and divide by mitosis to form an independent, haploid **gametophyte** generation. The gametophytes give rise to haploid gametes, the egg and sperm cells. Most brown algal gametophytes are multicellular structures only a few centimeters in diameter. Cells in the gametophyte, produced by mitosis, differentiate to form flagellated, swimming sperm cells or nonmotile eggs. Fusion of a sperm and an egg cell gives rise to a diploid zygote, which grows by mitotic divisions into the sporophyte generation. Other variations occur in smaller brown algae, including some life cycles in which the sporophytes and gametophytes are the same size and some in which the gametophyte is larger than the sporophyte.

The alginic acid in brown algal cell walls, called **algin** when extracted, is an essentially tasteless and nontoxic substance used to thicken such diverse prod-

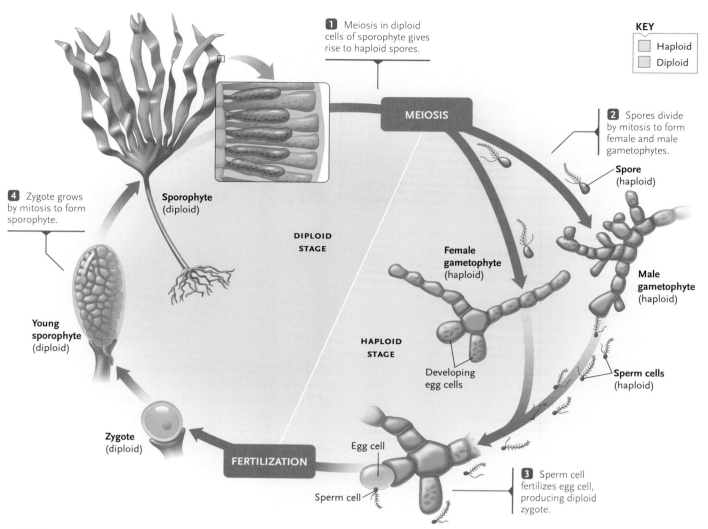

1 Meiosis in diploid cells of sporophyte gives rise to haploid spores.

MEIOSIS

KEY

| | Haploid |
| | Diploid |

2 Spores divide by mitosis to form female and male gametophytes.

Spore (haploid)

4 Zygote grows by mitosis to form sporophyte.

Sporophyte (diploid)

DIPLOID STAGE

Female gametophyte (haploid)

Male gametophyte (haploid)

Young sporophyte (diploid)

HAPLOID STAGE

Developing egg cells

Sperm cells (haploid)

Zygote (diploid)

Egg cell

FERTILIZATION

Sperm cell

3 Sperm cell fertilizes egg cell, producing diploid zygote.

Figure 26.13
The life cycle of the brown alga *Laminaria*, which alternates between a diploid sporophyte stage and a haploid gametophyte stage.

ucts as ice cream, pudding, salad dressing, jellybeans, cosmetics, paper, and floor polish. Brown algae are also harvested as food crops and fertilizers.

The Cercozoa Are Amoebas with Filamentous Pseudopods

Amoeba is a descriptive term for a single-celled protist that moves by means of temporary cellular projections called pseudopods. Several major groups of protists contain amoebas, which are similar in form but are not all closely related. The amoebas classified in cercozoa produce stiff, filamentous pseudopodia, and many produce hard outer shells, also called *tests*. We consider here two heterotrophic groups of cercozoan amoebae, the Radiolaria and the Foramin-

ifera, and a third, photosynthesizing group, the Chlorarachniophyta.

Radiolaria. Radiolarians are distinguished by axopods, slender, raylike strands of cytoplasm supported internally by long bundles of microtubules. They engulf prey organisms that stick to the axopods and digest them in food vacuoles.

Radiolarians live in marine environments. They secrete a glassy internal skeleton from which the axopods project (**Figure 26.14a** and **b**). Just outside the skeleton, the cytoplasm is crowded with frothy vacuoles and lipid droplets, which provide buoyancy.

The skeletons of dead radiolarians sink to the bottom and become part of the sediment. Over time, they harden into sedimentary rocks that form an important part of the geological record.

Foraminifera: Forams. Foraminifera, or forams, live in marine environments. Their shells consist of organic matter reinforced by calcium carbonate (**Figure**

a. Radiolarian skeletons

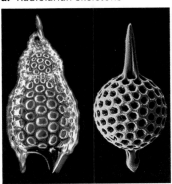

b. Living foram

Courtesy of Allen W. H. Be and David A. Caron

c. Foram shells

John Clegg/Ardea, London

d. Foram body plan

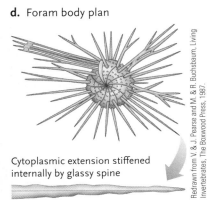

Cytoplasmic extension stiffened internally by glassy spine

Redrawn from V. & J. Pearse and M. & R. Buchsbaum, Living Invertebrates, The Boxwood Press, 1987.

Figure 26.14

Radiolarians and forams. **(a)** The internal skeletons of two radiolarian species, possibly *Pterocorys* and *Stylosphaera*. Bundles of microtubules support the cytoplasmic extensions of the radiolarians. **(b)** A living foram, showing the cytoplasmic strands extending from its shell. **(c)** Empty foram shells. **(d)** The body plan of a foram. Needlelike, glassy spines support the cytoplasmic extensions of the forams.

26.14c–d). Most foram shells are chambered, spiral structures that, although microscopic, resemble those of mollusks. Forams are identified and classified primarily by the form of the shell; about 250,000 species are known. Some species are planktonic, but they are most abundant on sandy bottoms and attached to rocks along the coasts. Their name comes from the perforations in their shells (*foramen* = little hole), through which extend long, slender strands of cytoplasm supported internally by a network of needlelike spines. The forams engulf prey that adhere to the strands and conduct them through the holes in the shell into the central cytoplasm, where they are digested in food vacuoles. Some forams have algal symbionts that carry out photosynthesis, allowing them to live as both heterotrophs and autotrophs.

Marine sediments are typically packed with the shells of dead forams. The sediments may be hundreds of feet thick; the White Cliffs of Dover in England are composed primarily of the shells of ancient forams. Most of the world's deposits of limestone and marble contain foram shells; the great pyramids and many other monuments of ancient Egypt are built from blocks cut from fossil foram deposits. Because distinct species lived during different geological periods, they are widely used to establish the age of sedimentary rocks containing their shells. They, along with radiolarian species, are also used as indicators by oil prospectors because layers of forams often overlie oil deposits.

Chlorarachniophyta. Chloroarachniophytes are green, photosynthetic amoebas that also engulf food. They contain chlorophylls *a* and *b*, but they are phylogenetically distinct from other chlorophyll *b*–containing eukaryotes. Many filamentous pseudopodia extend from the cell surface.

The Amoebozoa Includes Most Amoebas and Two Types of Slime Molds

Excavates | Discicristates | Alveolates | Heterokonts | Cercozoa | Amoebozoa | Opisthokonts | Archaeplastida

The Amoebozoa includes most of the amoebas (others are in the Cercozoa) as well as the cellular and plasmodial slime molds. All members of this group use pseudopods for locomotion and feeding for all or part of their life cycles.

Amoebas. Amoebas of the Amoebozoa are single-celled organisms that are abundant in marine and freshwater environments and in the soil. They use pseudopods for locomotion and feeding. The pseudopods extend and retract at any point on their body surface and are unsupported by any internal cellular organization. This type of pseudopod—called a *lobose* ("lobelike") *pseudopod*—distinguishes these amoebas from those in the Cercozoa, which have stiff, supported pseudopods. As a result of their pseudopod activity, and the ability to flatten or round up, these amoebas have no fixed body shape. A number of species are parasites, but most species feed on algae, bacteria, other protists, and bits of organic matter. The ingested matter is enclosed in food vacuoles and digested by enzymes secreted into the vacuoles. Any undigested matter is expelled to the outside by fusion of the vacuole with the plasma membrane. Their reproduction is entirely asexual, through mitotic divisions.

The most-studied amoebozoan is *Amoeba proteus* **(Figure 26.15)**. Its natural habitat is in freshwater ponds and streams. Another member, *Acanthamoeba*, which lives in the soil, is widely used as a source of actin and

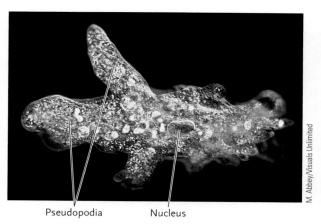

Pseudopodia   Nucleus

Figure 26.15
Amoeba proteus of the Amoebozoa is perhaps the most familiar protist of all.

myosin for scientific studies of amoeboid motion and cytoplasmic streaming.

The parasitic amoebas include some 45 species that infect the human digestive tract, one in the mouth and the rest in the intestine. One of the intestinal parasites, *Entamoeba histolytica,* causes amoebic dysentery. Cysts of this amoeba contaminate water supplies and soil in regions with inadequate sewage treatment. When ingested, a cyst breaks open to release an amoeba that feeds and divides rapidly in the digestive tract. Enzymes released by the amoebas destroy cells lining the intestine, producing the ulcerations, painful cramps, and debilitating diarrhea characteristic of the disease. Amoebic dysentery afflicts millions of people worldwide; in less-developed countries, it is a leading cause of death among infants and small children. Other parasitic amoebas cause less severe digestive upsets.

Slime Molds. Slime molds are heterotrophic protists that, at some stage of their life cycle, exist as individuals that move by amoeboid motion but the remainder of the time exist in more complex forms. They live on moist, rotting plant material such as decaying leaves and bark. The cells engulf particles of dead organic matter, and also bacteria, yeasts, and other microorganisms, and digest them internally. At one stage of their life cycles, they differentiate into a funguslike, stalked structure called a **fruiting body**, which forms spores by either asexual or sexual reproduction. Some species are brightly colored in hues of yellow, green, red, orange, brown, violet, or blue. The two major evolutionary lineages of slime molds, the cellular slime molds and the plasmodial slime molds, differ in cellular organization.

The Cellular Slime Molds. **Cellular slime molds** exist primarily as individual cells, either separately or as a coordinated mass. Among the 70 or so species of cellular slime molds, *Dictyostelium discoideum* is best known; its genome sequence was reported in May 2005. Its life cycle begins when a haploid spore lands in a suitably moist environment containing decaying organic matter **(Figure 26.16)**. The spore germinates

into an amoeboid cell that grows and divides mitotically into separate haploid cells as long as the food source lasts. When the food supply dwindles, some of the cells release a chemical signal (cyclic AMP; see Section 7.4) in pulses; in response, the amoebas move together and form a sausage-shaped mass that crawls in coordinated fashion like a slug. Some "slugs," although not much more than a millimeter in length, contain more than 100,000 individual cells. At some point the slug stops moving and differentiates into a stalked fruiting body, with cell walls reinforced by cellulose. When mature, the head of the fruiting body bursts, releasing spores that are carried by wind, water, or animals to new locations. Because the cells forming the slug and fruiting body are all products of mitosis, this pattern of reproduction is asexual.

Cellular slime molds also reproduce sexually by a pattern in which two haploid cells fuse to form a diploid zygote (also shown in Fig. 26.16) that enters a dormant stage. Eventually, the zygote undergoes meiosis, producing four haploid cells that may multiply inside the spore by mitosis. When conditions are favorable, the spore wall breaks down, releasing the cells. These grow and divide into separate amoeboid cells.

The Plasmodial Slime Molds. **Plasmodial slime molds** exist primarily as a large composite mass, the **plasmodium,** in which individual nuclei are suspended in a common cytoplasm surrounded by a single plasma membrane. (This is not to be confused with *Plasmodium,* the genus of apicomplexans that causes malaria.) There are about 500 known species of plasmodial slime molds. The main phase of the life cycle, the plasmodium (see Figure 26.1a), flows and feeds as a single huge amoeba—a single cell that contains thousands to millions or even billions of diploid nuclei surrounded by a single plasma membrane. Typically, a plasmodium, which may range in size from a few centimeters to more than a meter in diameter, moves in thick, branching strands connected by thin sheets. The movements occur by cytoplasmic streaming, driven by actin microfilaments and myosin (see Section 5.3). You may have seen one of these slimy masses crossing a lawn, moving over a mat of dead leaves, climbing a tree, or even in the movies—a slime mold in effect stars as a monster in the science fiction movie *The Blob.*

At some point, often in response to unfavorable environmental conditions, fruiting bodies form at sites on the plasmodium. At the tips of the fruiting bodies, nuclei become enclosed in separate cells, each surrounded by its own plasma membrane and cell wall. Depending on the species, either chitin or cellulose may reinforce the walls. These cells undergo meiosis, forming haploid, resistant spores that are released from the fruiting bodies and carried about by water or wind. If they reach a favorable environment, the spores germinate to form flagellated or unflagellated gametes, depending on the species, that fuse to form a diploid

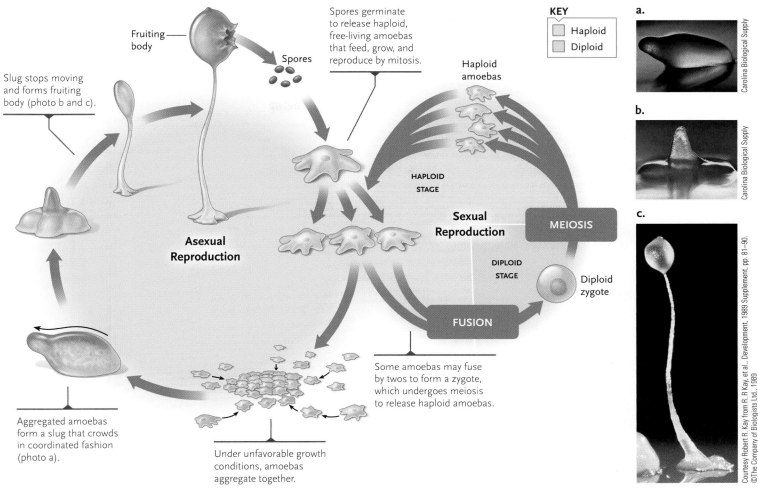

Fruiting body

Spores

Spores germinate to release haploid, free-living amoebas that feed, grow, and reproduce by mitosis.

Slug stops moving and forms fruiting body (photo b and c).

Haploid amoebas

HAPLOID STAGE

Asexual Reproduction

Sexual Reproduction

MEIOSIS

DIPLOID STAGE

Diploid zygote

FUSION

Some amoebas may fuse by twos to form a zygote, which undergoes meiosis to release haploid amoebas.

Aggregated amoebas form a slug that crowds in coordinated fashion (photo a).

Under unfavorable growth conditions, amoebas aggregate together.

KEY

Haploid
Diploid

a.

b.

c.

Figure 26.16
Life cycle of the cellular slime mold *Dictyostelium discoideum.* The light micrographs show **(a)** a migrating slug, **(b)** an early stage in fruiting body formation, and **(c)** a mature fruiting body.

zygote. The zygote nucleus then divides repeatedly without an accompanying division of the cytoplasm, forming many diploid nuclei suspended in the common cytoplasm of a new plasmodium.

The Slime Molds in Science. Both the cellular and plasmodial slime molds, particularly *Dictyostelium* (cellular) and *Physarum* (plasmodial; see Figure 26.1a), have been of great interest to scientists because of their ability to differentiate into fruiting bodies with stalks and spore-bearing structures. This differentiation is much simpler than the complex developmental pathways of other eukaryotes, providing a unique opportunity to study cell differentiation at its most fundamental level. One such study, examining the role of cyclic AMP in differentiation, is described in *Insights from the Molecular Revolution.*

The plasmodial slime molds are particularly useful in this kind of research because they become large enough to provide ample material for biochemical and molecular analyses. Actin and myosin extracted from *Physarum polycephalum,* for example, have been much used in studies of actin-based motility. A further advantage of plasmodial slime molds is that the many nuclei of a plasmodium usually replicate and pass

through mitosis in synchrony, making them useful in research tracking the changes that take place in the cell cycle.

The Archaeplastida Include the Red and Green Algae, and Land Plants

The Archaeplastida consist of the red and green algae, which are protists, and the land plants (the *viridaeplantae,* or "true plants"), which comprise the kingdom Plantae. These three groups have a common evolutionary origin, and they are all photosynthesizers. Here we describe the two types of algae; we discuss land plants in Chapter 27.

Rhodophyta: The Red Algae. Nearly all the 4000 known species of red algae, which are also known as the Rhodophyta (*rhodon* = rose), are small marine seaweeds **(Figure 26.17).** Fewer than 200 species are found in freshwater lakes and streams or in soils. Most red algae grow

Getting the Slime Mold Act Together

Development of differentiated structures can be followed at its simplest level in slime molds. In the cellular slime mold *Dictyostelium discoidium,* the aggregation of individual cells leading to differentiation begins when unfavorable living conditions induce some cells to secrete cyclic AMP (cAMP). Other *Dictyostelium* cells move toward the regions of highest cAMP concentration and aggregate into the slug stage. Further pulses of cAMP trigger differentiation into a stalk and spores.

Within the aggregating cells, the cAMP activates a cAMP-dependent protein kinase (PKA; see Section 7.4). The PKA, which is active only when cAMP is present, adds phosphate groups to target proteins in the cells. The target proteins, activated or deactivated by addition of the phosphate groups, trigger cellular developmental processes that lead to slug formation and differentiation of the stalk and spores.

These observations prompt several questions about development in *Dictyostelium.* Which is more important to the process, cAMP or the PKA? Is the PKA the only enzyme activated by the cAMP signal, or are other cAMP-activated pathways also essential to cell differentiation in the slime mold?

Adam Kuspa and his graduate student Bin Wang at Baylor College of Medicine in Houston, Texas, set out to answer these questions. They were aided by the availability of a mutant strain of *Dictyostelium* that lacks a normal form of *adenylyl cyclase,* the enzyme that converts ATP into cAMP.

Kuspa and Wang constructed an artificial gene by linking the promoter of an actin gene to the protein-encoding portion of a gene for the PKA. They chose the actin promoter because it is highly active and would induce essentially continuous transcription of the gene to which it is attached. The enzyme encoded in the artificial PKA gene was a modified form that does not require cAMP to be active, making it a *cAMP-independent* protein kinase. The researchers induced the mutant cells to take up the artificial gene by exposing them to Ca^{2+} ions (see Section 17.1). Once inside the cells, the actin promoter resulted in transcription of the artificial PKA gene, raising internal PKA concentration to levels about 1.6 times the amount in normal cAMP-activated cells.

Kuspa and Wang found that the cells with the artificial PKA gene aggregated into slugs when their cultures were deprived of food (in this case, bacteria). Moreover, the slugs differentiated normally into fruiting bodies. Tests for cAMP failed to detect the signal molecule, indicating that activated PKA by itself can trigger all the steps in the developmental pathway. Thus the requirement for cAMP in normal slime mold development is primarily or exclusively to stimulate the PKA. And, because development can proceed with active PKA alone, this protein kinase is probably more central to the growth and differentiation processes of the slime mold than is cAMP. Further, it appears from the results that no essential developmental pathways other than those involving PKA are triggered by cAMP.

These results are of more than passing interest because both cAMP and cAMP-dependent protein kinases are also active in animal development and intercellular signaling, including that of humans and other mammals. They also show that in *Dictyostelium,* a single molecule, the PKA normally activated by cAMP, can trigger all stages of development and differentiation.

---

attached to sandy or rocky substrates, but a few occur as plankton. Although most are free-living autotrophs, some are parasites that attach to other algae or plants.

Red algae are typically multicellular organisms, with plantlike bodies composed of interwoven filaments. The base of the body is differentiated into a holdfast, which anchors it to the bottom or other solid substrate, and into stalks with leaflike plates. Their cell walls contain cellulose and mucilaginous pectins that give them a slippery texture. In some species, the walls

Figure 26.17
Red algae.
(a) *Antithamnion plumula,* showing the filamentous and branched body form most common among red algae. **(b)** A sheetlike red alga growing on a tropical reef.

a. Filamentous red alga

Wim van Egmond

b. Sheetlike red alga

Douglas Faulkner/Sally Faulkner Collection

are hardened with stonelike deposits of calcium carbonate. Many of the red algae with stony cell walls resemble corals and occur with corals in reefs and banks.

Although most red algae are reddish in color, some are greenish purple or black. The color differences are produced by accessory pigments, mainly *phycobilins,* which mask the green color of their chlorophylls. The phycobilins are unusual photosynthetic pigments with structures related to the ring structure of hemoglobin. The accessory pigments of some red algae make them highly efficient in absorbing the shorter wavelengths of light that penetrate to the ocean depths, allowing them to grow at deeper levels than any other algae. Some red algae live at depths to 260 m if the water is clear enough to transmit light to these levels.

Red algae have complex reproductive cycles involving alternation between diploid sporophytes and haploid gametophytes. No flagellated cells occur in the red algae; instead, gametes are released into the water to be brought together by random collisions in currents.

Extracts containing the mucilaginous pectins of red algal cell walls are widely used in industry and science. Extracted **agar** is used as a moisture-preserving, inert agent in cosmetics and baked goods, as a setting agent for jellies and desserts, and as a culture medium in the laboratory. **Carrageenan,** extracted from the red alga *Eucheuma,* is used to thicken and stabilize paints, dairy products such as pudding and ice cream, and many other creams and emulsions.

Some red algae are harvested as food in Japan and China. *Porphyra,* one of these harvested algae, is used in sushi bars as the *nori* wrapped around fish and rice. Different *Porphyra* species have different flavors; all are nutritious.

Chlorophyta: The Green Algae. The green algae or Chlorophyta (*chloros* = green) are autotrophs that carry out photosynthesis using the same pigments as plants. They include single-celled, colonial, and multicellular species (**Figure 26.18;** see also Figure 26.1d). Most green algae are microscopic, but some range upward to the size of small seaweeds. Although the multicellular green algae have bodies that are filamentous, tubular, or leaflike, there is relatively little cellular differentiation as compared with the brown algae. However, the most complex green algae, such as the sea lettuce *Ulva* (see Figure 26.18c), have tissues differentiated into a leaflike body and a holdfast.

a. Single-celled green alga

b. Colonial green alga

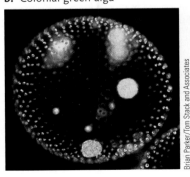

c. Multicellular green alga

Figure 26.18

Green algae. **(a)** A single-celled green alga, *Acetabularia,* which grows in marine environments. Each individual in the cluster is a large single cell with a rootlike base, stalk, and cap. **(b)** A colonial green alga, *Volvox.* Each green dot in the spherical wall of the colony is a potentially independent, flagellated cell. Daughter colonies can be seen within the parent colony. **(c)** A multicellular green alga, *Ulva,* common to shallow seas around the world.

With at least 16,000 species, green algae show more diversity than any other algal group. Most live in freshwater aquatic habitats, but some are marine, or live on rocks and soil surfaces, on tree bark, or even on snow. The green, slimy mat that grows profusely in stagnant pools and ponds, for example, consists of filaments of a green alga. A few species live as symbionts in other protists or in fungi and animals. Lichens (see Figure 28.14) are the primary example of a symbiotic relationship between green algae and fungi. Many animal phyla, including some marine snails and sea anemones, contain green algal chloroplasts, or entire green algae, as symbionts in their cells.

Life cycles among the green algae are as diverse as their body forms. Many can reproduce either sexually or asexually, and some alternate between haploid and diploid generations. Gametes in different species may be undifferentiated flagellated cells, or differentiated as a flagellated sperm cell and a nonmotile egg cell. Most common is a life cycle with a multicellular haploid phase and a single-celled diploid phase (**Figure 26.19**).

Among all the algae, the nucleic acid sequences of green algae are most closely related to those of land plants. In addition, as we have noted, green algae use the same photosynthetic pigments as plants, including chlorophylls *a* and *b*, and have the same complement of carotenoid accessory pigments. In some green algae, the thylakoid membranes within chloroplasts are arranged into stacks resembling the grana of plant chloroplasts (see Section 5.4). As storage reserves,

Figure 26.19
The life cycle of the green alga *Ulothrix*, in which the haploid stage is multicellular and the diploid stage is a single cell, the zygote. "+" and "−" are morphologically identical mating types ("sexes") of the alga.

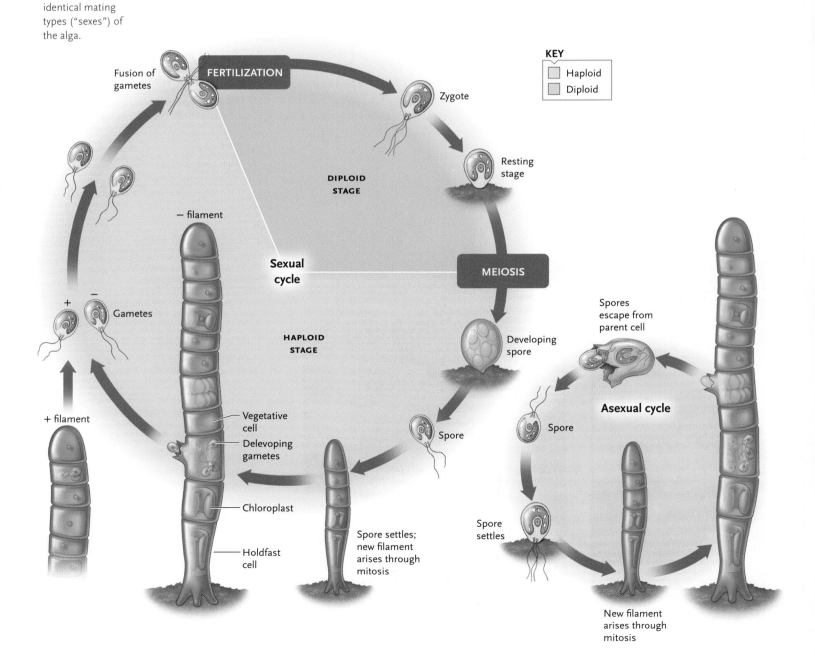

green algae contain starches of the same types as plants, and the cell walls of some green algal species contain cellulose, pectins, and other polysaccharides like those of plants. On the basis of these similarities, many biologists propose that some ancient green algae gave rise to the evolutionary ancestors of modern-day plants.

What green alga might have been the ancestor of modern land plants? Many biologists consider a group known as the **charophytes** to be most similar to the algal ancestors of land plants. These organisms, including *Chara* **(Figure 26.20)**, *Spirogyra, Nitella,* and *Coleochaete,* live in freshwater ponds and lakes. Their ribosomal RNA and chloroplast DNA sequences are more closely related to plant sequences than those of any other green alga. Further, the new cell wall separating daughter cells in charophytes is formed through development of a cell plate, by a mechanism closely similar to that of plants (see Section 10.2). The body form is distinctly plantlike, with a stemlike axis upon which whorls of leaflike blades occur at intervals.

The Opisthokonts Include the Choanoflagellates, Which May Be the Ancestors of Animals

Opisthokonts (*opistho* = posterior) are a broad group of eukaryotes that includes the choanoflagellates, protists thought to be the ancestors of fungi and animals. A single posterior flagellum is found at some stage in the life cycle of these organisms; sperm in animals is an example.

Choanoflagellata (*choanos* = collar) are named for a collar of closely packed microvilli that surrounds the single flagellum by which these protists move and take in food **(Figure 26.21).** The collar resembles an upside-down lampshade. There are about 150 species of choanoflagellates. They live in fresh and marine waters. Some species are mobile, with the flagellum pushing the cells along, as is the case with animal sperm, in contrast to most flagellates, which are pulled by their flagella. Most choanoflagellates, though, are *sessile;* that is, attached via a stalk to a surface. A number of species are colonial with a cluster of cells on a single stalk.

Choanoflagellates have the same basic structure as choanocytes (collar cells) of sponges, and they are similar to collared cells that act as excretory organs in organisms such a the flatworms and rotifers. These morphological similarities, as well as molecular sequence comparison data, indicate that a choanoflagel-

Reproductive structures

Dr. John Clayton, National Institute of Water and Atmospheric Research, New Zealand

Figure 26.20
The charophyte *Chara*, representative of a group of green algae that may have given rise to the plant kingdom.

late type of protist is likely to have been the ancestor of animals and, of course, of present-day choanoflagellates.

In Several Protist Groups, Plastids Evolved from Endosymbionts

We have encountered chloroplasts in a number of eukaryotic organisms in this chapter: red algae, green algae, land plants, euglenoids, dinoflagellates, heterokonts, and chlorarachniophytes. How did these chloroplasts evolve?

In Section 24.3 we discussed the endosymbiont hypothesis for the origin of eukaryotes. In brief, an anaerobic prokaryote ingested an aerobic prokaryote, which survived as an endosymbiont (see Figure 24.7). Over time, the endosymbiont became an organelle, the mitochondrion, which was incapable of free living, and the result was a true eukaryotic cell. Cells of animals,

Figure 26.21
A choanoflagellate.

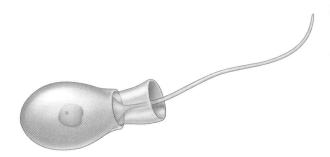

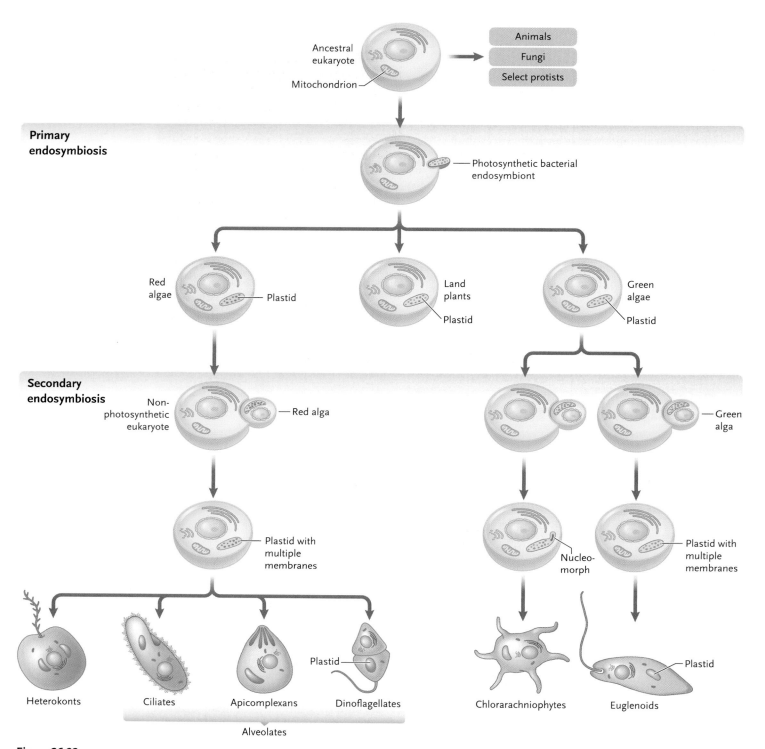

Figure 26.22
The origin and distribution of plastids among the eukaryotes by primary and secondary endosymbiosis.

fungi, and some protists derive from this ancestral eukaryote. The addition of plastids (the general term for chloroplasts and related organelles) through further endosymbiotic events produced the cells of all photosynthetic eukaryotes, including land plants, algae, and some other protists.

Figure 26.22 presents a model for the origin of plastids in eukaryotes through two major endosymbiosis events. First, in a single **primary endosymbiosis** event

perhaps 600 million years ago, a eukaryotic cell engulfed a photosynthetic cyanobacterium (a photosynthetic prokaryote, remember). In some such cells, the cyanobacterium was not digested, but instead formed a symbiotic relationship with the engulfing host cell—it become an endosymbiont. Over time the symbiont lost genes no longer required for independent existence, and most of the remaining genes migrated from the prokaryotic genome to the host's nuclear genome.

The symbiont had become an organelle—a chloroplast. All plastids subsequently evolved from this original chloroplast. Evidence for the single origin of plastids comes from a variety of sequence comparisons, including recent sequencing of the genomes of key protists, a red alga and a diatom.

The first photosynthesizing eukaryote was essentially an ancestral single-celled alga. The chloroplasts of the Archaeplastida—the red algae, green algae, and land plants—result from evolutionary divergence of this organism. Their chloroplasts, which originate from primary endosymbiosis, have two membranes, one from the plasma membrane of the engulfing eukaryote and the other from the plasma membrane of the cyanobacterium.

At least three **secondary endosymbiosis** events led to the plastids in other protists (see Figure 26.22). In each case, a nonphotosynthetic eukaryote engulfed a photosynthetic eukaryote, and new evolutionary lineages were produced. In one of these events, a red alga ancestor was engulfed and became an endosymbiont. In models accepted by a number of scientists, the transfer of functions that occurred over evolutionary time led to the chloroplasts of the heterokonts and the dinoflagellates. And, from the same photosynthetic ancestor, loss of chloroplast functions occurred

UNANSWERED QUESTIONS

What was the first eukaryote?

Since prokaryotes precede eukaryotes in the fossil record, we assume that eukaryotes arose after prokaryotes. The first eukaryote would have been some sort of protist—a single-celled organism with a nucleus and some rudimentary organelles, perhaps even a half-tamed mitochondrion. One approach to identifying which of the surviving protists is the most ancient has been to infer evolutionary trees from gene sequence data. To determine the earliest branching eukaryote, these trees need to include the prokaryotes. But herein lies the problem—prokaryotes are very distant, evolutionarily speaking, from even the simplest eukaryotes, and the mathematical models used to construct evolutionary trees are not yet up to the job. Initially, these models suggested that some protist parasites, like the excavates *Giardia* and *Trichomonas*, might be the most ancient eukaryotes, and this idea fit nicely with the fact that these protists lacked mitochondria. Indeed, for a time it was thought that the excavates might actually have diverged from the eukaryotic branch of life before the establishment of mitochondria. Nowadays, we know that *Giardia* and *Trichomonas* did initially have mitochondria. The latest research shows that they even have a tiny relic of the mitochondrion, though exactly what it does in these oxygen-shunning parasites remains to be figured out. Thus, trees depicting *Giardia* and *Trichomonas* at the base of the great expansion of eukaryotic life must be viewed with some caution—these protists might be the surviving representatives of the earliest cells with a nucleus, but they might not be. We simply need better methods for identifying just what the first eukaryotes were like.

How many times did plastids arise by endosymbioses?

For many years researchers thought that the green algae, plants, and red algae were the only organisms to have primary endosymbiosis-derived plastids. However, a second, independent primary endosymbiosis has been recently discovered in which a shelled amoeba has captured and partially domesticated a cyanobacterium. This organism, known as *Paulinella*, is a vital window into the process by which autotrophic eukaryotes first arose some 600 million years. *Paulinella* has tamed the cyanobacterium sufficiently to have it divide and segregate in coordination with host cell division, but the endosymbiont is still very much a cyanobacterium and has undergone little of the modification and streamlining we see in the red or green algal plastids.

After a primary endosymbiosis was established, the second chapter in plastid acquisition could take place. Secondary endosymbiosis involves a eukaryotic host engulfing and retaining a eukaryotic alga. Essentially, secondary endosymbiosis can convert a heterotrophic organism into an autotroph by hijacking a photosynthetic cell and putting it to work as a solar-powered food factory. Secondary endosymbiosis results in plastids with three or four membranes, and we know that it occurred at least three times—once for the euglenoids, once for the chlorarachniophytes, and once for the chromalveolates (a proposed grouping of heterokonts and alveolates). We can even tell what kind of endosymbiont was involved by the biochemistry and genetic makeup of the plastid: a green alga for euglenoids and chlorarachniophytes, and a red alga for chromalveolates. The number of secondary endosymbioses is hotly debated, largely because not all protistologists support the existence of chromalveolates. Some contend that there were multiple, independent enslavements of different red algae to produce the dinoflagellates, heterokonts, and apicomplexans. Understanding these events is crucial to confirming or refuting the proposed chromalveolate "supergroup."

A nice example of secondary endosymbiosis-in-action was recently discovered by Japanese scientists who found a flagellate, *Hatena*, with a green algal endosymbiont. *Hatena* hasn't yet assumed control of endosymbiont division and has to get new symbionts each time it divides, so it appears to be at a very early stage in establishing a relationship. We also want to know how secondary endosymbioses proceed because they have been a major driver in eukaryotic evolution. The heterokonts, for instance, are the most important ocean phytoplankton and are key to ocean productivity and global carbon cycling. Knowing exactly how they got to be autotrophs in the first place is fundamental to understanding the world we live in.

Dr. Geoff McFadden is a professor of botany at the University of Melbourne. He studies the early evolution of eukaryotes, especially the origin and evolution of plastids and mitochondria. You can learn more about his research by visiting http://homepage.mac.com/fad1/McFaddenLab.html.

in the lineage of the Apicomplexa, which have a remnant plastid. In an independent event, a nonphotosynthetic eukaryote engulfed a green alga ancestor. Subsequent evolution in this case produced the euglenoids. In a different event, a similar endosymbiosis involving a green alga led to the chlorarachniophytes. In these protists, the chloroplast is contained still within the remnants of the original symbiont cell, with a vestige of the original nucleus (the nucleomorph) also present.

Note that secondary endosymbiosis has produced plastids with additional membranes acquired from the new host, or series of hosts. For example, euglenoids have plastids with three membranes, while chlorarachniophytes have plastids with four membranes (see Figure 26.22). Sequencing the genomes of the chlorarachniophyte's nucleus, chloroplast, and vestigial nucleus is providing interesting information about the early endosymbiosis event that generated these organisms.

In sum, the protists are a highly diverse and ecologically important group of organisms. Their complex evolutionary relationships, which have long been the subject of contention, are now being revised as new information is discovered, including more complete genome sequences. A deeper understanding of protists is also contributing to a better understanding of their recent descendents, the fungi, plants, and animals. We turn to these descendents in the next four chapters, beginning with the fungi.

STUDY BREAK

1. What is the evidence that the Excavates, which lack mitochondria, derive from ancestors that had mitochondria rather than from ancestors that were in lineages that never contained mitochondria?
2. In primary endosymbiosis, a nonphotosynthetic eukaryotic cell engulfed a photosynthetic cyanobacterium. How many membranes surround the chloroplast that evolved?

Review

Go to ThomsonNOW™ at www.thomsonedu.com/login to access quizzing, animations, exercises, articles, and personalized homework help.

26.1 What Is a Protist?

- Protists are eukaryotes that differ from fungi in having motile stages in their life cycles and distinct cell wall molecules. Unlike plants, they lack true roots, stems, and leaves. Unlike animals, protists lack collagen, nerve cells, and an internal digestive tract, and they lack complex developmental stages (Figures 26.1 and 26.2).
- Protists are aerobic organisms that live as autotrophs or heterotrophs, or by a combination of both nutritional modes. Some are parasites or symbionts living in or among the cells of other organisms.
- Protists live in aquatic or moist terrestrial habitats, or as parasites within animals as single-celled, colonial, or multicellular organisms, and range in size from microscopic to some of Earth's largest organisms.
- Reproduction may be asexual by mitotic cell divisions, or sexual by meiosis and union of gametes in fertilization.
- Many protists have specialized cell structures including contractile vacuoles, food vacuoles, eyespots, and a pellicle, cell wall, or shell. Most are able to move by means of flagella, cilia, or pseudopodia (Figure 26.3).

26.2 The Protist Groups

- The Excavates, exemplified by the Diplomonadida and Parabasala are flagellated, single cells that lack mitochondria (Figure 26.4).
- The Discicristates are almost all single-celled, autotrophic or heterotrophic (some are both), motile protists that swim using flagella. The free-living, photosynthetic forms—the euglenoids—typically have complex cytoplasmic structures, including eyespots (Figures 26.5 and 26.6).
- Alveolates include the ciliates, dinoflagellates, and apicomplexans. The ciliates swim using cilia and have complex cytoplasmic structures and two types of nuclei, the micronucleus and macronucleus. The dinoflagellates swim using flagella and are primarily marine organisms; some are photosynthetic. The apicomplexans are nonmotile parasites of animals (Figures 26.7 and 26.8).
- Heterokonts include the funguslike Oomycota, which live as saprophytes or parasites, and three photosynthetic groups, the diatoms, golden algae, and brown algae. For most heterokonts, flagella occur only on reproductive cells. Many Oomycota grow as masses of microscopic hyphal filaments and secrete enzymes that digest organic matter in their surroundings. Diatoms are single-celled organisms covered by a glassy silica shell; golden algae are colonial forms; brown algae are primarily multicellular marine forms that include large seaweeds with extensive cell differentiation (Figures 26.9–26.13).
- Cercozoa are amoebas with filamentous pseudopods supported by internal cellular structures. Many produce hard outer shells. Radiolara (radiolarians) are primarily marine organisms that secrete a glassy internal skeleton. They feed by engulfing prey that adhere to their axopods. Foraminifera (forams) are marine, single-celled organisms that form chambered, spiral shells containing calcium. They engulf prey that adhere to the strands of cytoplasm extending from their shells. Chlorarachniophytes engulf food using their pseudopodia (Figure 26.14).
- The Amoebozoa includes most amoebas and two heterotrophic slime molds, cellular (which move as individual cells) and plasmodial (which move as large masses of nuclei sharing a common cytoplasm). The amoebas in this group are heterotrophs abundant in marine and freshwater environments and in the soil. They move by extending pseudopodia (Figures 26.15 and 26.16).
- The Archaeplastida include the red and green algae, as well as the land plants that comprise the kingdom Plantae. The red algae are typically multicellular, primarily photosynthetic organisms of marine environments, with plantlike bodies composed of interwoven filaments. They have complex life cycles including alternation of generations, with no flagellated cells at any stage. The green algae are single-celled, colonial, and multicel-

lular species that live primarily in freshwater habitats and carry out photosynthesis by mechanisms like those of plants; all produce flagellated gametes (Figures 26.17–26.20).

- The Opisthokonts are a broad group of eukaryotes that includes the choanoflagellates. These protists are characterized by a collar of microvilli surrounding a single flagellum. A choanoflagellate type of protist is considered likely to have been the ancestor of animals (Figure 26.21).

- Several groups of protists, as well as land plants, contain chloroplasts. Present-day chloroplasts and other plastids result from endosymbiosis events that took place millions of years ago: In a primary endosymbiosis event, a eukaryotic cell engulfed a cyanobacterium, which became an endosymbiont. Over time, the symbiont became an organelle, the chloroplast. This first photosynthesizing organism was a green alga. Evolutionary divergence produced the red algae, green algae, and land plants. By

secondary endosymbiosis, in which a nonphotosynthetic eukaryote engulfed a photosynthetic eukaryote, the various photosynthetic protists were produced (Figure 26.22).

Animation: Body plan of Euglena

Animation: Paramecium body plan

Animation: Ciliate conjugation

Animation: Apicomplexan life cycle

Animation: Red alga life cycle

Animation: Green alga life cycle

Animation: Amoeboid motion

Animation: Cellular slime mold life cycle

Questions

Self-Test Questions

1. Protists are characterized by:
   a. division by binary fission.
   b. multicellular structures.
   c. complex digestive systems.
   d. peptidoglycan cell walls.
   e. organelles and reproduction by meiosis/mitosis.

2. Which of the following is *not* found among the protist groups?
   a. life cycles
   b. contractile vacuoles
   c. pellicles
   d. collagen
   e. pseudopodia

3. Freely beating flagella buried in a fold of cytoplasm moving through viscous fluids of humans and commonly found as an infective agent in U.S. college health centers describes a member of:
   a. Ciliophora.
   b. Discicristates.
   c. Diplomonadida.
   d. Parabasala.
   e. Alveolates.

4. When *Paramecium* conjugate:
   a. cytoplasmic division produces four daughter cells, each having two micronuclei and two macronuclei.
   b. one haploid micronucleus in each cell remains intact; the other three degenerate. The micronucleus of each cell divides once, producing two nuclei, and each cell exchanges one nucleus with the other cell. In each partner the two micronuclei fuse, forming a diploid zygote micronucleus in each cell.
   c. and the partners disengage, the micronucleus of each divides meiotically. Macronuclei divide again in each cell and the original micronucleus breaks down. Each cell has two haploid micronuclei; one of the macronuclei develops into a micronucleus.
   d. the mating cells join together at opposite sites of their oral depression.
   e. the micronucleus in each cell undergoes mitosis. When mitosis is complete there are four diploid macronuclei; the micronucleus then breaks down.

5. The protist group Diplomonadida is characterized by:
   a. a mouthlike gullet and hairlike surface. *Paramecium* is an example.
   b. flagella and a lack of mitochondria. *Giardia* is an example.

   c. nonmotility, parasitism, and sporelike infective stages. *Toxoplasma* is an example.
   d. switching between autotrophic and heterotrophic life styles. *Euglena* is an example.
   e. large protein deposits. Movement is by two flagella, which are part of an undulating membrane. *Trypanosoma* is an example.

6. The greatest contributors to protist fossil deposits are:
   a. Oomycota.
   b. Chrysophyta.
   c. Bacillariophyta.
   d. Sporophyta.
   e. Alveolates.

7. The group with the distinguishing characteristic of gas-filled bladders and a cell wall composed of alginic acid is:
   a. Chrysophyta.
   b. Phaeophyta.
   c. Oomycota.
   d. Bacillariophyta.
   e. none of the preceding.

8. *Plasmodium* is transmitted to humans by the bite of a mosquito *(Anopheles)* and engages in a life cycle with infective spores, gametes, and cysts. This infective protist belongs to the group:
   a. Apicomplexa.
   b. Heterokonts.
   c. Dinoflagellata.
   d. Oomycota.
   e. Ciliophora.

9. Tripping on a rotten log, a hunter notices a mucus-looking mass moving slowly toward brightly colored fruiting bodies. The organisms in the mass are:
   a. amoebas in the group Cercozoa.
   b. slime molds.
   c. red algae.
   d. green algae.
   e. charophytes.

10. The latest stage for evolving the double membrane seen in modern day algal chloroplasts is thought to be the combining of:
    a. two ancestral nonphotosynthetic prokaryotes.
    b. two ancestral photosynthetic prokaryotes.
    c. a nonphotosynthetic eukaryote with a photosynthetic eukaryote.
    d. a photosynthetic prokaryote with a nonphotosynthetic eukaryote.
    e. mitochondria with an already established plastid.

Questions for Discussion

1. You decide to vacation in a developing country where sanitation practices and standards of personal hygiene are inadequate. Considering the information about protists covered in this chapter, what would you consider safe to drink in that country? What treatments could make the water safe to drink? What kinds of foods might be best avoided? What kinds of preparation might make foods safe to eat?

2. The overreproduction of dinoflagellates, producing red tides, is sometimes caused by fertilizer runoff into coastal waters. The red tides kill countless aquatic species, birds, and other wildlife. Would you consider drastic cutbacks in the use of fertilizers as a means to lessen the red tides? Why?

Experimental Analysis

Design an experiment to demonstrate whether the flagellated protist *Euglena* is phototropic, that is, is attracted to and moves toward light. Also propose a follow-up experiment (on the assumption of a positive result) to determine the wavelength range and light intensity range sufficient to cause phototropic movement.

Evolution Link

Use the Internet to research why studies of a molecular sensor, receptor tyrosine kinase (see Section 7.3), supports the hypothesis that a choanoflagellate type of protist is the ancestor of animals. Summarize your findings.

How Would You Vote?

The pathogen that causes sudden oak death has already infected 26 kinds of plants in California and Oregon. Some infected species are commonly sold as nursery stock. Should the states that are free of this pathogen be allowed to prohibit shipping of all plants from the states that are affected? Go to www.thomsonedu.com/login to investigate both sides of the issue and then vote.

A temperate forest with representatives of three major groups of land plants—mosses (bryophytes), conifers (gymnosperms), and flowering plants (angiosperms).

Animals, Animals—Earth Scenes

STUDY PLAN

27.1 The Transition to Life on Land

Early biochemical and structural adaptations enhanced plant survival on land

Vascular tissue was an innovation for transporting substances within a large plant body

Root and shoot systems were adaptations for nutrition and support

In the plant life cycle, the diploid phase became dominant

Some vascular plants evolved separate male and female gametophytes

27.2 Bryophytes: Nonvascular Land Plants

Liverworts may have been the first land plants

Hornworts have both plantlike and algalike features

Mosses most closely resemble vascular plants

27.3 Seedless Vascular Plants

Early seedless vascular plants flourished in moist environments

Modern lycophytes are small and have simple vascular tissues

Ferns, whisk ferns, horsetails, and their relatives make up the diverse phylum Pterophyta

27.4 Gymnosperms: The First Seed Plants

Major reproductive adaptations occurred as gymnosperms evolved

Modern gymnosperms include conifers and a few other groups

Cycads are restricted to warmer climates

Ginkgos are limited to a single living species

Gnetophytes include simple seed plants with intriguing features

Conifers are the most common gymnosperms

27.5 Angiosperms: Flowering Plants

The fossil record provides little information about the origin of flowering plants

Angiosperms are subdivided into several clades, including monocots and eudicots

Many factors contributed to the adaptive success of angiosperms

Angiosperms coevolved with animal pollinators

Current research focuses on genes underlying transitions in plant traits

27 Plants

WHY IT MATTERS

Ages ago, along the edges of the ancient supercontinent Laurentia, the only sound was the rhythmic muffled crash of waves breaking in the distance. There were no birds or other animals, no plants with leaves rustling in the breeze. In the preceding eons, oxygen-producing photosynthetic cells had come into being and had gradually changed the atmosphere. Solar radiation had converted much of the oxygen into a dense ozone layer—a shield against lethal doses of ultraviolet radiation, which had kept early organisms below the water's surface. Now, they could populate the land.

Cyanobacteria were probably the first to adapt to intertidal zones and then to spread into shallow, coastal streams. Later, green algae and fungi made the same journey. Seven to eight hundred million years ago, green algae living near the water's edge, or perhaps in a moist terrestrial environment, became the ancestors of modern plants. Several lines of evidence indicate that these algae were charophytes, a group discussed in Chapter 26. Today the **Kingdom Plantae** encompasses more than 300,000 living species, organized in this textbook into 10 phyla. These modern plants range from mosses, horsetails, and ferns to conifers and flowering plants **(Figure 27.1).** Most

575

a. Mosses growing on rocks

b. A ponderosa pine

Craig Wood/Visuals Unlimited

Robert Potts, California Academy of Sciences

c. An orchid

© Craig Allikas/www.orchidworks.com

Figure 27.1

Representatives of the Kingdom Plantae. **(a)** Mosses growing on rocks. Mosses evolved relatively soon after plants made the transition to land. **(b)** A ponderosa pine, *Pinus ponderosa*. This species and other conifers belonging to the phylum Coniferophyta represent the gymnosperms. **(c)** An orchid, *Cattalya rojo*, a showy example of a flowering plant.

Courtesy Microbial Culture Collection, National Institute for Environmental Studies, Japan

Figure 27.2

Chara, a stonewort. This representative of the charophyte lineage is known commonly as muskweed because of its skunky odor.

plants living today are terrestrial, and nearly all plants are multicellular autotrophs that use sunlight energy, water, carbon dioxide, and dissolved minerals to produce their own food. Together with photosynthetic bacteria and protists, plant tissues provide the nutritional foundation for nearly all communities of life. Humans also use plants as sources of medicinal drugs, wood for building, fibers used in paper and clothing, and a wealth of other products.

While the ancestors of land plants were making the transition to a fully terrestrial life, some remarkable adaptive changes unfolded. Eons of natural selection sorted out solutions to fundamental problems, among them avoiding desiccation, physically supporting the plant body in air, obtaining nutrients from soil, and reproducing sexually in environments where water would not be available for dispersal of eggs and sperm. With time, plants evolved features that not only addressed these problems but also provided access to a wide range of terrestrial environments. Those ecological opportunities opened the way for a dramatic radiation of varied plant species—and for the survival of plant-dependent organisms such as ourselves.

27.1 The Transition to Life on Land

Land plants and green algae share several fundamental traits: they have cellulose in their cell walls, they store energy captured during photosynthesis as starch, and their light-absorbing pigments include both chlorophyll *a* and chlorophyll *b*. Like other green algae, the charophyte lineage that produced the ancestor of land plants arose in water and has aquatic descendants today **(Figure 27.2)**. Yet because terrestrial environments pose very different challenges than aquatic environments, evolution in land plants produced a range of adaptations crucial to survival on dry land.

The algal ancestors of plants probably invaded land between 425 and 490 million years ago (mya). We say "probably" because the fossil record is sketchy in pinpointing when the first truly terrestrial plants appeared. A British and Arab research team working in the Middle East found fossilized tissue and spores from what appears to be a land plant in rocks dated to 475 mya. If the remains indeed are from a plant, they represent the earliest known plant fossils. Even in more recent deposits the most common finds of possible plant parts are microscopic bits and pieces. Obvious leaves, stems, roots, and reproductive parts seldom occur together, or if they do, it can be difficult to determine if the fossilized bits all belong to the same individual. Whole plants are extremely rare. Adding to the challenge, some chemical and structural adaptations to life on land arose independently in several plant lineages. Consequently, a fossil may have some but not all the features of modern land plants, leaving the puzzled paleobotanist to guess whether a given specimen was aquatic, terrestrial, or a transitional form. Despite these problems, botanists have been able to gain insight into several innovations and overall trends in plant evolution.

Early Biochemical and Structural Adaptations Enhanced Plant Survival on Land

To survive on land, plants had to have protection against drying out, a demand that had not been a problem for algae in their aquatic habitats. The earliest land plants may have benefited from an inherited ability to make **sporopollenin**, a resistant polymer that surrounds the zygotes of modern charophytes. In land plants, sporopollenin is a major component of the thick wall that protects reproductive spores from drying and other damage. Some of the first land plants also evolved an outer waxy layer called a **cuticle**, which slows water loss, helping to prevent desiccation **(Figure 27.3a)**. Another multifaceted adaptation was the presence of **stomata**, tiny passageways through the cuticle-covered surfaces **(Figure 27.3b)**. Stomata (singular, *stoma; stoma* = mouth), which can open and close, became the main route for plants to take up carbon di-

a. Cuticle on the surface of a leaf

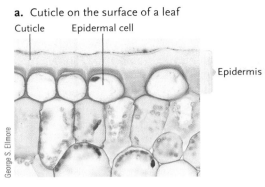

b. Stomata

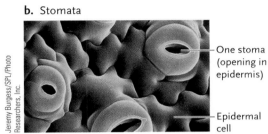

Figure 27.3
Land plant adaptations for limiting water loss. **(a)** A waxy cuticle, which covers the epidermis of land plants and helps reduce water loss. **(b)** Surface view of stomata in the epidermis (surface layer of cells) of a leaf. Stomata allow carbon dioxide and water to enter plant tissues and oxygen to leave.

oxide and control water loss by evaporation. The next unit describes these tissue specializations more fully.

By about 470 million years ago, land plants had split into two major groups, the **nonvascular plants**, or **bryophytes**, such as mosses, which lack internal transport vessels, and the **vascular plants**, or **tracheophytes**. This split correlates with the appearance of several fundamental adaptations in the vascular plant lineage **(Table 27.1)**. Transport vessels, which we describe shortly, was one adaptation. Another was **lignin**, a tough, rather inert polymer that strengthens the secondary walls of various plant cells and thus helps vascular plants to grow taller and stay erect on land, giving photosynthetic tissues better access to sunlight. Another was the **apical meristem**, a region of unspecialized dividing cells near the tips of shoots and roots. Descendants of such unspecialized cells differentiate and form all mature plant tissues. Meristem tissue is the foundation for a vascular plant's extensively branching stem parts, and is a central topic of Chapter 31.

Other land plant adaptations were related to the demands of reproduction in a dry environment. As described in more detail shortly, they included multicellular chambers that protect developing gametes, and a dependent, multicellular embryo that is sheltered inside tissues of a parent plant. Botanists use the term **embryophyte** (*phyton* = plant) as a synonym for land plants because all land plants produce an embryo during their reproductive cycle.

Vascular Tissue Was an Innovation for Transporting Substances within a Large Plant Body

The Latin *vas* means duct or vessel, and vascular plants have specialized tissues made up of cells arranged in lignified, tubelike structures that branch throughout the plant body, conducting water and solutes. One type

---

Table 27.1 | **Trends in Plant Evolution**

Traits derived from algal ancestor: cell walls with cellulose, energy stored in starch, two forms of chlorophyll (*a* and *b*); possibly, sporopollenin in spore wall

Bryophytes	Ferns and Their Relatives	Gymnosperms	Angiosperms	Functions in Land Plants
Cuticle ———————————————————————————————→				Protection against water loss, pathogens
Stomata ———————————————————————————————→				Regulation of water loss and gas exchange (CO_2 in, O_2 out)
Nonvascular ———→	Vascular ——————————————————————————→			Internal tubes that transport water, nutrients
	Lignin ——————————————————————————————→			Mechanical support for vertical growth
	Apical meristem ——————————————————————→			Branching shoot system
	Roots, stems, leaves ————————————————————→			Enhanced uptake, transport of nutrients and enhanced photosynthesis
Haploid phase dominant ———→	Diploid phase dominant ——————————————————→			Genetic diversity
One spore type (homospory) ———→	Two spore types (heterospory) ————————————→			Promotion of genetic diversity
Motile gametes ———————————————→		Nonmotile gametes ————————————→		Protection of gametes within parent body
Seedless ———————————————————→		Seeds ————————————————————→		Protection of embryo

Figure 27.4
Fossil of one of the earliest vascular plants, *Cooksonia*, which dates to about 420 mya. Cooksonia was small and, as this image shows, its stems lacked leaves and probably were less than 3 cm long. The cup-shaped structures at the top of the stems produced reproductive spores.

Reprinted with permission from Elsevier

of vascular tissue, called **xylem**, distributes water and dissolved mineral ions through plant parts. Another vascular tissue, **phloem**, distributes sugars manufactured during photosynthesis. Chapter 32 explains how xylem and phloem perform these key internal transport functions.

Ferns, conifers, and flowering plants—most of the plants you are familiar with—are vascular plants. Supported by lignin and with a well-developed vascular system, the body of a plant can grow large. Extreme examples are the giant redwood trees of the northern California coast, some of which are more than 300 feet tall. By contrast, nonvascular plants lack lignin, and have very simple internal transport systems, or none at all. As a result, modern nonvascular plants generally are small, as are the examples you will read about shortly.

Root and Shoot Systems Were Adaptations for Nutrition and Support

The body of a bryophyte is not differentiated into true roots and stems—structures that are fundamental adaptations for absorbing nutrients from soil and for support of an erect plant body. The evolution of sturdy stems—the basis of an aerial *shoot system*—went hand in hand with the capacity to synthesize lignin. To become large, land plants would also require a means of anchoring aerial parts in the soil, as well as effective strategies for obtaining soil nutrients. **Roots**—anchoring structures that also absorb water and nutrients—were the eventual solution to these problems. The earliest fossils showing clear evidence of roots are from vascular plants, although the exact timing of this change is uncertain. The first unquestioned fossils of a vascular plant, a small plant called *Cooksonia* **(Figure 27.4),** were found in deposits that date to about 420 mya. *Cooksonia* fossils have been unearthed in various locales but, frustratingly, none has ever included the lower portion of the plant—only its leafless, branching upper stems. *Cooksonia* probably was supported physically only by a **rhizome**—a horizontal, modified stem that can penetrate a substrate and anchor the plant. At some point, however, ancestral forms of vascular plants did come to have true roots. Ultimately, vascular plants developed specialized **root systems,** which generally consist of underground, cylindrical absorptive structures with a large surface area that favors the rapid uptake of soil water and dissolved mineral ions.

Above ground, the simple stems of early land plants also became more specialized, evolving into **shoot systems** in vascular plants. Shoot systems have stems and leaves that arise from apical meristems and that function in the absorption of light energy from the sun and carbon dioxide from the air. Stems grew larger and branched extensively after the evolution of lignin. The mechanical strength of lignified tissues almost certainly provided plants with several adaptive advantages. For instance, a strong, internal scaffold could support upright stems bearing leaves and other photosynthetic structures—and so help increase the surface area for intercepting sunlight. Also, reproductive structures borne on aerial stems might serve as platforms for more efficient launching of spores from the parent plant.

Structures we think of as "leaves" arose several times during plant evolution. In general, leaves represent modifications of stems, and **Figure 27.5** illustrates the basic steps of two main evolutionary pathways. In at least one early group of plants, the club mosses described in Section 27.3, leaflike parts evolved as outgrowths of the plant's main vertical axis (see Figure 27.5a). In other groups, leaves arose when small, neighboring stem branches became joined by thin, weblike tissue containing cells that had chloroplasts (see Figure 27.5b).

a. Leaf development as an offshoot of the main vertical axis

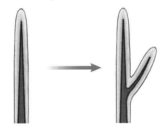

b. Development of leaves in a branching pattern

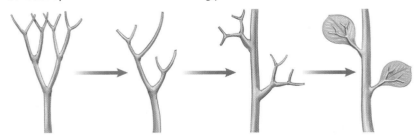

Figure 27.5
Evolution of leaves. **(a)** One type of early leaflike structure may have evolved as offshoots of the plant's main vertical axis; there was only one vein (transport vessel) in each leaf. Today, the seedless vascular plants known as lycophytes (club mosses) have this type of leaf. **(b)** In other groups of seedless vascular plants, leaves arose in a series of steps that began when the main stem evolved a branching growth pattern. Small side branches then fanned out and photosynthetic tissue filled the space between them, becoming the leaf blade. With time the small branches became modified into veins.

In the Plant Life Cycle, the Diploid Phase Became Dominant

As early plants moved into drier habitats, their life cycles also were modified considerably. You may recall that in sexually reproducing organisms, meiosis in diploid cells produces haploid (*n*) reproductive cells (see Chapter 11). These cells may be gametes—sperm or eggs—or they may be **spores**, which can give rise to a new haploid individual asexually, without mating.

As noted in Chapter 26, in green algae the haploid phase that starts at meiosis is usually the greater part of the life cycle, and the haploid alga spends much of its life producing and releasing gametes into the surrounding water. A much shorter diploid (2*n*) phase starts when gametes fuse at fertilization. Plants also cycle between haploid and diploid life phases, a phenomenon called **alternation of generations (Figure 27.6).** The diploid generation produces haploid spores and is called a **sporophyte** ("spore grower"). The haploid generation produces gametes and is called a **gametophyte** ("gamete grower"). As plants evolved on land, the haploid gametophyte phase became physically smaller and less complex and had a shorter life span while just the opposite occurred with the diploid sporophyte phase. In mosses and other nonvascular plants the sporophyte is a little larger and long-lived than in green algae, and in vascular plants the sporophyte clearly is larger and more complex and lives much longer than the gametophyte **(Figure 27.7).** When you look at a pine tree, for example, you see a large, long-lived sporophyte.

The sporophyte generation begins after fertilization, when the resulting zygote grows mitotically into a multicellular, diploid organism. Its body will eventually develop capsules called **sporangia** ("spore chambers"; singular, *sporangium*), which produce spores. Many botanists hypothesize that the trend toward "diploid dominance" in vascular plants reflects the advantages conferred by genetic diversity in land environments, where the supply of water and nutrients is inconsistent. Whereas haploid organisms are genetically identical to the parent, in a changeable environment the new combinations of parental alleles in a diploid organism may provide the genetic basis for adaptations to varying circumstances.

The haploid phase of the plant life cycle begins in the reproductive parts of the sporophyte. There, meiosis produces haploid spores in the sporangia. The spores then divide by mitosis and give rise to multicellular haploid gametophytes. A gametophyte's function is to nourish and protect the forthcoming generation. Unlike nonvascular plants, most groups of vascular plants retain spores and gametophytes until environmental conditions favor fertilization.

Some Vascular Plants Evolved Separate Male and Female Gametophytes

As already noted, during sexual reproduction in plants, meiosis produces spores. When a plant makes only one type of spore it is said to be **homosporous** ("same spore"). A gametophyte that develops from such a spore is bisexual—it can produce both sperm and eggs.

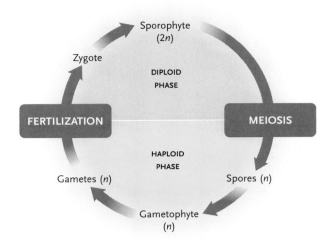

The sperm have flagella and are motile, for they must swim through liquid water in order to encounter female gametes. Other vascular plants, including gymnosperms and angiosperms, are **heterosporous.** They produce two types of spores in two different types of sporangia, and those spores develop into small, sexually different gametophytes. The smaller spore type develops into a male gametophyte—a *pollen grain.* The larger one develops into a female gametophyte, in which eggs form and fertilization occurs. The pollen grains of most vascular plants produce nonmotile sperm and also the structures required to deliver them to the egg.

Figure 27.6
Overview of the alternation of generations, the basic pattern of the plant life cycle. The relative dominance of haploid and diploid phases is different for different plant groups (compare with Figure 27.7).

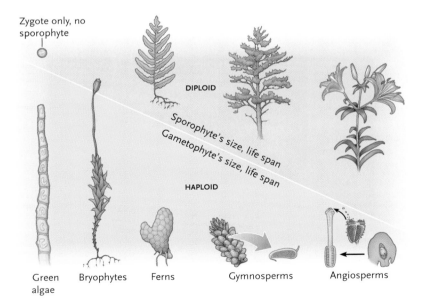

Figure 27.7
Evolutionary trend from dominance of the gametophyte (haploid) generation to dominance of the sporophyte (diploid) generation, represented here by existing species ranging from a green alga *(Ulothrix)* to a flowering plant. This trend developed as early plants were colonizing habitats on land. In general, the sporophytes of vascular plants are larger and more complex than those of bryophytes, and their gametophytes are reduced in size and complexity. In this diagram the fern represents seedless vascular plants.

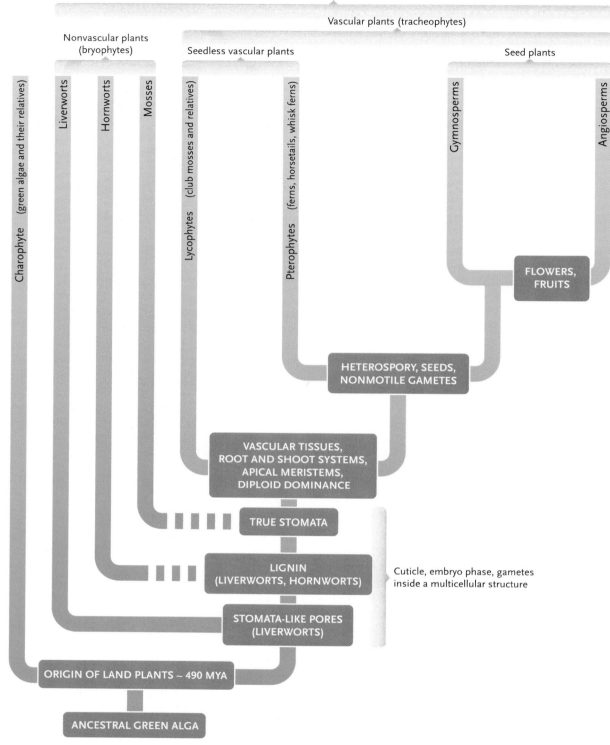

Figure 27.8
Overview of the possible phylogenetic relationships between major groups of land plants. Plant systematists do not agree on the relative place of bryophytes in this evolutionary history, hence the dashed lines. This diagram provides only a general picture of the points in land plant evolution where major adaptations took hold. For example, heterospory and seeds are shown as adaptations common to all seed plants, but some living fern species also are heterosporous. Fossil evidence indicates that certain ancient lycophytes and horsetails also produced two types of spores and some had seeds as well. Cycads and ginkgoes are unlike other gymnosperms in that they have motile sperm.

Labels within the figure:

Land plants
Vascular plants (tracheophytes)
Nonvascular plants (bryophytes)
Seedless vascular plants
Seed plants

Charophyte (green algae and their relatives)
Liverworts
Hornworts
Mosses
Lycophytes (club mosses and relatives)
Pterophytes (ferns, horsetails, whisk ferns)
Gymnosperms
Angiosperms

FLOWERS, FRUITS
HETEROSPORY, SEEDS, NONMOTILE GAMETES
VASCULAR TISSUES, ROOT AND SHOOT SYSTEMS, APICAL MERISTEMS, DIPLOID DOMINANCE
TRUE STOMATA
LIGNIN (LIVERWORTS, HORNWORTS)
Cuticle, embryo phase, gametes inside a multicellular structure
STOMATA-LIKE PORES (LIVERWORTS)
ORIGIN OF LAND PLANTS ~ 490 MYA
ANCESTRAL GREEN ALGA

As you will read in a later section, the evolution of pollen grains and pollination helped spark the rapid diversification of plants in the Devonian period, 408–360 mya. During this time another innovation, the seed, contributed to this diversification. In fact, so many new fossils appear in Devonian rocks that paleobotanists—scientists who specialize in the study of fossil plants—have thus far been unable to determine which fossil lineages gave rise to the modern plant phyla. Clearly, however, as each major lineage came into being, its characteristic adaptations included major modifications of existing structures and functions **(Figure 27.8)**. The next sections fill out this general picture, beginning with the plants that most clearly resemble the plant kingdom's algal ancestors.

STUDY BREAK

1. How did plant adaptations such as a root system, a shoot system, and a vascular system collectively influence the transition to terrestrial life?
2. Describe the difference between homospory and heterospory, and explain how heterospory paved the way for other reproductive adaptations in land plants.

27.2 Bryophytes: Nonvascular Land Plants

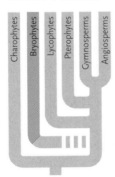

The **bryophytes** (*bryon* = moss)—liverworts, hornworts, and mosses—have a curious combination of traits that allow them to bridge aquatic and land environments. Because bryophytes lack a well-developed system for conducting water, it is not surprising that they commonly grow on wet sites along creek banks or on rocks just above running water; in bogs, swamps, or the dense shade of damp forests; and on moist tree trunks or rooftops. Some species are **epiphytes** (*epi* = upon)—they grow independently (that is, not as a parasite) on another organism and in a host of other moist places, ranging from the splash zone just above high tide on rocky shores, to edges of snowbanks, to coastal salt marshes.

In general, bryophytes are strikingly algalike. They produce flagellated sperm that must swim through water to reach eggs, and as noted they do not have a complex vascular system (although some have a primitive type of conducting tissue). Bryophytes have parts that are rootlike, stemlike, and leaflike. However, the

"roots" are rhizoids (slender rootlike structures), and bryophyte "stems" and "leaves" did not evolve from the same structures as vascular plant stems and leaves did. (Said another way, stems and leaves are not homologous in the two groups.) Also, as already mentioned, bryophyte tissues do not contain lignin. The absence of this strengthening material and the lack of internal pipelines for efficient nutrient transport partly account for bryophytes' small size—typically less than 20 cm long—and for their tendency to grow sprawled along surfaces instead of upright.

In other ways, bryophytes are clearly adapted to land. Along with their leaflike, stemlike, and fibrous, rootlike organs, sporophytes of some species have a water-conserving cuticle and stomata. Like most plants, bryophytes also have both sexual and asexual reproductive modes. And as is true of all plants, the life cycle has both gametophyte (*n*) and sporophyte (*2n*) phases, though the sporophyte is tiny and lives only a short time. **Figure 27.9** shows the green, leafy gametophyte of a moss plant, with miniscule diploid sporophytes attached to it by slender stalks. Bryophyte gametophytes produce gametes sheltered within a layer of protective cells called a **gametangium** (plural, *gametangia*). The gametangia in which bryophyte eggs form are flask-shaped structures called **archegonia** (*archi* = first, *gonos* = seed). Flagellated sperm form in rounded gametangia called **antheridia** (*antheros* = flowerlike). The sperm swim through a film of water to the archegonia and fertilize eggs. Each fertilized egg gives rise to a diploid embryo sporophyte, which stays attached to the gametophyte, produces spores—and the cycle repeats.

Despite these similarities with more complex plants, bryophytes are unique in several ways. Unlike in vascular species, the gametophyte is much larger than the sporophyte and obtains its nutrition independently of the sporophyte body. In fact, the comparatively tiny sporophyte remains attached to the gametophyte and depends on the gametophyte for much of its nutrition.

Because of bryophytes' mix of characteristics, their position in plant evolution is still an open question. The basic bryophyte body plan may be similar to the ancestral condition from which higher plants evolved, but it is also possible that bryophytes represent structurally simplified vascular plant lineages that evolved after vascular plants had already appeared. In another view, they are a side shoot of evolution, completely separate from the path that led to vascular plants. The fossil record provides little help in resolving the issue, because the first undisputed bryophyte fossils appear in late-Devonian rocks 350 mya, after vascular plants were already on the scene. (Fossil remains that may resemble liverworts, however, have recently been discovered in rocks that are 50 to 100 million years older.)

Despite questions raised by recent fossil finds, most current molecular, biochemical, cellular, and

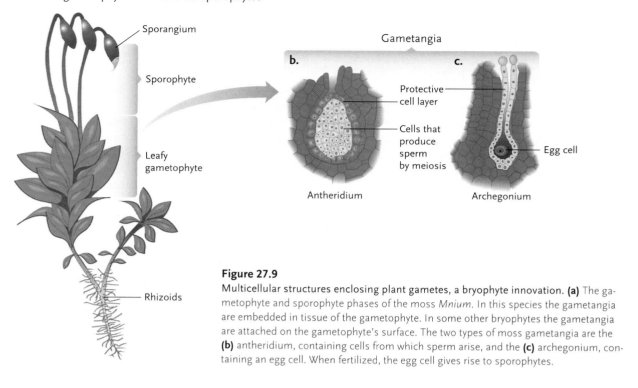

a. Moss gametophyte with attached sporophytes

Sporangium

Sporophyte

Leafy
gametophyte

Rhizoids

Gametangia

b.

c.

Protective
cell layer

Cells that
produce
sperm
by meiosis

Egg cell

Antheridium

Archegonium

Figure 27.9

Multicellular structures enclosing plant gametes, a bryophyte innovation. (a) The gametophyte and sporophyte phases of the moss *Mnium*. In this species the gametangia are embedded in tissue of the gametophyte. In some other bryophytes the gametangia are attached on the gametophyte's surface. The two types of moss gametangia are the **(b)** antheridium, containing cells from which sperm arise, and the **(c)** archegonium, containing an egg cell. When fertilized, the egg cell gives rise to sporophytes.

morphological evidence supports the view that bryophytes are not a monophyletic group. Instead, the various bryophytes evolved as separate lineages, in parallel with vascular plants. The relationships are far from resolved, however. For example, molecular evidence can be interpreted to mean that liverworts diverged early on from the lineage that led to all other land plants, with hornworts diverging later and mosses later still. Until new discoveries and interpretive work clarify this picture, the classification scheme in this chapter places liverworts, hornworts, and mosses in separate phyla. Our survey of nonvascular plants begins with the liverworts and hornworts, the simplest of the group, and concludes with mosses—plants that not only are more familiar to most of us, but whose structure and physiology more closely resemble that of vascular plants.

Liverworts May Have Been the First Land Plants

Liverworts make up the phylum **Hepatophyta**, and early herbalists thought that these small plants were shaped like the lobes of the human liver (*hepat* = liver; *wort* = herb). The resemblance might be a little vague to modern eyes: many of the 6000 species of liverworts consist of a flat, branching, ribbonlike plate of tissue closely pressed against damp soil. This simple body, called a **thallus** (plural, *thalli*) is the gametophyte generation. Threadlike rhizoids anchor the gametophytes to their substrate. About two-thirds of liverwort species have leaflike structures and some have stemlike parts.

None have true stomata, the openings that regulate gas exchange in most other land plants, although some species do have pores that open and close. Mitochondrial gene sequence data show that liverworts lack a few features (three introns) that are present in other bryophytes and in vascular plants. Taken together with liverwort morphology, this finding has led many researchers to conclude that liverworts were probably the first land plants.

In species of the liverwort genus *Marchantia* **(Figure 27.10)**, male and female gametophytes are separate plants. Male plants produce antheridia and female plants produce archegonia on specialized stalked organs (see Figure 27.10a–b). The motile sperm released from an antheridium swim through surface water, and some eventually encounter an egg inside an archegonium of a female gametophyte. After fertilization, a small, diploid sporophyte develops inside the archegonium, matures there, and produces haploid spores by meiosis. During meiosis, *Marchantia* sex chromosomes segregate, so some spores have the male genotype and others the female genotype. As in other liverworts, the spores develop inside jacketed sporangia that split open to release the spores. The capsules contain elongated cells twisted into a corkscrew shape. When certain regions of the cell wall absorb water and swell, the "corkscrews" rapidly unwind, helping to eject spores to the outside. A spore that is carried by air currents to a suitable location germinates and gives rise to a haploid gametophyte, which is either male or female. *Marchantia* also can reproduce asexually by way of **gemmae** (*gem* = bud), small cell masses that form in cuplike growths on a thallus (see Figure 27.10c). Gemmae can grow into new thalli when rainwater splashes them out of the cups and onto an appropriately moist substrate.

Hornworts Have Both Plantlike and Algalike Features

Roughly 100 species of hornworts make up the phylum **Anthocerophyta.** Many of them have cell features in common with green algae, including the presence in each cell of a single large chloroplast that contains algalike protein bodies called pyrenoids. No other group of land plants has this feature, and some biologists have speculated that the distinction of "first land plant" should be assigned not to liverworts but to hornworts instead. Like some liverworts, a hornwort gametophyte has a flat thallus, but the sporangium of the sporophyte phase is long and pointed, like a horn **(Figure 27.11).** The sporangia split into two or three ribbonlike sections when they release spores. Sexual reproduction occurs in basically the same way as in liverworts: free-swimming sperm fertilize eggs, which give rise to the sporophytes. Hornworts sometimes reproduce asexually by fragmentation as pieces of a thallus break off, form rhizoids, and develop into new individuals.

Mosses Most Closely Resemble Vascular Plants

Chances are that you have seen, touched, or sat upon at least several of the approximately 10,000 species of mosses, and the use of the name **Bryophyta** for this phylum underscores the fact that mosses are the best-known bryophytes. They also are structurally and functionally most similar to the vascular plants we will consider in following sections. Their spores, produced by the tens of millions in sporangia, give rise to thread-like, haploid gametophytes that grow into the familiar moss plants, which often form tufts or carpets of vegetation on the surface of rocks, soil, or bark.

The moss life cycle, diagrammed in **Figure 27.12,** begins when a haploid (*n*) spore lands on a wet soil surface. After the spore germinates it elongates and branches into a filamentous web of tissue called a **protonema** ("first thread"), which can become dense enough to color the surface of soil, rocks, or bark visibly green. After several weeks of growth, the bud-like cell masses on a protonema develop into leafy, green gametophytes anchored by rhizoids. A single protonema can be extremely prolific, producing bud after bud—and in this way giving rise to a dense clone of genetically identical gametophytes. Leafy mosses also may reproduce asexually by gemmae produced at the surface of rhizoids as well as on above-ground parts.

When a leafy moss is sexually mature, gametangia develop on its gametophytes and gametes form in

a. Male plant   **b.** Female plant   **c.** Asexual reproduction

Male gametophyte   Female gametophyte   Gemmae

Martin Hutten/National Park Service   *Paul Stehr-green/National Park Service*   *Wayne P. Armstrong, Professor of Biology and Botany, Palomar College, San Francisco, CA*

Figure 27.10

The bryophyte *Marchantia*, the only liverwort to produce **(a)** male and **(b)** female gametophytes on separate plants. *Marchantia* also reproduces asexually by way of **(c)** gemmae, multicellular vegetative bodies that develop in tiny cups on the plant body. Gemmae can grow into new plants when splashing raindrops transport them to suitable sites.

them. In some moss genera, plants are unisexual and produce male *or* female gametangia—antheridia at the tips of male gametophytes and archegonia at the tips of female gametophytes. In other genera, plants are bisexual and produce both antheridia and archegonia. Propelled by a pair of flagella, sperm released from antheridia swim through a film of dew or rainwater and down a channel in the neck of the archegonium, attracted by a chemical gradient secreted by each egg. Fertilization produces the new sporophyte generation inside the archegonium, in the form of diploid zygotes that develop into small, mature sporophytes, each consisting of a sporangium on a stalk. Moss sporophytes may eventually develop chloroplasts and nourish themselves photosynthetically, but initially they depend on the gametophytes for food. And even after a moss sporophyte begins photosynthesis, it still must obtain water, carbohydrates, and some other nutrients from the gametophyte.

Certain moss gametophytes are structurally complex, with features similar to those of higher plants. For example, some species have a central strand of primitive conducting tissue. One kind of tissue is made up of elongated, thin-walled, dead and empty cells that

Figure 27.11

The hornwort *Anthoceros.* The base of each long, slender sporophyte is embedded in the flattened, leafy gametophyte.

Clive@hiddenforest.co.nz (www.hiddenforest.com)

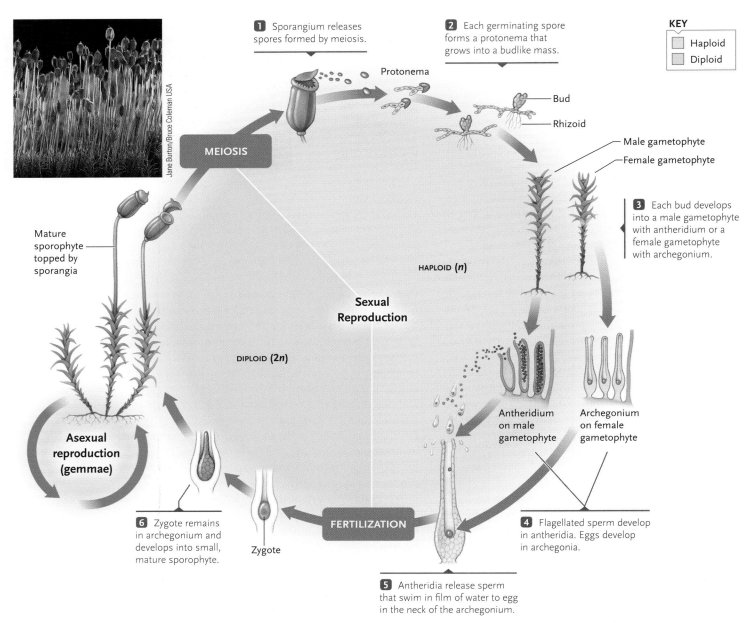

1 Sporangium releases spores formed by meiosis.

2 Each germinating spore forms a protonema that grows into a budlike mass.

Protonema

KEY
Haploid
Diploid

Bud
Rhizoid

Male gametophyte
Female gametophyte

MEIOSIS

3 Each bud develops into a male gametophyte with antheridium or a female gametophyte with archegonium.

Mature sporophyte topped by sporangia

HAPLOID (*n*)

Sexual Reproduction

DIPLOID (*2n*)

Asexual reproduction (gemmae)

Antheridium on male gametophyte

Archegonium on female gametophyte

6 Zygote remains in archegonium and develops into small, mature sporophyte.

FERTILIZATION

Zygote

4 Flagellated sperm develop in antheridia. Eggs develop in archegonia.

5 Antheridia release sperm that swim in film of water to egg in the neck of the archegonium.

Jane Burton/Bruce Coleman USA

Figure 27.12
Life cycle of a moss, *Polytrichum*.

conduct water. These specialized cells, called hydroids, have oblique end walls that sometimes are partly dissolved or perforated with pores. Experiments with dyes show that water moves through them, as it does in similar xylemlike arrangements in vascular plants (see Chapters 31 and 32). In a few mosses, the water-conducting cells are surrounded by sugar-conducting tissue resembling the phloem of vascular plants.

Mosses and other bryophytes are important both ecologically and economically. As colonizers of bare land, their small bodies trap particles of organic and inorganic matter, helping to build soil on bare rock and stabilizing soil surfaces with a biological crust in harsh places like coastal dunes, inland deserts, and embankments created by road construction. Some hornworts harbor mutualistic nitrogen-fixing cyanobacteria, and so increase the amount of nitrogen available to other plants. In arctic tundras, bryophytes constitute as

much as half the biomass, and they are crucial components of the food web that supports animals in that ecosystem. People have long used *Sphagnum* and other absorbent "peat" mosses (which typically grow in bogs) for everything from diapering babies and filtering whiskey to increasing the water-holding capacity of garden soil. Peat moss also has found use as a fuel; each day the Rhode generating station in Ireland, one of several in that nation, burns 2000 metric tons of peat to produce electricity.

In the next section we turn to the vascular plants, which have specialized tissues that can transport water, minerals, and sugars. Without the capacity to move these substances efficiently throughout the plant body, large sporophytes could not have survived on land. Unlike bryophytes, modern vascular plants are monophyletic—all groups are descended from a common ancestor.

27.3 Seedless Vascular Plants

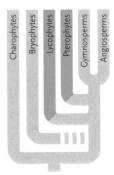

The first vascular plants, which did not "package" their embryos inside protective seeds, were the dominant plants on Earth for almost 200 million years, until seed plants became abundant. The fossil record shows that seedless vascular plants were well established by the late Silurian, some 428 mya, and they flourished until the end of the Carboniferous, about 250 mya. Some living seedless vascular plants have certain bryophyte-like traits, whereas others have some characteristics of seed plants. On one hand, like bryophytes, seedless vascular plants reproduce sexually by releasing spores, and they have swimming sperm that require free water to reach eggs. On the other hand, as in seed plants, the sporophyte of a seedless vascular plant separates from the gametophyte at a certain point in its development and has well-developed vascular tissues (xylem and phloem). Also, the sporophyte is the larger, longer-lived stage of the life cycle and the gametophytes are very small. Some bryophytes even lack chlorophyll. **Table 27.2** summarizes these characteristics and gives an overview of seedless vascular plant features within the larger context of modern plant phyla.

Seedless vascular plants once encompassed a huge number of diverse species of trees, shrubs, and herbs. In the late Paleozoic era, they were Earth's dominant vegetation. Some lineages have endured to the present, but collectively these survivors total fewer than 14,000 species. The taxonomic relationships between various lines are still under active investigation, and comparisons of gene sequences from the genomes in plastids, cell nuclei, and sometimes mitochondria are revealing previously unsuspected links between some of them. In this book we assign seedless vascular plants to two phyla, the Lycophyta (club mosses and their close relatives) and the Pterophyta (ferns, whisk ferns, and horsetails).

Early Seedless Vascular Plants Flourished in Moist Environments

The extinct plant genus *Cooksonia* (see Figure 27.4) probably was one of the earliest ancestors of modern seedless vascular plants. Like other members of its extinct phylum (Rhyniophyta) *Cooksonia* was small, rootless, and leafless, but its simple stems had a central core of xylem, an arrangement seen in many existing vascular plants. Mudflats and swamps of the damp Devonian period were dominated by plants like *Cooksonia* and *Rhynia* **(Figure 27.13)**. While these and other now-extinct phyla came and went, ancestral forms of both modern phyla of seedless vascular plants appeared. In botanical terms, the earliest seedless vascular plants were "herbs"—that is, they did not have woody, lignified tissue. By the start of the Carboniferous period, however, the small herbaceous Devonian plants had given rise to larger shrubby species and to trees with some woody tissue, bark, roots, leaves, and even seeds.

Carboniferous forests were swampy places dominated by members of the phylum **Lycophyta**, and fascinating fossil specimens of this group have been unearthed in North America and Europe. One example is *Lepidodendron,* which had broad, straplike leaves and sporangia near the ends of the branches **(Figure 27.14a)**. It also had xylem and several other types of tissues that are typical of all modern vascular plants (although probably not in the same proportions as seen today). Like trees growing in modern year-round tropical climates, the fossils do not exhibit growth rings. This observation implies that the continents of Europe and North America lay along the equator during the Carboniferous period. Also abundant at the time were representatives of the phylum **Pterophyta**, including ferns such as *Medullosa* and giants such as *Calamites*—huge horsetails that could have a trunk diameter of 30 cm. The sturdy, upright stems were attached to a system of rhizomes—horizontal underground stems. Ferns populated the forest understory. Some early seed plants also were present, including now-extinct fernlike plants, called seed ferns, which bore seeds at the tips of leaves **(Figure 27.14b)**.

Lepidodendron and *Calamites* dominated lush swamp forests in a subtropical climate. After leaves, branches, and old trees fell to the ground, they became buried in anaerobic sediments. Over geologic time, these buried remains became compressed and fossilized, and today they form much of the world's coal reserves. This is why coal is called a "fossil fuel," and the Carboniferous period is called the Coal Age. Characterized by a moist climate over much of the planet, and by the dominance of seedless vascular plants, the Carboniferous period continued for 150 million years, ending when climate patterns changed during the Paleozoic era.

Most modern seedless vascular plants are ferns, and like their ancestors they also are confined largely to wet or humid environments because they require external water for reproduction. Except for whisk ferns, their gametophytes have no vascular tissues for water transport, and male gametes must swim through water to reach eggs. The few vascular seedless plants that are

Table 27.2　Plant Phyla and Major Characteristics

Phylum	Common Name	Number of Species	Common General Characteristics
Bryophytes: Nonvascular plants. Gametophyte dominant, free water required for fertilization, cuticle and stomata present in some.			
Hepatophyta	Liverworts	6000	Leafy or simple flattened thallus, rhizoids; spores in capsules. Moist, humid habitats.
Anthocerophyta	Hornworts	100	Simple flattened thallus, rhizoids; hornlike sporangia. Moist, humid habitats.
Bryophyta	Mosses	10,000	Feathery or cushiony thallus, some have hydroids; spores in capsules. Moist, humid habitats; colonizes bare rock, soil, or bark.
Seedless vascular plants: Sporophyte dominant, free water required for fertilization, cuticle and stomata present.			
Lycophyta	Club mosses	1000	Simple leaves, true roots; most species have sporangia on sporophylls. Mostly wet or shady habitats.
Pterophyta	Ferns, whisk ferns, horsetails	13,000	*Ferns:* Finely divided leaves, woody stems in tree ferns; sporangia in sori. Habitats from wet to arid. *Whisk ferns:* Branching stem from rhizomes; sporangia on stem scales. Tropical to subtropical habitats. *Horsetails:* Hollow photosynthetic stem, scalelike leaves, sporangia in strobili. Swamps, disturbed habitats.
Gymnosperms: Vascular plants with "naked" seeds. Sporophyte dominant, fertilization by pollination, cuticle and stomata present.			
Cycadophyta	Cycads	185	Shrubby or treelike with palmlike leaves, pithy stems; male and female strobili on separate plants. Widespread distribution.
Ginkgophyta	Ginkgo	1	Woody-stemmed tree, deciduous fan-shaped leaves. Male, female structures on separate plants. Temperate areas of China.
Gnetophyta	Gnetophytes	70	Shrubs or woody vines; one has strappy leaves. Male and female strobili on separate plants. Limited to deserts, tropics.
Coniferophyta	Conifers	550	Mostly evergreen, woody trees and shrubs with needlelike or scalelike leaves; male and female cones usually on same plant.
Angiosperms: Plants with flowers and seeds protected inside fruits. Sporophyte dominant, fertilization by pollination, cuticle and stomata present. Major groups: Monocots, eudicots.			
Anthophyta	Flowering plants	268,500+ (including magnoliids, other basal angiosperms)	Wood and herbaceous plants. Nearly all land habitats, some aquatic.
Monocots	Grasses, palms, lilies, orchids, and others	60,000	Pollen grains have a single groove; one cotyledon. Parallel-veined leaves common.
Eudicots	Most fruit trees, roses, cabbages, melons, beans, potatoes, and others	200,000	Pollen grains have three grooves. Most species have two cotyledons; net-veined leaves common.

adapted to dry environments such as deserts can reproduce sexually only when adequate water is available, as during seasonal rains.

Modern Lycophytes Are Small and Have Simple Vascular Tissues

Lycophytes such as club mosses were highly diverse 350 mya, when some tree-sized forms inhabited lush swamp forests. Today, however, such giants are no more. The most familiar of the 1000 or so living species of lycophytes are club mosses, including members of genera such as *Lycopodium* and *Selaginella,* which grow on forest floors **(Figure 27.15a).** Other groups include

the spike mosses and quillworts. The sporophyte of a club moss has upright or horizontal stems that contain a small amount of xylem and bear small green leaves and roots—both of which have vascular tissue. Sporangia are clustered at the bases of specialized leaves, called **sporophylls,** that occur near stem tips. A cluster of sporophylls forms a **cone** or **strobilus** (plural, *strobili*). In some species the sporangia release haploid spores produced by meiosis **(Figure 27.15b).** If a spore eventually germinates (which can occur even several years after it is released), it forms a free-living gametophyte, but one that differs markedly from the sporophyte. Ranging in size from nearly invisible to several centimeters, the gametophyte easily becomes buried under decompos-

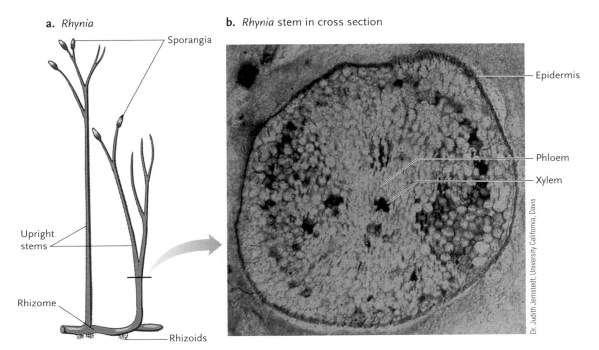

a. *Rhynia*

Sporangia

Upright stems

Rhizome

Rhizoids

b. *Rhynia* stem in cross section

Epidermis

Phloem

Xylem

Dr. Judith Jernstedt, University California, Davis

Figure 27.13
Rhynia, an early seedless vascular plant. **(a)** Fossil-based reconstruction of the entire plant, about 30 cm tall. **(b)** Cross section of the stem, approximately 3 mm in diameter. This fossil was embedded in chert approximately 400 million years ago. Still visible in it are traces of the transport tissues xylem and phloem, along with other specialized tissues.

ing plant litter. There rhizoids attach it to its substrate. It cannot photosynthesize, and instead obtains nutrients by way of mycorrhizae. Although all species of *Lycopodium* are homosporous—that is, one bisexual gametophyte produces both eggs and sperm—those of other genera are heterosporous. Regardless, as with ancestral lycophytes, the sperm require water in which they can swim to the eggs. After fertilization, the life cycle comes full circle as the zygote develops into a diploid embryo that grows into a sporophyte.

Figure 27.14
Reconstruction of the lycophyte tree *(Lepidodendron)* and its environment. **(a)** Fossil evidence suggests that *Lepidodendron* grew to be about 35 m tall with a trunk 1 m in diameter. **(b)** Artist's depiction of a Coal Age forest.

a. The lycophyte tree *(Lepidodendron)*

b. Artist's depiction of a Coal Age forest

Stem of a giant lycophyte *(Lepidodendron)*

Seed fern *(Medullosa)*; probably related to the progymnosperms, which may have been among the earliest seed-bearing plants

Stem of a giant horsetail *(Calamites)*

Field Museum of Natural History, Chicago

Figure 27.15

Lycophytes. **(a)** *Lycopodium* sporophyte, showing the conelike strobili in which spores are produced. **(b)** A fossilized lycophyte spore bearing a characteristic Y-shaped mark (arrow) called a trilete scar.

a. *Lycopodium* sporophyte

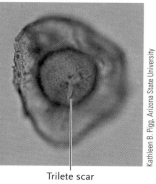

b. Fossilized lycophyte spore

Strobilus

Trilete scar

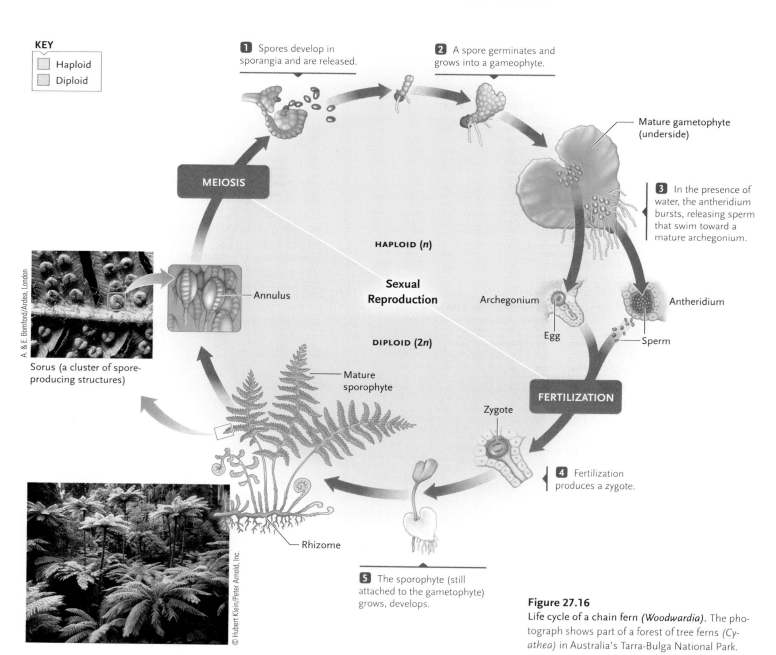

KEY

Haploid
Diploid

1 Spores develop in sporangia and are released.

2 A spore germinates and grows into a gameophyte.

MEIOSIS

Mature gametophyte (underside)

3 In the presence of water, the antheridium bursts, releasing sperm that swim toward a mature archegonium.

HAPLOID (*n*)

Sexual Reproduction

Archegonium

Antheridium

DIPLOID (2*n*)

Egg

Sperm

Annulus

Mature sporophyte

FERTILIZATION

Sorus (a cluster of spore-producing structures)

Zygote

4 Fertilization produces a zygote.

Rhizome

5 The sporophyte (still attached to the gametophyte) grows, develops.

Figure 27.16

Life cycle of a chain fern *(Woodwardia)*. The photograph shows part of a forest of tree ferns *(Cyathea)* in Australia's Tarra-Bulga National Park.

Ferns, Whisk Ferns, Horsetails, and Their Relatives Make Up the Diverse Phylum Pterophyta

Second only to the flowering plants, the phylum Pterophyta (*pteron* = wing) contains a large and diverse group of vascular plants—the 13,000 or so species of ferns, whisk ferns, and horsetails. Most ferns, including some that are poplar houseplants, are native to tropical and temperate regions. Some floating species are less than 1 cm across, while some tropical tree ferns grow to 25 m tall. Other species are adapted to life in arctic and alpine tundras, salty mangrove swamps, and semi-arid deserts.

Complex Anatomical Features in Ferns. The familiar plant body of a fern is the sporophyte phase (**Figure 27.16**). It produces an above-ground clump of fern leaves, called fronds. Often finely divided and featherlike, and containing multiple strands of vascular tissue, fronds are the most complex leaves of the plant kingdom. Young fronds are tightly coiled, and as they emerge above the soil these "fiddleheads" (so named because they resemble the scrolled pegheads of violins) unroll and expand. Before they unfurl, fiddleheads may be gathered by people who relish them as a gastronomic treat (albeit with care—the fiddleheads of some species contain a carcinogen). Leaves of some species last for only a single growing season, while in others they grow for several years. A typical frond has a well-developed epidermis with chloroplasts in the epidermal cells and stomata on the lower surface.

Except for tropical tree ferns, the stems of most ferns are underground rhizomes. The stem's vascular system is organized into a complex, interconnecting network of bundles, each having a central core of xylem surrounded by phloem. Roots descend along the length of the rhizomes. A rhizome can live for centuries, growing at its tip and extending outward horizontally through the soil, sometimes over a considerable area. In most ferns, the fronds arise from nodes positioned along the rhizome. A **node** is the point on a stem where one or more leaves are attached.

A fern sporophyte produces sporangia on the lower surface or margin of some leaves. Often, several sporangia are clustered into a rust-colored **sorus** (plural, *sori*). Sori may be exposed or they may be protected with a flap of tissue. Each sporangium is a delicate case, shaped rather like an old-fashioned pocket watch and covered by a layer of epidermal cells. In the layer, a row of thick-walled cells called the **annulus** ("ring") nearly encircles the sporangium.

Inside the sporangium, haploid spores arise by meiosis. Meanwhile, the sporangium slowly dries out, and as it does so the annulus steadily contracts. Eventually the force of the contracting annulus rips open the sporangium, which snaps back on itself, flinging out the mature spores. In this way fern spores can be dispersed up to 2 m away from the parent plant. Wind may carry them much farther: on board the *Beagle,* Charles Darwin collected fern spores hundreds of miles from shore.

A germinating spore develops into a gametophyte, which is typically a small, heart-shaped plant anchored to the soil by rhizoids. Both antheridia and archegonia are present on the lower surface of each gametophyte, where moisture is trapped. Inside an antheridium is a globular packet of haploid cells, each of which develops into a helical sperm with many flagella. When water is present, the antheridium bursts, releasing the sperm. If mature archegonia are nearby, the sperm swim toward them, drawn by a chemical attractant that diffuses from the neck of the archegonium, which is open when free water is present.

After a sperm fertilizes an egg, the diploid zygote begins dividing and developing into an embryo, which at this stage obtains nutrients from the gametophyte. In a short time, however, the embryo develops into a young sporophyte that is larger than the gametophyte and has its own green leaf and a root system. The sporophyte now is nutritionally independent and the parent gametophyte degenerates and dies.

Features of Early Vascular Plants in Whisk Ferns. The whisk ferns and their relatives are represented by two genera, *Psilotum* (pronounced si-lo'-tum) and *Tmesipteris* (may-sip'-ter-is), with only about 10 species in all. These rather uncommon plants grow in tropical and subtropical regions, often as epiphytes. In the United States the range for *Psilotum* species (**Figure 27.17**) includes Hawaii, Gulf Coast states such as Florida and Louisiana, and parts of the West.

The sporophytes of whisk ferns are up to 60 cm tall and resemble the extinct *Cooksonia* and *Rhynia.* Like those early vascular plants, they lack true roots and leaves. Instead, small leaflike scales adorn an upright, green, branching stem, which arises from a horizontal rhizome system anchored by rhizoids. The absorptive rhizoids have mycorrhizal fungi associated with them, which provide enhanced access to some nutrients.

A whisk fern's stem is structurally and functionally multifaceted. The stem's epidermal cells carry out photosynthesis, while its core has the transport tissues xylem and phloem and other anatomical features of more complex vascular plants. Sporangia rest atop some of the stem scales. Inside them, meiotic divisions of specialized cells produce haploid spores.

Horsetails, Possibly the Most Ancient Living Plant Species. The ancient relatives of modern-day horsetails included treelike forms taller than a two-story building. Only fifteen species in a single genus,

Figure 27.17
Sporophytes of a whisk fern *(Psilotum)*, a seedless vascular plant. Three-lobed sporangia occur at the ends of stubby branchlets; inside the sporangia, meiosis gives rise to haploid spores.

Kingsley R. Stern

a. Sporophyte stem

William Ferguson

b. Sporangia

Strobilus, an aggregation of sporangia at the tip of the horsetail sporophyte

c. Each petal shaped sporangium of a strobilus contains spores that formed by meiosis.

W. H. Hodges

Figure 27.18
A species of *Equisetum*, the horsetails. **(a)** Vegetative stem. **(b)** Strobili, which bear sporangia. **(c)** Close-up of sporangia and associated structures on a strobilus.

Equisetum, have survived to the present **(Figure 27.18)**. Horsetails grow in moist soil along streams and in disturbed habitats, such as roadsides and beds of railroad tracks. Their sporophytes typically have underground rhizomes and roots that anchor the rhizome to the soil. The scalelike leaves are arranged in whorls about a photosynthetic stem that is stiff and gritty because horsetails accumulate silica in their tissues. American pioneers used them to scrub out pots and pans, hence their other common name, "scouring rushes."

Equisetum sporangia are borne in strobili on highly specialized stem structures quite different from the sporophylls of club mosses. In most horsetails the strobili occur on ordinary vegetative shoots, but in a few species they occur only on special fertile shoots. Each stalked spore-bearing structure in a strobilus resembles an umbrella and is attached at right angles to a main axis. Haploid spores develop in sporangia attached near the edge of the "umbrella's" underside, and air currents disperse them. They must germinate within a few days to produce gametophytes, which are free-living plants about the size of a small pea.

STUDY BREAK

1. Compare and contrast the lycophyte and bryophyte life cycles with respect to the sizes and longevity of gametophyte and sporophyte phases.
2. In ferns, whisk ferns, and horsetails, what kinds of structures fulfill the roles of roots and leaves?
3. How does the life cycle of a horsetail differ from that of a fern?

27.4 Gymnosperms: The First Seed Plants

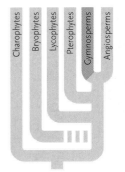

Kratz/Zefa

Gymnosperms are the conifers and their relatives. The earliest fossils identified as gymnosperms are found in Devonian rocks. By the Carboniferous, when nonvascular plants were dominant, many lines of gymnosperms had also evolved, and the first true conifers appeared. These radiated during the Permian period; the Mesozoic era that followed, 248 to 65 mya, was the age not only of the dinosaurs but of the gymnosperms as well.

The evolution of gymnosperms involved sweeping changes in plant structures related to reproduction. As a prelude to our survey of modern gymnosperms, we begin by considering some of these innovations, which opened new adaptive options for land plants.

Major Reproductive Adaptations Occurred as Gymnosperms Evolved

The word *gymnosperm* is derived from the Greek *gymnos,* meaning naked, and *sperma,* meaning seed. The evolution of gymnosperms included important reproductive adaptations—pollen and pollination, the ovule, and the seed. The fossil record has not revealed the sequence in which these changes arose, but all of them contributed to the radiation of gymnosperms into land environments. **Figure 27.19** shows an artist's rendering of *Archaeopteris,* which may have been one of the first true trees. Called a *progymnosperm,* it belonged to an evolutionary line that is thought to have given rise to modern seed plants.

Pollen and Ovules: Shelter for Spores. Unlike bryophytes and seedless vascular plants, gymnosperm sporophytes do not disperse their spores. The sporophyte produces haploid spores by meiosis, but it retains these spores inside reproductive structures where they give rise to multicellular haploid gametophytes. As noted briefly earlier, sperm arise inside a **pollen grain**, a male gametophyte that typically has walls reinforced with the polymer sporopollenin. All but a few gymnosperms have nonmotile sperm. Usually, two of these nonswimming sperm develop inside each pollen grain—very different from the flagellated, swimming sperm of algae and plants that do not produce seeds.

An **ovule** is a structure in a sporophyte in which a female gametophyte develops, complete with an egg. Physically connected to the sporophyte and surrounded by the ovule's protective layers, a female gametophyte no longer faces the same risks of predation or environmental assault that can threaten a free-living gametophyte.

Figure 27.19
Fossil-based reconstruction of *Archaeopteris*, a large Devonian progymnosperm. It could grow 25 m tall and may have been a seed-forming ancestor of modern gymnosperms.

Pollination is the transfer of pollen to female reproductive parts via air currents or on the bodies of animal pollinators. Pollen and pollination were enormously important adaptations for gymnosperms, because the shift to nonswimming sperm along with a means for delivering them to female gametes meant that reproduction no longer required liquid water. The only gymnosperms that have retained swimming sperm are the cycads and ginkgoes, described shortly, which have relatively few living species and are restricted to just a few native habitats.

Seeds: Protecting and Nourishing Plant Embryos. A **seed** is the structure that forms when an ovule matures after a pollen grain reaches it and a sperm fertilizes the egg. It consists of three basic parts: (1) the embryo sporophyte, (2) tissues around it containing carbohydrates, proteins, and lipids that nourish the embryo until it becomes established as a plantlet with leaves and roots, and (3) a tough, protective outer seed coat **(Figure 27.20).** This complex structure makes seeds ideal packages for sheltering an embryo from drought, cold, or other adverse conditions. As a result, seed-making plants enjoy a tremendous survival advantage over species that simply release spores to the environment. Encased in a seed, the embryo also can be transported far from its parent, as when ocean currents carry coconut seeds ("coconuts" protected in large, buoyant fruits) hundreds of kilometers across the sea. As discussed in Chapter 34, some plant embryos housed in seeds can remain dormant for months or years before environmental conditions finally prompt them to germinate and grow.

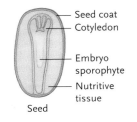

Figure 27.20
Generalized view of a seed—in this case, the seed of a pine, a gymnosperm.

Modern Gymnosperms Include Conifers and a Few Other Groups

Today there are about 800 gymnosperm species. The sporophytes of nearly all are large trees or shrubs, although a few are woody vines. The most widespread and familiar gymnosperms are the conifers (Coniferophyta). Others are the cycads (Cycadophyta), ginkgoes (Ginkgophyta), and gnetophytes (Gnetophyta).

Economically, gymnosperms, particularly conifers, are vital to human societies. They are sources of lumber, paper pulp, turpentine, and resins, among other products. They also have huge ecological importance. Their habitats range from tropical forests to deserts, but gymnosperms are most dominant in the cool-temperate zones of the northern and southern hemispheres. They flourish in poor soils where flowering plants don't compete as well. In North America, for example, gymnosperm forests cover more than one-third of the continent's landmass—although in some areas, logging has significantly reduced the once-lush forest cover. Our survey of gymnosperms begins, however, with the cycads, ginkgoes, and gnetophytes—the latter two groups remnants of lineages that have all but vanished from the modern scene.

Cycads Are Restricted to Warmer Climates

During the Mesozoic era, the **Cycadophyta** (*kykas* = palm), or cycads, flourished along with the dinosaurs. About 185 species have survived to the present, but they are confined to the tropics and subtropics.

At first glance, you might mistake a cycad for a small palm tree **(Figure 27.21).** Some cycads have massive, cone-shaped strobili (clusters of sporophylls) that bear either pollen or ovules. Air currents or crawling insects transfer pollen from male plants to the developing gametophyte on female plants. Poisonous alkaloids that may help deter insect predators occur in various cycad tissues. In tropical Asia, some people consume cycad seeds and flour made from cycad trunks, but only after the toxic compounds have been rinsed away. Much in demand from fanciers of unusual plants, cycads in some countries are uprooted and sold in what amounts to a black-market trade—greatly diminishing their numbers in the wild.

Figure 27.21
The cycad *Zamia*. Note the large, terminal female cone and fernlike leaves.

Ginkgoes Are Limited to a Single Living Species

The phylum **Ginkgophyta** has only one living species, the ginkgo (or maiden-hair) tree *(Ginkgo biloba)*, which grows wild today only in warm-temperate forests of central China. Ginkgo trees are large, diffusely branching trees with characteristic fan-shaped leaves **(Figure 27.22)** that turn a brilliant yellow in autumn. Nursery-propagated male trees often are planted in cities because they are resistant to insects, disease, and air pollutants. The female trees are equally pollution-resistant, but gardeners shy away from them—their fleshy fruits produce a notoriously foul odor.

Gnetophytes Include Simple Seed Plants with Intriguing Features

The phylum Gnetophyta contains three genera—*Gnetum, Ephedra,* and *Welwitschia*—that together include about 70 species. Moist, tropical regions are home to about 30 species of *Gnetum*, which includes both trees and leathery leafed vines (lianas). About 35 species of *Ephedra* grow in desert regions of the world **(Figure 27.23a–c)**.

Of all the gymnosperms, *Welwitschia* is the most bizarre. This seed-producing plant grows in the hot deserts of south and west Africa. The bulk of the plant is a deep-reaching taproot. The only exposed part is a woody disk-shaped stem that bears cone-shaped strobili and leaves. The plant never produces more than two strap-shaped leaves, which split lengthwise repeatedly as the plant grows older, producing a rather scraggly pile **(Figure 27.23d)**.

Although gnetophytes are structurally and functionally simpler than most other seed plants, recent studies of sexual reproduction mechanisms in *Gnetum* and *Ephedra* species uncovered a two-step process of fertilization—which is a hallmark of angiosperms, the most advanced seed plants. This discovery raised some provocative evolutionary questions, even leading some investigators to propose that ancient gnetophytes may have given rise to flowering plants. Complicating this picture, however, are molecular findings, such as those arrived at by a research team at the Academia Sinica in

a. Ginkgo tree

b. Fossil and modern gingko leaves

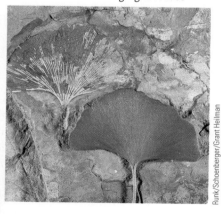

c. Male cone

d. Ginkgo seeds

Figure 27.22
Ginkgo biloba. **(a)** A ginkgo tree. **(b)** Fossilized ginkgo leaf compared with a leaf from a living tree. The fossil formed at the Cretaceous–Tertiary boundary. Even though 65 million years have passed, the leaf structure has not changed much. **(c)** Pollen-bearing cones and **(d)** fleshy-coated seeds of the *Ginkgo*.

Taiwan, People's Republic of China. When the team compared 65 nuclear rRNA sequences from ferns, gymnosperms, and angiosperms, their analysis supported the hypothesis that cycads and ginkgoes represent the earliest gymnosperm lineage, with a divergent lineage of gnetophytes and conifers arising later. The team found no molecular evidence for a link between the Gnetophyta and angiosperms.

Conifers Are the Most Common Gymnosperms

About 80% of all living gymnosperm species are members of one phylum, the **Coniferophyta**, or conifers ("cone-bearers"). Conifer trees and shrubs are longer-lived, and anatomically and morphologically more complex, than any sporophyte phase we have discussed so far. Characteristically, they form woody reproductive cones, and most of the 550 conifer species are woody trees or shrubs with needlelike or scalelike leaves, which are anatomically adapted to aridity. For instance, needles have a thick cuticle, sunken stomata, and a fibrous epidermis, all traits that reduce the loss of water vapor.

Most conifers are evergreens. That is, although they shed old leaves, often in autumn, they retain enough leaves so that they still look "green," unlike deciduous species like maples, which shed *all* their leaves as winter approaches. Familiar conifer examples are the pines, spruces, firs, hemlocks, junipers, cypresses, and redwoods. Like other seed plants, conifers are heterosporous, producing pollen in clusters of small strobili and eggs in larger, woody ones. Both of these structures are often referred to as cones. Seeds develop on the shelflike scales of the female cones.

Pines and many other gymnosperms produce resins, a mix of organic compounds that are by-products of metabolism. Resin accumulates and flows in long resin ducts through the wood, inhibiting the activity of wood-boring insects and certain microbes. Pine resin extracts are the raw material of turpentine and (minus the volatile terpenes) the sticky rosin used to treat violin bows.

We know a great deal about the pine life cycle **(Figure 27.24)**, so it is a convenient model for gymnosperms. All but 1 of the 93 pine species are trees (*Pinus mugo*, native to high elevations in Europe, is a shrub). The male cones (strobili) are relatively small and delicate, only about 1 cm long, and are borne on the lower branches. Each one consists of many small scales, which are specialized leaves (called sporophylls) attached to the cone's axis in a spiral. Two sporangia develop on the underside of each scale. Inside the sporangia, spore "mother cells" called microsporocytes undergo meiosis and give rise to haploid **microspores.** Each microspore then undergoes mitosis to develop into a winged pollen grain—an immature male gametophyte. At this stage the pollen grain consists of four

a. *Ephedra* plant

b. *Ephedra* male cone

William Ferguson

c. *Ephedra* female cone

Robert & Linda Mitchell Photography

Edward S. Ross

d. *Welwitschia* plant with female cones

© Fletcher and Baylis/Photo Researchers, Inc.

Figure 27.23
Gnetophytes. **(a)** Sporophyte of *Ephedra*, with close-ups of **(b)** its pollen-bearing cones and **(c)** a seed-bearing cone, which develop on separate plants. **(d)** Sporophyte of *Welwitschia mirabilis*, with seed-bearing cones.

cells, two that will degenerate and two that will function later in reproduction.

Young female cones develop higher in the tree, at the tips of upper branches. The cone scales bear ovules. Inside each ovule is a spore mother cell called a megasporocyte. Unlike microsporocytes, the megasporocyte in an ovule undergoes meiosis only when conditions are right and produces four haploid spores called **megaspores.** Only one megaspore survives, however, and it develops slowly, becoming a mature female gametophyte only when pollination is underway. In a pine, the process takes well over a year. The mature female gametophyte is a small oval mass of cells with several archegonia at one end, each containing an egg.

Each spring, air currents lift vast numbers of pollen grains off male cones—by some estimates, billions may be released from a single pine tree. The extravagant numbers assure that at least some pollen grains will land on female cones. The process is not as random as it might seem: studies have shown that the contours of female cones create air currents that

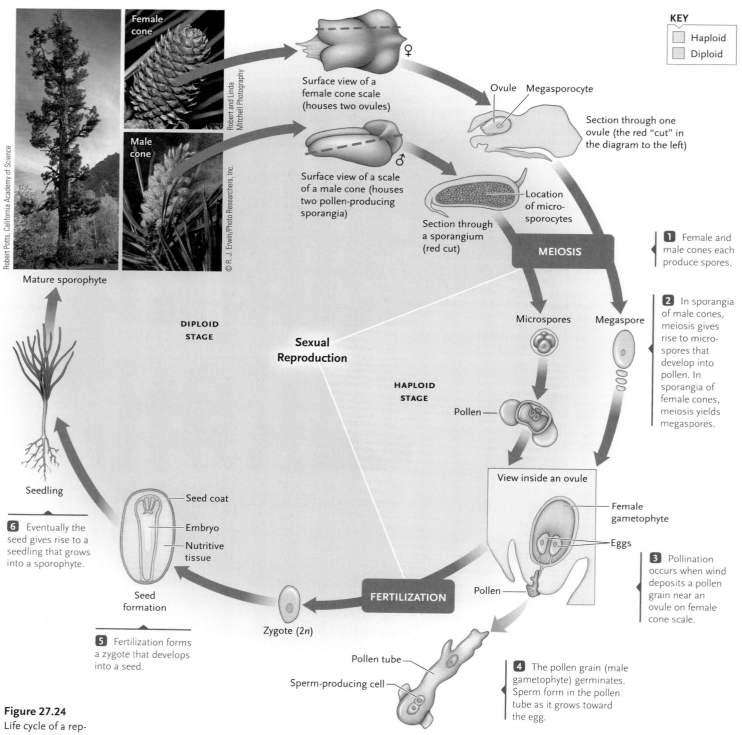

KEY
- Haploid
- Diploid

Female cone

Male cone

Robert Potts, California Academy of Science
Robert and Linda Mitchell Photography
© R. J. Erwin/Photo Researchers, Inc.

Surface view of a female cone scale (houses two ovules)

Surface view of a scale of a male cone (houses two pollen-producing sporangia)

Ovule   Megasporocyte

Section through one ovule (the red "cut" in the diagram to the left)

Location of micro-sporocytes

Section through a sporangium (red cut)

MEIOSIS

1 Female and male cones each produce spores.

Mature sporophyte

DIPLOID STAGE

Sexual Reproduction

HAPLOID STAGE

Microspores   Megaspore

2 In sporangia of male cones, meiosis gives rise to microspores that develop into pollen. In sporangia of female cones, meiosis yields megaspores.

Pollen

Seedling

6 Eventually the seed gives rise to a seedling that grows into a sporophyte.

View inside an ovule

Female gametophyte

Seed coat
Embryo
Nutritive tissue

Eggs

Seed formation

FERTILIZATION

Pollen

3 Pollination occurs when wind deposits a pollen grain near an ovule on female cone scale.

Zygote (2n)

5 Fertilization forms a zygote that develops into a seed.

Pollen tube
Sperm-producing cell

4 The pollen grain (male gametophyte) germinates. Sperm form in the pollen tube as it grows toward the egg.

Figure 27.24
Life cycle of a representative conifer, a ponderosa pine *(Pinus ponderosa).* Pines are the dominant conifers in the Northern Hemisphere and their large sporophytes provide a heavily exploited source of wood.

can favor the "delivery" of pollen grains near the cone scales. After pollination, the pollen grain develops into a *pollen tube* that grows toward the female spore mother cell. As it does, sperm form in the tube and stimulate maturation of the female gametophyte and the production of eggs. When a pollen tube reaches an egg, the stage is set for fertilization, the formation of a zygote, and early development of the plant em-

bryo. Often, fertilization occurs months to a year after pollination. Once an embryo forms, a pine seed—which, recall, includes the embryo, female gametophyte tissue, and seed coat—eventually is shed from the cone. The seed coat protects the embryo from drying out, and the female gametophyte tissue serves as its food reserve. This tissue makes up the bulk of a "pine nut."

1. What are the four major reproductive adaptations that evolved in gymnosperms?
2. What are the basic parts of a seed, and how is each one adaptive?
3. Describe some features that make conifers structurally more complex than other gymnosperms.

27.5 Angiosperms: Flowering Plants

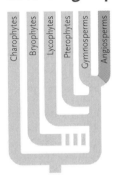

Of all plant phyla, the flowering plants, or **angiosperms**, are the most successful today. At least 260,000 species are known (**Figure 27.25** shows a few examples), and botanists regularly discover new ones in previously unexplored regions of the tropics. The word angiosperm is derived from the Greek *angeion* (meaning a case or vessel) and *sperma*

(seed). The "vessel" refers to the modified leaf, called a *carpel,* that surrounds and protects the ovules and later, the seeds of angiosperms. Carpels are **flowers**, reproductive structures that are a key defining feature of angiosperms. Another defining feature is the **fruit**—botanically speaking, a structure that surrounds the angiosperm embryo and aids seed dispersal.

In addition to having flowers and fruits, angiosperms are the most ecologically diverse plants on Earth, growing on dry land and in wetlands, fresh water, and the seas. They range in size from tiny duckweeds about 1 mm long to towering *Eucalyptus* trees more than 100 m tall. Most are free-living photosynthesizers. Others lack chloroplasts and feed on nonliving organic matter or are parasites that feed on living host organisms.

The Fossil Record Provides Little Information about the Origin of Flowering Plants

The evolutionary origin of angiosperms has confounded plant biologists for well over a hundred years. Charles Darwin called it the "abominable mystery," because

a. Flowering plants in a desert

George H. Huey/Corbis

b. Alpine angiosperms

Bill Coster/Peter Arnold, Inc.

c. Triticale, a grass

© John Mason/Ardea, London

d. A parasitic angiosperm

© Peter F. Zika/Visuals Unlimited

Figure 27.25

Flowering plants. Diverse photosynthetic species are adapted to nearly all environments, ranging from **(a)** deserts to **(b)** snowlines of high mountains. **(c)** Triticale, a hybrid grain derived from parental stocks of wheat *(Triticum)* and rye *(Secale)*, is one example of the various grasses utilized by humans. **(d)** The parasitic flowering plant Indian pipe *(Monotropa uniflora)* having no chlorophyll of its own, obtains food by associating with mycorrhizae, which are in turn associated with the roots of photosynthetic plants.

Archaefructus sinensis

Sketch of *Archaefructus sinensis*

© David Dilcher, Florida Museum of Natural History/Paleobotany Laboratory

Figure 27.26
Fossil of *Archaefructus sinensis*, thought to have been an early flowering plant. The sketch shows what this small, possibly aquatic plant may have looked like.

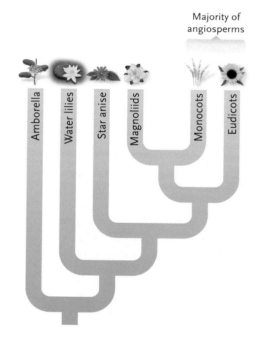

Majority of angiosperms

Amborella | Water lilies | Star anise | Magnoliids | Monocots | Eudicots

Figure 27.27
A hypothetical phylogenetic tree for flowering plants.

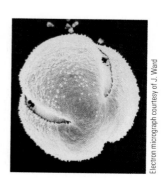

Electron micrograph courtesy of J. Ward

Figure 27.28
Eudicot pollen grain. Eudicot pollen grains have three slitlike grooves, only two of which are visible here. Pollen made by all other seed plants, including monocots, have just one groove.

flowering plants appear suddenly in the fossil record, without a fossil sequence that links them to any other plant groups. The oldest well-documented fossil specimens date back 125 million years. Discovered in China, these remarkable fossils show complex and strikingly modern-looking plants that have leaves, stems, fruits, and seeds **(Figure 27.26)**. Two species have been unearthed and have been assigned to the genus *Archaefructus*, representing a newly discovered, extinct angiosperm group.

The fossil record has yet to reveal obvious transitional organisms between flowering plants and either gymnosperms or seedless vascular plants. As with gymnosperms, attempts to reconstruct the earliest flowering plant lineages from morphological, developmental, and biochemical characteristics have produced several conflicting classifications and family trees. Some paleobotanists hypothesize that flowering plants arose in the Jurassic period; others propose they evolved in the Triassic from now-extinct gymnosperms or from seed ferns.

As the Mesozoic era ended and the modern Cenozoic era began, great extinctions occurred among both plant and animal kingdoms. Gymnosperms declined, and dinosaurs disappeared. Flowering plants, mammals, and social insects flourished, radiating into new environments. Today we live in what has been called "the age of flowering plants."

Angiosperms Are Subdivided into Several Clades, Including Monocots and Eudicots

Angiosperms are assigned to the phylum **Anthophyta**, a name that derives from the Greek *anthos*, meaning flower. **Figure 27.27** shows one current model of major clades within the phylum. The great majority of angiosperms are classified either as monocots or eudicots. **Monocots** are distinguished by the morphology of their embryos, which have a single seed leaf called a cotyledon ("cuplike hollow"). **Eudicots** ("true dicots"), which generally have two cotyledons, are set apart from other angiosperms by the structure of their pollen grains, which have three grooves **(Figure 27.28)**. By contrast, the pollen of monocots and all other seed plants (including more than 8500 species once lumped with eudicots under the term "dicots") have only a single groove. Paleobotanists use this clear structural difference not only to help establish the general type of plant that produced fossil pollen, but also what types of plants were present in fossil deposits of a particular age or geographic location.

While most angiosperms can fairly easily be categorized as either monocots or eudicots, figuring out the appropriate classification for other angiosperms is an ongoing challenge and an extremely active area of plant research. The diagram in Figure 27.27 reflects a synthesis of evidence from both morphological and molecular studies, an approach examined in this chapter's *Insights from the Molecular Revolution*. Along with eudicots and monocots, botanists currently recognize four other clades **(Figure 27.29)**. The **magnoliids**, a group that includes magnolias (see Figure 27.29a), laurels, and avocados, are more closely related to monocots than to eudicots. Some researchers also place plants that are the sources of spices such as peppercorns, nutmeg, and cinnamon in the magnoliid clade. The other three clades are considered to be **basal angiosperms** representing the earliest branches of the flowering plant lineage. They include the star anise group (see Figure 27.29b), water lilies (see Figure

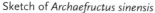

The Powerful Genetic Toolkit for Studying Plant Evolution

Unlike animals and most other eukaryotic organisms, plants have three distinct sets of genes—in the cell nucleus, in mitochondria, and in chloroplasts. Chloroplast DNA, or cpDNA, has been especially useful for evolutionary studies, particularly the chloroplast *rbcL* gene. Mutations of the gene have occurred slowly, at about one-fourth to one-fifth the rate of genes in the nucleus. As a result, the DNA sequences of *rbcL* genes of different species diverge less than those of most other plant genes. Further, there are no introns—noncoding sequences—interrupting the coding sequence of the *rbcL* gene. Researchers can compare *rbcL* DNA from different species base by base, with no need to subtract introns. At the same time, the *rbcL* genes of different species are different enough to allow researchers to assemble evolutionary trees based on the degree of sequence variation.

Studies using cpDNA have helped fuel several fundamental shifts in our understanding of branch points in plant evolution. For example, together with gene sequence data from nuclear DNA, analysis of *rbcL* genes provided the molecular foundation for the now widely accepted view that charophyte green algae were the evolutionary forerunners of land plants. Similarly, in the late 1990s an international research team led by Yin-Long Qiu at the University of Massachusetts at Amherst correlated the loss of introns from two mitochondrial genes with the hypothesis that the first land plants were liverworts. Qiu and his colleagues carried out a genetic survey of more than 350 land plants representing all major lineages. They discovered that the noncoding sequences were present in all other bryophytes and all major lines of vascular plants, but were absent in liverworts, green algae, and all other eukaryotes. The findings are supported by analysis of *rbcL* sequences in various plant groups. Data from cpDNA and mtDNA analyses also underlies the hypothesis that, as land plants evolved, the ancient relatives of club mosses (lycophytes) were the forerunners of other vascular plants. Clearly, these varied molecular tools, and cpDNA in particular, are helping plant scientists explore evolutionary relationships across the whole spectrum of the Kingdom Plantae.

27.29c), and an intriguing ancient line represented by a single shrub, *Amborella trichopoda* (see Figure 27.29d). Found only in cloud forests of the South Pacific island of New Caledonia, *Amborella*'s small white flowers and vascular system are structurally simpler than those of other angiosperms, and its female gametophyte differs as well. These morphological differences and a comparison of the nucleotide sequences of genes encoding the two angiosperm phytochromes (photoreceptors discussed in Chapter 35) suggest that *Amborella* is the closest living relative of the first flowering plants.

Figure 27.30a gives some idea of the variety of living monocots, which include grasses, palms, lilies, and orchids. The world's major crop plants (wheat, corn, rice, rye, sugarcane, and barley) are domesticated grasses, and all are monocots. There are at least 60,000 species of monocots, including 10,000 grasses and 20,000 orchids. Eudicots are even more diverse, with nearly 200,000 species **(Figure 27.30b)**. They include flowering shrubs and trees, most nonwoody (herbaceous) plants, and cacti. **Figure 27.31** shows the life cycle of a lily, a monocot. The life cycle of a typical eudicot is described in detail in the next unit, which focuses on the structure and function of flowering plants.

Many Factors Contributed to the Adaptive Success of Angiosperms

At this writing, molecular studies place the origin of flowering plants at least 140 mya. It took only about 40 million years—a short span in geological time—for angiosperms to eclipse gymnosperms as the prevailing form of plant life on land (see Figure 22.19). Several

Figure 27.29
Representatives of basal angiosperm clades. **(a)** Southern magnolia (*Magnolia grandiflora*), a magnoliid. **(b)** Star anise (*Illicium floridanum*). **(c)** Sacred lotus (*Nelumbo nucifera*), a water lily. **(d)** *Amborella trichopoda*.

a. Southern magnolia (*Magnolia grandiflora*), a magnoliid

b. Star anise (*Illicium floridanum*)

c. Sacred lotus (*Nelumbo nucifera*), a water lily

d. *Amborella*

D. Harms/Peter Arnold, Inc.

Rob & Ann Simpson/Visuals Unlimited

© Gregory C. Dimijian/Photo Researchers, Inc.

© Sangtae Kim, University of Florida

a. Representative monocots

Wheat *(Triticum)*

Tulips *(Tulipa)*

Eastern prairie fringed orchid
(Platanthera leucophaea)

b. Representative eudicots

Rose *(Rosa)*

Yellow bush lupine
(Lupinus arboreus)

Cherry *(Prunus)*

Claret cup cactus
(Echinocereus triglochidratus)

Figure 27.30
Examples of monocots and eudicots.

factors fueled this adaptive success. As with other seed plants, the large, diploid sporophyte phase dominates a flowering plant's life cycle, and the sporophyte retains and nourishes the much smaller gametophytes. But flowering plants also show some evolutionary innovations not seen in gymnosperms.

More Efficient Transport of Water and Nutrients. Where gymnosperms have only one type of water-conducting cell (tracheids), angiosperms have an additional, more specialized type (called vessel elements). As a result, an angiosperm's xylem vessels move water more rapidly from roots to shoot parts. Also, modifications in angiosperm phloem tissue allow it to more efficiently transport sugars produced in photosynthesis through the plant body.

Enhanced Nutrition and Physical Protection for Embryos. Other changes in angiosperms made it more likely that reproduction would succeed. For example, a two-step *double fertilization* process in the seeds of flowering plants gives rise to both an embryo and a unique nutritive tissue (called endosperm) that nour-

ishes the embryonic sporophyte. The ovule containing a female gametophyte is enclosed within an **ovary,** which develops from a carpel and shelters the ovule against desiccation and against attack by herbivores or pathogens. In turn, ovaries develop into the fruits that house angiosperm seeds. As noted earlier, a fruit not only protects seeds, but helps disperse them—for instance, when an animal eats a fruit, seeds may pass through the animal's gut none the worse for the journey and be released in a new location in the animal's feces. Above all, angiosperms have flowers, the unique reproductive organs that you will read much more about in the next unit.

Angiosperms Coevolved with Animal Pollinators

The evolutionary success of angiosperms correlates not only with the adaptations just described, but also with efficient mechanisms of transferring pollen to female reproductive parts. While a conifer depends on air currents to disperse its pollen, angiosperms coevolved with pollinators—insects, bats, birds, and other animals that

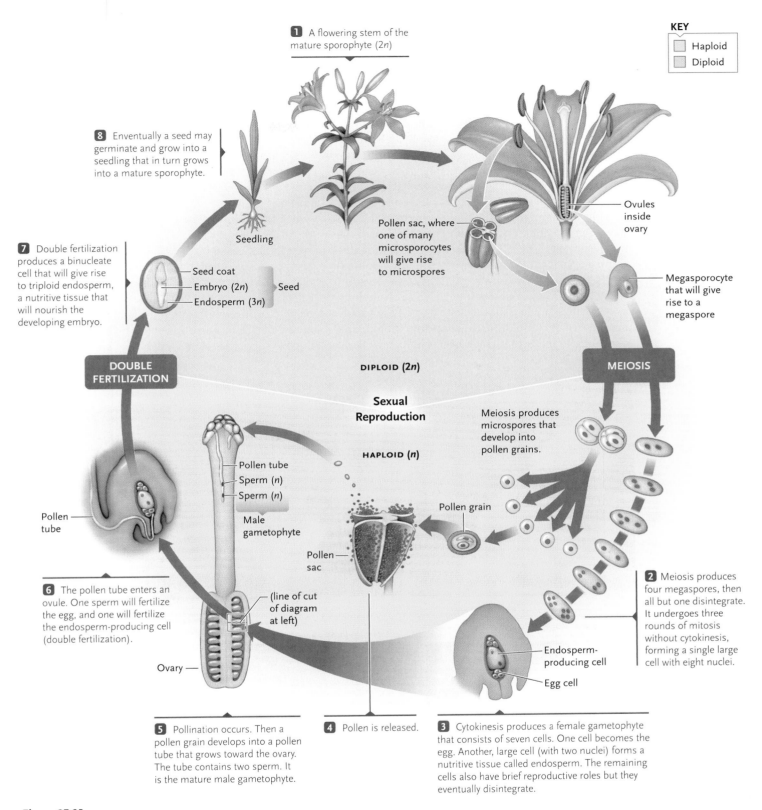

1 A flowering stem of the mature sporophyte (2*n*)

8 Enventually a seed may germinate and grow into a seedling that in turn grows into a mature sporophyte.

Seedling

Pollen sac, where one of many microsporocytes will give rise to microspores

Ovules inside ovary

7 Double fertilization produces a binucleate cell that will give rise to triploid endosperm, a nutritive tissue that will nourish the developing embryo.

Seed coat
Embryo (2*n*)
Endosperm (3*n*)
Seed

Megasporocyte that will give rise to a megaspore

DOUBLE FERTILIZATION

DIPLOID (2*n*)

MEIOSIS

KEY
Haploid
Diploid

Sexual Reproduction

Meiosis produces microspores that develop into pollen grains.

HAPLOID (*n*)

Pollen tube
Sperm (*n*)
Sperm (*n*)

Male gametophyte

Pollen grain

Pollen tube

Pollen sac

Pollen tube

(line of cut of diagram at left)

6 The pollen tube enters an ovule. One sperm will fertilize the egg, and one will fertilize the endosperm-producing cell (double fertilization).

Ovary

Endosperm-producing cell

Egg cell

2 Meiosis produces four megaspores, then all but one disintegrate. It undergoes three rounds of mitosis without cytokinesis, forming a single large cell with eight nuclei.

5 Pollination occurs. Then a pollen grain develops into a pollen tube that grows toward the ovary. The tube contains two sperm. It is the mature male gametophyte.

4 Pollen is released.

3 Cytokinesis produces a female gametophyte that consists of seven cells. One cell becomes the egg. Another, large cell (with two nuclei) forms a nutritive tissue called endosperm. The remaining cells also have brief reproductive roles but they eventually disintegrate.

Figure 27.31
Life cycle of a flowering plant, the monocot *Lilium.* Double fertilization is a notable feature of the cycle. The male gametophyte delivers two sperm to an ovule. One sperm fertilizes the egg, forming the embryo, and the other fertilizes the endosperm-producing cell, which nourishes the embryo.

withdraw pollen from male floral structures (often while obtaining nectar) and inadvertently transfer it to female reproductive parts. **Coevolution** occurs when two or more species interact closely in the same ecological setting. A heritable change in one species affects selection pressure operating between them, and the other species evolves as well. Over time, plants that came to have distinctive flowers, scents, and sugary nectar coevolved with animals that could take advantage of the rich food source.

In general, a flower's reproductive parts are positioned so that visiting pollinators will brush against them. In addition, many floral features correlate with specific pollinators. For example, reproductive parts may be located above nectar-filled floral tubes the same length as the feeding structure of a preferred pollinator. Nectar-sipping bats **(Figure 27.32a)** and moths forage by night. They pollinate intensely sweet-smelling flowers with white or pale petals that are more visible than colored petals in the dark. Long, thin mouthparts of moths and butterflies reach nectar in narrow floral tubes or flora spurs. The Madagascar hawkmoth uncoils a mouthpart the same length—an astonishing 22 cm—as a narrow floral spur of an orchid it pollinates, *Angraecum sesquipedale* **(Figure 27.32b)**. Red and yellow flowers attract birds **(Figure 27.32c)**, which have good daytime vision but a poor sense of smell. Hence bird-pollinated plants do not squander metabolic resources to make fragrances. By contrast, flowers of species that are pollinated by beetles or flies may smell like rotten meat, dung, or decaying matter. Daisies and other fragrant flowers with distinctive patterns, shapes, and red or orange components attract butterflies, which forage by day.

Bees see ultraviolet light and visit flowers with sweet odors and parts that appear to humans as yellow, blue, or purple **(Figure 27.32d)**. Produced by pigments that absorb ultraviolet light, the colors form patterns called "nectar guides" that attract bees—which may pick up or "drop off" pollen during the visit. Here, as in our other examples, flowers contribute to the reproductive success of plants that bear them.

a. Bat pollinating a giant saguaro

b. Hawk moth pollinating an orchid

c. Hummingbird visiting a hibiscus flower

d. Bee-attracting pattern of a marsh marigold

Visible light   UV light

Figure 27.32

Coevolution of flowering plants and animal pollinators. The colors and configurations of some flowers, and the production of nectar or odors, have coevolved with specific animal pollinators. **(a)** At night, nectar-feeding bats sip nectar from flowers of the giant saguaro *(Carnegia gigantea)*, transferring pollen from flower to flower in the process. **(b)** The hawkmoth *Xanthopan morgani praedicta* has a proboscis long enough to reach nectar at the base of the equally long floral spur of the orchid *Angraecum sesquipedale*. **(c)** A Bahama woodstar hummingbird *(Calliphlox evelynae)* sipping nectar from a hibiscus blossom *(Hibiscus)*. The long narrow bill of hummingbirds coevolved with long, narrow floral tubes. **(d)** Under ultraviolet light, the bee-attracting pattern of a gold-petaled marsh marigold becomes visible to human eyes.

Current Research Focuses on Genes Underlying Transitions in Plant Traits

Improvements in the ability of plant scientists to manipulate, analyze, and compare modern plant genomes, coupled with advances in the analysis of fossil plants, are having a profound impact on our understanding of the evolution of flowering plants. A case in point is research on the gene *LFY,* which encodes the regulatory protein LEAFY (Chapter 16 discusses regulatory proteins in detail). The LEAFY protein typically controls expression of several genes by binding to the genes' control sequences. All land plants carry the *LFY* gene, but its effects on phenotype vary markedly in different plant groups. In mosses, which arose almost 400 million years ago, the LEAFY protein regulates growth throughout the plant. In ferns and gymnosperms, which arose later, LEAFY controls growth in a subset of tissues. In angiosperms, LEAFY regulates gene expression only in the particular type of meristem

UNANSWERED QUESTIONS

Where did flowering plants come from?

Flowers are a unique feature of the angiosperms, yet botanists still understand little of their evolutionary origin. When flowering plants appear in the Cretaceous fossil record, they appear suddenly and diversify immediately, a situation Darwin famously referred to as an "abominable mystery." What did the first angiosperms and the first flower look like? And where did they arise?

As described in this chapter, recent molecular analyses have converged on *Amborella trichopoda* as the living representative of the most ancient lineage in the angiosperm family tree. This research has shed light on many questions. For example, *Amborella* flowers have some features considered evolutionarily primitive, such as petals and sepals that are not distinctly different in form. This observation supports the hypothesis that two other types of flower parts, the calyx and corolla, arose later in angiosperm evolution. But *Amborella* also has some features thought to have evolved much more recently, such as single-sex flowers that have either male or female reproductive parts (but never both). Should we be surprised to find both primitive and advanced traits in this ancient lineage? Not at all. *Amborella* has existed on Earth for millions of years and its flowers may have evolved new features over that time.

The puzzle of where angiosperms came from and what the first flowering plants looked like has not been solved by fossil studies, either. This chapter discusses the fossil species *Archaefructus,* which dates from the Jurassic and is thought to be the oldest known fossil flower. It consists of an elongated axis with what its discoverers described as stamens (male reproductive structures) toward the base and carpels (female reproductive structures) toward the apex, and no sepals and petals. This elongated flower is unlike the flowers of any modern angiosperm, and its structure suggests that the earliest flowers may have been very different from what we see today. However, some paleobotanists have reinterpreted the *Archefructus* "flower" as an inflorescence (a flower cluster), with male flowers at the base and female flowers toward the apex. In addition, radiometric dating places *Archaefructus* in the early-mid Cretaceous, a period from which other early angiosperm fossils are known. Thus, *Archaefructus* may not be the oldest flower, and the fossil specimen may represent a cluster of flowers instead of a single flower. This debate continues.

Botanists also disagree about the ancestors of angiosperms. Some gnetophytes—gymnosperms that include *Welwitschia* and *Ephedra* species (refer to Figure 27.23)—have features similar to angiosperms. Botanists long speculated that the two groups were closely related, with a common ancestor that had flowerlike features. However, recent analyses based on DNA sequence data suggest that gnetophytes are not closely related to angiosperms after all. There are also fossil gymnosperm taxa with features that might be forerunners of carpels or other flower parts, but paleobotanists disagree on these interpretations as well. Thus examinations of fossils and extant species have yet to resolve key questions about the evolution of angiosperms.

What, then, can molecular data tell us? Studies of the genetic mechanisms that guide the development of flower parts have provided a framework for understanding how genes control flower formation (a topic of Chapter 34). This research has also given us insight into what kinds of molecular changes may have led to the evolution of flowers. For instance, certain genes that encode transcription factors required for the formation of reproductive organs in flowers are found also in gymnosperms. This finding is not surprising, because gymnosperms also form male and female reproductive structures; the most logical hypothesis is that angiosperms retained the gymnosperm developmental program for these organs. Yet genes for other transcription factors active in flower formation are *not* found in gymnosperms. We know that transcription factors may turn on and off entire developmental pathways, such as those that cause undifferentiated tissue (called meristem tissue) to form a flower. One hypothesis is that in an ancient gymnosperm ancestor, duplications in a particular gene family gave rise to genes that in turn accumulated mutations allowing them to perform new functions that resulted in the formation of the first flowers.

As much insight as these molecular studies give us into events that might have resulted in the evolution of flowers, they have not brought us any closer to understanding the fundamental question of where angiosperms arose. Additional fossil data may help provide the answer, but it is also possible that the earliest angiosperms, or their direct ancestors, lived in habitats where fossils do not readily form. Additional molecular data may deepen our understanding of how changes in genes produced the first flower. But molecular data based on contemporary species will not help decipher what the first angiosperm and the first flower looked like. Thus, it is possible that the abominable mystery will live on.

Amy Litt is Director of Plant Genomics and Cullman Curator at the New York Botanical Garden, where she also earned her Ph.D. Her main interests lie in the evolution of plant form and how changes in gene function during the course of plant evolution have produced novel plant forms and functions—particularly new flower and fruit morphologies. Learn more about her work at http://sciweb.nybg.org/science2/Profile_106.asp.

tissue that gives rise to flowers (a topic of Chapter 34). Curious about the evolutionary shift from a general to a specific effect, Alexis Maizel and his team at the Max Planck Institute for Developmental Biology in Germany compared *LFY* sequences and their corresponding proteins in fourteen species, including a moss, ferns, gymnosperms, and the angiosperms *Arabidopsis* (thale cress) and snapdragon. Remarkably, they discovered that the evolutionary honing of the effects of the LEAFY protein correlated with only a handful of changes in the base sequence of the *LFY* gene. Each change affected how—or if—the LEAFY protein regulated the expression of a given gene. Over time, LEAFY took on its highly specific, crucial role in angiosperms, helping to direct the developmental events that produce flowers.

Today some of the most exciting research in all of biology involves studies exploring the connections between genetic changes and key evolutionary transitions in plant form and functioning. As the genes of many more plant species are sequenced and correlated with evidence from comparative morphology and the fossil record, we can expect a steady stream of new insights about the evolutionary journey of all major plant lineages. In Chapter 28 a very different group of organisms, the fungi, takes center stage. Although many fungal species seem superficially plantlike, biologists today are avidly exploring evolutionary links between fungi and animals.

STUDY BREAK

1. How has the relative lack of fossil early angiosperms affected our understanding of this group?
2. Describe two basic features that distinguish monocots from eudicots, and give some examples of species in each clade.
3. List at least three adaptations that have contributed to the evolutionary success of angiosperms as a group.

Review

27.1 The Transition to Life on Land

- Plants are thought to have evolved from charophyte green algae between 425 and 490 million years ago (Figure 27.2).
- Adaptations to terrestrial life in early land plants include an outer cuticle that helps prevent desiccation, lignified tissues, spores protected by a wall containing sporopollenin, multicellular chambers that protect developing gametes, and an embryo sheltered inside a parent plant (Figure 27.3).
- Other key evolutionary trends among land plants included the development of vascular tissues, root systems, and shoot systems, including lignified stems and leaves equipped with stomata; a shift from dominance by a long-lived, larger haploid gametophyte to dominance of a long-lived, larger diploid sporophyte, and a shift from homospory to heterospory with separate male and female gametophytes (Figures 27.5–27.7).
- Male gametophytes (pollen) became specialized for dispersal without liquid water, and female gametophytes became specialized for enclosing embryo sporophytes in seeds.

Animation: Milestones in plant evolution

Animation: Haploid to diploid dominance

Animation: Evolutionary tree for plants

Animation: The importance of alternation of generations

27.2 Bryophytes: Nonvascular Land Plants

- Existing nonvascular land plants, or bryophytes, include the liverworts (Hepatophyta), hornworts (Anthocerophyta), and mosses (Bryophyta). Liverworts may have been the first land plants.

- Bryophytes produce flagellated sperm that swim through free water to reach eggs. They lack a vascular system; true roots, stems, and leaves; and lignified tissue. A larger, dominant gametophyte (*n*) phase alternates with a small, fleeting sporophyte (2*n*) phase. Spores develop inside jacketed sporangia (Figures 27.9 and 27.12).

Animation: Moss life cycle

Animation: *Marchantia*, a liverwort

27.3 Seedless Vascular Plants

- Existing seedless vascular land plants include the lycophytes (club mosses), whisk ferns, horsetails, and ferns. Like bryophytes, they release spores and have swimming sperm. Unlike bryophytes, they have well-developed vascular tissues. The sporophyte is the larger, longer-lived stage of the life cycle and develops independently of the small gametophyte.
- Club mosses (Lycophyta) have sporangia clustered at the bases of specialized leaves called sporophylls. Each sporophylls cluster forms a strobilus (cone). Haploid spores dispersed from the sporangia germinate to form small, free-living gametophytes. Ferns, whisk ferns, and horsetails (Pterophyta) have a similar life cycle. Horsetail sporophytes typically have underground stems (rhizomes) anchored to the soil by roots.
- Ferns are the largest and most diverse group of seedless vascular plants. Most species do not have aboveground stems, only leaves that arise from nodes along an underground rhizome. Fern leaves typically have well-developed stomata, and the vascular system consists of bundles, each with xylem surrounded by phloem. Sporangia on the lower surface of sporophylls (fronds) release spores that develop into gametophytes. Sexual reproduction produces a much larger, long-lived sporophyte (Figure 27.16).

Animation: Seedless vascular plants

Animation: Fern life cycle

27.4 Gymnosperms: The First Seed Plants

- Gymnosperms (conifers and their relatives), together with angiosperms (flowering plants), are the seed-bearing vascular plants. Reproductive innovations include pollination, the ovule, and the seed. An ovule is a sporangium containing a female gametophyte, so the female gametophyte is attached to and protected by the sporophyte. The smaller spore type produces a male gametophyte. Since pollination takes place via air currents or animal pollinators, plants fertilized by pollination do not require liquid water to reproduce. A seed forms when an ovule matures following fertilization; in gymnosperms, its main function is to protect and help disperse the embryonic sporophyte (Figure 27.24).

- During the Mesozoic, gymnosperms were the dominant land plants. Today conifers are the primary vegetation of forests at higher latitudes and elevations and have important economic uses as sources of lumber, resins, and other products.

Animation: *Pinus* cones

Animation: Pine life cycle

27.5 Angiosperms: Flowering Plants

- Angiosperms (Anthophyta) have dominated the land for more than 100 million years and currently are the most diverse plant group. There are two main angiosperm clades: monocots and eudicots. Other clades are represented by magnolias and their relatives (magnoliids), water lilies, the star anise group, and *Amborella*, a single species thought to be the most basal living angiosperm (Figures 27.29 and 27.30).

- The angiosperm vascular system moves water from roots to shoots more efficiently than in gymnosperms, and the phloem tissue moves sugars more efficiently through the plant body. Reproductive adaptations include a protective ovary around the ovule, endosperm, fruits that aid seed dispersal, the complex organs called flowers, and the coevolution of flower characteristics with the structural and/or physiological characteristics of animal pollinators (Figures 27.31 and 27.32).

Animation: Flower parts

Animation: Monocot life cycle

Questions

Self-Test Questions

1. Which of the following is *not* an evolutionary trend among plants?
   a. developing vascular tissues
   b. becoming seedless
   c. having a dominant diploid generation
   d. producing nonmotile gametes
   e. producing two types of spores

2. As plants made the evolutionary transition to a terrestrial existence, they benefited from adaptations that:
   a. increased the motility of their gametes on dry land.
   b. flattened the plant body to expose it to the sun.
   c. reduced the number and distribution of roots to prevent drying.
   d. provided mechanisms for gaining access to nutrients in soil.
   e. allowed stems and leaves to absorb water from the atmosphere.

3. Land plants no longer required water as a medium for reproduction with the evolution of:
   a. fruits and roots.
   b. flowers and leaves.
   c. cell walls and rhizoids.
   d. lignified stems.
   e. seeds and pollen.

4. Which is the correct matching of phylum and plant group?
   a. Anthophyta: pines
   b. Bryophyta: gnetophytes
   c. Coniferophyta: angiosperms
   d. Hepatophyta: cycads
   e. Pterophyta: horsetails

5. A homeowner noticed moss growing between bricks on his patio. Closer examination revealed tiny brown stalks with cuplike tops emerging from green leaflets. These brown structures were:
   a. the sporophyte generation.
   b. the gametophyte generation.
   c. elongated haploid reproductive cells.
   d. archegonia.
   e. antheridia.

6. Horsetails are most closely related to:
   a. mosses and whisk ferns.
   b. liverworts and hornworts.
   c. cycads and ginkgos.
   d. club mosses and ferns.
   e. gnetophytes and gymnosperms.

7. Which feature(s) do ferns share with all other land plants?
   a. sporophyte and gametophyte life cycle stages
   b. gametophytes supported by a thallus
   c. dispersal of spores from a sorus
   d. asexual reproduction by way of gemmae
   e. water uptake by means of rhizoids

8. The evolution of true roots is first seen in:
   a. liverworts.
   b. seedless vascular plants.
   c. mosses.
   d. flowering plants.
   e. conifers.

9. Based solely on numbers of species, the most successful plants today are:
   a. angiosperms.
   b. ferns.
   c. gymnosperms.
   d. mosses.
   e. the bryophytes as a group.

10. Angiosperms and gymnosperms share the following characteristic(s):
    a. pollination by means of water.
    b. seeds protected within an ovary.
    c. embryonic cotyledons.
    d. a dominant sporophyte generation.
    e. a seasonal loss of all leaves.

Questions for Discussion

1. Suggest adjustments in the angiosperm life cycle that would better suit plants to some future world where environments were generally hotter and more arid. Do the same for a colder and wetter environment.

2. Working in the field, you discover a fossil of a previously undescribed plant species. The specimen is small and may not be complete; the parts you have do not include any floral organs. What sorts of observations would you need in order to classify the fossil as a seedless vascular plant with reasonable accuracy? What evidence would you need in order to distinguish between a fossil lycopod and a fern?

3. Modern humans emerged about 100,000 years ago. How accurate is it to state that our species has lived in the Age of Wood? Explain.

4. Compare the size, anatomical complexity, and degree of independence of a moss gametophyte, a fern gametophyte, a Douglas fir female gametophyte, and a dogwood female gametophyte. Which one is the most protected from the external environment? Which trends in plant evolution does your work on this question bring to mind?

Experimental Analysis

You are studying mechanisms that control the development of flowers, and your research to date has focused on eudicots, which tend to have showier flowers than monocots. A colleague has suggested that you broaden your analysis to include representative basal angiosperms. Outline the rationale for this expanded approach and indicate which additional species or group(s) you plan to include. Discuss the type(s) of data you plan to gather and why you feel the information will make your study more complete.

Evolution Link

Plant evolutionary biologist Spencer C. H. Barrett has written that the reproductive organs of angiosperms are more varied than the equivalent structures of any other group of organisms. Which angiosperm organs was Barrett talking about? Explain why you agree or disagree with his view.

How Would You Vote?

Demand for paper is a big factor in deforestation. However, using recycled paper can add to the cost of a product. Are you willing to pay more for papers, books, and magazines that are printed on recycled paper? Go to www.thomsonedu.com/login to investigate both sides of the issue and then vote.

The mushroom-forming fungus *Inocybe fastigiata*, a forest-dwelling species that commonly lives in close association with conifers and hardwood trees.

STUDY PLAN

28.1 General Characteristics of Fungi

Fungi may be single-celled or multicellular

Fungi obtain nutrients by extracellular digestion and absorption

All fungi reproduce by way of spores, but other aspects of reproduction vary

28.2 Major Groups of Fungi

Fungi were present on Earth by at least 500 million years ago

Once they appeared, fungi radiated into at least five major lineages

Chytrids produce motile spores that have flagella

Zygomycetes form zygospores for sexual reproduction

Glomeromycetes form spores at the ends of hyphae

Ascomycetes, the sac fungi, produce sexual spores in saclike asci

Basidiomycetes, the club fungi, form sexual spores in club-shaped basidia

Conidial fungi are species for which no sexual phase is known

Microsporidia are single-celled sporelike parasites

28.3 Fungal Associations

A lichen is an association between a fungus and a photosynthetic partner

Mycorrhizae are symbiotic associations of fungi and plant roots

28 Fungi

WHY IT MATTERS

In a forest, decay is everywhere—rotting leaves, moldering branches, perhaps the disintegrating carcass of an insect or a small mammal. Each year in most terrestrial ecosystems, an astounding amount of organic matter is produced, cast off, broken down, and its elements gradually recycled. This recycling has a huge impact on world ecosystems; for example, each year it returns at least 85 billion tons of carbon, in the form of carbon dioxide, to the atmosphere. Chief among the recyclers are the curious organisms of the **Kingdom Fungi**—about 60,000 described species of molds, mushroom-forming fungi, yeasts, and their relatives **(Figure 28.1),** and an estimated 1.6 million more that are yet to be described.

Fungi are eukaryotes, most are multicellular, and all are heterotrophs, obtaining their nutrients by breaking down organic molecules that other organisms have synthesized. Molecular evidence suggests that fungi were present on land at least 500 million years ago, and possibly much earlier. In the course of the intervening millennia, evolution equipped fungi with a remarkable ability to break down organic matter, ranging from living and dead organisms and animal wastes to your groceries, clothing, paper and wood, even photographic

605

Sulfur shelf fungus, *Polyporus*

Big laughing mushroom, *Gymnopilus*

Baker's yeast cells, *Saccharomyces cerevisiae*

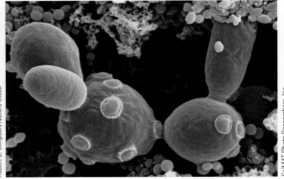

Figure 28.1
Examples of fungi that hint at the rich diversity within the Kingdom Fungi.

film. Along with heterotrophic bacteria, they have become Earth's premier decomposers.

Fungi collectively also are the single greatest cause of plant diseases, and a host of species cause disease in humans and other animals. Some even produce carcinogenic toxins. On the other hand, 90% of plants obtain needed minerals by way of a symbiotic relationship with a fungus. Humans have harnessed the metabolic activities of certain fungi to obtain substances ranging from flavorful cheeses and wine to therapeutic drugs such as penicillin and the immunosuppressant cyclosporin. And, as you know from previous chapters, species such as the yeast *Saccharomyces cerevisiae* and the mold *Neurospora crassa* have long been pivotal model organisms in studies of DNA structure and function, and the yeast has also been important in the development of genetic engineering methods.

Despite their profound impact on ecosystems and other life forms, most of us have only a passing acquaintance with the fungi—perhaps limited to the mushrooms on our pizza or the invisible but annoying types that cause skin infections like athlete's foot. This chapter provides you with an introduction to mycology, the study of fungi (*mykes* = mushroom; *logos* = knowledge). We begin with general characteristics of this kingdom and then discuss its major divisions.

28.1 General Characteristics of Fungi

We begin our survey of fungi by examining how fungi differ from other forms of life, how fungi obtain nutrients, and the adaptations for reproduction and growth that enable fungi to spread far and wide through the environment.

Fungi May Be Single-Celled or Multicellular

Two basic body forms, single-celled and multicellular, emerged as the lineages of fungi evolved. Some fungi are single cells, a form called **yeast**, while others exist in a multicellular form made up of threadlike filaments. Still others alternate between yeast and multicellular forms at different stages of the life cycle. Whether a fungus is single-celled or multicellular, a rigid wall usually surrounds the plasma membrane of its cells. Generally the polysaccharide **chitin** provides this rigidity, the same function it serves in the external skeletons of insects and other arthropods.

In a multicellular fungus, exploiting food sources is the province of a cottony mesh of tiny filaments that branch repeatedly as they grow over or into organic matter. Each filament is a **hypha** (*hyphe* = web; plural, hyphae); the combined mass of hyphae is a **mycelium** (plural, mycelia). Hyphae generally are tube-shaped **(Figure 28.2)**. In most multicellular fungi the hyphae are partitioned by cross walls called **septa** (*saeptum* = partition; singular, septum). The septa create cell-like compartments that contain organelles. However, in one group, the zygomycetes described shortly, most hyphae are *aseptate*—they lack cross walls—although septa do arise to separate reproductive structures from the rest of the hypha.

The unusual features of fungal hyphae have led many mycologists to question whether "multicellular" is really an accurate description for most fungal architecture. For instance, depending on the species, hyphal cells may have more than one nucleus, and septa have pores that permit nuclei and other organelles to move between hyphal cells. These passages also allow cytoplasm to extend from one hyphal cell to the next, throughout the whole mycelium. By a mechanism called **cytoplasmic streaming**, cytoplasm containing nutrients can flow unimpeded through the hyphae, from food-absorbing parts of the fungal body to other, nonabsorptive parts such as reproductive structures.

A multicellular fungus grows larger as its hyphae elongate and branch. Each hypha elongates at its tip as new wall polymers (delivered by vesicles) are incorporated and additional cytoplasm, including organelles, is synthesized. A hypha branches a few micrometers behind its tip, and as the new hyphae elongate, then branch themselves, an extensive mycelium can form quickly. Each forming branch fills with cytoplasm that includes new nuclei produced by mitosis. Although the rapid branching of hyphae is what allows multicellular fungi to grow aggressively—sometimes increasing in mass many times over within a few days—researchers have only recently gained the tools to explore the mechanisms that underlie this phenomenon. Studies spurred by the sequencing of the genome of *Neurospora crassa* suggest that multiple steps involving a variety of genes and their interacting protein products determine where and when a new branch arises. Given that the rapid growth of fungal mycelia has such a tremendous impact in nature, fungal diseases, and many other areas, this topic is a central focus of much mycological research.

Beyond their role in nutrient transport, aggregations of hyphae are the structural foundation for all other parts that arise as a multicellular fungus develops. For example, in many fungi a subset of hyphae interweave tightly, becoming prominent reproductive structures (sometimes called fruiting bodies). Grocery store mushrooms are examples. But while a mushroom or some analogous structure may be the most conspicuous part of a given fungus, it usually represents only a small fraction of the organism's total mass. The rest penetrates the food source the fungus is slowly digesting. In some fungi, modified hyphae called **rhizoids** anchor the fungus to its substrate. Most fungi

a. Multicellular fungus

— Mycelium —

b. Fungal hyphae

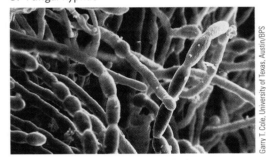

Garry T. Cole, University of Texas, Austin/BPS

Figure 28.2
Fungal mycelia. **(a)** Sketch of the mycelium of a mushroom-forming fungus, which consists of branching septate hyphae. **(b)** Micrograph of fungal hyphae.

that parasitize living plants produce hyphal branches called **haustoria** (*haustor* = drinker) that penetrate the walls of a host plant's cells and channel nutrients back to the fungal body.

Fungi Obtain Nutrients by Extracellular Digestion and Absorption

Some major challenges have shaped the adaptations by which fungi obtain nutrients. As heterotrophs, fungi must secure nutrients by breaking down organic substances formed by other organisms. Nearly all fungi are terrestrial, but unlike other land-dwelling heterotrophs (such as animals), fungi are not mobile. They also lack mouths or appendages for seizing, handling, and dismantling food items. Instead, fungi have a very different suite of adaptations for obtaining nutrients.

To begin with, most species of fungi can synthesize nearly all their required nutrients from a few raw materials, including water, some minerals and vitamins (especially B vitamins), and a sugar or some other organic carbon source. For many species, carbohydrates in dead organic matter are the carbon sources, and fungi with this mode of nutrition are called **saprobes** (*sapros* = rotten). Other fungi are parasites, which extract carbohydrates from tissues of a living host, harming it in the process. Parasitic fungi include those responsible for many devastating plant diseases, such as wheat rust and Dutch elm disease. Still other fungi are nourished by plants with which they have a mutually beneficial symbiotic association.

Regardless of their nutritional mode, all fungi gain the raw materials required to build and maintain their cells by absorption from the environment. Fungi can absorb many small molecules directly from their surroundings, and gain access to other nutrients through extracellular digestion. In this process, a fungus releases enzymes that digest nearby organic matter, breaking down larger molecules into absorbable fragments. Fungal species differ in the particular digestive enzymes they synthesize, so a substrate that is a suitable food source for one species may be unavailable to another. Although there are exceptions, fungi typically thrive only in moist environments where they can directly absorb water, dissolved ions, simple sugars, amino acids, and other small molecules. When some of a mycelium's hyphal filaments contact a source of food, growth is channeled in the direction of the food source. Nutrients are absorbed only at the porous tips of hyphae; small atoms and molecules pass readily through these tips, and then transport mechanisms move them through the underlying plasma membrane.

Large organic molecules, such as the carbohydrate cellulose (see Section 3.3), *cannot* directly enter any part of a fungus. To use such substances as a food source, a fungus must secrete hydrolytic enzymes that break down the large molecules into smaller, absorbable subunits. Depending on the size of the subunit, further digestion may occur inside cells.

With their adaptations for efficient extracellular digestion, fungi are masters of the decay so vital to terrestrial ecosystems. For instance, in a single autumn one elm tree can shed 400 pounds of withered leaves; and in a tropical forest, a year's worth of debris may total 60 tons per acre. Without the metabolic activities of saprobic fungi and other decomposers such as bacteria, natural communities would rapidly become buried in their own detritus. As fungi digest dead tissues of other organisms, they also make a major contribution to the recycling of chemical elements those tissues contain. For instance, over time the degradation of organic compounds by saprobic fungi helps return key nutrients such as nitrogen and phosphorus to ecosystems. But the prime example of this recycling virtuosity involves carbon. The respiring cells of fungi and other decomposers give off carbon dioxide, liberating carbon that would otherwise remain locked in the tissues of dead organisms. Each year this activity recycles a vast amount of carbon to plants, the primary producers of nearly all ecosystems on Earth.

All Fungi Reproduce by Way of Spores, but Other Aspects of Reproduction Vary

Biologists have observed a striking number of reproductive variations in fungi, differences that are part of what makes them fascinating to study. As you will learn in the next section, fungi have traditionally been classified on the basis of their reproductive characteristics, although today evidence from molecular analysis also plays a prominent role.

Overall, most fungi have the capacity to reproduce both sexually and asexually. Although no single diagram can depict all the variations, **Figure 28.3** gives an overview of the life cycle stages that mycologists have observed in several groups of fungi. The figure illustrates two general points. First, the life cycle of multicellular fungi typically involves a diploid stage ($2n$), a haploid stage (n), and a dikaryotic ("two nuclei") stage in which the fungus forms hyphae (and a mycelium) that are $n + n$—neither strictly haploid *nor* diploid. Depending on the type of fungus, this stage may be long lasting or extremely brief, and it is described more fully later in this section. Second, all fungi, whether they are multicellular or in a single-celled, yeast form, can reproduce via **fungal spores.** The spores are microscopic, and in all but one group they are not motile—that is, they are not propelled by flagella. Each spore is a walled single cell or multicellular structure that is dispersed from the parent body, often via wind or water. The spores of single-celled fungi form inside the parent cell, then escape when the wall breaks open. In multicellular fungi, spores arise in or on specialized hyphal structures and may develop thick walls that help them withstand cold or drying out after they are released.

Reproduction by way of spores is one of the crucial fungal adaptations. Most fungi are opportunists, obtaining energy by exploiting food sources that occur unpredictably in the environment. Having lightweight spores that are easily disseminated by air or water increases opportunities for finding food. And releasing vast numbers of spores, as some fungi do, improves the odds that at least a few spores will germinate and produce a new individual.

In nature generally, opportunistic organisms are adapted to reach new food sources quickly and utilize them rapidly. Fungi that are adept at degrading simple sugars and starches often are among the first decomposers to exploit a new source of food. They meet with keen competition from each other and from other decomposers. However, once fungal spores encounter potential food and favorable conditions, they can quickly develop into new individuals that simultaneously feed and rapidly make more spores.

Many opportunistic fungi develop rapidly, growing and reproducing before the food source is depleted. A common trade-off for speed, however, is small, even microscopic body size. Larger species of fungi are often adapted to move in later, exploiting food sources such as cellulose and lignin (a complex polymer in the walls of many plant cells), which their predecessors may have lacked the enzymatic machinery to digest efficiently. Some of these fungi may produce huge mycelia (and reproductive "fruiting bodies" such as mushrooms) by extracting nutrients from dead trees that

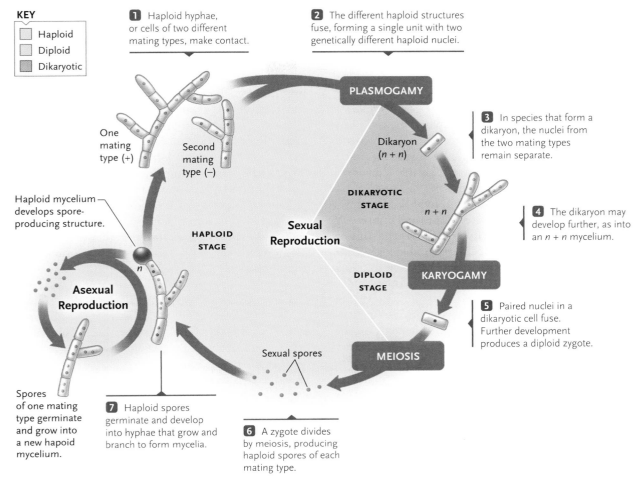

KEY

- Haploid
- Diploid
- Dikaryotic

1 Haploid hyphae, or cells of two different mating types, make contact.

2 The different haploid structures fuse, forming a single unit with two genetically different haploid nuclei.

PLASMOGAMY

One mating type (+)

Second mating type (−)

Dikaryon (*n* + *n*)

3 In species that form a dikaryon, the nuclei from the two mating types remain separate.

DIKARYOTIC STAGE

n + *n*

4 The dikaryon may develop further, as into an *n* + *n* mycelium.

Haploid mycelium develops spore-producing structure.

HAPLOID STAGE

Sexual Reproduction

DIPLOID STAGE

KARYOGAMY

n

Asexual Reproduction

5 Paired nuclei in a dikaryotic cell fuse. Further development produces a diploid zygote.

Sexual spores

MEIOSIS

Spores of one mating type germinate and grow into a new hapoid mycelium.

7 Haploid spores germinate and develop into hyphae that grow and branch to form mycelia.

6 A zygote divides by meiosis, producing haploid spores of each mating type.

Figure 28.3

Generalized life cycle for many fungi. Overall, fungi are diploid for only a short time. The duration of the dikaryon stage varies considerably, being lengthy for some species and extremely brief in others. Some types of fungi reproduce only asexually while in others shifts in environmental factors, such as the availability of key nutrients, can trigger a shift from asexual to sexual reproduction or vice versa. For still others sexual reproduction is the norm.

contain enough organic material to sustain an extended period of growth.

Features of Asexual Reproduction in Fungi. When a fungus produces spores asexually (see Figure 28.3), it may disperse billions of them into the environment. Some fungi (including many yeasts) also can reproduce asexually by budding or fission, or, in multicellular types, when fragments of hyphae break away from the mycelium and grow into separate individuals. In still others, environmental factors may determine whether the fungus produces hyphal fragments *or* asexual spores. These asexual reproductive strategies all result in new individuals that are essentially clones of the parent fungus. They can be viewed as another adaptation for speed, because the alternative—sexual reproduction—requires the presence of a suitable partner and generally involves several more steps.

The asexual stage of many multicellular fungi—including the pale gray fuzz you might see on berries or bread—is often called a **mold.** The term can be con-

fusing if you are attempting to keep track of taxonomic groupings; for example, the water molds and slime molds described in Chapter 26 are protists, although they were grouped with fungi until additional research revealed their true evolutionary standing. The mold visible on an overripe raspberry is actually a mycelium with aerial structures bearing sacs of haploid spores at their tips.

Features of Sexual Reproduction in Fungi. Although asexual reproduction is the norm, quite a few fungi shift to sexual reproduction when environmental conditions (such as a lack of nitrogen) or other influences dictate. As you may remember from Chapter 11, in sexual reproduction two haploid cells unite, and in most species fertilization—the fusion of two gamete nuclei to form a diploid zygote nucleus—soon follows. In fungi, however, the partners in sexual union can be two hyphae, two gametes, or other types of cells; the particular combination depends on the species involved. And in sharp contrast to other life forms, many days, months, or even years may pass between the time fertilization gets underway and when it is completed.

During the initial sexual stage, called **plasmogamy** (*plasma* = a formed thing; *gamos* = union), the cytoplasms of two genetically different partners fuse. The resulting new cell, a **dikaryon** (*di* = two; *karyon* = nucleus), contains two haploid nuclei, one from each parent. A dikaryon itself is not haploid (the condition of having one set of chromosomes) because it contains two nuclei. But neither is it diploid, because the nuclei are not fused. So, to be precise, we say that a dikaryon has an $n + n$ nuclear condition.

Plasmogamy can occur when hyphal cells of two different **mating type**, termed plus (+) and minus (−), fuse, a process that occurs in most fungi. The uniting hyphae belong to mycelia of different individuals of the same species that happen to grow near one another. The fusion of different mating types ensures genetic diversity in new individuals.

Once a dikaryon forms, the amount of time that elapses before the next stage begins depends on the type of fungus, as described in the next section. Sooner or later, however, a second phase of fertilization unfolds: The nuclei in the dikaryotic cell fuse to make a $2n$ zygote nucleus. This process is called **karyogamy** ("nuclear union"); in fungi that form mushrooms, it occurs in the tips of hyphae that end in the gills, which you may be able to see if you look closely at the underside of a mushroom cap (see Figure 28.1). In animals, a zygote is the first cell of a new individual, but in the world of fungi the zygote has a different fate. After it forms, meiosis converts the zygote nucleus into four haploid (*n*) nuclei. Those nuclei are packaged into haploid "sexual spores," which vary genetically from each parent. Then the spores are released to spread throughout the environment.

To sum up, in fungi both asexual and sexual spores are haploid, and both can germinate into haploid individuals. However, asexual spores are genetically identical products of asexual reproduction, while sexual spores are genetically varied products of sexual reproduction. We turn now to current ideas on the evolutionary history of fungi, and a survey of the major taxonomic groups in this kingdom.

STUDY BREAK

1. What features distinguish the two basic fungal body forms?
2. What is a fungal spore, and how does it function in reproduction?
3. Fungi reproduce sexually or asexually, but for many species the life cycle includes an unusual stage not seen in other organisms. What is this genetic condition, and what is its role in the life cycle?

28.2 Major Groups of Fungi

The evolutionary origins and lineages of fungi have been obscure ever since the first mycologists began puzzling over the characteristics of this group. With the advent of molecular techniques for research, these topics have become extremely active and exciting areas of biological research that may shed light on fundamental events in the evolution of all eukaryotes. Not surprisingly when so much new information is coming to light, mycologists hold a wide range of views on how different groups arose and may be related. Even so, there is wide agreement on five phyla of fungi, known formally as the Chytridiomycota, Zygomycota, Glomeromycota, Ascomycota, and Basidiomycota **(Figure 28.4).**

In a sixth group, termed conidial fungi, asexual reproduction produces spores called conidia. "Conidial" is not a true taxonomic classification, however. Rather, it serves as a holding station for fungal species that have not yet been assigned to one of the five phyla because no sexual reproductive phase has been observed. This is another instance in which the name for a fungal group can be confusing, because numerous species belonging to the Zygomycota, Ascomycota, and Basidiomycota also form conidia as part of their asexual reproductive cycle.

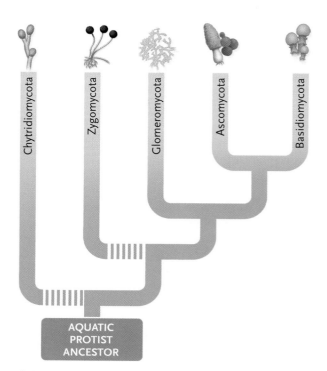

Figure 28.4

A phylogeny of fungi. This scheme represents a widely accepted view of the general relationships between major groups of fungi, but it may well be revised as new molecular findings provide more information. The dashed lines indicate that two groups, the chytrids and zygomycetes, are probably paraphyletic—they include subgroups that are not all descended from a single ancestor.

There Was Probably a Fungus among Us

The relationships of the fungi to protists, plants, and animals are buried so far back in evolutionary history that they have proved difficult to reconstruct. On the basis of morphological comparisons, for a long time taxonomists classified fungi as more closely related to the protists or plants than to animals. However, an investigation of ribosomal RNA (rRNA) sequences led to the conclusion that fungi and animals are more closely related to each other than either group is to protists or plants.

Patricia O. Wainwright, Gregory Hinkle, Mitchell L. Sogin, and Shawn K. Stickel of Rutgers University and the Woods Hole Marine Biology Laboratory carried out the analysis by comparing sequences of 18S rRNA, an rRNA molecule that forms part of the small ribosomal subunit in eukaryotes (see Section 15.4). The investigators began their work by sequencing the 18S rRNA molecules of species among the sponges, ctenophores, and cnidarians (see Chapter 29), which had never been sequenced before. These sequences were then compared with the

18S rRNA sequences of fungi, plants, and several protists, including protozoans and algae, which had been previously obtained by others. For the comparisons, the investigators used a computer program that sorts the rRNA sequences into related groups under the assumption that species with the greatest similarities in 18S rRNA sequence are most closely related. The sequence information was entered into the program in several different combinations; each time the analysis came up with the same family tree (see **figure**).

The family tree placed animals as the branch most closely related to fungi, and indicates that the two groups share a common ancestor not shared with any of the other groups. Other investigators have cited similarities in biochemical pathways in fungi and animals, such as pathways that make the amino acid hydroxyproline, the protein ferritin (which combines with iron atoms), and the polysaccharide chitin, which is a primary constituent of both fungal cell walls and arthropod exoskeletons. Studies of fungi called chytrids (p. 612) also are providing provocative insights on this topic.

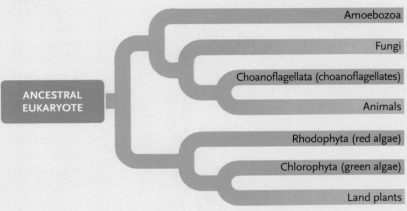

Finally, we will briefly consider a particular odd group of single-celled parasites called **microsporidia.** Based on genetic studies, many mycologists believe they make up a possible sixth phylum within the Kingdom Fungi.

ganism that does not fossilize well. Although traces of what may be fossil fungi exist in rock formations nearly 1 billion years old, the oldest fossils that we can confidently assign to the modern Kingdom Fungi appear in rock strata laid down about 500 million years ago.

Fungi Were Present on Earth by at Least 500 Million Years Ago

Many fungi look plantlike, and for many years fungi were classified as plants. As biologists learned more about the distinctive characteristics of fungi, however, it became clear that fungi merited a separate kingdom. The discovery of chitin in fungal cells, and recent comparisons of DNA and RNA sequences, all indicate that fungi and animals are more closely related to each other than they are to other eukaryotes (see *Insights from the Molecular Revolution*). Analysis of the sequences of several genes suggests that the lineages leading to animals and fungi may have diverged around 965 million years ago. Whenever the split developed, phylogenetic studies indicate that fungi first arose from a single-celled, flagellated protist—the sort of or-

Once They Appeared, Fungi Radiated into at Least Five Major Lineages

Most likely, the first fungi were aquatic. When other kinds of organisms began to colonize land, they may well have brought fungi along with them. For example, researchers have discovered what appear to be mycorrhizae—symbiotic associations of a fungus and a plant—in fossils of the some of the earliest known land plants (see Chapter 27). The final section of this chapter examines the nature of mycorrhizae more fully.

Over time, fungi diverged into the strikingly diverse lineages that we consider in the rest of this section **(Table 28.1).** As the lineages diversified, different adaptations associated with reproduction arose. For this reason, mycologists traditionally assigned fungi to

Table 28.1 Summary of Fungal Phyla

Phylum	Body Type	Key Feature	
Chytridiomycota (chytrids)	One to several cells	Motile spores propelled by flagella; usually asexual	
Zygomycota (zygomycetes)	Hyphal	Sexual stage in which a resistant zygospore forms for later germination	
Glomeromycota (glomero-mycetes)	Hyphal	Hyphae associated with plant roots, forming arbuscular mycorrhizae	
Ascomycota (ascomycetes)	Hyphal	Sexual spores produced in sacs called asci	
Basidiomycota (basidio-mycetes)	Hyphal	Sexual spores (basidiospores) form in basidia of a prominent fruiting body (basidiocarp)	

phyla according to the type of structure that houses the final stages of sexual reproduction and releases sexual spores. These features can still be useful indicators of the phylogenetic standing of a fungus, although now the powerful tools of molecular analysis are bringing many revisions to our understanding of the evolutionary journey of fungi. Our survey begins with chytrids, which probably most closely resemble the fungal kingdom's most ancient ancestors.

Chytrids Produce Motile Spores That Have Flagella

The phylum Chytridiomycota includes about a thousand species, referred to simply as chytrids. Chytrids are the only fungi that produce motile spores, which swim by way of flagella. Nearly all chytrids are microscopic **(Figure 28.5a),** and mycologists have recently begun paying significant attention to them, in part because their characteristics strongly suggest that the group arose near the beginning of fungal evolution. Another reason for research interest is the discovery that the chytrid *Batrachochytrium dendrobatis* is responsible for a disease epidemic that recently has wiped out an estimated two-thirds of the species of harlequin frogs *(Atelopus)* of the American tropics **(Figure 28.5b).** The epidemic has correlated with the rising average temperature in the frogs' habitats, an increase credited to global warming. Studies show that the warmer environment provides optimal growing temperatures for the chytrid pathogen.

Most chytrids are aquatic, although a few live as saprobes in soil, feeding on decaying plant and animal matter; as parasites on insects, plants, and some animals or even as symbiotic partners in the gut of cattle

a. *Chytriomyces hyalinus*   **b.** Chytridiomycosis in a frog   **c.** Harlequin frog

Skin surface

John Taylor/Visuals Unlimited

Centers for Disease Control and Prevention

Courtesy Ken Nemuras

Figure 28.5
Chytrids. **(a)** *Chytriomyces hyalinus,* one of the few chytrids that reproduces sexually. **(b)** Chytridiomycosis, a fungal infection, shown here in the skin of a frog. The two arrows point to flask-shaped cells of the parasitic chytrid *Batrachochytrium dendrobatis,* which has devastated populations of harlequin frogs **(c).**

and some other herbivores. Wherever a chytrid lives, reproduction requires at least a film of water through which the swimming spores can move.

A chytrid may advance though its entire life cycle within a matter of days, and for most species, much of this brief lifetime is spent in asexual reproduction. Although individuals initially exist as a vegetative (nonreproductive) phase, the fungus soon shifts into a reproductive mode. First, one or more spore-forming chambers called **sporangia** (*angeion* = vessel; singular, sporangium) develop, each containing one or more haploid nuclei. More developmental steps package the nuclei one by one in flagella-bearing spores. The spores are released to the environment through a pore or tube, and each swims briefly until it comes to rest on a substrate and a tough cyst forms around it. Under proper conditions, it will soon germinate and launch the life cycle anew.

A few chytrids reproduce sexually. Mycologists have observed a remarkable variety of sexual modes, but in all of them spores of different mating types unite. Karyogamy directly follows plasmogamy to produce a 2*n* zygote. This cell may form a mycelium that gives rise to sporangia, or it may directly give rise to either asexual or sexual spores.

Zygomycetes Form Zygospores for Sexual Reproduction

The phylum Zygomycota—fungi that reproduce sexually by way of structures called *zygospores*—contains fewer than a thousand species. What zygomycetes lack in numbers, however, they make up for in impact on other organisms. Many zygomycetes are saprobes that live in soil, feeding on plant detritus. There, their metabolic activities release mineral nutrients in forms that plant roots can take up. Some zygomycetes are parasites of insects (and even other zygomycetes), and some wreak havoc on human food supplies, spoiling stored grains, bread, fruits, and vegetables such as sweet potatoes. Others, however, have become major partners in commercial enterprises, where they are used in manufacturing products that range from industrial pigments to pharmaceuticals.

Most zygomycetes have aseptate hyphae, a feature that distinguishes them from the other multicellular fungi. Like other fungi, however, zygomycetes usually reproduce asexually, as shown at the lower left in **Figure 28.6.** When a haploid spore lands on a favorable substrate, it germinates and gives rise to a branching mycelium. Some of the hyphae grow upward, and saclike, thin-walled sporangia form at the tips of these aerial hyphae. Inside the sporangia the asexual cycle comes full circle as new haploid spores arise through mitosis and are released.

The black bread mold, *Rhizopus stolonifer,* may produce so many charcoal-colored sporangia **(Figure 28.7a)** that moldy bread looks black. The spores released are lightweight, dry, and readily wafted away by air currents. In fact, winds have dispersed *R. stolonifer* spores just about everywhere on Earth, including the Arctic. Another zygomycete, *Pilobolus* **(Figure 28.7b),** forcefully spews its sporangia away from the dung in which it grows. A grazing animal may eat a sporangium on a blade of grass; the spores then pass through the animal's gut unharmed and begin the life cycle again in a new dung pile.

Mycelia of many zygomycetes may occur in either the + or − mating type, and the nuclei of the different mating types are equivalent to gametes. Each strain secretes steroidlike hormones that can stimulate the development of sexual structures in the complementary strain and cause sexual hyphae to grow toward each other. When + and − hyphae come into close proximity, a septum forms behind the tip of each hypha, producing a terminal **gametangium** that contains several haploid nuclei (see Figure 28.6). When the gametangia of the two strains make contact, cellular enzymes digest the wall between them, yielding a single large, thin-walled cell that contains many nuclei from both parents. In other words, plasmogamy has occurred, and this new cell is a dikaryon. Gradually a second, inner wall forms, thickens, and hardens. This structure, with the multinucleate cell inside it, is a **zygospore,** the structure that gives this fungal group its scientific name. It becomes dormant and sometimes stays dormant for months or years.

Karyogamy follows plasmogamy, but the timing varies in different groups of zygomycetes. The exact trigger is unknown, but eventually the diploid zygospore ends its dormancy. The cell undergoes meiosis and produces a stalked sporangium (see Figure 28.6, step 5). The sporangium contains haploid spores of each mating type, which are released to the outside world. When a spore later germinates, it produces either a + or a − mycelium, and the sexual cycle can continue.

Zygomycetes that have aseptate hyphae are structurally simpler than the species in most other fungal groups. Although septa wall off the reproductive structures, in effect the branching mycelium of each fungus is a single, huge, multinucleate cell—the same body structure as found in some algae and certain protists. Because such zygomycetes have numerous nuclei in a common cytoplasm, these fungi are said to be **coenocytic,** which means "contained in a shared vessel." By contrast, in other fungal groups septa at least partially divide the hyphae into individual cells, which typically contain two or more nuclei.

Presumably, having hyphae that lack septa confers some selective advantages. One benefit may be that without septa to impede the flow, nutrients can move freely from the absorptive hyphal tips to other hyphae where reproductive parts develop. Hence the fungus may be able to reproduce faster.

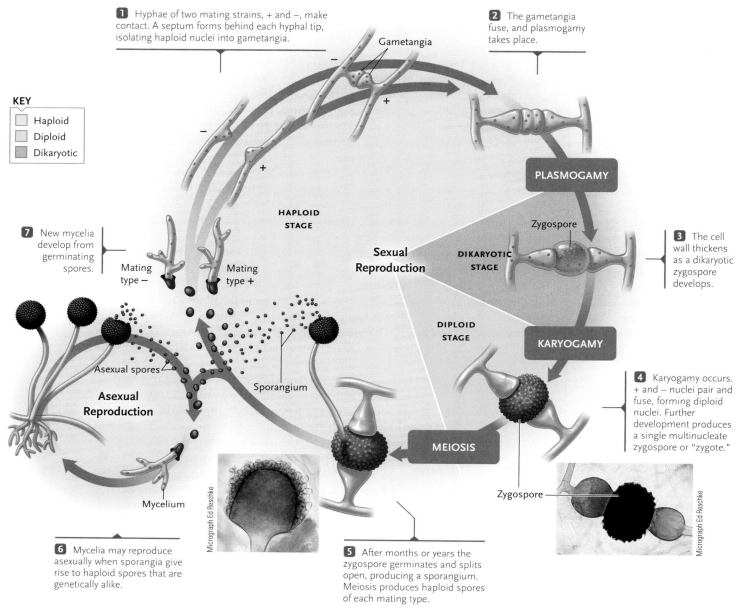

1 Hyphae of two mating strains, + and −, make contact. A septum forms behind each hyphal tip, isolating haploid nuclei into gametangia.

2 The gametangia fuse, and plasmogamy takes place.

Gametangia

KEY
- Haploid
- Diploid
- Dikaryotic

PLASMOGAMY

HAPLOID STAGE

Sexual Reproduction

DIKARYOTIC STAGE

Zygospore

3 The cell wall thickens as a dikaryotic zygospore develops.

7 New mycelia develop from germinating spores.

Mating type −

Mating type +

DIPLOID STAGE

KARYOGAMY

Asexual spores

Asexual Reproduction

Sporangium

4 Karyogamy occurs. + and − nuclei pair and fuse, forming diploid nuclei. Further development produces a single multinucleate zygospore or "zygote."

Zygospore

MEIOSIS

Mycelium

Micrograph Ed Reschke

Micrograph Ed Reschke

6 Mycelia may reproduce asexually when sporangia give rise to haploid spores that are genetically alike.

5 After months or years the zygospore germinates and splits open, producing a sporangium. Meiosis produces haploid spores of each mating type.

Figure 28.6
Life cycle of the bread mold *Rhizopus stolonifer*, a zygomycete. Asexual reproduction is common, but different mating types (+ and −) also reproduce sexually. In both cases, haploid spores form and give rise to new mycelia.

In zygomycetes, aggregations of "cooperating" hyphae may form body structures specialized for certain functions. However, such structures are more common in the three groups of more complex fungi that we consider next.

Glomeromycetes Form Spores at the Ends of Hyphae

The 160 known members of the phylum Glomeromycota are all specialized to form the associations called mycorrhizae with plant roots. It would be hard to overestimate their ecological impact, for Glomeromycota collectively make up roughly half of the fungi in soil and form mycorrhizae with an estimated 80% to 90% of all land plants. Virtually all glomeromycetes reproduce asexually, by way of spores that form at the tips of hyphae. The hyphae also secrete enzymes that

allow them to enter plant roots, where their tips branch into treelike clusters. As you will read in the next section, the clusters, called arbuscules, nourish the fungus by taking up sugars from the plant and in return supply the plant roots with a steady supply of dissolved minerals from the surrounding soil.

Ascomycetes, the Sac Fungi, Produce Sexual Spores in Saclike Asci

The phylum Ascomycota includes more than 30,000 species that produce reproductive structures called *asci* (**Figure 28.8**). A few ascomycetes prey upon various agricultural insect pests and thus have potential for use as "biological pesticides." Many more are destructive plant pathogens, including *Venturia inaequalis,* the fungus responsible for apple scab, and *Ophiostoma ulmi,* which causes Dutch elm disease. Several ascomycetes can be serious pathogens of humans. For example, *Claviceps purpurea*, a parasite on rye and other grains, causes ergotism, a disease marked by vomiting, hallucinations, convulsions, and in severe cases, gangrene and even death. Other ascomycetes cause nuisance infections such as athlete's foot and ringworm. Strains of *Aspergillus* grow in damp grain or peanuts; their metabolic wastes, known as aflatoxins, can cause cancer in humans who eat the poisoned food over an extended period. A few ascomycetes even show trapping behavior, ensnaring small worms that they then digest **(Figure 28.9a).** Yet some ascomycetes are valuable to humans: one species, the orange bread mold *Neurospora crassa*, has been important in genetic research, including the elucidation of the one gene–one enzyme hypothesis (see Section 15.1). And certain species of *Penicillium* **(Figure 28.9b)** are the source of the penicillin family of antibiotics, while others produce the aroma and distinctive flavors of Camembert and Roquefort cheeses. This multifaceted division also includes gourmet delicacies such as truffles *(Tuber melanosporum)* and the succulent true morel *Morchella esculenta*.

Although yeasts and filamentous fungi with a yeast stage in the life cycle occur in all fungal groups except chytrids, many of the best-known yeasts are ascomycetes. The yeast *Candida albicans* **(Figure 28.10)** infects mucous membranes, especially of the mouth (where it causes a disorder called thrush) and the vagina. *Saccharomyces cerevisiae*, which produces the ethanol in alcoholic beverages and the carbon dioxide

a. Sporangia of *Rhizopus stolonifer*

J. D. Cunningham/Visuals Unlimited

b. Sporangia (dark sacs) of *Pilobolus*

John Hodgin

500 μm

Figure 28.7
Two of the numerous strategies for spore dispersal by zygomycetes. **(a)** The sporangia of *Rhizopus stolonifer*, shown here on a slice of bread, release powdery spores that are easily dispersed by air currents. **(b)** In *Pilobolus*, the spores are contained in a sporangium (the dark sac) at the end of a stalked structure. When incoming rays of sunlight strike a light-sensitive portion of the stalk, turgor pressure (pressure against a cell wall due to the movement of water into the cell) inside a vacuole in the swollen portion becomes so great that the entire sporangium may be ejected outward as far as 2 m—a remarkable feat, given that the stalk is only 5 to 10 mm tall.

a. Ascocarp

Ascospore (sexual spore)

Ascus

Spore-bearing hypha of this ascocarp

b. Asci

© North Carolina State University, Department of Plant Pathology

c. Asci within ascocarp

© Michael Wood/mykob.com

d. Morel

© Fred Stevens/mykob.com

Figure 28.8
A few of the ascomycetes, or sac fungi. The examples shown are multicellular species that form mushrooms as reproductive structures. **(a)** A cup-shaped ascocarp, composed of tightly interwoven hyphae. The spore-producing asci occur inside the cup. **(b)** Asci on the inner surface of an ascocarp. **(c)** Scarlet cup fungus (*Sarcoscypha*). **(d)** A true morel (*Morchella esculenta*), a prized edible fungus.

a. A penicillium species   **b.** A trapping ascomycete

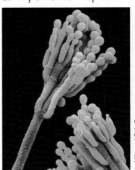

N. Allin and G. L. Barron

© Dennis Kunkel Microscopy, Inc.

Figure 28.9

Other ascomycete representatives. **(a)** *Eupenicillium*. Notice the rows of conidia (asexual spores) atop the structures that produce them. **(b)** Hyphae of *Arthrobotrys dactyloides*, a trapping ascomycete, form nooselike rings. When the fungus is stimulated by the presence of a prey organism, rapid changes in ion concentrations draw water into the hypha by osmosis. The increased turgor pressure shrinks the "hole" in the noose and captures this nematode. The hypha then releases digestive enzymes that break down the worm's tissues.

Yeast cells

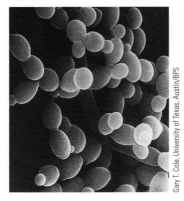

Gary T. Cole, University of Texas, Austin/BPS

Figure 28.10
Candida albicans, cause of yeast infections of the mouth and vagina.

that leavens bread, has also been a model organism for genetic research. By one estimate it has been the subject of more genetic experiments than any other eukaryotic microorganism. Yeasts commonly reproduce asexually by fission or budding from the parent cell, but many also can reproduce sexually after the fusion of two cells of different mating types (analogous to the mating types described earlier). Many ascomycete yeasts are found naturally in the nectar of flowers and on fruits and leaves. At least 1500 species have been described, and mycologists suspect that thousands more are yet to be identified.

Tens of thousands of ascomycetes, however, are not yeasts. They are multicellular, with tissues built up from septate hyphae. Although septa do slow the flow of nutrients (which, recall, can cross septa through pores), they also confer advantages. For example, septa present barriers to the loss of cytoplasm if a hypha is torn or punctured, whereas in an aseptate zygomycete, fluid pressure may force out a significant amount of cytoplasm before a breach can be sealed by congealing cytoplasm. In ways that are not well understood, septa can also limit the damage from toxins that are secreted by competing fungi.

As with zygomycetes, certain hyphae in ascomycetes are specialized for asexual reproduction. Instead of making spores inside sporangia, however, many ascomycetes produce asexual spores called **conidia** ("dust"; singular, conidium). In some of the species, the conidia form in chains that elongate from modified hyphal branches called **conidiophores**. In other ascomycetes, the conidia may pinch off from the hyphae in a series

of "bubbles," a bit like a string of detachable beads. Either way, an ascomycete can form and release spores much more quickly than a zygomycete can. Each newly formed conidium contains a haploid nucleus and some of the parent hypha's cytoplasm. Conidia and conidiophores of some ascomycete species are visible as the white powdery mildew that attacks grapes, roses, grasses, and the leaves of squash plants.

Ascomycetes can also reproduce sexually, and are commonly termed sac fungi because the meiotic divisions that generate haploid sexual spores occur in saclike cells called **asci** (*askos* = bladder; singular, ascus). In *Neurospora crassa* **(Figure 28.11)** and other complex ascomycetes, reproductive bodies called **ascocarps** bear or contain the asci. Some ascocarps resemble globes, others flasks or open dishes. An ascocarp begins to develop when two haploid mycelia of + and − mating types fuse (step 1). Plasmogamy then takes place, with the details differing from species to species. (In some species, hormonal signals cause the tip of one hypha to enlarge and form a "female" reproductive organ called an ascogonium, while the other hyphal tip develops into a "male" antheridium.) Paired nuclei, one from each mating type, migrate into the hyphae. During plasmogamy, the fused sexual structures give rise to dikaryotic hyphae, which develop inside the ascocarp. Asci form at the hyphal tips. Inside them, karyogamy takes place, producing a diploid zygote nucleus. It divides by meiosis, producing four haploid nuclei. In yeasts and some other ascomycetes cell division stops at this point, but in *N. crassa* and in many other species a round of mitosis ensues and results in eight nuclei. Regardless, the nuclei, other organelles, and a portion of cytoplasm then are incorporated into ascospores that may germinate on a suitable substrate and continue the life cycle.

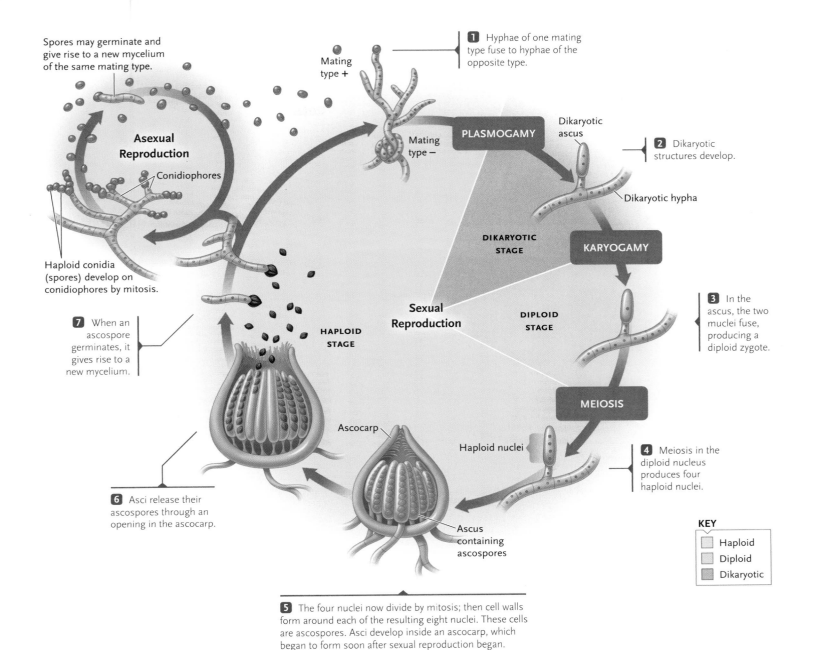

Spores may germinate and give rise to a new mycelium of the same mating type.

Mating type +

1 Hyphae of one mating type fuse to hyphae of the opposite type.

Asexual Reproduction

Conidiophores

PLASMOGAMY

Mating type −

Dikaryotic ascus

2 Dikaryotic structures develop.

Dikaryotic hypha

Haploid conidia (spores) develop on conidiophores by mitosis.

DIKARYOTIC STAGE

KARYOGAMY

Sexual Reproduction

DIPLOID STAGE

3 In the ascus, the two nuclei fuse, producing a diploid zygote.

7 When an ascospore germinates, it gives rise to a new mycelium.

HAPLOID STAGE

MEIOSIS

Ascocarp

Haploid nuclei

4 Meiosis in the diploid nucleus produces four haploid nuclei.

6 Asci release their ascospores through an opening in the ascocarp.

Ascus containing ascospores

KEY

	Haploid
	Diploid
	Dikaryotic

5 The four nuclei now divide by mitosis; then cell walls form around each of the resulting eight nuclei. These cells are ascospores. Asci develop inside an ascocarp, which began to form soon after sexual reproduction began.

Figure 28.11
Life cycle of the ascomycete *Neurospora crassa*.

Basidiomycetes, the Club Fungi, Form Sexual Spores in Club-Shaped Basidia

The 25,000 or so species of fungi in the phylum Basidiomycota include the mushroom-forming species, shelf fungi, coral fungi, bird's nest fungi, stinkhorns, smuts, rusts, and puffballs **(Figure 28.12)**. The common name for this group is club fungi, so named because the spore-producing cells, called **basidia** (meaning base or foundation), usually are club shaped. Some species have enzymes for digesting cellulose and lignin and are important decomposers of woody plant debris. A surprising number of basidiomycetes, including the prized edible oyster mushrooms (*Pleurotus ostreatus*), also can trap and consume bacteria and small animals such as rotifers and nematodes by secreting paralyzing toxins or gluey substances that immobilize the prey. This adaptation gives the fungus access to a rich source of molecular nitrogen, an essential nutrient that often is scarce in terrestrial habitats.

Many basidiomycetes take part in vital mutualistic associations with the roots of forest trees, as discussed later in this chapter. Others, the rusts and smuts, are parasites that cause serious diseases in wheat, rice, and other plants. Still others produce millions of dollars' worth of the reproductive structures commonly called mushrooms.

Amanita muscaria, the fly agaric mushroom (see Figure 28.12d), has been used as a fly poison, from

a. Coral fungus

b. Shelf fungus

c. White-egg bird's nest fungus

d. Fly agaric mushroom

e. Scarlet hood

Figure 28.12

Representative basidiomycetes, or club fungi. **(a)** The light red coral fungus *Ramaria*. **(b)** The shelf fungus *Polyporus*. **(c)** The white-egg bird's nest fungus *Crucibulum laeve*. Each tiny "egg" contains spores. Raindrops splashing into the "nest" can cause "eggs" to be ejected, thereby spreading spores into the surrounding environment. **(d)** The fly agaric mushroom *Amanita muscaria*, which causes hallucinations. **(e)** The scarlet hood *Hygrophorus*.

which it gets its common name. Due to its hallucinogenic effects, *A. muscaria* also is used in the religious rituals of ancient societies in Central America, Russia, and India. Other species of this genus, including the death cap mushroom *Amanita phalloides,* produce deadly toxins. The *A. phalloides* toxin, called α-amanitin, halts gene transcription, and hence protein synthesis, by inhibiting the activity of RNA polymerase. Within 8 to 24 hours of ingesting as little as 5 mg of the toxin, vomiting and diarrhea begin. Later, kidney and liver cells start to degenerate; without intensive medical care, death can follow within a few days.

A few basidiomycetes generally reproduce only by asexual means, by budding or shedding a fragment of a hypha. One is *Cryptococcus neoformans*, which causes a form of meningitis in humans. In general, however, basidiomycetes do reproduce sexually, producing large numbers of haploid sexual spores. **Figure 28.13** shows the life cycle of a typical basidiomycete.

Basidia typically develop on a **basidiocarp**, which is the reproductive body of the fungus. A basidiocarp consists of tight clusters of hyphae; the feeding mycelium is buried in the soil or decaying wood. The shelflike bracket fungi visible on trees are basidiocarps, and about

10,000 species of club fungi produce the basidiocarps we call mushrooms. Each is a short-lived reproductive body consisting of a stalk and a cap. Basidia develop on "gills," which are the sheets of tissue on the underside of the cap. The basidia undergo meiosis to produce microscopic, haploid **basidiospores** (Figure 28.13, inset) that disperse throughout the environment.

When a basidiospore lands on a suitable food source, it germinates and gives rise to a haploid mycelium. Two compatible mating types growing near each other may undergo plasmogamy. The resulting mycelium is dikaryotic, its cells containing one nucleus from each mating type. The dikaryotic stage of a basidiomycete is the feeding mycelium that can grow for years—a major departure from an ascomycete's short-lived dikaryotic stage. Accordingly, a basidiomycete has many more opportunities for producing sexual spores, and the mycelium can give rise to reproductive bodies many times.

After an extensive mycelium develops, and when environmental conditions such as moisture are favorable, basidiocarps grow from the mycelium and develop basidia. At first, each basidium in the mushroom or other reproductive body is dikaryotic, but then the two

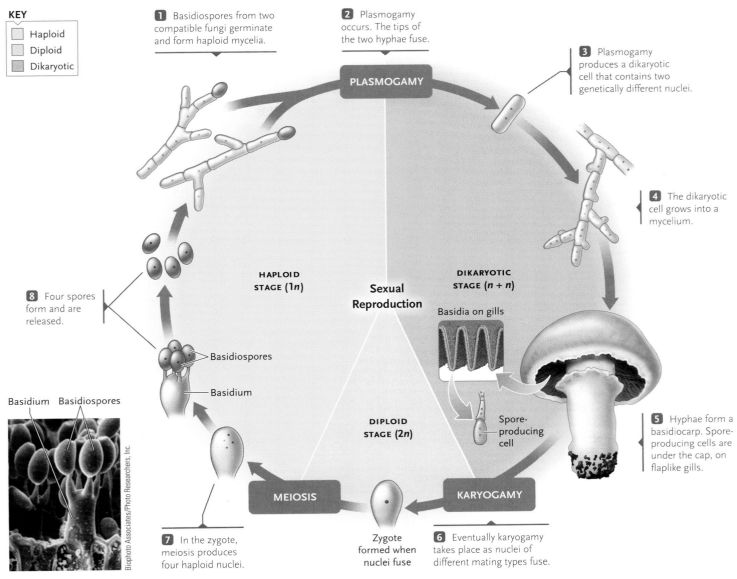

KEY

	Haploid
	Diploid
	Dikaryotic

1 Basidiospores from two compatible fungi germinate and form haploid mycelia.

2 Plasmogamy occurs. The tips of the two hyphae fuse.

PLASMOGAMY

3 Plasmogamy produces a dikaryotic cell that contains two genetically different nuclei.

4 The dikaryotic cell grows into a mycelium.

HAPLOID STAGE (1n)

Sexual Reproduction

DIKARYOTIC STAGE (n + n)

Basidia on gills

8 Four spores form and are released.

Basidiospores

Basidium

Basidium   Basidiospores

Biophoto Associates/Photo Researchers, Inc.

DIPLOID STAGE (2n)

Spore-producing cell

5 Hyphae form a basidiocarp. Spore-producing cells are under the cap, on flaplike gills.

MEIOSIS

KARYOGAMY

7 In the zygote, meiosis produces four haploid nuclei.

Zygote formed when nuclei fuse

6 Eventually karyogamy takes place as nuclei of different mating types fuse.

Figure 28.13
Generalized life cycle of the basidiomycete *Agaricus bisporus*, a species known commonly as the button mushroom. During the dikaryotic stage, cells contain two genetically different nuclei, shown here in different colors. Inset: Micrograph showing basidia and basidiospores.

nuclei undergo karyogamy, fusing to form a diploid zygote nucleus. The zygote exists only briefly; meiosis soon produces haploid basidiospores, which are wafted away from the basidium by air currents. Basidia can produce huge numbers of spores—for many species, estimates run as high as 100 million spores *per hour* during reproductive periods, day after day.

Squirrels and many other small animals may eat mushrooms almost as soon as they appear, but in some species the underlying mycelium can live for many years. For example, U.S. Forest Service scientists have found that the mycelium of a single individual of *Armillaria ostoyae* covers an area equivalent to 1665 football fields in an eastern Oregon forest. By one estimate, it measures an average of 1 m deep and nearly 6000 m across, making it perhaps one of the largest organisms

on Earth. As such a mycelium grows, specialized mechanisms during cell division maintain the dikaryotic condition and the paired nuclei in each hyphal cell.

Conidial Fungi Are Species for Which No Sexual Phase Is Known

As noted earlier, fungi generally are classified on the basis of their structures for sexual reproduction. When a sexual phase is absent or has not yet been detected, the fungal species is said to be anamorphic ("no related form") and is lumped into a convenience grouping, the conidial fungi (recall that conidia are asexual spores). This classification is the equivalent of "unidentified." Other names for this grouping are "imperfect fungi" and deuteromycetes.

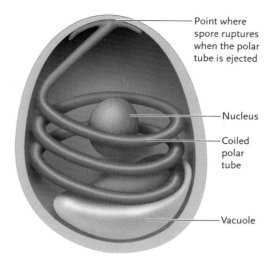

Point where spore ruptures when the polar tube is ejected

Nucleus

Coiled polar tube

Vacuole

Figure 28.14

Structure of microsporidia. When a spore germinates, its vacuole expands and forces the coiled "polar tube" outward and into a nearby, soon-to-be host cell. The nucleus and cytoplasm of the parasite enter the host through the tube, launching developmental steps that lead to the development of more microsporidia inside the host.

When researchers discover a sexual phase for a conidial fungus, or when molecular studies establish a clear relationship to a sexual species, the conidial fungus is reassigned to the appropriate phylum. Thus far, some have been classified as basidiomycetes, but most conidial fungi have turned out to be ascomycetes.

Microsporidia Are Single-Celled Sporelike Parasites

There are more than 1200 species of the single-celled parasites called **microsporidia.** They are known to infect insects including honeybees and grasshoppers, and vertebrates including fish and humans—especially individuals with compromised immune systems such as people with AIDS. Microsporidia are rather mysterious organisms. Physically they resemble spores **(Figure 28.14),** but they lack mitochondria and have several other puzzling characteristics. Molecular studies suggest that they are related to zygomycetes, and some researchers have proposed that the group may have lost many typical fungal features as it evolved a highly specialized parasitic lifestyle.

STUDY BREAK

1. Name the five phyla of the Kingdom Fungi and describe the reproductive adaptations that distinguish each one.
2. In terms of structure, which are the simplest fungal groups? The most complex?
3. Describe some ways, positive or negative, that members of each fungal phylum interact with other life forms.

28.3 Fungal Associations

Many fungi are partners in mutually beneficial interactions with photosynthetic organisms, and these associations play major roles in the functioning of ecosystems. A **symbiosis** is a state such as parasitism or mutualism in which two or more species live together in close association. Chapter 50 discusses general features of symbiotic associations more fully; here we are interested in some examples of the symbioses fungi form with photosynthetic partners—cyanobacteria, green algae, and plants.

A Lichen Is an Association between a Fungus and a Photosynthetic Partner

You may be familiar with one type of lichen, the leathery patches of various colors growing on certain rocks. Technically, a **lichen** is a single vegetative body that is the result of an association between a fungus and a photosynthetic partner. The fungal partner in a lichen, called the **mycobiont,** usually makes up only about 10% of the whole. The other 90% is the photosynthetic partner, called the **photobiont.** Most frequently, these are green algae of the genus *Trebouxia* or cyanobacteria of the genus *Nostoc.* Thousands of ascomycetes and a few basidiomycetes form this kind of symbiosis, but only about 100 photosynthetic species serve as photobionts.

Lichens often live in harsh, dry microenvironments, including on bare rock and wind-whipped tree trunks. Yet lichens have vital ecological roles and important human uses. Lichens secrete acid that eats away at rock, breaking it down and converting it to soil that can support larger plants. Some paleobiologists have suggested that lichens may have been some of the earliest land organisms, covering bare rocks during the Ordovician period (roughly 500 million to 425 million years ago). In this scenario, millennia of decaying lichens would have created the first soils in which the earliest land plants could grow. Today, lichens continue to enhance the survival of other life forms. For instance, in arctic tundra, where plants are scarce, reindeer and musk oxen can survive by eating lichens. Insects, slugs, and some other invertebrates also consume lichens, and they are nest-building materials for many birds and small mammals. People have derived dyes from lichens; they are even a component of garam masala, an ingredient in Indian cuisine. Some environmental chemists monitor air pollution by observing lichens, most of which cannot grow in heavily polluted air (see *Focus on Research*).

Because lichens are composite organisms, it may seem odd to talk of lichen "species." Biologists do give lichens binomial names, however, based on the characteristics of the mycobiont. More than 13,500 lichens are recognized, each one a unique combination of a

Applied Research: Lichens as Monitors of Air Pollution's Biological Damage

Lichens have become reliable pollution-monitoring devices all over the world—in some cases, replacing costly electronic monitoring stations. Different species are vulnerable to specific pollutants. For example, *Ramalina* lichens are damaged by nitrate and fluoride salts. Elevated levels of sulfur dioxide (a major component of acid rain) cause old man's beard (*Usnea trichodea*) to shrivel and die, but strongly promote the growth of a crusty European lichen, *Lecanora conizaeoides*. The sensitivity of yellow *Evernia* lichens to SO_2 enabled the scientist who discovered its damage at remote Isle Royale in Michigan to point the finger northward to coal-burning furnaces at Thunder Bay, Canada. Conversely, healthy lichens on damaged trees of Germany's Black Forest lifted suspicion from French

coal-burning power plants and allowed investigators to identify the true source of the tree damage: nitrogen oxides from automobile exhausts. The

result was Germany's first auto emission standards, which went into effect in the 1990s.

Usnea (old man's beard), a pendent (hanging) lichen.

Mark Mattock/Planet Earth Pictures

particular species of fungus and one or more species of photobiont. The relationship often begins when a fungal mycelium contacts a free-living cyanobacterium, algal cell, or both. The fungus parasitizes the photosynthetic host cell, sometimes killing it. If the host cell can survive, however, it multiplies in association with the fungal hyphae. The result is a tough, pliable body called a **thallus,** which can take a variety of forms **(Figure 28.15a)**. Short, specialized hyphae penetrate algal cells of the thallus, which become the fungus's sole source of nutrients. Often, the mycobiont of a lichen absorbs up to 80% of the carbohydrates the photobiont produces.

Benefits for the photobiont are less clear-cut, in part because the drain on nutrients hampers its growth and reproduction. In one view, many and possibly most lichens are parasitic symbioses in which the photobiont does not receive equal benefit. On the other hand, it is relatively rare to find a lichen's photobiont species living independently in the same conditions under which the lichen survives, whereas as part of a lichen it may eke out an enduring existence; some lichens have been dated as being more than 4000 years old! Studies have also revealed that at least some green algae do clearly benefit from the relationship. Such algae are sensitive to desiccation and intense ultraviolet radiation. Sheltered by a lichen's fungal tissues, a green alga can thrive in locales where alone it would perish. Clearly, we still have quite a bit

to learn about the physiological interactions between lichen partners.

As you might expect with such a communal life form, reproduction has its quirky aspects. In lichens that involve an ascomycete, the fungus produces ascospores that are dispersed by the wind. The spores germinate to form hyphae that may colonize new photosynthetic cells and so establish new symbioses. A lichen itself can also reproduce in at least two ways. In some types, a section of the thallus detaches and grows into a new lichen. In about one-third of lichens, specialized regions of the thallus give rise asexually to reproductive cell clusters called **soredia** (*soros* = heap; singular, soredium). Each cluster includes both algal and hyphal cells **(Figure 28.15b)**. As the lichen grows, the soredia detach and are dispersed by water, wind, or passing animals.

Mycorrhizae Are Symbiotic Associations of Fungi and Plant Roots

A **mycorrhiza** ("fungus-root") is a mutualistic symbiosis in which fungal hyphae associate intimately with plant roots. Mycorrhizae greatly enhance the plant's ability to extract various nutrients, especially phosphorus and nitrogen, from soil (see Chapter 33).

In **endomycorrhizae,** the fungal hyphae penetrate the cells of the root. This kind of association occurs on the roots of nearly all flowering plants, and in most

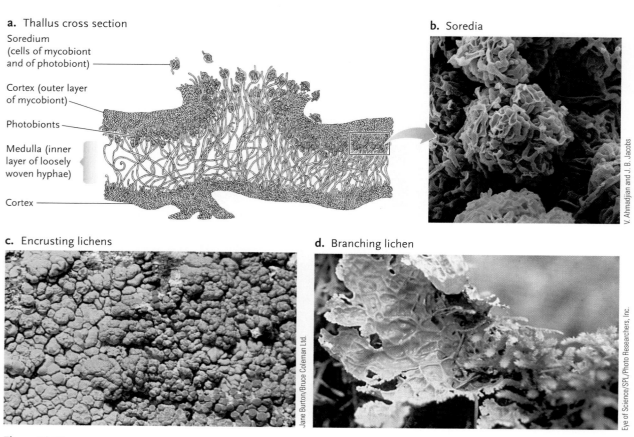

a. Thallus cross section

Soredium
(cells of mycobiont
and of photobiont)

Cortex (outer layer
of mycobiont)

Photobionts

Medulla (inner
layer of loosely
woven hyphae)

Cortex

b. Soredia

c. Encrusting lichens

d. Branching lichen

Figure 28.15

Lichens. **(a)** Sketch of a cross section through the thallus of the lichen *Lobaria verrucosa*. The soredia **(b),** which contain both hyphae and algal cells, are a type of dispersal fragment by which lichens reproduce asexually. **(c)** Encrusting lichens. **(d)** Erect, branching lichen, *Cladonia rangiferina*.

cases a glomeromycete is the fungal partner. The tree-like, branched hyphae of endomycorrhizae are called arbuscules **(Figure 28.16),** and glomeromycetes are sometimes referred to as arbuscular fungi.

Basidiomycetes are the usual fungal partners in **ectomycorrhizae (Figure 28.17),** in which hyphal tips grow between and around the young roots of trees and shrubs but never enter the root cells. Ectomycorrhizal associations—often several of them—are very common with trees. For instance, the extensive root system of a mature pine may be studded with ectomycorrhizae involving dozens of fungal species. The musky-flavored truffles *(Tuber melanosporum)* prized by gourmets are ascomycetes that form ectomycorrhizal associations with oak trees (genus *Quercus*).

Orchids are partners in a unique mycorrhizal relationship. The fungal partner, usually a basidiomycete, lives inside the orchid's tissues and provides the plant with a variety of nutrients. In fact, seeds of wild orchids germinate, and seedlings survive, only when such mycorrhizae are present.

In general, mycorrhizae represent a "win-win" situation for the partners. The fungal hyphae absorb carbohydrates synthesized by the plant, along with

some amino acids and perhaps growth factors as well. The growing plant in turn absorbs mineral ions made accessible to it by the fungus. Collectively, the fungal hyphae have a tremendous surface area for absorbing mineral ions from a large volume of the surrounding soil. Dissolved mineral ions accumulate in the hyphae when they are plentiful in the soil, and are released to the plant when they are scarce. This service is a survival boon to a great many plants, especially species that cannot readily absorb mineral ions, particularly phosphorus **(Figure 28.18).** For plants that inhabit soils poor in mineral ions, such as in tropical rain forests, mycorrhizal associations are crucial for survival. Likewise, in temperate forests, species of spruce, oak, pine, and some other trees die unless mycorrhizal fungi are present. Plants that live in dry habitats often rely on specialized mycorrhizal hyphae that serve as conduits for water into the root. Like lichens, mycorrhizae are highly vulnerable to damage from pollutants, especially acid rain.

Mycorrhizae have a long evolutionary history. Fossils show that endomycorrhizae were common among ancient land plants, and some biologists have speculated they might have been key for enhancing

Figure 28.16
Endomycorrhizae. **(a)** In this instance, the roots of leeks are growing in association with the glomeromycete *Glomus versiforme* (longitudinal section). Notice the arbuscules that have formed as the fungal hyphae branched after entering the leek root **(b)**.

a. Leek root with endomycorrhizae (black)

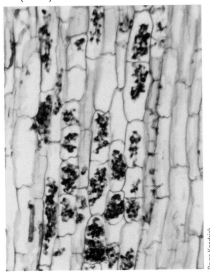

Bryce Kendrick

b. Arbuscule

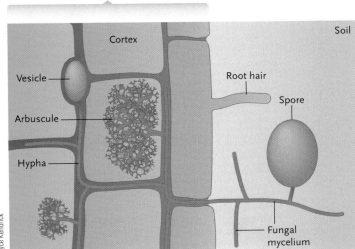

Root

Soil

Cortex

Vesicle

Root hair

Arbuscule

Spore

Hypha

Fungal mycelium

a. Lodgepole pine

Prof. D. J. Read, University of Sheffield

b. Mycorrhiza

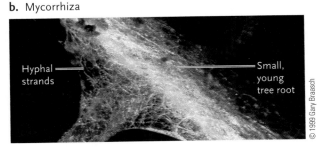

Hyphal strands

Small, young tree root

© 1999 Gary Braasch

Figure 28.17
Ectomycorrhizae. **(a)** Lodgepole pine, *Pinus contorta*, seedling, longitudinal section. Notice the extent of the mycorrhiza compared with the above-ground portion of the seedling, which is only about 4 cm tall. **(b)** Mycorrhiza of a hemlock tree.

F. B. Reeves

Figure 28.18
Effect of mycorrhizal fungi on plant growth. The six-month-old juniper seedlings on the left were grown in sterilized low-phosphorus soil inoculated with a mycorrhizal fungus. The seedlings on the right were grown under the same conditions but without the fungus.

the transport of water and minerals to the plants. In that scenario, endomycorrhizae may have played a crucial role in allowing plants to make the transition to life on land.

STUDY BREAK

1. Explain what a lichen is, and how each partner contributes to the whole.
2. Describe the biological and ecological roles of mycorrhizae.
3. How do endomycorrhizae and ectomycorrhizae differ?

How do plant pathogenic fungi invade plants?

Many species of fungi are pathogenic to plants. Of particular interest to humans are the pathogenic fungi that invade crop plants. In general, to invade a plant the pathogenic fungus must first break down any form of natural resistance that the plant has, and then establish an infection. Moreover, each pathogenic fungus has specificity—it invades only a particular set of plants, not all plants. For a number of pathogenic fungi, scientists are beginning to gain an understanding of the cellular and molecular events involved in invasion and the spread of the infection through the plant. A complete understanding of these processes will open the way to developing approaches that protect crop plants from fungal invasion, or at least reduce the extent of damage to the plants.

One example of the research being done in this area concerns the ascomycete fungus *Cochliobolus carbonum*. This fungus is pathogenic to maize, causing leaf blight (early drying of the leaves) and ear rot disease. *C. carbonum* secretes a toxin called HC-toxin to infect maize hosts. Guri Johal of Purdue University is studying the infection process, in particular investigating the molecular mechanisms by which HC-toxin leads to fungal colonization of maize tissues. Currently, little is known about those mechanisms.

Another example of research with pathogenic fungi concerns the ascomycete *Magnaporthe grisea*, the fungus that causes rice blast (lesions on leaves and other parts of the plant). The genome of this fungus has been sequenced, making possible the use of genomic/proteomic tools and approaches for studying pathogenesis. Dan Ebbole at Texas A&M University is using those tools and approaches to analyze proteins secreted by *M. grisea* with the aim of understanding their roles in the interaction of the pathogen with rice plants. Specifically, Ebbole and his group are looking at 300 proteins that, based on analysis of the genome, are predicted to be secreted. They produce tagged versions of the proteins by expressing the genes for them in fungal cultures. Then they test the purified proteins directly on plants one by one to see if any elicits a specific response by the host plant. They anticipate that this approach will serve as a screen to identify proteins that play significant roles in the pathogen–plant interaction. Those proteins will then be analyzed more completely, with the objective of developing cellular and molecular models for pathogenesis.

What are the interactions between all the molecular components of a fungus?

As you learned in Section 18.3, the study of the interactions between all of the molecular components of a cell or organism is systems biology. Over the years, significant advances have been made toward a molecular understanding of many processes in fungi, particularly in model fungi such as the yeast *Saccharomyces cerevisiae* and the mold *Neurospora crassa*. In addition, genome sequences have been obtained for a number of fungi, including the two species just mentioned as well as some pathogenic species. For a number of fungi, then, researchers are poised for systems biology studies. To that end, scientists from around the world have established the Yeast Systems Biology Network (YSBN) to coordinate research efforts in the systems biology of *S. cerevisiae*. The researchers argue that this yeast is a particularly appropriate model system for a concentrated effort to obtain a systems-level understanding of biological processes. Indeed, yeast has been a model system for eukaryotic cell structure and function, and for a number of aspects of fungal biology (see Chapter 10's *Focus on Research*). It was also one of the original model eukaryotes chosen for genome sequencing in the Human Genome Project (see Chapter 18).

Peter J. Russell

Review

Go to **ThomsonNOW** at www.thomsonedu.com/login to access quizzing, animations, exercises, articles, and personalized homework help.

28.1 General Characteristics of Fungi

- Fungi are key decomposers contributing to the recycling of carbon and some other nutrients. They occur as single-celled yeasts or multicellular filamentous organisms.

- The fungal mycelium consists of filamentous hyphae that grow throughout the substrate the fungus feeds upon (Figure 28.2). A wall containing chitin surrounds the plasma membrane, and in most species septa partition the hyphae into cell-like compartments. Pores in septa permit cytoplasm and organelles to move between hyphal cells. Aggregations of hyphae form all other tissues and organs of a multicellular fungus.

- Fungi gain nutrients by extracellular digestion and absorption. Saprobic species feed on nonliving organic matter. Parasitic types obtain nutrients from tissues of living organisms. Many fungi are partners in symbiotic relationships with plants.

- All fungi may reproduce via spores generated either asexually or sexually (Figure 28.3). Some types also may reproduce asexually by budding or fragmentation of the parent body. Sexual reproduction usually has two stages. First, in plasmogamy, the cytoplasms of two haploid cells fuse to become a dikaryon containing a haploid nucleus from each parent. Later, in karyogamy, the nuclei fuse and form a diploid zygote. Meiosis then generates haploid spores.

Animation: Mycelium

28.2 Major Groups of Fungi

- The main phyla of fungi are the Chytridiomycota (which have motile spores), Zygomycota (zygospore-forming fungi), Glomeromycota, Ascomycota (sac fungi), and Basidiomycota (club fungi) (Figure 28.4). The phyla traditionally have been distinguished mainly on the basis of the structures that arise as part of sexual reproduction. When a sexual phase cannot be detected or is absent from the life cycle, the specimen is assigned to an informal grouping, the conidial fungi.

- Chytrids usually are microscopic. They are the only fungi that produce motile, flagellated spores. Many are parasites (Figure 28.5).

- Zygomycetes have aseptate hyphae and are coenocytic, with many nuclei in a common cytoplasm. They sometimes reproduce sexually by way of hyphae that occur in + and − mating types; haploid nuclei in the hyphae function as gametes. Further development produces the zygospore, which may go dormant for a time. When the zygospore breaks dormancy it produces a stalked sporangium containing haploid spores of each mating type, which are released (Figures 28.6 and 28.7).

- Glomerulomycetes form a distinct type of endomycorrhizae in association with plant roots. They reproduce asexually, by way of spores that form at the tips of hyphae.

- Most ascomycetes are multicellular (Figure 28.9). In asexual reproduction, chains of haploid asexual spores called conidia elongate or pinch off from the tips of conidiophores (modified aerial hyphae; Figure 28.10). In sexual reproduction, haploid sexual spores called ascospores arise in saclike cells called asci. In the most complex species, reproductive bodies called ascocarps bear or contain the asci. Ascospores can give rise to a new haploid mycelium (Figures 28.8 and 28.11).

- Most basidiomycete species reproduce only sexually. Club-shaped basidia develop on a basidiocarp and bear sexual spores on their surface. When dispersed, these basidiospores may germinate and give rise to a haploid mycelium (Figure 28.13).

- Microsporidia are single-celled sporelike parasites of arthropods, fish, and humans (Figure 28.14).

Animation: Zygomycete life cycle

Animation: Sac fungi

Animation: Club fungus life cycle

28.3 Fungal Associations

- Many ascomycetes and a few basidiomycetes enter into symbioses with cyanobacteria or green algae to produce the communal life form called a lichen, which has a spongy body called a thallus. The algal cells supply the lichen's carbohydrates, most of which are absorbed by the fungus. In some lichens a section of the thallus may detach and grow into a new individual. In others, specialized regions of the thallus give rise asexually to reproductive soredia that include both algal and hyphal cells (Figure 28.15).

- In the symbiosis called a mycorrhiza, fungal hyphae make mineral ions and sometimes water available to the roots of a plant partner. The fungus in turn absorbs carbohydrates, amino acids, and possibly other growth-enhancing substances from the plant (Figures 28.16–28.18). In endomycorrhizae, the fungal hyphae (usually of a glomeromycete) penetrate the cells of the root. With ectomycorrhizae, hyphal tips grow close to young roots but do not enter roots cells; the usual fungal partner is a basidiomycete.

Animation: Lichens

Animation: Mycorrhiza

Questions

Self-Test Questions

1. Which of the following attributes best exemplifies a filamentous saprobic fungus?
   a. reproduction by spores on week-old bread
   b. metabolic by-products that make bread rise
   c. extracellular digestion of tissues in a fallen log
   d. extracellular digestion of a living leaf's cellulose with hydrolytic enzymes
   e. aggressive expansion of the fungal mycelium into the tissues of a living elm tree

2. Which of the following events is/are *not* part of a typical fungal life cycle involving asexual reproduction?
   a. formation of a dikaryon
   b. hyphae developing into a mycelium
   c. formation of a diploid zygote
   d. plasmogamy, which occurs when hyphae fuse at their tips
   e. production and release of large numbers of spores

3. A trait common to all fungi is:
   a. reproduction via spores.
   b. parasitism.
   c. septate hyphae.
   d. a dikaryotic phase inside a zygospore.
   e. plasmogamy after an antheridium and ascogonium come into contact.

4. The chief characteristic used to classify fungi into the major fungal phyla is:
   a. nutritional dependence on nonliving organic matter.
   b. recycling of nutrients in terrestrial ecosystems.
   c. adaptations for obtaining water.
   d. features of reproduction.
   e. cell wall metabolism.

5. At lunch George ate a mushroom, some truffles, a little Camembert cheese, and a bit of moldy bread. Which of the following groups was *not* represented in the meal?
   a. Basidiomycota
   b. Ascomycota
   c. conidial fungi
   d. chytrids
   e. Zygomycota

6. Which of the following fungal reproductive structures is diploid?
   a. basidiocarps
   b. ascospores
   c. conidia
   d. gametangia
   e. zygospores

7. A mushroom is:
   a. the food-absorbing region of an ascomycete.
   b. the food-absorbing region of a basidiomycete.
   c. a reproductive structure formed only by basidiomycetes.
   d. a specialized form of mycelium not constructed of hyphae.
   e. a collection of saclike cells called asci.

8. A zygomycete is characterized by:
   a. aseptate hyphae.
   b. mostly sexual reproduction.
   c. absence of + and − mating types.
   d. the tendency to form mycorrhizal associations with plant roots.
   e. a life cycle in which karyogamy does not occur.

9. Which best describes a lichen?
   a. It is a fungus that breaks down rock to provide nutrients for an alga.
   b. It colonizes bare rocks and slowly degrades them to small particles.
   c. It spends part of the life cycle as a mycobiont and part as a fungus.
   d. It is an association between a basidiomycete and an ascomycete.
   e. It is an association between a photobiont and a fungus.

10. In a college greenhouse a new employee observes fuzzy mycorrhizae in the roots of all the plants. Destroying no part of the plants, he carefully removes the mycorrhizae. The most immediate result of this "cleaning" is that the plants cannot:
    a. carry out photosynthesis.
    b. absorb water through their roots.
    c. transport water up their stems.
    d. extract phosphorus and nitrogen from water.
    e. store carbohydrates in their roots.

Questions for Discussion

1. A mycologist wants to classify a specimen that appears to be a new species of fungus. To begin the classification process, what kinds of information on body structures and/or functions must the researcher obtain in order to assign the fungus to one of the major fungal groups?

2. In a natural setting—a pile of horse manure in a field, for example—the sequence in which various fungi appear illustrates ecological succession, the replacement of one species by another in a community (see Chapter 50). The earliest fungi are the most efficient opportunists, for they can form and disperse spores most rapidly. In what order would you expect representatives from each division of fungi to appear on the manure pile? Why?

3. As the text noted, conifers, orchids, and some other types of plants cannot grow properly if their roots do not form associations with fungi, which provide the plant with minerals such as phosphate and in return receive carbohydrates and other nutrients synthesized by the plant. In some instances, however, the plant receives proportionately more nutrients than the fungus does. Even so, biologists still consider this to be a mycorrhizal association. Explain why you agree or disagree.

4. Humans are fundamentally diploid organisms. Explain how this state of affairs compares with the fungal life cycle, then compare the two general life cycles in light of the two groups' overall reproductive strategies.

Experimental Analysis

Experiments on the orange bread mold *Neurospora crassa*, an ascomycete, were pivotal in elucidating the concept that each gene encodes a single enzyme. As *N. crassa* ascospores arise through meiosis and then mitosis in an ascus, each ascospore occupies a particular position in the final string of eight spores the ascus contains:

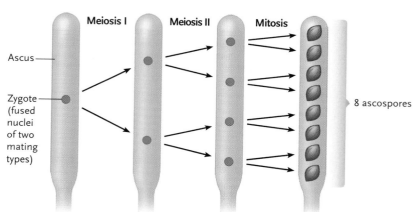

This quirk of ascospore development was extremely useful to early geneticists, because it vastly simplified the task of figuring out which alleles ended up in particular ascospores following meiosis. Recalling genetics topics discussed in Chapter 11, why was the analysis easier?

Evolution Link

The hypothesis that fungi are more closely related to animals than to plants has receive support from studies of fungus genomes. For instance, scientists have documented striking similarities in the structure of many fungal and human genes—similarities that may be especially important in medicine. One mycologist, John Taylor of the University of California at Berkeley, suggests that a close biochemical relationship between fungi and animals may explain why fungal infections are typically so resistant to treatment, and why it has proven rather difficult to develop drugs that kill fungi without damaging their human or other animal hosts. About 100 fungal genomes have been or soon will be sequenced, including genomes of several medically important species. If you are a researcher working to develop new antifungal drugs, how could you make use of this growing genetic understanding? Using Web resources, can you find examples of antifungal drugs that exploit biochemical differences between animals and fungi?

How Would You Vote?

The disappearance of lichens and soil fungi may be an early indication that coal-fired power plants are emitting pollutants that also can endanger human health. Controlling emissions raises the cost of energy for consumers. Should pollution standards for these power plants be tightened? Go to www.thomsonedu.com/login to investigate both sides of the issue and then vote.

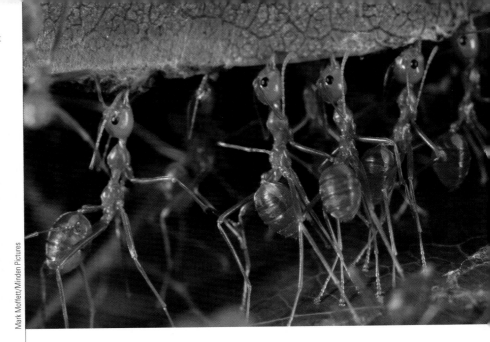

Weaver ants *(Oecophylla longinoda)* carry a leaf to repair their nest in Papua New Guinea.

Mark Moffett/Minden Pictures

STUDY PLAN

29.1 What Is an Animal?

All animals share certain structural and behavioral characteristics

The animal lineage probably arose from a colonial choanoflagellate ancestor

29.2 Key Innovations in Animal Evolution

Tissues and tissue layers appeared early in animal evolution

Most animals exhibit either radial or bilateral symmetry

Many animals have body cavities that surround their internal organs

Developmental patterns mark a major divergence in animal ancestry

Segmentation divides the bodies of some animals into repeating units

29.3 An Overview of Animal Phylogeny and Classification

Molecular analyses have refined our understanding of animal phylogeny

The molecular phylogeny reveals surprising patterns in the evolution of key morphological innovations

29.4 Animals without Tissues: Parazoa

Sponges have simple body plans and lack tissues

29.5 Eumetazoans with Radial Symmetry

Cnidarians use nematocysts to stun or kill prey

Ctenophores use tentacles to feed on microscopic plankton

29.6 Lophotrochozoan Protostomes

The lophophorate phyla share a distinctive feeding structure

Flatworms have digestive, excretory, nervous, and reproductive systems, but lack a coelom

Rotifers are tiny pseudocoelomates with a jawlike feeding apparatus

Ribbon worms use a proboscis to capture food

Mollusks have a muscular foot and a mantle that secretes a shell or aids in locomotion

Annelids exhibit a serial division of the body wall and some organ systems

29.7 Ecdysozoan Protostomes

Nematodes are unsegmented worms covered by a flexible cuticle

Velvet worms have segmented bodies and numerous unjointed legs

Arthropods are segmented animals with a hard exoskeleton and jointed appendages

29 Animal Phylogeny, Acoelomates, and Protostomes

WHY IT MATTERS

In 1909, a lucky fossil hunter named Charles Wolcott tripped over a rock on a mountain path in British Columbia, Canada. Under the force of his hammer, the rock split apart, revealing the discovery of a lifetime. Wolcott and other workers soon found fossils of more than 120 species of previously undescribed animals from the Cambrian period. These creatures had lived on the muddy sediments of a shallow ocean basin. About 530 million years ago, an underwater avalanche buried them in a rain of silt that was eventually compacted into finely stratified shale. Over millions of years, the shale was uplifted by tectonic activity and incorporated into the mountains of western Canada. It is now known as the Burgess Shale formation.

Some animals in the Burgess Shale were truly bizarre (**Figure 29.1**). For example, *Opabinia* was about as long as a tube of lipstick; it had five eyes on its head and a grasping organ that it may have used to capture prey. No living animals even remotely resemble *Opabinia*. The smaller *Hallucigenia* sported seven pairs of large spines on one side and seven pairs of soft organs on the other. Recent research suggests that *Hallucigenia* may belong in the phylum Onychophora, described in Section 29.7. Nevertheless, most species of the Burgess

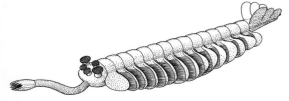
Opabinia

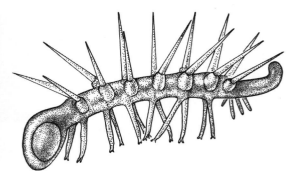

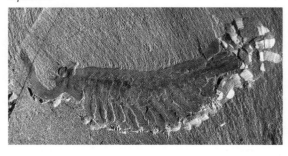
Hallucigenia

Figure 29.1
Animals of the Burgess Shale. *Opabinia* had five eyes and a grasping organ on its head. *Hallucigenia* had seven pairs of spines and soft protuberances. (Images: Dr. Chip Clark, National Museum of Natural History, Smithsonian Institution.)

Shale left no descendants that are still alive today. Thus, this remarkable assemblage of fossils provides a glimpse of some evolutionary novelties that—whether through the action of natural selection or just plain bad luck—were ultimately unsuccessful.

Other animal lineages have shown much greater longevity. Zoologists have described nearly 2 million living species in the kingdom **Animalia**, about five times as many as in all the other kingdoms combined. The familiar **vertebrates**, animals with a backbone, encompass only a small fraction (about 47,000 species) of the total. The overwhelming majority of animals fall within the descriptive grouping of **invertebrates**, animals without a backbone.

The remarkable evolutionary diversification of animals resulted from their ability to consume other organisms as food and, for most groups, their ability to move from one place to another. Today animals are important consumers in nearly every environment on Earth. Their diversification has been accompanied by the evolution of specialized tissues and organ systems as well as complex behaviors.

In this chapter, we introduce the general characteristics of animals and a phylogenetic hypothesis about their evolutionary history and classification. We also survey some of the major invertebrate phyla; a *phylum* is an ancient monophyletic lineage with a distinctive body plan. In Chapter 30 we examine the deuterostome lineage, which includes the vertebrates and their nearest invertebrate relatives.

29.1 What Is an Animal?

Biologists recognize the Kingdom Animalia as a monophyletic group that is easily distinguished from the other kingdoms.

All Animals Share Certain Structural and Behavioral Characteristics

Animals are eukaryotic, multicellular organisms. Their cells lack cell walls, a trait that differentiates them from plants and fungi. The individual cells of most animals are similar in size, so that very large animals like elephants have many more cells than small ones like fleas. In large animals, most cells are far from the body surface, but specialized tissues and organ systems deliver nutrients and oxygen to them and carry wastes away.

All animals are **heterotrophs**: they acquire energy and nutrients by eating other organisms. Food is ingested (eaten) and then digested (broken down) and absorbed by specialized tissues. Animals use oxygen to metabolize the food they eat through the biochemical pathways of aerobic respiration, and most store excess energy as glycogen, oil, or fat.

All animals are **motile**—able to move from place to place—at some time in their lives. They travel through the environment to find food or shelter and to interact with other animals. Most familiar animals are motile as adults. However, in some species, such as mussels and barnacles, only the young are motile; they eventually settle down as **sessile**—unable to move from one place to another—adults. The advantages of motility have fostered the evolution of locomotor structures, including fins, legs, and wings. And in many animals, locomotion results from the action of muscles, specialized contractile tissues that move individual body parts. Most animals also have sensory and nervous systems that allow them to receive, process, and respond to information about the environment.

Animals reproduce either asexually or sexually; in many groups they switch from one mode to the other. Sexually reproducing species produce short-lived, haploid **gametes** (eggs and sperm), which fuse to form diploid **zygotes** (fertilized eggs). Animal life cycles generally include a period of development during which mitosis transforms the zygote into a multicelled **embryo**, which develops into a sexually immature juvenile or a free-living **larva**, which becomes a sexually mature adult. Larvae often differ markedly from adults, and they may occupy different habitats and consume different foods.

The Animal Lineage Probably Arose from a Colonial Choanoflagellate Ancestor

An overwhelming body of morphological and molecular evidence indicates that all animal phyla had a common ancestor. For example, all animals share similarities in their cell-to-cell junctions and the molecules in their extracellular matrices (see Section 5.5) as well as similarities in the structure of their ribosomal RNAs.

Most biologists agree that the common ancestor of all animals was probably a colonial, flagellated protist that lived at least 700 million years ago, during the Precambrian era. It may have resembled the minute, sessile choanoflagellates that live in both freshwater and marine habitats today (see Figure 26.21). In 1874 the German embryologist Ernst Haeckel proposed a colonial, flagellated ancestor, suggesting that it was a hollow, ball-shaped organism with unspecialized cells. According to his hypothesis, its cells became specialized for particular functions, and a developmental re-

organization produced a double-layered, sac-within-a-sac body plan **(Figure 29.2)**. As you will see in Chapter 48, the embryonic development of many living animals roughly parallels this hypothetical evolutionary transformation.

STUDY BREAK

1. What characteristics distinguish animals from plants?
2. How does the ability of animals to move through the environment relate to their acquisition of nutrients and energy?

29.2 Key Innovations in Animal Evolution

Once established, the animal lineage diversified quickly into an amazing array of body plans. Before the development of molecular sequencing techniques, biologists used several key morphological innovations to unravel the evolutionary relationships of the major animal groups.

Tissues and Tissue Layers Appeared Early in Animal Evolution

The presence or absence of **tissues**, groups of cells that share a common structure and function, divides the animal kingdom into two distinct branches. One branch, the sponges, or Parazoa (*para* = alongside; *zoon* = animal), lacks tissues. All other animals, collectively grouped in the Eumetazoa (*eu* = true; *meta* = later), have tissues.

During the development of eumetazoans, embryonic tissues form as either two or three concentric **primary cell layers.** The innermost layer, the **endoderm,** eventually develops into the lining of the gut (digestive system) and, in some animals, respiratory organs. The outermost layer, the **ectoderm,** forms the external cover-

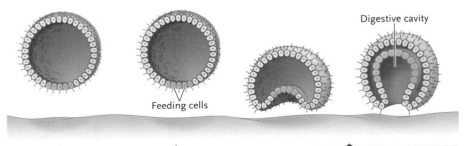

Digestive cavity

1 Colonial flagellated protist with unspecialized cells

Feeding cells

2 Certain cells became specialized for feeding and other functions.

3 A developmental reorganization produced a two-layered animal with a sac-within-a-sac body plan.

Figure 29.2

Animal origins. Many biologists believe that animals arose from a colonial, flagellated protist in which cells became specialized for specific functions and a developmental reorganization produced two cell layers. The cell movements illustrated here are similar to those that occur during the development of many animals, as described in Chapter 48.

ing and nervous system. Between the two, the **mesoderm** forms the muscles of the body wall and most other structures between the gut and the external covering. Some simple animals have a **diploblastic** body plan that includes only two layers, endoderm and ectoderm. However, most animals are **triploblastic**, having all three primary cell layers.

Most Animals Exhibit either Radial or Bilateral Symmetry

The most obvious feature of an animal's body plan is its shape **(Figure 29.3)**. Most animals are **symmetrical;** in other words, their bodies can be divided by a plane into mirror-image halves. By contrast, most sponges have irregular shapes and are therefore **asymmetrical.**

Most eumetazoans exhibit one of two body symmetry patterns. The Radiata includes two phyla, Cnidaria (hydras, jellyfishes, and sea anemones) and Ctenophora (comb jellies), which have **radial symmetry.** Their body parts are arranged regularly around a central axis, like the spokes on a wheel. Thus, any cut down the long axis of a hydra divides it into matching halves. Radially symmetrical animals are usually sessile or slow moving and receive sensory input from all directions.

All other eumetazoan phyla fall within the Bilateria, animals that have **bilateral symmetry.** In other words, only a cut along the midline from head to tail divides them into left and right sides that are essentially mirror images of each other. Bilaterally symmetrical animals also have **anterior** (front) and **posterior** (back) ends as well as **dorsal** (upper) and **ventral** (lower) surfaces. As these animals move through the environment, the anterior end encounters food, shelter, or enemies first. Thus, in bilaterally symmetrical animals, natural selection also favored **cephalization,** the development of an anterior head where sensory organs and nervous system tissue are concentrated.

Many Animals Have Body Cavities That Surround Their Internal Organs

The body plans of many bilaterally symmetrical animals include a body cavity that separates the gut from the muscles of the body wall **(Figure 29.4)**. **Acoelomate** animals (*a* = not; *koilos* = hollow), such as flatworms (phylum Platyhelminthes), do not have such a cavity; a continuous mass of tissue, derived largely from mesoderm, packs the region between the gut and the body wall (see Figure 29.4a). **Pseudocoelomate** animals (*pseudo* = false), including the roundworms (phylum Nematoda) and wheel animals (phylum Rotifera), have a **pseudocoelom,** a fluid- or organ-filled space between the gut and the muscles of the body wall (see Figure 29.4b). Internal organs lie within the pseudocoelom and are bathed by its fluid. **Coelomate** animals have a true **coelom,** a fluid-filled body cavity completely lined by the **peritoneum,** a thin tissue derived from mesoderm (see Figure 29.4c). Membranous extensions of the inner and outer layers of the peritoneum, the **mesenteries,** surround the internal organs and suspend them within the coelom.

Biologists describe the body plan of pseudocoelomate and coelomate animals as a "tube within a tube"; the digestive system forms the inner tube, the body wall forms the outer tube, and the body cavity lies between them. The body cavity separates internal organs from the body wall, allowing them to function independently of whole-body movements. The fluid within the cavity also protects delicate organs from mechanical damage. And, because the volume of the body cavity is fixed, the incompressible fluid within it serves as a **hydrostatic skeleton,** which provides support; in some animals muscle contractions can shift the fluid, changing the animals' shape and allowing them to move from place to place (see Section 41.2).

Developmental Patterns Mark a Major Divergence in Animal Ancestry

Embryological and molecular evidence suggests that bilaterally symmetrical animals are divided into two lineages: the protostomes, which includes most phyla of invertebrates, and the deuterostomes, which includes the vertebrates and their nearest invertebrate relatives. Protostomes and deuterostomes differ in several developmental characteristics **(Figure 29.5)**.

Shortly after fertilization, an egg undergoes a series of mitotic divisions called **cleavage** (see Section 48.1). The first two cell divisions divide a zygote as you might slice an apple, cutting it into four wedges from top to bottom. In many protostomes, subsequent cell divisions produce daughter cells that lie *between* the pairs of cells below them; this pattern is called **spiral cleavage** (left side of Figure 29.5a). In deuterostomes, by contrast, subsequent cell divisions produce a mass

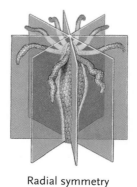
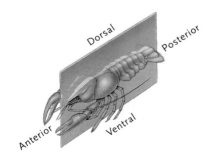

Radial symmetry   Bilateral symmetry

Figure 29.3
Patterns of body symmetry. Most animals have either radial or bilateral symmetry.

a. In acoelomate animals, no body cavity separates the gut and body wall.

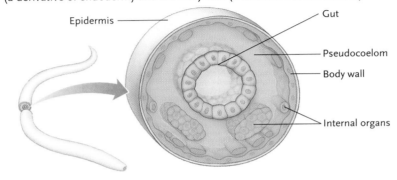

Epidermis

Gut

Internal organs

Body wall

Figure 29.4
Body plans of triploblastic animals.

b. In pseudocoelomate animals, the pseudocoelom forms between the gut (a derivative of endoderm) and the body wall (a derivative of mesoderm).

Epidermis

Gut

Pseudocoelom

Body wall

Internal organs

c. In coelomate animals, the coelom is completely lined by peritoneum (a derivative of mesoderm).

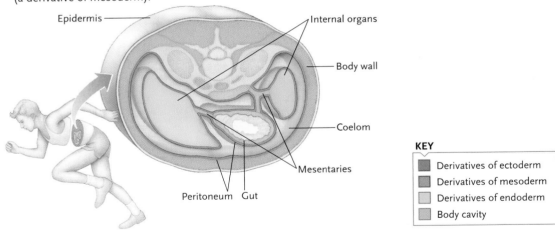

Epidermis

Internal organs

Body wall

Coelom

Mesenteries

Peritoneum   Gut

KEY

■	Derivatives of ectoderm
■	Derivatives of mesoderm
□	Derivatives of endoderm
▨	Body cavity

of cells that are stacked directly above and below one another; this pattern is called **radial cleavage** (right side of Figure 29.5a).

Protostomes and deuterostomes often differ in the timing of important developmental events. During cleavage, certain genes are activated at specific times, determining a cell's developmental path and ultimate fate. Many protostomes undergo **determinate cleavage:** each cell's developmental path is determined as the cell is produced. Thus, one cell isolated from a two- or four-cell protostome embryo cannot develop into a functional embryo or larva. By contrast, many deuterostomes have **indeterminate cleavage:** the developmental fates of cells are determined later. A cell isolated from a four-cell deuterostome embryo will develop into a functional, although smaller than usual, embryo or larva. Like other deuterostomes, humans have indeterminate cleavage; thus, the cells produced by the first few cleavage divisions sometimes separate and develop into identical twins.

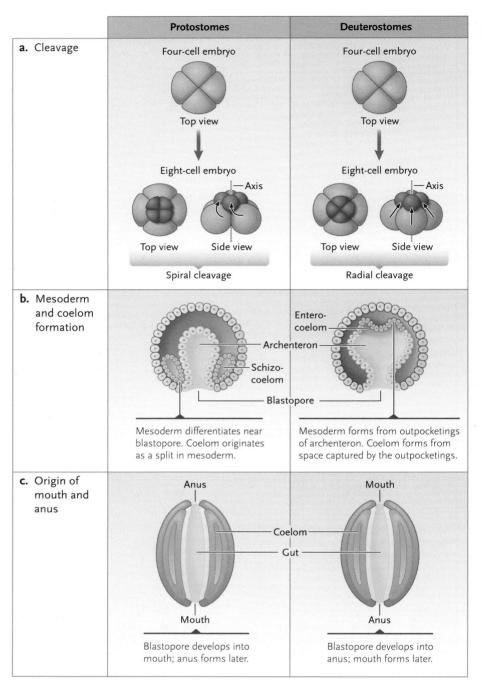

	Protostomes	Deuterostomes
a. Cleavage	Four-cell embryo Top view Eight-cell embryo Top view — Side view — Axis Spiral cleavage	Four-cell embryo Top view Eight-cell embryo Top view — Side view — Axis Radial cleavage
b. Mesoderm and coelom formation	Archenteron Schizo-coelom Blastopore Mesoderm differentiates near blastopore. Coelom originates as a split in mesoderm.	Entero-coelom Archenteron Blastopore Mesoderm forms from outpocketings of archenteron. Coelom forms from space captured by the outpocketings.
c. Origin of mouth and anus	Anus Coelom Gut Mouth Blastopore develops into mouth; anus forms later.	Mouth Coelom Gut Anus Blastopore develops into anus; mouth forms later.

KEY
- Derivatives of ectoderm
- Derivatives of mesoderm
- Derivatives of endoderm
- Body cavity

Figure 29.5
Protostomes and deuterostomes. The two lineages of coelomate animals differ in **(a)** cleavage patterns, **(b)** the origin of mesoderm and the coelom, and **(c)** the polarity of the digestive system.

As development proceeds, an opening on the surface of the embryo connects the developing gut, called the **archenteron,** to the outside environment. This initial opening is called the **blastopore** (see Figure 29.5b). Later in development, a second opening at the opposite end of the embryo transforms the pouchlike gut into a digestive tube (see Figure 29.5c). In protostomes (*protos* = first; *stoma* = mouth), the blastopore develops into the mouth, and the second opening forms the anus. In some deuterostomes (*deuteros* = second), the blastopore develops into the anus, and the second opening becomes the mouth.

Protostomes and deuterostomes also differ in the origin of mesoderm and the coelom (see Figure 29.5b). In most protostomes, mesoderm originates from a few specific cells near the blastopore. As the mesoderm grows and develops, it splits into inner and outer layers. The space between the layers is called a **schizocoelom** (*schizo* = split). In deuterostomes, mesoderm often forms from outpocketings of the archenteron. The space pinched off by the outpocketings is called an **enterocoelom** (*enteron* = intestine).

Several other characteristics also differ in adult protostomes and deuterostomes. For example, the central nervous system of protostomes is generally positioned on the ventral side of the body, and their brain surrounds the opening of the digestive tract. By contrast, the nervous system and brain of deuterostomes lie on the dorsal side of the body.

Segmentation Divides the Bodies of Some Animals into Repeating Units

Some phyla in both the protostome and deuterostome lineages exhibit varying degrees of **segmentation,** the production of body parts as repeating units. During development, segmentation first arises in the mesoderm, the middle tissue layer that produces most of the body's bulk. In humans and other vertebrates, we see evidence of segmentation in the vertebral column (backbone), ribs, and associated muscles, such as the "six-pack abs" that sit-ups accentuate. In some animals, segmentation is also reflected in structures derived from the endoderm and ectoderm. For example, the ringlike pattern on an earthworm or a caterpillar matches the underlying segments.

Segmentation provides several advantages. In markedly segmented animals, such as earthworms and their relatives, each segment may include a complete set of important organs, including respiratory surfaces and parts of the nervous, circulatory, and excretory systems. Thus, a segmented animal may survive damage to the organs in one segment, because those in other segments perform the same functions. Segmentation also improves control over movement, especially in wormlike animals. Each segment has its own set of muscles, which can act independently of those in other segments. Thus, an earthworm can move its anterior end to the left while it swings its posterior end to the right. The segmented backbone and body wall musculature of vertebrates allow greater flexibility of movement than would unsegmented structures.

STUDY BREAK

1. What is a tissue, and what three primary tissue layers are present in the embryos of most animals?
2. What type of body symmetry do humans have?
3. What is the functional significance of the coelom?
4. What are some advantages of having a segmented body?

29.3 An Overview of Animal Phylogeny and Classification

For many years, biologists used the morphological innovations and embryological patterns described earlier to trace the phylogenetic history of animals. These efforts were sometimes hampered by the difficulty of identifying homologous structures in different phyla

and by morphological data that led to contradictory interpretations. Recently, biologists have used molecular sequence data to reanalyze animal relationships. Although biologists now recognize nearly 40 animal phyla, we focus primarily on the phyla that include substantial numbers of species.

Molecular Analyses Have Refined Our Understanding of Animal Phylogeny

Molecular analyses of animal relationships are often based on nucleotide sequences in small subunit ribosomal RNA and mitochondrial DNA (see Chapter 15). Recent analyses of *Hox* gene sequences provide similar results. (*Hox* genes are described in Sections 22.6 and 48.6.) These molecular analyses are still reasonably new, and they include studies of relatively few genes. Thus, the phylogenetic tree based on molecular sequences **(Figure 29.6)** represents a working hypothesis; its details will likely change as researchers accumulate more data.

The phylogenetic tree based upon molecular characters includes the major lineages that biologists had defined using the morphological innovations and embryological characters just described. For example, molecular data confirm the distinctions between the Parazoa and the Eumetazoa and between the Radiata and the Bilateria. They also confirm the separation of the deuterostome phyla from all others within the Bilateria.

However, the molecular phylogeny groups many other phyla—including the acoelomate animals, pseudocoelomate animals, protostomes, and a few others—into one taxon, the Protostomia. This group is, in turn, subdivided into two major lineages, the Lophotrochozoa and the Ecdysozoa, groups that were not previously recognized. The name Lophotrochozoa (*lophos* = crest; *trochos* = wheel) refers to both the lophophore, a feeding structure found in three phyla (illustrated in Figure 29.15), and the trochophore, a type of larva found in annelids and mollusks (illustrated in Figure 29.23). The name Ecdysozoa (*ekdysis* = escape) refers to the cuticle or external skeleton that these species secrete and periodically molt (or "escape from") when they experience a growth spurt or begin a different stage of the life cycle (illustrated in Figure 29.34); the molting process is called **ecdysis.**

The Molecular Phylogeny Reveals Surprising Patterns in the Evolution of Key Morphological Innovations

Phylogenetic trees contain explicit hypotheses about evolutionary change, and the molecular phylogeny has forced biologists to reevaluate the evolution of several important morphological innovations. For example, traditional phylogenies based upon morphology and embryology usually inferred that the absence of a body cavity, the acoelomate condition, was ancestral and that the

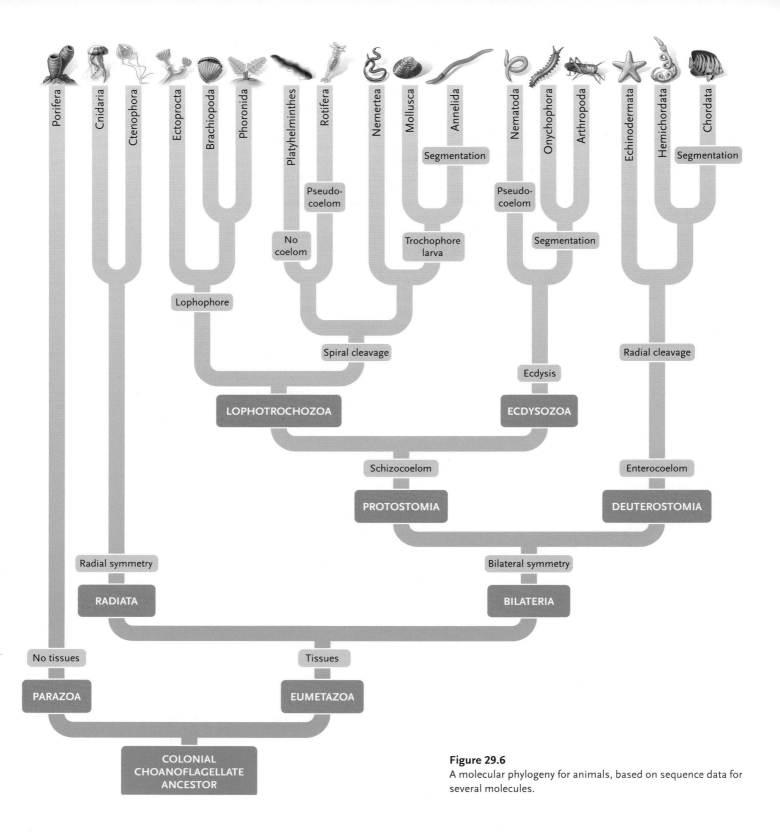

Porifera
Cnidaria
Ctenophora
Ectoprocta
Brachiopoda
Phoronida
Platyhelminthes
Rotifera
Nemertea
Mollusca
Annelida
Nematoda
Onychophora
Arthropoda
Echinodermata
Hemichordata
Chordata

Segmentation
Segmentation

Pseudo-coelom

Pseudo-coelom

No coelom

Trochophore larva

Segmentation

Lophophore

Spiral cleavage

Radial cleavage

Ecdysis

LOPHOTROCHOZOA
ECDYSOZOA

Schizocoelom

Enterocoelom

PROTOSTOMIA
DEUTEROSTOMIA

Radial symmetry
Bilateral symmetry

RADIATA
BILATERIA

No tissues
Tissues

PARAZOA
EUMETAZOA

COLONIAL CHOANOFLAGELLATE ANCESTOR

Figure 29.6
A molecular phylogeny for animals, based on sequence data for several molecules.

presence of a body cavity, the pseudocoelomate or coelomate condition, was derived. But the molecular tree provides a very different view. Because most protostome phyla have a schizocoelom, the molecular tree suggests that this trait is ancestral within the lineage, having evolved in the common ancestor of the lineage. If that hypothesis is correct, then the acoelomate condition of flatworms (phylum Platyhelminthes) represents the evolutionary *loss* of the schizocoelom, *not* an ancestral condition. Similarly, the molecular tree hypothesizes that the pseudocoelom evolved independently in rotifers (Lophotrochozoa, phylum Rotifera) and in roundworms (Ecdysozoa, phylum Nematoda) as modifications of the ancestral schizocoelom. Thus, according to the molecular tree, the pseudocoelomate condition of these organisms is the product of convergent evolution.

Traditional phylogenies also suggested that the segmented body plan of several protostome phyla was inherited from a segmented common ancestor and that segmentation arose independently in the chordates by convergent evolution. The molecular tree, by contrast, suggests that segmentation evolved independently in *three* lineages—segmented worms (Lophotrochozoa, phylum Annelida), arthropods and velvet worms (Ecdysozoa, phyla Arthropoda and Onychophora), and chordates (Deuterostomia, phylum Chordata)—rather than in just two lineages.

Despite these surprising findings about morphological evolution, most biologists now embrace the hypothesis provided by molecular sequence studies. In the future, new data will undoubtedly foster active discussions, heated disputes, and revisions of the phylogeny. Students may be understandably frustrated by the lack of consensus among experts and an ever-changing phylogeny for animals. But these disputes highlight the uniqueness of science as a way of knowing the natural world through the process of collecting evidence and rigorously challenging accepted hypotheses. The phylogenetic tree based on molecular sequence data is truly revolutionary, and the dust has yet to settle on the disagreements that these new analyses have provoked.

STUDY BREAK

1. Which major groupings of animals defined on the basis of morphological characters have been confirmed by molecular sequence studies?
2. What type of body cavity is ancestral within the Protostomes?

29.4 Animals without Tissues: Parazoa

The Parazoa is a lineage that includes just one group of animals, the sponges.

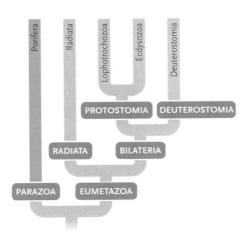

Sponges Have Simple Body Plans and Lack Tissues

Sponges, phylum Porifera (meaning "pore bearers"), lack true tissues: during development, their cells do not form the layers typical of other phyla. Mature sponges are sessile, and their shapes are less fixed than those of other animals, because mobile cells allow them to change shape in response to local conditions **(Figure 29.7)**. Sponges have been abundant since the Cambrian, especially in shallow coastal areas. Most of the 8000 living species are marine. Mature sponges range in size from 1 cm to 2 m.

Sponges have simple body plans **(Figure 29.8)**. Flattened cells form an outer layer, the **pinacoderm**. The inner surface of saclike sponges is lined by collar cells, called **choanocytes,** each equipped with a beating flagellum and a surrounding "collar" of modified microvilli. (Choanocytes resemble the cells of choanoflagellates, the hypothesized ancestor of all animals.) Amoeboid cells wander through the gelatinous **mesohyl** between the two layers; they secrete a supporting skeleton of a fibrous protein and *spicules,* small needlelike structures of calcium carbonate or silica (see Figure 29.8d). The natural sponges that we use are the fibrous remains of the bath sponge *(Spongia),* which lacks mineralized parts.

A sponge's body is an elaborate system for filtering food particles from the surrounding water. Water flows through pores in the pinacoderm into a central chamber, the **spongocoel,** and then out of the sponge through one or more openings called **oscula** (singular, *osculum*). The beating flagellae of the choanocytes maintain a constant flow, and contractile pore cells *(porocytes)* adjust the flow rate. Even a small sponge may filter as much as 20 liters of water per day. The choanocytes capture suspended parti-

Figure 29.7
Asymmetry in sponges. The shapes of sponges vary with their habitats. Those that occupy calm waters, such as this stinker vase sponge (*Ircinia campana*), may be lobed, tubular, cuplike, or vaselike.

Marty Snyderman/Planet Earth Pictures

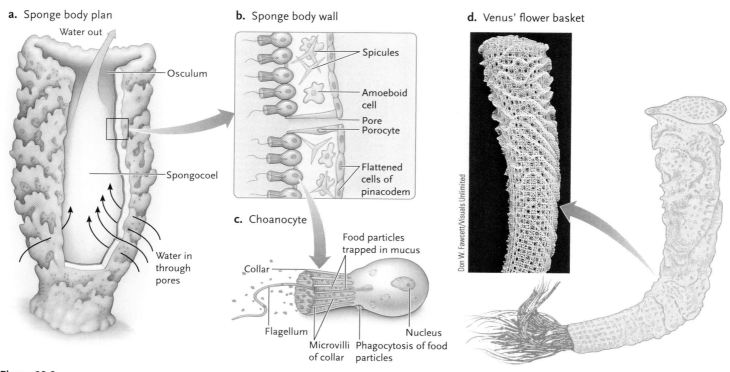

a. Sponge body plan

Water out
Osculum
Spongocoel
Water in through pores

b. Sponge body wall

Spicules
Amoeboid cell
Pore
Porocyte
Flattened cells of pinacodem

c. Choanocyte

Food particles trapped in mucus
Collar
Flagellum
Microvilli of collar
Phagocytosis of food particles
Nucleus

d. Venus' flower basket

Don W. Fawcett/Visuals Unlimited

Figure 29.8
The body plan of sponges. Most sponges have **(a)** simple body plans and **(b)** relatively few cell types. **(c)** Beating flagella on the choanocytes create a flow of water through incurrent pores, into the spongocoel, and out through the osculum. **(d)** Venus' flower basket (*Euplectella* species), a marine sponge, has spicules of silica fused into a rigid framework.

cles and microorganisms from the water and pass this food to mobile amoeboid cells, which carry nutrients to cells of the pinacoderm.

Most sponges are **hermaphroditic**: individuals produce both sperm and eggs. Sperm are released into the spongocoel and then out into the environment through oscula; eggs remain in the mesohyl, where sperm from other sponges, drawn in with water, fertilize them. Zygotes develop into flagellated larvae that are expelled to fend for themselves. Surviving larvae attach to substrates and undergo **metamorphosis** (a reorganization of form) into sessile adults. Some sponges also reproduce asexually; small fragments break off an adult and grow into new sponges. Many species also produce *gemmules,* clusters of cells with a resistant covering that allows them to survive unfavorable conditions; gemmules germinate into new sponges when conditions improve.

STUDY BREAK

1. What type of body symmetry do sponges exhibit?
2. How does a sponge gather food from its environment?

29.5 Eumetazoans with Radial Symmetry

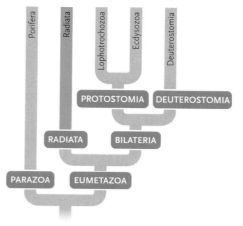

Unlike sponges, eumetazoans have true tissues, which develop from distinct layers in the embryo. Working together, the cells of a tissue perform complex functions beyond the capacity of individual cells. For example, nerve tissue transmits information rapidly through an animal's body, and epithelial tissue forms barriers that surround the body and line body cavities. In this section, we describe eumetazoans with radial symmetry, which enables them to sense stimuli from all directions, an effective adaptation for life in open water.

Two phyla of soft-bodied organisms, Cnidaria and Ctenophora, have radial symmetry and tissues, but they lack organ systems and a coelom. Species in both phyla possess a **gastrovascular cavity** that serves both digestive and circulatory functions. It has a single opening, the mouth. Gas exchange and excretion occur by diffusion because no cell is far from a body surface.

The radiate phyla have a diploblastic body plan with only two tissue layers, the inner *gastrodermis* (an endoderm derivative) and the outer *epidermis* (an ectoderm derivative). Most species also possess a gelatinous *mesoglea* (*mesos* = middle; *glia* = glue) between the two layers. The mesoglea contains widely dispersed fibrous and amoeboid cells.

Cnidarians Use Nematocysts to Stun or Kill Prey

Nearly all of the 8900 species in the phylum Cnidaria (*knide* = stinging nettle, a plant with irritating surface hairs) live in the sea **(Figure 29.9)**. Their body plan is organized around a saclike gastrovascular cavity; the mouth is ringed with tentacles, which push food into it. Cnidarians may be vase-shaped, upward-pointing **polyps** or bell-shaped, downward-pointing **medusae** (see Figure 29.9a). Most polyps attach to a substrate at the *aboral* (opposite the mouth) end; medusae are unattached and float.

Cnidarians are the simplest animals that exhibit a division of labor among specialized tissues (see Figure 29.9b, c). (Sponges have specialized cells, but no tissues.) The gastrodermis includes gland cells and phagocytic nutritive cells. Gland cells secrete enzymes for the extracellular digestion of food, which is then engulfed by nutritive cells and exposed to intracellular digestion. The epidermis includes nerve cells, sensory cells, contractile cells, and cells specialized for prey capture. A layer of acellular mesoglea separates the gastrodermis from the epidermis.

Cnidarians prey on crustaceans, fishes, and other animals. The epidermis includes unique cells, **cnidocytes**, each armed with a stinging **nematocyst (Figure 29.10)**. The nematocyst is an encapsulated, coiled thread that is fired at prey or predators, sometimes releasing a toxin through its tip. Discharge of nematocysts may be triggered by touch, vibrations, or chemical stimuli. The toxin can paralyze small prey by disrupting nerve cell membranes. The painful stings of some jellyfishes and corals result from the discharge of nematocysts.

Cnidarians engage in directed movements by contracting specialized cells in the epidermis. In medusae, the mesogleal jelly serves as a deformable skeleton against which contractile cells act. Rapid contractions narrow the bell, forcing out jets of water that propel the animal. Polyps use their water-filled gastrovascular cavity as a hydrostatic skeleton. When some cells contract, fluid within the chamber is shunted about,

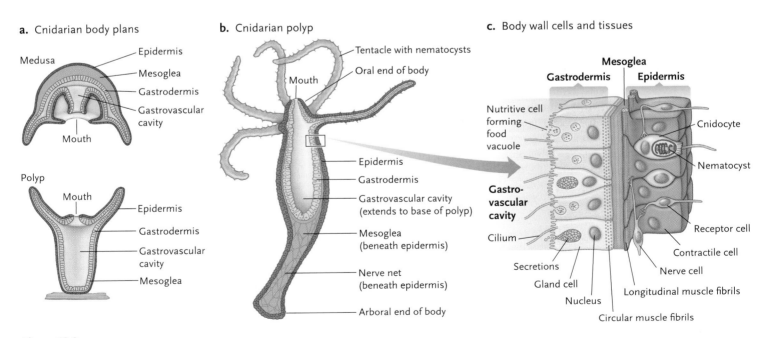

a. Cnidarian body plans

Medusa
- Epidermis
- Mesoglea
- Gastrodermis
- Gastrovascular cavity
- Mouth

Polyp
- Mouth
- Epidermis
- Gastrodermis
- Gastrovascular cavity
- Mesoglea

b. Cnidarian polyp

- Tentacle with nematocysts
- Oral end of body
- Mouth
- Epidermis
- Gastrodermis
- Gastrovascular cavity (extends to base of polyp)
- Mesoglea (beneath epidermis)
- Nerve net (beneath epidermis)
- Arboral end of body

c. Body wall cells and tissues

Mesoglea
Gastrodermis   Epidermis
- Nutritive cell forming food vacuole
- Cnidocyte
- Nematocyst
- Gastrovascular cavity
- Cilium
- Receptor cell
- Contractile cell
- Secretions
- Gland cell
- Nerve cell
- Nucleus
- Longitudinal muscle fibrils
- Circular muscle fibrils

Figure 29.9

The cnidarian body plan. **(a)** Cnidarians exist as either polyps or medusae. **(b)** The body of both forms is organized around a gastrovascular cavity, which extends all the way to the aboral end of the animal. **(c)** The two tissue layers in the body wall, the gastrodermis and the epidermis, include a variety of cell types.

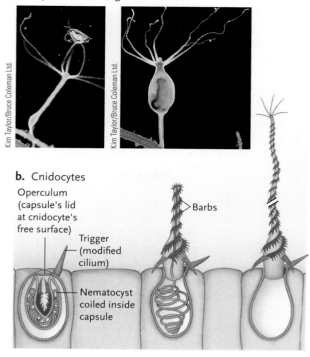

a. *Hydra* consuming a crustacean

Kim Taylor/Bruce Coleman Ltd.

Kim Taylor/Bruce Coleman Ltd.

b. Cnidocytes

Operculum
(capsule's lid
at cnidocyte's
free surface)

Trigger
(modified
cilium)

Nematocyst
coiled inside
capsule

Barbs

Figure 29.10

Predation by cnidarians. **(a)** A polyp of a freshwater *Hydra* captures a small crustacean with its tentacles and swallows it whole. **(b)** Cnidocytes, special cells on the tentacles, encapsulate nematocysts, which are discharged at prey.

changing the body's shape and moving it in a particular direction.

The **nerve net**, which threads through both tissue layers, is a simple nervous system that coordinates responses to stimuli but has no central control organ or brain. Impulses initiated by sensory cells are transmitted in all directions from the site of stimulation.

Many cnidarians exist in only the polyp or the medusa form, but some have a life cycle that alternates between them **(Figure 29.11).** In the latter type, the polyp often produces new individuals asexually from buds that break free of the parent (see Figure 47.2). The medusa is often the sexual stage, producing sperm and eggs, which are released into the water. The four lineages of Cnidaria differ in the form that predominates in the life cycle.

Hydrozoa. Most of the 2700 species in the Hydrozoa have both polyp and medusa stages in their life cycles (see Figure 29.11). The polyps form sessile colonies that develop asexually from one individual. A colony can include thousands of polyps, which may be specialized for feeding, defense, or reproduction. They share food through their connected gastrovascular cavities.

Unlike most hydrozoans, freshwater species of *Hydra* (see Figure 29.10a) live as solitary polyps that attach temporarily to rocks, twigs, and leaves. Under favorable conditions hydras reproduce by budding. Under adverse conditions they produce eggs and sperm; the zygotes, which are encapsulated in a protective coating, develop and grow when conditions improve.

Scyphozoa. The medusa stage predominates in the 200 species of the Scyphozoa **(Figure 29.12a),** or jellyfishes. They range from 2 cm to more than 2 m in diameter. Nerve cells near the margin of the bell control their tentacles and coordinate the rhythmic activity of contractile cells, which move the animal. Specialized sensory cells are clustered at the edge of the bell: **statocysts** sense gravity and **ocelli** are sensitive to light. Scyphozoan medusae are either male or female; they release gametes into the water where fertilization takes place.

Cubozoa. The 20 known species of box jellyfish, the Cubozoa **(Figure 29.12b),** exist primarily as cube-shaped medusas only a few centimeters tall; the largest species grows to 25 cm in height. Nematocyst-rich tentacles grow in clusters from the four corners of the boxlike medusa, and groups of light receptors and image-forming eyes occur on the four sides of the bell. Unlike the true jellyfish, cubozoans are active swimmers. They feed on small fishes and invertebrates, immobilizing their prey with one of the deadliest toxins produced by animals. Cubozoans live in tropical and subtropical coastal waters, where they sometimes pose a serious threat to swimmers.

Anthozoa. The Anthozoa includes 6000 species of corals and sea anemones **(Figure 29.13).** Anthozoans exist only as polyps, and often reproduce by budding or fission; most also reproduce sexually. Corals are always sessile and usually colonial. Most species build calcium carbonate skeletons, which sometimes accumulate into gigantic underwater reefs. The energy needs of many corals are partly fulfilled by the photosynthetic activity of symbiotic protists that live within the corals' cells. For this reason, corals are restricted to shallow water where sunlight can penetrate. Sea anemones, by contrast, are soft-bodied, solitary polyps, ranging from 1 cm to 10 cm in diameter. They occupy shallow coastal waters. Most species are essentially sessile, but some move by crawling slowly or by using the gastrovascular cavity as a hydrostatic skeleton.

Ctenophores Use Tentacles to Feed on Microscopic Plankton

Like the cnidarians, the 100 species of comb jellies in the phylum Ctenophora (*kteis* = comb; *-phoros* = bearing) have radial symmetry, mesoglea, and feeding tentacles. However, they differ from cnidarians in signifi-

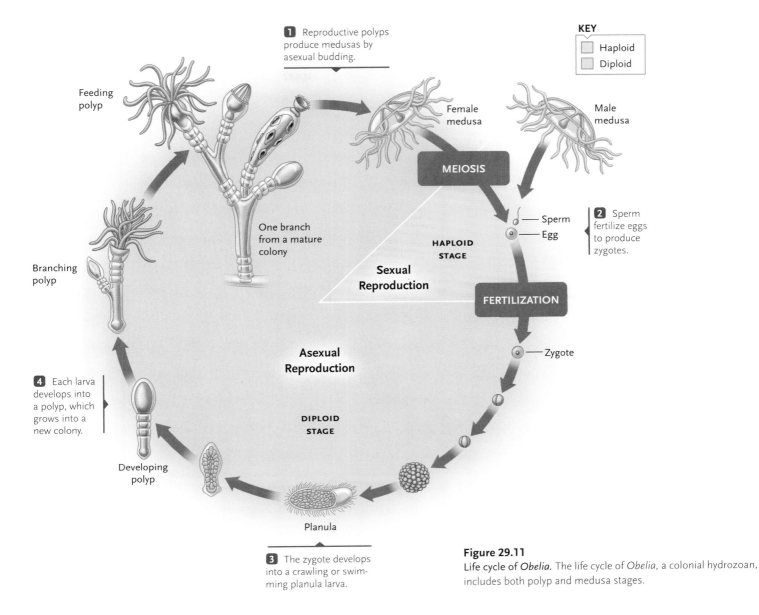

1 Reproductive polyps produce medusas by asexual budding.

KEY
Haploid
Diploid

Feeding polyp

Female medusa

Male medusa

MEIOSIS

One branch from a mature colony

Branching polyp

Sperm

Egg

2 Sperm fertilize eggs to produce zygotes.

HAPLOID STAGE

Sexual Reproduction

FERTILIZATION

Asexual Reproduction

Zygote

4 Each larva develops into a polyp, which grows into a new colony.

DIPLOID STAGE

Developing polyp

Planula

3 The zygote develops into a crawling or swimming planula larva.

Figure 29.11
Life cycle of *Obelia*. The life cycle of *Obelia*, a colonial hydrozoan, includes both polyp and medusa stages.

cant ways: they lack nematocysts; they expel some waste through anal pores opposite the mouth; and certain tissues appear to be of mesodermal origin. These transparent, and often luminescent, animals range in size from a few millimeters to a few meters **(Figure 29.14).** They live primarily in coastal regions of the oceans.

Ctenophores move by beating cilia arranged on eight longitudinal plates that resemble combs. They are the largest animals to use cilia for locomotion, but they are feeble swimmers. Nerve cells connected to the cilia coordinate the animals' movements, and a gravity-sensing statocyst helps them maintain an upright position. They capture microscopic plankton with their two tentacles, which have specialized cells that discharge sticky filaments; the food-laden tenta-

cles are drawn across the mouth. Ctenophores are hermaphroditic, producing gametes in cells that line the gastrovascular cavity. Eggs and sperm are expelled through the mouth or from special pores, and fertilization occurs in the open water.

STUDY BREAK

1. How do cnidarians capture, consume, and digest their prey?
2. Which group of cnidarians has only a polyp stage in its life cycle?
3. What do ctenophores eat, and how do they collect their food?

a. Scyphozoan

b. Cubozoan

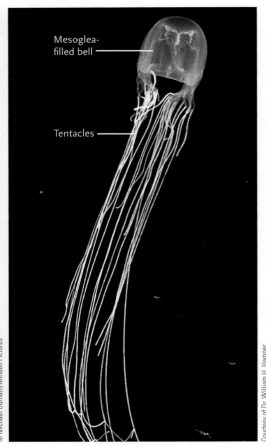

Mesoglea-filled bell

Tentacles

© Michael Durham/Minden Pictures

Courtesy of Dr. William H. Hamner

Figure 29.12

Scyphozoans and cubozoans. **(a)** Most scyphozoans, like the sea nettle *(Chrysaora quinquecirrha)*, live as floating medusae. Their tentacles trap prey, and the long oral arms transfer it to the mouth on the underside of the bell. **(b)** Cubozoans, like the sea wasp *(Chironex fleckeri)*, are strong swimmers that actively pursue small fishes and invertebrates.

a. Coral

b. Sea anemone escape behavior

Tentacle of one polyp

Christian DellaCorte

F. S. Westmorland

F. S. Westmorland

F. S. Westmorland

Interconnected skeletons of polyps of a colonial coral

Figure 29.13

Anthozoans. **(a)** Many corals, like the staghorn coral *(Acropora cervicornis)*, are colonial, and their polyps build a hard skeleton of calcium carbonate. The skeletons accumulate to form coral reefs in shallow tropical waters (see Figure 52.28b). **(b)** A white-spotted sea anemone *(Urticina lofotensis)* detaches from its substrate to escape from a predatory sea star.

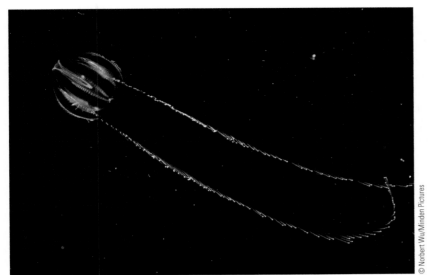

Figure 29.14
Ctenophores. The comb jelly *Mertensia ovum* collects micro-scopic prey on its two long sticky tentacles and then wipes the food-laden tentacles across its mouth.

© Norbert Wu/Minden Pictures

29.6 Lophotrochozoan Protostomes

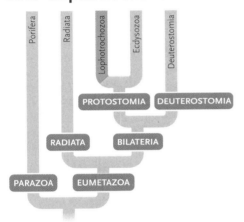

The remaining organisms described in this chapter fall within the group Bilateria: they have bilateral symmetry and a greater variety of tissues than do the Radiata. Bilaterians also have organ systems, structures that include two or more tissue types organized to perform specific functions; most of them also possess a coelom or pseudocoelom. With bilateral symmetry and sensory organs concentrated at the anterior end of the body, most bilaterians engage in highly directed, often rapid movements in pursuit of food or mates or to escape danger. And their complex organ systems accomplish tasks more efficiently than simple tissues. For example, animals that have a tubular digestive system surrounded by a space (the coelom or pseudocoelom) use muscular contractions of the digestive system to move ingested food past specialized epithelia that break it down and absorb the breakdown products.

Molecular analyses group eight of the phyla that we consider into the Lophotrochozoa, one of the two main protostome lineages (see Figure 29.6).

The Lophophorate Phyla Share a Distinctive Feeding Structure

Three small groups of aquatic, coelomate animals—the phyla Ectoprocta, Brachiopoda, and Phoronida—possess a **lophophore**, a circular or U-shaped fold with one or two rows of hollow, ciliated tentacles surrounding the mouth **(Figure 29.15)**. Molecular sequence data as well as the lophophore suggest that these phyla share a common ancestry.

a. Ectoprocta *(Plumatella repens)*

b. Brachiopoda *(Terebraulina septentrionalis)*

c. Phoronida *(Phoronopsis californica)*

© blickwinkel/Hecker/Alamy

© Andrew J. Martinez/Photo Researchers, Inc.

© Lawrence Naylor/Photo Researchers, Inc.

Figure 29.15
Lophophorate animals. Although the lophophorate animals differ markedly in appearance, they all use a lophophore—the feathery structures in the photos—to acquire food.

The lophophore, which looks like a crown of tentacles at the anterior end of the animal, serves as a site for gas exchange and waste elimination as well as for food capture. Most lophophorates are sessile suspension-feeders as adults: movement of cilia on the tentacles brings food-laden water toward the lophophore, the tentacles capture small organisms and debris, and the cilia transport them to the mouth. The lophophorates have a complete digestive system, which is U-shaped in most species, with the anus lying outside the ring of tentacles.

Phylum Ectoprocta. The Ectoprocta (sometimes called Bryozoa) are tiny colonial animals that mainly occupy marine habitats (see Figure 29.15a). They secrete a hard covering over their soft bodies and feed by extending the lophophore through a hole. Each colony, which may include more than a million individuals, is produced asexually by a single animal. Ectoproct colonies are permanently attached to solid substrates, where they form encrusting mats, bushy upright growths, or jellylike blobs. Nearly 4000 living species are known.

Phylum Brachiopoda. The Brachiopoda, or lampshells, have two calcified shells that develop on the animal's dorsal and ventral sides (see Figure 29.15b). Most species attach to substrates with a stalk that protrudes through one of the shells. The lophophore is held within the two shells, and the animal feeds by opening its shell and drawing water over its tentacles. Although only 250 species of brachiopods live today, more than 30,000 extinct species are known as fossils, mostly from Paleozoic seas.

Phylum Phoronida. The 18 or so species of phoronid worms vary in length from a few millimeters to 25 cm (see Figure 29.15c). They usually build tubes in soft ocean sediments or on hard substrates, and feed by protruding the lophophore from the top of the tube. The animal can withdraw into the tube when disturbed. Phoronida reproduce both sexually and by budding.

Flatworms Have Digestive, Excretory, Nervous, and Reproductive Systems, but Lack a Coelom

The 13,000 flatworm species in the phylum Platyhelminthes (*platys* = flat; *helmis* = worm) live in aquatic and moist terrestrial habitats. Like cnidarians, flatworms can swim or float in water, but they are also able to crawl over surfaces. They range from less than 1 mm to more than 20 m in length, but most are just a few millimeters thick. Free-living species eat live prey or decomposing carcasses, whereas parasitic species consume the tissues of living hosts.

Like the radiate phyla, flatworms are acoelomate, but they have a complex structural organization that

Digestive system
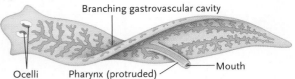

Nervous system
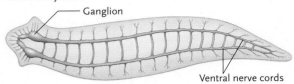

Reproductive system
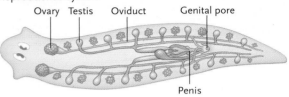

Excretory system

Figure 29.16

Flatworms. The phylum Platyhelminthes, exemplified by a freshwater planarian, have well-developed digestive, excretory, nervous, and reproductive systems. Because flatworms are acoelomate, their organ systems are embedded in a solid mass of tissue between the gut and the epidermis.

reflects their triploblastic construction **(Figure 29.16).** Endoderm lines the digestive cavity with cells specialized for the chemical breakdown and absorption of ingested food. Mesoderm, the middle tissue layer, produces muscles and reproductive organs. Ectoderm produces a ciliated epidermis, the nervous system, and the *flame cell system,* a simple excretory system; flame cells regulate the concentrations of salts and water within body fluids, allowing free-living flatworms to live in freshwater habitats. Flatworms do not have circulatory or respiratory systems, but, because all cells of their dorsoventrally (top-to-bottom) flattened bodies are near an interior or exterior surface, diffusion supplies them with nutrients and oxygen.

The flatworm nervous system includes two or more longitudinal ventral nerve cords interconnected by numerous smaller nerve fibers, like rungs on a ladder. An anterior **ganglion,** a concentration of nervous system tissue that serves as a primitive brain, integrates their behavior. Most free-living species have *ocelli,* or eye spots, that distinguish light from dark and tiny chemoreceptor organs that sense chemical cues.

Figure 29.17
Turbellaria. A few turbellarians, such as *Pseudoceros dimidiatus*, are colorful marine worms.

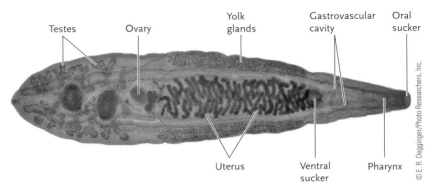

Figure 29.18
Trematoda. The hermaphroditic Chinese liver fluke *(Opisthorchis sinensis)* uses a well-developed reproductive system to produce thousands of eggs.

The phylum Platyhelminthes includes four lineages, defined largely by their anatomical adaptations to free-living or parasitic habits.

Turbellaria. Most free-living flatworms (Turbellaria) live in the sea **(Figure 29.17),** but the familiar planarians and a few others live in fresh water or on land. Turbellarians swim by undulating the body wall musculature, or they crawl across surfaces by using muscles and cilia to glide on mucous trails produced by the ventral epidermis.

The gastrovascular cavity in free-living flatworms is similar to that in cnidarians. Food is ingested and wastes eliminated through a single opening, the mouth, located on the ventral surface. Most turbellarians also acquire food with a muscular **pharynx** that connects the mouth to the digestive cavity (see Figure 29.16). Chemicals secreted into the saclike cavity digest the food into particles; then cells throughout the gastrovascular surface engulf the particles and subject them to intracellular digestion. In some species, the digestive cavity is highly branched, increasing the surface area for digestion and absorption.

Nearly all turbellarians are hermaphroditic, with complex reproductive systems (see Figure 29.16). When they mate, each partner functions simultaneously as a male and a female. Many free-living species also reproduce asexually by simply separating the anterior half of the animal from the posterior half. Both halves subsequently regenerate the missing parts.

Trematoda and Monogenoidea. Flukes (Trematoda and Monogenoidea) are parasites that obtain nutrients from host tissues **(Figure 29.18).** Most adult trematodes are **endoparasites,** living in the gut, liver, lungs, bladder, or blood vessels of vertebrates. Monogenes are **ectoparasites,** attaching to the gills or skin of aquatic vertebrates. Flukes are structurally specialized for a parasitic existence. They use suckers or hooks to attach to hosts, and a tough outer covering protects them from chemical attack. They produce large numbers of eggs that can readily infect new hosts. Monogene flukes usually have simple life cycles with a single host species. Trematodes, by contrast, have complex life cycles and multiple hosts. Humans suffer potentially fatal infections by many flukes, as discussed in *Focus on Applied Research.*

Cestoda. Tapeworms (Cestoda) develop, grow, and reproduce within the intestines of vertebrates **(Figure 29.19).** Through evolution, they have lost their mouths

a. Tapeworm   **b.** Scolex

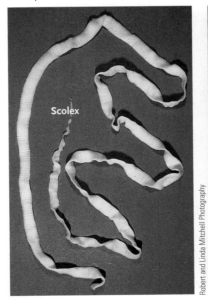
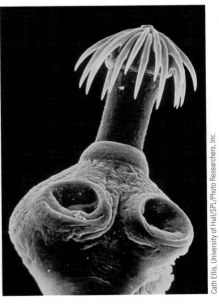

Figure 29.19
Cestoda. **(a)** Tapeworms have long bodies comprised of a series of proglottids that each produce thousands of fertilized eggs. **(b)** The anterior end is a scolex with hooks and suckers that attach to the host's intestinal wall.

Applied Research: A Rogue's Gallery of Parasitic Worms

Many parasitic flatworms (phylum Platyhelminthes) and roundworms (phylum Nematoda) call the human body home, frequently causing disfiguring or life-threatening infections. The effort to control or eliminate these infections often begins when parasitologists or public health researchers study the worms' life cycles and ecology. This approach is successful because the worms often have more than one host: humans may be the *primary host,* harboring the sexually mature stage of the parasites' life cycles, but other animals serve as *intermediate hosts* to the larval stages. If researchers can learn the details of a parasite's life cycle, they can identify ways to cut it short before the parasite infects a human host.

More than 200 million people in tropical and subtropical regions suffer from *schistosomiasis,* a disease caused by three species of flatworms called blood flukes (Trematoda). Japanese blood flukes *(Schistosoma japonicum)* mature and mate in blood vessels of the human intestine **(Figure a).** Sharp spines on their eggs rupture the blood vessels, releasing the eggs into the lumen of the gut, from which they are passed with feces. When the infected human waste enters standing fresh water, the eggs hatch into ciliated larvae, which burrow into certain aquatic snails. The larvae feed on the snail's tissues and reproduce asexually. Their tailed offspring leave the snail and, when they contact human skin, bore inward to a blood vessel. They eventually reach the intestine, where they complete their complex life cycle and produce fertilized eggs. Infected humans mount an immune response against flukes, but it is always a losing battle. Severe infections cause coughs, rashes, pain, and eventually diarrhea, anemia, and permanent damage to the intestines, liver, spleen, bladder, and kidneys. Death often results. Drug therapy can reduce the symptoms, but schistosomiasis is most common in less-developed countries with limited access to medical care. Research has demonstrated that the disease is best controlled by proper sanitation and the elimination of snails that serve as intermediate hosts.

Tapeworms (Cestoda), another group of flatworms, rely primarily on vertebrate hosts. Fishes, hogs, or cattle become intermediate hosts by inadvertently eating tapeworm eggs **(Figure b).** When the host's digestive enzymes dissolve the protective covering on the eggs, the newly freed larvae bore through the intestinal wall and travel through the bloodstream to muscles or other tissues, where they *encyst* (produce a protective covering and enter a resting stage). Humans and other carnivores ingest living cysts when they eat the undercooked flesh of these animals. The scolex of the tapeworm larva then attaches to the primary host's intestinal wall, and the worm begins to grow. When mature, its proglottids produce huge numbers of eggs, which are released with feces to begin the next generation. Tapeworm infection can result in malnutrition of the host, because tapeworms consume much of the nutrients that the host ingests. They can also grow large enough to cause intestinal blockage. A full understanding of the tapeworm life cycle suggests that careful inspection and adequate cooking of meat can prevent infection.

The roundworm *Trichinella spiralis* causes the painful and sometimes fatal symptoms of *trichinosis.* Adult trichinas breed in the small intestine of their hosts, including hogs and some game animals. Female worms release juveniles, which burrow into blood vessels and travel to various organs, where they become encysted **(Figure c).** Humans become infected when they consume insufficiently cooked pork or other meat that contains encysted larvae. Once in the human digestive tract, the encysted worms complete their development, mate, and produce larvae. Most of the

Figure a
Life cycle of the Japanese blood fluke *(Schistosoma japonicum).*

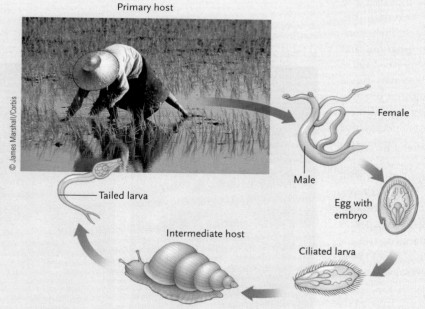

Primary host

Female

Male

Tailed larva

Egg with embryo

Intermediate host

Ciliated larva

© James Marshall/Corbis

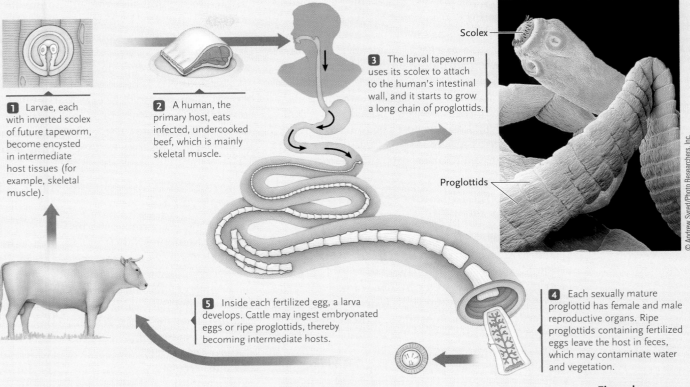

1 Larvae, each with inverted scolex of future tapeworm, become encysted in intermediate host tissues (for example, skeletal muscle).

2 A human, the primary host, eats infected, undercooked beef, which is mainly skeletal muscle.

3 The larval tapeworm uses its scolex to attach to the human's intestinal wall, and it starts to grow a long chain of proglottids.

Scolex

Proglottids

© Andrew Syred/Photo Researchers, Inc.

5 Inside each fertilized egg, a larva develops. Cattle may ingest embryonated eggs or ripe proglottids, thereby becoming intermediate hosts.

4 Each sexually mature proglottid has female and male reproductive organs. Ripe proglottids containing fertilized eggs leave the host in feces, which may contaminate water and vegetation.

Figure b
Life cycle of the beef tapeworm.

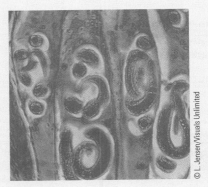

© L. Jensen/Visuals Unlimited

Figure c
Trichinella spiralis juveniles in muscle tissue.

awful symptoms of trichinosis, including severe pain, high fever, and debilitating anemia, are produced by the migration of millions of larvae throughout the primary host's tissues. Once begun, the infection is difficult or impossible to control, but, as with tapeworms, thorough cooking of meat can prevent infection.

Filarial worms (*Wuchereria* and *Brugia*) cause another debilitating nematode infection. These large roundworms (up to 10 cm long) live in the lymphatic system, where they obstruct the normal flow of lymphatic fluid to the bloodstream. Female worms release first-stage larvae, which are acquired by mosquitoes, the intermediate host, when they feed on human blood. The larvae develop into second-stage larvae in mosquitoes, and when a mosquito bites another human, it may transmit those larvae to a new host. If victims experience severe filarial worm infection, their lymphatic vessels can be so obstructed that surrounding tissues swell grotesquely, a condition know as *elephantiasis* **(Figure d).** Pub- lic health programs that reduce or eliminate mosquito populations lower the incidence of this disease.

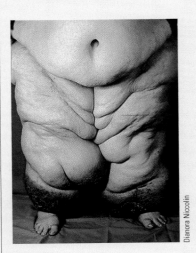

Dianora Niccolin

Figure d
Elephantiasis caused by *Wuchereria bancrofti.*

and digestive systems and absorb nutrients through their body wall. Tapeworms have a specialized structure, the *scolex,* with hooks and suckers that attach to the host's intestine; like the flukes, they also have a protective covering resistant to digestive enzymes.

Most of a tapeworm's body, which may be up to 20 m long, consists of a series of identical structures, *proglottids,* that contain little more than male and female reproductive systems. New proglottids are generated near the scolex; older proglottids, each carrying as many as 80,000 fertilized eggs, break off from the tapeworm's posterior end and leave the host's body in feces. As described in *Focus on Applied Research,* tapeworms are important parasites on humans, pets, and livestock.

Rotifers Are Tiny Pseudocoelomates with a Jawlike Feeding Apparatus

Most of the 1800 species in the phylum Rotifera (*rota* = wheel; *ferre* = to carry) live in fresh water **(Figure 29.20).** All are microscopic—about the size of a ciliate protist—but they have well-developed digestive, reproductive, excretory, and nervous systems as well as a pseudocoelom. In some habitats, rotifers make up a large part of the zooplankton, tiny animals that float in open water.

Rotifers use coordinated movements of cilia, arranged in a wheel-like *corona* around the head, to propel themselves in the environment. Cilia also bring food-laden water to their mouths. Ingested microorganisms are conveyed to the *mastax,* a toothed grinding organ, and then passed to the stomach and intestine. Rotifers have a **complete digestive tract:** food enters through the mouth, and undigested waste is voided through the anus.

The life history patterns of some rotifers are adapted to the ever-changing environments in small bodies of water. During most months, rotifer populations include only females that reproduce by **parthenogenesis,** a form of asexual reproduction in which unfertilized eggs develop into diploid females (see Section 47.1). When environmental conditions deteriorate, females produce eggs that develop into haploid males. The males fertilize haploid eggs to produce diploid female zygotes. The fertilized eggs have durable shells and food reserves to survive drying or freezing.

Ribbon Worms Use a Proboscis to Capture Food

The 650 species of ribbon worms or proboscis worms (phylum Nemertea) vary from less than 1 cm to 30 m in length **(Figure 29.21).** Most species are marine, but a few occupy moist terrestrial habitats. Although the often brightly colored ribbon worms superficially resemble free-living flatworms, their body plans are more complex. First, they possess both a mouth and an anus; thus, they have a complete digestive tract. Second, nemerteans have a circulatory system in which fluid flows through **circulatory vessels** that carry nutrients and oxygen to tissues and remove wastes (see Section 42.1). Finally, they have a muscular, mucus-covered proboscis, a tube that can be everted (turned inside out) through the mouth or a separate pore to capture prey. The proboscis is housed within a chamber, the *rhynchocoel,* which is unique to this phylum.

Figure 29.20

Phylum Rotifera. **(a)** Despite their small size, rotifers such as *Philodina roseola* have complex body plans and organ systems. **(b)** This rotifer, another *Philodina* species, is laying eggs.

a. Rotifer body plan

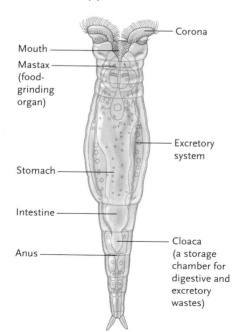

Corona
Mouth
Mastax (food-grinding organ)
Excretory system
Stomach
Intestine
Anus
Cloaca (a storage chamber for digestive and excretory wastes)

b. Rotifer laying eggs

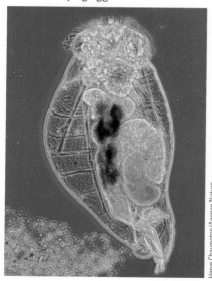

Herve Chaumeton/Agence Nature

a. Ribbon worm

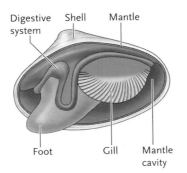

b. Ribbon worm anatomy

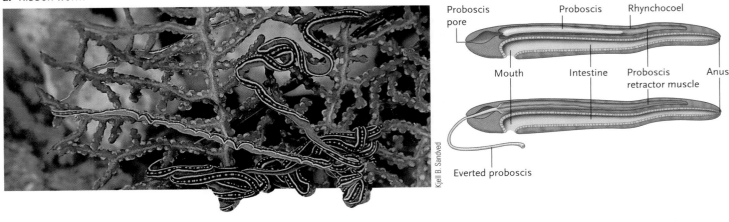

Kjell B. Sandved

Figure 29.21
Phylum Nemertea. (a) The flattened, elongated bodies of ribbon worms, such as this species in the genus *Lineus*, are often brightly colored. **(b)** Ribbon worms have a complete digestive system and a specialized cavity, the rhynchocoel, that houses a protrusible proboscis.

Mollusks Have a Muscular Foot and a Mantle That Secretes a Shell or Aids in Locomotion

Most of the 100,000 species in the phylum Mollusca (*mollis* = soft)—including clams, snails, octopuses, and their relatives—are marine. However, many clams and snails occupy freshwater habitats, and some snails live on land. Mollusks vary in length from clams less than 1 mm to the giant squids, which can exceed 18 m.

In mollusks, the body is divided into three regions: the visceral mass, the head–foot, and the mantle **(Figure 29.22)**. The **visceral mass** contains the digestive, excretory, reproductive systems, and heart. The muscular **head–foot** often provides the major means of locomotion. In the more active groups, the head region is well

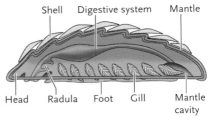

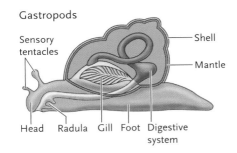

Figure 29.22
Molluskan body plans. The bilaterally symmetrical body plans of mollusks include a muscular head–foot, a visceral mass, and a mantle.

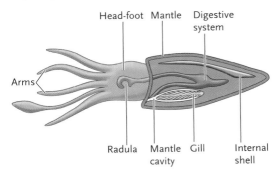

KEY
▢ Head-foot
▢ Visceral mass
▢ Mantle

Figure 29.23
Trochophore larva. At the conclusion of their embryological development, both mollusks and annelids typically pass through a trochophore stage. The top-shaped trochophore larva has a band of cilia just anterior to its mouth.

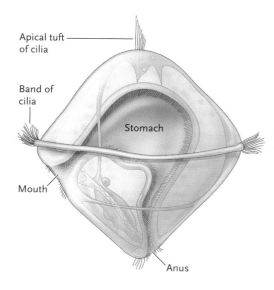

Apical tuft of cilia

Band of cilia

Stomach

Mouth

Anus

defined and carries sensory organs and a brain. The mouth often includes a toothed **radula**, which scrapes food into small particles or drills through the shells of prey. Many mollusks are covered by a protective shell of calcium carbonate secreted by the **mantle**, one or two folds of the body wall that often enclose the visceral mass. The mantle also defines a space, the *mantle cavity*, that houses the *gills*, delicate respiratory structures with enormous surface area. In most mollusks, cilia on the mantle and gills generate a steady flow of water into the mantle cavity.

The large size of mollusks—as well as their possession of a true coelom—requires a circulatory system to maintain cells that are far from the body surface. Most mollusks have an **open circulatory system** in which **hemolymph**, a bloodlike fluid, leaves the circulatory vessels and bathes tissues directly (see Figure 42.3a). Hemolymph pools in spaces called *sinuses*, and then drains into vessels that carry it back to the heart.

The sexes are usually separate, although many snails are hermaphroditic. Fertilization may be internal or external. The zygotes of marine species often develop into free-swimming, ciliated **trochophore** larvae **(Figure 29.23)**, typical of both this phylum and the phylum Annelida, which we describe next. In some

mollusks, the trochophore develops into a second larval stage, called a *veliger,* before metamorphosing into an adult. Some snails as well as octopuses and squids have direct development: embryos develop into miniature replicas of the adults.

Mollusca includes eight lineages. In the following sections, we examine the four that are most commonly encountered.

Polyplacophora. The 600 species of chitons (Polyplacophora: *poly* = many; *plax* = flat surface) are sedentary mollusks that graze on algae along rocky marine coasts. The oval, bilaterally symmetrical body has a dorsal shell divided into eight plates that allow it to conform to irregularly shaped surfaces **(Figure 29.24)**. When a chiton is disturbed or exposed to strong wave action, the muscles of its broad foot maintain a tenacious grip, and the mantle's edge functions like a suction cup to hold fast to the substrate.

Gastropoda. Snails and slugs (Gastropoda: *gaster* = belly; *pod* = foot) are the largest molluskan group, numbering 40,000 species **(Figure 29.25)**. Aquatic species use gills to acquire oxygen, but in terrestrial species a modified mantle cavity functions as an air-breathing lung. Gastropods feed on algae, vascular plants, or animal prey. Some are scavengers, and a few are parasites.

The visceral mass of most snails is housed in a coiled or cone-shaped shell that is balanced above the rest of the body, much as you balance a backpack full of books (see Figure 29.25a, b). Most shelled snails undergo **torsion**, a curious realignment of body parts that is independent of shell coiling. Muscle contractions and differential growth twist the visceral mass and mantle 180° relative to the head–foot. This rearrangement moves the mantle cavity forward so that the head can be withdrawn into the shell in times of danger. But it also brings the gills, anus, and excretory openings above the head—a potentially messy configuration, were it not for cilia that sweep away wastes.

Some gastropods, including terrestrial slugs and colorful nudibranchs (sea slugs), are shell-less, a condition that leaves them somewhat vulnerable to predators (see Figure 29.25c). Some nudibranchs consume cnidarians and then transfer undischarged nematocysts to projections on their dorsal surface, where these "borrowed" stinging capsules provide protection.

The nervous and sensory systems of gastropods are well developed. Tentacles on the head include chemical and touch receptors; the eyes detect changes in light intensity but don't form images.

Bivalvia. The 8000 species of clams, scallops, oysters, and mussels (Bivalvia: *bi* = two; *valva* = folding door) are restricted to aquatic habitats. They are enclosed within a pair of shells, hinged together dorsally by a ligament **(Figure 29.26)**. Contraction of the ligament

Figure 29.24
Polyplacophora. Chitons live on rocky shores, where they use their foot and mantle to grip rocks and other hard substrates. This bristled chiton (*Mopalia ciliata*) lives in Monterey Bay, California.

Jeff Foott/Tom Stack & Associates

a. Gastropod body plan

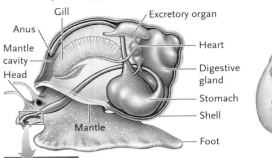

Gill
Anus
Mantle cavity
Head
Mantle
Radula

Excretory organ
Heart
Digestive gland
Stomach
Shell
Foot

Mouth
Anus

b. Terrestrial snail

J. Kottmann/Peter Arnold, Inc.

c. Marine nudibranchs

© Joe McDonald/Corbis

Figure 29.25

Gastropoda. **(a)** Most gastropods have a coiled shell that houses the visceral mass. A developmental process called torsion causes the digestive and excretory systems to eliminate wastes into the mantle cavity, near the animal's head. **(b)** The edible snail *(Helix pomatia)* is a terrestrial gastropod. **(c)** Nudibranchs, like these Spanish shawl nudibranchs *(Flabellina iodinea)*, are shell-less marine snails. (Micrograph: Danielle C. Zacherl with John McNulty.)

opens the shell by pulling the two sides apart, and contraction of one or two **adductor muscles** closes it by pulling them together (see Figure 29.26a). Although some bivalves are tiny, giant clams of the South Pacific can be more than 1 m across and weigh 225 kg.

Adult mussels and oysters are sessile and permanently attached to hard substrates, but many clams are mobile and use their muscular foot to burrow in sand or mud. Some bivalves, such as young scallops, swim by rhythmically clapping their valves together, forcing a current of water out of the mantle cavity (see Figure 29.26b). The "scallops" that we eat are their well-developed adductor muscles.

Bivalves have a reduced head, and they lack a radula. Part of the mantle forms two tubes called *siphons* (see Figure 29.26c). Beating of cilia on the gills and mantle carry water into the mantle cavity through the incurrent siphon and out through the excurrent siphon. Incurrent water carries dissolved oxygen and food particles to the gills, where oxygen is absorbed. Mucus strands on the gills trap the food, which is then transported by cilia to *palps*, where final sorting takes place; acceptable bits are carried to the mouth. The excurrent water carries away metabolic wastes and feces.

Despite their sedentary existence, bivalves have moderately well developed nervous systems; sensory organs that detect chemicals, touch, and light; and statocysts to sense their orientation. When they encounter pollutants, many bivalves stop pumping water and close their shells. When confronted by a predator, some burrow into sediments or swim away.

a. Bivalve body plan

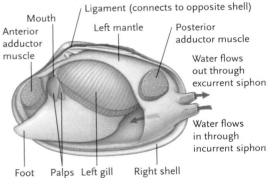

Mouth
Anterior adductor muscle
Ligament (connects to opposite shell)
Left mantle
Posterior adductor muscle
Water flows out through excurrent siphon
Water flows in through incurrent siphon
Foot Palps Left gill Right shell

b. Bivalve locomotion

Herve Chaumeton/Agence Nature

c. Geoduck

© Tom McHugh/Photo Researchers, Inc.

Figure 29.26

Bivalvia. **(a)** Bivalves are enclosed in a hinged two-part shell. Part of the mantle forms a pair of water-transporting siphons. **(b)** When threatened by a predator (in this case a sea star), some scallops clap their shells together rapidly, propelling the animal away from danger. **(c)** The geoduck *(Panope generosa)* is a clam with enormous muscular siphons.

Cephalopoda. The 600 species of octopuses, squids, and nautiluses (Cephalopoda: *kephale* = head) are active marine predators, including the fastest and most intelligent invertebrates **(Figure 29.27)**. They vary in length from a few centimeters to 18 m. Giant squids, the largest invertebrates known, may be the source of "sea monster" stories.

The cephalopod body has a fused head and foot (see Figure 29.27d). The head comprises the mouth and eyes. The ancestral "foot" forms a set of arms and tentacles, equipped with suction pads, adhesive structures, or hooks. Cephalopods use these structures to capture prey and a pair of beaklike jaws to bite or crush it. Venomous secretions often speed the captive's death. Some species use their radula to drill through the shells of other mollusks.

Cephalopods have a highly modified shell. Octopuses have no remnant of a shell at all. In squids and cuttlefishes, it is reduced to a stiff internal support. Only the chambered nautilus and its relatives retain an external shell.

Squids are the most mobile cephalopods, moving rapidly by a kind of jet propulsion. When muscles in the mantle relax, water is drawn into the mantle cavity. When they contract, a jet of water is squeezed out through a funnel-shaped excurrent siphon. By manipulating the position of the mantle and siphon, the animal can control the rate and direction of its locomotion. When threatened, a squid can make a speedy escape. Many species simultaneously release a dark fluid, commonly called "ink," that obscures their direction of movement. Being highly active, cephalopods need lots of oxygen, and they alone among the mollusks have a **closed circulatory system**: hemolymph is confined within the walls of hearts and blood vessels, providing increased pressure to the vascular fluid. Moreover, accessory hearts speed the flow of hemolymph through the gills, enhancing the uptake of oxygen and release of carbon dioxide.

Compared with other mollusks, cephalopods have large and complex brains. Giant nerve fibers connect the brain with the muscles of the mantle, enabling quick responses to food or danger. Their image-forming eyes are similar to those of vertebrates. Cephalopods are also highly intelligent. Octopuses, for example, learn to recognize objects with distinctive shapes or colors, and they can be trained to approach or avoid them.

Cephalopods have separate sexes and elaborate courtship rituals. Males store sperm within the mantle cavity and use a specialized tentacle to transfer packets of sperm into the female's mantle cavity, where fertilization occurs. Fertilized eggs, wrapped in a protective jelly, are attached to objects in the environment. The young hatch with an adult body form.

Annelids Exhibit a Serial Division of the Body Wall and Some Organ Systems

The 15,000 species of segmented worms (phylum Annelida: *anellus* = ring) occupy marine, freshwater, and moist terrestrial habitats. Terrestrial annelids eat organic debris; aquatic species consume algae, microscopic organisms, detritus, or other animals. They range from a few millimeters to several meters in length.

The annelid body is highly segmented: the body wall muscles and some organs—including respiratory surfaces, parts of the nervous, circulatory, and excretory systems, and the coelom itself—are divided into similar repeating units **(Figure 29.28)**. Body segments are separated by transverse partitions called **septa**. The digestive system and major blood vessels are not segmented and run the length of the animal.

The body wall muscles of annelids have both circular and longitudinal layers. Alternate contractions of these muscle groups allow annelids to make directed movements, using the pressure of the fluid in the coe-

a. Squid

b. Octopus

c. Chambered nautilus

Eye

Eye

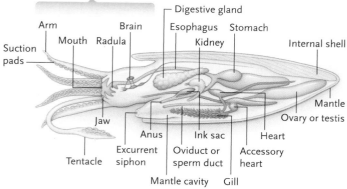

d. Internal anatomy of squid

Fused head and foot

Arm · Mouth · Radula · Brain · Digestive gland · Esophagus · Stomach · Kidney · Internal shell · Suction pads · Mantle · Ovary or testis · Jaw · Anus · Ink sac · Heart · Tentacle · Excurrent siphon · Oviduct or sperm duct · Accessory heart · Mantle cavity · Gill

Figure 29.27

Cephalopoda. **(a)** Squids, such as *Dosidicus gigas*, and **(b)** octopuses, such as *Octopus vulgaris*, are the most familiar cephalopods. **(c)** The chambered nautilus (*Nautilus macromphalus*) and its relatives retain an external shell. **(d)** Like other cephalopods, the squid body includes a fused head and foot; most organ systems are enclosed by the mantle.

a. Digestive system

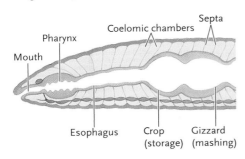

Pharynx
Mouth
Coelomic chambers
Septa
Esophagus
Crop (storage)
Gizzard (mashing)

b. Circulatory system

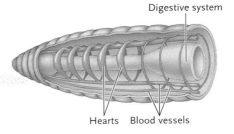

Digestive system
Hearts   Blood vessels

c. Nervous system

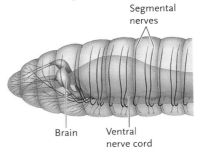

Segmental nerves
Brain
Ventral nerve cord

d. Excretory system (metanephridium)

Bladderlike storage region of excretory organ
Excretory organ's thin loop reabsorbs some solutes, returns them to blood
Septum
Blood vessels
Body wall
Coelomic fluid with waste enters funnel
Wastes discharged at external pore

e. Cross section

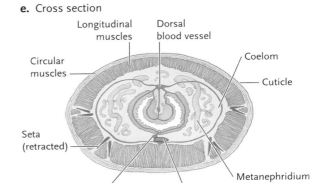

Longitudinal muscles
Dorsal blood vessel
Circular muscles
Coelom
Cuticle
Seta (retracted)
Metanephridium
Ventral blood vessel
Ventral nerve cord

Figure 29.28
Segmentation in the phylum Annelida. Although the digestive system **(a)**, the longitudinal blood vessels **(b)**, and the ventral nerve cord **(c)** are not segmented, the coelom **(a)**, blood vessels **(b)**, nerves **(c)**, and excretory organs **(d)** appear as repeating structures in most segments. The body musculature **(e)** includes both circular and longitudinal layers that allow these animals to use the coelomic chambers as a hydrostatic skeleton.

lom as a hydrostatic skeleton. All annelids except leeches also have chitin-reinforced bristles, called **setae** (sometimes written *chaetae*), which protrude outward from the body wall. Setae anchor the worm against the substrate, providing traction.

Annelids have a complete digestive system and a closed circulatory system. However, they lack a discrete respiratory system; oxygen and carbon dioxide diffuse through the skin. The excretory system is composed of paired **metanephridia**, which usually occur in all body segments posterior to the head. The nervous system is highly developed, with ganglia (local control centers) in every segment, a simple brain in the head, and sensory organs that detect chemicals, moisture, light, and touch.

Most freshwater and terrestrial annelids are hermaphroditic, and worms exchange sperm when they mate. Newly hatched worms have an adult morphology. Some terrestrial annelids also reproduce asexually by fragmenting and regenerating missing parts. Marine annelids usually have separate sexes, and release gametes into the sea for fertilization. The zygotes develop into trochophore larvae that add segments, gradually assuming an adult form.

Annelida includes three lineages, each of which is largely restricted to one environment.

Polychaeta. The 10,000 species of bristle worms (Polychaeta: *chaite* = bristles) are primarily marine **(Figure 29.29)**. Many live under rocks or in tubes constructed from mucus, calcium carbonate secretions, grains of sand, and small shell fragments. Their setae project from well-developed **parapodia** (singular, *parapodium* = closely resembling a foot), fleshy lateral extensions of the body wall used for locomotion and gas exchange. Sense organs are concentrated on a well-developed head.

Crawling or swimming polychaetes are often predatory; they use sharp jaws in a protrusible muscular pharynx to grab small invertebrate prey. Other species graze on algae or scavenge organic matter. A few tube dwellers draw food-laden water into the tube by beating their parapodia; most others collect food by extending feathery, ciliated, mucus-coated tentacles.

Oligochaeta. Most of the 3500 species of oligochaete worms (Oligochaeta: *oligos* = few) are terrestrial **(Figure 29.30)**, but they are restricted to moist habitats because

a. Fan worm

b. Polychaete feeding structures

- Jaws
- Teeth
- Everted pharynx
- Antenna
- Palp (food-handling appendage)
- Tentacle
- Eyes
- Chemical-sensing organs
- Parapodium

c. Polychaete setae

- Setae

Figure 29.29

Polychaeta. **(a)** The tube-dwelling fan worm *(Sabella melanostigma)* has mucus-covered tentacles that trap small food particles. **(b)** Some polychaetes, like *Nereis*, actively seek food; when they encounter a suitable tidbit, they evert their pharynx, exposing sharp jaws that grab the prey and pull it into the digestive system. **(c)** Many marine polychaetes (such as *Proceraea cornuta*, shown here) have numerous setae, which they use for locomotion.

they quickly dehydrate in dry air or soil. They range in length from a few millimeters to more than 3 m. Terrestrial oligochaetes, the earthworms, are nocturnal, spending their days in burrows that they excavate. Aquatic species live in mud or detritus at the bottom of lakes, rivers, and estuaries.

Earthworms are scavengers on decomposing organic matter. In his book *The Formation of Vegetable Mould through the Action of Worms,* Darwin noted that earthworms can ingest their own weight in soil every day. He calculated that a typical population of 16,000 worms per hectare (10,000 sq m) consumes more than 20 tons of soil in a year. This impressive activity aerates soil and makes nutrients available to plants by mixing

Figure 29.30

Oligochaeta. Earthworms *(Lumbricus terrestris)* generally move across the ground surface at night.

the subsoil with the topsoil. Earthworms have complex organ systems (see Figure 29.28), and they sense light and touch at both ends of the body. In addition, they have moisture receptors, an important adaptation in organisms that must stay wet to allow gas exchange across the skin.

Hirudinea. The 500 species of leeches (Hirudinea: *hirudo* = leech) are mostly freshwater parasites. They have dorsoventrally flattened, tapered bodies with a sucker at each end. Although the body wall is segmented, the coelom is reduced and not partitioned. Many leeches are ectoparasites of vertebrates, but some attack small invertebrate prey.

Parasitic leeches feed on the blood of their hosts. Most attach to the host with the posterior sucker, and then use their sharp jaws to make a small, often painless, triangular incision. A sucking apparatus draws blood from the prey, while a special secretion, *hirudin,* maintains the flow by preventing the host's blood from coagulating. Leeches have a highly branched gut that allows them to consume huge blood meals **(Figure 29.31).** For centuries, doctors used medicinal leeches *(Hirudo medicinalis)* to "bleed" patients; today, surgeons use them to drain excess fluid from tissues after recon-

Leech before feeding   Leech after feeding

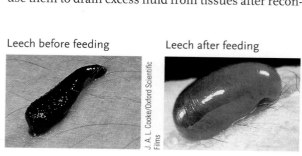

Figure 29.31

Hirudinea. Parasitic leeches consume huge blood meals, as shown by these before and after photos of a medicinal leech *(Hirudo medicinalis)*. Because suitable hosts are often hard to locate, gorging allows a leech to take advantage of any host it finds.

structive surgery, reducing swelling until the patient's blood vessels regenerate and resume this function.

STUDY BREAK

1. What organs systems are present in free-living flatworms (Turbellaria)? Which of these organ systems is absent in tapeworms (Cestoda)?
2. What characteristic reveals the close evolutionary relationship of ectoprocts, brachiopods, and phoronid worms?
3. What anatomical structures and physiological systems allow squids and other cephalopods to be much more active than other types of mollusks?
4. Which organs systems exhibit segmentation in most annelid worms?

29.7 Ecdysozoan Protostomes

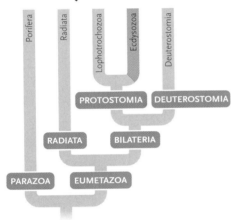

The three phyla in the protostome group Ecdysozoa all have an external covering that they shed periodically. The outer covering protects these animals from harsh environmental conditions, and it helps parasitic species resist host defenses. Although many of these animals live in either aquatic or moist terrestrial habitats, a tough exoskeleton allows one group, the insects, to thrive on dry land and in the air.

Nematodes Are Unsegmented Worms Covered by a Flexible Cuticle

Roundworms (phylum Nematoda: *nema* = thread) are perhaps the most abundant animals on Earth. A cupful of rich soil, a dead earthworm, or a rotting fruit may contain thousands of them. Although 80,000 species have been described, experts estimate that more than half a million exist. Many nematodes are almost microscopically small, but some species are a meter or more long. They occupy nearly every freshwater, ma-

rine, and terrestrial habitat on Earth, consuming detritus, microorganisms, plants, or animals.

The roundworm body is cylindrical and usually tapered at both ends **(Figure 29.32)**. None of the cells have cilia or flagella. Roundworms are covered in a tough but flexible, water-resistant **cuticle**, which is replaced by the underlying epidermis as the worm grows. The cuticle prevents the animal from dehydrating in dry environments, and, in parasitic species, it resists attack by acids and enzymes in a host's digestive system. Beneath the cuticle and epidermis, a layer of longitudinal muscles extends the length of the body. Nematodes use alternating muscle contractions to push against the substrate and propel themselves forward, usually with a thrashing motion.

The adults of one soil-dwelling species, *Caenorhabditis elegans,* are transparent and contain fewer than 1000 cells. As *Focus on Model Research Organisms* explains, biologists have studied this worm inside and out.

Nematodes reproduce sexually, and the sexes are separate in most species. In some, internal fertilization produces many thousands of fertile eggs per day. The eggs of many species can remain dormant if environmental conditions are unsuitable.

Because of their great numbers, nematodes have enormous ecological, agricultural, and medical significance. Free-living species are responsible for decomposition and nutrient recycling in many habitats. Parasitic nematodes attack the roots of plants, causing tremendous crop damage. Roundworms also parasitize animals, including humans. Although some, like pinworms *(Enterobius),* are more of a nuisance than a danger, others, like trichinas or filarial worms, can cause serious disease, disfigurement, or even death (see *Focus on Applied Research* on pages 644–645). More than 1 billion people worldwide suffer from debilitating and life-threatening nematode infections.

Figure 29.32

Phylum Nematoda. Some roundworms, like these *Anguillicola crassus* in the swim bladder of an eel, are parasites of plants or animals. Others are important consumers of dead organisms in most ecosystems.

Model Research Organisms: *Caenorhabditis elegans*

Researchers studying the tiny, free-living nematode *Caenorhabditis elegans* have made many recent advances in molecular genetics, animal development, and neurobiology. It is so popular as a model research organisms that most workers simply refer to it as "the worm."

Several attributes make *C. elegans* a model research organism. It has an adult size of about 1 mm and thrives on cultures of *E. coli* or other bacteria; thus, thousands can be raised in a culture dish. It is hermaphroditic and often self-fertilizing, which allows researchers to maintain pure genetic strains. It completes its life cycle from egg to reproductive adult within 3 days at room temperature. Furthermore, stock cultures can be kept alive indefinitely by freezing them in liquid nitrogen or in an ultra-cold freezer set to −80°C. Researchers can therefore store new mutants for later research without having to clean, feed, and maintain active cultures.

Best of all, the worm is anatomically simple (see **figure**); an adult contains just 959 cells (excluding the gonads). Having a fixed cell number is relatively uncommon among animals, and developmental biologists have made good use of this trait. The eggs, juveniles, and adults of the worm are completely transparent, and researchers can observe cell divisions and cell movements in living animals with straightforward microscopy techniques. There is no need to kill, fix, and stain specimens for study. And virtually every cell in the worm's body is accessible for manipulation by laser microsurgery, microinjection, and similar approaches.

The genome of *C. elegans*, which was sequenced in 1998, is also simple, consisting of 100 million base pairs organized into roughly 17,000 genes on six pairs of chromosomes. The genome, which is about the same size as one human chromosome, specifies the amino acid sequences of about 10,000 protein molecules, far fewer than are found in more complex animals.

The worm entered the biological limelight in 1965 when Sidney Brenner of the Medical Research Council's Laboratory of Molecular Biology in Cambridge, England, identified it as an ideal organism for research on the genetic control of development. By 1983 numerous researchers had collectively identified the cell lineage for every one of its 959 cells. We now know the exact patterns of cell divisions, cell migrations, and programmed cell death that generate an adult from a fertilized egg.

Research on *C. elegans* has generated interesting results, some of which contradict old assumptions about animal development. For example, researchers once believed that a particular tissue always arises from the same embryonic germ layer, but in *C. elegans* some muscle cells do not develop from mesoderm and some nervous system cells do not arise from ectoderm. Moreover, the bilaterally symmetrical body of the adult worm does not arise from symmetrical events; matching cells on the left and right sides of the worm sometimes arise from different developmental pathways. Finally, researchers were surprised by the important role of cell death in this organism. Developmental events in the embryo produce 671 cells, but 113 of them are programmed to die before the larva hatches from the egg.

Studies of the worm's nervous system have been equally fruitful. Researchers have mapped the development of all 302 neurons and their 7000 connections. Now they are working to identify the molecules that function in sensory recognition and cell-to-cell signaling. Molecular analyses reveal that some of the proteins that carry messages between nerve cells in the worm are similar to those found in vertebrates.

The knowledge gained from research on *C. elegans* is highly relevant to studies of larger and more complex organisms, including vertebrates. Recent research demonstrates some striking similarities among nematodes, fruit flies, and mice in the genetic control of development, in some of the proteins that govern important events like cell death, and in the molecular signals used for cell-to-cell communication. Using a relatively simple model like *C. elegans*, researchers can answer research questions more quickly and more efficiently than they could if they studied larger and more complex animals.

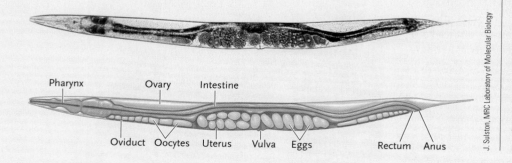

Pharynx   Ovary   Intestine   Oviduct   Oocytes   Uterus   Vulva   Eggs   Rectum   Anus

J. Sulston, MRC Laboratory of Molecular Biology

Velvet Worms Have Segmented Bodies and Numerous Unjointed Legs

The 65 living species of velvet worms (phylum Onychophora: *onyx* = claw) live under stones, logs, and forest litter in the tropics and in moist temperate habitats in the southern hemisphere. They range in size from 15 mm to 15 cm and feed on small invertebrates.

Onychophorans have a flexible cuticle, segmented bodies, and numerous pairs of unjointed legs (**Figure 29.33**). Like the annelids, they have pairs of excretory organs in most segments. But unlike annelids, they

Figure 29.33
Phylum Onychophora. Members of the small phylum Ony-chophora, such as species in the genus *Dnycophor*, have seg-mented bodies and unjointed appendages.

have an open circulatory system, a specialized respira-tory system, relatively large brains, jaws, and tiny claws on their feet. Many produce live young, which, in some species, are nourished within a uterus. Fossil evidence indicates that the onychophoran body plan has not changed much over the last 500 million years.

Arthropods Are Segmented Animals with a Hard Exoskeleton and Jointed Appendages

The more than 1 million known species of arthropods (phylum Arthropoda: *arthron* = joint) include more than half the animal species on Earth—and only a frac-tion of the living arthropods have been described. This huge lineage includes insects, spiders, crustaceans, millipedes, centipedes, the extinct trilobites, and their relatives.

Arthropods have a segmented body encased in a rigid **exoskeleton**. This external covering is made of chitin, a mix of polysaccharide fibers glued together with glycoproteins, as well as waxes and lipids that block the passage of water. In some marine groups, such as crabs and lobsters, it is hardened with cal-cium carbonate. The exoskeleton probably first evolved in marine species, providing protection against predators. In terrestrial habitats, it provides support against gravity and protection from dehydra-tion, contributing to the success of insects in even the driest places on Earth. The exoskeleton is especially thin and flexible at the joints between body segments and in the appendages. Contractions of muscles at-tached to the inside of the exoskeleton move individ-ual body parts like levers, allowing highly coordinated movements and patterns of locomotion that are more precise than those in soft-bodied animals with hydro-static skeletons.

Although the exoskeleton has obvious advantages, it is nonexpandable and therefore could limit growth of the animal. But, like other Ecdysozoa, arthropods grow and periodically develop a soft, new exoskeleton beneath the old one, which they shed in the complex process of ecdysis **(Figure 29.34).** After shedding the old exoskeleton, aquatic species swell with water and ter-restrial species swell with air before the new one hard-ens. They are especially vulnerable to predators at these times. "Soft-shelled" crabs, prized as food in many countries, are ones that have recently molted.

As arthropods evolved, body segments became fused in various ways, reducing their overall number. Each region, along with its highly modified paired ap-pendages, is specialized, but the structure and function vary greatly among groups. In insects (see Figure 29.43), which have three body regions, the **head** in-cludes a brain, sensory structures, and some sort of feeding apparatus. The segments of the **thorax** bear walking legs and, in some insects, wings. The **abdomen** includes much of the digestive system and sometimes part of the reproductive system.

The coelom of arthropods is greatly reduced, but another cavity, the *hemocoel,* is filled with bloodlike he-molymph. The heart pumps the hemolymph through an open circulatory system, bathing tissues directly.

Arthropods are active animals and require sub-stantial quantities of oxygen. Different groups have distinctive mechanisms for gas exchange, because oxy-gen cannot cross the impermeable exoskeleton. Ma-rine and freshwater species, like crabs and lobsters, rely on diffusion across gills. The terrestrial groups—

Figure 29.34
Ecdysis in insects. Like all other arthropods, this cicada (*Graptopsaltsia nigrofusca*) sheds its old exoskeleton as it grows and when it undergoes metamorphosis into a winged adult.

Unscrambling the Arthropods

This book follows a classification that divides arthropods into five subphyla—trilobites (now extinct), chelicerates, crustaceans, myriapods, and hexapods. In the past, biologists grouped hexapods and myriapods together **(figure, left)** because they share certain morphological characteristics, which may indicate a common ancestry: unbranched appendages, a tracheal system for gas exchange, Malpighian tubules for excretion, and one pair of antennae. Some biologists have argued, however, that these traits may have been produced by convergent evolution. Furthermore, the comparison of other morphological traits suggests that hexapods are more closely related to crustaceans than to myriapods. For example, both hexapods and crustaceans have jawlike mandibles on the fourth head segment, similar compound eyes, comparable development of the nervous system, and similarities in the structure of thoracic appendages.

A molecular study by Jeffry L. Boore and his colleagues at the University of Michigan lends support to the hypothesis that hexapods may indeed be more closely related to crustaceans than to myriapods. For their study, Boore and his coworkers compared the arrangement of genes in the mitochondrial DNA of representative arthropod species. Mitochondrial DNA (mtDNA) was chosen for study because it has been subject to many random rearrangements during the evolutionary history of the arthropods, producing wide variations in the order and placement of the 36 or 37 genes typically present. As long as the rearrangements do not disrupt gene function, they may have little or no effect on fitness; therefore, they may not be affected by natural selection. As Boore and his colleagues noted, the large number of possible rearrangements of

the mtDNA genes makes it unlikely that the same gene order would appear by chance in any two groups; thus, groups with the same arrangement are likely to share a common ancestor in which the rearrangement first appeared.

The investigators sequenced the mtDNA of a chelicerate, two crustaceans, and a myriapod (listed in the figure caption). They also sequenced the mtDNAs of a mollusk and an annelid, animals that are only distantly related to arthropods, to provide outgroup comparisons. They compared these sequences with those in 14 other invertebrates, including four hexapods and one crustacean, already obtained by other workers.

The locations of two genes encoding leucine-carrying tRNAs provide the strongest clue to the evolutionary relationships of these organisms. These genes are positioned next to each other in the mtDNA of the chelicerate, the mollusk, and the annelid. This arrangement appears to reflect the ancient gene order present in the common ancestor of annelids, mollusks, and arthropods. However, in the hexapods and the crustaceans studied, one of the leucine tRNA genes is located at a different position, between two genes

coding for electron transfer proteins. Moreover, this rearrangement does not appear in the myriapod examined. It is extremely unlikely that the same translocation of the leucine tRNA gene occurred independently in hexapods and crustaceans. Instead, this gene rearrangement probably occurred in an ancestor shared by hexapods and crustaceans after this ancestor diverged from the lineage that includes the chelicerates and myriapods.

The shared gene arrangement in insects and crustaceans and the absence of a matching arrangement in myriapods suggests that hexapods and crustaceans are closely related, and that hexapods and myriapods are not, in spite of their morphological similarities **(figure, right).** Other investigators, who compared the sequences of ribosomal RNA genes in a variety of arthropod species, reached the same conclusions independently.

Because the molecular research studied a relatively few characteristics, it cannot yet be accepted as the definitive answer to questions about arthropod lineages. However, the results give strong support to the idea that hexapods and crustaceans are more closely related than the traditional family tree suggested.

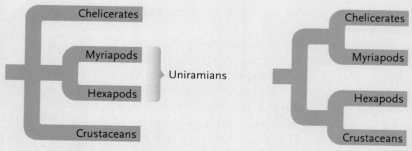

The traditional phylogenetic tree for arthropods (left) differs from that suggested by the research described in this box. The DNA sequences used to construct the new phylogenetic tree (right) were obtained from the chelicerate *Limulus*, the myriapod *Thyrophygus*, the hexapod *Drosophila*, and the crustaceans *Daphnia* and *Homarus*.

insects and spiders—have developed unique and specialized respiratory systems.

High levels of activity also require intricate sensory structures. Many arthropods are equipped with a highly organized central nervous system, touch receptors, chemical sensors, **compound eyes** that include multiple image-forming units, and in some, hearing organs.

Arthropod systematics is an active area of research, and scientists are currently using molecular, morpho-

Figure 29.35
Subphylum Trilobita. Trilobites, like *Olenellus gilberti*, bore many pairs of relatively undifferentiated appendages.

Dr. Chip Clark

mored, with two deep longitudinal grooves that divided the body into the one median and two lateral lobes for which the group is named **(Figure 29.35)**. Their segmented bodies were organized into a head, which included a pair of sensory **antennae** (chemosensory organs) and two compound eyes, and a thorax and an abdomen, both of which bore pairs of walking legs.

The position of trilobites in the fossil record indicates that they were among the earliest arthropods. Thus, biologists are confident that their three body regions and numerous unspecialized appendages—one pair per segment—represent ancestral traits within the phylum. As you will learn as you read about the other four subphyla, the subsequent evolution of the different arthropod groups included dramatic remodeling of the major body regions as well as modifications of the ancestral, unspecialized paired appendages into structures specialized for different functions.

logical, and developmental data to reexamine relationships within this immense group. As *Insights from the Molecular Revolution* explains, hypotheses about the phylogeny and classification of arthropods are in a state of flux. Some researchers even argue for splitting them into four or more phyla. We follow the traditional definition of five *subphyla,* partly because this classification adequately reflects arthropod diversity and partly because no alternative hypothesis has been widely adopted by experts.

Subphylum Trilobita. The trilobites (subphylum Trilobita: *tri* = three; *lobos* = lobe), now extinct, were among the most numerous animals in shallow Paleozoic seas. They disappeared in the Permian mass extinction, but the cause of their demise is unknown. Most trilobites were ovoid, dorsoventrally flattened, and heavily ar-

Subphylum Chelicerata. In spiders, ticks, mites, scorpions, and horseshoe crabs (subphylum Chelicerata: *chela* = claw; *keras* = horn), the first pair of appendages, the **chelicerae,** are fanglike structures used for biting prey. The second pair of appendages, the *pedipalps,* serve as grasping organs, sensory organs, or walking legs. All chelicerates have two major body regions, the **cephalothorax** (a fused head and thorax) and the abdomen. The group originated in shallow Paleozoic seas, but most living species are terrestrial. They vary in size from less than a millimeter to 20 cm; all are predators or parasites.

The 60,000 species of spiders, scorpions, mites, and ticks (Arachnida) represent the vast majority of chelicerates **(Figure 29.36)**. Arachnids have four pairs

a. Wolf spider

b. Spider anatomy

P. J. Bryand, University of California–Irvine/BPS

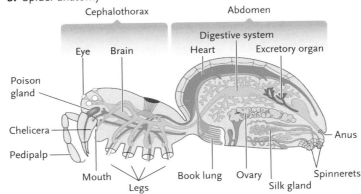

c. Scorpion

P. J. Bryand, Univ. of California–Irvine/BPS

d. House dust mite

© Andrew Syred/Photo Researchers, Inc.

Figure 29.36
Subphylum Chelicerata, subgroup Arachnida. **(a)** The wolf spider (*Lycosa* species) is harmless to humans. **(b)** The arachnid body plan includes a cephalothorax and abdomen. **(c)** Scorpions have a stinger at the tip of the segmented abdomen. Many, like *Centruroides sculpuratus*, protect their eggs and young. **(d)** House dust mites (*Dermatophagoides pteronyssinus*), shown in a scanning electron micrograph, feed on microscopic debris.

of walking legs on the cephalothorax and highly modified chelicerae and pedipalps. In some spiders, males use their pedipalps to transfer packets of sperm to females. Scorpions use them to shred food and to grasp one another during courtship. Many predatory arachnids have excellent vision, provided by simple eyes on the cephalothorax. Scorpions and some spiders also have unique pocketlike respiratory organs called **book lungs,** derived from abdominal appendages.

Spiders, like most other arachnids, subsist on a liquid diet. They use their chelicerae to inject paralyzing poisons and digestive enzymes into prey and then suck up the partly digested tissues. Many spiders are economically important predators, helping to control insect pests. Only a few are a threat to humans. The toxin of a black widow *(Latrodectus mactans)* causes paralysis, and the toxin of the brown recluse *(Loxosceles reclusa)* destroys tissues around the site of the bite.

Although many spiders hunt actively, others capture prey on silken threads secreted by **spinnerets,** which are modified abdominal appendages. Some species weave the threads into complex, netlike webs. The silk is secreted as a liquid, but quickly hardens on contact with air. Spiders also use silk to make nests, to protect their egg masses, as a safety line when moving through the environment, and to wrap prey for later consumption.

Most mites are tiny, but they have a big impact. Some are serious agricultural pests that feed on the sap of plants. Others cause mange (patchy hair loss) or painful and itchy welts on animals. House dust mites cause allergic reactions in many people. Ticks, which are generally larger than mites, are blood-feeding ectoparasites that often transmit pathogens, such as those causing Rocky Mountain spotted fever and Lyme disease.

The subphylum Chelicerata also includes five species of horseshoe crabs (Merostomata), an ancient lineage that has not changed much during its 350 million year history **(Figure 29.37).** Horseshoe crabs are carnivorous bottom feeders in shallow coastal waters. Beneath their characteristic shell, they have one pair

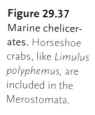

Figure 29.37
Marine chelicerates. Horseshoe crabs, like *Limulus polyphemus,* are included in the Merostomata.

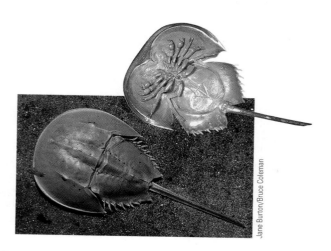

Jane Burton/Bruce Coleman

of chelicerae, a pair of pedipalps, four pairs of walking legs, and a set of paperlike gills, derived from ancestral walking legs.

Subphylum Crustacea. The 35,000 species of shrimps, lobsters, crabs, and their relatives (subphylum Crustacea, meaning "encrusted") represent a lineage that emerged more than 500 million years ago. They are abundant in marine and freshwater habitats. A few species, such as sowbugs and pillbugs, live in moist, sheltered terrestrial environments. In many crustaceans two, or even all three, of the arthropod body regions—head, thorax, and abdomen—are fused; a fused cephalothorax and a separate abdomen is an especially common pattern. The edible "tail" of a lobster or crayfish is actually a highly muscularized abdomen. In some, the exoskeleton includes a **carapace,** a protective covering that extends backward from the head. Crustaceans vary in size from water fleas less than 1 mm long to lobsters that can grow to 60 cm in length and weigh as much as 20 kg.

Crustaceans generally have five characteristic pairs of appendages on the head **(Figure 29.38).** Most have two pairs of sensory antennae and three pairs of mouthparts. The latter include one pair of *mandibles,* which move laterally to bite and chew, and two pairs of *maxillae,* which hold and manipulate food. Numerous paired appendages posterior to the mouthparts vary among groups.

Most crustaceans are active animals that exhibit complex patterns of movement. Their activities are coordinated by elaborate sensory and nervous systems, including chemical and touch receptors in the antennae, compound eyes, statocysts on the head, and sensory hairs embedded in the exoskeleton throughout the body. The nervous system is similar to that in annelids, but the ganglia are larger and more complex, allowing a finer level of motor control. High levels of activity require substantial oxygen, and larger species have complex, feathery gills tucked beneath the carapace. Activity also produces abundant metabolic wastes that are excreted by diffusion across the gills or, in larger species, by glands located in the head.

The sexes are typically separate, and courtship rituals are often complex. Eggs are usually brooded on the surface of the female's body or beneath the carapace. Many have free-swimming larvae that, after undergoing a series of molts, gradually assume an adult form.

The subphylum includes so many different body plans that it is usually divided into six major groups with numerous subgroups. The crabs, lobsters, and shrimps (Decapoda, meaning "10 feet," a subgroup of the Malacostraca) number more than 10,000 species. The vast majority of decapods are marine, but a few shrimps, crabs, and crayfishes occupy freshwater habitats. Some crabs also live in moist terrestrial habitats, where they scavenge dead vegetation, clearing the forest floor of debris.

a. Crab

b. Lobster

c. Lobster anatomy

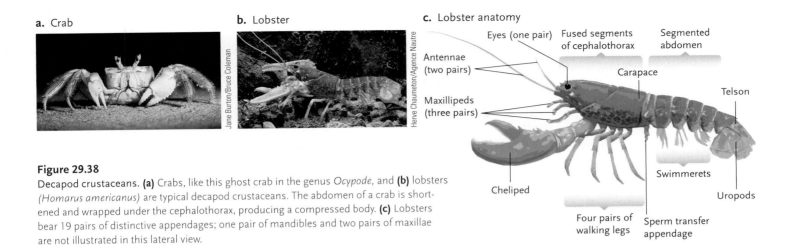

Figure 29.38
Decapod crustaceans. **(a)** Crabs, like this ghost crab in the genus *Ocypode*, and **(b)** lobsters (*Homarus americanus*) are typical decapod crustaceans. The abdomen of a crab is shortened and wrapped under the cephalothorax, producing a compressed body. **(c)** Lobsters bear 19 pairs of distinctive appendages; one pair of mandibles and two pairs of maxillae are not illustrated in this lateral view.

All decapods exhibit extreme specialization of their appendages. In the American lobster, for example, each of the 19 pairs of appendages is different (see Figure 29.38c). Behind the antennae, mandibles, and maxillae, the thoracic segments have three pairs of *maxillipeds*, which shred food and pass it up to the mouth, a pair of large *chelipeds* (pinching claws), and four pairs of walking legs. The abdominal appendages include a pair specialized for sperm transfer (in males only), *swimmerets* for locomotion and for brooding eggs, and *uropods,* a pair of appendages that, combined with the *telson,* the tip of the abdomen, form a fan-shaped tail. If any appendage is damaged, the animal can autotomize (drop) it and begin growing a new one before its next molt.

Representatives of several crustacean groups—fairy shrimps, amphipods, water fleas, krill, ostracods, and copepods **(Figure 29.39)**—live as plankton in the upper waters of oceans and lakes. Most are only a few millimeters long, but are present in huge numbers. They feed on microscopic algae or detritus and are themselves food for larger invertebrates, fishes, and some suspension-feeding marine mammals like the baleen whales. Planktonic crustaceans are among the most abundant animals on Earth.

Adult barnacles (Cirripedia, meaning "hairy footed," a subgroup of the Maxillopoda) are sessile, marine crustaceans that live within a strong, calcified cup-shaped shell **(Figure 29.40)**. Their free-swimming larvae attach permanently to substrates—rocks, wooden pilings, the hulls of ships, the shells of mollusks, even the skin of whales—and secrete the shell, which is actually a modified exoskeleton. To feed, barnacles open the shell and extend six pairs of feathery legs. The beating legs capture microscopic plankton and transfer it the mouth. Unlike most crustaceans, barnacles are hermaphroditic.

Subphylum Myriapoda. The 3000 species of centipedes (Chilopoda) and 10,000 species of millipedes (Diplopoda) are classified together in the subphylum Myriap-

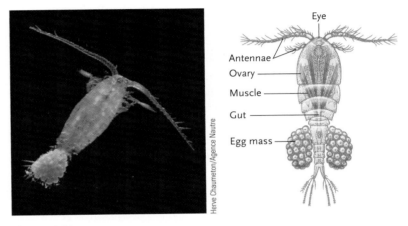

Figure 29.39
Copepods. Tiny crustaceans, like these copepods (*Calanus* species on the left, *Cyclops* species on the right), occur by the billions in freshwater and marine plankton.

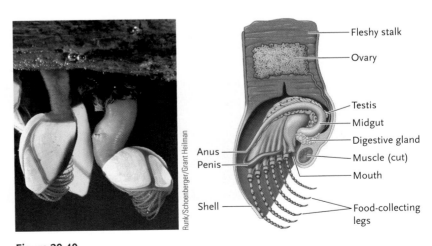

Figure 29.40
Barnacles. Gooseneck barnacles (*Lepas anatifera*) attach to the underside of floating debris. Like other barnacles, they open their shells and extend their feathery legs to collect particulate food from seawater.

a.

Steve Martin/Tom Stack & Associates

b.

Z. Leszczynski/Animals, Animals

Figure 29.41

Millipedes and centipedes. **(a)** Millipedes, like *Spirobolus* species, feed on living and decaying vegetation. They have two pairs of walking legs on most segments. **(b)** Like all centipedes, this Southeast Asian species *(Scolopendra subspinipes)*, shown feeding on a small frog, is a voracious predator. Centipedes have one pair of walking legs per segment.

Figure 29.42

Insect diversity. Insects are grouped into about 30 orders, 8 of which are illustrated here.

oda (meaning "countless feet"). Myriapods have two body regions, a head and a segmented trunk **(Figure 29.41)**. The head bears one pair of antennae, and the trunk bears one (centipedes) or two (millipedes) pairs of walking legs on most of its many segments. Myriapods are terrestrial, and many species live under rocks or dead leaves. Centipedes are fast and voracious predators, using powerful toxins to kill their prey; they generally feed on invertebrates, but some eat small vertebrates. The bite of some species is harmful to humans. Although most species are less than 10 cm long, some grow to 25 cm. The millipedes are slow but powerful

© Arthur Evans/Animals, Animals—Earth Scenes

a. Silverfish (Thysanura, *Ctenolepisma longicaudata*) are primitive wingless insects.

Edward S. Ross

b. Dragonflies, like the flame skimmer (Odonata, *Libellula saturata*), have aquatic larvae that are active predators; adults capture other insects in mid-air.

© BIOS Borrell Bartomeu/Peter Arnold, Inc.

c. Male praying mantids (Mantodea, *Mantis religiosa*) are often eaten by the larger females during or immediately after mating.

Edward S. Ross

d. This rhinoceros beetle (Coleoptera, *Dynastes granti*) is one of more than 250,000 beetle species that have been described.

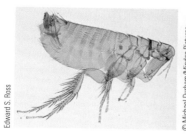

Edward S. Ross

e. Fleas (Siphonoptera, *Hystrichopsylla dippiei*) have strong legs with an elastic ligament that allows these parasites to jump on and off their animal hosts.

© Michael Durham/Minden Pictures

f. Crane flies (Diptera, *Tipula* species) look like giant mosquitoes, but their mouthparts are not useful for biting other animals; the adults of most species live only a few days and do not feed at all.

C. P. Hickman, Jr.

g. The luna moth (Lepidoptera, *Actias luna*), like other butterflies and moths, has wings that are covered with colorful microscopic scales.

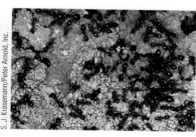

S. J. Krasemann/Peter Arnold, Inc.

h. Like many other ant species, fire ants (Hymenoptera, *Solenopsis invicta*) live in large cooperative colonies. Fire ants—named for their painful sting—were introduced into southeastern North America, where they are now serious pests.

herbivores or scavengers. The largest species attain a length of nearly 30 cm. Although they lack a poisonous bite, they curl into a ball and exude noxious liquids when disturbed.

Subphylum Hexapoda. In terms of sheer numbers, diversity, and the range of habitats they occupy, the 1,000,000 or more species of insects and their closest relatives (subphylum Hexapoda, meaning "six feet") are the most successful animals on Earth. They were among the first animals to colonize terrestrial habitats, where most species still live. The oldest insect fossils date from the Devonian, 380 million years ago. Insects have one pair of antennae on the head, a pair of mandibles for feeding, and unbranched appendages. Insects are generally small, ranging from 0.1 mm to 30 cm in length. The group is divided into about 30 subgroups **(Figure 29.42)**. The insect body plan always includes a head, thorax, and abdomen **(Figure 29.43)**. The head is equipped with multiple mouthparts, a pair of compound eyes, and one pair of sensory antennae. The thorax has three pairs of walking legs and often one or two pairs of wings. Insects are the only invertebrates capable of flight. Their wings, which are made of lightweight but durable sheets of chitin and sclerotin, arise embryonically from the body wall; unlike the wings of birds and bats, insect wings are not derived from ancestral appendages.

Studies in evolutionary developmental biology have begun to unravel the genetic changes that fostered certain aspects of the insect body plan. For example, the *Distal-less* gene (*Dll* for short) is a highly conserved tool-kit gene (see Section 22.6) that triggers the development of appendages in all sorts of animals—the legs of chickens, the fins of fishes, the parapodia of polychaete worms, and the diverse appendages of arthropods. All arthropods also have a gene called *Ultrabithorax* (*Ubx* for short). It is one of the *Hox* genes that control the overall body plans of animals. In insects, the *Ubx* gene contains a unique mutation, not found in other arthropods, that causes the protein for which it codes to repress *Dll*, thereby preventing the formation of appendages wherever *Ubx* is expressed. And because insects express *Ubx* in their abdomen, they do not grow abdominal appendages. All other arthropods, which have the ancestral, nonrepressing form of the *Ubx* gene, have appendages in the posterior region of their body. Thus, one mutation in a *Hox*-family gene has fostered the evolution of a highly distinctive morphological trait in insects—having legs on the thorax, but not on the abdomen.

Insects exchange gases through a specialized **tracheal system**, a branching network of tubes that carry oxygen from small openings in the exoskeleton to tissues throughout the body. Insects excrete nitrogenous wastes through specialized **Malpighian tubules** that transport wastes to the digestive system for disposal with feces. Both of these organ systems are unique among animals. Insect sensory systems are diverse and complex. Besides image-forming compound eyes, many insects have light-sensing ocelli on their heads. Many also have hairs, sensitive to touch, on their antennae, legs, and other regions of the body. Chemical receptors are particularly common on the legs, allowing the identification of food. And many groups of insects have hearing organs to detect predators and potential mates. The familiar chirping of crickets, for example, is a mating call emitted by males that may repel other males and attract females.

As a group, insects feed in every conceivable way and on most other organisms **(Figure 29.44)**. Species that eat plants, such as grasshoppers, have a pair of rigid mandibles, which chew food before it is ingested. Behind the mandibles is a pair of maxillae, which may also aid in food acquisition. Insects also have inflexible upper and lower lips, the *labrum* and *labium,* respectively. A tonguelike structure just dorsal to the labium in chewing insects houses the openings of salivary glands. But evolution has modified this ancestral mandibulate pattern in numerous ways. In some biting

External anatomy of a grasshopper

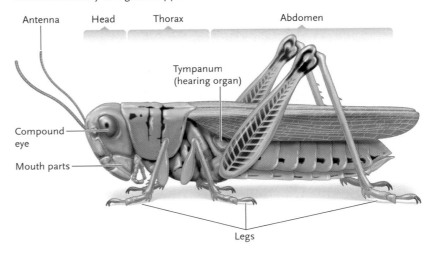

Internal anatomy of a female grasshopper

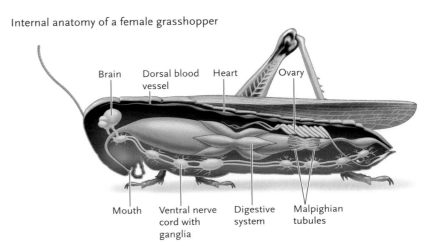

Figure 29.43
The insect body plan. Insects have distinct head, thorax, and abdomen. Of all the internal organ systems, only the dorsal blood vessel, ventral nerve cord, and some muscles are strongly segmented.

Figure 29.44

Specialized insect mouth-parts. The **(a)** ancestral chewing mouthparts have been modified by evolution, allowing different insects to **(b)** sponge up food, **(c)** drink nectar, and **(d)** pierce skin to drink blood.

a. Grasshopper

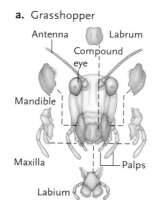

Antenna · Labrum · Compound eye · Mandible · Maxilla · Palps · Labium

b. Housefly

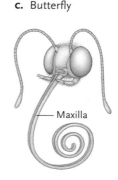

Labium

c. Butterfly

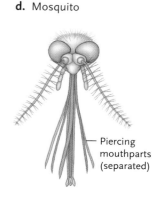

Maxilla

d. Mosquito

Piercing mouthparts (separated)

flies, like mosquitoes and blackflies, the mouthparts have evolved into piercing structures. In butterflies and moths the mouthparts include a long proboscis to drink nectar. And in houseflies, the mouthparts are adapted for sopping up liquid food.

After it hatches from an egg, an insect passes through a series of developmental stages called *instars*. Several hormones control development and ecdysis, which marks the passage from one instar to the next. Insects exhibit one of three basic patterns of postem-

Unanswered Questions

What are the evolutionary relationships among the invertebrate lineages?

If we step back from our vertebrate and terrestrial biases, invertebrates are the dominant form of life on Earth. Many students of natural history are overwhelmed by the sheer numbers of organisms that one encounters, particularly in the aquatic environment, and the challenge of categorizing them taxonomically. Sorting them out in terms of their ecological roles is equally daunting. The problem we face is that invertebrates do it all—they are predators, herbivores, parasites, detritivores, and the primary symbiotic organisms on Earth. This adaptability of form and function is a fascinating hallmark of the animal way of life, but it poses a legion of questions, many still unanswered. Recent advances in genetic technology have advanced our understanding, but there is much left to do.

In the past two decades, our ability to compare genetic information from various invertebrate groups has led to a remarkable reshuffling of the long-established categories used to classify these organisms. The first categories to be eliminated were groups based on superficial phenotypic resemblances, such as the "pseudocoelomates," which had plagued student understanding of diversity. Today, we have a much deeper knowledge of the evolutionary relationships of these organisms. Perhaps the most exciting discovery is that much of the diversity we see is not the product of slow changes in protein-coding gene sequences, but rather the result of variations in the timing and location of the expression of genes that affect development. Evo-devo, the melding of evolutionary and developmental biology—mostly made possible by intensive studies of two model invertebrates, *Caenorhabditis elegans* and *Drosophila melanogaster*—has revealed that changes in the expression of relatively simple sets of genes have brought about the myriad forms of life we see among the invertebrates. As systematists incorporate these new discoveries in their analyses, our understanding of the evolutionary relationships among the invertebrates will surely change.

What is the genetic basis of the diversity of form and function observed among invertebrates?

Because of advances in genetics research, we are on the cusp of being able to answer some fundamental questions. How does an animal's body develop either radial or bilateral symmetry? How does an organism develop a head with a concentration of nervous system tissue, and how do all the exquisite sensory systems associated with a big brain develop? From where do the respiratory pigments, which increase the capacity of the hemolymph or blood to carry oxygen, increasing an organism's capacity for activity, arise? How does the immune system develop, and what genetic change fosters the quantum leap from a nonspecific defense system to one that responds specifically to foreign invaders? More practically, are there genetic switches that we can manipulate? Can we make blood-feeding invertebrates, such as mosquitoes, or parasitic species, like tapeworms or filarial worms, innocuous? Is it possible to use genetic engineering to reduce the ability of the mosquito *Anopheles gambiae* (the "deadliest organism on Earth") to transmit malaria? Will it be possible to forestall or reverse the global decline of coral reefs, one of the richest habitats on Earth?

The answers to these and many more basic and applied questions lie within a deep knowledge and understanding of the invertebrates and the roles they play on our planet. If one looks at invertebrates as dynamic systems, as rich sources of clues to life on Earth, the questions they pose easily provide a lifetime of investigation and reward.

William S. Irby is an associate professor in the Department of Biology at Georgia Southern University in Statesboro. His research focuses on the ecology and evolution of blood-feeding behavior in mosquitoes. To learn more about his work, go to http://www.bio.georgiasouthern.edu/amain/fac-list.html.

a. No metamorphosis

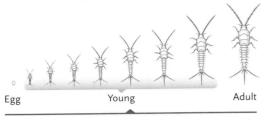

Egg   Young   Adult

Some wingless insects, like silverfish (order Thysanura), do not undergo a dramatic change in form as they grow.

b. Incomplete metamorphosis

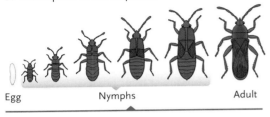

Egg   Nymphs   Adult

Other insects, such as true bugs (order Hemiptera), have incomplete metamorphosis; they develop from nymphs into adults with relatively minor changes in form.

c. Complete metamorphosis

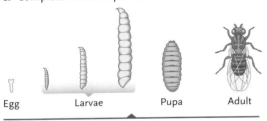

Egg   Larvae   Pupa   Adult

Fruit flies (order Diptera) and many other insects have complete metamorphosis; they undergo a total reorganization of their internal and external anatomy when they pass through the pupal stage of the life cycle.

Figure 29.45
Patterns of postembryonic development in insects.

bryonic development **(Figure 29.45)**. Primitive, wingless species simply grow and shed their exoskeleton without undergoing major changes in morphology. Other species undergo **incomplete metamorphosis**; they hatch from the egg as a *nymph,* which lacks functional wings. In many species, such as grasshoppers (order Orthoptera), the nymphs resemble the adults. In other insects, such as dragonflies (order Odonata), the aquatic nymphs are morphologically very different from the adults.

Most insects undergo **complete metamorphosis:** the larva that hatches from the egg differs greatly from the adult. Larvae and adults often occupy different habitats and consume different food. The larvae (cat-

erpillars, grubs, or maggots) are often worm-shaped, with chewing mouthparts. They grow and molt several times, retaining their larval morphology. Before they transform into sexually mature adults, they spend a period of time as a sessile **pupa**. During this stage, the larval tissues are drastically reorganized. The adult that emerges is so different from the larva that it is often hard to believe that they are of the same species. Moths, butterflies, beetles, and flies are examples of insects with complete metamorphosis. Their larval stages specialize in feeding and growth, whereas the adults are adapted for dispersal and reproduction. In some species, the adults never feed, relying on the energy stores accumulated during the larval stage.

The 240-million-year-history of insects has been characterized by innovations in morphology, life cycle patterns, locomotion, feeding, and habitat use. Their well-developed nervous systems govern exceptionally complex patterns of behavior, including parental care, a habit that reaches its zenith in the colonial social insects, the ants, bees, and wasps (see Chapter 55). The factors that contribute to the insects' success also make them our most aggressive competitors. They destroy vegetable crops, stored food, wool, paper, and timber. They feed on blood from humans and domesticated animals, sometimes transmitting disease-causing pathogens as they do so. Nevertheless, insects are essential members of terrestrial ecological communities. Many species pollinate flowering plants, including important crops. Many others attack or parasitize species that are harmful to human activities. And most insects are a primary source of food for other animals. Some even make useful products like honey, beeswax, and silk.

In the next chapter we consider the lineage of deuterostomes, which includes the vertebrates and their closest invertebrate relatives.

STUDY BREAK

1. What part of a parasitic nematode's anatomy protects it from the digestive enzymes of its host?
2. If an arthropod's rigid exoskeleton cannot be expanded, how does the animal grow?
3. How do the number of body regions and the appendages on the head differ among the four subphyla of living arthropods?
4. How do the life stages differ between insects that have incomplete metamorphosis and those that have complete metamorphosis?

Review

Go to **ThomsonNOW** at www.thomsonedu.com/login to access quizzing, animations, exercises, articles, and personalized homework help.

29.1 What Is an Animal?

- Animals are eukaryotic, multicellular heterotrophs that are motile at some time in their lives.
- Animals probably arose from a colonial flagellated ancestor during the Precambrian era (Figure 29.2).

29.2 Key Innovations in Animal Evolution

- All animals except sponges have tissues, organized into either two or three tissue layers.
- Although some animals exhibit radial symmetry, most exhibit bilateral symmetry (Figure 29.3).
- Acoelomate animals have no body cavity. Pseudocoelomate animals have a body cavity between the derivatives of endoderm and mesoderm. Coelomate animals have a body cavity that is entirely lined by derivatives of mesoderm (Figure 29.4).
- Two lineages of animals differ in developmental patterns (Figure 29.5). Most protostomes exhibit spiral, determinate cleavage, and their coelom forms from a split in a solid mass of mesoderm. Most deuterostomes have radial, indeterminate cleavage, and the coelom usually forms within outpocketings of the primitive gut.
- Four animal phyla exhibit segmentation.

  Animation: Types of body symmetry

  Animation: Types of body cavities

  Animation: Developmental differences between protostomes and deuterostomes

29.3 An Overview of Animal Phylogeny and Classification

- Analyses of molecular sequence data have refined our view of animal evolutionary history (Figure 29.6). The molecular phylogeny recognizes some major lineages that had been identified on the basis of morphological and embryological characters. Sponges are grouped in the Parazoa. All other lineages are grouped in the Eumetazoa. Among the Eumetazoa, the Radiata includes animals with two tissue layers and radial symmetry, and the Bilateria includes animals with three tissue layers and bilateral symmetry.
- Bilateria is further subdivided into Protostomia and Deuterostomia. The new phylogeny divides the Protostomia into the Lophotrochozoa and the Ecdysozoa.
- The molecular phylogeny suggests that ancestral protostomes had a coelom and that segmentation arose independently in three lineages.

29.4 Animals without Tissues: Parazoa

- Sponges (phylum Porifera) are asymmetrical animals with limited integration of cells in their bodies (Figure 29.7).
- The body of many sponges is a water-filtering system (Figure 29.8). Flagellated choanocytes draw water into the body and capture particulate food.

  Animation: Body plan of a sponge

29.5 Eumetazoans with Radial Symmetry

- The two major radiate phyla, Cnidaria and Ctenophora, have two well-developed tissue layers with a gelatinous mesoglea between (Figure 29.9). They lack organ systems. All are aquatic.
- Cnidarians capture prey with tentacles and stinging nematocysts (Figures 29.10, 29.12, and 29.13). Their life cycles may include polyps, medusae, or both (Figure 29.11).
- Ctenophores use long tentacles to capture particulate food and use rows of cilia for locomotion (Figure 29.14).

  Animation: Cnidarian body plans

  Animation: Nematocyst action

  Animation: Cnidarian life cycle

29.6 Lophotrochozoan Protostomes

- The taxon Lophotrochozoa includes eight phyla.
- Three small phyla (Ectoprocta, Brachiopoda, and Phoronida) use a lophophore to feed on particulate matter (Figure 29.15).
- Free-living flatworm species (phylum Platyhelminthes) have well-developed digestive, excretory, reproductive, and nervous systems (Figures 29.16 and 29.17). Parasitic species attach to their animal hosts with suckers or hooks (Figures 29.18 and 29.19).
- The rotifers (phylum Rotifera) are tiny and abundant inhabitants of freshwater and marine ecosystems (Figure 29.20). Movements of cilia in the corona control their locomotion and bring food to their mouths.
- The ribbon worms (phylum Nemertea) are elongate and often colorful animals with a proboscis housed in a rhynchocoel (Figure 29.21).
- Mollusks (phylum Mollusca) have fleshy bodies that are often enclosed in a hard shell. The molluskan body plan includes a head–foot, visceral mass, and mantle (Figures 29.22, 29.24–29.27).
- Segmented worms (phylum Annelida) generally exhibit segmentation of the coelom and of the muscular, circulatory, excretory, respiratory, and nervous systems. They use the coelom as a hydrostatic skeleton for locomotion (Figures 29.28–29.31).

  Animation: Planarian organ systems

  Animation: Blood fluke life cycle

  Animation: Tapeworm life cycle

  Animation: Earthworm body plan

  Animation: Molluscan groups

  Animation: Snail body plan

  Animation: Torsion in gastropods

  Animation: Clam body plan

  Animation: Cuttlefish body plan

29.7 Ecdysozoan Protostomes

- The taxon Ecdysozoa includes three phyla that periodically shed their cuticle or exoskeleton.
- Roundworms (phylum Nematoda) feed on decaying organic matter or parasitize plants or animals (Figure 29.32). They move by contracting longitudinal muscles of the body wall.

- The velvet worms (phylum Onychophora) have segmented bodies and unjointed legs (Figure 29.33). Some species bear live young, which develop in a uterus.
- The segmented bodies of the arthropods (phylum Arthropoda) have specialized appendages for feeding, locomotion, or reproduction. Arthropods shed their exoskeleton as they grow or enter a new stage of the life cycle (Figure 29.34). They have an open circulatory system, a complex nervous system, and, in some groups, highly specialized respiratory and excretory systems.
- Arthropods are divided into five subphyla. The extinct trilobites (subphylum Trilobita), with three-lobed bodies and relatively undifferentiated appendages, were abundant in Paleozoic seas (Figure 29.35). Chelicerates have a cephalothorax and abdomen; appendages on the head serve as pincers or fangs and pedipalps

(Figures 29.36 and 29.37). Crustaceans have a carapace that covers the cephalothorax as well as highly modified appendages, including antennae and mandibles (Figures 29.38–29.40). Myriapods have a head and an elongate, segmented trunk (Figure 29.41). Hexapods have three body regions, three pairs of walking legs on the thorax, and three pairs of feeding appendages on the head (Figures 29.42–29.44). Insects exhibit three patterns of postembryonic development (Figure 29.45).

Animation: Roundworm body plan

Animation: Crab life cycle

Animation: Chelicerates

Animation: Insect head parts

Animation: Insect development

Questions

Self-Test Questions

1. Which of the following characteristics is *not* typical of most animals?
   a. heterotrophic
   b. sessile
   c. bilaterally symmetrical
   d. multicellular
   e. motile at some stage of life cycle

2. A body cavity that separates the digestive system from the body wall but is *not* completely lined with mesoderm is called a:
   a. schizocoelom.
   b. mesentery.
   c. peritoneum.
   d. pseudocoelom.
   e. hydrostatic skeleton.

3. Protostomes and deuterostomes typically differ in:
   a. their patterns of body symmetry.
   b. the number of germ layers during development.
   c. their cleavage patterns.
   d. the size of their sperm.
   e. the size of their digestive systems.

4. The nematocysts of cnidarians are used primarily for:
   a. capturing prey.
   b. detecting light and dark.
   c. courtship.
   d. sensing chemicals.
   e. gas exchange.

5. Which organ system is absent in flatworms (phylum Platyhelminthes)?
   a. nervous system
   b. reproductive system
   c. circulatory system
   d. digestive system
   e. excretory system

6. Which part of a mollusk secretes the shell?
   a. visceral mass
   b. radula
   c. trochophore
   d. head–foot
   e. mantle

7. What is the major morphological innovation seen in annelid worms?
   a. a complete digestive system
   b. image-forming eyes
   c. a respiratory system
   d. an open circulatory system
   e. body segmentation

8. Which phylum includes the most abundant animals in soil?
   a. Nematoda
   b. Rotifera
   c. Mollusca
   d. Annelida
   e. Brachiopoda

9. Which body region of an insect bears the walking legs?
   a. head
   b. carapace
   c. abdomen
   d. thorax
   e. trunk

10. Ecdysis refers to a process in which:
    a. bivalves use siphons to pass water across their gills.
    b. arthropods shed their old exoskeletons.
    c. cnidarians build skeletons of calcium carbonate.
    d. rotifers produce unfertilized eggs.
    e. squids escape from predators in a cloud of ink.

Questions for Discussion

1. Many invertebrate species are hermaphroditic. What selective advantages might this characteristic offer? In what kinds of environments might it be most useful?

2. People who eat raw clams and oysters harvested from sewage-polluted waters often develop mild to severe gastrointestinal infections. These mollusks are suspension feeders. Develop a hypothesis about why people who eat them raw may be at risk.

3. On a voyage to the ocean bottom, a biologist discovers a worm that appears to be new to science. What characteristics of this animal should the biologist examine to determine whether or not she has discovered a previously undescribed phylum?

4. The phylogenetic tree and classification based on molecular sequence data suggests that segmentation evolved independently in Lophotrochozoa (phylum Annelida), Ecdysozoa (phyla Onychophora and Arthropoda), and Deuterostomia (phylum Chordata). What morphological evidence would you try to collect to confirm that segmentation is not homologous in these three groups?

5. What are the relative advantages and disadvantages of radially symmetrical and bilaterally symmetrical body plans?

Experimental Analysis

Design an experiment to test the hypothesis that the cuticle of parasitic nematodes protects them from the acids and enzymes present in the digestive systems of their hosts. Your design must include both experimental and control treatments.

Evolution Link

Many insects have a larval stage that is morphologically different from the adult and that feeds on different foods. What selection pressures may have fostered the evolution of a life cycle with such distinctive life stages? Your answer should address the different biological activities that characterize each life cycle stage.

How Would You Vote?

Cone snails are diverse, but most kinds have a limited geographic range, which makes them highly vulnerable to extinction. We do not know how many are harvested, because no one monitors the trade. Should the United States push to extend regulations on trade in endangered species to cover any species captured from the wild? Go to www.thomsonedu.com/login to investigate both sides of the issue and then vote.

Snow monkeys *(Macaca fuscata)*. These snow monkeys, which have the northernmost distribution of any nonhuman primate, are soaking in a hot spring in Japan.

© Ingo Arndt/Foto Natura/Minden Pictures

STUDY PLAN

30.1 Invertebrate Deuterostomes

Echinoderms have secondary radial symmetry and an internal skeleton

Acorn worms use gill slits and a pharynx to acquire food and oxygen

30.2 Overview of the Phylum Chordata

Key morphological innovations distinguish chordates from other deuterostome phyla

Invertebrate chordates are small, marine suspension feeders

Vertebrates possess several unique tissues, including bone and neural crest

30.3 The Origin and Diversification of Vertebrates

Vertebrates probably arose from a cephalochordate-like ancestor through the duplication of genes that regulate development

Early vertebrates diversified into numerous lineages with distinctive adaptations

30.4 Agnathans: Hagfishes and Lampreys, Conodonts and Ostracoderms

Hagfishes and lampreys are the living descendants of ancient agnathan lineages

Conodonts and ostracoderms were early jawless vertebrates with bony structures

30.5 Jawed Fishes

Jawed fishes first appeared in the Paleozoic era

Chondrichthyes includes fishes with cartilaginous endoskeletons

The Actinopterygii and Sarcopterygii are fishes with bony endoskeletons

30.6 Early Tetrapods and Modern Amphibians

Key adaptations facilitated the transition to land

Modern amphibians are very different from their Paleozoic ancestors

30.7 The Origin and Mesozoic Radiations of Amniotes

Key adaptations allow amniotes to live a fully terrestrial life

Amniotes diversified into three main lineages

Diapsids diversified wildly during the Mesozoic era

30.8 Testudines: Turtles

Turtles have bodies encased in a bony shell

Continued on next page

30 Deuterostomes: Vertebrates and Their Closest Relatives

WHY IT MATTERS

In 1798, naturalists at the British Museum skeptically probed a curious specimen that had been sent from Australia. The furry creature—about the size of a housecat—had webbed front feet, a ducklike bill, and a flat, paddlelike tail **(Figure 30.1)**. The scientists eagerly searched for evidence that a prankster had stitched together parts from wildly different animals, but they found no signs of trickery and soon accepted the duck-billed platypus *(Ornithorhynchus anatinus)* as a genuine zoological novelty.

Further study has revealed that the platypus is even stranger than those scientists could have imagined. Like other mammals, the platypus is covered with fur, and females produce milk that the offspring lick off the fur on their mother's belly. But like turtles and birds, a platypus has no teeth, and it reproduces by laying eggs instead of giving birth to its offspring. And like turtles, birds, lizards, snakes, and crocodilians, it has a cloaca, a multipurpose chamber through which it releases feces, urine, and eggs. Scientists had never before seen such a weird combination of traits, and they didn't quite know what to make of them.

Continued from previous page

30.9 Living Nonfeathered Diapsids: Sphenodontids, Squamates, and Crocodilians

Living sphenodontids are remnants of a diverse Mesozoic lineage

Squamates—lizards and snakes—are covered by overlapping, keratinized scales

Crocodilians are semiaquatic, predatory archosaurs

30.10 Aves: Birds

Key adaptations reduce body weight and provide power for flight

Flying birds were abundant by the Cretaceous period

Modern birds vary in morphology, diet, habits, and patterns of flight

30.11 Mammalia: Monotremes, Marsupials, and Placentals

Mammals exhibit key adaptations in anatomy, physiology, and behavior

The major groups of modern mammals differ in their reproductive adaptations

30.12 Nonhuman Primates

Key derived traits enabled primates to become arboreal, diurnal, and highly social

Living primates include two major lineages

30.13 The Evolution of Humans

Hominids first walked upright in East Africa about 6 million years ago

Homo habilis was probably the first hominid to manufacture stone tools

Homo erectus dispersed from Africa to other continents

Modern humans are the only surviving descendants of *Homo erectus*

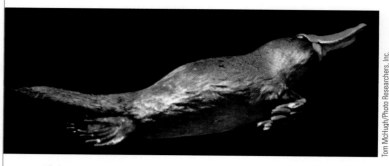

Figure 30.1

A puzzling animal. Because of its strange mixture of traits, the platypus (*Ornithorhynchus anatinus*) amazed the first European zoologists who saw it.

Studies of the platypus under natural conditions have helped biologists make sense of its characteristics. The platypus inhabits streams and lagoons in Australia and Tasmania. It rests in streamside burrows during the day, but at night it slips into the water to hunt for invertebrates. Its dense fur keeps its body warm and dry under water, and its tail serves both as a rudder and as a storehouse for energy-rich fat. It uses its bill to scoop up food and the horny pads that line its jaws to grind up prey. While underwater, the platypus clamps shut its eyes, ears, and nostrils, relying on roughly 800,000 sensory receptors in its bill to detect the movements and weak electrical discharges of nearby prey.

The platypus, with its strange combination of characteristics, illustrates the remarkable diversity of adaptations that enable vertebrates—animals with backbones—to occupy nearly every habitat on Earth. Despite the platypus's mixed characteristics, biologists eventually classified it as a member of the mammal lineage because, like all other mammals, it has hair on its body and produces milk to nourish its offspring. Today biologists know that it is one of just a few remaining survivors of an early lineage of egg-laying mammals.

In this chapter, we survey the Deuterostomia, a monophyletic lineage of animals that dates to the Paleozoic. The deuterostomes are defined by features of early embryological development and molecular sequence data (see Chapter 29). There are three living phyla of deuterostomes; we briefly consider two phyla of invertebrate deuterostomes before focusing on the Phylum Chordata, which includes a few thousand species of invertebrates as well as nearly 50,000 living species of vertebrates.

30.1 Invertebrate Deuterostomes

Deuterostome body plans have been so modified by evolution that a casual observer would not readily group the two phyla of invertebrate deuterostomes—Echinodermata and Hemichordata—together with the Phylum Chordata. However, embryological and molecular analyses agree that all three are indeed closely related.

Echinoderms Have Secondary Radial Symmetry and an Internal Skeleton

The phylum Echinodermata (*echinos* = spiny; *derma* = skin) includes 6600 species of sea stars, sea urchins, sea cucumbers, brittle stars, and sea lilies. These slow moving or sessile, bottom-dwelling animals are important herbivores and predators in shallow coastal waters and, paradoxically, the ocean depths. The phylum was diverse in the Paleozoic, but only a remnant of that fauna remains. Living species vary in size from less than 1 cm to more than 50 cm in diameter.

Echinoderms develop from a bilaterally symmetrical, free-swimming larva. As a larva develops, it assumes a secondary radial symmetry, often organized around five rays or "arms" **(Figure 30.2)**. Many echinoderms have an *oral surface,* with the mouth facing the substrate, and an *aboral surface* facing in the opposite direction. Virtually all echinoderms have an internal skeleton made of calcium-stiffened *ossicles* that develop from mesoderm. In some groups, fused ossicles form a rigid container called a *test*. In most, spines or bumps project from the ossicles.

The internal anatomy of echinoderms is unique among animals **(Figure 30.3)**. They have a well-defined coelom and a complete digestive system (see Figure 30.3a), but no excretory or respiratory systems, and most have only a minimal circulatory system. In many, gases are exchanged and metabolic wastes eliminated through projections of the epidermis and peritoneum near the base of the spines. Given their radial symmetry, there is no head or central brain; the nervous system is organized around nerve cords that encircle the mouth and branch into the rays. Sensory cells are abundant in the skin.

Echinoderms move using a unique system of fluid-filled canals, the *water vascular system* (see Figure 30.3b). In a sea star, for example, water enters the system through the *madreporite,* a sievelike plate on the aboral surface. A short tube connects it to the *ring canal,* which surrounds the esophagus. The ring canal branches into five *radial canals* that extend into the arms. Each radial canal is connected to numerous *tube feet* that protrude through holes in the plates. Each tube foot has a mucus-covered, suckerlike tip and a small muscular bulb, the *ampulla,* that lies inside the body. When an ampulla contracts, fluid is forced into the tube foot, causing it to lengthen and attach to the substrate (see Figure 30.3c). The tube foot then contracts, pulling the animal along. As the tube foot shortens, water is forced back into the ampulla, and the tube foot releases its grip on the substrate. The tube foot can then take another step forward, reattaching to the substrate. Although each tube foot has limited strength, the coordinated action of hundreds or even thousands of them is so strong that they can hold an echinoderm to a substrate even against strong wave action.

Echinoderms have separate sexes, and most reproduce by releasing gametes into the water. Radial cleavage is so clearly apparent in the transparent eggs of some sea urchins that they are commonly used for demonstrations of cleavage in introductory biology laboratories. A few echinoderms also reproduce asexually by splitting in half and regenerating the missing parts; some can regenerate body parts lost to predators.

Echinoderms are divided into six groups, one of which, the sea daisies (Concentricycloidea) was discovered only in 1986. These small, medusa-shaped animals occupy sunken, waterlogged wood in the deep sea. In the following sections, we describe the five other groups, which are more diverse and better known.

Asteroidea. The 1500 species of sea stars (Asteroidea, from *asteroeides* = starlike) live from rocky shorelines to depths of 10,000 m. The body consists of a central disk surrounded by 5 to 20 radiating "arms" (see Figure 30.2a), with the mouth centered on the oral surface. The ossicles of the endoskeleton are not fused, permitting flexibility of the arms and disk. Small pincers, **pedicellariae**, at the base of short spines remove debris that falls onto the animal's surface (see Figure 30.3c). Many sea stars feed on invertebrates and small fishes. Species that consume bivalve mollusks pry apart the two valves using their tube feet and slip their everted stomachs between the bivalve's shells. The stomach secretes digestive enzymes that dissolve the mollusk's tissues. Some sea stars are destructive predators of corals, endangering many reefs.

Ophiuroidea. The 2000 species of brittle stars and basket stars (Ophiuroidea, from *ophioneos* = snakelike) occupy roughly the same range of habitats as sea stars. Their bodies have a well-defined central disk and slender, elongate arms that are sometimes branched (see Figure 30.2b). Ophiuroids can crawl fairly swiftly across substrates by moving their arms in coordinated fashion. As their common name implies, the arms are delicate and easily broken, an adaptation that allows them to escape from predators with only minor losses. Brittle stars feed on small prey, suspended plankton, or detritus that they extract from muddy deposits.

Echinoidea. The 950 species of sea urchins and sand dollars (Echinoidea, "having spines") lack arms altogether (see Figure 30.2c). Their ossicles are fused into solid tests, which provide excellent protection but restrict flexibility. The test is spherical in sea urchins and flattened in sand dollars. Five rows of tube feet, used primarily for locomotion, emerge through pores in the test. Most echinoids have movable spines, some with poison glands; a jab from certain tropical species can cause severe pain and inflammation to a careless swimmer. Echinoids graze on algae and other organisms that cling to marine surfaces. In the center of an

Figure 30.2
Echinoderm diversity. Echinoderms exhibit secondary radial symmetry, usually organized as five rays around an oral-aboral axis.

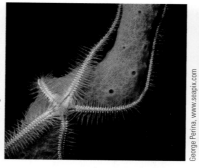

a. Asteroidea: This sea star *(Fromia milleporella)* lives in the intertidal zone.

b. Ophiuroidea: A brittle star *(Ophiothrix swensonii)* perches on a coral branch.

c. Echinoidea: A sea urchin *(Strongylocentrotus purpuratus)* grazes on algae.

d. Holothuroidea: A sea cucumber *(Cucumaria miniata)* extends its tentacles, which are modified tube feet, to trap particulate food.

e. Crinoidea: A feather star *(Himerometra robustipinna)* feeds by catching small particles with its numerous arms.

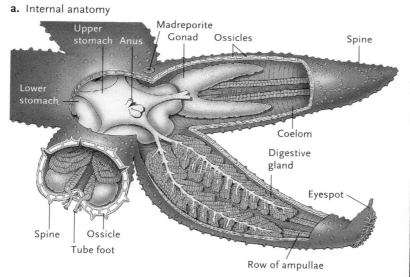

a. Internal anatomy

Upper stomach · Anus · Madreporite · Gonad · Ossicles · Spine · Lower stomach · Coelom · Digestive gland · Eyespot · Spine · Ossicle · Tube foot · Row of ampullae

b. Water vascular system

Madreporite · Radial canal · Ring canal · Ampulla

c. Tube feet

Aboral surface · Pedicellaria · Oral surface · Tube foot

Figure 30.3
Internal anatomy of a sea star. **(a)** The coelom is well developed in echinoderms, as illustrated by this cutaway diagram of a sea star. **(b)** The water vascular system, unique in the animal kingdom, operates the tube feet **(c)**, which are responsible for locomotion. Note the pedicellariae on the upper surface of the sea star's arm **(c)**.

urchin's oral surface is a five-part nipping jaw that is controlled by powerful muscles. Some species damage kelp beds, disrupting the habitat of young lobsters and other crustaceans. But echinoid ovaries are a delicacy in many countries, making these animals a prized natural resource.

Holothuroidea. The 1500 species of sea cucumbers (Holothuroidea, from *holothourion* = water polyp) are elongate animals that lie on their sides on the ocean bottom (see Figure 30.2d). Although they have five rows of tube feet, their endoskeleton is reduced to widely separated microscopic plates. The body, which is elongated along the oral-aboral axis, is soft and fleshy, with a tough, leathery covering. Modified tube feet form a ring of tentacles around the mouth, which points to the side or upward. Some species secrete a mucous net that traps plankton or other food particles. The net and tentacles are inserted into the mouth where the net and trapped food are ingested. Other species extract food from bottom sediments. Many sea cucumbers exchange gases through an extensively branched *respiratory tree* that arises from the rectum, the part of the digestive system just inside the anus at the aboral end of the animal. A well-developed circulatory system distributes oxygen and nutrients to tissues throughout the body.

Crinoidea. The 600 living species of sea lilies and feather stars (Crinoidea, from *krinon* = lily) are the surviving remnants of a fauna that was diverse and abundant 500 million years ago (see Figure 30.2e). Most species occupy marine waters of medium depth. The central disk and mouth point upward rather than toward the substrate. Between five and several hundred branched arms surround the disk; new arms are added as a crinoid grows larger. The branches of the arms are covered with tiny mucus-coated tube feet, which trap suspended microscopic organisms. The sessile sea lilies have the central disk attached to a flexible stalk that can reach a meter in length. Adult feather stars can swim or crawl weakly, attaching temporarily to substrates.

Acorn Worms Use Gill Slits and a Pharynx to Acquire Food and Oxygen

The 80 species of acorn worms (phylum Hemichordata, from *hemi* = half and *chorda* referring to the phylum Chordata) are sedentary marine animals that live in U-shaped tubes or burrows in coastal sand or mud. Their soft bodies, which range from 2 cm to 2 m in length, are organized into an anterior proboscis, a tentacled collar, and an elongate trunk **(Figure 30.4).** They use their muscular, mucus-coated proboscis to construct burrows and trap food particles. Acorn worms also have pairs of **gill slits** in the pharynx, the part of the digestive system just posterior to the mouth. Beating cilia create a flow of water, which enters the phar-

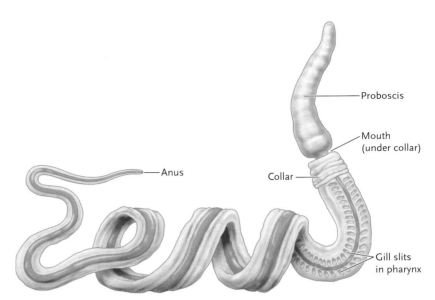

Figure 30.4
Phylum Hemichordata. Acorn worms draw food- and oxygen-laden water in through the mouth and expel it through gill slits in the anterior region of the trunk.

ynx through the mouth and exits through the gill slits. As water passes through, suspended food is trapped and shunted into the digestive system, and gases are exchanged across the partitions between gill slits. The coupling of feeding and respiration—as well as a dorsal nerve cord—reflects the close evolutionary relationship between hemichordates and chordates, the phylum that we consider next.

STUDY BREAK

1. What organ system is unique to echinoderms, and what is its function?
2. How does a perforated pharynx enable hemichordates to acquire food and oxygen from seawater?

30.2 Overview of the Phylum Chordata

The phylum Chordata contains three subphyla: two lineages of invertebrates, Urochordata and Cephalochordata, and a diverse lineage of vertebrates, Vertebrata.

Key Morphological Innovations Distinguish Chordates from Other Deuterostome Phyla

Chordates are distinguished from other deuterostomes by a set of key morphological innovations: a *notochord, segmental muscles in the body wall and tail,* a *dorsal hollow nerve chord,* and a *perforated pharynx* **(Figure 30.5).** These structures foster higher levels of activity, unique modes of aquatic locomotion, and more efficient feeding and oxygen acquisition.

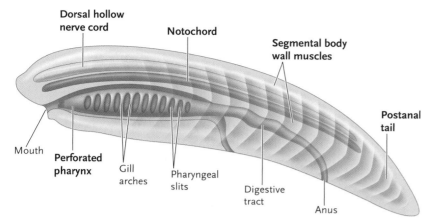

Figure 30.5

Diagnostic chordate characters. Chordates have a notochord; a muscular postanal tail; a segmental body wall and tail muscles; a dorsal, hollow nerve cord; and a perforated pharynx.

Segmental Body Wall and Tail Muscles. Chordates evolved in water, and they swim by contracting segmentally arranged blocks of muscles in the body wall and tail. The chordate tail, which is posterior to the anus, provides much of the propulsion in aquatic species. Segmentation allows each muscle block to contract independently; waves of contractions pass down one side of the animal and then down the other, sweeping the body and tail back and forth in a smooth and continuous movement.

Dorsal Hollow Nerve Cord. The central nervous system of chordates is a hollow nerve cord on the dorsal side of the animal (see Section 48.3). By contrast, most nonchordate invertebrates have solid nerve cords on the ventral side. In vertebrates, an anterior enlargement of the nerve cord forms a brain; in invertebrates, an anterior concentration of nervous system tissue is usually described as a *ganglion*.

Notochord. Early in chordate development, mesoderm that is dorsal to the developing digestive system forms a **notochord** (*noton* = the back; *chorda* = string). This flexible rod, constructed of fluid-filled cells surrounded by tough connective tissue, supports the embryo from head to tail. The notochord forms the skeleton of invertebrate chordates. Their body wall muscles are anchored to the notochord, and when these muscles contract, the notochord bends, but it does not shorten. As a result, the chordate body swings left and right during locomotion, propelling the animal forward; unlike annelids and other nonchordate invertebrates, the chordate body does not shorten when the animal is moving. Remnants of the notochord persist as gelatinous disks in the backbones of adult vertebrates.

Perforated Pharynx. The chordate pharynx, the part of the digestive system just posterior to the mouth, typically contains perforations or slits during some stage of the life cycle. These paired openings originated as exit holes for water that carried particulate food into the mouth, allowing chordates to gather large quantities of food. Invertebrate chordates also collect oxygen and release carbon dioxide across the walls of the pharynx. In fishes, **gill arches**, the supporting structures between the slits in the pharynx, are often sites of gas exchange, allowing animals to extract oxygen efficiently from the water. Invertebrate chordates and fishes retain a perforated pharynx throughout their lives. In most air-breathing vertebrates, the slits are present only during embryonic development and in larvae.

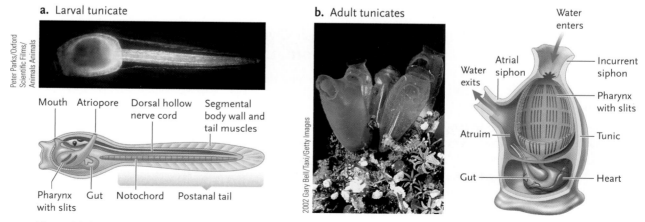

Figure 30.6

Urochordates. **(a)** A tadpolelike tunicate larva metamorphoses into **(b)** a sessile adult; shown here is *Rhopalaea crassa*. After a larva attaches to a substrate at its anterior end, the tail, notochord, and most of the nervous system are recycled to form new tissues. Slits in the pharynx multiply, the mouth becomes the incurrent siphon, and the atriopore becomes the atrial siphon.

Invertebrate Chordates Are Small, Marine Suspension Feeders

Two subphyla of invertebrate chordates exhibit the basic chordate body plan in its simplest form.

Subphylum Urochordata. The 2500 species of tunicates, sometimes called sea squirts (subphylum Urochordata, from *oura* = tail), float in surface waters or attach to substrates in shallow marine habitats. The sessile adults of many species secrete a gelatinous or leathery "tunic" around their bodies and squirt water through a siphon when disturbed; adults grow to several centimeters **(Figure 30.6)**. In the most common group of sea squirts (Ascidiacea), the swimming larvae possess the defining chordate features. Larvae eventually attach to substrates and transform into sessile adults. During metamorphosis, they lose most traces of the notochord, dorsal nerve cord, and tail, and their basketlike pharynx enlarges. In adults, beating cilia pull water into the pharynx through an **incurrent siphon.** A mucous net traps particulate food, which is carried, with the mucus, to the gut. Water passes through the pharyngeal slits, enters a chamber called the **atrium,** and is expelled—along with digestive wastes and carbon dioxide—through the **atrial siphon.** Oxygen is absorbed across the walls of the pharynx.

Subphylum Cephalochordata. The 28 lancelet species (subphylum Cephalochordata, from *kephale* = head) occupy warm, shallow marine habitats where they lie mostly buried in sand **(Figure 30.7)**. Although generally sedentary, they have well-developed body wall muscles and a prominent notochord. Most species are included in the genus *Branchiostoma* (formerly *Amphioxus*). Lancelet bodies, which are 5 to 10 cm long, are pointed at both ends like the double-edged surgical tools for which they are named. Adults have light receptors on the head as well as chemical sense organs on tentacles that grow from the **oral hood.** Lancelets use cilia to draw food-laden water through hundreds of pharyngeal slits; water flows into the atrium and is expelled through the **atriopore.** Most gas exchange occurs across the skin.

Vertebrates Possess Several Unique Tissues, Including Bone and Neural Crest

The most distinctive anatomical characteristic of the subphylum Vertebrata is an internal skeleton that provides structural support for muscles and protection for the nervous system and other organs. The skeleton and the muscles attached to it enable most vertebrates to move rapidly through the environment. A vertebrate's skeleton is composed of many separate, bony elements. Indeed, vertebrates are the only animals that have **bone,** a connective tissue in which living cells secrete the mineralized matrix that surrounds them (see Figure 36.5d). The **vertebral column,** made up of individual **vertebrae,** surrounds and protects the dorsal nerve cord, and a bony **cranium** surrounds the brain. The cranium, vertebral column, ribs, and sternum (breastbone) make up the **axial skeleton.** Most vertebrates also have a **pectoral girdle** anteriorly and a **pelvic girdle** posteriorly that attach bones in the fins or limbs to the axial skeleton. Bones of the two girdles and the appendages constitute the **appendicular skeleton.** One vertebrate lineage, Chondrichthyes, has lost its bone over evolutionary time; its skeleton is made of cartilage, a dense but flexible connective tissue that is often a developmental precursor of bone (see Section 36.2).

Vertebrates also possess a unique cell type, **neural crest,** which is distinct from endoderm, mesoderm, and ectoderm. Neural crest cells arise next to the developing nervous system, but later migrate throughout a vertebrate's body. They ultimately contribute to many uniquely vertebrate structures, including parts of the cranium, teeth, sensory organs, cranial nerves, and the medulla (that is, the interior part) of the adrenal glands.

Finally, the brains of vertebrates are much larger and more complex than those of invertebrate chordates.

Figure 30.7
Cephalochordates. **(a)** The unpigmented skin of adult lancelets reveals their segmental body wall muscles. A cutaway view **(b)** illustrates their internal anatomy.

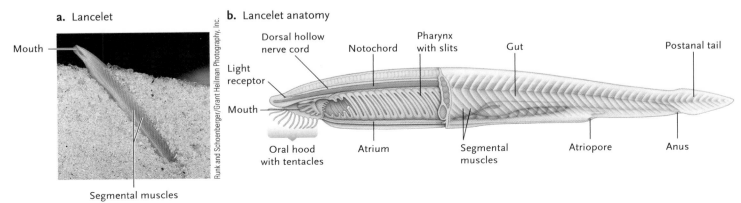

a. Lancelet

Mouth

Runk and Schoenberger/Grant Heilman Photography, Inc.

Segmental muscles

b. Lancelet anatomy

Dorsal hollow nerve cord — Pharynx with slits — Notochord — Gut — Postanal tail

Light receptor

Mouth

Oral hood with tentacles — Atrium — Segmental muscles — Atriopore — Anus

Moreover, the vertebrate brain is divided into three regions—the forebrain, midbrain, and hindbrain—each of which governs distinct nervous system functions (see Section 38.1).

STUDY BREAK

1. On a field trip to a lake, a college student captures a worm-shaped animal with segmental body wall muscles. While examining the specimen in laboratory the following day, she determines that the main nerve cord runs along the ventral side of the animal. Is this animal a chordate?
2. What structures distinguish vertebrates from invertebrate chordates?

30.3 The Origin and Diversification of Vertebrates

Biologists use embryological, molecular, and fossil evidence to trace the origin of vertebrates and to chronicle their evolutionary diversification.

Vertebrates Probably Arose from a Cephalochordate-Like Ancestor through the Duplication of Genes That Regulate Development

Molecular sequence studies suggest that vertebrates are more closely related to cephalochordates than to urochordates. The evolution of vertebrates from a cephalochordate-like ancestor was marked by the emergence of neural crest, bone, and other typically vertebrate traits. What genetic changes were responsible for these remarkable developments? Biologists now hypothesize that an increase in the number of homeotic—structure determining—genes may have made the development of more complex anatomy possible. (Homeotic genes are described further in Sections 22.6 and 48.6.)

In animals, one group of homeotic genes, the *Hox* genes, influences the three-dimensional shape of the animal and the locations of important structures—such as eyes, wings, and legs—particularly along the head to tail axis of the body. *Hox* genes are arranged on the chromosomes in a particular order, forming what biologists call the *Hox* gene complex (see Section 22.6). Each gene in the complex governs the development of particular structures. Animal groups with the simplest structure,

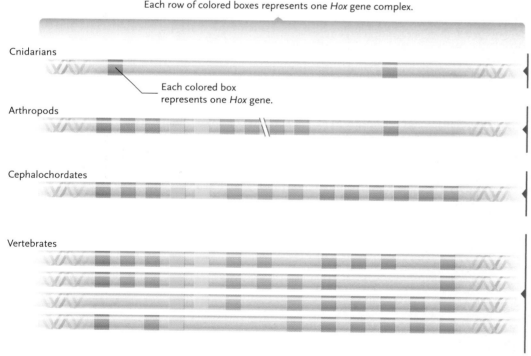

Each row of colored boxes represents one *Hox* gene complex.

Cnidarians

Each colored box represents one *Hox* gene.

Arthropods

Cephalochordates

Vertebrates

a. Invertebrates with simple anatomy, such as cnidarians, have a single *Hox* gene complex that includes just a few *Hox* genes.

b. Invertebrates with more complicated anatomy, such as arthropods, have a single *Hox* gene complex, but with a larger number of *Hox* genes.

c. Invertebrate chordates, such as cephalochordates, also have a single *Hox* gene complex, but with even more *Hox* genes than are found in nonchordate invertebrates.

d. Vertebrates, such as the laboratory mouse, have numerous *Hox* genes, arranged in two to seven *Hox* gene complexes. The additional *Hox* gene complexes are products of wholesale duplications of the ancestral *Hox* gene complex. The additional copies of *Hox* genes specify the development of uniquely vertebrate characteristics, such as the cranium, vertebral column, and neural crest cells.

Figure 30.8

Hox genes and the evolution of vertebrates. The *Hox* genes in different animals appear to be homologous, indicated here by their color and position in the complex. Vertebrates have many more individual *Hox* genes than most invertebrates do, and the entire *Hox* gene complex was duplicated in the vertebrate lineage.

such as cnidarians, have two *Hox* genes. Those with more complex anatomy, such as insects, have 10. Chordates have as many as 13 or 14. Thus, lineages with many *Hox* genes generally have more complex anatomy than do those with fewer *Hox* genes.

Molecular analyses also reveal that the entire *Hox* gene complex was duplicated several times in the evolution of vertebrates, producing multiple copies of all its genes **(Figure 30.8)**. The cephalochordate *Branchiostoma* has just one *Hox* gene complex, but the most primitive living vertebrates, the jawless hagfishes described later, have two. All vertebrates that possess jaws, a derived characteristic, have at least four sets, and some fishes have as many as seven. Evolutionary developmental biologists hypothesize that the duplication of *Hox* genes and other tool-kit genes allowed the evolution of new structures: while the original copies of these genes maintained their ancestral functions, the duplicate copies assumed *new* functions, directing the development of novel structures, such as the vertebral column and jaws.

Early Vertebrates Diversified into Numerous Lineages with Distinctive Adaptations

The oldest known vertebrate fossils were discovered in the late 1990s, when scientists in China described several species from the early Cambrian period, about 550 million years ago. Both *Myllokunmingia* and *Haikouichthys* were fish-shaped animals about 3 cm long **(Figure 30.9)**. In both species the brain was surrounded by a cranium, which, in these cases, was formed of fibrous connective tissue or cartilage. They also had segmental body wall muscles and fairly well-developed fins, but neither shows any evidence of bone.

The early vertebrates gave rise to numerous descendants, which varied greatly in anatomy, physiology, and ecology. New feeding mechanisms and locomotor structures were often crucial to their success. Today, vertebrates occupy nearly every habitat and feed on virtually all other organisms. Here we briefly introduce the major vertebrate lineages **(Figure 30.10)**.

Although biologists use four key morphological innovations—a cranium, vertebrae, bone, and neural crest cells—to identify vertebrates, these structures did not arise all at once. Instead, they appeared somewhat independently of one another as new groups arose. Some researchers and textbooks present a phylogeny and classification that places the "vertebrates" (animals that have vertebrae) within a larger lineage called the "craniates" (animals that have a cranium). But only one small group, the hagfishes (Myxinoidea, described later), has a cranium but no vertebrae, and some recent molecular analyses do not support its separation from the other vertebrates. Thus, for the sake of simplicity, we describe organisms that possessed any of the four key innovations as "vertebrates."

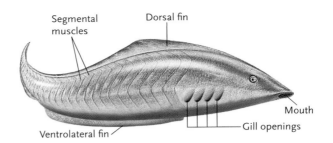

Labels: Segmental muscles; Dorsal fin; Mouth; Gill openings; Ventrolateral fin

Figure 30.9
An early vertebrate. *Myllokunmingia*, one of the earliest vertebrates yet discovered, had no bones; it was about 3 cm long.

Several groups of early jawless vertebrates are described as *agnathans* (*a* = not; *gnathos* = jawed), but they do not form a monophyletic group. Although most became extinct in the Paleozoic era, two ancient lineages, Myxinoidea and Petromyzontoidea, still live today. All other vertebrates possess moveable jaws; they are members of the monophyletic lineage **Gnathostomata** ("jawed mouth"). The first jawed fishes, the Acanthodii and Placodermi, are now extinct, but several other lineages of jawed fishes are still abundant in aquatic habitats: the Chondrichthyes includes fishes with cartilaginous skeletons, such as sharks and skates; the Actinopterygians and Sarcopterygians comprise the bony fishes, which have bony endoskeletons. All jawless vertebrates and jawed fishes are restricted to aquatic habitats, and they use gills to extract oxygen from the water that surrounds them.

The Gnathostomata also includes the monophyletic lineage **Tetrapoda** (*tetra* = four; *pod* = foot); most tetrapods use four limbs for locomotion. Many tetrapods are semiterrestrial or terrestrial, although some, like sea turtles and porpoises, have secondarily returned to aquatic habitats. Adult tetrapods generally use air-breathing lungs for gas exchange. Within the Tetrapoda, one lineage, the amphibians, includes animals, such as frogs and salamanders, that typically need standing water to complete their life cycles. Another lineage, the **Amniota**, comprises animals with specialized eggs that can develop on land. Shortly after their appearance, the amniotes diversified into three lineages: one is ancestral to living mammals; another to the living turtles; and a third to lizards, snakes, alligators, and birds. We consider the detailed evolutionary history of the amniotes in Section 30.7.

STUDY BREAK

1. How do the *Hox* genes of vertebrates differ from those of cephalochordates?
2. Which of the taxonomic groups Amniota, Gnathostomata, and Tetrapoda includes the largest number of species? Which includes the fewest?

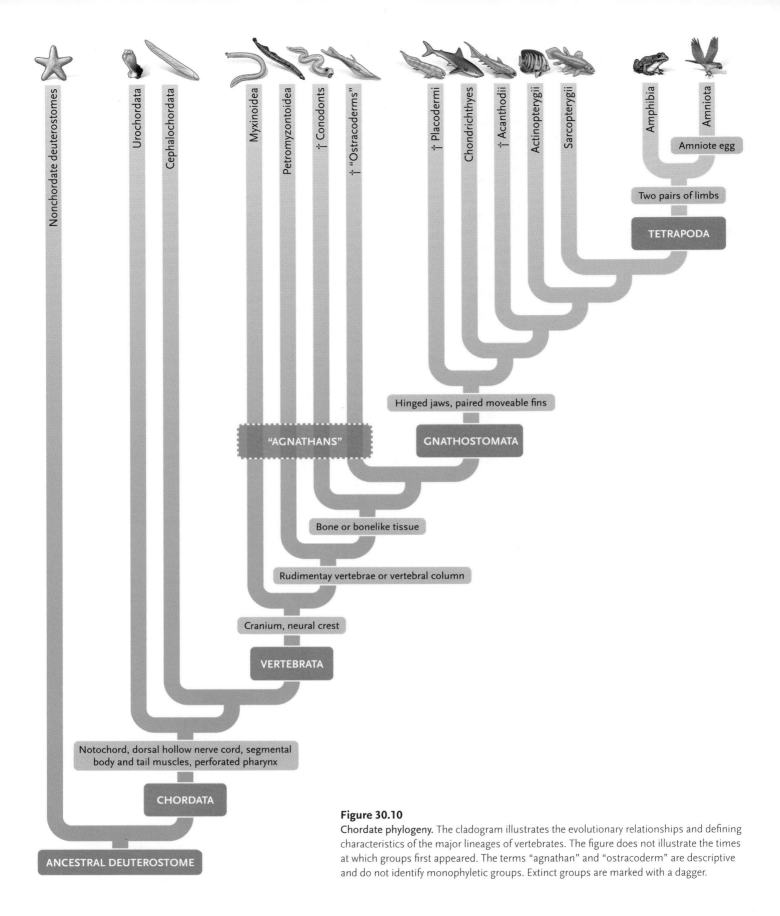

Figure 30.10

Chordate phylogeny. The cladogram illustrates the evolutionary relationships and defining characteristics of the major lineages of vertebrates. The figure does not illustrate the times at which groups first appeared. The terms "agnathan" and "ostracoderm" are descriptive and do not identify monophyletic groups. Extinct groups are marked with a dagger.

30.4 Agnathans: Hagfishes and Lampreys, Conodonts and Ostracoderms

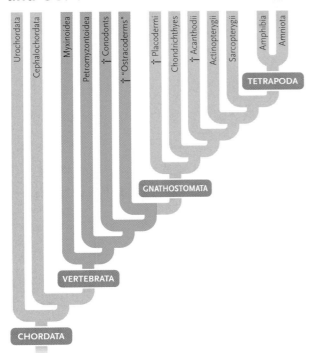

Lacking jaws, most of the earliest vertebrates used a muscular pharynx to suck edible tidbits into their mouths. The two living groups of agnathans as well as species that flourished in the Paleozoic vary greatly in size and shape as well as in the number of vertebrate characters they possess.

Hagfishes and Lampreys Are the Living Descendants of Ancient Agnathan Lineages

Two apparently separate lineages of jawless vertebrates, hagfishes (Myxinoidea) and lampreys (Petromyzontoidea), still live today. Both have skeletons composed entirely of cartilage. Although scientists have found no fossilized hagfishes or lampreys from the early Paleozoic era, the absence of jaws and bone in their living descendants suggests that their lineages arose early in vertebrate history, before the evolution of bone. Hagfishes and lampreys have a well-developed notochord, but no true vertebrae or paired fins, and their skin has no scales. Individuals grow to a maximum length of about 1 m **(Figure 30.11)**.

The axial skeleton of the 60 living species of hagfishes includes only a cranium and a notochord; it has no specialized structures surrounding the dorsal nerve cord. Some biologists do not even include hagfishes among the Vertebrata, because they lack any sign of vertebrae. Hagfishes are marine scavengers that burrow in sediments on continental shelves. They feed on inverte-

brate prey and on dead or dying fishes. In response to predators, they secrete an immense quantity of sticky, noxious slime; when no longer threatened, a hagfish ties itself into a knot and wipes the slime from its body. Hagfish life cycles are simple and lack a larval stage.

The 40 or so living species of lampreys have traces of an axial skeleton. Their notochord is surrounded by dorsally pointing cartilages that partially cover the nerve cord; many biologists suspect that this arrangement may reflect an early stage in the evolution of the vertebral column. Most lamprey species are parasitic as adults. They have a circular mouth surrounded by a sucking disk with which they attach to a fish or other vertebrate host; they feed on a host's body fluids after rasping through its skin. In most species, sexually mature adults migrate from the ocean or a lake to the headwaters of a stream, where they lay eggs and then die. Their suspension-feeding larvae, which resemble adult cephalochordates, burrow into mud and develop for as long as seven years before undergoing metamorphosis and migrating to the sea or a lake to live as parasitic adults.

Conodonts and Ostracoderms Were Early Jawless Vertebrates with Bony Structures

Mysterious bonelike fossils, most less than 1 mm long, have long been known in oceanic rocks dating from the early Paleozoic era through the early Mesozoic era. Called **conodont** ("cone tooth") elements, these abun-

a. Living jawless fishes

Hagfish

Tentacles   Gill slits   Slime glands

Lamprey

Oral disk   Gill slits

b. Mouth of a lamprey

Heather Angel

Figure 30.11
Living agnathans. **(a)** Two groups of jawless fishes, hagfishes and lampreys, survive to-day. **(b)** Lampreys use a toothed oral disk to attach to a host and feed on its blood and soft tissues.

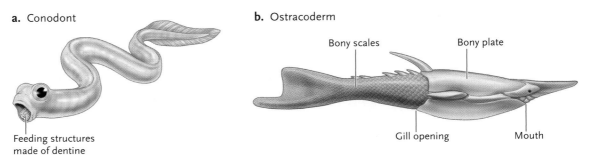

a. Conodont

Feeding structures
made of dentine

b. Ostracoderm

Bony scales

Bony plate

Gill opening

Mouth

Figure 30.12

Extinct agnathans. **(a)** Conodonts were elongate, soft-bodied animals with bonelike feeding structures in the mouth and pharynx. **(b)** *Pteraspis*, an ostracoderm, had large bony plates on its head and small bony scales on the rest of its body; it was about 6 cm long.

dant fossils were once described as the support structures of marine algae or the feeding structures of ancient invertebrates. However, recent analyses of their mineral composition reveal that they were made of dentine, a bonelike component of vertebrate teeth. In the 1980s and 1990s, several research teams discovered fossils of intact conodont animals with these elements in place.

Conodonts were elongate, soft-bodied animals; most were 3 to 10 cm long. They had a notochord, cranium, segmental body wall muscles, and large, moveable eyes **(Figure 30.12a)**. The conodont elements at the front of the mouth were forward pointing, hook-shaped structures that apparently functioned to collect food; those in the pharynx were stouter, suitable for crushing items that had been consumed. Paleontologists now classify conodonts as vertebrates—the earliest vertebrates with bonelike structures.

An assortment of jawless fishes, representing several evolutionary lineages and collectively called **ostracoderms** (*ostrakon* = shell), were abundant from the Ordovician through the Devonian periods **(Figure 30.12b)**. Like their invertebrate chordate ancestors, ostracoderms used their pharynx to extract small food particles from mud and water. However, the ostracoderms' muscular pharynx enabled them to *suck* mud and water into their mouths, providing a much stronger flow than the cilia-driven currents of invertebrate chordates. The greater flow rate allowed ostracoderms to collect food more rapidly. It also supported a larger body size: although most were much smaller, some ostracoderms reached a length of 2 m.

The skin of ostracoderms was heavily armored with plates and scales formed of bone. Although some ostracoderms had paired lateral extensions of their bony armor, they could not move them the way living fishes move their paired fins. Ostracoderms lacked a true vertebral column, but they had rudimentary support structures surrounding the nerve cord. They also had other distinctly vertebrate-like characteristics. For example, imprints in the head shields indicate that their brains had the three regions—forebrain, mid-

brain, and hindbrain—typical of all later vertebrates (see Section 38.1).

STUDY BREAK

1. What characteristics of the living hagfishes and lampreys suggest that their lineages arose very early in vertebrate evolution?
2. What traits in conodonts and ostracoderms are derived relative to those in hagfishes and lampreys?

30.5 Jawed Fishes

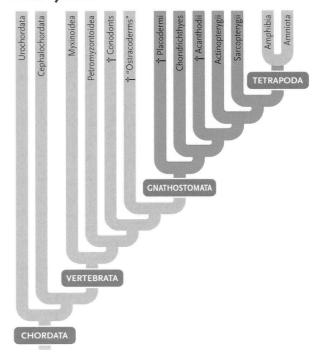

The first gnathostomes were jawed fishes. Key derived traits made their feeding and locomotion more efficient than those of their ancestors.

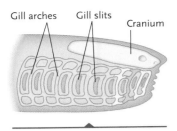

Gill arches   Gill slits   Cranium

a. Jaws evolved from gill arches in the pharynx of jawless fishes.

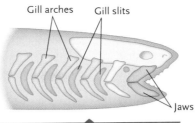

Gill arches   Gill slits

Jaws

b. In early jawed fishes, the upper jaw was firmly attached to the cranium.

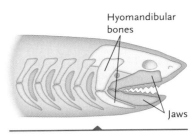

Hyomandibular bones

Jaws

c. In later jawed fishes, the jaws were supported by the hyomandibular bones, which were derived from a second pair of gill arches.

Figure 30.13
The evolution of jaws.

Jawed Fishes First Appeared in the Paleozoic Era

The renowned anatomist and paleontologist Alfred Sherwood Romer of Harvard University described the evolution of jaws as "perhaps the greatest of all advances in vertebrate history." Hinged jaws allow vertebrates to grasp, kill, shred, and crush large food items. Some species also use their jaws for defense, for grooming, to construct nests, and to transport their young.

The Origin of Jaws and Fins. Embryological evidence suggests that jaws evolved from paired gill arches in the pharynx of a jawless ancestor **(Figure 30.13)**. One pair of ancestral gill arches formed bones in the upper and lower jaws, while a second pair was transformed into the hyomandibular bones that braced the jaws against the cranium. Nerves and muscles of the ancestral suspension-feeding pharynx control and move the jaws.

Innovative locomotor mechanisms have often appeared at roughly the same time as innovative feeding mechanisms in the vertebrate lineage, and many early jawed fishes also had fins. The earliest fins were folds of skin and moveable spines that stabilized locomotion and deterred predators. Moveable fins appeared independently in several lineages, and by the Devonian period, most fishes had unpaired (dorsal, anal, and caudal) and paired (pectoral and pelvic) fins **(Figure 30.14)**.

Early Jawed Fishes. In two early lineages of jawed fishes, spiny sharks and placoderms, the upper jaw was firmly attached to the cranium (see Figure 30.13b); their inflexible mouths simply snapped open and shut **(Figure 30.15)**. Both groups also show evidence of an internal skeleton.

Spiny sharks (Acanthodii, from *akantha* = thorn), which persisted from the late Ordovician through the Permian periods, were less than 20 cm long. Their small, light scales; streamlined bodies; well-developed eyes; large jaws; and numerous teeth suggest that they were fast swimmers and efficient predators. Most had a row of ventral spines and fins with internal skeletal support on each side of the body. Acanthodian anatomy suggests that they are closely related to the bony fishes alive today.

Placoderms (Placodermi, from *plax* = flat surface) appeared in the Silurian and diversified in the Devonian and Carboniferous periods, but they left no direct descendants. Some, like *Dunkleosteus,* reached a length of 10 m. Their bodies were covered with large, heavy plates of bone anteriorly and smaller scales posteriorly. Their jaws had sharp cutting edges, but not separate teeth, and their paired fins had internal skeletons and powerful muscles.

a. Spiny shark

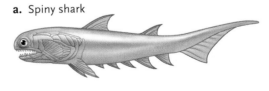

b. Placoderm

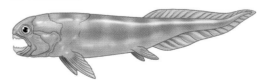

Figure 30.15
Early gnathostomes. **(a)** *Climatius,* an acanthodian, was small, reaching a total length of about 8 cm. **(b)** The placoderm *Dunkleosteus* was gigantic, sometimes growing to 10 m in length. Some acanthodians had teeth on their jaws, but placoderms had only sharp, cutting edges.

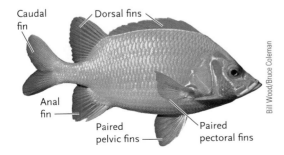

Caudal fin   Dorsal fins

Anal fin

Paired pelvic fins   Paired pectoral fins

Bill Wood/Bruce Coleman

Figure 30.14
Fish fins. Most fishes have both paired and unpaired fins.

Chondrichthyes Includes Fishes with Cartilaginous Endoskeletons

The 850 living species in the Chondrichthyes (*chondros* = cartilage; *ichthys* = fish) have skeletons composed entirely of cartilage, which is much lighter than bone. The absence of bone in Chondrichthyes is a derived trait, however, because all earlier fishes had bony armor or bony endoskeletons.

Most living chondrichthyans are grouped in the Elasmobranchii, which includes the skates, rays, and sharks; nearly all are marine predators **(Figure 30.16).** Skates and rays are dorsoventrally flattened (see Figure 30.16a). They swim by undulating their enlarged pectoral fins. Most are bottom dwellers that often lie partly buried in sand. They feed on hard-shelled invertebrates, which they crush with massive, flattened teeth. The largest species, the manta ray *(Manta birostris),* which measures 6 m across, feeds on plankton in the open ocean. Some rays have electric organs that stun prey with as much as 200 volts.

Sharks (see Figure 30.16b) are among the ocean's dominant predators. Flexible fins, lightweight skeletons, streamlined bodies, and the absence of heavy body armor allow most sharks to pursue prey rapidly. Their livers often contain **squalene,** an oil that is lighter than water, which increases their buoyancy. The great white shark *(Carcharodon carcharias)* is the largest predatory species, attaining a length of 10 m. The whale shark *(Rhincodon typus),* which grows to 18 m, is the largest fish; it feeds on plankton.

Elasmobranchs—including sharks, skates, and rays—exhibit remarkable adaptations for acquiring and processing food. Their teeth develop in whorls under the fleshy parts of the mouth. New teeth migrate forward as old, worn teeth break free. In many sharks, the upper jaw is loosely attached to the cranium, and it swings down during feeding. As the jaws open, the mouth spreads widely, sucking in large, hard-to-digest chunks of prey, which are swallowed intact. Although the elasmobranch digestive system is short, it includes a corkscrew-shaped **spiral valve,** which slows the passage of material and increases the surface area available for digestion and absorption.

Elasmobranchs also have well-developed sensory systems. In addition to vision and olfaction, they use **electroreceptors** to detect weak electric currents produced by other animals. And their **lateral-line system,** a row of tiny sensors in canals along both sides of the body, detects vibrations in water (see Figure 39.4).

Chondrichthyans exhibit numerous reproductive specializations. Males have a pair of organs, the **claspers,** on the pelvic fins, which help transfer sperm into the female's reproductive tract. Fertilization occurs internally. In many species, females produce yolky eggs with tough leathery shells (see Figure 30.16c).

a. Manta ray

b. Galápagos shark

c. Swell shark egg case

Figure 30.16

Chondrichthyes. **(a)** Skates and rays, like the manta ray *(Manta birostris),* as well as **(b)** sharks, like the Galápagos shark *(Carcharhinus galapagensis),* are grouped in the Elasmobranchii. **(c)** Many shark egg cases, like that of the swell shark *(Cephaloscylium ventricosum),* include a large yolk that nourishes the developing embryo.

Others retain the eggs within the oviduct until the young hatch. A few species nourish young within a uterus.

The Actinopterygii and Sarcopterygii Are Fishes with Bony Endoskeletons

In terms of diversity and sheer numbers, the fishes with bony endoskeletons—a cranium, vertebral column with ribs, and bones supporting their moveable fins—are the most successful of all vertebrates. The endoskeleton provides lightweight support, particularly compared with the heavy bony armor of ostracoderms and placoderms, and enhances their locomotor efficiency. Bony fishes first appeared in the Silurian period and rapidly diversified into two lineages. The ray-finned fishes (Actinopterygii, from *aktis* = ray and *pteron* = wing) have fins that are supported by thin and flexible bony rays. The fleshy-finned fishes (Sarcopterygii, from *sarco* = flesh) have fins that are supported by muscles and an internal bony skeleton. Ray-finned fishes have always been more diverse, and they vastly outnumber the fleshy-finned fishes today. The 21,000 living species of bony fishes occupy nearly every aquatic habitat and represent more than 95% of living fish species. Adults range from 1 cm to more than 6 m in length.

Bony fishes have numerous adaptations that increase their swimming efficiency. In many modern ray-finned fishes, a gas-filled **swim bladder** serves as a hydrostatic organ that increases buoyancy (see Figure 30.18a). The swim bladder is derived from an ancestral air-breathing lung that allowed early actinopterygians to gulp air, supplementing their gill respiration in aquatic habitats where dissolved oxygen concentration was low. The scales of most bony fishes are small, smooth, and lightweight. And their bodies are covered with a protective coat of mucus, which retards bacterial growth and smoothes the flow of water.

Actinopterygii. The most primitive living actinopterygians, sturgeons and paddlefishes, have mostly cartilaginous skeletons **(Figure 30.17a)**. These large fishes live in rivers and lakes of the northern hemisphere. Sturgeons feed on detritus and invertebrates; paddlefish consume plankton. Gars and bowfins are remnants of a more recent radiation **(Figure 30.17b)**. They occur only in the eastern half of North America, where they feed on fishes and other prey. Gars are protected from predators by a heavy coat of bony scales.

Teleosts, the latest radiation of Actinopterygii, are the most diverse, successful, and familiar bony fishes. Evolution has produced a wide range of body forms **(Figure 30.18)**. Teleosts have an internal skeleton made almost entirely of bone. On either side of the head, a flap of the body wall, the **operculum,** covers a chamber that houses the gills. Sensory systems generally include large eyes, a lateral-line system, sound receptors, chemoreceptive nostrils, and taste buds. Variations in jaw structure allow different teleosts to consume plankton, seaweed, invertebrates, or other vertebrates.

Teleosts exhibit remarkable feeding and locomotor adaptations. When some teleosts open their mouths, bones at the front of the jaws swing forward to create a circular opening. Folds of skin extend backward, forming a tube through which they suck food (see Figure 30.18f). Many also have symmetrical caudal fins, posterior to the vertebral column, which provide power for locomotion. And their pectoral fins lie high on the sides of the body, providing fine control over swimming. Some species use their pectoral fins for acquiring food, for courtship, and for care of eggs and young. Some teleosts even use them for crawling on land or gliding in air.

Most marine species produce small eggs that hatch into planktonic larvae. Eggs of freshwater teleosts are generally larger and hatch into tiny versions of the adults. Parents often care for their eggs and young, fanning oxygen-rich water over them, removing fungal growths, and protecting them from predators. Some freshwater species, such as guppies, give birth to live young.

Sarcopterygii. Two groups of fleshy-finned fishes (Sarcopterygii), lobe-finned fishes and lungfishes, are now represented by only eight living species **(Figure 30.19)**.

a. Lake sturgeon

b. Long-nosed gar

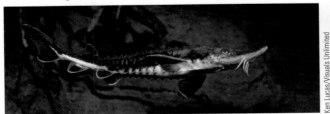
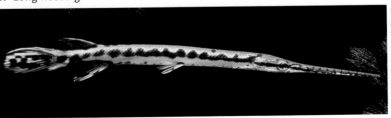

Ken Lucas/Visuals Unlimited

Patrice Ceisel/© 1986 John G. Shedd Aquarium

Figure 30.17
Primitive actinopterygians. **(a)** A lake sturgeon *(Accipenser fulvescens)* and **(b)** a long-nosed gar *(Lepisosteus osseus)* are living representatives of early actinopterygian radiations.

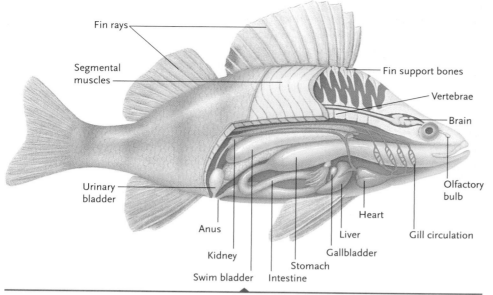

Fin rays

Segmental muscles

Fin support bones

Vertebrae

Brain

Olfactory bulb

Urinary bladder

Anus

Kidney

Swim bladder

Intestine

Stomach

Gallbladder

Liver

Heart

Gill circulation

a. Teleost internal anatomy

Kit Kittle/Corbis

c. The long, flexible body of a spotted moray eel *(Gymnothorax moringa)* can wiggle through the nooks and crannies of a reef.

F. Graner/Peter Arnold, Inc.

d. Flatfishes, like this European flounder *(Platichthys flesus)*, lie on one side and leap at passing prey.

Digital Vision/Getty Images, Inc.

b. Sea horses, like the northern sea horse *(Hippocampus hudsonius),* use a prehensile tail to hold on to substrates; they are weak swimmers.

Operculum

Brandon Cole/Visuals Unlimited

e. Open ocean predators, like the yellowfin tuna *(Thunnus albacares),* have strong, torpedo-shaped bodies and powerful caudal fins.

Arthur W. Ambler/Photo Researchers, Inc.

f. Kissing Gouramis *(Helostoma temmincki)* extend their jaws into a tube that sucks food into the mouth.

Figure 30.18

Teleost diversity. Although all teleosts share similar internal features, their diverse shapes adapt them to different diets and types of swimming.

a. Coelacanth

Norbert Wu/Peter Arnold, Inc.

b. Australian lungfish

Wernher Krutein/photovault.com

Figure 30.19

Sarcopterygians. **(a)** The coelacanth *(Latimeria chalumnae)* is one of two living species of lobe-finned fishes. **(b)** The Australian lungfish *(Neoceratodus forsteri)* is one of only six living lungfish species.

Although lobe-finned fishes were once thought to have been extinct for 65 million years, a living coelacanth *(Latimeria chalumnae)* was discovered in 1938 near the Comoros Islands, off the southeastern coast of Africa. We now know that a population of this meter-long fish lives at depths of 70 to 600 m, feeding on fishes and squid. Remarkably, a second population of coelacanths was discovered in 1998, when a specimen was found in an Indonesian fish market, 10,000 km east of the Comoros population. Based on analyses of its DNA, it is a distinct species *(Latimeria menadoensis)*.

Lungfishes have changed relatively little over the last 200 million years. Six living species are distributed on southern continents. The Australian lungfishes, which live in rivers and pools, use their lungs to supplement gill respiration when dissolved oxygen concentration is low. The South American and African species, which live in swamps, use their lungs to collect oxygen during the annual dry season, which they spend encased in a mucus-lined burrow in the dry mud. When the rains begin, water fills the burrow and the fishes awaken from dormancy.

STUDY BREAK

1. What characteristics of sharks and rays make them more efficient predators than the acanthodians or placoderms?
2. How do the air bladder and fins of ray-finned bony fishes increase their locomotor abilities?
3. How do the lungs of lungfishes allow them to survive in stressful environments?

30.6 Early Tetrapods and Modern Amphibians

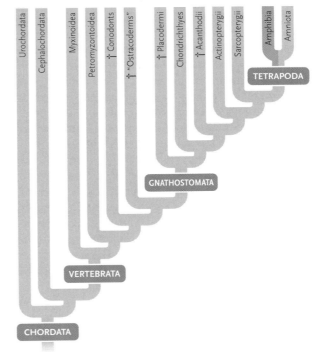

The fossil record suggests that tetrapods arose from a group of fleshy-finned fishes, the *osteolepiforms,* in the late Devonian period. Osteolepiforms and early tetrapods shared several derived characteristics: both had curious infoldings of their tooth surfaces, a trait with unknown function; and the shapes and positions of bones on the dorsal side of their crania and in their appendages were similar **(Figure 30.20).**

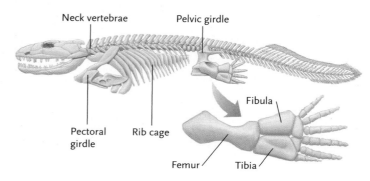

a. *Eusthenopteron*, an osteolepiform fish

b. *Ichthyostega*, an early tetrapod

Pectoral girdle   Pectoral fin   Pelvic fin   Fibula   Femur   Tibia

Neck vertebrae   Pelvic girdle   Pectoral girdle   Rib cage   Fibula   Femur   Tibia

Figure 30.20

Evolution of tetrapod limbs. The limb skeleton of osteolepiform fishes, such as **(a)** *Eusthenopteron*, is homologous to that of early tetrapods, such as **(b)** *Ichthyostega*. Although *Ichthyostega* retained many fishlike characteristics, its pectoral girdle was completely freed from the cranium and it had a heavy rib cage. Fossils of its forefoot have not yet been discovered.

Key Adaptations Facilitated the Transition to Land

Fishes are not adapted to live on land, and the first tetrapods faced serious environmental challenges. First, because air is less dense than water, it provides less support for an animal's body. Second, animals exposed to air inevitably lose body water by evaporation. Third, the sensory systems of fishes, which work well under water, do not function well on land. However, swampy late Devonian habitats also offered distinct advantages. Land plants, worms, and arthropods provided abundant food; oxygen was more readily available in air than in water; and no predators lived in these new habitats.

In some ways, osteolepiforms were preadapted for terrestrial life (see Figure 30.20a). Most had strong, stout fins that enabled them to crawl on the muddy bottom of shallow pools, and their vertebral column included crescent-shaped bones that provided good support. They had nostrils leading to sensory pits that housed olfactory (smell) receptors. And they almost certainly had lungs to augment gill respiration in the swampy, oxygen-poor waters where they lived.

The earliest tetrapod for which we have nearly complete skeletal data is the semiterrestrial, meter-long *Ichthyostega* (see Figure 30.20b). Compared with its fleshy-finned ancestors, *Ichthyostega* had a stronger vertebral column, sturdier girdles and appendages, a rib cage that protected its internal organs (including lungs), and a neck. Fishes have no neck: the pectoral girdle is fused to the cranium. But several vertebrae separated these structures in *Ichthyostega,* allowing it to move its head to scan the environment and to capture food. However, *Ichthyostega* retained a fishlike lateral-line system, caudal fin, and scaly covering on its body.

Life on land also required changes in sensory systems. In fishes, for example, the body wall picks up sound vibrations and transfers them to sensory receptors directly. But sound waves are harder to detect in air. Early tetrapods developed a **tympanum,** a specialized membrane on either side of the head that is vibrated by airborne sounds. The tympanum connects to the **stapes,** a bone that is homologous to the hyomandibula, which had supported the jaws of fishes (see Figure 23.4). The stapes, in turn, transfers vibrations to the sensory cells of an inner ear.

Modern Amphibians Are Very Different from Their Paleozoic Ancestors

Most of the more than 6000 species of living amphibians—including frogs, salamanders, and caecelians—are small, and their skeletons contain fewer bones than those of Paleozoic tetrapods like *Ichthyostega.* All living amphibians are carnivorous as adults, but the aquatic larvae of some species are herbivores.

Most living amphibians have a thin, scaleless skin, well supplied with blood vessels, that is a major site of gas exchange. Because gases must enter the body across a thin layer of water, the skin of most amphibians must remain moist, restricting them to aquatic or wet terrestrial habitats. Adults of some species also acquire oxygen through saclike lungs. The evolution of lungs was accompanied by modifications of the heart and circulatory system that increase the efficiency with which oxygen is delivered to body tissues (see Section 42.1).

The life cycles of many amphibians (*amphi* = both; *bios* = life) include both larval and adult stages. Eggs are laid and fertilized in water, where they hatch into larvae, such as the tadpoles of frogs, that eventu-

a. A frog

Stephen Dalton/Photo Researchers, Inc.

b. A salamander

Bill M. Campbell, MD

c. A caecelian

Juan M. Renjifo/Animals Animals

Figure 30.21

Living amphibians. **(a)** Anurans, like the northern leopard frog *(Rana pipiens)*, have compact bodies and long hind legs. **(b)** Urodeles, such as the red-spotted newt *(Notophthalmus viridescens)*, have an elongate body and four legs. **(c)** Caecelians, like *Caecelia nigricans* from Colombia, are legless burrowing animals.

ally metamorphose into adults (see Figure 40.9). Although the larvae of most species are aquatic, adults may be aquatic, amphibious, or terrestrial. Some salamanders are paedomorphic: the larval stage attains sexual maturity without changing its form or moving to land. By contrast, some frogs and salamanders reproduce on land, skipping the larval stage altogether. But even though they are terrestrial breeders, their eggs dry out quickly unless they are laid in moist places.

Modern amphibians are represented by three lineages **(Figure 30.21)**. Populations of practically all amphibians have declined rapidly in recent years, probably because of exposure to acid rain, high levels of ultraviolet B radiation, or parasitic infections.

Anura. The 3700 species of frogs and toads (Anura, from *an* = without; *oura* = tail) have short, compact bodies, and adults lack tails. Their elongate hind legs and webbed feet allow them to hop on land or swim. A few species are adapted to dry habitats, withstanding periods of drought by encasing themselves in mucous cocoons.

Urodela. Most of the 400 species of salamanders (Urodela, from *oura* = tail; *delos* = visible) have an elongate, tailed body and four legs. They walk by alternately contracting muscles on either side of the body much the way fishes swim. Species in the most diverse group, the lungless salamanders, are fully terrestrial throughout their lives, using their skin and the lining of the throat for gas exchange.

Gymnophiona. The 200 species of caecelians (Gymnophiona, from *gymnos* = naked; *ophioneos* = snakelike) are legless burrowing animals with wormlike bodies. They occupy tropical habitats throughout the world. Unlike other modern amphibians, caecelians have small bony scales embedded in their skin. Fertilization is internal, and females give birth to live young.

STUDY BREAK

1. For the first tetrapods, what were the advantages and disadvantages of moving onto the land?
2. What parts of the life cycle in most modern amphibians are dependent on water or very moist habitats?

30.7 The Origin and Mesozoic Radiations of Amniotes

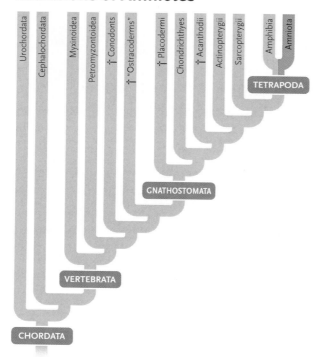

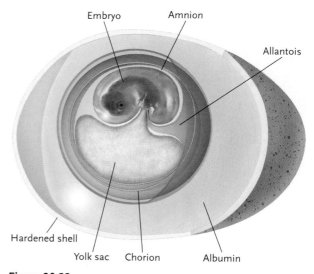

Figure 30.22

The amniote egg. A water-retaining egg with four specialized membranes (the amnion, allantois, chorion, and yolk sac) and a hard or leathery shell allowed amniotes and their descendants to reproduce in dry environments.

The amniote lineage arose during the Carboniferous period, when seed plants and insects, which served as excellent food resources, began to occupy higher ground. The lineage is named for the amnion, a fluid-filled sac that surrounds the embryo during development.

Key Adaptations Allow Amniotes to Live a Fully Terrestrial Life

Although the fossil record includes abundant skeletons of early amniotes, it provides little direct information about their soft body parts and physiology. For amniotes living today, three key adaptations allow them to live in dry habitats, freeing them from a dependency on moist surroundings and standing water. First, they have a tough, dry skin. Its cells are filled with keratin and lipids, which are relatively impermeable to water. Thus, amniotes do not dehydrate in air as quickly as amphibians do.

Second, many amniotes produce an **amniote egg**, which can survive and develop on dry land. The eggs of modern reptiles and birds have four specialized membranes and a hard or leathery shell perforated by microscopic pores **(Figure 30.22)**. The membranes protect the developing embryo and facilitate gas exchange and excretion; the shell mediates the exchange of air and water between the egg and its environment. The egg also includes generous supplies of **yolk**, the embryo's main energy source, and **albumin**, a source of nutrients and water. Compared with those of amphib-

ians, amniote eggs are large; and lacking a larval stage, the young hatch as miniature versions of the adult. By contrast to reptiles and birds, the eggs of virtually all mammals lack a shell; embryos, with the same four membranes, implant in the wall of the mother's uterus and receive nutrients and oxygen directly from her.

Third, some amniotes produce uric acid as a waste product of nitrogen metabolism (see Chapter 46). By contrast, fishes and amphibians produce ammonium ions or urea, toxic materials that require lots of water to flush them from body tissues. Because uric acid is less toxic than these other compounds, it can be excreted as a semisolid paste, conserving body water.

Amniotes Diversified into Three Main Lineages

Based on the abundance and diversity of their fossils, amniotes were extremely successful; they quickly replaced many nonamniote species in terrestrial habitats. During the Carboniferous and Permian periods, amniotes produced three radiations: synapsids, anapsids, and diapsids **(Figure 30.23)**. Differences in skull structure—specifically, the number of bony arches in the temporal region of the skull—distinguish the three groups. In those animals that have temporal arches, the openings between the arches provide space for

Figure 30.23

Amniote ancestry. The early amniotes gave rise to three lineages (anapsids, synapsids, and diapsids) and numerous descendants. The lineages are distinguished by the number of bony arches in the temporal region of the skull (indicated on the small icons).

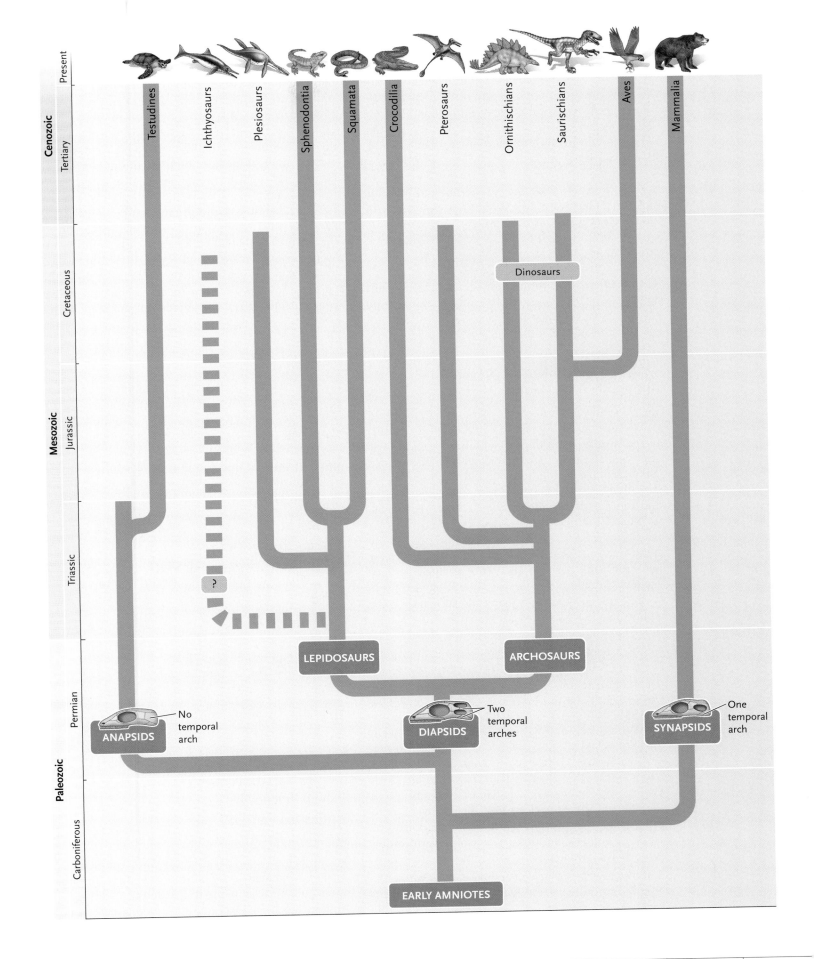

Present

Cenozoic

Tertiary

Testudines

Ichthyosaurs

Plesiosaurs

Sphenodontia

Squamata

Crocodilia

Pterosaurs

Ornithischians

Saurischians

Aves

Mammalia

Cretaceous

Dinosaurs

Mesozoic

Jurassic

Triassic

?

LEPIDOSAURS

ARCHOSAURS

Permian

No temporal arch

ANAPSIDS

Two temporal arches

DIAPSIDS

One temporal arch

SYNAPSIDS

Paleozoic

Carboniferous

EARLY AMNIOTES

large and powerful jaw muscles to bunch up when they contract.

Synapsida. The first offshoot from the ancestral amniotes was a group of small terrestrial predators, the **synapsids** (*syn* = with; *apsis* = arch), which had one temporal arch on each side of the head. Synapsids emerged late in the Permian period; mammals are their living descendants.

Anapsida. A second lineage to emerge was the **anapsids** ("no arch"), which had no temporal arches and no spaces on the sides of the skull. Many biologists believe that turtles are living representatives of this group.

Diapsida. Most Mesozoic amniotes belong to the third lineage, **diapsids** ("two arches"), which had two temporal arches. Their living descendants include lizards and snakes, crocodilians, and birds.

Diapsids Diversified Wildly during the Mesozoic Era

The early diapsids differentiated into two lineages, the **Archosauromorpha** (*archos* = ruler; *saurus* = lizard; *morphe* = form) and the **Lepidosauromorpha** (*lepis* = scale), which differed in many skeletal characteristics. The archosauromorphs (commonly called archosaurs), or "ruling reptiles," include crocodilians, pterosaurs, and dinosaurs. Crocodilians, which first appeared during the Triassic period, have bony armor and a laterally flattened tail that propels them through water. Pterosaurs, now extinct, were flying predators of the Jurassic and Cretaceous periods. Their wings, which spanned as much as 13 m, were composed of thin sheets of skin attached to the sides of the body and supported by an elongate finger. Small pterosaurs may have been active fliers, but large ones probably soared on air currents as vultures do today.

Two lineages of dinosaurs, "lizard-hipped" saurischians and "bird-hipped" ornithischians proliferated in the Triassic and Jurassic periods. As their names imply, they differed in the anatomy of their pelvic girdles. The saurischian lineage included bipedal carnivores and quadrupedal herbivores. Most carnivorous saurischians were swift runners. Their forelimbs, however, were often ridiculously short. *Tyrannosaurus*, which was 15 m long and stood 6 m high, is the most familiar, but most species were much smaller. One group of small carnivorous saurischians was ancestral to birds. By the Cretaceous period, some herbivorous saurischians had also attained gigantic size, and many had long, flexible necks. For example, *Apatosaurus* (previously known as *Brontosaurus*) was 25 m long and may have weighed 50,000 kg.

The largely herbivorous ornithischian dinosaurs had enormous, chunky bodies. This lineage included the armored or plated dinosaurs (*Ankylosaurus* and *Stegosaurus*), the duck-billed dinosaurs (*Hadrosaurus*), horned dinosaurs (*Styracosaurus*), and some with remarkably thick skulls (*Pachycephalosaurus*). The ornithischians were most abundant in the Jurassic and Cretaceous periods.

The second major lineage of diapsids was the lepidosauromorphs (commonly called lepidosaurs), a diverse group that included both marine and terrestrial animals. Plesiosaurs were marine, fish-eating creatures that used long, paddlelike limbs to row through the water. Ichthyosaurs were porpoiselike animals with laterally flattened tails. They were so highly specialized for marine life that they could not venture onto land, even to reproduce. Instead, they gave birth to live young, as porpoises and whales do today. A third important group within this lineage is the squamates, which includes the living lizards and snakes.

STUDY BREAK

1. How did the evolution of the amniote egg free amniotes from a dependence on standing water?
2. What groups of animals are included in each of the three amniote lineages?
3. Based upon the evolutionary history of the diapsid amniotes, are crocodilians more closely related to lizards or to birds?

30.8 Testudines: Turtles

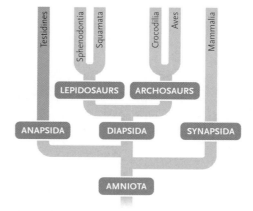

Turtles Have Bodies Encased in a Bony Shell

The turtle body plan, largely defined by a bony, boxlike shell, has changed little since the group first appeared during the Triassic period (**Figure 30.24**). The shell includes a dorsal **carapace** and a ventral **plastron.** A turtle's ribs are fused to the inside of the carapace, and, in contrast to other tetrapods, the pectoral and pelvic girdles lie within the ribcage. Large keratinized scales cover the bony plates that form the shell.

The 250 living species of turtles occupy terrestrial, freshwater, and marine habitats. They range from 8 cm to 2 m in length. All species lack teeth, but they use a keratinized beak and powerful jaw muscles to feed on plants or animal prey. When threatened, most species retract into their shells. Many species are now highly endangered because adults are hunted for meat, their eggs are consumed by humans and other predators, and their young are collected for the pet trade.

STUDY BREAK

How does the overall structure of turtles distinguish them from other amniotes?

30.9 Living Nonfeathered Diapsids: Sphenodontids, Squamates, and Crocodilians

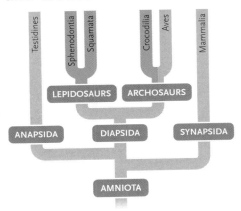

Biologists once grouped living turtles, lizards and snakes, and crocodilians in the class Reptilia, because they all have a dry, scaly skin, produce amniote eggs, and have low metabolic rates and variable body temperatures. But as you read in Section 23.5, these animals do not represent a monophyletic lineage. Turtles are probably not closely related to the other groups. Moreover, the "class Reptilia" excludes birds, which, like crocodilians, are part of the archosaur lineage. In this section, we describe all of the living diapsids except birds, which we consider separately in Section 30.10 because of their unique derived traits and their conspicuous evolutionary success.

Living Sphenodontids Are Remnants of a Diverse Mesozoic Lineage

The tuatara (*Sphenodon punctatus*) is one of two living representatives of the sphenodontids (*sphen* = wedge; *odont* = tooth), a diverse Mesozoic lineage (**Figure 30.25a**). These lizardlike animals survive on a few is-

a. The turtle skeleton

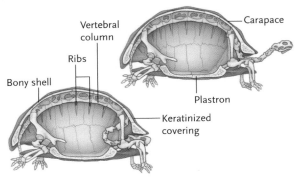

b. An aquatic turtle

Paul J. Fusco/Photo Researchers, Inc.

Figure 30.24
Testudines. **(a)** Most turtles can withdraw their heads and legs into a bony shell. **(b)** Aquatic turtles, like the Eastern painted turtle (*Chrysemys picta*), often bask in the sun to warm up. The sunlight may also help eliminate parasites that cling to the turtle's skin.

lands off the coast of New Zealand. Adults are about 60 cm long. They live in dense colonies, where males and females defend small territories using vocal and visual displays. They often share underground burrows with seabirds, feeding mainly on invertebrates and small vertebrates. They are primarily nocturnal and maintain low body temperatures during periods of activity.

Squamates—Lizards and Snakes—Are Covered by Overlapping, Keratinized Scales

The skin of lizards and snakes (Squamata, from *squama* = scale) is composed of overlapping, keratinized scales that protect against dehydration. Squamates periodically shed their skin as they grow, much the way arthropods shed their old exoskeletons (see Section 29.7). Most squamates regulate their body temperature behaviorally: they are active only when weather conditions are favorable, and they shuttle between sunny and shady places when they need to warm up or cool down (see Section 46.7).

Most of the 3700 lizard species are less than 15 cm long (**Figure 30.25b**). However, the Komodo dragon (*Varanus komodoensis*) grows to nearly 3 m. Lizards occupy a wide range of habitats, but they are especially

Pete & Judy Morrin/Ardea London

© Stephen Dalton/Photo Researchers, Inc.

a. Sphenodontia includes the tuatara *(Sphenodon punctatus)* and one other species.

b. Basilisk lizards *(Basiliscus basiliscus)* escape from predators by running across the surface of streams.

Venom gland

Hollow fang

Andrew Dennis/A.N.T. Photo Library

Mary Ann McDonald/Corbis

c. A western diamondback rattlesnake *(Crotalus atrox)* of the American southwest bares its fangs with which it injects a powerful toxin into prey.

d. Crocodilia includes semiaquatic predators, like this resting African Nile crocodile *(Crocodylus niloticus)*, that frequently bask in the sun.

Figure 30.25
Living nonfeathered diapsids.

common in deserts and the tropics; one species *(Zootoca vivipara)* occurs within the Arctic Circle. Most lizards feed on insects, although some eat leaves or meat. The diverse tropical genus *Anolis* has become a frequent subject of research, as described in *Focus on Research*.

The 2300 species of snakes evolved from a lineage of lizards that lost their legs over evolutionary time **(Figure 30.25c).** Streamlined bodies make snakes efficient burrowers or climbers. Many subterranean species are only 10 or 15 cm long, but the giant constrictors may grow to 10 m. Unlike lizards, all snakes are predators that swallow prey whole. Snakes have smaller skull bones than their lizard ancestors did, and the bones are connected to each other by elastic ligaments that stretch remarkably, allowing some snakes to swallow food that is larger than their head. Snakes also have well-developed sensory systems for detecting prey. The flicking tongue carries airborne molecules to sensory receptors in the roof of the mouth. Most snakes can detect vibrations on the ground, and some, like rattlesnakes, have heat-sensing organs (see Figure 39.22). Many snakes kill by constriction, which suffocates

prey, and several groups produce toxins that immobilize, kill, and partially digest it.

Crocodilians Are Semiaquatic, Predatory Archosaurs

The 21 species of alligators and crocodiles (Crocodilia, from *crocodilus* = crocodile), along with the birds, are the remnants of the once-diverse archosaur lineage **(Figure 30.25d).** The largest species, the Australian saltwater crocodile *(Crocodylus porosus)*, grows to 7 m. Crocodilians are aquatic predators that consume other vertebrates. Striking anatomical adaptations distinguish them from living lepidosaurs, including a four-chambered heart that is homologous to the heart in birds.

American alligators *(Alligator mississippiensis)* exhibit strong maternal behavior, which also reflects their relationship to birds. Females guard their nests ferociously and free their offspring from the nest after they hatch. Young stay close to the mother for about a year, feeding on scraps that fall from her mouth and

Model Research Organisms: *Anolis* Lizards of the Caribbean

The lizard genus *Anolis* has been a model system for studies in ecology and evolutionary biology since the 1960s, when Ernest E. Williams of Harvard University's Museum of Comparative Zoology first began studying it. With more than 400 known species—and new ones being described all the time—*Anolis* is one of the most diverse vertebrate genera. Most anoles are less than 10 cm long, not including the tail, and many occur at high densities, making it easy to collect lots of data in a relatively short time. Male anoles defend territories, and their displays make them conspicuous even in dense forests.

Anolis species are widely distributed in South America and Central America, but nearly 40% occupy Caribbean islands. The number of species on an island is generally proportional to the island's size. Cuba, the largest island, has more than 50 species, whereas small islands have just one or two.

Studies by Williams and others suggest that the anoles on some large islands are the products of independent adaptive radiations. Eight of the 10 *Anolis* species now found on Puerto Rico probably evolved on that island from a common ancestor. Similarly, the seven *Anolis* species on Jamaica shared a common ancestor, which was different from the ancestor of the Puerto Rican species. The anole faunas on Cuba and Hispaniola are the products of several independent radiations on each island.

Williams discovered that these independent radiations had produced similar-looking species on different islands. He developed the concept of the *ecomorph,* a group of species that have similar morphological, behavioral, and ecological characteristics even though they are not closely related within the genus. Williams named the ecomorphs after the vegetation that they commonly used (see **figure**). For example, grass anoles are small, slender species that usually perch on low, thin vegetation. Trunk-ground anoles have chunky bodies and large heads, and they perch low on tree trunks, frequently jumping to the ground to feed. Although the grass anoles or the trunk-ground anoles on different islands are similar in many ways, they are not closely related to each other. Their resemblances are the products of convergent evolution.

Ecomorphs exist because evolutionary processes have accentuated the morphological differences among species that occupy different types of vegetation. Jon Losos of Harvard University has demonstrated that trunk-ground anoles, which have relatively long legs and tails, can run faster on wide surfaces and jump farther than species with relatively short legs. And in nature the trunk-ground anoles run and jump more frequently than the other ecomorphs do.

Different ecomorphs on an island use different parts of their habitats by choosing different perch sites (grass, tree trunks, rocks). When two or more species of the same ecomorph inhabit the same island, they occupy habitats with different temperature and shade conditions (see the figure). For example, in Puerto Rico, one species of trunk-ground anole *(Anolis gundlachi)* occupies cool, shady uplands; another *(Anolis cristatellus)* lives in warm, fairly open lowland habitats; and a third species *(Anolis cooki)* lives in desert habitats. Other species in Puerto Rico exhibit similar differences in their distributions. These differences in geographical distribution and habitat use presumably allow the different species to avoid competition with each other and gain access to the resources they need to survive and reproduce.

Evolutionary processes have also fostered physiological differences that reinforce the ecological separation established by the lizards' use of different habitats. For example, *A. cristatellus* maintains higher body temperatures than *A. gundlachi,* and neither is physiologically adapted to the environment of the other: *A. cristatellus* dies in the high altitude forests where *A. gundlachi* thrives, while *A. gundlachi* suffers heat stress at body temperatures that are typical for *A. cristatellus.*

Researchers throughout the Americas continue to explore the ecology and evolution of anoles. Some unravel their biogeography and systematic relationships; others focus on the ecology of populations and communities; still others study their social behavior or sensory physiology. With so many species distributed across hundreds of Caribbean islands, the lizard genus *Anolis* provides fertile ground for testing hypotheses about nearly every aspect of vertebrate biology.

A. cooki

Manuel Leal, Duke University

A. poncensis

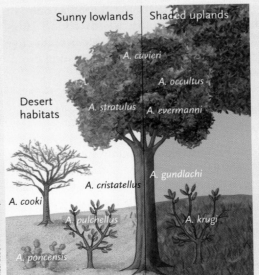

Sunny lowlands   Shaded uplands

Desert habitats

A. cuvieri
A. occultus
A. stratulus   A. evermanni
A. cristatellus   A. gundlachi
A. cooki
A. pulchellus   A. krugi
A. poncensis

Manuel Leal, Duke University

A. gundlachi

Manuel Leal, Duke University

A. krugi

Manuel Leal, Duke University

living under her watchful protection. Most alligator and crocodile species are highly endangered. Their habitats have been disrupted by human activities, and they have been hunted for meat and leather. Protection efforts have been extremely successful, however. American alligators, for example, recently recovered from the brink of extinction.

30.10 Aves: Birds

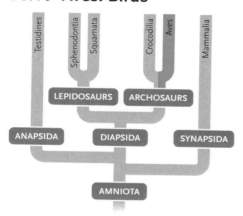

Birds (Aves) appeared in the Jurassic period as descendants of carnivorous, bipedal dinosaurs. Thus, they are full-fledged members of the archosaur lineage. Their evolutionary relationship to dinosaurs is evident in their skeletal anatomy, the scales on their legs and feet, and their posture when walking. However, powered flight gave birds access to new adaptive zones, setting the stage for their astounding evolutionary success (Figure 30.26a).

Key Adaptations Reduce Body Weight and Provide Power for Flight

The skeletons of birds are both lightweight and strong (Figure 30.26b). For example, the endoskeleton of the frigate bird, which has a 1.5 m wingspan, weighs little more than 100 g, far less than its feathers. Most birds have hollow limb bones with small supporting struts that crisscross the internal cavities. Evolution has also reduced the number of separate bony elements in the wings, skull, and vertebral column (especially the tail),

making the skeleton light and rigid. And all modern birds lack teeth, which are dense and heavy; they acquire food with a lightweight, keratinized bill. Many species have a long, flexible neck, which allows them to use their bills for feeding, grooming, nest-building, and social interactions.

The bones associated with flight are generally large. The forelimb and forefoot are elongate, forming the structural support for the wing. And most modern birds possess a **keeled sternum** (breastbone) to which massive flight muscles attach (Figure 30.26c). Not all birds are strong fliers, however; ostriches and other bipedal runners have strong, muscular legs but small wings and flight muscles (see Figure 19.2).

Like the skeleton, soft internal organs are modified in ways that reduce weight. Most birds lack a urinary bladder; uric acid paste is eliminated with digestive wastes. Females have only one ovary and never carry more than one mature egg; eggs are laid as soon as they are shelled.

All birds also possess **feathers** (Figure 30.26d), sturdy, lightweight structures derived from scales in the skin of their ancestors. Each feather has numerous barbs and barbules with tiny hooks and grooves that maintain the feathers' structure, even during vigorous activity. Flight feathers on the wings provide lift; contour feathers streamline the surface of the body; and down feathers form an insulating cover close to the skin. Worn feathers are replaced once or twice each year.

Other adaptations for flight allow birds to harness the energy needed to power their flight muscles. Their metabolic rates are eight to ten times higher than those of other comparably sized diapsids, and they process energy-rich food rapidly. A complex and efficient respiratory system (see Figure 44.7) and four-chambered heart (see Figure 42.5d) enable them to consume and distribute oxygen efficiently. As a consequence of high rates of metabolic heat production, birds maintain a high and constant body temperature (see Section 46.8).

Flying Birds Were Abundant by the Cretaceous Period

Although the earliest known bird, the pigeon-sized *Archaeopteryx,* had feathers, its skeleton was essentially that of a small dinosaur (see Figure 19.12). It had digits and claws on the forelimbs, teeth on its jaws, many bones in its wings and vertebral column, and only a poorly developed sternum. How could flight evolve in so unbirdlike an animal? Biologists hypothesize that *Archaeopteryx* ran after prey, using its feathered wings like fly swatters. Larger wings would have provided extra lift when they jumped at prey, and gradual evolutionary modifications of the wing bones and muscles could have led to powered flight.

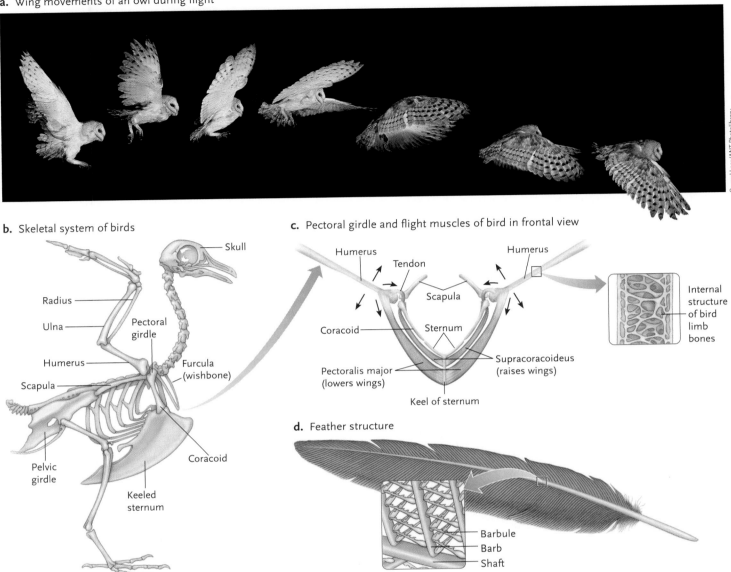

a. Wing movements of an owl during flight

Gerard Lacz/ANT Photolibrary

b. Skeletal system of birds

- Skull
- Radius
- Ulna
- Pectoral girdle
- Humerus
- Furcula (wishbone)
- Scapula
- Pelvic girdle
- Coracoid
- Keeled sternum

c. Pectoral girdle and flight muscles of bird in frontal view

- Humerus
- Tendon
- Humerus
- Scapula
- Coracoid
- Sternum
- Pectoralis major (lowers wings)
- Supracoracoideus (raises wings)
- Keel of sternum
- Internal structure of bird limb bones

d. Feather structure

- Barbule
- Barb
- Shaft

Figure 30.26
Adaptations for flight in birds. **(a)** The flapping movements of a bird's wing provide *thrust* for forward movement and *lift* to counteract gravity. **(b)** The bird skeleton includes a boxlike trunk, short tail, long neck, lightweight skull and beak, and well-developed limbs. In large birds, limb bones are hollow. **(c)** Two sets of flight muscles attach to a keeled sternum; one set raises the wings, and the other lowers it. **(d)** Flexible feathers form an airfoil on the wing surface.

Crow-sized birds with full flight capability appeared by the early Cretaceous period. They had a keeled sternum and other modern skeletal features. The modern groups of wading birds and seabirds first appear in late Cretaceous rocks; fossils of other modern groups are found in slightly later deposits. Woodpeckers, perching birds, birds of prey, pigeons, swifts, the flightless ratites, penguins, and some other groups were all present by the end of the Oligocene; birds continued to diversify through the Miocene (see Table 22.1).

Modern Birds Vary in Morphology, Diet, Habits, and Patterns of Flight

The 9000 living bird species show extraordinary ecological specializations, but they share the same overall body plan. Living birds are traditionally classified into nearly 30 groups **(Figure 30.27)**. They vary in size from the bee hummingbird *(Mellisuga helenae)* of Cuba, which weighs little more than 1 g, to the ostrich *(Struthio camelus),* which can weigh as much as 150 kg.

a. The Laysan albatross (Procellariiformes, *Phoebastria immutabilis*) has the long thin wings typical of birds that fly great distances.

b. The roseate spoonbill (Ciconiformes, *Ajaia ajaja*) uses its bill to strain food particles from water.

c. The bald eagle (Falconiformes, *Haliaeetus leucocephalus*) uses its sharp bill and talons to capture and tear apart prey.

d. A European nightjar (Caprimulgiformes, *Caprimulgus europaeus*) uses its wide mouth to capture flying insects.

e. A Bahama woodstar hummingbird (Apodiformes, *Calliphlox evelynae*) hovers before a hibiscus blossom to drink nectar from the base of the flower.

f. The chestnut-backed chickadee (Passeriformes, *Parus rufescens*) uses its thin bill to probe for insects in dense vegetation.

Figure 30.27
Bird diversity.

The structure of the bill usually reflects a bird's diet. Seed and nut eaters, such as finches and parrots, have deep, stout bills that crack hard shells. Carnivorous hawks and carrion-eating vultures have sharp beaks to rip flesh, and nectar-feeding hummingbirds have slender bills to reach into flowers. The bills of ducks are modified to extract particulate matter from water, and many perching birds have slender bills to feed on insects.

Birds also differ in the structure of their feet and wings. Predators have large, strong talons (claws), whereas ducks and other swimming birds have webbed feet that serve as paddles. Long-distance fliers like albatrosses have narrow wings; those that hover at flowers, such as hummingbirds, have wide ones. The wings of some species, like penguins, are so specialized for swimming that they are incapable of aerial flight.

All birds have well-developed sensory and nervous systems, and their brains are proportionately larger than those of other diapsids of comparable size. Large eyes provide sharp vision, and most species also have good hearing, which nocturnal hunters like owls use to locate prey. Some vultures and other species have a good sense of smell, which they use to find food. Migrating birds use polarized light, changes in air pressure, and Earth's magnetic field for orientation.

Most birds exhibit complex social behavior, including courtship, territoriality, and parental care. Many species communicate with vocalizations and visual displays to challenge other individuals or attract mates. Most raise their young in a nest, using body heat to incubate eggs. The nest may be a simple depression on a gravely beach, a cup woven from twigs and grasses, or a feather-lined hole in a tree.

Many bird species embark on a semiannual long-distance migration (see Section 55.1). The golden plover *(Pluvialis dominica)*, for example, migrates 20,000 km twice each year. Migrations are a response to seasonal changes in climate. Birds travel toward the tropics as winter approaches; in spring, they return to high latitudes using seasonally available food sources.

STUDY BREAK

1. What specific adaptations allow birds to fly?
2. How do the structures of a bird's bill, wings, and feet reflect its dietary and habitat specializations?

30.11 Mammalia: Monotremes, Marsupials, and Placentals

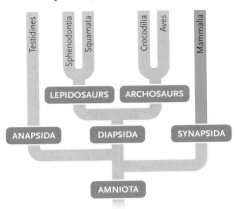

The synapsid lineage, which includes the living mammals, was the first group of amniotes to diversify broadly on land. Indeed, during the late Paleozoic era, medium- to large-sized synapsids were the most abundant predators in terrestrial habitats. One particularly successful and persistent branch of the synapsid lineage, the *therapsids,* exhibited many mammal-like characteristics in their legs, skulls, jaws, and teeth. And by the end of the Triassic period, the earliest mammals—most of them no bigger than a rat—had emerged from therapsid ancestors. Several mammalian lineages coexisted with dinosaurs and other diapsids throughout the Mesozoic, but paleontologists hypothesize that most Mesozoic mammals were active only at night to avoid predatory dinosaurs, which were active during the day. Two mammalian lineages, the egg-laying Prototheria (or monotremes) and the live-bearing Theria (marsupials and placentals), survived the mass extinction that eliminated most dinosaurs at the end of the Mesozoic. The Theria diversified into the mammalian groups that are most familiar today.

Mammals Exhibit Key Adaptations in Anatomy, Physiology, and Behavior

Four sets of key adaptations fostered the success of mammals.

High Metabolic Rate and Body Temperature. Like birds, mammals have high metabolic rates that release enough energy to maintain high activity levels and enough heat to maintain high body temperatures (see Section 46.8). An outer covering of fur and a layer of subcutaneous fat help retain body heat. Using metabolic heat to stay warm requires lots of oxygen, and mammals have a muscular organ, the diaphragm, that fills their lungs with air (see Figure 44.8). Four-chambered hearts and complex circulatory systems deliver oxygen to active tissues (see Figure 42.5d).

Specializations of the Teeth and Jaws. Mammals also have anatomical features that allow them to feed efficiently. Ancestrally, mammals have four types of teeth (see Figure 45.17): flattened **incisors** nip and cut food; pointed **canines** pierce and kill prey; and two sets of cheek teeth, **premolars** and **molars**, grind and crush food. Moreover, teeth in the upper and lower jaws occlude (that is, fit together) tightly as the mouth is closed; thus, mammals can use their large jaw muscles to chew food thoroughly.

Parental Care. Mammals provide more parental care to their young than any other animals. In most species, young complete development within a female's uterus, deriving nourishment through the **placenta**, a specialized organ that mediates the delivery of oxygen and nutrients (see Section 47.2). Females also have **mammary glands**, specialized structures that produce energy-rich milk, a watery mixture of fats, sugars, proteins, vitamins, and minerals. This perfectly balanced diet is the sole source of nutrients for newborn offspring.

Complex Brains. Finally, mammals have larger brains than other tetrapods of equivalent body size; the difference lies primarily in the **cortex**, the part of the forebrain responsible for information processing and learning (see Figure 38.6). Extensive postnatal care provides opportunities for offspring to learn from older individuals. Thus, mammalian behavior is strongly influenced by past experience as well as by genetically programmed instincts.

The Major Groups of Modern Mammals Differ in Their Reproductive Adaptations

Biologists recognize a primary distinction between two lineages of modern mammals: the egg-laying Prototheria (*protos* = first; *therion* = wild beast), or monotremes, and the live-bearing Theria. The Theria, in turn, diversified into two sublineages, the Metatheria (*meta* = between), or marsupials, and the Eutheria (*eu* = true), or placentals, which also differ in their reproductive adaptations.

Monotremes. The three living species of monotremes (Prototheria), which are limited to the Australian region, reproduce with a leathery-shelled egg **(Figure 30.28)**. Newly hatched young lap up milk secreted by modified sweat glands on the mother's belly. The duck-billed platypus *(Ornithorhynchus anatinus)* lives in burrows along riverbanks and feeds on aquatic invertebrates. The two species of echidnas, or spiny anteaters (*Tachyglossus aculeatus* and *Zaglossus bruijni*), feed on ants or termites.

Marsupials. The 240 species of marsupials (Metatheria) have short gestation: the young are nourished through a placenta very briefly—sometimes only for

a. Short-nosed echidna

D. & V. Blagden/ANT Photo Library

b. Duck-billed platypus

Jean Phillipe Varin/Jacana/Photo Researchers, Inc.

Figure 30.28

Monotremes. **(a)** The short-nosed echidna *(Tachyglossus aculeatus)* is terrestrial. **(b)** The duck-billed platypus *(Ornithorhynchus anatinus)* raises its young in a streamside burrow.

8 to 10 days—before birth. Newborns use their forelimbs to drag themselves across the mother's belly fur and enter her abdominal pouch, the **marsupium**, where they complete development attached to a teat. Marsupials are the dominant native mammals of Australia and a minor component of the South American fauna **(Figure 30.29);** only one species, the opossum *(Didelphis virginiana),* occurs in North America. South America once had a diverse marsupial fauna, but it declined after the Isthmus of Panama bridged the sea-

Milse, T./Arco Images/Peter Arnold, Inc.

Figure 30.29

Marsupials. An Eastern gray kangaroo *(Macropus giganteus)* carries her "joey" in her pouch.

way between North and South America (see *Focus on Research* in Chapter 22).

Placentals. The 4000 species of placental mammals (Eutheria) are the dominant mammals today. They complete embryonic development in the mother's uterus, nourished through a placenta until they reach a fairly advanced stage of development. Some species, like humans, are helpless at birth, but others, such as horses, are quickly mobile.

Biologists divide eutherians into about 18 groups, only eight of which contain more than 50 living species **(Figure 30.30).** Rodents (Rodentia) make up about 45% of eutherian species, and bats (Chiroptera) comprise another 22%. Our own group, Primates, is represented by fewer than 170 living species (less than 5% of all mammalian species), many of which are highly endangered. Researchers still do not agree on the details of eutherian evolution. *Insights from the Molecular Revolution* describes the use of molecular techniques to resolve one question about their relationships.

Some eutherians have highly specialized locomotor structures. Whales and dolphins (Cetacea) and manatees and dugongs (Sirenia) are descended from terrestrial ancestors, but their appendages do not function on land, and they are now restricted to aquatic habitats. By contrast, seals and walruses (Carnivora) feed under water but rest and breed on land. Bats (Chiroptera) use wings for powered flight.

Although early mammals appear to have been insectivorous, the diets of modern eutherians are diverse. Odd-toed ungulates *(ungula* = hoof) like horses and rhinoceroses (Perissodactyla), even-toed ungulates like cows and camels (Artiodactyla), and rabbits and hares (Lagomorpha) feed on vegetation. Carnivores (Carnivora) consume other animals. Most insectivores (Insectivora) and bats eat insects, but some feed on flowers,

a. The capybara (Rodentia, *Hydrochoerus hydrochaeris*), the largest rodent, feeds on vegetation in South American wetlands.

b. Most bats, like the Yuma Myotis (Chiroptera, *Myotis yumanensis*), are nocturnal predators on insects.

c. Walruses (Carnivora, *Obodenus rosmarus*) feed primarily on marine invertebrates in frigid arctic waters.

d. The black rhinoceros (Perissodactyla, *Diceros bicornis*) feeds on grass in sub-Saharan Africa.

e. Arabian camels (Artiodactyla, *Camelus dromedarius*) use enlarged foot pads to cross hot desert sands.

f. Antillean manatees (Sirenia, *Trichechus manatus*) are herbivores that live in warm coastal marshes and rivers from Florida to northern South America.

Figure 30.30
Eutherian diversity.

fruit, and nectar. Many whales and dolphins prey on fishes and other animals, but some eat plankton. And some groups, including rodents and primates, feed opportunistically on both plant and animal matter.

STUDY BREAK

1. During the Mesozoic era, why were most mammals active only at night?
2. Which key adaptations in mammals allow them to be active under many types of environmental conditions?
3. On what basis are the major groups of living mammals distinguished?

30.12 Nonhuman Primates

We now focus our attention on Primates, the mammalian lineage that includes humans, apes, monkeys, and their close relatives. The first Primates appeared early in the Eocene epoch, about 55 million years ago, in forested habitats in North America, Europe, Asia, and North Africa.

Key Derived Traits Enabled Primates to Become Arboreal, Diurnal, and Highly Social

Several derived traits allow primates to be arboreal (to live in trees rather than on the ground). For example, most primates have a more erect posture than

INSIGHTS FROM THE MOLECULAR REVOLUTION

The Guinea Pig Is Not a Rat

Using the Linnaean system of taxonomy, the Rodentia has traditionally included more than 1800 species distributed among 29 families, including squirrels, rats and mice, guinea pigs, and porcupines. Their placement in the same order implies that they have a common evolutionary ancestor not shared by any other groups within the mammals. Biologists commonly accepted this interpretation until a molecular study compared the amino acid sequences of 15 proteins encoded in the nuclear DNA of various rodents. The comparisons revealed differences suggesting that guinea pigs should be placed in a separate order. Since then, further molecular evaluations of nuclear genes have produced contradictory results, with some studies supporting the tradi-

tional classification of guinea pigs as rodents and others placing them outside the Rodentia.

A cooperative study by Anna Maria D'Erchia and her colleagues at universities and institutes in Italy and Sweden now adds molecular weight to the conclusion that guinea pigs belong in an order of their own. The research team used mitochondrial DNA (mtDNA) sequences because they are easy to isolate and purify, and typically undergo many random mutations and rearrangements that have no apparent effect on gene function. Thus the changes observed in mtDNA are expected to reflect more faithfully the ticking of the molecular clock that tracks the time course of evolutionary events.

For their study, the researchers sequenced mtDNA of the guinea pig and

another mammal considered by some biologists to be closely related to rodents, rabbits (Lagomorpha). Other workers had previously sequenced the mtDNAs of 14 other mammals in eight orders, including primates (Primates), seals (Carnivora), cows (Artiodactyla), whales (Cetacea), horses (Perissodactyla), mice and rats (Rodentia), hedgehogs (Insectivora), and opossums (Marsupialia).

The researchers evaluated these sequences with three different statistical programs that use similarities and differences in mtDNA sequences to construct evolutionary trees. They also conducted analyses using nuclear DNA, including separate evaluations of the entire nuclear genome, the protein-encoding sequences, and the DNA encoding ribosomal RNA. Significantly, all the methods produced essentially the same family tree (shown in **Figure a**).

The tree places guinea pigs in a group of their own, sharing a more recent common ancestor with all of the mammalian orders examined except those represented by rodents, hedgehogs, and opossums. The lineage that includes guinea pigs and most other mammals shared a common ancestor with the lineage leading to rodents at a point further back in evolutionary time. And the lineage that includes rodents, guinea pigs, and most other mammals split off from the ancestors of hedgehogs and opossums even earlier. Thus, guinea pigs merit placement in a separate group from rodents. The results for rabbits also indicate that they are more closely related to other mammals than they are to mice and rats. Incidentally, the tree also supports the conclusion from other molecular studies that cows and whales are more closely related to each other than to other mammals (see *Insights from the Molecular Revolution* in Chapter 23).

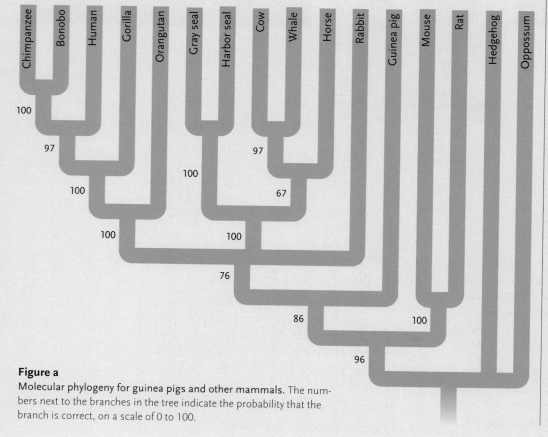

Figure a

Molecular phylogeny for guinea pigs and other mammals. The numbers next to the branches in the tree indicate the probability that the branch is correct, on a scale of 0 to 100.

other mammals, and they have flexible hip and shoulder joints, which allow a variety of locomotor activities. They can grasp objects with their hands and feet, because they have nails, not claws, on their fingers and toes; their fingertips are well endowed with sensory nerves that enhance the sense of touch. Unlike other mammals, most primates have an opposable big toe, which can touch the tips of other digits and the sole of the foot; many species also have an opposable thumb.

Most primates are diurnal (active during daylight hours), and, unlike most mammals, they rely more on vision than on their sense of smell. Thus, they generally have short snouts and small olfactory lobes of the brain. Most species have forward-facing eyes with overlapping fields of vision, providing excellent depth perception, which comes in handy when moving through trees. Many species have color vision.

Primate brains—especially the regions that integrate information—are large and complex. As a result, they have an exceptional capacity to learn. Most species live in social groups; thus, young primates, which mature slowly, can interact with and learn from their elders and peers during an extended period of parental care. Females give birth to only one or two young at a time, allowing them to devote substantial attention to each offspring.

Living Primates Include Two Major Lineages

Primatologists recognize two lineages within the Primates **(Figure 30.31)**, the Strepsirhini (*streptos* = twisted or turned, *rhin* = nose) and the Haplorhini (*haploos* = single or simple).

Strepsirhini. The 36 living species of Strepsirhini—lemurs, lorises, and galagos—possess many ancestral morphological traits, including moist, fleshy noses and eyes that are positioned somewhat laterally on their heads **(Figure 30.32)**. Strepsirhines generally have short gestation periods and rapid maturation. Today, 22 lemur species survive on Madagascar, a large island off the east coast of Africa; they are ecologically diverse and range in size from 40 g to 7 kg; some lemurs are arboreal, whereas others spend substantial time on the ground. The 12 species of lorises and galagos occupy tropical forest and subtropical woodlands in Africa, India, and Southeast Asia; they are all arboreal and nocturnal.

Haplorhini. Most species in the Haplorhini—the familiar monkeys and apes—have many derived primate characteristics, including compact, dry noses, and forward-facing eyes.

However, five species of tarsiers, which are restricted to tropical forests on the islands of Southeast Asia, exhibit several ancestral traits: small body size

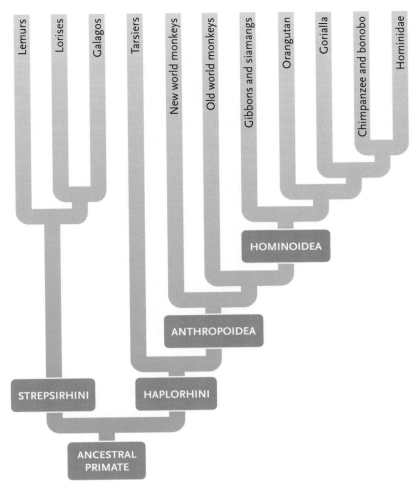

Figure 30.31

Primate phylogeny. A phylogenetic tree for the Primates illustrates the two main lineages: Strepsirhini and Haplorhini. Note that chimpanzees are the closest living relatives of humans.

(about 100 g), large eyes and ears, and two grooming claws on each foot **(Figure 30.33)**. But they share the derived characteristics of dry noses and forward-facing eyes with the other haplorhines; and DNA sequence data link them to the monkeys and apes and not to the strepsirhines described earlier.

The 130 or so species of monkeys, 13 species of apes, and humans constitute the monophyletic haplorhine lineage **Anthropoidea,** which probably arose in Africa; fossils of a diverse and abundant radiation of forest-dwelling anthropoids, dating from the late Eocene epoch, have been discovered in northern Egypt. Continental drift then established long-term geographical and evolutionary separation of anthropoids in the New World and Old World **(Figure 30.34)**.

By the middle of the Oligocene epoch, about 30 million years ago, the ancestors of the New World monkeys had arrived in South America and begun to diversify there. They probably rafted across the

Figure 30.32

Strepsirhines. The ring-tailed lemur *(Lemur catta)* of Madagascar has ancestral primate characteristics, such as a long snout and wet nose.

Figure 30.33

Tarsiers. Tarsiers *(Tarsius bancanus)* are classified as haplorhines, but they retain many ancestral characteristics.

a. Spider monkey

b. Hamadryas baboon

Figure 30.34

New World and Old World monkeys. (a) Many New World monkeys, such as the spider monkey *(Ateles geoffroyi)*, have prehensile tails, which they use as a fifth limb. Old World monkeys lack prehensile tails, and many, such as **(b)** the Hamadryas baboon *(Papio hamadryas)*, are largely terrestrial.

Atlantic, which was narrower at that time, on trees or other storm debris. New World monkeys now live in Central and South America (see Figure 30.34a). They range in size from tiny marmosets and tamarins (350 g) to hefty howler monkeys (10 kg). Most are exclusively arboreal and diurnal. The larger species may hang below branches by their arms, and some use a prehensile (grasping) tail as a fifth limb.

Anthropoids diversified most spectacularly in the Old World, however, eventually giving rise to two lineages—one ancestral to Old World monkeys and the other to apes and humans. Although many people assume that the apes are descended from Old World monkeys, the fossil record contradicts that impression. The earliest hominoid (ape) fossils date to the early Miocene, roughly 23 million years ago, but the oldest known Old World monkeys appeared several million years later.

Old World monkeys, which occupy habitats ranging from tropical rain forests to deserts in Africa and Asia, may grow as large as 35 kg (see Figure 30.34b). Many species are sexually dimorphic; in other words, males and females attain different adult sizes (see Section 20.3). Arboreal species use all four limbs for locomotion, but none has a prehensile tail. Some species, such as baboons, often walk or run on the ground.

Within the anthropoid lineage, the **Hominoidea** ("humanlike") is a monophyletic group that includes apes and humans. The climate of the early Miocene was wetter than it is today, and eastern Africa, where many early hominoid fossils are found, was covered with extensive forests. A climate shift in the middle Miocene, around 14 million years ago, converted dense forests into woodlands and grasslands. Hominoids probably adopted a more terrestrial existence and shifted their diets. Miocene hominoids ranged in size from 4 kg to 80 kg. They occupied both forest and open woodland habitats; some were probably ground dwelling.

Although hominoids are closely related to Old World monkeys, several characteristics distinguish them. Apes lack a tail, and great apes (orangutans, gorillas, chimpanzees, and bonobos) are much larger than monkeys. Moreover, the posterior region of the vertebral column is shorter and more stable in apes. Apes also show more complex behavior.

The gibbons and siamangs, which live in tropical forests in Southeast Asia, are the smallest of the apes, ranging in weight from 6 to 11 kg. With extremely long arms and strong shoulders, they hang below branches by their arms and swing themselves forward, a pattern of locomotion called **brachiation (Figure 30.35a).** The much larger orangutan *(Pongo pygmaeus)*, now restricted to forested areas on the islands of Borneo and Sumatra, can grow to 90 kg. Orangutans use both hands and feet to climb trees; they sometimes venture onto the ground on all fours.

Gorillas *(Gorilla gorilla)*, which are currently restricted to two large central African forests, are the largest of the living primates. Males can weigh 180 kg; females are about half that size. Because of their size, gorillas spend most of their time on the ground. They often use "knuckle-walking" locomotion, leaning forward and supporting part of their weight on the backs of their hands. Gorillas are almost exclusively vegetarian.

Chimpanzees *(Pan troglodytes)* are also forest dwellers, weighing up to 45 kg **(Figure 30.35b).** Like gorillas, they spend most of their waking hours on the ground; they often knuckle-walk, but sometimes adopt a **bipedal** (two-legged) stance and swagger short distances. Groups of related males form loosely defined communities of up to 50 individuals, which may cooperate in hunts and foraging. Bonobos *(Pan paniscus),* sometimes called pygmy chimpanzees, are restricted to a small area in central Africa. Somewhat smaller than chimps, they have longer legs and smaller heads.

The Primates also includes humans *(Homo sapiens),* which occupy virtually all terrestrial habitats. Humans have adaptations that allow an upright posture and bipedal locomotion. They are ground-dwelling animals with extremely broad diets and complex social behavior.

a. Black-handed gibbon   **b.** Chimpanzee

Figure 30.35

Apes. **(a)** Small-bodied apes, such as the black-handed gibbon *(Hylobates agilis)* are agile brachiators that swing through the trees with ease. **(b)** Among the large-bodied apes, chimpanzees *(Pan troglodytes)* have opposable thumbs and big toes.

Figure 30.36

Adaptations for bipedal locomotion. Differences in the posture, skeleton, and muscles of monkeys, great apes, and humans illustrate the anatomical changes that accompanied upright, bipedal locomotion. Evolutionary changes in the spine, pelvis, hip, knee, ankle, and foot were accompanied by changes in the sizes of leg muscles and their points of attachment to the bones they move.

Monkey   Ape   Human

STUDY BREAK

1. What characteristics of primates allow them to spend a great deal of time in trees?
2. What is the lowest taxonomic group that includes monkeys, apes, and humans? What is the lowest taxonomic group that includes only apes and human?
3. Which species of ape spend most of the time on the ground?

30.13 The Evolution of Humans

Genetic analyses of living hominoid species indicate that African hominoids diverged into several lineages between 10 million and 5 million years ago; one lineage, the **hominids**, includes modern humans and our bipedal ancestors.

Hominids First Walked Upright in East Africa about 6 Million Years Ago

Upright posture and bipedal locomotion are key adaptations that distinguish hominids from apes. Researchers identify early hominid fossils from features of the skull, spine, pelvis, knees, ankles, and feet that make bipedal locomotion possible **(Figure 30.36)**. As a conse-

quence of bipedal locomotion, the hands were no longer used for locomotor functions, allowing them to become specialized for other activities, such as tool use. Evolutionary refinements in grasping ability allow hominids to hold objects tightly with a *power grip* or manipulate them precisely with a *precision grip* **(Figure 30.37)**. Hominids also developed larger brains.

Paleontologists have uncovered fossil of numerous hominids that lived in East Africa and South Africa from roughly 6 million to 1 million years ago **(Figure 30.38)**. In 2000, researchers found 13 fossils of *Orrorin tugenensis* ("first man" in a local African language), a species that lived about 6 million years ago in East Af-

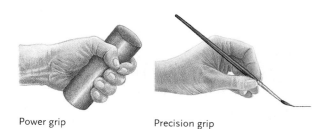

Power grip   Precision grip

Figure 30.37

Power grip versus precision grip. Hominids grasp objects in two distinct ways. **(a)** The power grip allows us to grasp an object firmly. **(b)** The precision grip allows us to manipulate objects with fine movements.

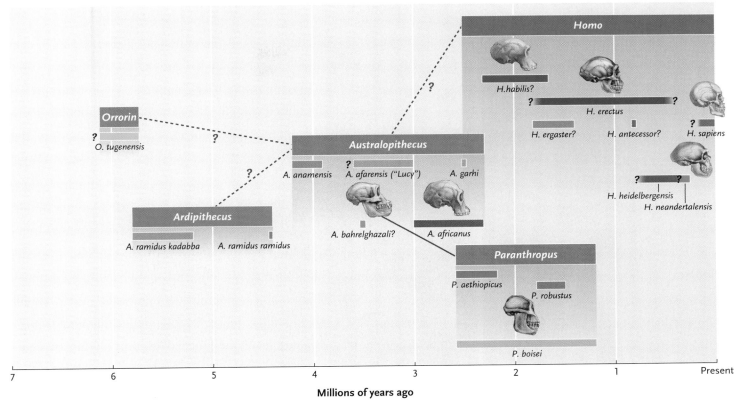

Figure 30.38

Hominid time line. Several species of hominids lived simultaneously at sites in eastern and southern Africa. The time line for each species and genus reflects the ages of known fossils. The numerous question marks indicate researchers' uncertainty about the classification, the ages of fossils, and the evolutionary relationships among the species. Some of the skulls pictured are reconstructed from fragmentary fossils.

rican forests. However, these remains are fragmentary, and experts are still evaluating them. The best-studied early hominid fossils, the remains of 50 individuals discovered in the East African Rift Valley, date from about 5 million years ago. Named *Ardipithecus ramidus*, these hominids stood 120 cm tall and had apelike teeth. Other *Ardipithecus* fossils, recently discovered at a different site, appear to be much older (5.8 million years) and show evidence of bipedal locomotion.

Hominid fossils from 4.2 million to 1.2 million years ago are known from many sites in East, Central, and South Africa. They are currently assigned to the genera *Australopithecus* (*australis* = southern; *pithekos* = ape) and *Paranthropus* (*para* = beside; and *anthropos* = human being). With their large faces, protruding jaws, and small skulls and brains, most of these hominids had an apelike appearance (see Figure 30.38). *Australopithecus anamensis*, which lived in East Africa around 4 million years ago, is the oldest known species. It had thick enamel covering on its teeth, a derived hominid characteristic; the structure of a fossilized leg bone suggests that it was bipedal.

Specimens of more than 60 individuals of *Australopithecus afarensis* have been found in northern Ethiopia, including about 40% of a female's skeleton, named "Lucy" by its discoverers **(Figure 30.39)**. *A. afarensis* lived 3.5 million to 3 million years ago, but it retained several ancestral characteristics. For example, it had moderately large and pointed canine teeth, and a relatively small brain. Males and females were 150 cm and 120 cm tall, respectively. Skeletal analyses suggest that *A. afarensis* was fully bipedal, a conclusion supported by fossilized footprints preserved in a layer of volcanic ash.

Other species of *Australopithecus* and *Paranthropus* lived in East Africa or South Africa between 3.7 million and 1 million years ago. Adult males ranged from 40 to 50 kg in weight and from 130 to 150 cm in height; females were smaller. Most species had deep jaws and large molars. Several species had a crest of bone along the midline of the skull, providing a large surface for the attachment of jaw muscles. These anatomical features suggest that they fed on hard food, such as nuts, seeds, and other vegetable products. One species,

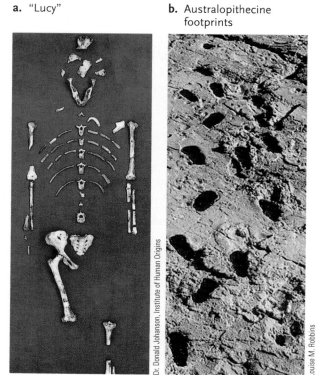

a. "Lucy"   **b.** Australopithecine footprints

Dr. Donald Johanson, Institute of Human Origins

Louise M. Robbins

Figure 30.39
Australopithecines. **(a)** Researchers named the most complete fossil of *Australopithecus afarensis* "Lucy." **(b)** Mary Leakey discovered australopithecine footprints, made in soft, damp volcanic ash about 3.7 million years ago. The footprints indicate that australopithecines were fully bipedal.

Australopithecus africanus, known only from South Africa, had small jaws and teeth, indicating that it probably consumed a softer diet. The phylogenetic relationships of the species classified as *Australopithecus* and *Paranthropus*—and their exact relationships to later hominids—are not yet fully understood. But most scientists agree that *Australopithecus* was ancestral to humans, which are classified in the genus *Homo.*

Homo habilis Was Probably the First Hominid to Manufacture Stone Tools

Pliocene fossils of the earliest humans, which may have included several species, are fragmentary. They are also widely distributed in space and time, complicating analyses of their relationships. For the sake of simplicity, we describe them as belonging to one species, *Homo habilis* ("handy man").

From 2.3 million to 1.7 million years ago, *H. habilis* occupied the woodlands and savannas of eastern and southern Africa, sharing these habitats with various species of *Paranthropus.* The two genera are easy to tell apart because the brains of *H. habilis* were at least 20% larger, and they had larger incisors and smaller molars than their hominid cousins. Their diet

included hard-shelled nuts and seeds as well as soft fruits, tubers, leaves, and insects. They may also have hunted small prey or scavenged carcasses left by large predators.

Researchers have found numerous tools dating to the time of *H. habilis,* but they are not sure which species used them. Many of the hominid species alive at the time probably cracked marrowbones with rocks or scraped flesh from bones with sharp stones. Paleoanthropologist Louis Leakey was the first to discover evidence of tool *making* at East Africa's Olduvai Gorge, which cuts through a great sequence of sedimentary rock layers. The oldest tools at this site are crudely chipped pebbles, which were probably manufactured by *H. habilis.*

Homo erectus Dispersed from Africa to Other Continents

Early in the Pleistocene epoch, about 1.8 million years ago, new species of humans appeared in East Africa. Most fossils are fragmentary. For convenience, we describe them all as *Homo erectus* ("upright man"), recognizing that they probably represent several species. One nearly complete skeleton suggests that *H. erectus* was taller than its ancestors, had a much larger brain, a thicker skull, and protruding brow ridges **(Figure 30.40).**

H. erectus made fairly sophisticated tools, including the hand axe (see Figure 30.40b), which they apparently used to cut food and other materials, to scrape meat from bones, and to dig for roots. *H. erectus* probably fed on both plants and animals; they may have hunted and scavenged

a. *Homo erectus*   **b.** Hand axe

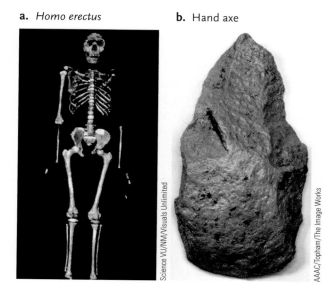

Science VU/NM/Visuals Unlimited

AAAC/Topham/The Image Works

Figure 30.40
Homo erectus. **(a)** A nearly complete skeleton of *Homo erectus* was discovered in Kenya. **(b)** Hand axes are frequently found at *H. erectus* sites.

animal prey. Archaeological data points to their use of fire to process food and to keep themselves warm.

The pressure of growing populations apparently forced groups of *H. erectus* out of Africa about 1.5 million years ago. They dispersed northward from East Africa into both northwestern Africa and Eurasia. Some moved eastward through Asia as far as the island of Java. Recent discoveries in Spain indicate that *H. erectus* also occupied parts of Western Europe.

Modern Humans Are the Only Surviving Descendants of *Homo erectus*

Judging from its geographical distribution, *Homo erectus* was successful in many environments. It produced several descendant species, of which modern humans (*Homo sapiens*, "wise man") are the only survivors.

Fossils from Africa, Asia, and Europe indicate that archaic humans, the now-extinct descendants of *H. erectus*, first appeared at least 400,000 years ago. They generally had a larger brain, rounder skull, and smaller molars than *H. erectus*.

Neanderthals. The Neanderthals (*Homo neanderthalensis*), who occupied Europe and western Asia from 150,000 to 28,000 years ago, are the best-known archaic humans. Compared with modern humans, they had a heavier build, more-pronounced brow ridges, and slightly larger brains (see Figure 30.38). Neanderthals were culturally and technologically sophisticated. They made complex tools, including wooden spears, stone axes, and flint scrapers and knives. At some sites they built shelters of stones, branches, and animal hides, and they routinely used fire. They were successful hunters and probably consumed nuts, berries, fishes, and bird eggs. Some groups buried their dead, and they may have had rudimentary speech.

Researchers once classified Neanderthals as a subspecies of *H. sapiens,* but most scientists now believe that they were a separate species. In 1997 two teams of researchers, Matthias Kring and Svante Pääbo of the University of Munich and Anne Stone and Mark Stoneking of Pennsylvania State University, independently analyzed short segments of mitochondrial DNA (mtDNA) extracted from the fossilized arm bone of a Neanderthal. Unlike nuclear DNA, which individuals inherit from both parents, only mothers pass mtDNA to offspring. It does not undergo genetic recombination (see Section 13.5), and it has a high mutation rate, making it useful for phylogenetic analyses. Many scientists believe that mutation rates in mtDNA are fairly constant, allowing this molecule to serve as a molecular clock (see Section 23.6). Comparing the Neanderthal sequence with mtDNA from 986 living humans, the researchers discovered three times more differences between the Neanderthals and modern humans than between pairs of modern humans in their sample. These results suggest that Neanderthals and modern humans are different species that diverged from a common ancestor 690,000 to 550,000 years ago—hundreds of thousands of years before modern humans appeared.

Modern Humans. Modern humans (*Homo sapiens*) differ from Neanderthals and other archaic humans in having a slighter build, less-protruding brow ridges, and a more prominent chin. The earliest fossils of modern humans found in Africa and Asia are 150,000 years old; those from the Middle East are 100,000 years old. Fossils from about 20,000 years ago are known from Western Europe, the most famous being those of the Cro-Magnon deposits in southern France. The widespread appearance of modern humans roughly coincided with the demise of Neanderthals in Western Europe and the Middle East 40,000 to 28,000 years ago. Although the two species apparently coexisted in some regions for thousands of years, we have little concrete evidence that they interacted.

One Origin or Many? *Homo erectus* apparently left Africa in waves between 1.5 million and 500,000 years ago. But when and where did modern humans first arise? Researchers use fossils and genetic data from contemporary human populations to address two competing hypotheses about this question.

According to the **African Emergence Hypothesis,** a population of *H. erectus* gave rise to several descendant species between 1.5 million and 0.5 million years ago. The early descendants, archaic humans, left Africa and established populations in the Middle East, Asia, and Europe. Some time later, 200,000 to 100,000 years ago, *H. sapiens* arose in Africa. These modern humans also migrated into Europe and Asia, and eventually drove the archaic humans to extinction. Thus, the African Emergence Hypothesis suggests that all modern humans are descended from a fairly recent African ancestor.

According to the **Multiregional Hypothesis,** populations of *H. erectus* and archaic humans had spread through much of Europe and Asia by 0.5 million years ago. Modern humans then evolved from archaic humans in many regions simultaneously. Although these geographically separated populations may have experienced some evolutionary differentiation (see Section 21.3), gene flow between them prevented reproductive isolation and maintained them as a single, but variable, species, *H. sapiens*.

Paleontological data do not clearly support either hypothesis. Some scientists argue that human remains with a mixture of archaic and modern characteristics confirm the Multiregional Hypothesis. In late 1998, for example, researchers in Portugal discovered a fossilized child that had been buried only 24,000 years ago, when only modern humans are thought to have occupied Europe. This fossil shows a surprising mix of Neanderthal and modern human traits, possibly indicating that the two groups interbred. On the other hand, recent finds in the Mideast indicate that Neanderthals and modern humans coexisted without interbreeding for 50,000 years. Thus, Neanderthals could not have been the ancestors of those modern humans.

Scientists also use DNA sequences from modern humans to evaluate the two hypotheses. In 1987, Rebecca Cann, Mark Stoneking, and Allan Wilson of the University of California at Berkeley and their colleagues published an analysis of mtDNA sequences from more than 100 ethnically diverse humans on four continents. They found that contemporary African populations contain the greatest variation in mtDNA. One explanation for this observation is that neutral mutations have been accumulating in African

UNANSWERED QUESTIONS

What causes the evolution of diversity?

In this chapter, you have read about the extensive diversity—in size, shape, color, structure, lifestyle, and habitats—of the vertebrate animals. Their mechanisms for maintaining themselves—including behaviors such as moving, feeding, reproducing, hiding, fighting, and sleeping—are equally varied. But nearly all vertebrates share some fundamental features of life. Recent research shows us that this generality is especially true of early development, but it is also apparent in many of the basic homeostatic mechanisms that vertebrates share—features of digestion and metabolism, respiration, and other characteristics regulated by the products of genetic networks and cascades. Biologists had long thought that once we understood the genetics of a variety of species, we would also understand the basis for their evolution and the maintenance of diversity. But now that a number of animal genomes have been sequenced completely, we have actually learned more about the genetic information that is shared by all animal species than we have about the genes that promote diversification. One of the great unanswered questions in biology, therefore, is how diversity arises and how it is maintained, given that so much genetic information is shared among even distantly related species. Equally important is why the same kinds of features evolved almost identically time after time in many unrelated lineages.

The ever-increasing body of genetic information is opening the "black box" of why there are so many species, and how they came to be. For example, we've long known that limbs evolved from fins, based on fossil evidence (which continues to accumulate, providing additional supportive evidence). But we haven't known what the *mechanisms* for forming a fin or a limb are, and how selection works to modify such structures. Now, with our knowledge of *Hox* and other tool-kit genes that control embryonic development, we understand how fins and limbs are formed. We can also experimentally manipulate development and perform selection experiments to test possible pathways of evolution—the "how they came to be."

Why do we see the recurrent, independent evolution of common themes?

A phenomenon that we often see, but don't yet fully understand, is why certain themes—such as body elongation, limblessness, and tooth modification—recur among distantly related vertebrates. The evolution of viviparity, live-bearing reproduction, is one such theme. In the vast

majority of animals, reproduction occurs when a female lays her eggs in water and a male sprays sperm over them; typically both parents then abandon the fertilized eggs. But some species in many separate vertebrate lineages have evolved forms of viviparity. Cartilaginous and bony fishes exhibit diverse modes of embryonic nutrition, and in one group, the sea horses, it is the males that become pregnant. Amphibians also exhibit diverse patterns of viviparity: some have pregnant fathers and others have mothers that brood embryos in the skin of their backs, in their stomachs, or in their oviducts. And many squamate reptiles grow placentas similar to those of mammals. All mammals except monotremes are live-bearers with maternal nutrition. In fact, among the living vertebrate groups, birds are the only group that has not evolved the live-bearing habit. Can you think of some reasons why?

In some fishes, amphibians, and squamates, the mother supplies all the nutrition for the developing young through her investment in the egg; but in other species, females resorb their yolks and provide nutrients directly to offspring that are born as fully metamorphosed juveniles. Given that viviparity has evolved independently in approximately 200 vertebrate groups, it is not surprising that we see such a variety of reproductive patterns. Biologists are just beginning to understand the evolution of viviparity and to determine which patterns are shared among different evolutionary lineages and which are not. It appears that the hormonal basis for viviparity is similar in many groups, although the timing, the receptors involved, and the physiological responses vary. Researchers are now identifying candidate genes in the hope of unraveling the genetic networks that have fostered the different modes of viviparity. Ecological studies are revealing the interactions of potential selection regimes that influence reproductive modes. However, we still have much to learn about how genetics, development, physiology, and ecology interact in the evolution of diversity and common recurring themes.

Marvalee H. Wake is a professor of the graduate school in the Department of Integrative Biology at the University of California at Berkeley. She studies vertebrate evolutionary morphology, development, and reproductive biology, with the goal of understanding evolutionary patterns and processes. To learn more about her research, go to http://ib.berkeley.edu/research/interests/research_profile.php?person=236.

populations longer than in others, marking the African populations as the oldest on Earth. They also found that all human populations contain at least one mtDNA sequence of African origin, suggesting an African ancestry for all modern humans. Cann and her colleagues named the ancestral population, which lived approximately 200,000 years ago, the "mitochondrial Eve."

Proponents of the Multiregional Hypothesis criticized the statistical techniques used by Cann and her colleagues. Some also noted that if ancient populations in Africa were larger than those in other parts of the world, population size, rather than age, could account for high genetic variability (see Section 20.4). Moreover, recent research suggests that mutation rates in mtDNA might not be constant and that natural selection can influence mtDNA sequences, calling into question mtDNA's usefulness as a molecular clock.

Other researchers have examined genetic material that males inherit only from their fathers. In 1995, L. Simon Whitfield and his colleagues at Cambridge University published a study on an 18,000-base-pair sequence from the Y chromosome, which does not undergo recombination with the X chromosome. Because the sequence contains no genes, it should not be subject to natural selection. Thus, sequence variations should result only from random mutations, which can serve as a molecular clock. The researchers discovered only three sequence mutations among the five subjects examined—a surprising result, given that the sample included a European, a Melanesian, a South American, and two Africans. By contrast, a chimpanzee exhibited 207 differences from the human version of the sequence. Using a sophisticated statistical analysis, the Whitfield team calculated that the common ancestor of these five diverse humans, dubbed the "African Adam," lived between 37,000 and 49,000 years ago. The limited genetic diversity and relatively recent origin of a common ancestor clearly support the African Emergence Hypothesis. Follow-up studies on the Y chromosomes of thousands of men from Africa, Europe, Asia, Australia, and the Americas have confirmed that all modern humans are the descendants of a single migration out of Africa.

Some controversies about human evolution arise because researchers who use genetic data must make assumptions about the sizes and geographical ranges of ancient populations, the amount of gene flow they experienced, and how natural selection may have affected them. And one of a scientist's important responsibilities is to question the assumptions made by other workers. Controversies also surround the particular statistical tests that researchers use to analyze these data. But intellectual disputes are routine in science, and they challenge researchers to refine their hypotheses and to test them in new ways. Questions about the details of human origins are at the center of one of the liveliest debates in evolutionary biology today; additional research will surely clarify the evolutionary history of our species.

STUDY BREAK

1. What trait allows researchers to distinguish between apes and humans?
2. What evidence suggests that Neanderthals and modern humans represent two distinct species?
3. How do the African Emergence Hypothesis and the Multiregional Hypothesis differ in what they say about the origin of modern humans?

Review

Go to **Thomson**NOW™ at www.thomsonedu.com/login to access quizzing, animations, exercises, articles, and personalized homework help.

30.1 Invertebrate Deuterostomes

- Echinoderms have secondary radial symmetry, a five-part body plan, and a unique water vascular system that is used in locomotion and feeding (Figures 30.2 and 30.3).
- Hemichordates collect oxygen and particulate food from seawater that enters the mouth and exits the pharynx through the gill slits (Figure 30.4).

30.2 Overview of the Phylum Chordata

- Chordates share several derived characteristics: a notochord; postanal tail; segmentation of body wall and tail muscles; a dorsal, hollow nerve cord; and a perforated pharynx at some stage of the life cycle (Figure 30.5).
- Two subphyla of invertebrate chordates use their perforated pharynx to collect particulate food (Figures 30.6 and 30.7).
- The subphylum Vertebrata includes animals with a bony endoskeleton, structures derived from neural crest cells, and a brain divided into three regions.

Animation: Tunicate body plan

Animation: Lancelet body plan

30.3 The Origin and Diversification of Vertebrates

- Vertebrates evolved from an invertebrate chordate ancestor, probably through duplication of the *Hox* gene complex (Figure 30.8).

- Vertebrates diversified into numerous lineages (Figure 30.10). Gnathostomata includes all jawed vertebrates. Tetrapoda includes all lineages that ancestrally had four legs. Amniota includes groups descended from animals that produced an amniote egg.

Animation: Vertebrate evolution

30.4 Agnathans: Hagfishes and Lampreys, Conodonts and Ostracoderms

- Living agnathans—hagfishes and lampreys—are jawless fishes that lack vertebrae and paired fins (Figure 30.11).
- Conodonts had bonelike elements in the pharynx. Ostracoderms were heavily armored, jawless fishes that sucked particulate food into their mouths (Figure 30.12).

Animation: Jawless fishes

30.5 Jawed Fishes

- Jaws arose through the evolutionary modification of gill arches, which supported the pharynx of ostracoderms (Figure 30.13). Fins arose at the same time (Figure 30.14).
- The first jawed fishes, Acanthodii and Placodermi, are now extinct (Figure 30.15). Chondrichthyans have a skeleton composed of cartilage (Figure 30.16). Actinopterygians and Sarcopterygians have a skeleton composed of bone. Actinopterygians have fins supported by bony rays (Figures 30.17 and 30.18). Sarcopterygians have fins supported by muscles and a bony endoskeleton (Figure 30.19).

Animation: Evolution of jaws

Animation: Cartilaginous fishes

Animation: Bony fish body plan

30.6 Early Tetrapods and Modern Amphibians

- Tetrapods arose in the late Devonian (Figure 30.20). Key tetrapod adaptations include a strong vertebral column, girdles, limbs, and modified sensory systems.
- Modern amphibians are generally restricted to moist habitats. Their life cycles often include larval and adult stages. Urodeles (salamanders) are elongate, tailed amphibians. Anurans (frogs) have compact bodies, long legs, and no tails. Caecelians are legless burrowers (Figure 30.21).

Animation: Evolution of limb bones

Animation: Salamander locomotion

30.7 The Origin and Mesozoic Radiations of Amniotes

- Key adaptations in amniotes, the first fully terrestrial vertebrates, included a water-resistant skin, amniote eggs (Figure 30.22), and the excretion of nitrogen wastes as uric acid.
- Early amniotes diversified into anapsids, synapsids, and diapsids, distinguished by the number of temporal arches on the skull (Figure 30.23).
- Diapsids split into two lineages, archosaurs and lepidosaurs.

Animation: Amniote egg

30.8 Testudines: Turtles

- The turtle body plan includes a bony shell, with a dorsal carapace and ventral plastron (Figure 30.24).

Animation: Tortoise shell and skeleton

30.9 Living Nonfeathered Diapsids: Sphenodontids, Squamates, and Crocodilians

- Sphenodontids are remnants of a once-diverse lineage (Figure 30.25a).
- Squamates have skin composed of overlapping, keratinized scales (Figure 30.25b and c).
- Crocodilians are semiaquatic predators (Figure 30.25d).

Animation: Crocodile body plan

30.10 Aves: Birds

- Birds have adaptations that reduce their weight and generate power for flight (Figure 30.26).
- Modern birds exhibit adaptations of their bills, feet, wings, and behavior (Figure 30.27).

Animation: Feather development

Animation: Avian bone and muscle structure

30.11 Mammalia: Monotremes, Marsupials, and Placentals

- Key adaptations of mammals include endothermy, which allows high levels of activity; modification of the teeth and jaws; extensive parental care of offspring; and large and complex brains.
- Three major groups of mammals, Prototheria, the monotremes; and two lineages of Theria, the marsupials and the placentals, differ in their reproductive patterns.
- Monotremes are restricted to Australia and New Guinea (Figure 30.28). Marsupials are abundant in Australia and occur in South America and North America (Figure 30.29). Most living mammals are placentals, occupying nearly all terrestrial and aquatic habitats (Figure 30.30).

Animation: Mammalian dentition

Animation: Mammalian radiations

Animation: Structure of the placenta

30.12 Nonhuman Primates

- Key adaptations allow Primates to be arboreal and diurnal: upright posture and flexible limbs; good depth perception; and a large and complex brain.
- The Strepsirhini have many ancestral primate characteristics (Figure 30.32). The Haplorhini have many derived primate characteristics (Figures 30.33 and 30.34).
- Primates arose in forests about 55 mya. The hominoid lineage, which includes apes and humans, arose in Africa about 23 mya (Figure 30.35).

Animation: Primate skeletons

Animation: Skulls of extinct primates

30.13 The Evolution of Humans

- Hominids, the lineage that includes humans, arose in Africa between 10 mya and 5 mya. Hominid anatomy permits bipedal locomotion (Figure 30.36). Over time, hominids developed larger brains, improved grasping ability in the hands (Figure 30.37), and tool-making behavior. Several genera of hominids occupied sub-Saharan Africa for several million years (Figures 30.38 and 30.39).
- *Homo habilis* was the first hominid species to make stone tools. *Homo erectus*, which arose in East Africa about 1.8 mya, made sophisticated stone tools (Figure 30.40).

- The early descendants of *H. erectus* left Africa in waves, populating Asia and Europe. Neanderthals, the best known of these groups, became extinct about 30,000 years ago.
- Modern humans, *Homo sapiens*, arose approximately 150,000 years ago and migrated out of Africa, eventually replacing archaic humans in Europe and Asia.

Animation: Fossils of australopiths

Animation: *Homo* skulls

Animation: Primate phylogenetic tree

Animation: Genetic distance between human groups

Questions

Self-Test Questions

1. Which phylum includes animals that have a water vascular system?
   a. Echinodermata
   b. Hemichordata
   c. Chordata
   d. Tetrapoda
   e. Amniota

2. Which of the following is *not* a characteristic of all chordates?
   a. notochord with postanal tail
   b. segmental body wall and tail muscles
   c. segmented nervous system
   d. dorsal hollow nerve cord
   e. perforated pharynx

3. Which group of vertebrates has adaptations that allow it to reproduce on land?
   a. agnathans
   b. tetrapods
   c. gnathostomes
   d. amniotes
   e. ichthyosaurs

4. Which group of fishes has the most living species today?
   a. sarcopterygians
   b. actinopterygians
   c. chondrichthyans
   d. acanthodians
   e. ostracoderms

5. Modern amphibians:
   a. closely resemble their Paleozoic ancestors.
   b. always occupy terrestrial habitats as adults.
   c. never occupy terrestrial habitats as adults.
   d. are generally larger than their Paleozoic ancestors.
   e. are generally smaller than their Paleozoic ancestors.

6. Which of the following key adaptations allows amniotes to occupy terrestrial habitats?
   a. the production of carbon dioxide as a metabolic waste product
   b. an unshelled egg that is protected by jellylike material
   c. a dry skin that is largely impermeable to water
   d. a lightweight skeleton with hollow bones
   e. feathers or fur that provide insulation against cold weather

7. Which of the following characteristics does *not* contribute to powered flight in birds?
   a. a lightweight skeleton
   b. efficient respiratory and circulatory systems
   c. enlarged forelimbs and a keeled sternum
   d. a high metabolic rate that releases energy from food rapidly
   e. scaly skin on the legs and feet

8. Which of the following characteristics did *not* contribute to the evolutionary success of mammals?
   a. extended parental care of young
   b. an erect posture and flexible hip and shoulder joints
   c. specializations of the teeth and jaws
   d. enlargement of the brain
   e. high metabolic rate and high body temperature

9. The Hominoidea is a monophyletic group that includes:
   a. apes and monkeys
   b. apes only
   c. humans and human ancestors
   d. apes and humans
   e. monkeys, apes, and humans

10. Which of the following hominid species was the earliest?
   a. *Ardipithecus ramidus*
   b. *Australopithecus afarensis*
   c. *Homo habilis*
   d. *Homo erectus*
   e. *Homo neanderthalensis*

Questions for Discussion

1. Most sharks and rays are predatory, but the largest species feed on plankton. Construct a hypothesis to explain this observation. How would you test your hypothesis?

2. When tetrapods first ventured onto the land, what new selection pressures did they face? What characteristics might have fostered the success of these animals as they made the transition from aquatic to terrestrial habitats?

3. Use a pair of binoculars to observe several species of birds that live in different types of environments, such as lakes and forests. How are their beaks and feet adapted to their habitats and food habits?

4. Imagine that you unearthed the complete fossilized remains of a mammal. How would you determine the food habits of this now extinct animal?

5. Many myths about human evolution are embraced by popular culture. Using the information you have learned about human evolution, argue against each of the following myths.
   a. Humans evolved from chimpanzees.
   b. Evolution occurred in a steady linear progression from primitive primate to anatomically modern humans.
   c. All human characteristics, such as bipedal locomotion and an enlarged brain, evolved simultaneously and at the same rate.

Experimental Analysis

Walking along a rocky coast one day, you discover two small creatures—one lumpy and the other worm-shaped. What anatomical studies would you conduct to determine whether or not they are chordates? What genetic studies might provide supplementary evidence?

Evolution Link

Birds and crocodiles are both descended from an ancestral archosaur. What shared anatomical and behavioral characteristics reflect this common ancestry? Explain why dinosaurs, which were also members of the archosaur lineage, may have shared these traits as well. Review Figure 30.23 before formulating your answer.

How Would You Vote?

Private collectors find and protect fossils, but a private market for rare vertebrate fossils raises the cost of museums and encourages theft from protected fossil beds. Should private collecting of vertebrate fossils be banned? Go to www.thomsonedu.com/login to investigate both sides of the issue and then vote.

A population of Caribbean flamingos *(Phaenicopterus ruber)*. Each pair of flamingos in this breeding colony incubates a single egg in a mud nest.

© Gerry Ellis/Minden Pictures

STUDY PLAN

49.1 The Science of Ecology

Ecologists study levels of organization ranging from individual organisms to the biosphere

Ecologists test hypotheses with observational and experimental data

49.2 Population Characteristics

A population's size and density determine the amount of resources it uses

Populations differ in how they are distributed in space

A population's age structure, generation time, and sex ratio influence how quickly it will grow

49.3 Demography

Life tables summarize a population's survival and reproductive rates

Survivorship curves depict changes in survival rate over the life span

49.4 The Evolution of Life Histories

Organisms face trade-offs in their allocation of resources

Life history patterns vary dramatically among species

Ecologists analyze the individual components of life histories

49.5 Models of Population Growth

Models of exponential population growth describe growth without limitation

The logistic model describes population growth when resources are limited

49.6 Population Regulation

Density-dependent factors often regulate population size

Density-independent factors can limit population size

Interacting environmental factors often limit population sizes

The life history characteristics of a species govern fluctuations in its population size through time

Some species exhibit regular cycles in population size

49.7 Human Population Growth

Human populations have sidestepped the usual density-dependent controls

Age structure and economic development may now control our population growth

49 Population Ecology

WHY IT MATTERS

When humans immigrate to new places, they often transport familiar plants and animals from home, introducing them into their new gardens, fields, and forests. Some organisms fail to survive in the new environments. But other species—like the European starlings *(Sturnus vulgaris)* and house sparrows *(Passer domesticus)* that are now so common in North America—flourish and sometimes become pests.

In 1859, an Australian rancher released a few pairs of European rabbits *(Oryctolagus cuniculus)* for sport hunting in the state of Victoria. The rabbits bred rapidly, sometimes producing litters of four or five offspring every month. They had no natural predators in Australia, and by 1900, an estimated 20 million rabbits had overrun much of the continent; their advance was limited only by extreme climates, clay soil, and lack of food or water. The rabbits destroyed natural vegetation and the pastures that supported a large sheep industry. The government tried in vain to poison the rabbits. Ranchers introduced predators, hoping that they would eat rabbits faster than the rabbits could reproduce. But the rabbits continued to multiply. Eventually, the government built a "rabbit-proof fence" that stretched more than

1125

Figure 49.1
Introduced organisms. European rabbits multiplied so rapidly and destroyed so much vegetation in Australia that the government built a fence across the country to prevent their spread.

3200 km (2000 miles) to keep the rabbits out of the rich pasture lands in Western Australia (**Figure 49.1**).

In 1950, scientists tackled the devastating problems caused by the introduced rabbits. Biologists collected myxoma virus (a relative of smallpox) from infected rabbits in South America and released it among the European rabbits in Australia. The virus was lethal to European rabbits, which had never evolved resistance to it. The first epidemic of myxomatosis killed more than 99% of infected rabbits. But in the following season, the virus killed only 90% of infected rabbits, and within a few years, the virus was killing only half the rabbits it infected. Clearly, some rabbits were becoming more resistant to the virus. Resistant rabbits survived and reproduced, comprising a larger percentage of the population over time (see Section 20.3 to review natural selection). Subsequent research showed that the virus had also become less virulent. Today, wildlife-control agents develop and release more deadly viruses to control the rabbit population.

This brief history of an environmental disruption introduces our unit on **ecology**, the study of interactions between organisms and their environments. All environments have both **abiotic** (nonbiological) and **biotic** (biological) components. The abiotic environment includes temperature, moisture, soil chemistry, and other physical factors; the biotic environment includes all the organisms found in a particular place.

This story also identifies several ecological phenomena that we consider in this chapter. For example, some species produce large numbers of young at frequent intervals; their numbers may increase rapidly for a period of time and then drop precipitously. Moreover, a species' abundance is often governed by the presence of other species—its food, predators, parasites, and disease-causing microorganisms. Finally, over time, ecological interactions foster adaptation and evolutionary change.

49.1 The Science of Ecology

The subject matter of ecology is so vast and so diverse that research in ecology is often linked to work in genetics, physiology, anatomy, behavior, paleontology, and evolution as well as in geology, geography, and environmental science. Many ecological phenomena occur over huge areas and long time spans. Ecologists

must devise ways to determine how environments influence organisms and how organisms change the environments in which they live. Today, the science of ecology encompasses two related disciplines. The major research questions of *basic ecology* relate to the distribution and abundance of species and how they interact with each other and the physical environment. Using these data as a baseline, workers in *applied ecology* develop conservation plans and amelioration programs to limit, repair, and mitigate ecological damage caused by human activities.

Ecologists Study Levels of Organization Ranging from Individual Organisms to the Biosphere

Ecology can be divided into five increasingly complex and inclusive levels of organization. In **organismal ecology** researchers study the genetic, biochemical, physiological, morphological, and behavioral adaptations of organisms to the abiotic environment. We have described many such adaptations in Units V and VI; we describe the evolution of animal behavior in Chapter 55.

Population ecology, the subject of this chapter, focuses on **populations**, groups of individuals of the same species that live together. Population ecologists study how the size and other characteristics of populations change in space and time. Research in **community ecology** examines groups of populations that occur together in one area. Community ecologists study interactions between species, analyzing how predation, competition, and environmental disturbances influence a community's development, organization, and structure. We address major issues in community ecology in Chapter 50. Ecologists studying **ecosystem ecology** explore the cycling of nutrients and the flow of energy between the biotic components of an ecological community and the abiotic environment. We consider this topic in Chapter 51. Finally, some ecological studies focus on the **biosphere**, the total of all ecosystems on Earth. In Chapter 52, we examine global patterns in abiotic factors and their effects on populations, communities, and ecosystems. We discuss biodiversity and conservation biology in Chapter 53.

Ecologists Test Hypotheses with Observational and Experimental Data

Ecology has its roots in descriptive natural-history studies that date back to the ancient Greeks. Modern ecology was born in 1870 when the German biologist Ernst Haeckel coined the term (*oikos* = house). Contemporary researchers still gather descriptive information about ecological relationships, but these observations are only a starting point for more rigorous studies.

Most ecologists create hypotheses about ecological relationships and how they change through time or differ from place to place. Like other scientists, some ecologists formalize these ideas in mathematical models that express clearly defined, but hypothetical, relationships among important variables in a system. Manipulation of a model, usually with the help of a computer, can allow researchers to ask what would happen if some of the variables or their relationships change. Thus, researchers can simulate natural events and large-scale experiments before investing time, energy, and money in field and laboratory work. Bear in mind, however, that mathematical models are no better than the ideas and assumptions they embody, and useful models are constructed only after basic observations define the relevant variables.

Ecologists often conduct field or laboratory studies to test the predictions of their hypotheses. In controlled experiments, researchers compare data from an experimental treatment (in which one or more variables are artificially manipulated) with data from a control (in which nothing is changed). Sometimes the distributions of species create "natural experiments," eliminating the need to manipulate variables. For example, two species of fishes, cutthroat trout (*Oncorhynchus clarki*) and Dolly Varden char (*Salvelinus malma*), live in coastal lakes of British Columbia, Canada. Some lakes have either trout or char, but other lakes contain both species. The natural distributions of these fishes allowed researchers to measure the effect of each species on the other. In lakes where both species live, each restricts its activities to fewer areas and feeds on a smaller variety of prey than it does in lakes where it occurs alone.

STUDY BREAK

1. Why are studies of ecosystems more "inclusive" than studies of populations?
2. In what ways are mathematical models useful in ecological research?

49.2 Population Characteristics

Populations have characteristics that transcend those of the individuals they comprise. For example, every population has a **geographical range**, the overall spatial boundaries within which it lives. Geographical ranges vary enormously. A population of snails might inhabit a small tidepool, whereas a population of marine phytoplankton might occupy an area orders of magnitude larger. Every population also occupies a **habitat**, the specific environment in which it lives, as characterized by its biotic and abiotic features. Ecologists also mea-sure other population characteristics, such as size, distribution in space, and age structure.

A Population's Size and Density Determine the Amount of Resources It Uses

Population size is simply the number of individuals in a population at a specified time. **Population density** is the number of individuals per unit area or per unit volume of habitat. Species with large body size generally have lower population densities than those with small body size **(Figure 49.2)**. Although population size and density are related measures, knowing a population's density provides more information about its relationship to the resources it uses. For example, if a population of 200 oak trees occupies 1 hectare (10,000 m²), its population density is 200/10,000 m² or one tree per 50 m². But if a population of 200 oaks is spread over 5 hectares, its density is one tree per 250 m². Clearly, the second population is less dense than the first, and its members will have greater access to sunlight, water, and other resources.

Ecologists measure population size and density to monitor and manage populations of endangered species, economically important species, and agricultural pests. For large-bodied species, a simple head count provides accurate information. For example, ecologists survey the size and density of African elephant (*Loxodonta africana*) populations by flying over herds and counting individuals. Researchers use a variation on that technique to estimate population size in tiny organisms that live at high population densities. To estimate the density of aquatic phytoplankton, for example, you might collect water samples of known volume from representative areas in a lake and use a microscope to count the organisms; you could then extrapolate their population size and density based on the estimated volume of the entire lake. In other cases, researchers use the mark-release-recapture sampling technique **(Figure 49.3)**.

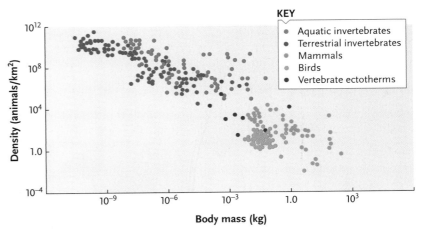

KEY
- Aquatic invertebrates
- Terrestrial invertebrates
- Mammals
- Birds
- Vertebrate ectotherms

Figure 49.2
Population density and body size. Population density generally declines with increasing body size among animal species. Similar trends exist for other types of organisms.

Figure 49.3   Research Method

Using Mark-Release-Recapture to Estimate Population Size

PURPOSE: Ecologists use the mark-release-recapture technique to estimate the population size of mobile animals that live within a restricted geographic range.

PROTOCOL: A sample of organisms is captured, marked in some permanent but harmless way, and released. Insects and reptiles are marked with ink or paint, birds with rings on their legs, and mammals with ear tags or collars. Some time later, a second sample of organisms is captured, and the researcher notes what proportion of the second sample carries the mark. That proportion tells us what percentage of the total population was captured and marked at the first sampling. The total population size is estimated as (number marked) × (number in the second sample/number of marked recaptures).

Michael C. Singer, University of Texas

INTERPRETING THE RESULTS: Imagine that you capture 120 butterflies, mark each with a black spot on its wing, and release them. A week later, you capture a second sample of 150 butterflies and find that 30 of them have the black mark. Thus, you had marked one out of every five butterflies (30/150) on your first field trip. Because you captured 120 individuals on that first excursion, you would estimate that the total population size is 120 × (150/30) = 600 butterflies.

The technique is based on several assumptions that are critical to its accuracy: (1) that being marked has no effect on survival; (2) that marked and unmarked animals mix randomly in the population; (3) that no migration into or out of the population takes place during the estimating period; and (4) that marked individuals are just as likely to be captured as unmarked individuals. (Sometimes animals become "trap shy" or "trap happy," a violation of the fourth assumption.)

Populations Differ in How They Are Distributed in Space

Populations also vary in their **dispersion**, the spatial distribution of individuals within the geographical range. Ecologists define three theoretical patterns of dispersion: *clumped, uniform,* and *random* **(Figure 49.4).**

Three reasons explain why a **clumped dispersion**—with individuals grouped together—is extremely common in nature. First, suitable conditions often have a patchy distribution. For example, certain pasture plants may be clumped in small, scattered areas where cowpats fell months before, locally enriching the soil. Second, some animals live in social groups (see Section 55.5). Mates are easy to locate within groups, and individuals may cooperate in rearing offspring, feeding, or defending themselves from predators. Third, some organisms are clumped because of their reproductive pattern. Plants and animals that produce asexual clones, such as aspen trees and sea anemones, often occur in large aggregations (see Chapters 34 and 47).

In other species, seeds, eggs, or larvae lack dispersal mechanisms, and offspring grow near their parents.

Organisms are evenly spaced in their habitat, a pattern called **uniform dispersion**, when individuals repel each other because resources are in short supply. For example, creosote bushes *(Larrea tridentata)* are uniformly distributed in the dry scrub deserts of the American Southwest. Mature bushes deplete the surrounding soil of water and secrete toxic chemicals, making it impossible for seedlings to grow. Moreover, seed-eating ants and rodents that live at the base of mature bushes consume any seeds that fall nearby. Territorial behavior, the defense of an area and its resources, produces uniform dispersion in animals (see Section 55.2).

For some populations, environmental conditions don't vary much within a habitat, and individuals are neither attracted to nor repelled by others of their species. These populations exhibit **random dispersion**, which has a formal statistical definition that serves as a theoretical baseline for assessing whether organisms are clumped or uniformly distributed. In cases of random dispersion, individuals are distributed unpredictably. Some spiders, burrowing clams, and rainforest trees exhibit random dispersion.

Whether the spatial distribution of a population appears to be clumped, uniform, or random depends partly on how large an area an ecologist studies. Oak seedlings may be randomly dispersed on a spatial scale of a few square meters, but over an entire mixed hardwood forest, they are clumped under the parent trees.

In addition, the dispersion of animal populations often varies through time in response to natural environmental rhythms. Few habitats provide a constant supply of resources throughout the year, and many animals move from one habitat to another on a seasonal cycle. For example, tropical birds and mammals are often widely dispersed in deciduous forests during the wet season. But during the dry season, they crowd into narrow "gallery forests" along watercourses where evergreen trees provide food and shelter.

A Population's Age Structure, Generation Time, and Sex Ratio Influence How Quickly It Will Grow

All populations have an **age structure**, a statistical description of the relative numbers of individuals in each age class (see Section 49.7). Individuals can be roughly categorized as prereproductive (younger than the age of sexual maturity), reproductive, or postreproductive (older than the maximum age of reproduction). A population's age structure reflects its recent growth history and predicts its future growth potential. Populations that include many prereproductive individuals grew rapidly in the recent past and will continue to grow larger as the young individuals mature and reproduce.

Another characteristic that influences a population's growth is its **generation time**, the average time between the birth of an organism and the birth of its offspring. Generation time is usually short in species that reach sexual maturity at a small body size **(Figure 49.5)**. Their populations often grow rapidly because of the speedy accumulation of reproductive individuals.

Populations also vary in their **sex ratio**, the relative proportions of males and females. In general, the number of females in a population has a bigger impact on population growth than the number of males because only females actually produce offspring. Moreover, in many species, one male can mate with several females, and the number of males may have little effect on the population's reproductive output. In northern elephant seals *(Mirounga angustirostris)*, for example, mature bulls fight for dominance on the beaches where the seals mate, and only a few males may ultimately inseminate a hundred or more females. Thus, the presence of other males in the group has little effect on the size of future generations. In animals that form lifelong pair bonds, such as geese and swans, the number of males does influence reproduction in the population.

Population ecologists often try to determine the proportion of individuals in a population that are reproducing. This issue is particularly relevant to the conservation of any species in which individuals are rare or widely dispersed in the habitat (see Section 53.4). As *Insights from the Molecular Revolution* describes, ecologists now use DNA analysis to address this question. In the next section we consider factors that influence the age structure of a population and its potential for future growth.

STUDY BREAK

1. What is the difference between a population's size and its density?
2. What do the three patterns of dispersion imply about the relationships between individuals in a population?

49.3 Demography

Populations grow larger through the birth of individuals and the **immigration** (movement into the population) of organisms from neighboring populations. Conversely, death and **emigration** (movement out of the population) reduce population size. **Demography** is the statistical study of the processes that change a population's size and density through time.

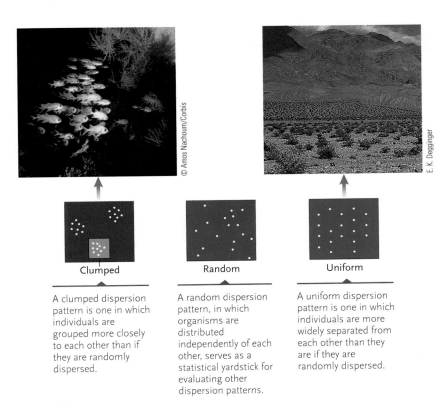

Clumped	Random	Uniform
A clumped dispersion pattern is one in which individuals are grouped more closely to each other than if they are randomly dispersed.	A random dispersion pattern, in which organisms are distributed independently of each other, serves as a statistical yardstick for evaluating other dispersion patterns.	A uniform dispersion pattern is one in which individuals are more widely separated from each other than they are if they are randomly dispersed.

Figure 49.4

Dispersion patterns. Schooling fishes, like these sabre squirrelfish *(Sargocentron spiniferum)* from the Maldives in the Indian Ocean, exhibit a clumped pattern of dispersion. A random pattern of dispersion, which is fairly rare in nature, occurs in organisms that are neither attracted to nor repelled by each other. Creosote bushes *(Larrea tridentata)* near Death Valley, California, exhibit uniform dispersion.

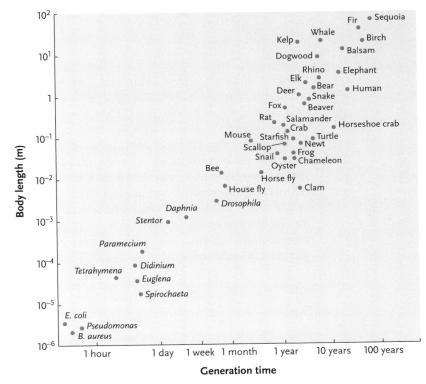

Figure 49.5

Generation time and body size. Generation time increases with body size among bacteria, protists, plants, and animals. The logarithmic scale on both axes compresses the data into a straight line.

INSIGHTS FROM THE MOLECULAR REVOLUTION

Tracing Armadillo Paternity and Migration

Nine-banded armadillos (*Dasypus novemcinctus*) are slow moving and solitary animals. Given their almost completely asocial behavior, what proportion of an armadillo population successfully mates? And, given their slow movements, how have they spread from Mexico and southern Texas through most of the southern United States in only 100 years?

Paulo A. Prodöhl and his colleagues at the University of Georgia, University of Washington, and Valdosta State University used molecular techniques to answer these questions. Their previous work had identified seven short tandem repeat (STR) loci in the armadillo genome. An STR locus consists of a segment of a chromosome with a short sequence repeated in series (see Section 18.2). Here the loci are approximately 200 bp long with 2- and 4-bp tandemly repeated sequences. Alleles of each locus vary in the number of copies of the repeated sequences, resulting in variable lengths for the locus. Because alleles of STR loci are inherited in the same way as alleles of genes, the allelic variations of the

STR loci made it possible for the researchers to trace parentage and migration patterns, in a manner analogous to human DNA fingerprinting.

Prodöhl and his coworkers collected tissue samples by clipping small pieces of tissue from the ears of 290 armadillos living in the Tall Timbers Research Station near Tallahassee, Florida. They extracted genomic DNA from the tissue sample and used the polymerase chain reaction (PCR) to amplify alleles for each of the seven STR loci. They determined the sizes of the fragments amplified by PCR using gel electrophoresis. (These molecular techniques are discussed in Chapter 18.)

The investigators used statistical methods to compare the gel patterns produced by DNA from adults with those from juveniles to determine their relatedness. Adult males and females with gel patterns most similar to those of a given juvenile were considered to be its parents. When the data identified more than two possible parents, the male and female living closest to a juvenile were scored as the most likely candidates. These techniques allowed

the investigators to assign parents to 69 sets of juveniles. Only seven juveniles could not be assigned parents, possibly because the parents had died, avoided capture, or emigrated from the population.

The results from 4 years of study suggest that 36% to 46% of adult armadillos reproduced at least once, despite their asocial habits—a moderately successful reproductive rate. In general, parents and offspring lived between 800 and 1500 m of each other, and individuals were usually captured within a 200-m radius of the same spot from season to season and from one year to the next. Thus, migration appears to be very limited, leaving the basis of their rapid spread unexplained.

© Fred Whitehead/Animals, Animals-Earth Scenes

Ecologists use demographic analysis to predict a population's future population growth. For human populations, these data help governments anticipate the need for social services such as schools and hospitals. Demographic data also allow conservation ecologists to develop plans to protect endangered species. For example, demographic data on the northern spotted owl (*Strix occidentalis caurina*) helped convince the courts to restrict logging in the owl's primary habitat, the old growth forests of the Pacific Northwest. *Life tables* and *survivorship curves* are among the tools ecologists use to analyze demographic data.

Life Tables Summarize a Population's Survival and Reproductive Rates

Although every species has a characteristic life span, few individuals survive to the maximum age possible. Mortality results from starvation, disease, accidents, predation, or the inability to find a suitable habitat. Life insurance companies first developed techniques for measuring mortality rates, but ecologists adapted these approaches to the study of nonhuman populations.

A **life table** summarizes the demographic characteristics of a population **(Table 49.1).** To collect life-table data for short-lived organisms, demographers typically mark a **cohort**, a group of individuals of similar age, at birth and monitor their survival until all members of the cohort die. For organisms that live more than a few years, a researcher might sample the population for 1 or 2 years, recording the ages at which individuals die, and then extrapolate those results over the species' life span.

In any life table, the life span of the organisms is divided into age intervals of convenient length: days, weeks, or months for short-lived species; years or groups of years for longer-lived organisms. Mortality can be expressed in two complementary ways. **Age-specific mortality** is the proportion of individuals alive at the start of an age interval that died during that age interval. Its more cheerful reflection, **age-specific survivorship,** is the proportion of individuals alive at the start of an age interval that survived until the start of the next age interval. Thus, for the data shown in Table 49.1, the age-specific mortality rate during the 3-to-6-month age interval is 195/722 = 0.270, and the

Table 49.1   Life Table for a Cohort of 843 Individuals of the Grass *Poa annua* (Annual Bluegrass)

Age Interval (in months)	Number Alive at Start of Age Interval	Number Dying During Age Interval	Age-Specific Mortality Rate	Age-Specific Survivorship Rate	Proportion of Original Cohort Alive at Start of Age Interval	Age-Specific Fecundity (Seed Production)
0–3	843	121	0.144	0.856	1.000	0
3–6	722	195	0.270	0.730	0.856	300
6–9	527	211	0.400	0.600	0.625	620
9–12	316	172	0.544	0.456	0.375	430
12–15	144	90	0.625	0.375	0.171	210
15–18	54	39	0.722	0.278	0.064	60
18–21	15	12	0.800	0.200	0.018	30
21–24	3	3	1.000	0.000	0.004	10
24–	0	—	—	—	—	—

Source: Begon, M., and M. Mortimer. *Population Ecology.* Sunderland, MA: Sinauer Associates, 1981. Adapted from R. Law. 1975.

age-specific survivorship rate is 527/722 = 0.730. For any age interval, the sum of age-specific mortality and age-specific survivorship always equals 1. Life tables also summarize the proportion of the cohort that survived to a particular age, a statistic that identifies the probability that any randomly selected newborn will still be alive at that age. For the 3-to-6-month age interval in Table 49.1, this probability is 722/843 = 0.856.

Life tables also include data on **age-specific fecundity**, the average number of offspring produced by surviving females during each age interval. Table 49.1 shows, for example, that plants in the 3-to-6-month age interval each produced an average of 300 seeds. In some species, including humans, fecundity is highest in individuals of intermediate age. Younger individuals have not yet reached sexual matu-

rity, and older individuals are past their reproductive prime. However, in some plants and animals fecundity increases steadily with age.

Survivorship Curves Depict Changes in Survival Rate over the Life Span

Survivorship data are depicted graphically in a **survivorship curve**, which displays the rate of survival for individuals over the species' average life span. Ecologists have identified three generalized survivorship curves (blue lines in **Figure 49.6**), although most organisms exhibit survivorship patterns that fall between these idealized patterns.

Type I curves reflect high survivorship until late in life, when mortality takes a great toll. Type I curves are

Figure 49.6
Survivorship curves. The survivorship curves of many organisms (pink data points) roughly match one of three idealized patterns (blue curves).

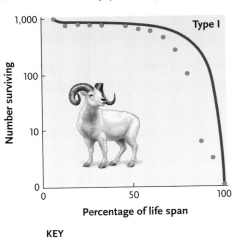

Dall mountain sheep *(Ovis dalli)*

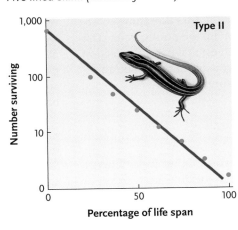

Five-lined skink *(Eumeces fasciatus)*

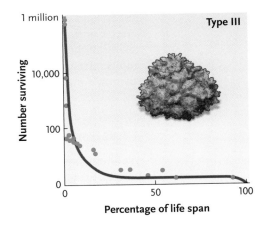

Perennial desert shrub *(Cleome droserifolia)*

KEY
— Theoretical   • Data

typical of large animals that produce few young and provide them with extended care, which reduces juvenile mortality. For example, large mammals, such as the Dall mountain sheep *(Ovis dalli),* produce only one or two offspring at a time and nurture them through their vulnerable first year.

Type II curves reflect a relatively constant rate of mortality in all age classes, a pattern that produces steadily declining survivorship. Many lizards, such as the five-lined skink *(Eumeces fasciatus),* as well as songbirds and small mammals, face a constant probability of mortality from predation, disease, and starvation.

Type III curves reflect high juvenile mortality, followed by a period of low mortality once the offspring reach a critical age and size. For example, *Cleome droserifolia,* a desert shrub from the Middle East, experiences extraordinarily high mortality in its seed and seedling stages. Researchers estimate that for every 1 million seeds produced, fewer than 1000 germinate, and only about 40 individuals survive their first year. Once a plant becomes established, however, its likelihood of future survival is higher, and the survivorship curve flattens out. Many plants, insects, marine invertebrates, and fishes exhibit type III survivorship.

STUDY BREAK

1. What statistics are usually included in a life table?
2. Which type of survivorship curve is characteristic of humans in industrialized countries? Explain your answer.

49.4 The Evolution of Life Histories

The analysis of life tables reveals how natural selection affects the **life histories**—the lifetime patterns of growth, maturation, and reproduction—that characterize different species. Ecologists study life histories to understand how trade-offs in the allocation of resources influence the evolution of specific traits. The results of their research suggest that natural selection adjusts the allocation of resources to maximize an individual's number of surviving offspring.

Organisms Face Trade-Offs in Their Allocation of Resources

Every organism is constrained by a finite **energy budget,** the total amount of energy that it can accumulate and use to fuel its activities. An organism's energy budget is like a savings account. When the individual accumulates more energy than it needs, it makes deposits to this account—energy is stored as starch, glycogen, or fat. When it expends more energy than it

harvests, it makes withdrawals from its energy stores. But unlike a bank account, an organism's energy budget cannot be overdrawn, and no loans against future "earnings" are possible.

Organisms use the energy they harvest for three broadly defined functions: maintenance (the preservation of good physiological condition), growth, and reproduction. And when an organism devotes energy to any one of these functions, the balance in its energy budget is reduced, leaving less energy for the other functions.

Life History Patterns Vary Dramatically among Species

A fish, a deciduous tree, and a mammal illustrate the dramatic variations that exist in life history patterns. Larval coho salmon *(Oncorhynchus kisutch)* hatch in the headwaters of a stream, where they feed and grow for about a year before assuming their adult body form. After swimming downstream to the ocean, they remain at sea for a year or two, feeding voraciously and growing rapidly. Eventually, salmon use sun-compass, geomagnetic, and chemical cues to return to the rivers and streams where they hatched. The fishes swim upstream, and each female lays hundreds or thousands of relatively small eggs. After spending all of their energy reserves on the upstream journey and reproduction, their condition deteriorates and they die.

Most deciduous trees in the temperate zone, such as oaks (genus *Quercus*) and maples (genus *Acer*), begin their lives as seeds in late summer. The seeds remain metabolically inactive until the following spring or a later year. After germinating, trees collect nutrients and energy and continue to grow throughout their lives. Once they achieve a critical size, they may produce thousands of seeds annually for many years. Thus, growth and reproduction occur simultaneously through much of the trees' life.

European red deer *(Cervus elaphus)* are born in spring, and young remain with their mothers for an extended period, nursing and growing rapidly. After weaning, they feed on their own. Female red deer begin to breed after reaching adult size in their third year, producing one or two offspring annually until they are about 16 years old, when they reach their maximum life span and die.

How can we summarize the similarities and differences in the life histories of these organisms? All three species harvest energy throughout their lives. Salmon and deciduous trees continue to grow until old age, whereas deer reach adult size fairly early in life. Salmon produce many offspring in a single reproductive episode, whereas deciduous trees and deer reproduce repeatedly. However, most trees produce thousands of seeds annually, whereas deer produce only one or two young each spring.

What factors have produced these variations in life history patterns? Life history traits—like all population characteristics—are modified by natural selection. Thus, organisms exhibit evolutionary adaptations that increase the fitness of individuals. Each species' life history is, in fact, a highly integrated "strategy"—not in the human sense of planning ahead, but as a suite of selection-driven adaptations.

Ecologists Analyze the Individual Components of Life Histories

In analyzing life histories, ecologists compare the number of offspring with the amount of care provided to each by the parents; they look at the number of reproductive episodes in the organism's lifetime; and they look at the timing of first reproduction. Because these characteristics evolve together, a change in one trait is likely to influence the success of the others.

Fecundity versus Parental Care. If a female has a fixed amount of energy for reproduction, she can package that energy in various ways. By way of illustration, a female duck with 1000 units of energy for reproduction might lay 10 eggs that each contain 100 units of energy per egg. A salmon, which has higher fecundity, might lay 1000 eggs, each endowed with 1 unit of energy. The amount of energy invested in each offspring *before* it is born represents the **passive parental care** that the female provides. Passive parental care is provided through yolk in an egg, endosperm in a seed, or, in mammals, nutrients that cross the placenta.

Many animals, especially birds and mammals, also provide **active parental care** to offspring *after* their birth. In general, species that produce many offspring in a reproductive episode—such as the coho salmon—provide relatively little active parental care *to each offspring*. In fact, female coho salmon, which produce 2400 to 4500 eggs, die before their eggs even hatch. Conversely, species that produce only a few offspring at a time—such as the European red deer—provide a lot of care to each. A red deer doe nurses its single fawn for up to 8 months before weaning it.

One Reproductive Episode versus Several. A second life history characteristic adjusted by natural selection is the number of reproductive episodes in an organism's lifetime. Some organisms, like the coho salmon, devote all of their stored energy to a single reproductive event. Any adult that survives the upstream migration is likely to leave some surviving offspring. Other species, like deciduous trees and red deer, reproduce multiple times. In contrast to salmon, individuals of these species devote only some of their energy budget to reproduction at any time, with the balance allocated to maintenance and growth. Moreover, in some plants, invertebrates, fishes, and reptiles, larger individuals produce more offspring than small ones do. Thus, one advantage of using only part of the energy budget for reproduction is that continued growth may result in greater fecundity at a later age. However, if an organism does not survive until the next breeding season, the potential advantage of putting energy into maintenance and growth is lost.

Early Reproduction versus Late Reproduction. Individuals that first reproduce at the earliest possible age may stand a good chance of leaving some surviving offspring. But the energy devoted to reproduction is no longer available for maintenance and growth. Thus, early reproducers may be smaller and less healthy than individuals that delay reproduction in favor of these other functions. Conversely, an individual that delays reproduction may increase its chance of survival and its future fecundity by becoming larger or more experienced. But there is always some chance that it will die before the next breeding season, leaving no offspring at all. Thus, a finite energy budget and the risk of mortality establish a trade-off in the timing of first reproduction. Mathematical models suggest that delayed reproduction will be favored by natural selection if a sexually mature individual has a good chance of surviving to an older age, if organisms grow larger as they age, and if larger organisms have higher fecundity. Early reproduction will be favored if adult survival rates are low, if animals don't grow larger as they age, or if larger size does not increase fecundity.

Life history characteristics not only vary from one species to another, but they also vary among populations of a single species. *Focus on Research* describes how predation influences life history characteristics in natural populations of guppies (*Poecilia reticulata*) in Trinidad.

STUDY BREAK

1. To what two broad categories of activities do children devote their energy budget?
2. Why do fecundity and the amount of parental care devoted to each offspring exhibit an inverse relationship?

49.5 Models of Population Growth

We now examine mathematical models of population growth that describe very different responses to changes in a population's density. *Exponential* models apply when populations experience unlimited growth. The *logistic* model applies when population growth is limited, often because available resources are finite. These simple models are tools that help ecologists refine their hypotheses, but neither provides entirely accurate predictions of population growth in nature. In the simplest versions of these models, ecologists define births as the production of offspring by any form

Basic Research: The Evolution of Life History Traits in Guppies

Some years ago, drenched with sweat and with fishnets in hand, two ecologists were engaged in fieldwork on the Caribbean island of Trinidad. They were after guppies (*Poecilia reticulata*)— small fish that bear live young in shallow mountain streams **(Figure a)**. John Endler and David Reznick, then of the University of California at Santa Barbara, were studying the environmental variables that influence the evolution of life history patterns in guppies.

Male guppies are easy to distinguish from females. Males, which stop growing at sexual maturity, are smaller, and their scales have bright colors that serve as visual signals in intricate courtship displays. The drably colored females continue to grow larger throughout their lives.

In the mountains of Trinidad, guppies living in different streams—and even in different parts of the same stream—are eaten by one of two other fish species **(Figure b)**. In some streams, a large pike-cichlid (*Crenicichla alta*) prefers mature guppies and tends not to spend time hunting small, immature ones. In other streams, a small killifish (*Rivulus hartii*) preys on immature guppies but does not have much success with the larger adults.

Reznick and Endler found that the life history patterns of guppies vary among streams with different predators. In streams with pike-cichlids, both male and female guppies mature faster and begin to reproduce at a smaller size and a younger age than their counterparts in streams where killifish live **(Figure c)**. In addition, female guppies from pike-cichlid streams reproduce more often and produce smaller and more numerous young **(Figure d)**. These differences allow guppies to avoid some predation. Those in pike-cichlid streams begin to reproduce when they are smaller than the size preferred by that predator. And those from killifish streams grow

Male guppy (right) that shared a stream with pike-cichlids (below)

Male guppy (right) that shared a stream with killifish (below)

Figure a
David Reznick surveys a shallow stream in the mountains of Trinidad.

Figure b
Male guppies from streams where pike-cichlids live (top) are smaller, more streamlined, and have duller colors than those from streams where killifish live (bottom). The pike-cichlid prefers to eat large guppies, and the killifish feeds on small guppies. Guppies are shown approximately life-size, adult pike-cichlids grow to 16 cm in length, and adult killifish grow to 10 cm. (Guppy photos: David Reznick/University of California, Riverside: computer enhanced by Lisa Starr; predator photos: Hippocampus Bildarchiv.)

of reproduction, and ignore the effects of immigration and emigration.

Models of Exponential Population Growth Describe Growth without Limitation

Sometimes populations increase in size for a period of time with no apparent limits on their growth. In models of exponential growth, population size increases steadily by a constant ratio. Bacterial populations provide the most obvious examples, but multicellular organisms also sometimes exhibit exponential population growth.

Bacterial Population Growth. Bacteria reproduce by binary fission. A parent cell divides in half, producing two daughter cells, which each divide to produce two granddaughter cells. Generation time in a bacterial population is simply the time between successive cell divisions. And if no bacteria in the population die, the population doubles in size each generation.

Bacterial populations grow quickly under ideal temperatures and with unlimited space and food. Consider a population of the human intestinal bacterium *Escherichia coli,* for which the generation time can be as short as 20 minutes. If we start with a population of one bacterium, the population doubles to two cells after one generation, to four cells after two generations, and to eight cells after three generations **(Figure 49.7)**. After only 8 hours, or 24 generations, the population will number more than 16 million. And after a single day, or 72 generations, the population will number nearly 5×10^{21} cells. Although other bacteria grow more slowly than *E. coli,* it is no wonder that patho-

quickly to a size that is too large to be consumed by killifish.

Although these life history differences were correlated with the distributions of the two predatory fishes, they might result from some other, unknown differences between the streams. Endler and Reznick investigated this possibility with controlled laboratory experiments. They shipped groups of guppies to California, where they bred guppies from each kind of stream for two generations. Both types of experimental populations were raised under identical conditions in the absence of predation. Even when predators were absent, the two types of experimental populations retained their life history differences. These results provided evidence of a heritable genetic basis for the observed life history differences.

Endler and Reznick also examined the role of predators in the *evolution* of the size differences. They raised guppies for many generations in the laboratory under three experimental conditions—some alone, some with killifish, and some with pike-cichlids. As predicted, the guppy lineage that was subjected to predation by killifish became larger at maturity. Individuals that were small at maturity were frequently eaten, and their reproduction was limited. The lineage that was

raised with pike-cichlids showed a trend toward earlier maturity. Individuals that matured at a larger size faced a greater likelihood of being eaten before they had reproduced.

Finally, when they first visited Trinidad, Endler and Reznick had introduced guppies from a pike-cichlid stream to another stream that contained killifish but no pike-cichlids or guppies. Eleven years later, the guppy populations had changed. As the researchers predicted, the guppies had become larger in size and reproduced more slowly, characteristics that are typical of natural guppy populations that live and die with killifish.

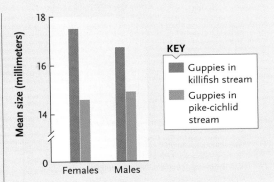

Figure c
Guppies in streams occupied by pike-cichlids are smaller than those in streams occupied by killifish.

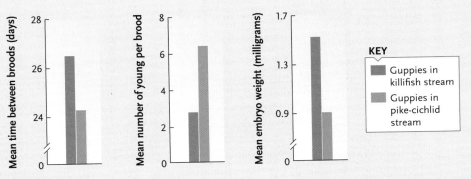

Figure d
Female guppies from streams occupied by pike-cichlids reproduce more often (shorter time between broods) and produce more young per brood and smaller young (lower embryo weight) than females living in streams occupied by killifish.

genic bacteria, such as those causing cholera or plague, can quickly overtake the defenses of an infected animal.

Exponential Population Growth in Other Organisms.
By contrast to bacteria, many plants and animals live side-by-side with their offspring. In these populations, births increase a population's size and deaths decrease it. Over a given time period:

change in population size =
number of births − number of deaths

We express this relationship mathematically by defining N as the population size; ΔN (pronounced "delta N") as the change in population size; Δt as the time period during which the change occurs; and B and D as the numbers of births and deaths, respectively, dur-

ing that time period. Thus, $\Delta N/\Delta t$ symbolizes the change in population size over time, and

$$\Delta N/\Delta t = B - D$$

The preceding equation applies to any population for which we know the exact numbers of births and deaths.

Ecologists usually express births and deaths as *per capita* (per individual) rates, allowing them to apply the model to a population of any size. The per capita birth rate, symbolized b, is simply the number of births in the population during the specified time period divided by the population size: $b = (B/N)$. Similarly, the per capita death rate, d, is the number of deaths divided by the population size: $d = (D/N)$. If, for example, in a population of 2000 field mice, 1000 mice are born and 200 mice die during 1 month's time, $b = 1000/2000 =$

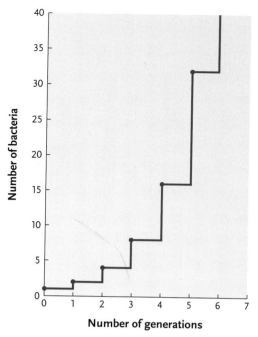

Figure 49.7

Bacterial population growth. If all members of a bacterial population divide simultaneously, a plot of population size over time forms a stair-stepped curve in which the steps get larger as the number of dividing cells increases.

0.5 births per individual per month, and $d = 200/2000 = 0.1$ deaths per individual per month. Of course, no mouse can give birth to half an offspring, and no individual can die one-tenth of a death. But these rates tell us the per capita birth and death rates *averaged over all mice in the population*. Per capita birth and death rates are always expressed over a specified time period. For long-lived organisms, such as humans, time is measured in years; for short-lived organisms, such as fruit flies, time is measured in days. We can calculate per capita birth and death rates from data in a life table.

We can now revise the population growth equation to use per capita birth and death rates instead of the actual numbers of births and deaths. The change in a population's size during a given time period ($\Delta N/\Delta t$) depends on the per capita birth and death rates, as well as on the number of individuals in the population. Mathematically, we can write

$$\Delta N/\Delta t = B - D = bN - dN = (b - d)N$$

or, in the notation of calculus,

$$dN/dt = (b - d)N$$

This equation describes the **exponential model of population growth.** (Note that in calculus, dN/dt is the notation for the population growth rate; the "d" in dN/dt is *not* the same "d" that we use to symbolize the per capita death rate.)

The difference between the per capita birth rate and the per capita death rate, $b - d$, is the **per capita growth rate** of the population, symbolized by r. Like b and d, r is always expressed per individual per unit time. Using the per capita growth rate, r, in place of ($b - d$), the exponential growth equation is written

$$dN/dt = rN$$

If the birth rate exceeds the death rate, r has a positive value ($r > 0$), and the population is growing. In our example with field mice, $r = 0.5 - 0.1 = 0.4$ mice per mouse per month. If, on the other hand, the birth rate is lower than the death rate, r has a negative value ($r < 0$), and the population is getting smaller. In populations where the birth rate equals the death rate, r is exactly zero, and the population's size is not changing—a situation known as **zero population**

Figure 49.8

Exponential population growth. Exponential population growth produces a J-shaped curve of population size plotted against time. Although the per capita growth rate *(r)* remains constant, the increase in population size gets larger every month, because more individuals are reproducing.

Month	Old Population Size		Net Monthly Increase		New Population Size
1	2,000	+	800	=	2,800
2	2,800	+	1,120	=	3,920
3	3,920	+	1,568	=	5,488
4	5,488	+	2,195	=	7,683
5	7,683	+	3,073	=	10,756
6	10,756	+	4,302	=	15,058
7	15,058	+	6,023	=	21,081
8	21,081	+	8,432	=	29,513
9	29,513	+	11,805	=	41,318
10	41,318	+	16,527	=	57,845
11	57,845	+	23,138	=	80,983
12	80,983	+	32,393	=	113,376
13	113,376	+	45,350	=	158,726
14	158,726	+	63,490	=	222,216
15	222,216	+	88,887	=	311,103
16	311,103	+	124,441	=	435,544
17	435,544	+	174,218	=	609,762
18	609,762	+	243,905	=	853,667
19	853,677	+	341,467	=	1,195,134

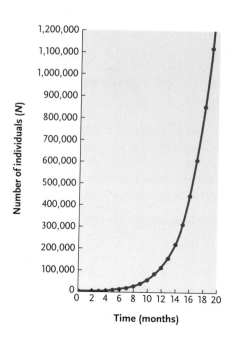

growth, or ZPG. Even under conditions of ZPG, births and deaths still occur, but the numbers of births and deaths cancel out.

As long as a population's per capita growth rate is positive ($r > 0$), the population will increase in size. In our hypothetical population of field mice, we started with $N = 2000$ mice, and calculated a per capita growth rate of 0.4 mice per individual per month. In the first month, the population grows by $0.4 \times 2000 = 800$ mice **(Figure 49.8)**. At the start of the second month, $N = 2800$ and r still $= 0.4$. Thus, in the second month, the population grows by $0.4 \times 2800 = 1120$ mice. Notice that even though r remains constant, the *increase* in population size gets larger each month simply because more individuals are reproducing. In less than 2 years, the mouse population will grow to more than 1 million! A graph of exponential population growth has a characteristic J shape, getting steeper through time. The population grows at an ever-increasing pace because the change in a population's size depends on the number of individuals in the population as well as its per capita growth rate.

Population Growth under Ideal Conditions. Imagine a hypothetical population living in an ideal environment—one with unlimited food and shelter; no predators, parasites, or disease; and a comfortable abiotic environment. Under such circumstances, which are admittedly unrealistic, the per capita birth rate is very high, the per capita death rate is very low, and the per capita growth rate, r, is as high as it can possibly be. This maximum per capita growth rate, symbolized r_{max}, is the population's **intrinsic rate of increase.** Under these ideal conditions, our exponential growth equation is

$$dN/dt = r_{max}N$$

When populations are growing at their intrinsic rate of increase, population size increases very rapidly. Across a wide variety of protists and animals r_{max} varies inversely with generation time: species with short generation time have higher intrinsic rates of increase than those with long generation time **(Figure 49.9).**

The Logistic Model Describes Population Growth When Resources Are Limited

The exponential model predicts unlimited population growth. But we know from even casual observations that the population sizes of most species are somehow limited—we are not knee-deep in bacteria, rosebushes, or garter snakes. What factors limit the growth of populations? As a population gets larger, it uses more vital resources, and a shortage of resources may eventually develop. As a result, individuals may have less energy available for maintenance and reproduction, causing per capita birth rates to decrease and per capita death rates to increase. Changes in these rates reduce the

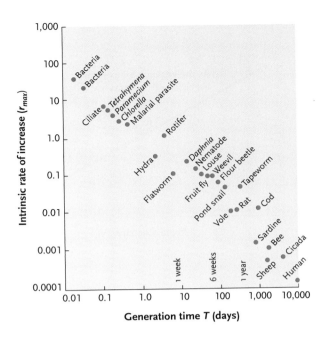

population's per capita growth rate, causing population growth to slow down or to stop altogether.

The Logistic Model. Environments provide enough resources to sustain only a finite population of any species. The maximum number of individuals that an environment can support indefinitely is termed its **carrying capacity,** symbolized K. The carrying capacity, which is defined for each population, is a property of the environment, and it varies from one habitat to another and in a single habitat through time. For example, the spring and summer flush of insects in temperate habitats supports large populations of insectivorous birds. But fewer insects are available in autumn and winter, causing a seasonal decline in the carrying capacity *for birds.* Many birds then migrate to habitats that provide more food and better weather.

The **logistic model of population growth** assumes that a population's per capita growth rate, r, decreases as the population gets larger **(Figure 49.10a).** In other words, population growth slows as the population size approaches the carrying capacity. The mathematical expression ($K - N$) tells us how many individuals can be added to a population before it reaches carrying capacity. And the expression ($K - N$)/K indicates what *percentage* of the carrying capacity is still available.

To create the logistic model, we factor the impact of carrying capacity into the exponential model by letting $r = r_{max}(K - N)/K$. This calculation reduces the per capita growth rate (r) from its maximum value (r_{max}) as N increases:

$$dN/dt = r_{max}N(K - N)/K$$

The calculation of how r varies with population size is straightforward **(Table 49.2).** In a very small population (N much smaller than K), plenty of resources are still

Figure 49.9
Generation time and r_{max}. The intrinsic rate of increase (r_{max}) is high for protists and animals with short generation time and low for those with long generation time. The logarithmic scale on both axes compresses the data into a straight line.

a. The predicted effect of *N* on *r*

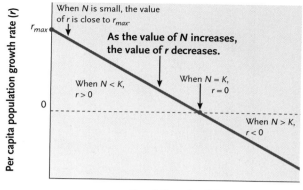

When *N* is small, the value of *r* is close to r_{max}.

r_{max}

As the value of *N* increases, the value of *r* decreases.

When *N* < *K*, *r* > 0

When *N* = *K*, *r* = 0

0

When *N* > *K*, *r* < 0

Per capita population growth rate (*r*)

Population size (*N*)

b. Population size through time

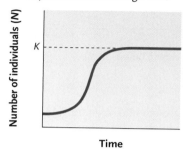

Number of individuals (*N*)

K

Time

Figure 49.10
The logistic model of population growth. **(a)** The logistic model assumes that the per capita population growth rate (*r*) decreases linearly as population size (*N*) increases. **(b)** The logistic model predicts that population size increases quickly at first, but then slowly approaches the carrying capacity (*K*).

Table 49.2	**The Effect of *N* on *r* and ΔN^* in a Hypothetical Population Exhibiting Logistic Growth in which $K = 2000$ and $r_{max} = 0.04$ per capita per year**		
N (population size)	(*K* − *N*)/*K* (% of *K* available)	$r = r_{max}(K - N/K)$ (per capita growth rate)	$\Delta N = rN$ (change in *N*)
50	0.990	0.0396	2
100	0.950	0.0380	4
250	0.875	0.0350	9
500	0.750	0.0300	15
750	0.625	0.0250	19
1000	0.500	0.0200	20
1250	0.375	0.0150	19
1500	0.250	0.0100	15
1750	0.125	0.0050	9
1900	0.050	0.0020	4
1950	0.025	0.0010	2
2000	0.000	0.0000	0

*ΔN rounded to the nearest whole number.

available; the value of (*K* − *N*)/*K* is close to 1, and the per capita growth rate (*r*) is therefore close to the maximum possible (r_{max}). Under these conditions, population growth is close to exponential. If a population is large (*N* close to *K*), few additional resources are available; the value of (*K* − *N*)/*K* is small, and the per capita growth rate (*r*) is very low. When the size of the population exactly equals the carrying capacity, (*K* − *N*)/*K* becomes zero, and so does the population growth rate—the situation defined as ZPG.

The logistic model of population growth predicts an S-shaped graph of population size over time, with the population slowly approaching its carrying capacity and remaining at that level **(Figure 49.10b)**. According to this model, the population grows slowly when the population size is small, because there are few individuals reproducing. It also grows slowly when the population size is large because the per capita population growth rate is low. The population grows quickly (*dN/dt* is highest) at intermediate population sizes, when a sizable number of individuals are breeding and the per capita population growth rate (*r*) is still fairly high (see Table 49.2).

Intraspecific Competition. The logistic model assumes that vital resources become increasingly limited as a population grows larger. Thus, the model is a mathematical portrait of **intraspecific** (within species) **competition**, the dependence of two or more individuals in a population on the same limiting resource. For animals, limiting resources can be food, water, nesting sites, refuges from predators, and, for sessile species (those permanently attached to a surface), space. For plants, sunlight, water, inorganic nutrients, and growing space can be limiting. The pattern of uniform dispersion described earlier often reflects intraspecific competition for limited resources.

In some very dense populations, the accumulation of poisonous waste products may also reduce survivorship and reproduction. Most natural populations live in open systems where wastes are consumed by other organisms or flushed away. But the buildup of toxic wastes is common in laboratory cultures of microorganisms. For example, yeast cells ferment sugar and produce ethanol as a waste product. Thus, the alcohol content of wine rarely exceeds 13% by volume, the ethanol concentration that poisons winemaking yeasts.

Logistic Growth in the Laboratory and in Nature. How well do species conform to the predictions of the logistic model? In simple laboratory cultures, relatively small organisms, such as *Paramecium*, some crustaceans, and flour beetles, often show an S-shaped pattern of population growth **(Figure 49.11)**. Moreover, large animals that have been introduced into new environments sometimes exhibit a pattern of population

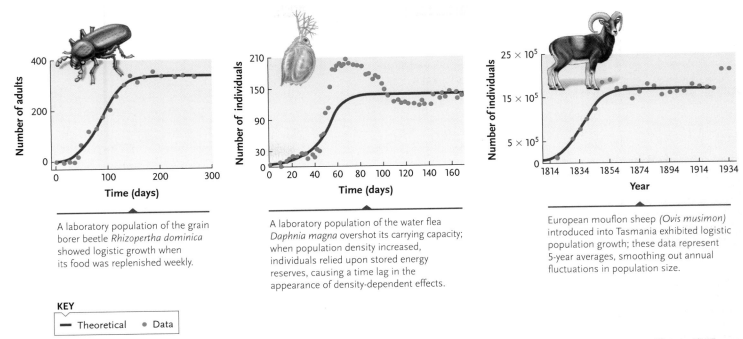

A laboratory population of the grain borer beetle *Rhizopertha dominica* showed logistic growth when its food was replenished weekly.

A laboratory population of the water flea *Daphnia magna* overshot its carrying capacity; when population density increased, individuals relied upon stored energy reserves, causing a time lag in the appearance of density-dependent effects.

European mouflon sheep *(Ovis musimon)* introduced into Tasmania exhibited logistic population growth; these data represent 5-year averages, smoothing out annual fluctuations in population size.

KEY

— Theoretical   • Data

Figure 49.11
Examples of logistic population growth.

growth that matches the predictions of the logistic model (see Figure 49.11).

Nevertheless, some assumptions of the logistic model are unrealistic. For example, the model predicts that survivorship and fecundity respond immediately to changes in a population's density. But many organisms exhibit a delayed response, called a **time lag.** Some time lags occur because fecundity is usually determined by the availability of resources at some time in the past, when individuals were adding yolk to eggs or endosperm to seeds. Moreover, when food resources become scarce, individuals may use stored energy reserves to survive and reproduce, and the effects of crowding may not be felt until those reserves are depleted. As a result, the population size may temporarily overshoot its carrying capacity (see Figure 49.11b). Deaths may then outnumber births, causing the population size to drop below the carrying capacity, at least temporarily. Time lags often cause a population to oscillate around its carrying capacity.

Another unrealistic assumption of the logistic model is that the addition of new individuals to a population always decreases survivorship and fecundity, no matter how small the population is. But in small populations, modest population growth probably doesn't have much effect on these processes. In fact, most organisms probably require a minimum population density to survive and reproduce. For example, some plants flourish in small clumps that buffer them from physical stresses, whereas a single individual living in the open would suffer adverse effects. And in some animal populations, a minimum population density is necessary for individuals to find mates—an important issue in conservation biology (see Chapter 53).

STUDY BREAK

1. How does the prediction of the exponential model of population growth differ from that of the logistic model?
2. What is carrying capacity? Is it a property of a habitat or of a population?
3. What is a time lag?

49.6 Population Regulation

As you have seen, the population sizes of some species change from month to month or from year to year, whereas others remain fairly stable. What environmental factors influence population growth rates and control fluctuations in population size?

Density-Dependent Factors Often Regulate Population Size

Some factors that affect population size are **density-dependent:** their influence increases or decreases with the density of the population. Examples of density-dependent environmental factors include intraspecific competition and predation. The logistic model includes the effects of density-dependence in its assumption that per capita birth and death rates change with a population's density.

The Effects of Crowding. Numerous laboratory and field studies show that crowding (high population density) decreases the individual growth rate, adult size,

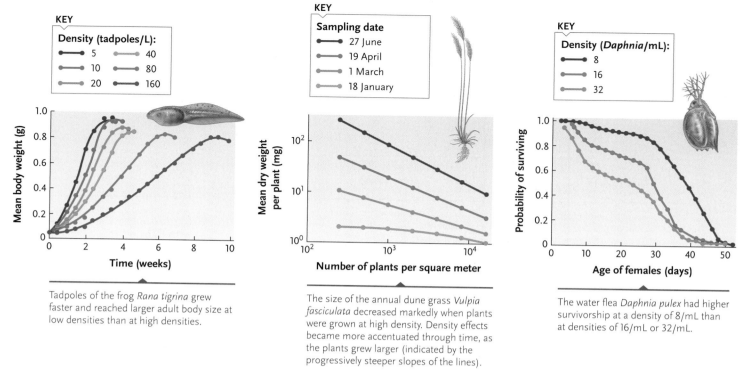

KEY
Density (tadpoles/L):
● 5   ● 40
● 10   ● 80
● 20   ● 160

Tadpoles of the frog *Rana tigrina* grew faster and reached larger adult body size at low densities than at high densities.

KEY
Sampling date
● 27 June
● 19 April
● 1 March
● 18 January

The size of the annual dune grass *Vulpia fasciculata* decreased markedly when plants were grown at high density. Density effects became more accentuated through time, as the plants grew larger (indicated by the progressively steeper slopes of the lines).

KEY
Density (*Daphnia*/mL):
● 8
● 16
● 32

The water flea *Daphnia pulex* had higher survivorship at a density of 8/mL than at densities of 16/mL or 32/mL.

Figure 49.12
Effects of crowding on individual growth, size, and survival.

Figure 49.13
Effects of crowding on fecundity.

The number of seeds produced by shepherd's purse (*Capsella bursa-pastoris*) decreased dramatically with increasing density in experimental plots.

The mean number of eggs produced by the great tit (*Parus major*), a woodland bird, declined as the number of breeding pairs in Marley Wood increased.

and survival of plants and animals **(Figure 49.12)**. Organisms living in extremely dense populations are unable to harvest enough resources; they grow slowly and tend to be small, weak, and less likely to survive. Gardeners understand this relationship, thinning their plants to a density that maximizes the number of vigorous individuals.

Crowding also has a negative effect on reproduction **(Figure 49.13)**. When resources are in short supply, each individual has less energy available for reproduction after meeting its basic maintenance needs. Hence, females in crowded populations produce either fewer offspring or smaller offspring that are less likely to survive.

In some species, crowding stimulates developmental and behavioral changes that may influence the density of a population. For example, migratory locusts (*Locusta migratoria*) can develop into either solitary or migratory forms in the same population. Migratory individuals have longer wings and more body fat, characteristics that allow them to disperse great distances. High population density increases the frequency of the migratory form, and huge numbers of locusts move away from the area of high density **(Figure 49.14)**, reducing the size of the original population.

These studies confirm the assumptions of the logistic equation, but they don't prove that natural populations are regulated by density-dependent factors. A convincing demonstration requires experimental evi-

Figure 49.14

A swarm of locusts. Migratory locusts (*Locusta migratoria*), moving across an African landscape, can devour their own weight in plant material every day.

dence that an increase in population density causes population size to decrease, and that a decrease in density causes it to increase. In one study conducted in the 1960s, Robert Eisenberg of the University of Michigan experimentally increased the numbers of aquatic snails in some ponds, decreased them in others, and maintained natural densities in control ponds. Although adult survivorship did not differ between experimental and control treatments, snails in the high-density ponds produced fewer eggs, and those in the low-density ponds produced more eggs, than those living at the control density. In addition, the survival rates of young snails declined as density increased. After 4 months, the densities in the two experimental groups converged on those in the control, providing strong evidence of density-dependent population regulation.

Other Density-Dependent Factors. Our discussion of the logistic equation described intraspecific competition as the primary density-dependent factor regulating population size. Competition between populations of different species also exerts density-dependent effects on population growth, a topic we consider in Section 50.1.

Predation can also cause density-dependent population regulation. As a particular prey species becomes more numerous, predators may consume more of it because it is easier to find and catch. Once a prey species has exceeded a threshold density, predators may consume a *larger percentage* of the prey population, which is a density-dependent effect (see Figure 20.16). For example, on rocky shores in California, sea stars concentrate their feeding on the most abundant of several invertebrate species. When one prey species becomes common, predators feed on it disproportionately, drastically reducing its numbers.

Like predation, parasitism and disease cause density-dependent regulation of plant and animal populations. Infectious microorganisms spread quickly in a crowded population. In addition, if crowded individuals are weak or malnourished, they are more susceptible to infection and may die from diseases that healthy organisms would survive.

Density-Independent Factors Can Limit Population Size

Some populations are affected by **density-independent** factors that reduce population size regardless of its density. If an insect population is not physiologically adapted to high temperature, a sudden hot spell may kill 80% of the insects whether they number 100 or 100,000. Fires, earthquakes, storms, and other natural disturbances may contribute directly or indirectly to density-independent mortality. But because such factors do not cause a population to fluctuate around its carrying capacity, density-independent factors do not *regulate* population size, although they may reduce it.

Density-independent factors have a particularly strong effect on populations of small-bodied species that cannot buffer themselves against environmental change. Their populations grow exponentially for a time, but shifts in climate or random events cause high mortality before populations reach a size at which density-dependent factors regulate their numbers. When conditions improve, populations grow exponentially—at least until another density-independent factor causes them to crash again. For example, a small Australian insect, *Thrips imaginis*, feeds on pollen and flowers of plants in the rose family; they are frequently abundant enough to damage the blooms. *Thrips* populations grow exponentially in spring, when many flowers are available and the weather is warm and moist **(Figure 49.15)**. But populations crash predictably during summer because *Thrips* do not tolerate extremely hot

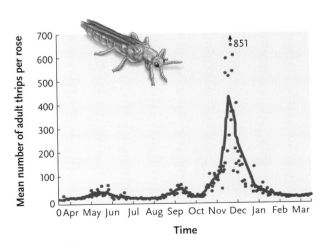

Figure 49.15

Booms and busts in a *Thrips* population. Populations of the Australian insect *Thrips imaginis* grow exponentially when conditions are favorable during spring (which begins in September in the southern hemisphere). The populations crash in summer, however, when hot and dry conditions cause high mortality.

Figure 49.16 Experimental Research

Evaluating Density-Dependent Interactions between Species

QUESTION: Does the population density of lizards on Caribbean islands have any effect on the population density of spiders?

EXPERIMENT: Spiller and Schoener built fences around a series of study plots on a small island in the Bahamas. They excluded all individuals of three lizard species from the experimental plots, but left resident lizards undisturbed in the control plots. They then made monthly measurements of population densities of the web-building spider *Metepeira datona* in both experimental plots and control plots.

RESULTS: Over the 20-month course of the experiment, spider densities were as much as five times higher in the experimental plots than in the control plots.

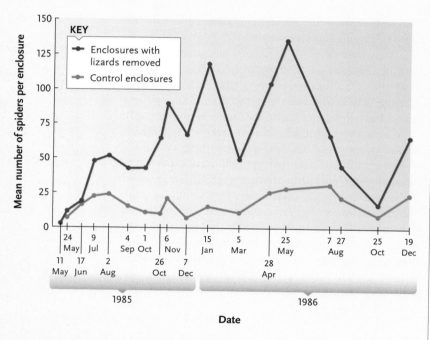

CONCLUSION: Spiller and Schoener concluded that the presence of lizards has a large impact on spider populations. The lizards not only compete with the spiders for insect food, but they also appear to prey on the spiders.

and dry conditions. After the crash, a few individuals survive in remaining flowers, forming the stock from which the population grows exponentially the following spring.

Interacting Environmental Factors Often Limit Population Sizes

Sometimes several density-dependent factors influence a population at the same time. For example, on small islands in the West Indies, the spider *Metepeira datona* is rare wherever lizards (*Ameiva festiva, Anolis*

carolinensis, and *Anolis sagrei*) are abundant, but common where the lizards are rare or absent. To test whether the presence of lizards limits the abundance of spiders, David Spiller and Tom Schoener of the University of California, Davis, built fences around plots on islands where these species occur. They eliminated lizards from experimental plots, but left them in control plots. After 2 years, spider populations in some experimental plots were five times denser than those in control plots **(Figure 49.16)**. In this case, lizards had two density-dependent effects on spider populations: they preyed upon spiders, and they competed with them for food.

Density-dependent factors can also interact with density-independent factors, limiting population growth. For example, food shortage caused by high population density (a density-dependent factor) may lead to malnourishment; in turn, malnourished individuals may be more likely to succumb to the stress of extreme weather (a density-independent factor).

Populations can also be affected by density-independent factors in a density-dependent manner. For example, animals often retreat into shelters to escape environmental stresses, such as floods or severe heat. If a population is small, most individuals can fit into a limited number of available refuges. But if a population is large, only a small proportion will find suitable shelter; and the larger the population is, the greater the percentage of individuals that will experience the stress. Thus, although the density-independent effects of weather limit *Thrips* populations, it is the availability of flowers in summer—clearly a density-dependent factor—that regulates the size of the *Thrips* starting stock the following spring. Hence, both types of factors influence the size of *Thrips* populations.

The Life History Characteristics of a Species Govern Fluctuations in Its Population Size through Time

Even casual observation reveals tremendous variation in how rapidly population size changes in different species. For example, new weeds often appear in a vegetable garden overnight, whereas the number of oak trees in a forest may remain relatively stable for years. Why do some species have the potential for explosive population growth, but others do not? The answer lies in how natural selection has molded life history strategies that are adapted to different ecological conditions. Ecologists describe two divergent life history patterns—**r-selected** species and **K-selected** species—with very different characteristics **(Table 49.3, Figure 49.17)**. These strategies represent extremes on a continuum of possible patterns, and the life histories of most species actually fall somewhere between them.

Species with an *r*-selected life history are adapted to function well in rapidly changing environments. They are generally small, have short generation times, and produce numerous, tiny offspring, often in a single reproductive event. The offspring receive little or no parental care of any kind. Because species with short generation times tend to have high r_{max} (see Figure 49.9), their populations grow exponentially when environmental conditions are favorable—hence the name *r*-selected. Although their numerous offspring disperse and rapidly colonize available habitats, most die before reaching sexual maturity (Type III survivorship). Thus, the success of an *r*-selected life history depends on flooding the environment with a *large quantity* of young, only a few of which may be successful.

Because they have small body size, *r*-selected species lack physiological mechanisms to buffer them from environmental variation. Thus, as described earlier for the Australian thrips living in roses, survivorship and fecundity are often greatly influenced by density-independent factors, and population size fluctuates markedly. In good years, survivorship and fecundity may be high, and the population explodes. In bad years, survivorship and reproduction may be low, and the population crashes. Populations of *r*-selected species are often so greatly reduced by changes in abiotic environmental factors, such as temperature or moisture, that they never grow large enough to face a shortage of limiting resources; thus, their carrying capacity cannot be estimated, and changes in their population size cannot be described by the logistic model of population growth.

By contrast, *K*-selected species thrive in more stable environments. They are generally large, have long generation times, and produce offspring repeatedly during their lifetimes. Their offspring receive substantial parental care, either as energy reserves in an egg or seed or as active care, ensuring that most survive the early stages of life (Type I or Type II survivorship). Because *K*-selected species typically have a low r_{max}, their populations often grow slowly. The success of a *K*-selected life history therefore depends on the production of a relatively small number of *high quality* offspring that join an already well-established population.

The large body size of *K*-selected species allows them to use behavioral and physiological mechanisms to buffer themselves against environmental change, so that survivorship and fecundity do not fluctuate wildly in response to environmental variations. Instead, their populations are often affected by density-dependent factors, which regulate population size near their carrying capacity—hence the name *K*-selected. For these species, natural selection has favored life history characteristics that result in stable population sizes: the production of relatively few offspring, extensive parental care, good competitive ability, a long life span, and repeated reproductions. Many large terrestrial vertebrates are examples of *K*-selected species.

Table 49.3	Characteristics of *r*-Selected and *K*-Selected Species	
Characteristic	**r-Selected Species**	**K-Selected Species**
Maturation time	Short	Long
Life span	Short	Long
Mortality rate	Usually high	Usually low
Reproductive episodes	Usually one	Usually several
Time of first reproduction	Early	Late
Clutch or brood size	Usually large	Usually small
Size of offspring	Small	Large
Active parental care	Little or none	Often extensive
Population size	Fluctuating	Relatively stable
Tolerance of environmental change	Generally poor	Generally good

Some Species Exhibit Regular Cycles in Population Size

The population densities of many insects, birds, and mammals in the northern hemisphere fluctuate between species-specific lows and highs in a multiyear cycle. Arctic populations of small rodents vary in size over a 4-year cycle, whereas snowshoe hares, ruffed grouse, and lynxes have 10-year cycles. Ecologists documented such cyclic fluctuations more than a century ago, but none of the general hypotheses so far proposed explains the cycles in all species. The availability and quality of food, the abundance of predators, the prevalence of disease-causing microorganisms, and variations in weather may influence population growth. Furthermore, a cycling population's food supply and

a. An *r*-selected species

Nigel Cattlin/Holt/Holt Studios International, Ltd.

b. A *K*-selected species

© Goodshoot/Corbis

Figure 49.17
Life history differences. **(a)** An *r*-selected species, like quinoa (*Chenopodium quinoa*), matures in one growing season and produces many tiny seeds, which were a traditional food staple for the indigenous people of North and South America. **(b)** A *K*-selected species, like the coconut palm (*Cocos nucifera*), grows slowly and produces a few large seeds repeatedly during its long life.

predators are themselves influenced by the population's size.

Theories of *intrinsic control* suggest that as an animal population grows, individuals undergo hormonal changes that increase aggressiveness, reduce reproduction, and foster dispersal to other areas. The dispersal phase of the cycle may be dramatic. For example, when populations of the Norway lemming *(Lemmus lemmus),* a rodent that lives in the Scandinavian arctic, reach their peak density, aggressive interactions drive younger and weaker individuals away from their place of birth. The exodus of many thousands of lemmings, scrambling over rocks and even cliffs, is sometimes incorrectly portrayed in nature films as a suicidal mass migration. Researchers do not yet know how widespread these hormonal and behavioral changes are among different species or exactly what regulates them.

Other explanations focus on *extrinsic control,* such as the relationship between a cycling species and its food or predators. A dense population may exhaust its food supply, increasing mortality and decreasing reproduction. But experimental food supplementation does not always prevent the decline in mammal populations, indicating that other factors are also at work.

Some researchers have suggested that the cycles of predators and their prey are induced by time lags in each population's response to changes in density of the other **(Figure 49.18).** The 10-year cycles of snowshoe hares *(Lepus americanus)* and their feline predators, Canada lynxes *(Lynx canadensis),* were often cited as a classic example of such an interaction. But recent research has cast doubt on this straightforward explanation. Hare populations exhibit a 10-year fluctuation even on islands where lynxes are absent. Thus, the lynx cannot be solely responsible for the hare's cycle, although cycles in the hare populations may trigger cycles in populations of their predators.

Charles Krebs and his colleagues at the University of British Columbia studied hare and lynx interactions with a large-scale, multiyear experiment in the southern Yukon. They fenced experimental areas where they added food for the hares, excluded mammalian predators, or applied both experimental treatments; unmanipulated plots served as controls. Where mammalian

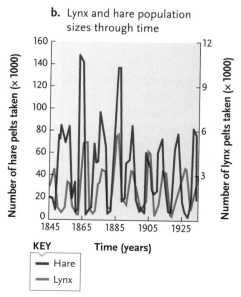

Figure 49.18

The predator-prey model. Predator-prey interactions may contribute to density-dependent regulation of both populations. **(a)** A mathematical model predicts cycles in the numbers of predators and prey because of time lags in each species' responses to changes in the density of the other. (Predator population size is exaggerated in this graph: predators are usually less common than prey.) **(b)** The interaction between the Canada lynx *(Lynx canadensis)* and the snowshoe hare *(Lepus americanus)* was often described as a cyclic predator-prey interaction. The abundances of lynx and hare are based on counts of pelts that trappers sold to Hudson's Bay Company over a 90-year period. Recent research has shown that population cycles in snowshoe hares are caused by complex interactions between the hare, its food plants, and its predators.

© Ed Cesar/Photo Researchers, Inc.

a. Predictions of a predator-prey model

When prey density is low, predators go hungry and their population size declines.

With fewer predators feeding on them, the prey population grows.

Predators capture the abundant prey, and their population starts to grow.

As prey density decreases, predators go hungry and their population size starts to decline.

The increasing number of predators causes the prey population to decline.

A new cycle begins.

KEY
— Prey
— Predator

Time

b. Lynx and hare population sizes through time

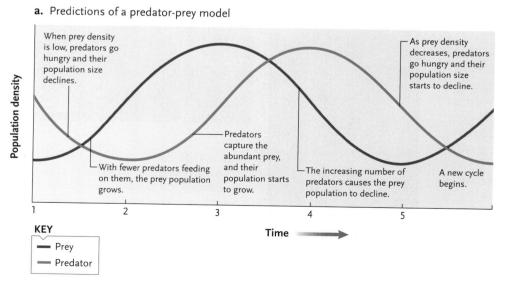

KEY
— Hare
— Lynx

Time (years)

predators were excluded, hare densities approximately doubled relative to the controls. Where food was added, hare densities tripled. But in plots where food was added *and* predators were excluded, the hare densities increased 11-fold. Krebs and his colleagues concluded that neither food availability nor predation alone is solely responsible for arctic hare population cycles; instead, complex interactions between the hares, their food plants, and their predators create the cyclic fluctuations in hare population size.

STUDY BREAK

1. How can you tell whether an environmental factor causes density-dependent or density-independent effects on a population?
2. Are the effects of infectious diseases on populations more likely to be density-dependent or density-independent?

49.7 Human Population Growth

How do human populations compare with those of other species we have studied? The worldwide human population surpassed 6 billion on October 12, 1999. Like many other species, humans live in somewhat isolated populations, which vary in their demographic traits and access to resources. Although many of us live comfortably, at least a billion people are malnourished or starving, lack clean drinking water, and live without adequate shelter or health care. Even if it were possible to double the food supply, increased agricultural production would inevitably increase pollution and contribute to spoiled croplands, deforestation, and desertification, which are described in Chapter 53.

Human Populations Have Sidestepped the Usual Density-Dependent Controls

For most of human history, our population grew slowly; but over the past two centuries, the worldwide human population has grown exponentially **(Figure 49.19)**. Demographers have identified three ways in which humans have avoided the effects of density-dependent regulating factors.

First, humans have expanded their geographical range into virtually every terrestrial habitat. Our early ancestors lived in tropical and subtropical grasslands, but by 40,000 years ago, they had dispersed through much of the world (see Section 30.13). Their success resulted from their ability to solve ecological problems by building fires, assembling shelters, making clothing and tools, and planning community hunts. Vital survival skills spread from generation to generation and from one population to another because language allowed the communication of complex ideas and knowledge.

Second, humans have increased the carrying capacities of habitats they occupy. About 11,000 years ago, many populations shifted from hunting and gathering to agriculture. They cultivated wild grasses, diverted water to irrigate crops, and used domesticated animals for food and labor. Such innovations increased the availability of food, raising both the carrying capacity and the population growth rates. In the mid-eighteenth century, people harnessed the energy in fossil fuels, and industrialization began in Western Europe and North America. Food supplies and the carrying capacity increased again, at least in the industrialized countries, through the use of synthetic fertilizers, pesticides, and efficient methods of transportation and food distribution.

Third, advances in public health have reduced the effects of critical population-limiting factors such as malnutrition, contagious diseases, and poor hygiene. Over the past 300 years, modern plumbing and sewage

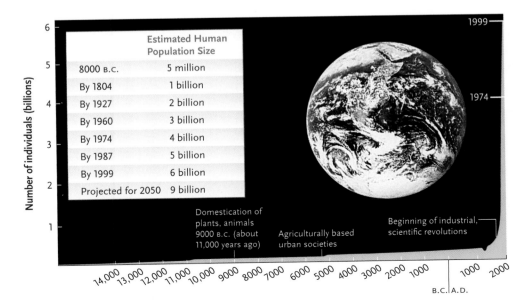

Figure 49.19

Human population growth. The worldwide human population grew slowly until 200 years ago, when it began to increase explosively. The dip in the mid-fourteenth century represents the death of 60 million Asians and Europeans from the bubonic plague. The table shows the years when the human population reached each additional billion people. (Photo: NASA.)

Estimated Human Population Size	
8000 B.C.	5 million
By 1804	1 billion
By 1927	2 billion
By 1960	3 billion
By 1974	4 billion
By 1987	5 billion
By 1999	6 billion
Projected for 2050	9 billion

treatment, improvements in food handling and processing, and medical discoveries have reduced death rates sharply. Births now greatly exceed deaths, especially in less industrialized countries, resulting in rapid population growth.

Age Structure and Economic Development May Now Control Our Population Growth

Where have our migrations and technological developments taken us? It took about 2.5 *million* years for the human population to reach 1 billion, 123 years to reach the second billion, and only 13 years to jump from 5 billion to 6 billion (see the inset table in Figure 49.19). Rapid population growth may now be an inevitable consequence of our age structure and economic development.

Population Growth and Age Structure. On a worldwide scale, the annual growth rate for the human population averaged nearly 1.2% ($r = 0.012$ new individuals per individual per year) between 2000 and 2005. Population experts expect that rate to decline, but even so, the human population will probably exceed 9 *billion* by 2050.

The population growth rates of individual nations vary widely, however, ranging from much less than 1% to more than 3% in 2001 **(Figure 49.20a).** The industrialized countries of Western Europe have achieved nearly zero population growth, but other countries—notably those in Africa, Latin America, and Asia—will experience huge increases over the next 20 or 25 years **(Figure 49.20b).**

For all long-lived species, differences in age structure are a major determinant of differences in population growth rates **(Figure 49.21).** The uniform age structure of countries with zero growth—with approximately equal numbers of people of reproductive and prereproductive ages—suggests that individuals have just been replacing themselves and that these populations will not experience a growth spurt when today's children mature. By contrast, the narrow-based age structure of countries with negative growth illustrates a continuing decrease in population size. Reproductives have been producing very few offspring, and the small group of prereproductives may not even replace themselves. Countries with rapid growth have a broad-based age structure, with many youngsters born during the previous 15 years. Worldwide, more than one-third of the human population falls within this prereproductive base. This age class will soon reach sexual maturity. Even if each woman produces only two offspring, populations will continue to grow rapidly because so many individuals are reproducing.

The age structure of the United States falls between those for countries with zero growth and countries with rapid growth. The average number of children per family has declined to the two that are necessary to replace their parents in the population. Nevertheless, the U.S. population will continue to grow slowly for the next couple of generations largely because of continued immigration.

Population Growth and Economic Development. The relationship between a country's population growth and its economic development can be depicted by the **demographic transition model (Figure 49.22).** This model describes historical changes in demographic patterns in the industrialized countries of Western Europe; we do not know if it accurately predicts the future for developing nations today.

a. Mean annual population growth rates

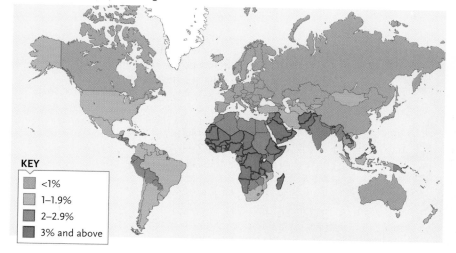

KEY
- <1%
- 1–1.9%
- 2–2.9%
- 3% and above

b. Projected population sizes for 2025

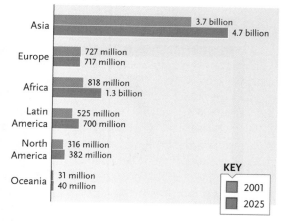

Asia	3.7 billion / 4.7 billion
Europe	727 million / 717 million
Africa	818 million / 1.3 billion
Latin America	525 million / 700 million
North America	316 million / 382 million
Oceania	31 million / 40 million

KEY
- 2001
- 2025

Figure 49.20

Local variation in human population growth rates. **(a)** Average annual population growth rates varied among countries and continents in 2001. **(b)** In some regions, the population is projected to increase greatly by 2025 (red) as compared with the population size in 2001 (orange); the population of Europe will likely decline.

a. Hypothetical age distributions for populations with different growth rates

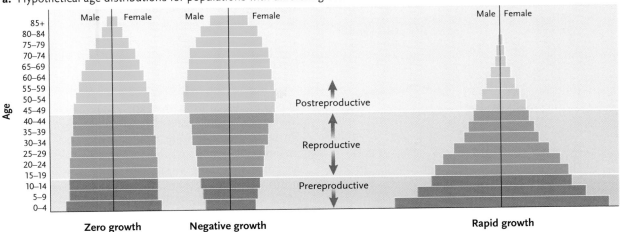

b. Age pyramids for the United States and Mexico in 2000

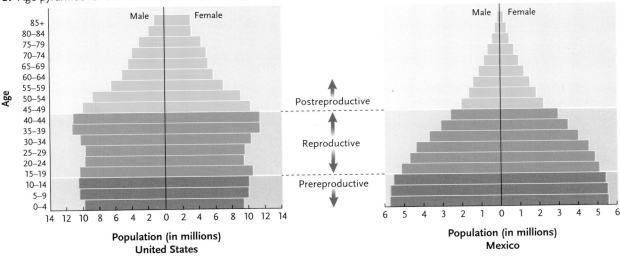

Figure 49.21
Age-structure diagrams. (a) Age-structure diagrams differ for countries with zero, negative, and rapid population growth rates. The width of each bar represents the proportion of the population in each age class. **(b)** Age-structure diagrams for the United States and Mexico in 2000 (measured in millions of people) suggest that these countries would experience different population growth rates.

According to this model, during a country's *preindustrial* stage, birth and death rates are high, and the population grows slowly. Industrialization begins a *transitional* stage, when food production rises, and health care and sanitation improve. The death rate declines, resulting in an increased rate of population growth. Later, as living conditions improve, the birth rate also declines, causing the population growth rate to drop. When the *industrial* stage is in full swing, population growth slows dramatically. People move from the countryside to cities, and urban couples often choose to accumulate material goods instead of having large families. Zero population growth is reached in the *postindustrial* stage. Eventually, the birth rate falls below the death rate, *r* falls below zero, and population size begins to decrease.

Today, the United States, Canada, Australia, Japan, Russia, and most of Western Europe are in the industrial stage. Their growth rates are slowly decreasing. In Bulgaria, Germany, Hungary, and Sweden, birth rates are lower than death rates, and populations are getting smaller, indicating their entry into the postindustrial stage. Kenya and other less industrialized countries are in the transitional stage, but they may not have enough skilled workers or enough capital to make the transition to an industrialized economy. Thus, many poorer nations may be stuck in the transitional stage.

Limiting Population Growth. Most governments realize that increased population size is now the major factor causing resource depletion, excessive pollution, and an overall decline in the quality of life. The principles of population ecology demonstrate that a slowing of population growth—or an actual decline in population size—can be achieved only by decreasing the birth rate or increasing the death rate. And because increasing mortality is neither a rational nor humane means

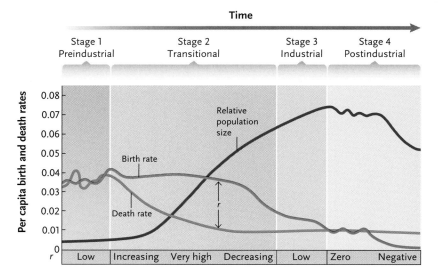

Stage 1
Preindustrial

Stage 2
Transitional

Stage 3
Industrial

Stage 4
Postindustrial

Figure 49.22

The demographic transition. The demographic transition model describes changes in the birth and death rates and relative population size as a country passes through four stages of economic development. The bottom bar describes the net population growth rate, *r*.

of population control, most governments are attempting to lower birth rates with **family planning programs.** These programs educate people about ways to produce an optimal family size on an economically feasible schedule. Programs vary in their details, but all provide information on methods of birth control (see Section 47.6). When thoughtfully developed and carefully administered, family planning programs cause birth rates to decline significantly.

All species face limits to their population growth. We have postponed the action of most factors that limit

UNANSWERED QUESTIONS

Are there universal governing principles in population ecology, similar to the laws of physical sciences? Or is the natural world so complex that each population must be considered individually, leaving us with just a series of case studies?

These types of broad questions motivated the founders of modern ecological studies, such as G. Evelyn Hutchinson and Robert MacArthur. Many ecologists have attempted to codify aspects of population ecology in terms of specific principles, sometimes imposing artificial dichotomies in the process. For example, this chapter considered whether or not natural populations are subject to either density-dependent or density-independent regulation. Ecologists have also attempted to uncover basic patterns in community ecology, which are described in the next chapter. We do know that some general principles are often important in governing the structure of populations or natural communities but, as yet, we cannot apply any of them to a specific system without also including a detailed study of that system.

What is the importance of scale in ecology?

Although an individual population can be a meaningful object of study, ecologists often collect data from multiple populations of the same species to compare results among the "replicates." However, although the separate populations may appear to be replicates, they are often quite different in appearance, age structure, life history, or other characteristics. Ecologists confronted with such variation might seek explanations in the differences between the populations' environments, including both abiotic and biotic factors. However, such local variation may also be attributable to the larger context, such as the landscape or surrounding communities.

One important manifestation of this question applies to how the populations of a species are distributed in space. In many cases, discrete populations are widely separated from one another. Such separation is easy to imagine in terms of fish that live in lakes or organisms that live on islands, but it also applies in a diversity of other organisms. For example, many plants and animals are found only on chemically distinct patches of soil that are distributed like islands across a terrestrial environment. Many

lizards prefer to live in rock outcrops that dot the landscape. Other organisms live on cool, wet mountaintops surrounded by desert. Such isolation is often exaggerated by human modification of the landscape, which progressively fragments and isolates habitable environments from one another. How does such subdivision change the dynamics of the individual populations? How does it change the way they evolve? These are questions that have challenged population and evolutionary biologists for decades. The effects of humans on the environment are making the answers to these questions more than a theoretical concern.

What is the importance of evolution in ecological interactions?

Most research in ecology, ranging from formal models of population growth and regulation to empirical studies, treats populations as if they were unchanging—as if they were not evolving. This implicit perspective does not deny that evolution is happening, but treats it as if it happens on such a long time scale that it need not be considered in contemporary studies. However, many recent studies have shown that populations may evolve quickly, often on a year-to-year basis. If this observation is generally true, then ecological studies that do not include evolutionary change may be compromised. For example, the monitoring and management of commercially exploited fish populations are based entirely on models of population growth, demography, and life histories similar to those considered in this chapter. Commercial fisheries often capture a large proportion of a population every year, focusing on the largest adults. Although research has clearly shown that these practices are likely to select for earlier maturity at a smaller size, these findings have not yet been incorporated into fisheries-management policy. More generally, ecologists have not yet included sufficient emphasis on the interaction between evolution as it occurs on the scale of our day-to-day existence and the modeling and empirical study of ecological processes.

David Reznick is Professor of Biology at the University of California, Riverside. He studies natural selection both from an experimental perspective and by testing evolutionary theory in natural populations. He works primarily with guppies on the island of Trinidad. Learn more about his work at http://www.biology.ucr.edu/people/faculty/Reznick.html.

population growth, but no amount of invention can expand the ultimate limits set by resource depletion and a damaged environment. We now face two options for limiting human population growth: we can make a global effort to limit our population growth, or we can wait until the environment does it for us.

STUDY BREAK

1. How have humans sidestepped the controls that regulate populations of other organisms?
2. How does the age structure of a population influence its future population growth?

Review

Go to **Thomson**NOW™ at www.thomsonedu.com/login to access quizzing, animations, exercises, articles, and personalized homework help.

49.1 The Science of Ecology

- Ecology is the study of the interactions between organisms and their environments. Basic ecology focuses on undisturbed natural systems, whereas applied ecology considers the effects of human disturbance (Figure 49.1).
- Ecologists do research at five levels of organization: organisms, populations, communities, ecosystems, and the biosphere.
- Ecologists test hypotheses about ecological relationships with experimental or observational data. They sometimes frame hypotheses in mathematical models.

49.2 Population Characteristics

- A population's size and density can be measured directly or with sampling techniques (Figures 49.2 and 49.3).
- Organisms within a population may be clumped, uniformly distributed, or randomly distributed within their habitat (Figure 49.4). Clumped dispersion is the most common, but animals may change their dispersion pattern seasonally.
- The relative numbers of individuals of different ages determines a population's age structure. Generation time is the average time between an individual's birth and the birth of its offspring (Figure 49.5). A population's sex ratio is the relative proportion of males and females.

Animation: Distribution patterns

Animation: Mark-recapture method

Animation: Age structure diagrams

49.3 Demography

- Demography is the study of the survivorship, reproduction, immigration, and emigration patterns that influence population characteristics.
- Life tables summarize age-specific mortality, survivorship, and age-specific fecundity of surviving individuals (Table 49.1).
- Survivorship curves depict a population's survival pattern over its life span. Ecologists define three general patterns of survivorship: high survivorship until late in life, a constant mortality level at all ages, and high juvenile mortality (Figure 49.6).

49.4 The Evolution of Life Histories

- An organism's energy budget mandates trade-offs in the allocation of energy to maintenance, growth, and reproduction.
- Natural selection has molded several interacting components of life history variation based upon the allocation of resources to growth, maintenance, and reproduction: the trade-off between fecundity and parental care; whether to reproduce once versus multiple times; and the age of first reproduction.

Animation: Life history patterns

Animation: Guppy characteristics

49.5 Models of Population Growth

- Bacteria reproduce by binary fission, and their populations double in size each generation (Figure 49.7).
- The exponential growth model, $dN/dt = rN$, describes unlimited population growth. A graph of exponential growth is J-shaped (Figure 49.8).
- The logistic model, $dN/dt = r_{max}N(K - N/K)$, includes the effects of resource limitation. The carrying capacity, K, is the maximum population size that an environment can sustain. The per capita population growth rate, r, decreases as N approaches K. A graph of logistic growth is S-shaped (Figures 49.9 and 49.10, Table 49.2).
- Some populations exhibit logistic growth in the laboratory and in nature, but time lags in responses to increased density may cause N to oscillate around K (Figure 49.11).

Animation: Exponential growth

Animation: Effect of death on growth

Practice: Comparison of exponential and logistic population growth

49.6 Population Regulation

- Density-dependent factors regulate population size by reducing individual growth rates, adult size, survivorship, and fecundity (Figures 49.12 and 49.13). Competition within populations or between species, predator–prey interactions, parasites, and infectious diseases can cause density-dependent population regulation (Figure 49.14).
- Abiotic environmental factors, which affect a population regardless of its size, cause density-independent limitation of population size (Figure 49.15).
- Interactions between density-dependent and density-independent factors often influence population size (Figure 49.16).
- The life history patterns of most organisms fall between two extremes: r-selected species and K-selected species (Figure 49.17). They differ in many life history characteristics (Table 49.3).
- Some animal populations exhibit cyclic fluctuations in size (Figure 49.18). No general model has successfully explained all population cycles.

49.7 Human Population Growth

- Human populations have sidestepped density-dependent population regulation by expanding into most terrestrial habitats, increasing carrying capacity, and reducing death rates with improved medical care and sanitation (Figures 49.19 and 49.20).
- Age structure may now control human population growth rates (Figure 49.21). In countries with large numbers of young people, populations will continue to grow rapidly as they reach sex-

ual maturity. The populations of countries with a uniform age structure will not experience much growth in the foreseeable future.

- The demographic transition model describes the influence of economic development on population growth (Figure 49.22).

- Many governments encourage population control through family planning programs.

Animation: Current and projected population sizes by region

Animation: U.S. age structure

Animation: Demographic transition model

Questions

Self-Test Questions

1. Ecologists sometimes use mathematical models to:
   a. avoid conducting laboratory studies or field work.
   b. simulate natural events before conducting detailed field studies.
   c. make basic observations about ecological relationships in nature.
   d. collect survivorship and fecundity data to construct life tables.
   e. determine the geographical ranges of populations.

2. The number of individuals per unit area or volume of habitat is called the population's:
   a. geographical range.   d. size.
   b. dispersion pattern.   e. age structure.
   c. density.

3. One day you caught and marked 90 butterflies in a population. A week later, you returned to the population and caught 80 butterflies, including 16 that had been marked previously. What is the size of the butterfly population?
   a. 170   b. 450   c. 154   d. 186   e. 106

4. A uniform dispersion pattern implies that members of a population:
   a. cooperate in rearing their offspring.
   b. work together to escape from predators.
   c. use resources that are patchily distributed.
   d. may experience intraspecific competition for vital resources.
   e. have no ecological interactions with each other.

5. The model of exponential population growth predicts that the per capita population growth rate (r):
   a. does not change as a population gets larger.
   b. gets larger as a population gets larger.
   c. gets smaller as a population gets larger.
   d. is always at its maximum level (r_{max}).
   e. fluctuates on a regular cycle.

6. A population of 1000 individuals experiences 462 births and 380 deaths in 1 year. What is the value of r for this population?
   a. 0.842/individual/year   d. 0.820/individual/year
   b. 0.462/individual/year   e. 0.082/individual/year
   c. 0.380/individual/year

7. According to the logistic model of population growth, the absolute number of individuals by which a population grows during a given time period:
   a. gets steadily larger as the population size increases.
   b. gets steadily smaller as the population size increases.
   c. remains constant as the population size increases.
   d. is highest when the population is at an intermediate size.
   e. fluctuates on a regular cycle.

8. Which example might reflect density-dependent regulation of population size?
   a. An exterminator uses a pesticide to eliminate carpenter ants from a home.
   b. Mosquitoes disappear from an area after the first frost.
   c. The lawn dies after a month-long drought.
   d. Storms blow over and kill all willow trees along a lake.
   e. The size of a clam population declines as the number of predatory herring gulls explodes.

9. A K-selected species is likely to exhibit:
   a. a Type I survivorship curve and a short generation time.
   b. a Type II survivorship curve and a short generation time.
   c. a Type III survivorship curve and a short generation time.
   d. a Type I survivorship curve and a long generation time.
   e. a Type II survivorship curve and a long generation time.

10. One reason that human populations have sidestepped factors that usually control population growth is:
    a. The carrying capacity for humans has remained constant since humans first evolved.
    b. Agriculture and industrialization have increased the carrying capacity for our species.
    c. The population growth rate (r) for the human population has always been small.
    d. The age structure of human populations has no impact on its population growth.
    e. Plagues have killed off large numbers of humans at certain times in the past.

Questions for Discussion

1. Choose an animal or plant species that lives in your environment and identify the density-dependent and density-independent factors that might influence its population size. How could you demonstrate conclusively that the factors work in either a density-dependent or density-independent fashion?

2. Many city-dwellers have noted that the density of cockroaches in apartment kitchens appears to vary with the habits of the occupants: people who wrap food carefully and clean their kitchen frequently tend to have fewer arthropod roommates than those who leave food on kitchen counters and clean less often. Interpret these observations from the viewpoint of a population ecologist.

3. How could you define the worldwide carrying capacity for humans? What factors would you have to take into account?

Experimental Analysis

Design an experiment using fruit flies or some other small laboratory animal to test the hypothesis that delaying the age of first reproduction will decrease a population's per capita birth rate. Your experimental design should include experimental and control groups as well as details about your experimental methods and the data you would collect.

Evolution Link

Many animals, including humans and other primates, live long beyond their reproductive years. Develop an evolutionary hypothesis to explain this observation, and design a study that might test it.

How Would You Vote?

Some people oppose any deer hunting, whereas others see hunters as a logical substitute for an absence of natural predators. Do you support encouraging hunting in areas where the presence of too many deer is harming the habitat? Go to www.thomsonedu.com/login to investigate both sides of the issue and then vote.

Three interacting populations. Ladybird beetles *(Coccinella septempunctata)* feed on aphids (order Hemiptera), which consume the sap of plants.

© Claude Nuridsany & Marie Perennou/SPL/Photo Researchers, Inc.

STUDY PLAN

50.1   Population Interactions

Coevolution produces reciprocal adaptations in species that interact ecologically

Predation and herbivory define many relationships in ecological communities

Interspecific competition occurs when different species depend on the same limiting resources

In symbiotic associations, the lives of two or more species are closely intertwined

50.2   The Nature of Ecological Communities

Most ecological communities blend into neighboring communities

50.3   Community Characteristics

The growth forms of plants establish a community's overall appearance

Communities differ in species richness and the relative abundance of species they contain

Feeding relationships within a community determine its trophic structure

50.4   Effects of Population Interactions on Community Characteristics

Interspecific competition can reduce species richness within communities

Predators can boost species richness by stabilizing competitive interactions among their prey

Herbivores may counteract or reinforce competition among their food plants

50.5   Effects of Disturbance on Community Characteristics

Frequent disturbances keep some communities in a constant state of flux

Moderate levels of disturbance may foster high species richness

50.6   Ecological Succession: Responses to Disturbance

Succession begins after disturbance alters a landscape or changes the species composition of an existing community

Community characteristics change during succession

Several hypotheses help to explain the processes underlying succession

50.7   Variations in Species Richness among Communities

Many types of organisms exhibit latitudinal gradients in species richness

The theory of island biogeography explains variations in species richness

50 Population Interactions and Community Ecology

WHY IT MATTERS

In some open woodlands in Central America, flocks of chestnut-headed oropendolas *(Zarhynchus wagleri)*, members of the blackbird family, build hanging nests in isolated trees **(Figure 50.1).** Female giant cowbirds *(Scaphidura oryzivora)* often bully their way into a colony, laying an egg or two in each oropendola nest. Cowbirds are *brood parasites* on oropendolas, tricking them into caring for cowbird young. The cowbird chicks grow faster than oropendola chicks, and they consume much of the food that the oropendolas bring to their own offspring. Because cowbird chicks take food away from their oropendola nest mates, we might expect adult oropendolas to eject cowbird eggs and chicks from their nests—but often they don't.

Why do some oropendolas care for offspring that are not their own? In an ingenious study conducted in the 1960s, Neal Smith of the Smithsonian Tropical Research Institute determined that cowbird chicks could actually increase the number of offspring that some oropendolas raise. Oropendola chicks are frequently parasitized by botfly larvae, which feed on their flesh. The aggressive cowbird chicks snap at adult botflies and pick fly larvae off their nest mates. Although cowbird chicks eat food meant for oropendola chicks, they also protect

Figure 50.1
Potential victims of brood parasitism. Chestnut-headed oropendolas *(Zarhynchus wagleri)* rear their young in elaborate hanging nests. Some populations of oropendolas are subject to brood parasitism by giant cowbirds *(Scaphidura oryzivora).*

Cortez C. Austin

them from potentially lethal parasites; twice as many young oropendolas survive in nests with cowbird chicks as in nests without them.

In other areas of Central America, oropendolas build nests near the hives of bees or wasps. These oropendolas chase cowbirds from their colonies, and when a cowbird does manage to sneak an egg into one of their nests, the oropendolas frequently eject it. Why do oropendolas in these colonies reject cowbird eggs when others do not? Smith determined that the swarms of bees and wasps keep botflies away from the oropendola colonies. At these sites, twice as many oropendola chicks survive in nests without cowbirds as in those that include them. Thus the oropendolas derive no benefit from having cowbird chicks in their nests, and natural selection has favored discriminating behavior in oropendolas that nest near bees and wasps.

The story of the oropendolas, cowbirds, botflies, bees, and wasps provides an example of the population interactions that characterize life in an **ecological community,** an assemblage of species living in the same place. And as this story reveals, the presence or absence of certain species may alter the effects of such interactions in almost unimaginably complex ways. We begin this chapter with a description of some of the many ways that populations in a community interact. We then examine how population interactions and other factors, such as the kinds of species present and

the relative numbers of each species, influence a community's characteristics.

50.1 Population Interactions

Population interactions usually provide benefits or cause harm to the organisms engaged in the interaction **(Table 50.1).** And because interactions with other species often affect the survival and reproduction of individuals, many of the relationships that we witness today are the products of long-term evolutionary modification. Before examining several general types of population interactions, we briefly consider how natural selection has shaped the relationships between interacting species.

Coevolution Produces Reciprocal Adaptations in Species That Interact Ecologically

Population interactions change constantly. New adaptations that evolve in one species exert selection pressure on another, which then evolves adaptations that exert selection pressure on the first. The evolution of genetically based, reciprocal adaptations in two or more interacting species is described as **coevolution.**

Some coevolutionary relationships are straightforward. For example, ecologists describe the coevolutionary interactions between some predators and their prey as a race in which each species evolves adaptations that temporarily allow it to outpace the other. When antelope populations suffer predation by cheetahs, natural selection fosters the evolution of faster speed in the antelopes. Cheetahs then experience selection for increased speed so that they can overtake and capture antelopes. Other coevolved interactions provide bene-

Table 50.1	**Population Interactions and Their Effects**	
Interaction		Effects on Interacting Populations
Predation	+/−	Predators gain nutrients and energy; prey are killed or injured.
Herbivory	+/−	Herbivores gain nutrients and energy; plants are killed or injured.
Competition	−/−	Both competing populations lose access to some resources.
Commensalism	+/0	One population benefits; the other population is unaffected.
Mutualism	+/+	Both populations benefit.
Parasitism	+/−	Parasites gain nutrients and energy; hosts are injured or killed.

fits to both partners. For example, the flower structures of different monkey-flower species have evolved characteristics that allow them to be visited by either bees or hummingbirds (see Figure 21.7).

Although one can hypothesize a coevolutionary relationship between any two interacting species, documenting the evolution of reciprocal adaptations is difficult. As our introductory story about oropendolas and their parasites illustrated, coevolutionary interactions often involve more than two species. Indeed, most organisms experience complex interactions with numerous other species in their communities, and the simple portrayal of coevolution as taking place between two species rarely does justice to the complexity of these relationships.

Predation and Herbivory Define Many Relationships in Ecological Communities

Because animals acquire nutrients and energy by consuming other organisms, **predation** (the interaction between predatory animals and the animal prey they consume) and **herbivory** (the interaction between herbivorous animals and the plants they eat) are often the most conspicuous relationships in ecological communities.

Adaptations for Feeding. Both predators and herbivores have evolved remarkable characteristics that allow them to feed effectively. Carnivores use sensory systems to locate animal prey and specialized behaviors and anatomical structures to capture and consume it. For example, a rattlesnake (genus *Crotalus*) uses heat sensors on its head (see Figure 39.22) and chemical sensors in the roof of its mouth to find rats or other endothermic prey. Its hollow fangs inject toxins that kill the prey and begin to digest its tissues even before the snake consumes it. And elastic ligaments connecting the bones of its jaws and skull allow a snake to swallow prey that is larger than its head. Herbivores have comparable adaptations for locating and processing their food plants. Insects use chemical sensors on their legs and heads to identify edible plants and sharp mandibles or sucking mouthparts to consume plant tissues or sap. Herbivorous mammals have specialized teeth to harvest and grind tough vegetation (see Section 45.5).

All animals must select their diets from a variety of potential food items. Some species, described as *specialists,* feed on one or just a few types of food. Among birds, for example, the Everglades kite (*Rostrhamus sociabilis*) consumes just one prey species, the apple snail (*Pomacea paludosa*). Other species, described as *generalists,* have broader tastes. Crows (genus *Corvus*) consume food ranging from grain to insects to carrion.

How does an animal select what type of food to eat? Some mathematical models, collectively described as **optimal foraging theory,** predict that an animal's diet is a compromise between the costs and benefits associated with different types of food. Assuming that animals try to maximize their energy intake in a given feeding time, their diets should be determined by the time and energy it takes to pursue, capture, and consume a particular kind of food compared with the energy that food provides. For example, a cougar (*Puma concolor*) will invest more time and energy hunting a mountain goat (*Oreamnos americanus*) than a jackrabbit (*Lepus townsendii*), but the payoff for the cougar is a bigger meal.

Food abundance also affects food choice. When prey are scarce, animals often take what they can get, settling for food that has a low benefit-to-cost ratio. But when food is abundant, they may specialize, selecting types that provide the largest energetic return. Bluegill sunfish (*Lepomis macrochirus*), for example, feed on *Daphnia* and other small crustaceans. When crustacean density is high, the fish hunt mostly large *Daphnia*, which provide more energy for their effort; but when prey density is low, bluegills feed on *Daphnia* of all sizes **(Figure 50.2)**.

Defenses against Herbivory and Predation. Because herbivory and predation have a negative impact on the organisms being consumed, plants and animals have evolved mechanisms to avoid being eaten. Some plants

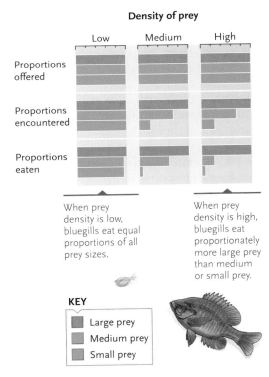

Figure 50.2

An experiment demonstrating that prey density affects predator food choice. Researchers tested the food size preferences of captive bluegill sunfish (*Lepomis macrochirus*) by offering them equal numbers of small, medium, and large-sized prey (*Daphnia magna*) at three different prey densities. Because large prey are the easiest to find, bluegills encountered them more frequently than small or medium-sized prey, especially at the highest prey density. The bluegills' selection of prey varied with prey density; they strongly preferred large prey when prey of all sizes were abundant.

Figure 50.3

Hiding in plain sight. Some animals, such as **(a)** giant swallowtail butterfly *(Papilio cresphontes)* larvae that resemble bird droppings and **(b)** some katydids *(Mimetica* species) that resemble insect-damaged leaves, do not attract the attention of predators.

a. Bird dropping mimic

Edward S. Ross

b. Damaged leaf mimic

Frans Lanting/Minden Pictures

use spines, thorns, and irritating hairs to protect themselves from herbivores. Many plant tissues also contain poisonous chemicals that deter herbivores from feeding. For example, plants in the milkweed family (Asclepiadeceae) exude a milky, irritating sap that contains cardiac glycosides, even small amounts of which are toxic to vertebrate heart muscle. Other compounds mimic the structure of insect hormones, disrupting the development of insects that consume them. Most of these poisonous compounds are volatile, giving plants their typical aromas; some herbivores have co-evolved the ability to recognize these odors and avoid the toxic plants. Recent research indicates that some plants increase their production of toxic compounds in response to herbivore feeding. For example, potato and tomato plants that have been damaged by herbivores produce higher levels of protease-inhibiting chemicals; these compounds prevent herbivores from digesting proteins they have just consumed, reducing the food value of these plant tissues.

Many animals have evolved an appearance that provides a passive defense against predation **(Figure 50.3).** Caterpillars that look like bird droppings, for example, may not attract much attention from a hungry predator. And as you learned in Chapter 1 (see Figure 1.9), **cryptic coloration** helps some prey (as well as some predators) to blend in with their surroundings.

Once discovered by a predator, many animals first try to run away. When cornered, they may try to startle or intimidate the predator with a display that increases their apparent size or ferocity **(Figure 50.4).** Such a display might confuse the predator just long enough to allow the potential victim to escape. Other species seek shelter in protected sites. For example, flexible-shelled African pancake tortoises *(Malacochersus tornieri)* retreat into rocky crevices and puff themselves up with air, becoming so tightly wedged between rocks that predators cannot extract them.

Other animals defend themselves actively. North American porcupines (genus *Erethizon*) release hairs modified into sharp, barbed quills that stick in a predator's mouth, causing severe pain and swelling. Other species fight back by biting, charging, or kicking an attacking predator. Chemical defenses also provide effective protection. Skunks release a noxious spray when threatened, and some frogs and toads produce neurotoxic skin secretions that paralyze and kill mammals. Some insects even protect themselves with poisons acquired from plants. The caterpillars of monarch butterflies *(Danaus plexippus)* are immune to the cardiac glycosides in the milkweed leaves they eat. They store these chemicals at high concentration, even through metamorphosis, making adult monarchs poisonous to vertebrate predators.

Poisonous or repellant species often advertise their unpalatability with bright, contrasting patterns, called **aposematic coloration (Figure 50.5).** Although a predator might attack a black-and-white skunk, a

Photo Researchers, Inc.

Figure 50.4

Startle defenses. A short-eared owl *(Asio flammeus)* increases its apparent size when threatened by a predator.

Courtesy of Ken Nemuras

Figure 50.5

Aposematic coloration. Poisonous animals, like the harlequin toad *(Atelopus varius)* from Central America often have bright warning coloration.

yellow-banded wasp, or an orange monarch butterfly once, it quickly learns to associate the gaudy color pattern with pain, illness, or severe indigestion—and rarely attacks these easily recognized animals again.

Mimicry, in which one species evolves an appearance resembling that of another **(Figure 50.6)**, is also a form of defense. In **Batesian mimicry**, named for English naturalist Henry W. Bates, a palatable or harmless species, the **mimic**, resembles an unpalatable or poisonous one, the **model**. Any predator that eats the poisonous model will subsequently avoid other organisms that resemble it. In **Müllerian mimicry**, named for German zoologist Fritz Müller, two or more unpalatable species share a similar appearance, which reinforces the lesson learned by a predator that attacks any species in the mimicry complex.

Despite the effectiveness of many antipredator defenses, coevolution has often molded the responses of predators to overcome them. For example, when threatened by a predator, the beetle *Eleodes longicollis* raises its rear end and sprays a noxious chemical from a gland at the tip of its abdomen. Although this behavior deters many would-be predators, grasshopper mice (genus *Onychiomys*) of the American southwest circumvent this defense: they grab the beetles and shove their abdomens into the ground, rendering the beetle's spray ineffective **(Figure 50.7)**.

Interspecific Competition Occurs When Different Species Depend on the Same Limiting Resources

Populations of different species often use the same limiting resources, causing **interspecific competition** (competition between species). The competing populations may experience increased mortality and decreased reproduction, responses that are similar to the effects of intraspecific competition (see Section 49.5). Interspecific competition reduces the size and population growth rate of one or more of the competing populations.

Community ecologists identify two main forms of interspecific competition. In **interference competition**, individuals of one species harm individuals of another species directly. Animals may fight for access to resources, as when lions chase smaller scavengers like hyenas and jackals from their kills. Similarly, many plant species, including creosote bushes (see Figure 49.4), release toxic chemicals, which prevent other plants from growing nearby. In **exploitative competition**, two or more populations use ("exploit") the same limiting resource; the presence of one species reduces resource availability for the others, even in the absence of snout-to-snout or root-to-root confrontations. For example, in the deserts of the American Southwest, many bird and ant species feed largely on seeds. Thus, each seed-eating species may deplete the food supply available to others.

a. Batesian mimicry

Drone fly *(Eristalis tenax)*, the mimic

Honeybee *(Apis mellifera)*, the model

b. Müllerian mimicry

Heliconius erato

Heliconius melpone

Figure 50.6

Mimicry. **(a)** Batesian mimics are harmless animals that mimic a dangerous one. The harmless drone fly *(Eristalis tenax)* is a Batesian mimic of the stinging honeybee *(Apis mellifera)*. **(b)** Müllerian mimics are poisonous species that share a similar appearance. Two distantly related species of butterfly, *Heliconius erato* and *Heliconius melpone*, have nearly identical patterns on their wings.

Competitive Exclusion and the Niche Concept. In the 1920s, the Russian mathematician Alfred J. Lotka and the Italian biologist Vito Volterra independently proposed a model of interspecific competition, modifying the logistic equation (see Section 49.5) to describe the effects of competition between two species. In their model, an increase in the size of one population reduces the population growth rate of the other.

a. *Eleodes* beetle   **b.** Grasshopper mouse

Figure 50.7

Coevolution of predators and prey. **(a)** When disturbed by a predator, the beetle *Eleodes longicollis* sprays a noxious chemical from its posterior end. **(b)** Grasshopper mice (genus *Onychiomys*) overcome this defense by shoving a beetle's rear end into the soil and dining on it headfirst.

Figure 50.8 Experimental Research

Gause's Experiments on Interspecific Competition in Paramecium

QUESTION: Can two species of *Paramecium* coexist in a simple laboratory environment?

EXPERIMENT: Gause grew populations of two *Paramecium* species, *Paramecium aurelia* and *Paramecium caudatum*, alone (single species cultures) or together (mixed culture) in small bottles in his laboratory. To determine whether the growth of these populations followed the predictions of the logistic equation, Gause had to maintain a reasonably constant carrying capacity in each culture. Thus, he fed the cultures a broth of bacteria, and he eliminated their waste products (by centrifuging the cultures and removing some of the culture medium) on a regular schedule. He then monitored their population sizes through time.

RESULTS: When grown separately, *P. caudatum* **(a)** and *P. aurelia* **(b)** each exhibited logistic population growth. But when the two species were grown together in a mixed culture **(c)**, *P. aurelia* persisted and *P. caudatum* was nearly eliminated from the culture.

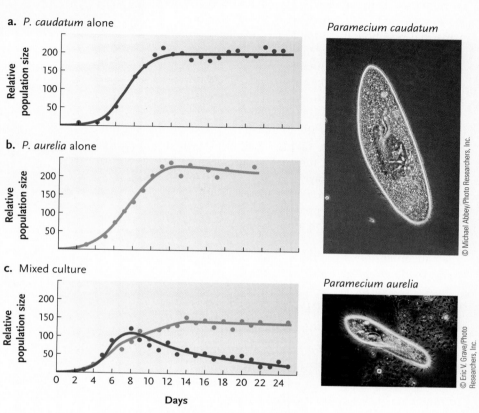

CONCLUSION: Because one species was almost always eliminated from mixed species cultures, Gause formulated the competitive exclusion principle: populations of two or more species cannot coexist indefinitely if they rely on the same limiting resources and exploit them in the same way.

A Russian biologist, G. F. Gause, tested the model experimentally in the 1930s. He grew cultures of two *Paramecium* species (ciliate protozoans) under constant laboratory conditions, regularly renewing food and removing wastes. Both species feed on bacteria suspended in the culture medium. When grown alone, each species exhibited logistic growth; but when grown together in the same dish, *Paramecium aurelia* persisted at high density, but *Paramecium caudatum* was nearly eliminated **(Figure 50.8)**. These results inspired Gause to define the **competitive exclusion principle:** populations of two or more species cannot coexist indefinitely if they rely on the same limiting resources and exploit them in the same way. One species inevitably harvests resources more efficiently and produces more offspring than the other.

Ecologists developed the concept of the **ecological niche** as a tool for visualizing resource use and the potential for interspecific competition in nature. We define a population's niche by the resources it uses and the environmental conditions it requires over its lifetime. In this context, the niche includes food, shelter, and nutrients as well as abiotic conditions, such as light intensity and temperature, which cannot be depleted. In theory, one could identify an almost infinite variety of conditions and resources that contribute to a population's niche. In practice, ecologists usually analyze a few critical resources for which populations might compete. Sunlight, soil moisture, and inorganic nutrients are important resources for plants. Food type, food size, and nesting sites are important for animals.

Ecologists distinguish the **fundamental niche** of a population, the range of conditions and resources that it can possibly tolerate and use, from its **realized niche**, the range of conditions and resources that it actually uses in nature. Realized niches are smaller than fundamental niches, partly because all tolerable conditions are not always present in a habitat, and partly because some resources are used by other species. We can visualize competition between two populations by plotting their fundamental and realized niches with respect to one or more resources **(Figure 50.9)**. If the fundamental niches of two populations overlap, they *might* compete in nature.

Evaluating Competition in Nature. The observation that several populations use the same resource does not demonstrate that competition occurs. For example, all terrestrial animals consume oxygen, but they don't compete for oxygen because it is usually plentiful. Nevertheless, two general observations provide *indirect* evidence that interspecific competition may have important effects. The first is the extremely common observation of **resource partitioning**, the use of different resources or the use of resources in different ways, by species living in the same place. For example, weedy plants might compete for water and dissolved nutrients in abandoned fields. But they avoid competition by partitioning these resources, collecting them from different depths in the soil **(Figure 50.10)**.

A second phenomenon that suggests the importance of competition is observed in comparisons of species that are sometimes sympatric (that is, living in the same place) and sometimes allopatric (that is, living in different places). In several studies of animals, researchers have documented **character displacement**: allopatric populations are morphologically similar and use similar resources, but sympatric populations are morphologically different and use different resources. The differences between the sympatric populations allow them to coexist without competing. Differences in bill size among sympatric finch species on the Galápagos Islands (see Sections 19.2 and 20.3) may be the product of character displacement **(Figure 50.11)**.

Data on resource partitioning and character displacement merely suggest the possible importance of interspecific competition in nature. To demonstrate *conclusively* that interspecific competition limits natural populations, one must show that the presence of one population reduces the population size or distribution of its presumed competitor. In a classic field experiment, Joseph Connell of the University of California, Santa Barbara, determined that competition between two barnacle species caused the realized niche of one species to be smaller than its fundamental niche

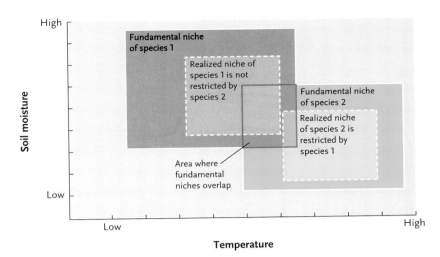

Figure 50.9
Fundamental versus realized niches. In this hypothetical example, both species 1 and species 2 can survive intermediate temperature and soil moisture conditions, as indicated by the shading where their fundamental niches overlap. Because species 1 actually occupies most of this overlap zone, its realized niche is not much affected by the presence of species 2. By contrast, the realized niche of species 2 is restricted by the presence of species 1, and species 2 occupies warmer and dryer parts of the habitat.

(Figure 50.12). Connell first observed the distributions of barnacles in undisturbed habitats. *Chthamalus stellatus* is generally found in shallow water on rocky coasts, where it is periodically exposed to air. *Balanus balanoides* typically lives in deeper water, where it is usually submerged.

Connell determined the fundamental niche of each species by removing either *Chthamalus* or *Balanus* from rocks and monitoring the distribution of each

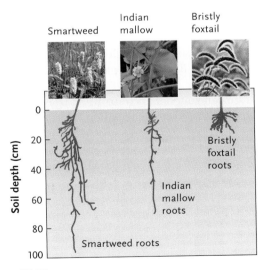

Figure 50.10
Resource partitioning. The root systems of three plant species that grow in abandoned fields partition water and nutrient resources in soil. Bristly foxtail grass *(Setaria faberii)* has a shallow root system; Indian mallow *(Abutilon theophraste)* has a moderately deep taproot; and smartweed *(Polygonum pensylvanicum)* has a deep taproot that branches at many depths.
(Photos: left, © Tony Wharton, Frank Lane Picture Agency/Corbis; middle, © Hal Horwitz/Corbis; right, © Joe McDonald/Corbis.)

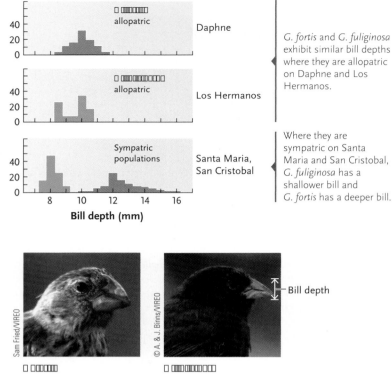

G. fortis and G. fuliginosa exhibit similar bill depths where they are allopatric on Daphne and Los Hermanos.

Where they are sympatric on Santa Maria and San Cristobal, G. fuliginosa has a shallower bill and G. fortis has a deeper bill.

Bill depth

Figure 50.11

Character displacement. *Geospiza fortis* and *Geospiza fuliginosa* exhibit character displacement in the depth of their bills, a trait that is correlated with the sizes of seeds they eat.

species in the absence of the other. When Connell removed *Balanus* from rocks in deep water, larval *Chthamalus* colonized the area and produced a flourishing population of adults. Connell observed that *Balanus* physically displaced *Chthamalus* from these rocks. Thus, interference competition from *Balanus* prevents *Chthamalus* from occupying areas where it would otherwise live. By contrast, the removal of *Chthamalus* from rocks in shallow water did not result in colonization by *Balanus. Balanus* is apparently unable to live in habitats that are frequently exposed to air. Connell therefore concluded that competition from *Chthamalus* does not affect the distribution of *Balanus*. Thus, the competitive interaction between these two species is asymmetrical: *Balanus* has a substantial effect on *Chthamalus*, but *Chthamalus* has virtually no effect on *Balanus*.

In Symbiotic Associations, the Lives of Two or More Species Are Closely Intertwined

Some species have a physically close ecological association called **symbiosis** (*sym* = together; *bio* = life; *sis* = process). Biologists define three types of symbiotic interactions—*commensalism, mutualism,* and *parasitism*—that differ in their effects.

Commensalism, in which one species benefits and the other is unaffected, is rare in nature, because few species are unaffected by their interactions with another. One possible example is the relationship between cattle egrets (*Bubulcus ibis*), birds in the heron family, and the large grazing mammals with which they associate **(Figure 50.13)**. Cattle egrets feed on insects and other small animals that their commensal partners flush from grass. Feeding rates of egrets are higher when they associate with large grazers than when they do not. The birds clearly benefit from this interaction, but the presence of birds has no apparent positive or negative impact on the mammals.

Mutualism, in which both partners benefit, is extremely common. The coevolved relationships between flowering plants and animal pollinators are largely mutualistic. Animals that feed on a plant's nectar or pollen carry its gametes from one flower to another **(Figure 50.14)**. Similarly, animals that eat the fruits of flowering plants disperse the seeds, "planting" them in a pile of nutrient-rich feces. These mutualistic relationships between plants and animals do not require active cooperation. Each species simply exploits the other for its own benefit.

Some associations between bacteria and plants are also mutualistic. One of the most important of these associations is between *Rhizobium* and leguminous plants, such as peas, beans, and clover (see Section 33.3). *Insights from the Molecular Revolution* describes how the genes responsible for the association were identified and their possible evolutionary origin.

Mutualistic relationships between animal species are also common. For example, some small marine fishes feed on parasites that attach to the mouths and gills of large predatory fishes **(Figure 50.15)**. Parasitized fishes hover motionless while the "cleaners" scour their tissues. The relationship is mutualistic because the cleaner fishes get a meal, and the larger fishes are relieved of parasites.

The relationship between the bull's horn acacia tree (*Acacia cornigera*) of Central America and a small ant species (*Pseudomyrmex ferruginea*) is one of the most highly coevolved mutualisms known **(Figure 50.16)**. Each acacia is inhabited by an ant colony that lives in the tree's swollen thorns. The ants swarm out of the thorns to sting—and sometimes kill—herbivores that touch the tree. The ants also clip any vegetation that grows nearby. Thus, acacia trees that are colonized by ants grow in a space free of herbivores and competitors, and occupied trees grow faster and produce more seeds than unoccupied trees. In return, the plants produce sugar-rich nectar consumed by adult ants and protein-rich structures that the ants feed to their larvae. Ecologists describe the coevolved mutualism between these species as *obligatory,* at least for the ants; they cannot subsist on any other food sources.

Figure 50.12 Experimental Research

Demonstration of Competition between Two Species of Barnacles

QUESTION: Do two barnacle species limit one another's realized niche in habitats where they coexist?

EXPERIMENT: Connell observed a difference in the distributions of two barnacle species on a rocky coast: *Chthamalus stellatus* occupies shallow water, and *Balanus balanoides* lives in deeper water. He then determined the fundamental niche of each species by removing either *Chthamalus* or *Balanus* from rocks and monitoring the distribution of each species in the absence of the other.

RESULTS: When Connell removed *Balanus* from rocks in deep water, larval *Chthamalus* colonized the area and produced a flourishing population of adults. By contrast, the removal of *Chthamalus* from rocks in shallow water did not result in colonization by *Balanus*.

Realized niches before experimental treatments

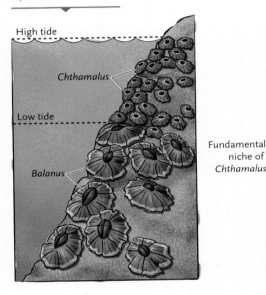

Treatment 1: Remove *Balanus*
In the absence of *Balanus*, *Chthamalus* occupies both shallow water and deep water.

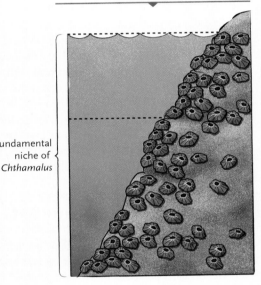

Treatment 2: Remove *Chthamalus*
In the absence of *Chthamalus*, *Balanus* still occupies only deep water.

CONCLUSION: In habitats where *Balanus* and *Chthamalus* coexist, the realized niche of *Chthamalus* is smaller than its fundamental niche because of competition from *Balanus*. The realized niche of *Balanus* is similar to its fundamental niche because it is not affected by the competitive interaction.

Parasitism is a type of interaction in which one species, the **parasite**, uses another, the **host**, in a way that is harmful to the host. Parasite–host relationships are like predator–prey relationships: one population of organisms feeds on another. But parasites rarely kill their hosts quickly because a dead host is useless as a continuing source of nourishment.

Tapeworms and other parasites that live *within* a host are **endoparasites**. Many endoparasites acquire their hosts passively, when a host accidentally ingests the parasite's eggs or larvae (see *Focus on Research*, Chapter 29). Endoparasites generally complete their

Figure 50.13
Commensalism. Cattle egrets *(Bubulcus ibis)* feed on insects and other small animals flushed by the movements of large grazing mammals, like this African buffalo *(Sycerus coffer)*.

Fritz Polking/Frank Lane Picture Agency

Figure 50.14

Mutualism between plants and animals. Several species of yucca plants (*Yucca* species) are each pollinated exclusively by one species of yucca moth (*Tegeticula* species). The adult stage of each moth appears at the time of year when its yucca plant flowers. These species are so mutually interdependent that the larvae of each moth species can feed on only one type of yucca, and the flowers of each yucca can be fertilized by only one species of moth. Most plant-pollinator mutualisms are much less specific.

a. Flowering yucca plant

b. Female yucca moth

c. Yucca moth larva

A female yucca moth uses highly modified mouthparts to gather the sticky pollen and roll it into a ball. She carries the pollen to another flower, and after piercing its ovary wall, she lays her eggs. She then places the pollen ball into the opening of the stigma.

When moth larvae hatch from the eggs, they eat some of the yucca seeds and gnaw their way out of the ovary to complete their life cycle. Enough seeds remain undamaged to produce a new generation of yuccas.

life cycle in one or two host individuals. By contrast, leeches, aphids, mosquitoes, and other parasites that feed on the *exterior* of a host are **ectoparasites.** Most animal ectoparasites have elaborate sensory and behavioral mechanisms that allow them to locate specific hosts, and they feed on numerous host individuals during their lifetimes. Some plants, such as mistletoes (genus *Phoradendron*), live as ectoparasites on the trunks and branches of trees; their roots penetrate the host's xylem and extract water and nutrients.

Not all parasites feed directly on a host's tissues. The giant cowbirds described earlier are brood parasites, as are other species of cowbirds and cuckoos. Although oropendolas sometimes benefit from the presence of cowbirds, most brood parasites have negative effects on their hosts. For example, brood parasitism by the brown-headed cowbird (*Molothrus ater*) has

played a large role in the near-extinction of Kirtland's warbler (*Dendroica kirtlandii*).

The feeding habits of some insects, called **parasitoids,** fall somewhere between true parasitism and predation. A female parasitoid lays eggs in the larva or pupa of another insect species, and her young consume the tissues of the living host. Because the hosts chosen by most parasitoids are highly specific, agricultural ecologists often release parasitoids to control populations of insect pests.

STUDY BREAK

1. Why are some carnivores willing to spend more time and energy capturing large prey than small prey?
2. What are the differences between cryptic coloration, aposematic coloration, and mimicry? Can a mimic ever have aposematic coloration?
3. How can field experiments demonstrate conclusively that two species compete for limiting resources?

— Cleaner wrasse

Figure 50.15

Mutualism between animal species. A large potato cod (*Epinephelus tukula*) from the Great Barrier Reef in Australia remains nearly motionless in the water while a striped cleaner wrasse (*Labroides dimidiatus*) carefully removes and eats ectoparasites attached to its lip. The potato cod is a predator, and the striped cleaner wrasse is a potential prey—but their mutualistic interaction supersedes a possible predator–prey interaction.

50.2 The Nature of Ecological Communities

Ecologists have often debated the nature of ecological communities, asking if they have emergent properties that transcend the interactions among the populations they contain.

Most Ecological Communities Blend into Neighboring Communities

How do complex population interactions affect the organization and functioning of ecological communities? In the 1920s, ecologists in the United States de-

Finding a Molecular Passport to Mutualism

The mutualistic association between *Rhizobium* bacteria and leguminous plants is established through a complex signaling process. When roots of one of these plants are invaded by *Rhizobium*, the plants respond by developing *root nodules* that house the bacteria and supply them with carbohydrates. In return, the bacteria fix atmospheric nitrogen into ammonia, which the plants use as a nitrogen source. This mutualistic association fixes about 120 million metric tons of nitrogen annually into ammonia, and greatly reduces farmers' need to use nitrogen-containing chemical fertilizers.

Proteins encoded in several sets of *Rhizobium* genes (called *nod*, *nif*, and *fix*) promote the mutualistic association with legumes. Enzymes encoded in the *nod* genes catalyze the synthesis of polysaccharides stimulating growth of a tubelike *infection thread*, which admits the bacteria to the root tissue. Once inside, the same polysaccharides promote development of the root nodule. The *nif* and *fix* genes encode enzymes involved in nitrogen fixation.

Most of these genes are carried on a single plasmid in *Rhizobium*. (Plasmids are small circles of DNA located outside the main bacterial chromosome.) The DNA sequence of the plasmid carrying the *nod*, *nif*, and *fix* genes was revealed by Cristoph Freiberg and his colleagues at the Institute for Molecular Biotechnology in Jena, Germany, and the University of Geneva in Switzerland.

The investigators studied the large plasmid of the *Rhizobium* species designated NGR234, which can invade an unusually large selection of legumes—more than 110 genera—and even one nonleguminous plant. They isolated the bacterium from root nodules and extracted the plasmid DNA. Sequencing showed that the plasmid contains an astounding 416 coding sequences. A computer search identified 277 sequences as relatives of known genes with established functions, including relatives of *nod*, *nif*, and *fix*. The remaining 139 genes have no known counterparts in any other living organism.

Among the known genes, close similarities were found to genes of a plasmid in *Rhizobium radiobacter*, another bacterium able to invade plant hosts. *Rhizobium radiobacter* invades various deciduous plants and promotes growth of large masses of tissue called *crown gall tumors* (see Figure 18.14). The similarities between the *Rhizobium* NGR234 and *Rhizobium radiobacter* plasmids suggest that the mechanisms by which they invade their host plants may have originated in a common evolutionary ancestor.

This research sequencing the *Rhizobium* NGR234 plasmid may help to reveal the molecular and biochemical basis of the mutualistic relationship between nodule-inducing *Rhizobium* and legumes. As a practical matter, the plasmid and its genes may provide a "genetic passport" that could be adapted to allow nodule-inducing *Rhizobium* to invade nonleguminous plants. If successful, this adaptation might allow the equivalent of nitrogen-fixing root nodules to be developed in many nonleguminous crops, eliminating their need for nitrogenous fertilizers and reducing both the cost of growing food crops and pollution by fertilizer runoff.

---

veloped two extreme hypotheses about the nature of ecological communities. Frederic Clements of the University of Minnesota championed an *interactive* view of communities. He described communities as "superorganisms," assemblages of species bound together by complex population interactions. According to this view, each species in a community requires interactions with a set of ecologically different species, just as every cell in an organism requires services that other types of cells provide. Clements believed that once a mature community was established, its **species composition**—the particular combination of species that occupy the site—was at *equilibrium*. If a fire or some other environmental factor disturbed the community, it would return to its predisturbance state.

Henry A. Gleason of the University of Michigan proposed an alternative, *individualistic* view of ecological communities. He believed that population interactions do not always determine species composition. Instead, a community is just an assemblage of species that are individually adapted to similar environmental conditions. According to Gleason's hypothesis, communities do not achieve equilibrium; rather, they constantly change in response to disturbance and environmental variation.

In the 1960s, Robert Whittaker of Cornell University suggested that ecologists could determine which

a. Ants patrolling an acacia

b. Cleared area around an acacia

Figure 50.16

A highly coevolved mutualism. **(a)** Bull's horn acacia trees (*Acacia cornigera*) provide colonies of small ants (*Pseudomyrmex ferruginea*) with homes in hollow enlarged thorns as well as other resources. Although individual ants are small, they are numerous and aggressive. **(b)** Because the ants attack herbivores and remove vegetation near their tree, acacias occupied by ants grow in a space that is free of herbivores and competitors.

a. Interactive hypothesis

The interactive hypothesis predicts that species within communities exhibit similar distributions along environmental gradients (indicated by the close alignment of several curves over each section of the gradient) and that boundaries between communities (indicated by arrows) are sharp.

b. Individualistic hypothesis

The individualistic hypothesis predicts that species distributions along the gradient are independent (indicated by the lack of alignment of the curves) and that sharp boundaries do not separate communities.

c. Siskiyou Mountains

d. Santa Catalina Mountains

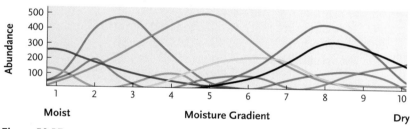

Most gradient analyses support the individualistic hypothesis, as illustrated by distributions of tree species along moisture gradients in Oregon's Siskiyou Mountains and Arizona's Santa Catalina Mountains.

Figure 50.17
Two views of ecological communities.

Jasper Ridge Biological Preserve

Figure 50.18

Sharp community boundaries. Soils derived from serpentine rock have high magnesium and heavy metal content, which many plants cannot tolerate. Although native California wildflowers (bright yellow in this photograph) thrive on serpentine soil at the Jasper Ridge Preserve of Stanford University, introduced European grasses (green in this photograph) competitively exclude them from adjacent soils derived from sandstone.

hypothesis was correct by analyzing communities along environmental gradients, such as temperature or moisture **(Figure 50.17)**. According to Clements' interactive hypothesis, species that typically occupy the same communities should always occur together. Thus, their distributions along the gradient would be clustered in discrete groups with sharp boundaries between groups (see Figure 50.17a). According to Gleason's individualistic hypothesis, each species is distributed over the section of an environmental gradient to which it is adapted. Different species would have unique distributions, and species composition would change continuously along the gradient. In other words, communities would not be separated by sharp boundaries (see Figure 50.17b).

Most gradient analyses support Gleason's individualistic view of ecological communities. Environmental conditions vary continuously in space, and most plant distributions match these patterns (see Figure 50.17c, d). Species occur together in assemblages because they are adapted to similar conditions, and the species compositions of the assemblages change gradually across environmental gradients.

Nevertheless, the individualistic view does not fully explain all patterns observed in nature. Ecologists recognize certain assemblages of species as distinctive communities and name them accordingly—redwood forests and coral reefs are good examples. But the borders between adjacent communities are often wide transition zones, called **ecotones**. Ecotones are generally rich with species because they include plants and animals from both neighboring communities as well as some species that thrive only under transitional conditions. In some places, however, a discontinuity in a critical resource or some important abiotic factor produces a sharp community boundary. For example, chemical differences between soils derived from serpentine rock and sandstone establish sharp boundaries between communities of native California wildflowers and introduced European grasses **(Figure 50.18)**.

STUDY BREAK

1. Which view of communities suggests that they are just chance assemblages of species that happen to be adapted to similar abiotic environmental conditions?
2. Why would you often find more species living in an ecotone than you would in the communities on either side of it?

50.3 Community Characteristics

Although the species composition of an ecological community may vary somewhat over geographical gradients, every community has certain characteristics that define its overall appearance and structure.

The Growth Forms of Plants Establish a Community's Overall Appearance

The growth forms—sizes and shapes—of plants vary markedly in different environments. Warm, moist environments support complex vegetation with multiple vertical layers. For example, tropical forests include a canopy, formed by the tallest trees; an understory of shorter trees and shrubs; an herb layer under openings in the canopy; vinelike lianas; and epiphytes, which grow on the trunks and branches of trees **(Figure 50.19)**. By contrast, physically harsh environments are occupied by low vegetation with simple structure. For example, trees on mountaintops buffeted by cold winds are short, and the plants below them cling to rocks and soil. Other environments support growth forms between these extremes (see Chapter 52).

Communities Differ in Species Richness and the Relative Abundance of Species They Contain

Communities differ greatly in their **species richness**, the number of species that live within them. For example, the harsh environment on a low desert island may support just a few species of microorganisms, fungi, algae, plants, and arthropods. By contrast, tropical forests, which grow under milder physical conditions, include many thousands of species. Ecologists have studied global patterns of species richness (described below in Section 50.7) for decades. Today, as human disturbance of natural communities has reached a crisis point, conservation biologists focus on such studies to determine which regions of Earth are most in need of preservation (see Chapter 53).

Within every community, populations differ in their commonness or the **relative abundance** of individuals. Some communities have just one or two abundant species and a number of rare species; in other communities, species are represented by more equal numbers of individuals. For example, in a temperate deciduous forest in West Virginia, tulip poplar *(Liriodendron tulipifera)* and sassafras *(Sassafras albidum)*

Figure 50.19

Layered forests. Tropical forests include a canopy of tall trees and an understory of short trees and shrubs. Huge vines (lianas) climb through the trees, eventually reaching sunlight in the canopy; and epiphytic plants grow on trunks and branches, increasing the structural complexity of the habitat.

Canopy

Epiphyte

Understory

Liana

Herb layer

Buttress

Forest A   Forest B   Forest C

Figure 50.20

Species diversity. In this hypothetical example, each of three forests contains 50 trees. Forest A and forest B each include 10 tree species, but forest C includes only two tree species. Because forest A is dominated by one tree species, but forest B is not, ecologists would say that forest B is more diverse. Forest C, with only two tree species, is less diverse than the others.

might together account for nearly 85% of the trees. By contrast, a tropical forest in Costa Rica may include more than 200 tree species, each making up only a small percentage of the total.

Species richness and relative abundance together contribute to a community characteristic that ecologists call **species diversity**. To demonstrate species diversity, we will compare two hypothetical forest communities, each with 50 trees distributed among 10 species **(Figure 50.20)**. In Forest A, the dominant species is represented by 39 individuals, two species by two individuals each, and seven species by one individual each. In Forest B, each of the 10 species is represented by five individuals. Although both communities have the same species richness (10 species), Forest A is less diverse than Forest B, because most of its trees are of the same species. A forest with only two tree species (Forest C in Figure 50.20) would be less diverse than either of the others.

Feeding Relationships within a Community Determine Its Trophic Structure

All ecological communities, regardless of their species richness, also have a trophic structure (*troph* = nourishment) that comprises all of the plant–herbivore, predator–prey, host–parasite, and potential competitive interactions **(Figure 50.21)**.

Trophic Levels. We can visualize the trophic structure of a community as a hierarchy of **trophic levels**, defined by the feeding relationships among its species (see Figure 50.21a). Photosynthetic organisms are the **primary producers**, the first trophic level. Primary producers are often described as **autotrophs** (*auto* = self) because they capture sunlight and convert it into chemical energy, using simple inorganic molecules acquired from the environment to build larger organic molecules that

other organisms can use. Plants are the dominant primary producers in terrestrial communities. Multicellular algae and plants are the major primary producers in shallow freshwater and marine environments, but photosynthetic protists and cyanobacteria play that role in deep, open water.

Animals, by contrast, are **consumers**. Herbivores, which feed directly on plants, form the second trophic level, the **primary consumers**. Carnivores that feed on herbivores are the third trophic level, or **secondary consumers**; and carnivores that feed on other carnivores form the fourth trophic level, the **tertiary consumers**. For example, songbirds feeding on herbivorous insects are secondary consumers, and falcons feeding on songbirds are tertiary consumers. Some organisms, like humans and some bears, are **omnivores**, feeding at several trophic levels simultaneously.

A separate and distinct trophic level includes organisms that extract energy from the organic detritus (refuse) produced at other trophic levels. Scavengers, or **detritivores**, are animals such as earthworms and vultures that ingest dead organisms, digestive wastes, and cast-off body parts such as leaves and exoskeletons. **Decomposers** are small organisms, such as bacteria and fungi, that feed on dead or dying organic material. As described in Chapter 51, detritivores and decomposers serve a critical ecological function because their activity reduces organic material to small inorganic molecules that producers can assimilate.

All of the consumers in a community—the animals, fungi, and diverse microorganisms—are described as **heterotrophs** (*hetero* = other) because they acquire energy and nutrients by eating other organisms or their remains.

Food Chains and Webs. Ecologists depict the trophic structure of a community in a **food chain**, a portrait

a. Trophic levels

b. Marine food web

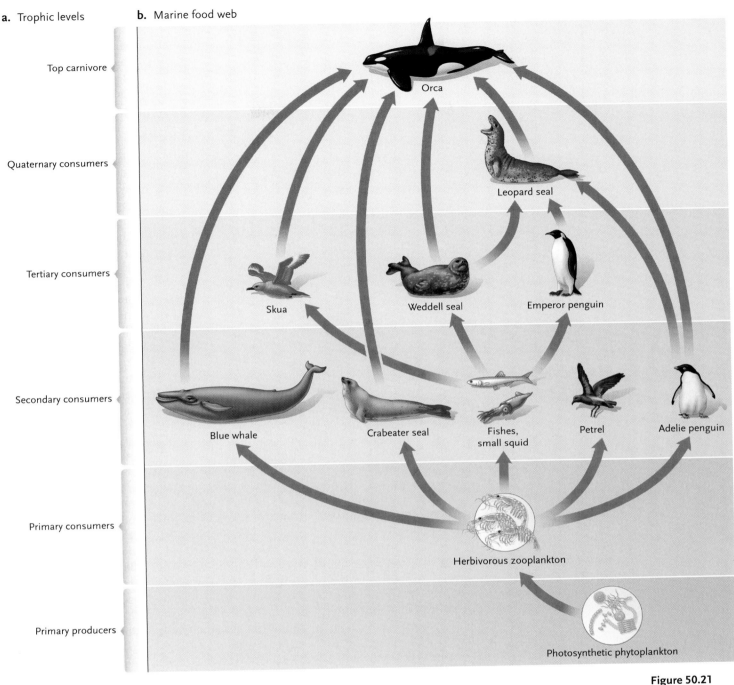

Top carnivore

Quaternary consumers

Tertiary consumers

Secondary consumers

Primary consumers

Primary producers

Orca

Leopard seal

Skua

Weddell seal

Emperor penguin

Blue whale

Crabeater seal

Fishes, small squid

Petrel

Adelie penguin

Herbivorous zooplankton

Photosynthetic phytoplankton

Figure 50.21
The marine food web off the coast of Antarctica.

of who eats whom. Each link in a food chain is represented by an arrow pointing from the food to the consumer. Simple, straight-line food chains are rare in nature because most consumers feed on more than one type of food, and because most organisms are eaten by more than one type of consumer. These complex relationships are portrayed as a **food web**, a set of interconnected food chains with multiple links.

In the food web for the waters off the coast of Antarctica (see Figure 50.21b), the primary producers and primary consumers are small organisms that occur in

vast numbers. Microscopic diatoms (phytoplankton) are responsible for most photosynthesis, and small shrimplike krill (zooplankton) are the major primary consumers. These tiny organisms are eaten by larger species, such as fishes, seabirds, and suspension-feeding baleen whales. Some of the secondary consumers are themselves eaten by birds and mammals at higher trophic levels. The top carnivore in this ecosystem, the orca *(Orcinus orca)*, feeds on carnivorous birds and mammals.

Ideally, depictions of food webs would include all species in a community, from microorganisms to the top consumer. But most ecologists simply cannot collect data on every species, particularly those that are rare or very small. Instead, they study the links between the most important species and simplify the analysis by grouping together trophically similar species. For example, Figure 50.21b categorizes the many different species of primary producers and primary consumers as phytoplankton and zooplankton, respectively.

Food-Web Analysis. In the late 1950s, Robert MacArthur of Princeton University pioneered the analysis of food webs to determine how the many links between trophic levels may contribute to a community's **stability**—its ability to maintain its species composition and relative abundances when environmental disturbances eliminate some species from the community. MacArthur hypothesized that in species-rich communities, where animals feed on many food sources, the absence of one or two species would have only minor effects on the structure and stability of the community as a whole. He therefore proposed a connection between species diversity, food-web complexity, and community stability.

Recent research has confirmed MacArthur's reasoning. For example, the average number of links per species generally increases with increasing species richness. Comparative food-web analysis also reveals that the relative proportions of species at the highest, middle, and lowest trophic levels are reasonably constant across communities. When researchers compared the number of prey species to the number of predator species in food webs from 92 communities of freshwater invertebrates, they discovered that, regardless of species richness, a community includes between two and three prey species for every predator species.

Interactions among species in a food web are often complex, indirect, and hard to unravel. In desert communities of the American Southwest, for example, rodents and ants potentially compete for seeds, their main food source. And the plants that produce the seeds compete for water, nutrients, and space. Rodents generally prefer to eat large seeds, but ants prefer small seeds. Thus, feeding by rodents reduces the potential population sizes of plants that produce large seeds. As a result, the population sizes of plants that produce small seeds may increase, ultimately providing more food for ants.

Some analyses of food webs focus on interactions in which predators or prey have significant influence on the growth rates and sizes of other populations in the community; these *strong interactions* can affect overall community structure. In the next section we provide examples of strong interactions when we describe how consumers influence the competitive interactions among populations of their prey.

STUDY BREAK

1. What plant growth forms are common in tropical forests?
2. What is the difference between species richness and relative abundance?
3. Peregrine falcons are predatory birds that have been introduced into many North American cities, where they feed primarily on pigeons. The pigeons eat mostly vegetable matter. To what trophic level do pigeons and peregrine falcons belong?

50.4 Effects of Population Interactions on Community Characteristics

Numerous studies have shown that interspecific competition and predation can influence a community's species composition.

Interspecific Competition Can Reduce Species Richness within Communities

Interspecific competition can cause the local extinction of species or prevent new species from becoming established in a community, thus reducing its species richness. During the 1960s and early 1970s, ecologists emphasized competition as the primary factor structuring communities. Observations of resource partitioning and character displacement suggested that some process had fostered differences in resource use among coexisting species, and competition provided the most straightforward explanation of these patterns.

Seeking to uncover direct evidence of competition, ecologists undertook many field experiments on competition in natural populations. The experiment on barnacles depicted in Figure 50.12 is typical of this approach, in which researchers determine whether adding or removing a species changes the distribution or population size of its presumed competitors. In the early 1980s, two independent reviews of the literature on these field experiments, one by Joseph Connell and the other by Thomas W. Schoener of the University of California at Davis, suggested that competition is sometimes a potent force. Connell's survey, which included 527 published experiments on 215 species, identified competition in roughly 40% of the experiments and more than 50% of the species. Schoener's review, which used different criteria to evaluate 164 experiments on approximately 400 species, found that competition affected more than 75% of the species.

Although these reviews confirm the importance of competition, the ecological literature upon which they were based probably contains several significant biases. First, ecologists who set out to study competition are more likely to study interactions in which they think competition occurs, and they are more likely to publish research that documents its importance. Thus, the literature includes more studies of competition in *K*-selected species than in *r*-selected species. Recall that populations of *r*-selected species, such as herbivorous insects, rarely reach carrying capacity, and competition may not limit their population sizes (review Section 49.6). Thus, the Connell and Schoener surveys may *overestimate* the importance of competition. (Nevertheless, a more recent survey suggests that interspecific competition may be common even among populations of herbivorous insects.) Another bias, which Connell called "the ghost of competition past," *underestimates* the importance of competition. If, as many ecologists believe, resource partitioning and character displacement are the results of past competition, we are unlikely to witness much competition today, even though it was once important in structuring those population interactions.

Ecologists have not yet reached consensus about whether interspecific competition strongly influences the species composition and structure of most communities. Plant ecologists and vertebrate ecologists, who often study *K*-selected species, generally believe that competition has a profound effect on species distributions and resource use. Insect ecologists and marine ecologists, who often study *r*-selected species, argue that competition is not the major force governing community structure, pointing instead to predation or parasitism and physical disturbance.

Predators Can Boost Species Richness by Stabilizing Competitive Interactions among Their Prey

Predators can influence the species richness and structure of communities by reducing the population sizes of their prey. On the rocky coast of the American Northwest, for example, algae and sessile invertebrates compete for attachment sites on rocks, a requirement for life on a wave-swept shore. Mussels *(Mytilus californianus)* are the strongest competitors for space, eliminating other species from the community. But at some sites, predatory sea stars *(Pisaster ochraceus)* preferentially feed on mussels, reducing their numbers and creating space for other species to grow. Because the interaction between *Pisaster* and *Mytilus* affects other species as well, it qualifies as a strong interaction.

In the 1960s, Robert Paine of the University of Washington conducted removal experiments to evaluate the effects of *Pisaster* predation **(Figure 50.22)**. In predator-free experimental plots, mussels outcompeted barnacles, chitons, limpets, and other invertebrate herbivores, reducing species richness from 18 species to 2 or 3. In control plots that contained preda-

tors, however, all 18 species persisted. Ecologists describe predators like *Pisaster* as **keystone species,** species that have a greater effect on community structure than their numbers might suggest.

Herbivores May Counteract or Reinforce Competition among Their Food Plants

Herbivores also exert complex effects on communities. In the 1970s, Jane Lubchenco, then of Harvard University, studied herbivory in a periwinkle snail *(Littorina littorea)*, a keystone species on rocky shores in Massachusetts **(Figure 50.23)**. Periwinkles preferentially graze on the tender green alga *Enteromorpha*. In tidepools, which are usually submerged, *Enteromorpha* outcompetes other algae. Moderate feeding by periwinkles, however, eliminates some *Enteromorpha,* allowing less competitive algal species to grow. Moderate herbivory by periwinkles therefore increases algal species richness in tidepools. But on high rocks, which are exposed to air during low tide, the dehydration-resistant red alga *Chondrus* is competitively dominant. Periwinkles don't eat the tough *Chondrus,* however, feeding instead on the less abundant and competitively inferior *Enteromorpha.* Thus, on exposed rocks, feeding by the snails reduces algal species richness.

STUDY BREAK

1. How is the scientific literature on interspecific competition biased?
2. What are keystone species, and how do they influence species richness in communities?

50.5 Effects of Disturbance on Community Characteristics

Recent research tends to support the individualistic view that many communities are not in equilibrium and that their species composition changes frequently. Environmental disturbances—storms, landslides, fires, floods, and cold spells—often eliminate some species, providing opportunities for others to become established.

Frequent Disturbances Keep Some Communities in a Constant State of Flux

Physical disturbances are common in some environments. For example, lightning-induced fires commonly sweep through grasslands, powerful hurricanes routinely demolish patches of forest, and waves wash over communities that live at the edge of the sea.

Joseph Connell and his colleagues conducted an ambitious long-term study of the effects of disturbance

Figure 50.22 Experimental Research

Effect of a Predator on the Species Richness of Its Prey

QUESTION: Does feeding by a predator influence the species richness and relative abundances of the species on which it feeds?

EXPERIMENT: The predatory sea star *Pisaster ochraceus* preferentially feeds on mussels *(Mytilus californianus)*, which is the strongest competitor for space in rocky intertidal habitats in Washington State. Paine removed *Pisaster* from caged experimental study plots, but left control study plots undisturbed. He then monitored the species richness of *Pisaster*'s invertebrate prey over many years.

RESULTS: Paine documented an increase in mussel populations in the experimental plots as well as complex changes in the feeding relationships among species in the intertidal food web. The overall effect of removing *Pisaster*, the top predator in this food web, was a rapid decrease in the species richness of invertebrates and algae. By contrast, control plots maintained their species richness over the course of the experiment.

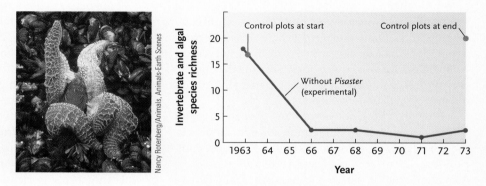

CONCLUSION: Predation by the sea star *Pisaster ochraceus* maintains the species richness of its prey by preventing mussels from outcompeting other invertebrates and algae on rocky shores.

on coral reefs, shallow tropical marine habitats that are among the most species-rich communities on Earth. In some parts of the world, reefs are routinely battered by violent storms, which wash corals off the substrate, creating bare patches in the reef. The scouring action of storms creates opportunities for coral larvae to settle on bare substrates and start a new colony; ecologists use the word *recruitment* to describe the process in which young individuals join a population.

From 1963 to 1992, Connell and his colleagues tracked the fate of the Heron Island Reef at the south end of Australia's Great Barrier Reef **(Figure 50.24).** The inner flat and protected crests of the reef are sheltered from severe wave action during storms, whereas some pools and crests are routinely exposed to physical disturbance. Because corals live in colonies of variable size, the researchers monitored coral abundance by measuring the percentage of the substrate (that is, the seafloor) that colonies covered. They revisited marked study plots at intervals, photographing and identifying individual coral colonies.

Five major cyclones crossed the reef during the 30-year study period. Coral communities in the exposed areas of the reef were in a nearly continual state of flux. In exposed pools, four of the five cyclones re-

duced the percentage of cover, often drastically. On exposed crests, the cyclone of 1972 eliminated virtually all of the corals, and subsequent storms slowed the recovery of these areas for more than 20 years. By contrast, corals in sheltered areas suffered much less storm damage. Nevertheless, their coverage also declined steadily during the study as a natural consequence of the corals' growth. As colonies grew taller and closer to the ocean's surface, their increased exposure to air resulted in substantial mortality.

Connell and his colleagues also documented recruitment, the growth of new colonies from settling larvae, in their study plots. They discovered that the rate at which new colonies developed was almost always higher in sheltered areas than in exposed areas. However, recruitment rates were extremely variable, depending in part on the amount of space that storms or coral growth had made available.

This long-term study of coral reefs illustrates that frequent disturbances prevent some communities from reaching an equilibrium determined by interspecific interactions. Changes in the coral reef community at Heron Island result from the combined effects of external disturbances that remove coral colonies from the reef and internal processes (growth and recruit-

Figure 50.23 Experimental Research

The Complex Effects of an Herbivorous Snail on Algal Species Richness

QUESTION: How does feeding by periwinkle snails (*Littorina littorea*) influence the species richness of algae in intertidal communities?

EXPERIMENT: Lubchenco manipulated the densities of periwinkle snails in tidepools and on exposed rocks in a rocky intertidal habitat by creating enclosures that prevented snails from either entering or leaving her study plots. She then monitored the species composition of algae in the study plots and examined those data by plotting them against periwinkle density.

RESULTS: The effects of periwinkle density on algal species richness varied dramatically between study plots in tidepools and on exposed rocks.

Periwinkle snails *(Littorina littorea)*

Enteromorpha growing in tidepools

Chondrus growing on exposed rocks

In tidepools

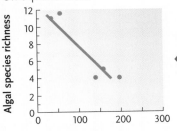

Periwinkles per square meter

In tidepools, snails at low densities eat little algae and *Enteromorpha* competitively excludes other algal species, reducing species richness. At high snail densities, heavy feeding on all species reduces algal species richness. At intermediate snail densities, grazing eliminates some *Enteromorpha*, allowing other species to grow.

On exposed rocks

Periwinkles per square meter

On exposed rocks, periwinkles never eat much *Chondrus*, but they consume the tender, less successful competitors. Thus, feeding by periwinkles reinforces the competitive superiority of *Chondrus*: as periwinkle density increases, algal species richness declines.

CONCLUSION: Grazing by periwinkle snails has complex effects on the species richness of competing algae. In tidepools, where periwinkle snails preferentially feed on *Enteromorpha*, the competitively dominant alga, snails at an intermediate density remove some *Enteromorpha*, which allows weakly competitive algae to grow, increasing species richness. Feeding by snails at either low or high densities reduces algal species richness. On exposed rocks, where periwinkle snails rarely eat the competitively dominant alga *Chondrus*, feeding by snails reduces algal species richness.

---

ment) that either eliminate colonies or establish new ones. In this community, growth and recruitment are slow processes, and disturbances are frequent. Thus, the community never attains equilibrium.

Moderate Levels of Disturbance May Foster High Species Richness

According to the **intermediate disturbance hypothesis,** proposed by Connell in 1978, species richness is greatest in communities that experience fairly frequent disturbances of moderate intensity. Moderate distur-

bances create some openings for *r*-selected species to arrive and join the community, but they allow *K*-selected species to survive. Thus, communities that experience intermediate levels of disturbance contain a rich mixture of species. Where disturbances are severe and frequent, communities include only *r*-selected species that complete their life cycles between catastrophes. Where disturbances are mild and rare, communities are dominated by long-lived *K*-selected species that competitively exclude other species from the community.

Several studies in diverse habitats have confirmed the predictions of the intermediate disturbance hy-

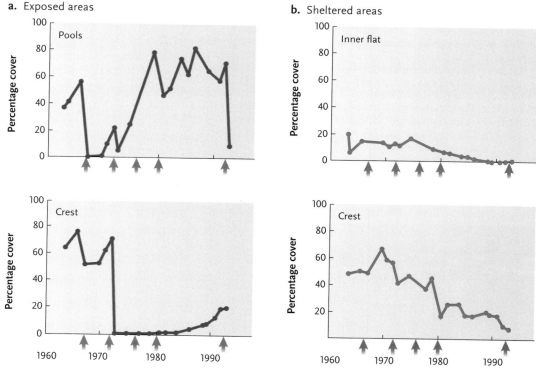

Figure 50.24

The effects of storms on corals. Five tropical cyclones (marked by gray arrows) damaged corals on the Heron Island Reef during a 30-year period. Storms reduced the percentage cover of corals in exposed parts of the reef **(a)** much more than in sheltered parts of the reef **(b)**.

pothesis. For example, Colin R. Townsend and his colleagues at the University of Otago studied the effects of disturbance at 54 stream sites in the Taieri River system in New Zealand. Disturbance occurs in these communities when water flow from heavy rains moves the rocks, soil, and sand in the streambed, disrupting the habitats where animals live. Townsend and his colleagues measured how much of the substrate moved in different streambeds to index the intensity of the disturbance. Their results indicate that species richness is highest in areas that experience intermediate levels of disturbance **(Figure 50.25)**.

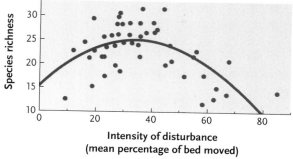

Figure 50.25

An observational study that supports the intermediate disturbance hypothesis. In the Taieri River system in New Zealand, species richness was highest in stream communities that experienced an intermediate level of disturbance.

Some ecologists have also suggested that species-rich communities recover from disturbances more readily than do less diverse communities. For example, David Tilman and his colleagues at the University of Minnesota conducted large-scale experiments in midwestern grasslands on the relationship between species number and the ability of communities to recover from disturbance. Their results demonstrate that grassland plots with high species richness recover from drought faster than plots with fewer species.

STUDY BREAK

1. How might disturbances from storms allow coral reefs to be rejuvenated by the recruitment of young individuals?
2. How do moderately severe and moderately frequent disturbances influence a community's species richness?

50.6 Ecological Succession: Responses to Disturbance

In response to disturbance, communities undergo **ecological succession**, a somewhat predictable series of changes in species composition over time.

Succession Begins after Disturbance Alters a Landscape or Changes the Species Composition of an Existing Community

Primary succession begins when organisms first colonize habitats without soil, such as those created by erupting volcanoes and retreating glaciers **(Figure 50.26)**. Lichens (see Section 28.3), which derive nutrients from rain and bare rock, are usually the first visible colonizers of such inhospitable habitats. They secrete mild acids that erode rock surfaces, initiating the slow development of soil, which is enriched by the organic material lichens produce. After lichens modify a site, mosses (see Section 27.2) colonize patches of soil and grow quickly.

As soil accumulates, hardy opportunistic plants—grasses, ferns, and broad-leaved herbs—colonize the site from surrounding areas. Their roots break up rock, and as they die, their decaying remains enrich the soil. Detritivores and decomposers facilitate these processes. As the soil gets deeper and richer, increased moisture and nutrients support bushes and, eventually, trees. Late successional stages are often dominated by K-selected species with woody trunks and branches that position leaves in sunlight and large root systems that acquire water and nutrients from soil.

In the classical view of ecological succession, long-lived species, which replace themselves over time, eventually dominate a community, and new species join it only rarely. This relatively stable, late successional stage is called a **climax community** because the dominant vegetation replaces itself and persists until an environmental disturbance eliminates it, allowing other species to invade. Local climate and soil conditions, the surrounding communities where colonizing species originate, and chance events determine the species composition of climax communities. However, recent research suggests that even "climax communities" change slowly in response to environmental fluctuations, as described below.

Secondary succession occurs after existing vegetation is destroyed or disrupted by an environmental disturbance, such as a fire, a storm, or human activity. The presence of soil makes the disturbed sites ripe for colonization. Moreover, the soil may contain numerous seeds that germinate after the disturbance. The early stages of secondary succession proceed rapidly, but later stages parallel those of primary succession.

Secondary succession in the North Temperate Zone is well studied in abandoned farms, called "old fields," where forests were cleared centuries earlier. Because the transformation from old field back to forest takes at least a hundred years, ecologists use historical records to find the age of different stands of vegetation and reconstruct the successional sequence by comparing stands of different ages. In the Piedmont region of southeastern North America, an abandoned field is covered by crabgrass (genus *Digitaria*), an an-

nual plant, during the first growing season. The following year, crabgrass is replaced by horseweed (*Conyza canadensis*), which cannot persist because it secretes substances that inhibit the germination of its own seeds. Ragweed (*Ambrosia artemisiifolia*), another annual, dominates during the third year, but it is gradually replaced by perennial asters (genus *Erigeron*) and broomsedges (genus *Andropogon*), which are, in turn, replaced by shrubs. Ten to fifteen years after the field was abandoned, pine (genus *Pinus*) seedlings germinate. Growing pines cast substantial shade and their fallen needles acidify the soil, making the site unsuitable for the plants from earlier successional stages. Because pines are intolerant of shade, pine seedlings don't flourish under mature pine trees. Thus, after 50 to 100 years, pines are replaced by a taller mixed hardwood forest of oaks (genus *Quercus*) and hickories (genus *Carya*), which develops in the thick, moist soil. The hardwood forest forms the climax community after more than a century of successional change.

Similar climax communities sometimes arise from alternative successional sequences. For example, hardwood forests also develop in sites that were once ponds. During **aquatic succession**, debris from rivers and run-off accumulates in a body of water, causing it to fill in at its margins. The pond is transformed into a swamp, inhabited by plants adapted to a semisolid substrate. As larger plants get established, their high transpiration rates dry the soil, allowing other plant species to colonize. Given enough time, the site may become a meadow or forest, where an area of moist, low-lying ground is the only remnant of the original pond.

Community Characteristics Change during Succession

Several characteristics undergo directional change as succession proceeds. First, because r-selected species are short-lived and K-selected species long-lived, species composition changes rapidly in the early stages, but slowly in the late stages of succession. Second, species richness increases rapidly during the early stages because new species join the community faster than resident species become extinct; as succession proceeds, however, species richness stabilizes or may even decline. Third, in terrestrial communities that receive sufficient rainfall, the maximum height and total mass of the vegetation increase steadily as large species replace small ones, creating the complex structure of the climax.

Because plants influence the physical environment below them, the community itself increasingly moderates the microclimate. The shade cast by a forest canopy retains soil moisture and reduces temperature fluctuations. The trunks and canopy also reduce wind speed. By contrast, the short vegetation in an early successional stage does not effectively shelter the space below it.

Roger K. Burnard

1 The glacier has retreated about 8 m per year since 1794.

Roger K. Burnard

2 This site was covered with ice less than 10 years before this photo was taken. When a glacier retreats, a constant flow of melt water leaches minerals, especially nitrogen, from the newly exposed substrate.

Roger K. Burnard

3 Once lichens and mosses have established themselves, mountain avens (genus *Dryas*) grows on the nutrient-poor soil. This pioneer species benefits from the activity of mutualistic nitrogen-fixing bacteria, spreading rapidly over glacial till.

Roger K. Burnard

4 Within 20 years, shrubby willows (genus *Salix*), cottonwoods (genus *Populus*), and alders (genus *Alnus*) take hold in drainage channels. These species are also symbiotic with nitrogen-fixing microorganisms.

Roger K. Burnard

5 In time, young conifers, mostly hemlocks (genus *Tsuga*) and spruce (genus *Picea*), join the community.

Ed Degginger

6 After 80 to 100 years, dense forests of Sitka spruce *(Picea sichensis)* and western hemlock *(Tsuga heterophylla)* have crowded out the other species.

Figure 50.26
Primary succession following glacial retreat. The retreat of glaciers at Glacier Bay, Alaska, has allowed ecologists to document primary succession on newly exposed rocks and soil.

Although ecologists usually describe succession in terms of vegetation, animals undergo succession, too. As the vegetation shifts, new resources become available, and animal species replace each other over time. Herbivorous insects, which often have strict food preferences, undergo succession along with their food plants. And as the herbivores change, so do their predators, parasites, and parasitoids. In old-field succession in eastern North America, different vegetation stages harbor a changing assortment of bird species (Figure 50.27).

Several Hypotheses Help to Explain the Processes Underlying Succession

Differences in dispersal abilities, maturation rates, and life spans among species are at least partly responsible for ecological succession. Early successional stages harbor many r-selected species because they produce numerous small seeds that colonize open habitats and grow quickly. Mature successional stages are dominated by K-selected species because they are long-lived. Nevertheless, coexisting populations inevitably affect one another. Although the role of population interactions in succession is generally acknowledged, ecologists debate the relative importance of processes that either facilitate or inhibit the turnover of species in a community.

The **facilitation hypothesis** suggests that species modify the local environment in ways that make it less suitable for themselves but more suitable for colonization by species typical of the next successional stage. For example, when lichens first colonize bare rock, they produce a small quantity of soil, which is required by mosses and grasses that grow there later. According to this hypothesis, changes in species composition are both orderly and predictable because the presence of each stage facilitates the success of the next. Facilitation is very important in primary succession, but it may not be the best model of interactions that influence secondary succession.

The **inhibition hypothesis** suggests that new species are prevented from occupying a community by whatever species are already present. According to this hypothesis, succession is neither orderly nor predictable because each stage is dominated by whichever species happen to colonize the site first. Species replacements occur only when individuals of the dominant species die of old age or when an environmental disturbance reduces their numbers. Eventually, long-lived species replace short-lived species, but the precise species composition of a mature community is up for grabs. Inhibition appears to play a role in some secondary successions. For example, the interactions among early successional species in an old field are highly competitive. Horseweed inhibits the growth of asters, which follow them in succession, by shading the aster seedlings and by releasing toxic substances from their roots. The experimental removal of horseweed enhances the growth of asters, confirming the inhibitory effect.

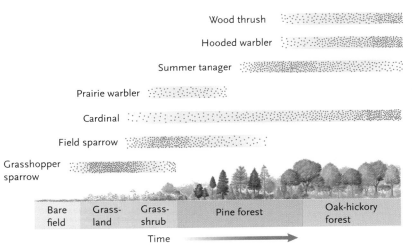

Figure 50.27

Succession in animals. Successional changes in bird species composition in an abandoned agricultural field in eastern North America parallel the changes in plant species composition. Residence times of several representative species are illustrated. The density of stippling inside each bar illustrates the density of each species through time.

The **tolerance hypothesis** asserts that succession proceeds because competitively superior species replace competitively inferior ones. According to this model, early-stage species neither facilitate nor inhibit the growth of later-stage species. Instead, as more species arrive at a site and resources become limiting, competition eliminates species that cannot harvest scarce resources successfully. In the Piedmont region of North America, for example, hardwood trees are more tolerant of shade than pine trees are, and hardwoods gradually replace pines during succession. Thus, the climax community includes only strong competitors. Tolerance may explain the species composition of many transitional and mature communities.

At most sites, succession probably results from a combination of facilitation, inhibition, and tolerance, coupled with interspecific differences in dispersal, growth, and maturation rates. Moreover, within a community, the patchiness of abiotic factors also strongly influences plant distributions and species composition. In the deciduous forests of eastern North America, maples (genus *Acer*) predominate on wet, low-lying ground, but oaks (genus *Quercus*) are more abundant at higher and drier sites. Thus, a mature deciduous forest is often a mosaic of species and not a uniform stand of trees.

Disturbance and density-independent factors also play important roles, in some cases speeding successional change. In northern forests, for example, moose prefer to feed on deciduous shrubs, accelerating the rate at which conifers replace them. In other cases, disturbance inhibits successional change, establishing a *disturbance climax* or **disclimax community.** In many grassland communities, grazing by large mammals and periodic fires kill the seedlings of trees that would otherwise become established. Thus, disturbance prevents the succession from grassland to forest, and grassland persists as a disclimax community.

On a local scale, disturbances often destroy small patches of vegetation, returning them to an earlier successional stage. A hurricane may knock over trees in a forest, creating small, sunny patches of open ground. Locally occurring *r*-selected species take advantage of the resources that are suddenly available and quickly colonize the openings. These local patches then undergo succession that is out of step with the immediately surrounding forest. Thus, moderate disturbance, accompanied by succession in local patches, can increase species richness in many communities.

50.7 Variations in Species Richness among Communities

Species richness often varies among communities according to a recognizable pattern. Two large-scale patterns of species richness—latitudinal trends and island patterns—have captured the attention of ecologists for more than a century.

Many Types of Organisms Exhibit Latitudinal Gradients in Species Richness

Ever since Darwin and Wallace traveled the globe (see Section 19.2), ecologists have recognized broad latitudinal trends in species richness. For many, but not all, plant and animal groups, species richness follows a latitudinal gradient, with the most species in the tropics and a steady decline in numbers toward the poles **(Figure 50.28)**. Several general hypotheses may explain these striking patterns.

Some hypotheses propose historical explanations for the *origin* of high species richness in the tropics. The benign climate in tropical regions allows some tropical organisms to have more generations per year than their temperate counterparts. And, given the small seasonal changes in temperature, tropical species may be less likely than temperate species to migrate from one habitat to another, thus reducing gene flow between geographically isolated populations (see Section 21.3). These factors may have fostered higher speciation rates in the tropics, accelerating the accumulation of species. Tropical communities may also have experienced severe disturbance less often than communities at higher latitudes, where periodic glaciations have caused repeated extinctions. Thus, new species may have accumulated in the tropics over longer periods of time.

Other hypotheses focus on ecological explanations for the *maintenance* of high species richness in the tropics. Some resources are more abundant, predictable, and diverse in tropical communities. Tropical regions experience more intense sunlight, warmer temperatures in most months, and higher annual rainfall than temperate and polar regions (see Chapter 52). These factors provide a long and predictable growing season for the lush tropical vegetation, which supports a rich assemblage of herbivores, and through them many carnivores and parasites. Furthermore, the abundance, predictability, and year-round availability of resources allow some tropical animals to have specialized diets. For example, tropical forests support many species of fruit-eating bats and birds, which could not survive in temperate forests where fruits are not available year-round.

Species richness may therefore be a self-reinforcing phenomenon in tropical communities. Complex webs of population interactions and interdependency

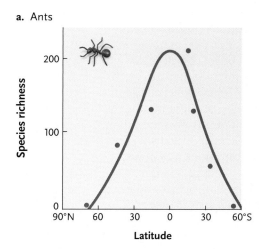

a. Ants

b. Birds

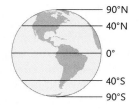

Figure 50.28
Latitudinal trends in species richness. The species richness of many animals and plants varies with latitude, as illustrated here **(a)** for ants and **(b)** for birds of North and Central America. The species-richness data for birds is based on records of where the species breed.

have coevolved in relatively stable and predictable tropical climates. Predator–prey, competitive, and symbiotic interactions may prevent individual species from dominating communities and reducing species richness.

The Theory of Island Biogeography Explains Variations in Species Richness

Although the species richness of communities may be stable over time, species composition is often in flux as new species join a community and others drop out. In the 1960s, Robert MacArthur of Princeton University and Edward O. Wilson of Harvard University addressed the question of why communities vary in species richness, using islands as model systems. Islands provide natural laboratories for studying ecological phenomena, just as they do for evolution (see *Focus on Research* in Chapter 21). Island communities are often small, have well-defined boundaries, and are isolated from surrounding communities.

In developing the **equilibrium theory of island biogeography**, MacArthur and Wilson sought to explain variations in species richness on islands of different size and different levels of isolation from other landmasses **(Figure 50.29)**. They hypothesized that the number of species on any island was governed by a give and take between two processes: the immigration of new species to an island and the extinction of species already there (see Figure 50.29a).

According to the MacArthur–Wilson model, the mainland harbors a *species pool* from which species immigrate to offshore islands. Seeds and small arthropods are carried by wind or floating debris; some animals, such as birds, arrive under their own power. When few species are already on an island, the rate at which new species immigrate to the island is high. But as more species inhabit the island over time, the immigration rate declines because there are fewer species left in the mainland pool that can still arrive on the island as *new* colonizers.

Once a species immigrates to an island, its population grows and persists for some time. But as the number of species on the island increases, the rate at which those species go extinct also rises. The extinction rate increases through time partly because there are more species that can go extinct there. In addition, as the number of species on the island increases, competition and predator–prey interactions can reduce the population sizes of some species and drive them to extinction.

According to MacArthur and Wilson's theory, an equilibrium between immigration and extinction determines the number of species that ultimately occupy an island. In other words, once equilibrium is reached, the number of species remains relatively constant because one species already on the island goes extinct in about the same time it takes a new species to immigrate to the island. The model does not specify which species immigrate to the island or which ones already on the island go extinct. It simply predicts that the number of species on the island is in equilibrium, although species composition is not. The ongoing processes of immigration and extinction establish a constant turnover in the roster of species that live on any island.

The MacArthur–Wilson model explains why some islands harbor more species than others. Large islands have higher immigration rates than small islands do because they present a larger target for dispersing organisms. Moreover, large islands have lower extinction rates because they can support larger populations and provide a greater range of habitats and resources. Thus, at equilibrium, large islands have more species than small islands (see Figure 50.29b). Similarly, islands

Figure 50.29
Predictions of the theory of island biogeography.

a. Immigration and extinction rates

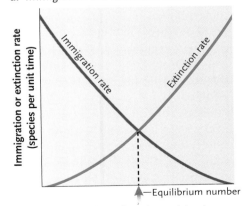

The number of species on an island at equilibrium (indicated by the arrow) is determined by the rate at which new species immigrate and the rate at which species already on the island go extinct.

b. Effect of island size

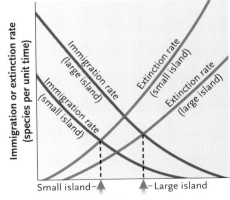

Immigration rates are higher and extinction rates lower on large islands than on small islands. Thus, at equilibrium, large islands have more species.

c. Effect of distance from mainland

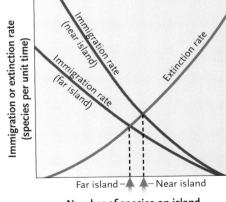

Organisms leaving the mainland locate nearby islands more easily than distant islands, causing higher immigration rates on near islands. Thus, near islands support more species than far ones.

a. Distance effect

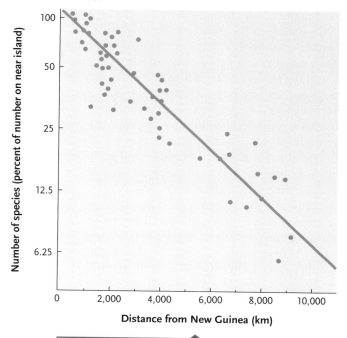

The number of lowland bird species on islands of the South Pacific declines with the islands' distance from the species source, the large island of New Guinea. Data in this graph were corrected for differences in the sizes of the islands. The number of bird species on each island is expressed as a percentage of the number of bird species on an island of equivalent size close to New Guinea.

b. Area effect

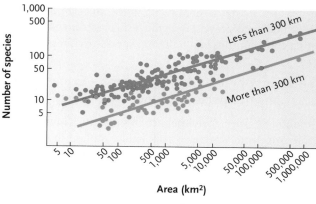

The number of bird species on tropical and subtropical islands throughout the world increases dramatically with island area. The data for islands near to a source and islands far from a mainland source are presented separately to minimize the effect of distance. Notice that the "distance effect" reduces the number of bird species on islands that are more than 300 km from a mainland source.

Figure 50.30
Factors that influence bird species richness on islands.
(a) Fewer bird species colonize islands that are distant from the mainland source. **(b)** More bird species colonize large islands than small ones.

near the mainland have higher immigration rates than distant islands do, because dispersing organisms are more likely to locate islands that are close to their point of departure. Distance does not affect extinction rates. Thus, at equilibrium, near islands have more species than far islands (see Figure 50.29c).

The equilibrium theory's predictions about the effects of area and distance are generally supported by data on plants and animals **(Figure 50.30)**. Experimental work has also verified some of its basic assumptions. For example, Amy Schoener of the University of Washington found that, within 30 days, more than 200 species of marine organisms colonized tiny artificial "islands" (plastic kitchen scrubbers) that she placed in a Bahaman lagoon. Her research confirmed that immigration rate increases with island size. In another ambitious study, Daniel Simberloff and Edward O. Wilson exterminated insects on small islands in the Florida Keys and monitored subsequent immigration and extinction (see the *Focus on Research*). Their research also confirmed the equilibrium theory's predictions that an island's size and distance from the mainland influence how many species will occupy it.

The equilibrial view of species richness also applies to mainland communities, which exist as islands in a metaphorical sea of dissimilar habitat. Lakes are "islands"

in a "sea" of dry land, and mountaintops are habitat "islands" in a "sea" of low terrain. Species richness in these communities is partly governed by the immigration of new species from distant sources and the extinction of species already present. As human activities disrupt environments across the globe, undisturbed sites function as islandlike refuges for threatened and endangered species. Conservation biologists now apply the general lessons of MacArthur and Wilson's theory to the design of nature preserves (see Chapter 53).

In the next chapter we examine ecosystems, which include ecological communities interacting with their abiotic environments, focusing on the movements of energy and nutrients.

STUDY BREAK

1. What factors may foster the maintenance of high species richness in tropical communities?
2. According to the equilibrium theory of island biogeography, what are the effects of an island's size and its distance from the mainland on the number of species that can occupy it?

Basic Research: Testing the Theory of Island Biogeography

Shortly after Robert MacArthur and Edward O. Wilson published the equilibrium theory of island biogeography in the 1960s, Daniel Simberloff, one of Wilson's graduate students at Harvard University, and Wilson himself undertook one of the most ambitious experiments ever attempted in community ecology. Simberloff reasoned that the best way to test the theory's predictions was to monitor immigration and extinction on barren islands.

Simberloff and Wilson devised a system for removing all the animals from individual red mangrove trees in the Florida Keys. The trees, with canopies that spread from 11 to 18 m in diameter, grow in shallow water and are isolated from their neighbors; thus, each tree is an island that harbors an arthropod community. The species pool on the Florida mainland includes about 1000 arthropod species, but each mangrove island contains no more than 40 species at one time.

After cataloging the species on each island, Simberloff and Wilson hired an extermination company to erect large tents and fumigate the islands to eliminate all arthropods on them **(Figure a).** The exterminators used methyl bromide, a pesticide that doesn't harm trees or leave any residue. Simberloff then monitored both the immigration of arthropods to the islands and the extinction of species that became established on them. He surveyed six islands regularly for 2 years and at intervals thereafter.

The results of this experiment confirm several predictions of MacArthur and Wilson's theory **(Figure b).** Arthropods recolonized the islands rapidly, and within 8 or 9 months the number of species living on each island had reached an equilibrium that was near the original species number. In addition, the island nearest to the mainland had more species than the most distant island. However, immigration

Figure a

After cataloguing the arthropods, Simberloff and Wilson hired an exterminating company to erect a tent over each mangrove island. Once the islands were fully covered, exterminators used methyl bromide to eliminate all living arthropods.

and extinction were incredibly rapid, and Simberloff and Wilson suspected that some species went extinct even before they had noted their presence. The researchers also discovered that 3 years after the experimental treatments, the species composition of the islands was still changing constantly and did not remotely resemble the species composition in the islands before they were defaunated.

Simberloff and Wilson's research was a landmark study in ecology because it tested the predictions of an important theory using a field experiment. Although such efforts are now almost routine in ecological studies, this project was one of the first to demonstrate that large-scale experimental manipulations of natural systems are feasible and that they often produce clear results.

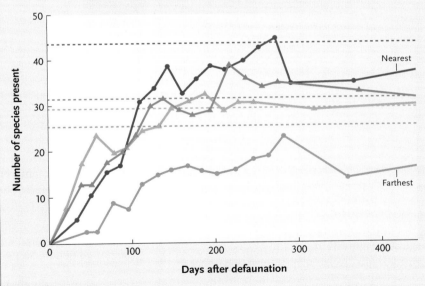

Figure b

On three of four islands, species richness gradually returned to the predefaunation level (indicated by color-coded dashed lines on the graph). The most distant island had not reached its predefaunation species richness after 2 years.

Do species interactions change predictably across environments?

As we learned in this chapter, the population interactions that occur between species range from mutualistic to parasitic. Some biologists have suggested that we should expect more competitive interactions between species in some kinds of environments, but more positive interactions in others. Community ecology will become a more quantitative and predictive discipline if researchers focus on how abiotic and biotic environmental factors—such as the presence of particular community members, environmental gradients, or global climate change—influence the strength of the interactions between species. For example, as physical environments become more stressful, the abundance and distribution of species should be determined less by resource limitation and more by the stress itself. Accordingly, plants tend to compete far less with each other in stressful environments than they do under ideal growing conditions. Scientists are now engaged in the intellectual feedback of theory development and experimental testing aimed at generating a predictive framework for particular types of interactions and their consequences for community structure.

What is the relative importance of positive versus negative interactions for community structure?

It was once suggested that ecologists in capitalist societies, like the United States, tend to more often study competition and predation, but ecologists in socialist societies tend to study mutualism. Although the truth of this anecdote is unclear, it is remarkable that ecologists still do not agree on the relative importance of positive interactions (for example, mutualism or commensalism) versus negative interactions (such as predation or competition) in generating community structure. Advances in this area of study may result from "factorial" experiments, in which two or more types of interactions are manipulated. For example, one might examine the relative effects of excluding pollinators versus excluding herbivores on the success of a plant population. In factorial experiments, the researcher can conclude that one factor has a bigger effect than the other, because all other factors were controlled. These sorts of experiments may eventually lead to an emerging picture of the relative importance of positive versus negative population interactions.

How does the evolutionary history of a species influence its ecology today?

The great evolutionary biologist Theodosius Dobzhansky once noted that "Nothing in biology makes sense except in the light of evolution." Although we know a great deal about both ecology and evolutionary biology, researchers are only beginning to explore the impact of an organism's evolutionary history on its ecology. This very active area of research includes the use of phylogenetic information (see Chapter 23), selection experiments (see Chapter 21), and a knowledge of the genetic basis of particular traits (see Chapter 12). For example, are closely related species more likely to compete with each other than more distantly related species are? Do organisms that are well adapted to particular environments fare poorly in other environments? Why do some organisms specialize in their resource use? Are the population dynamics that species experience shaped by past evolutionary events? These questions are currently being addressed, and the answers uncovered by researchers may unravel many current mysteries about the ecology of populations and communities.

Anurag Agrawal is an associate professor in the Departments of Entomology and Ecology and Environmental Biology at Cornell University. He studies the evolutionary and community ecology of plant–insect interactions. To learn more about Dr. Agrawal's research go to http://www.herbivory.com.

Review

Go to **Thomson**NOW™ at www.thomsonedu.com/login to access quizzing, animations, exercises, articles, and personalized homework help.

50.1 Population Interactions

- Coevolution is the evolution of reciprocal adaptations in species that interact ecologically (Figure 50.1).

- Predators and herbivores use diverse adaptations to select, locate, capture, and ingest an appropriate diet (Figure 50.2). Plants have both structural and chemical defenses against herbivores. Animal prey may try to hide or escape from predators, defend themselves actively, or advertise their unpalatability (Figures 50.3–50.5); some species mimic the appearance of poisonous species (Figure 50.6). Predators may evolve adaptations to counter prey defenses (Figure 50.7).

- Interspecific competition results if two or more populations use the same limiting resources; competition may lead to the extinction of one competitor (Figure 50.8). Ecologists use the ecological niche concept to visualize a population's resource use (Figure 50.9). Observations of resource partitioning (Figure 50.10) and character displacement (Figure 50.11) suggest that competition may be important, but only field experiments can demonstrate that competition occurs (Figure 50.12).

- Symbiosis is a close ecological association between species. In commensal interactions, one species benefits and the other is unaffected (Figure 50.13). In mutualistic interactions, both partners benefit (Figures 50.14–50.16). In parasitic interactions, one species benefits and the other is harmed.

Animation: Predator–prey interactions

Animation: Competitive exclusion

Animation: Hairston's experiment

Animation: Resource partitioning

Animation: Wasp and mimics

Practice: Understanding the major types of species interactions: competition, predation, parasitism, and mutualism

50.2 The Nature of Ecological Communities

- An interactive view suggests that species in a community are bound together in a complex web of necessary biotic interactions; an individualistic view recognizes communities as loose assemblages of organisms that have similar physical requirements (Figure 50.17).

- Ecotones occur where adjacent communities grade into one another; sharp boundaries occur between communities where a critical resource or an important abiotic factor is discontinuous (Figure 50.18).

50.3 Community Characteristics

- In benign environments, vegetation is tall and has a complex physical structure (Figure 50.19). In stressful environments, vegetation is short and has a simple physical structure.

- Communities differ in species richness and the relative abundances of species. Both characteristics contribute to a community's species diversity (Figure 50.20).

- Organisms are classified as producers, consumers, detritivores, or decomposers. Ecologists depict the trophic structure (feeding relationships) of communities in food webs (Figure 50.21). Food-web analyses seek to identify generalities about trophic structure and its relationship to community stability.

Animation: Trophic levels in a simple food chain

Animation: Rain forest food web

50.4 Effects of Population Interactions on Community Characteristics

- Interspecific competition often affects the species composition and structure of communities.

- Predators may increase species richness by reducing the population size of the competitively most successful prey, thus allowing other prey species to occupy the community (Figure 50.22).

- Herbivores sometimes increase species richness and sometimes decrease it (Figure 50.23).

Animation: Effect of keystone species on diversity

50.5 Effects of Disturbance on Community Characteristics

- Environmental disturbances may eliminate populations from a community. Some communities, such as coral reefs, experience such frequent disturbance that their species composition is never at equilibrium (Figure 50.24).

- Disturbances of intermediate intensity and frequency allow both r-selected and K-selected species to occupy a site, increasing species richness (Figure 50.25).

50.6 Ecological Succession: Responses to Disturbance

- Ecological succession is a somewhat predictable change in species composition over time.

- Primary succession occurs on bare ground or rock (Figure 50.26). Secondary succession occurs where a community existed in the past (Figure 50.27).

- Species composition changes quickly and species richness rises rapidly during early successional stages. Early stages include short-lived r-selected species; later stages include long-lived K-selected species. Some communities eventually achieve a relatively stable climax state.

- Most communities include a mosaic of species that reflect patchiness in environmental conditions and the mixture of relatively undisturbed and recently disturbed sites.

Animation: Succession

50.7 Variations in Species Richness among Communities

- Communities near the equator have higher species richness than those near the poles (Figure 50.28). Explanations for this latitudinal gradient focus on either the origin or the maintenance of high species richness in the tropics.

- The equilibrium theory of island biogeography predicts that the number of species on an island represents a balance between the immigration of new species and the extinction of species already present (Figure 50.29). Studies show that large islands harbor more species than small islands and that islands near a mainland source have more species than distant islands (Figure 50.30).

Animation: Species diversity by latitude

Animation: Area and distance effects

Questions

Self-Test Questions

1. According to optimal foraging theory, predators:
   a. always feed on the largest prey possible.
   b. always feed on the prey that are easiest to catch.
   c. choose prey based on the costs of capturing and consuming it compared with the energy it provides.
   d. feed on plants when animal prey are scarce.
   e. have coevolved mechanisms to overcome prey defenses.

2. The use of the same limiting resource by two species is called:
   a. brood parasitism.
   b. interference competition.
   c. exploitative competition.
   d. mutualism.
   e. optimal foraging.

3. The range of resources that a population can possibly use is called:
   a. its fundamental niche.
   b. its realized niche.
   c. character displacement.
   d. resource partitioning.
   e. its relative abundance.

4. Differences in bill size of finch species living on the same island in the Galápagos may be caused by:
   a. predation.
   b. character displacement.
   c. mimicry.
   d. interference competition.
   e. cryptic coloration.

5. Bacteria that live in the human intestine assist digestion and feed on nutrients the human consumed. This relationship might best be described as:
   a. commensalism.
   b. mutualism.
   c. endoparasitism.
   d. ectoparasitism.
   e. predation.

6. The table below shows how many individuals were recorded for each of five species in five separate communities (a–e). Which community has the highest species diversity?

Community	Species 1	Species 2	Species 3	Species 4	Species 5
a.	90	10	0	0	0
b.	80	10	10	0	0
c.	25	25	25	25	0
d.	2	4	6	8	80
e.	20	20	20	20	20

7. A keystone species:
   a. is usually a primary producer.
   b. has a critically important role in determining the species composition of its community.
   c. is always a predator.
   d. usually reduces the species diversity in a community.
   e. usually exhibits aposematic coloration.

8. Species richness is often highest in communities where disturbances are:
   a. very frequent and severe.
   b. very frequent and of moderate intensity.
   c. very rare and severe.
   d. of intermediate frequency and moderate intensity.
   e. very rare and mild.

9. The change in the species composition of a community from bare and lifeless rock to climax vegetation is called:
   a. disturbance.
   b. competition.
   c. secondary succession.
   d. primary succession.
   e. facilitation.

10. The equilibrium theory of island biogeography predicts that the number of species found on an island:
    a. increases steadily until it equals the number in the mainland species pool.
    b. is greater on large islands than on small ones.
    c. is smaller on islands near the mainland than on distant islands.
    d. can never reach an equilibrium number.
    e. is greater for islands near the equator than for islands near the poles.

Questions for Discussion

1. Using the terms and concepts introduced in this chapter, describe the interactions that humans have with ten other species. Try to pick at least eight species that we do not eat.

2. After reading about the two potential biases in the scientific literature on competition, describe how future studies of competition might avoid such biases.

3. Humans are destroying natural communities at an ever-increasing pace. Using the predictions of the theory of island biogeography, develop hypotheses about what might happen as patches of natural habitats get smaller and smaller. How would you test these hypotheses?

Experimental Analysis

Chaparral, a community of woody shrubs that is fairly common in California, often grows adjacent to grassland. The two communities are consistently separated by a "bare zone," usually less than 1 m wide, where no vegetation of either type grows. Ecologists have proposed two possible explanations for this strip of bare soil: (1) that the leaves of chaparral shrubs release harmful, water-soluble chemicals that keep the grass seeds from germinating in the adjacent soil; and (2) that small mammals living in the dense cover provided by chaparral consume the grass seeds before they germinate; the animals don't venture very far from the shrubs because they would be easy targets for predatory hawks. Design a set of field experiments to test the two hypotheses.

Evolution Link

Five processes can foster microevolutionary change: gene flow, genetic drift, mutation, natural selection, and nonrandom mating (see Section 20.3). Which of those processes might contribute to the evolution of Batesian mimicry in two butterfly species? Would the same processes affect both the mimic and the model similarly? Which processes might have contributed to the evolution of the mutualistic relationship between ants and acacia trees, and how would their action on the two mutualists differ?

How Would You Vote?

Currently, only a fraction of the crates being imported into the United States are inspected for the inadvertent or deliberate presence of exotic species. Would the cost of added inspections be worth it? Go to www.thomsonedu.com/login to investigate both sides of the issue and then vote.

Silver Springs, Florida. This small river was the site of one of the earliest comprehensive studies of ecosystem structure and function.

© Mark J. Barrett 2005 www.markjbarrett.com

51 Ecosystems

STUDY PLAN

51.1 Energy Flow and Ecosystem Energetics

Sunlight provides the energy input for practically all ecosystems

Primary productivity varies greatly on global and local scales

Some stored energy is always lost before it is transferred from one trophic level to the next

Ecological pyramids illustrate the effects of energy losses

Consumers sometimes regulate ecosystem processes

51.2 Nutrient Cycling in Ecosystems

Ecologists describe nutrient cycling with a generalized compartment model

The hydrologic cycle recirculates all the water on Earth

The carbon cycle includes a large atmospheric reservoir

The nitrogen cycle depends upon the activity of diverse microorganisms

The phosphorus cycle includes a large sedimentary reservoir

51.3 Ecosystem Modeling

Ecologists use conceptual models and simulation models to understand ecosystem dynamics

WHY IT MATTERS

Poor Lake Erie, the shallowest of the Great Lakes. Several major industrial cities, including Toledo, Cleveland, Erie, and Buffalo, sprawl along its shoreline. Most of its water comes from the Detroit River, which flows past Detroit; the other rivers that flow into Lake Erie carry runoff from agricultural fields in Canada and the United States.

When Europeans first settled along its shores roughly 300 years ago, Lake Erie was a wetland paradise. Fishes and waterfowl reproduced in marshes and bays. Even after steel mills and oil refineries were built nearby in the 1860s and 1870s, the lake supported a busy fishing industry and was famous as a recreation area.

By 1970, wetlands had been filled for building; bays had been dredged for shipping lanes; and the shoreline had been converted to beaches. Worst of all, household sewage, industrial effluent, and agricultural runoff had so polluted the lake that it no longer supported the activities that had made it famous **(Figure 51.1).** The water was murky with algae and cyanobacteria; dead fishes washed up on the shore; local health departments closed beaches; and the fishing industry collapsed.

Figure 51.1
Pollution of Lake Erie. A steel mill in Lackawanna, New York, discharged industrial wastes into Lake Erie until 1983, when the mill was closed.

How can a vibrant natural resource become a foul smelling dump? The answer lies in the human activities that disrupt an **ecosystem**, a biological community and the physical environment with which it interacts. Between the 1930s and the 1970s, Lake Erie's concentration of phosphorus, which had been a limiting nutrient, tripled, largely from household detergents and agricultural fertilizers. High phosphorus concentrations encouraged the growth of photosynthetic algae, changing the phytoplankton community. The density of coliform bacteria, which originate in the human gut and serve as indicators of organic pollution, also skyrocketed as a result of the surge in sewage and nutrients entering the lake.

Increased phytoplankton and bacterial populations depleted oxygen in the lake's waters, contributing to changes elsewhere in the lake. Mayflies (*Hexagenia* species), whose larvae live in well-oxygenated bottom sediments, had once been so abundant that their aerial breeding swarms were a public nuisance. But they became nearly extinct in the polluted lake, replaced by oligochaete worms, snails, and other invertebrates. Along with overfishing, changes in the bottom fauna shifted the composition of the fish community; the catch of desirable food fishes declined to almost zero by the mid-1960s.

In 1972, Canada and the United States began efforts to restore the lake. They spent billions of dollars to reduce the influx of phosphates and limited fishing of the most vulnerable native species. Nonnative salmon (*Onchorhynchus* species) and other predatory fishes were introduced in the hope that they could bring the lake back to its original condition. Even the accidental introduction of zebra mussels *(Dreissina polymorpha)*, an aquatic pest, inadvertently helped the effort because they feed on phytoplankton.

But, although somewhat improved, Lake Erie will never return to its former glory. Some native species

are now extinct there, and the introduced species that replaced them function differently within the ecosystem. The lake still suffers periods of uncontrolled algal growth, fish kills, and high levels of harmful bacteria.

This story of an ecological disaster and partial recovery introduces ecosystem ecology, the branch of ecology that analyzes the flow of energy and the cycling of materials between an ecosystem's living and nonliving components. These processes make the resident organisms highly dependent on each other and on their physical surroundings. Ultimately, the Lake Erie ecosystem unraveled because human activities disrupted the flow of energy and the cycling of materials upon which the organisms depended.

51.1 Energy Flow and Ecosystem Energetics

Ecosystems receive a steady input of energy from an external source, which in virtually all cases is the sun. Energy flows through an ecosystem, but, as dictated by the laws of thermodynamics (see Section 4.1), much of it is lost without being used by organisms.

Food webs define the pathways by which energy moves through an ecosystem's biotic components (see Section 50.3). In most ecosystems, energy moves simultaneously through a *grazing food web* and a *detrital food web* **(Figure 51.2).** The grazing food web includes the producer, herbivore, and carnivore trophic levels. The detrital food web includes detritivores and decomposers. Because detritivores and decomposers subsist on the remains and waste products of organisms at every trophic level, the two food webs are closely interconnected. Detritivores also contribute to the grazing food web when carnivores eat them.

All of the organisms in a trophic level are the same number of energy transfers from the ecosystem's ultimate energy source. Plants are one energy transfer removed from sunlight; herbivores are two transfers away; carnivores feeding on herbivores are three transfers away; and carnivores feeding on other carnivores are four transfers away. In this section, we consider the details of energy flow and the efficiency of energy transfer from one trophic level to another.

Sunlight Provides the Energy Input for Practically All Ecosystems

Virtually all life on Earth depends on the input of solar energy. Every minute of every day, the atmosphere intercepts roughly 19 kcal of energy per square meter. (Recall from Chapter 2 that 1 kcal = 1000 calories.) About half that energy is absorbed, scattered, or reflected by gases, dust, water vapor, and clouds without ever reaching the planet's surface (see Chapter 52). Most energy that reaches the surface falls on bodies of water or bare ground, where it is absorbed as heat or

Laurence Lowry/Photo Researchers, Inc.

reflected back into the atmosphere; reflected energy warms the atmosphere, as we discuss later in this chapter. Only a small percentage contacts primary producers, and most of that energy evaporates water, driving transpiration in plants (see Section 32.3).

Ultimately, photosynthesis converts less than 1% of the solar energy that arrives at Earth's surface into chemical energy. But primary producers capture enough energy to create an average of several kilograms of dry plant material per square meter per year. On a global scale, they produce more than 150 billion metric tons of new biological material annually. Some of the solar energy that producers convert into chemical energy is transferred to consumers at higher trophic levels.

The rate at which producers convert solar energy into chemical energy is an ecosystem's **gross primary productivity.** But like all other organisms, producers also use energy for their own maintenance. After deducting the energy used for these functions, which are collectively called *cellular respiration* (see Section 8.1), whatever chemical energy remains is the ecosystem's **net primary productivity.** In most ecosystems, net primary productivity is between 50% and 90% of gross primary productivity. In other words, producers use between 10% and 50% of the energy they capture for their own respiration.

Ecologists generally measure primary productivity in units of energy captured ($kcal/m^2/yr$) or in units of biomass created ($g/m^2/yr$). *Biomass* is the dry weight of biological material per unit area or volume of habitat. (We measure biomass as the *dry* weight of organisms because their water content, which fluctuates with water uptake or loss, has no energetic or nutritional value.) You should not confuse an ecosystem's productivity with its **standing crop biomass,** the total dry weight of plants present at a given time. Net primary productivity is the *rate* at which the standing crop produces *new* biomass.

The energy captured by plants is stored in biological molecules—mostly carbohydrates, lipids, and proteins. Ecologists can convert units of biomass into

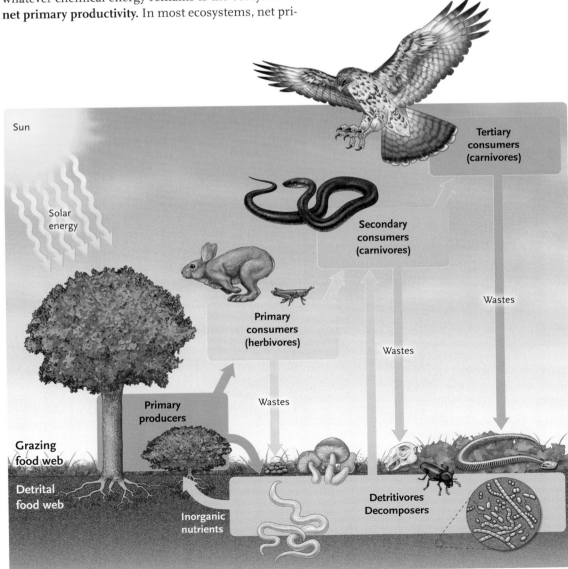

Figure 51.2
Grazing and detrital food webs. Energy and nutrients move through two parallel food webs in most ecosystems. The grazing food web includes producers, herbivores, and carnivores. The detrital food web includes detritivores and decomposers. Each box in this diagram represents many species, and each arrow represents many arrows.

Sun

Solar energy

Tertiary consumers (carnivores)

Secondary consumers (carnivores)

Wastes

Primary consumers (herbivores)

Wastes

Primary producers

Wastes

Grazing food web

Detrital food web

Inorganic nutrients

Detritivores Decomposers

Global variation in primary productivity. Satellite data for 1997–2000 provide a visual portrait of net primary productivity across Earth's surface. High-productivity regions on land are dark green; low-productivity regions are yellow. For the oceans, the highest productivity is red, down through orange, yellow, and green, with blue the lowest.

SeaWiFS Project, NASA/Goddard Space Flight Center and ORBIMAGE

units of energy or vice versa as long as they know how much carbohydrate, protein, and lipid a sample of biological material contains (4.2 kcal/g of carbohydrate; nearly 4.1 kcal/g of protein; and 9.5 kcal/g of lipid). Thus, net primary productivity is a measure of the rate at which producers accumulate energy as well as the rate at which new biomass is added to an ecosystem. Because it is far easier to measure biomass than energy content, ecologists usually measure changes in biomass to estimate productivity. New biomass takes several forms: the growth of existing producers; the creation of new producers by reproduction; and the storage of energy as carbohydrates. Because herbivores eat all three forms of new biomass, net primary productivity also measures how much new energy is available for primary consumers.

Primary Productivity Varies Greatly on Global and Local Scales

The potential rate of photosynthesis in any ecosystem is proportional to the intensity and the duration of sunlight, which vary geographically and seasonally (see

Chapter 52). Sunlight is most intense and day length least variable near the equator. By contrast, light intensity is weakest and day length most variable near the poles. Thus, producers at the equator can photosynthesize nearly 12 hours a day, every day of the year. Near the poles, photosynthesis is virtually impossible during the long, dark winter; in summer, however, plants can photosynthesize around the clock.

Sunlight is not the only factor that influences the rate of primary productivity, however; temperature as well as the availability of water and nutrients also have big effects. For example, many of the world's deserts receive plenty of sunshine but have low rates of productivity because water is in short supply and the soil is nutrient-poor. Thus, mean annual net primary productivity varies greatly on a global scale **(Figure 51.3)**, reflecting variations in these environmental factors (see Chapter 52).

On a finer geographical scale, within a particular terrestrial ecosystem, mean annual net productivity often increases with the availability of water **(Figure 51.4)**. In systems with sufficient water, a shortage of mineral nutrients may be limiting. All plants need specific ratios of macronutrients and micronutrients for maintenance and photosynthesis (see Section 33.1). But plants withdraw nutrients from soil, and if nutrient concentration drops below a critical level, photosynthesis may decrease or stop altogether. In every ecosystem, one nutrient inevitably runs out before the supplies of other nutrients are exhausted. The element in short supply is called a **limiting nutrient** because its absence limits productivity. Productivity in agricultural fields is subject to the same constraints as productivity in natural ecosystems. Farmers increase productivity by irrigating (adding water to) and fertilizing (adding nutrients to) their crops.

In freshwater and marine ecosystems, where water is always readily available, the depth of the water and the combined availability of sunlight *and* nutrients govern the rate of primary productivity. Productivity is high in near-shore ecosystems where sunlight pene-

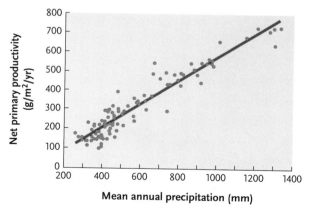

Figure 51.4

Water and net primary productivity. Mean annual net primary productivity increases with mean annual precipitation among 100 sites in the Great Plains of North America. These data include only aboveground productivity.

trates shallow, nutrient-rich waters. Kelp beds and coral reefs, for example, which occur along temperate and tropical coastlines respectively, are among the most productive ecosystems on Earth **(Table 51.1)**. By contrast, productivity is low in the open waters of a large lake or ocean: sunlight penetrates only the upper layers, and nutrients sink to the bottom. Thus, the two requirements for photosynthesis, sunlight and nutrients, are available in different places.

Although ecosystems vary in their net primary productivity, the differences are not always proportional to variations in their standing crop biomass (see Table 51.1). For example, biomass in temperate deciduous forests and temperate grasslands differs by a factor of 20, but the difference in their rates of net primary productivity is only twofold. Most biomass in trees is present in non-photosynthetic tissues such as wood. As a result, their ratio of productivity to biomass is low (1200 g/m^2 ÷ 30,000 g/m^2 = 0.040). By contrast, grasslands don't accumulate much biomass because annual mortality, herbivores, and fires remove plant material as it is produced; and their productivity to biomass ratio is much higher (600 g/m^2 ÷ 1600 g/m^2 = 0.375).

Some ecosystems contribute more than others to overall net primary productivity **(Figure 51.5)**. Ecosystems that cover large areas make substantial contributions, even if their productivity is low. Conversely, geographically restricted ecosystems make large contributions if their productivity is high. For example, the open ocean and tropical rain forests contribute about equally to total global productivity, but for different reasons. Open oceans have low productivity, but they cover nearly two-thirds of Earth's surface. Tropical rain forests cover only a small area, but they are highly productive.

Table 51.1	Standing Crop Biomass and Net Primary Productivity of Different Ecosystems	
Ecosystem	Mean Standing Crop Biomass (g/m^2)	Mean Net Primary Productivity (g/m^2/yr)
Terrestrial Ecosystems		
Tropical rain forest	45,000	2,200
Tropical deciduous forest	35,000	1,600
Temperate rain forest	35,000	1,300
Temperate deciduous forest	30,000	1,200
Savanna	4,000	900
Boreal forest (taiga)	20,000	800
Woodland and shrubland	6,000	700
Agricultural land	1,000	650
Temperate grassland	1,600	600
Tundra and alpine tundra	600	140
Desert and thornwoods	700	90
Extreme desert, rock, sand, ice	20	3
Freshwater Ecosystems		
Swamp and marsh	15,000	2,000
Lake and stream	20	250
Marine Ecosystems		
Open ocean	3	125
Upwelling zones	20	500
Continental shelf	10	360
Kelp beds and reefs	2,000	2,500
Estuaries	1,000	1,500
World Average	**3,600**	**333**

From Whittaker, R.H. 1975. *Communities and Ecosystems*. 2nd ed. Macmillan.

Some Stored Energy Is Always Lost before It Is Transferred from One Trophic Level to the Next

Net primary productivity ultimately supports all the consumers in grazing and detrital food webs. Consumers in the grazing food web eat some of the biomass at every trophic level except the highest; uneaten biomass eventually dies and passes into detrital food webs. However, consumers assimilate only a portion of the material they ingest, and unassimilated material is passed as feces, which also supports detritivores and decomposers.

As energy is transferred from producers to consumers, some is stored in new consumer biomass, called **secondary productivity**. Nevertheless, two factors cause energy to be lost from the ecosystem every time it flows from one trophic level to another. First, animals use much of the energy they assimilate for maintenance or locomotion rather than the production of new biomass. Second, as dictated by the second law of thermodynamics, no biochemical reaction is 100% effi-

cient; thus, some of the chemical energy liberated by cellular respiration is always converted to heat, which most organisms do not use.

Ecological efficiency is the ratio of net productivity at one trophic level to net productivity at the trophic level below it. For example, if the plants in an ecosystem have a net primary productivity of 100 g/m^2/year of new tissue and the herbivores that eat those plants produce 10 g/m^2/year, the ecological efficiency of the herbivores is 10%. The efficiencies of three processes—harvesting food, assimilating ingested energy, and producing new biomass—determine the ecological efficiencies of consumers.

Harvesting efficiency is the ratio of the energy content of food consumed to the energy content of food available. Predators harvest food efficiently when prey are abundant and easy to capture (see Section 50.1).

Assimilation efficiency is the ratio of the energy absorbed from consumed food to the food's total energy content. Because animal prey is relatively easy to digest,

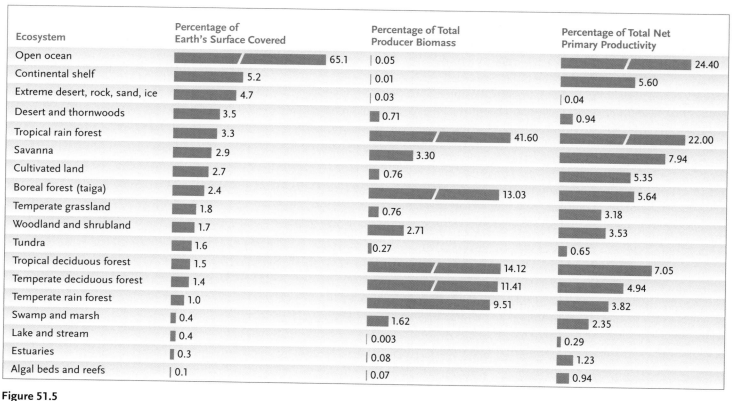

Ecosystem	Percentage of Earth's Surface Covered	Percentage of Total Producer Biomass	Percentage of Total Net Primary Productivity
Open ocean	65.1	0.05	24.40
Continental shelf	5.2	0.01	5.60
Extreme desert, rock, sand, ice	4.7	0.03	0.04
Desert and thornwoods	3.5	0.71	0.94
Tropical rain forest	3.3	41.60	22.00
Savanna	2.9	3.30	7.94
Cultivated land	2.7	0.76	5.35
Boreal forest (taiga)	2.4	13.03	5.64
Temperate grassland	1.8	0.76	3.18
Woodland and shrubland	1.7	2.71	3.53
Tundra	1.6	0.27	0.65
Tropical deciduous forest	1.5	14.12	7.05
Temperate deciduous forest	1.4	11.41	4.94
Temperate rain forest	1.0	9.51	3.82
Swamp and marsh	0.4	1.62	2.35
Lake and stream	0.4	0.003	0.29
Estuaries	0.3	0.08	1.23
Algal beds and reefs	0.1	0.07	0.94

Figure 51.5

Biomass and net primary productivity. The percentage of Earth's surface that an ecosystem covers is not proportional to its contribution to the total biomass of producers or its contribution to the total net primary productivity.

carnivores absorb between 60% and 90% of the energy in their food; assimilation efficiency is lower for prey with indigestible parts like bones or exoskeletons. Herbivores assimilate only 15% to 80% of the energy they consume because cellulose is not very digestible.

Production efficiency is the ratio of the energy content of new tissue produced to the energy assimilated from food. Production efficiency varies with maintenance costs. For example, endothermic animals often use less than 10% of their assimilated energy for growth and reproduction, because they use energy to generate body heat (see Section 46.8). Ectothermic animals, by contrast, channel more than 50% of their assimilated energy into new biomass.

The overall ecological efficiency of most organisms is between 5% and 20%. As a rule of thumb, only about 10% of the energy accumulated at one trophic level is converted into biomass at the next higher trophic level, as illustrated by energy transfers at Silver Springs, Florida **(Figure 51.6)**. Producers in the Silver Springs ecosystem convert 1.2% of the solar energy they intercept into chemical energy (represented by 20,810 kcal of gross primary productivity). However, they use about two-thirds of this energy for respiration, leaving only one-third to be included in new plant biomass, the net primary productivity. All consumers in the grazing food web (on the right in Figure 51.6) ultimately depend on this energy source, which dwindles with each transfer between trophic levels. Energy is lost to respiration and export (that is, the transport of energy-containing materials out of the ecosystem by flowing water) at each trophic level. In addition, substantial energy, represented in organic wastes and uneaten biomass, flows into the detrital food web (on the left in Figure 51.6). To determine the ecological efficiency of any trophic level, we divide its productivity by the productivity of the level below it. For example, the ecological efficiency of midlevel carnivores at Silver Springs is 111 kcal/yr ÷ 1103 kcal/yr =10.06%.

As energy works its way up a food web, energy losses are multiplied in successive energy transfers, greatly reducing the energy available to support the highest trophic levels. Consider a hypothetical example in which ecological efficiency is 10% for all consumers. Assume that the plants in a small field annually produce new tissues containing 100 kcal of energy. Because only 10% of that energy is transferred to new herbivore biomass, the 100 kcal in plants produces only 10 kcal of new herbivorous insects; only 1 kcal of new songbirds, which feed on insects; and only 0.1 kcal of new falcons, which feed on songbirds. Thus, after three energy transfers, only 0.1% of the energy from primary productivity remains at the highest trophic levels. If the energy available to each trophic level is depicted graphically, the result is a **pyramid of energy** with primary producers on the bottom and higher-level consumers on the top. We discuss ecological pyramids in detail in the next section.

The low ecological efficiencies that characterize most energy transfers illustrate one advantage of eating "lower on the food chain." Even though humans digest and assimilate meat more efficiently than vegetables, we might be able to feed more people if we all ate more vegetables directly instead of first passing

Figure 51.6 Observational Research

Energy Flow in the Silver Springs Ecosystem

METHOD: Howard T. Odum and his research team analyzed energy flow in an aquatic ecosystem at Silver Springs, Florida. The producers in this small spring are mostly aquatic plants. The herbivores include snails, shrimp, insects, fishes, and turtles. The carnivores include a variety of invertebrates and fishes. The top carnivores are large fish. Sunlight is available as an energy source all year round. After defining the food web in this ecosystem, researchers estimated the biomass and energy content (kcal/g) of each trophic level. They then constructed a diagram that illustrates how much energy is present at each trophic level and how much energy is lost as it works its way through the food web.

HYPOTHESIS: Only a small percentage of the energy present in a trophic level is transferred to the next higher trophic level in the ecosystem.

PREDICTION: The energy content of the organisms present in each trophic level will decline steadily from the lowest to highest trophic levels.

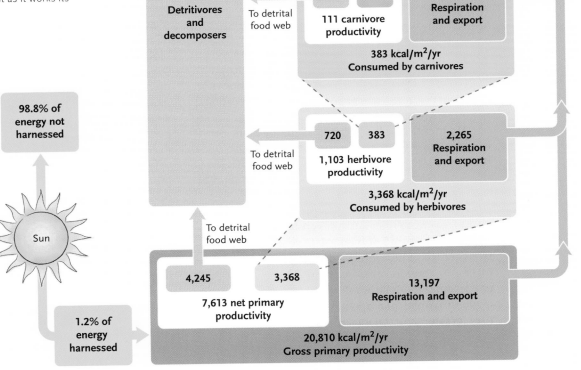

RESULTS: The diagram illustrates annual energy flow for the spring ecosystem at Silver Springs, Florida. Numbers on the diagram indicate the quantity of energy (kcal/m^2/yr). Because the ecosystem is based on flowing water, small quantities of energy arrive from other ecosystems and small quantities are exported in material carried away by stream flow.

CONCLUSION: The study confirmed the hypothesis that only a small proportion of the energy present at a trophic level is transferred to the next higher trophic level. Ultimately, all of the energy that passes through the grazing and detrital food webs is released as metabolically generated heat.

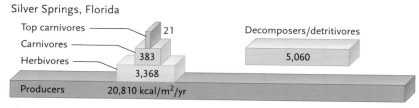

Silver Springs, Florida

Figure 51.7
Pyramids of energy. The pyramid of energy for Silver Springs, Florida, shows that the amount of energy passing through each trophic level decreases as it moves up the food web.

these crops through another trophic level, such as cattle or chickens, to produce meat. The production of animal protein is costly because much of the energy fed to livestock is used for their own maintenance rather than the production of new biomass. But despite the economic—not to mention health-related—logic of a more vegetarian diet, a change in our eating habits alone won't eliminate food shortages or the frequency of malnutrition. Many regions of Africa, Australia, North America, and South America support vegetation that is suitable only for grazing by large herbivores. These areas could not produce significant quantities of edible grains and vegetables.

Ecological Pyramids Illustrate the Effects of Energy Losses

The inefficiency of energy transfer from one trophic level to the next has profound effects on ecosystem structure. Ecologists illustrate these effects in diagrams called **ecological pyramids.** Trophic levels are drawn as stacked blocks, with the size of each block proportional to the energy, biomass, or numbers of organisms present. We mentioned the pyramid of energy in the previous section. Pyramids of energy typically have wide bases and narrow tops **(Figure 51.7)** because each trophic level contains only about 10% as much energy as the trophic level below it.

The progressive reduction in productivity at higher trophic levels, as illustrated in Figure 51.6, usually establishes a **pyramid of biomass (Figure 51.8).** The biomass at each trophic level is proportional to the chemical energy temporarily stored there. Thus, in terrestrial ecosystems, the total mass of producers is generally greater than the total mass of herbivores, which is, in turn, greater than the total mass of predators (see Fig-

ure 51.8a). Populations of top predators—animals like mountain lions or alligators—contain too little biomass and energy to support another trophic level; thus, they have no nonhuman predators.

Freshwater and marine ecosystems sometimes exhibit inverted pyramids of biomass (see Figure 51.8b). In the open waters of a lake or ocean, primary consumers (zooplankton) eat the primary producers (phytoplankton) almost as soon as they are produced. As a result, the standing crop of primary consumers at any moment in time is actually larger than the standing crop of primary producers. Food webs in these ecosystems are stable, however, because the producers have exceptionally high **turnover rates.** In other words, the producers divide and their populations grow so quickly that feeding by zooplankton doesn't endanger their populations or reduce their productivity. And on an annual basis, the *cumulative total* biomass of primary producers far outweighs that of primary consumers.

The reduction of energy and biomass also affects the population sizes of organisms at the top of a food web. Top predators are often relatively large animals. Thus, the limited biomass present in the highest trophic levels is concentrated in relatively few animals **(Figure 51.9).** The extremely narrow top of this **pyramid of numbers** has grave implications for conservation biology. Top predators tend to be large animals with small population sizes. And because each individual must patrol a large area to find sufficient food, the members of a population are often widely dispersed within their habitats. As a result, they are subject to genetic drift (see Section 20.3) and are highly sensitive to hunting, habitat destruction, and random events, which can lead to extinction (see Chapter 53). Top predators may also suffer from the accumulation of poisonous materials that move through food webs (see *Focus on Research* on biological magnification in Chapter 53). Even predators that feed below the top trophic level often suffer the ill effects of human activities. *Insights from the Molecular Revolution* describes how researchers determined that fishing diminishes fragile populations of loggerhead sea turtles *(Caretta caretta),* a predator that routinely travels from one ecosystem to another.

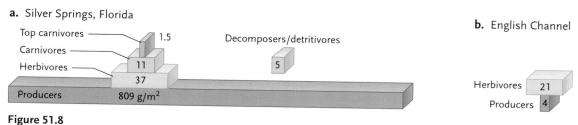

a. Silver Springs, Florida

b. English Channel

Figure 51.8
Pyramids of biomass. **(a)** The pyramid of standing crop biomass for Silver Springs is bottom heavy, as it is for most ecosystems. **(b)** Some marine ecosystems, such as that in the English Channel, have an inverted pyramid of biomass because producers are quickly eaten by primary consumers. Only the producer and herbivore trophic levels are illustrated here. The data for both pyramids are given in grams of dry biomass per square meter.

a. Grassland (summer)

1 | 90,000 | 200,000 | 1,500,000

b. Temperate forest (summer)

Top carnivores — 2
Carnivores — 120,000
Herbivores — 150,000
Producers — 200

Figure 51.9
Pyramids of numbers. **(a)** The pyramid of numbers (number of individuals per 1000 m²) for temperate grasslands is bottom-heavy because individual producers are small and very numerous. **(b)** The pyramid of numbers for forests may have a narrow base because herbivorous insects often outnumber the producers, which are large trees. Data for both pyramids were collected in summer. Detritivores and decomposers (soil animals and microorganisms) are not included because they are difficult to count.

Consumers Sometimes Regulate Ecosystem Processes

As you know from the preceding discussion, numerous abiotic factors—the intensity and duration of sunlight, rainfall, temperature, and the availability of nutrients—have significant effects on primary productivity. Primary productivity, in turn, has profound effects on populations of herbivores and the predators that feed on them. But what effect does feeding by these consumers have on primary productivity?

Research conducted in the 1990s suggests that consumers may sometimes influence rates of primary productivity, especially in ecosystems with low species diversity and relatively few trophic levels. For example, food webs in lake ecosystems depend primarily on the productivity of phytoplankton **(Figure 51.10)**. These producers are consumed by herbivorous zooplankton, which are in turn eaten by predatory invertebrates and fishes. The top nonhuman carnivore in these food webs is usually a predatory fish.

Herbivorous zooplankton play a central role in the regulation of lake ecosystems. Small zooplankton species consume only small phytoplankton. Thus, when small zooplankton are especially abundant, the large phytoplankton escape predation and survive, and the lake's primary productivity is high. By contrast, large zooplankton are voracious, eating both small and large phytoplankton. When *large* zooplankton are especially abundant, they reduce the overall biomass of phytoplankton, lowering the ecosystem's primary productivity.

In what has been termed a **trophic cascade**—predator–prey effects that reverberate through the population interactions at two or more trophic levels in an ecosystem—feeding by plankton-eating invertebrates and fishes has a *direct* impact on herbivorous zooplankton populations and an *indirect* impact on phytoplankton populations and the ecosystem's primary productivity. Invertebrate predators prefer small zooplankton. And when the invertebrates that eat small zooplankton are the dominant carnivores in the ecosystem, large zooplankton become more abundant;

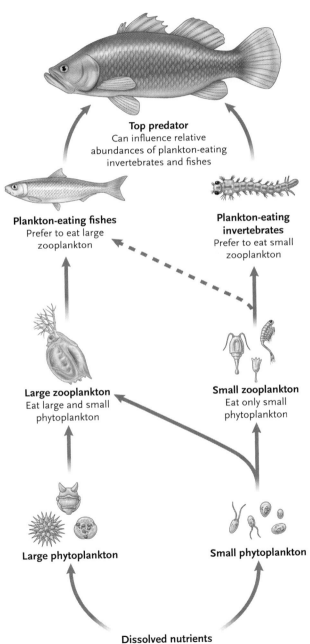

Top predator
Can influence relative abundances of plankton-eating invertebrates and fishes

Plankton-eating fishes
Prefer to eat large zooplankton

Plankton-eating invertebrates
Prefer to eat small zooplankton

Large zooplankton
Eat large and small phytoplankton

Small zooplankton
Eat only small phytoplankton

Large phytoplankton

Small phytoplankton

Dissolved nutrients

Figure 51.10
Consumer regulation of primary productivity. A simplified food web illustrates that lake ecosystems have relatively few trophic levels. The effects of feeding by top carnivores can cascade downward, exerting an indirect effect on phytoplankton and, thus, on primary productivity.

Fishing Fleets at Loggerheads with Sea Turtles

Populations of loggerhead sea turtles (*Caretta caretta*) that nest on Western Pacific beaches in Australia and Japan have been in decline. A surprising recent discovery indicates that the explanation may lie many thousands of miles away. Loggerhead sea turtles hatch from eggs that females bury on sandy beaches. The hatchlings then scurry to the surf and migrate to distant feeding grounds. The turtles mature at the feeding grounds, and eventually return to their hatching beaches to lay eggs.

Recently, a population of loggerhead sea turtles was discovered feeding along the coast of Baja California. Nesting grounds for these turtles are known only in the western Pacific, in Australia and Japan; none had been identified in the eastern Pacific. Did these turtles really migrate across 10,000 km of open ocean from Japan and Australia to Baja California? If so, the trip would be the longest open ocean migration known for any marine animal.

In addition, this long journey might explain the decline of the turtles in Japan and Australia. Scientists know that as many as 4000 loggerhead turtles drown in fishing nets in the north Pacific each year. Are these turtles intercepted on their way to Baja California from the Australian or Japanese feeding grounds? If so, the large numbers caught in fishing nets may contribute to the decline of loggerhead populations in the western Pacific.

Brian W. Bowen and his colleagues at the University of Florida, Gainesville, used mitochondrial DNA (mtDNA) sequences to answer these questions. One 350-base-pair segment of mtDNA was particularly useful because it includes sequence variations that are characteristic of different loggerhead populations.

The investigators took DNA samples from nesting populations in Australia and Japan, from feeding populations in Baja California, and from turtles drowned in fishing nets in the north Pacific. They used the polymerase chain reaction (PCR) to amplify the mtDNA segment from the DNA samples. Sequencing of the amplified segments revealed three major variants of mtDNA, which the researchers designated sequences A, B, and C. The sequences were distributed among loggerhead turtles as shown in the accompanying table.

The mtDNA of most turtles found in Baja California and in fishing nets in the north Pacific match that of turtles from the Japanese nesting areas, supporting the idea that loggerhead turtles hatched in Japan make the long migration across the north Pacific to Baja California. The data also indicate that a few turtles hatched in Australia may follow the same migratory route.

The investigators propose that the North Pacific Current, which moves from west to east, aids the migration. The return trip from Baja to Japan could be made via the North Equatorial Current, which runs from east to west just north of the equator. Loggerhead turtles have been found in this current; further tests will reveal whether they have the mtDNA sequence characteristic of the individuals nesting in Japan and feeding in Baja California.

Because only 2000 to 3000 female loggerhead turtles nest in Japan, it is uncertain whether the Japanese nesting population can survive the loss of thousands of offspring to fishing in the north Pacific. The number of female loggerhead turtles nesting in Australia has declined by 50% to 80% in the last decade; the loss of only a few individuals in fishing nets could have a drastic impact on this population as well. To save the loggerhead turtles, wildlife managers and international agencies must establish and enforce limits on the number of migrating individuals trapped and killed in the ocean fisheries.

Location	Sequence A	Sequence B	Sequence C
Australian nesting areas	26 turtles	0 turtles	0 turtles
Japanese nesting areas	0 turtles	23 turtles	3 turtles
Baja California feeding grounds	2 turtles	19 turtles	5 turtles
North Pacific	1 turtle	28 turtles	5 turtles

they consume many phytoplankton, causing productivity to decrease. By contrast, zooplankton-eating fishes prefer to eat large zooplankton (see Figure 50.2). Thus, when plankton-eating fishes are abundant, small zooplankton become the dominant herbivores. As a result, large phytoplankton become more numerous and the lake's productivity rises.

Large predatory fishes may add an additional level of control to the system because they feed on and regulate the population sizes of plankton-eating invertebrates and fishes. Thus, the effects of feeding by the top predator can cascade downward through the food web, affecting the densities of plankton-eating invertebrates and fishes, herbivorous zooplankton, and phytoplankton.

STUDY BREAK

1. What is the difference between gross primary productivity and net primary productivity?
2. What environmental factors influence rates of primary productivity in terrestrial and aquatic ecosystems?
3. Why is energy lost from an ecosystem at every transfer from one trophic level to the trophic level above it?
4. How can the presence of a top predator influence the interactions of organisms at lower trophic levels and an ecosystem's productivity?

51.2 Nutrient Cycling in Ecosystems

The availability of nutrients is as important to ecosystem function as the input of energy. Photosynthesis—the conversion of solar energy into chemical energy—requires carbon, hydrogen, and oxygen, which producers acquire from water and air. Producers also need nitrogen, phosphorus, and other minerals (see Table 33.1). A deficiency in any of these minerals can reduce primary productivity.

Earth is essentially a closed system with respect to matter. Thus, unlike energy, for which there is a constant cosmic input, virtually all the nutrients that will ever be available for biological systems are already present. Nutrient ions or molecules constantly circulate between the abiotic environment and living organisms in what ecologists describe as **biogeochemical cycles.** And unlike energy, which flows through ecosystems and is gradually lost as heat, matter is conserved in biogeochemical cycles. Although there may be local shortages of specific nutrients, Earth's overall supplies of these chemical elements are never depleted.

Ecologists Describe Nutrient Cycling with a Generalized Compartment Model

Nutrients take various forms as they pass through biogeochemical cycles. Some materials, such as carbon, nitrogen, and oxygen, form gases, which move through global *atmospheric cycles.* Geological processes move other materials, such as phosphorus, through local *sedimentary cycles,* carrying them between dry land and the seafloor. Rocks, soil, water, and air are the reservoirs where mineral nutrients accumulate, sometimes for many years.

Ecologists use a **generalized compartment model** to describe nutrient cycling **(Figure 51.11).** Two criteria divide ecosystems into four compartments where nutrients accumulate. First, nutrient molecules and ions are described as either *available* or *unavailable,* depending upon whether or not they can be assimilated by organisms. Second, nutrients are present either in *organic* material, the living or dead tissues of organisms, or in *inorganic* material, such as rocks and soil. For example, minerals in dead leaves on the forest floor are in the available-organic compartment because they are in the remains of organisms that can be eaten by detritivores. But calcium ions in limestone rocks are in the unavailable-inorganic compartment because they exist in a nonbiological form that producers cannot assimilate.

Nutrients move rapidly within and between the available compartments. Living organisms are in the available-organic compartment, and whenever heterotrophs consume food, they recycle nutrients within that reservoir (indicated by the oval arrow in the upper left of Figure 51.11). Producers acquire nutrients from the air, soil, and water of the available-inorganic compartment. Consumers also acquire nutrients from the available-inorganic compartment when they drink water or absorb mineral ions through the body surface. Several processes routinely transfer nutrients from organisms to the available-inorganic compartment. As one example, respiration releases carbon dioxide, moving both carbon and oxygen from the available-organic compartment to the available-inorganic compartment.

By contrast, the movement of materials into and out of the unavailable compartments is generally slow. Sedimentation, a long-term geological process, converts ions and particles of the available-inorganic compartment into rocks of the unavailable-inorganic compartment. Materials are gradually returned to the available-inorganic compartment when rocks are uplifted and eroded or weathered. Similarly, over millions of years, the remains of organisms in the available-organic compartment were converted into coal, oil, and peat of the unavailable-organic compartment.

Except for the input of solar energy, we have described energy flow and nutrient cycling as though ecosystems were closed systems. In fact, most ecosystems exchange energy and nutrients with neighboring ecosystems. For example, rainfall carries nutrients into a forest ecosystem, and runoff carries nutrients from a forest into a lake or river. Ecologists have mapped the biogeochemical cycles of important elements, often by

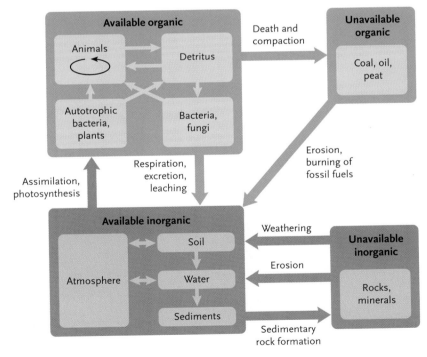

Figure 51.11

A generalized compartment model of nutrient cycling. Nutrients cycle through four major compartments within ecosystems. Processes that move nutrients from one compartment to another are indicated on the arrows. The oval arrow in the upper left corner of the figure represents animal predation on other animals.

using radioactively labeled molecules that they can follow in the environment. As you study the details of the four biogeochemical cycles described below, try to understand them in terms of the generalized compartment model of nutrient cycling.

The Hydrologic Cycle Recirculates All the Water on Earth

Although it is not a mineral nutrient, water is the universal intracellular solvent for biochemical reactions. Nevertheless, only a fraction of 1% of Earth's total water is present in biological systems at any time.

The cycling of water, called the **hydrologic cycle**, is global, with water molecules moving from the ocean into the atmosphere, to the land, through freshwater ecosystems, and back to the ocean **(Figure 51.12).** Solar energy causes water to evaporate from oceans, lakes, rivers, soil, and living organisms, entering the atmosphere as a vapor and remaining aloft as a gas, as droplets in clouds, or as ice crystals. It falls as precipitation, mostly in the form of rain and snow. When precipitation falls on land, water flows across the surface or percolates to great depth in the soil, eventually reentering the ocean reservoir through the flow of streams and rivers.

The hydrologic cycle maintains its global balance because the total amount of water that enters the atmosphere is equal to the amount that falls as precipitation. Most water that enters the atmosphere evaporates from the ocean, which represents the largest reservoir on the planet. A much smaller fraction evaporates from terrestrial ecosystems, and most of that results from transpiration in green plants.

The constant recirculation provides fresh water to terrestrial organisms and maintains freshwater ecosystems such as lakes and rivers. Water also serves as a transport medium that moves nutrients within and between ecosystems, as demonstrated in a series of classic experiments in the Hubbard Brook Experimental Forest, described in *Focus on Basic Research*.

The Carbon Cycle Includes a Large Atmospheric Reservoir

Carbon atoms provide the backbone of most biological molecules, and carbon compounds store the energy captured by photosynthesis (see Section 9.1). Carbon enters food webs when producers convert atmospheric carbon dioxide (CO_2) into carbohydrates. Heterotrophs acquire carbon by eating other organisms or detritus.

a. The water cycle

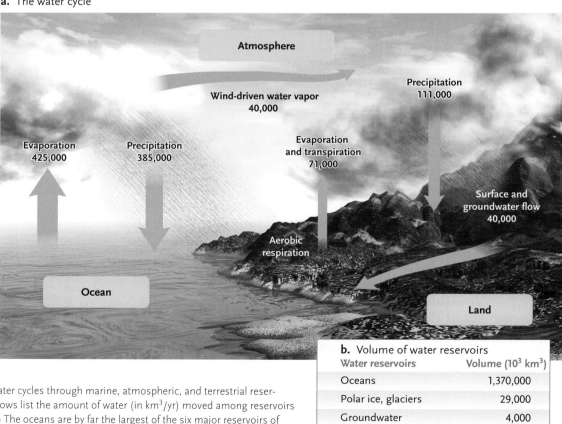

Figure 51.12
The hydrologic cycle. Water cycles through marine, atmospheric, and terrestrial reservoirs. **(a)** Data on the arrows list the amount of water (in km³/yr) moved among reservoirs by various processes. **(b)** The oceans are by far the largest of the six major reservoirs of water on Earth.

b. Volume of water reservoirs

Water reservoirs	Volume (10³ km³)
Oceans	1,370,000
Polar ice, glaciers	29,000
Groundwater	4,000
Lakes, rivers	230
Soil moisture	67
Atmosphere (water vapor)	14

Basic Research: Studies of the Hubbard Brook Watershed

Because water always flows downhill, local topography affects the movement of dissolved nutrients in terrestrial ecosystems. A **watershed** is an area of land from which precipitation drains into a stream or river system. Thus, each watershed represents a part of an ecosystem from which nutrients exit through a single outlet, much the way a bathtub empties through a single drain. When several streams join to form a river, the watershed drained by the river encompasses all of the smaller watersheds drained by the streams. For example, the Mississippi River watershed covers roughly one-third of the United States, and it includes watersheds drained by the Illinois, Missouri, and Tennessee Rivers as well as many other watersheds drained by smaller streams and rivers.

Because watersheds are relatively self-contained units, they are ideal for large-scale field experiments about nutrient flow in ecosystems. Herbert Bormann of Yale University and Gene Likens of Cornell University have conducted a classic experiment on this topic since the 1960s. Bormann and Likens manipulated small watersheds of temperate deciduous forest in the Hubbard Brook Experimental Forest in the White Mountain National Forest of New Hampshire. They measured precipitation and nutrient input into the watersheds, the uptake of nutrients by vegetation, and the amount of nutrients leaving the watershed via streamflow. Nutrients exported in streamflow were monitored in water samples collected from V-shaped concrete weirs built into bedrock below the streams that drained the watersheds **(Figure a).** Impermeable bedrock underlies the soil, preventing water from leaving the system by deep seepage.

After collecting several years of baseline data on six undisturbed watersheds, the researchers cut all the trees in one small watershed in 1965 and 1966. They also applied herbicides to prevent regrowth. After establishing this experimental treatment, they monitored the output of nutrients in streams that drained experimental and control watersheds. They attributed differences in nutrient export between undisturbed watersheds (controls) and the clear-cut watershed (experimental treatment) to the effects of deforestation.

Bormann and Likens determined that vegetation absorbed substantial water and conserved nutrients in undisturbed watersheds. Plants used about 40% of the precipitation for transpiration. The rest contributed to runoff and groundwater. Control watersheds lost only about 8–10 kg of calcium per hectare each year, an amount that was replaced by the erosion of bedrock and input from rain. Moreover, control watersheds actually accumulated about 2 kg of nitrogen per hectare per year and slightly smaller amounts of potassium.

By contrast, the experimentally deforested watershed experienced a 40% annual increase in runoff. During a 4-month period in the summer, runoff increased 300%. Some mineral losses were similarly large. The net loss of calcium was 10 times higher than in the control watersheds **(Figure b)** and the loss of potassium 21 times higher. Phosphorus losses did not increase;

Figure a
Weir used to measure the volume and nutrient content of water leaving a watershed by streamflow.

this mineral was apparently retained by the soil. However, the loss of nitrogen was an astronomical 120 kg per hectare per year. So much nitrogen entered the stream draining the experimental watershed that the stream became choked with algae and cyanobacteria. Thus, the results of the Hubbard Brook experiment suggest that deforestation increases flooding and decreases the fertility of ecosystems.

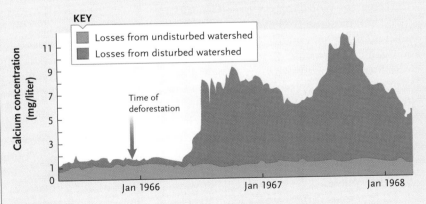

KEY
Losses from undisturbed watershed
Losses from disturbed watershed

Figure b
Calcium losses from a deforested watershed were much greater than those from controls. The arrow indicates the time of deforestation in early winter. Mineral losses did not increase until after the ground thawed the following spring; increased runoff also caused large water losses from the watershed.

c. The global carbon cycle

a. Amount of carbon in major reservoirs

Carbon reservoirs	Mass (10^{15} g)
Sediments and rocks	77,000,000
Ocean (dissolved forms)	39,700
Soil	1,500
Atmosphere	750
Biomass on land	715

b. Annual global carbon movement between reservoirs

Direction of movement	Mass (10^{15} g)
From atmosphere to plants (carbon fixation)	120
From atmosphere to ocean	107
To atmosphere from ocean	105
To atmosphere from plants	60
To atmosphere from soil	60
To atmosphere from burning fossil fuel	5
To atmosphere from burning plants	2
To ocean from runoff	0.4
Burial in ocean sediments	0.1

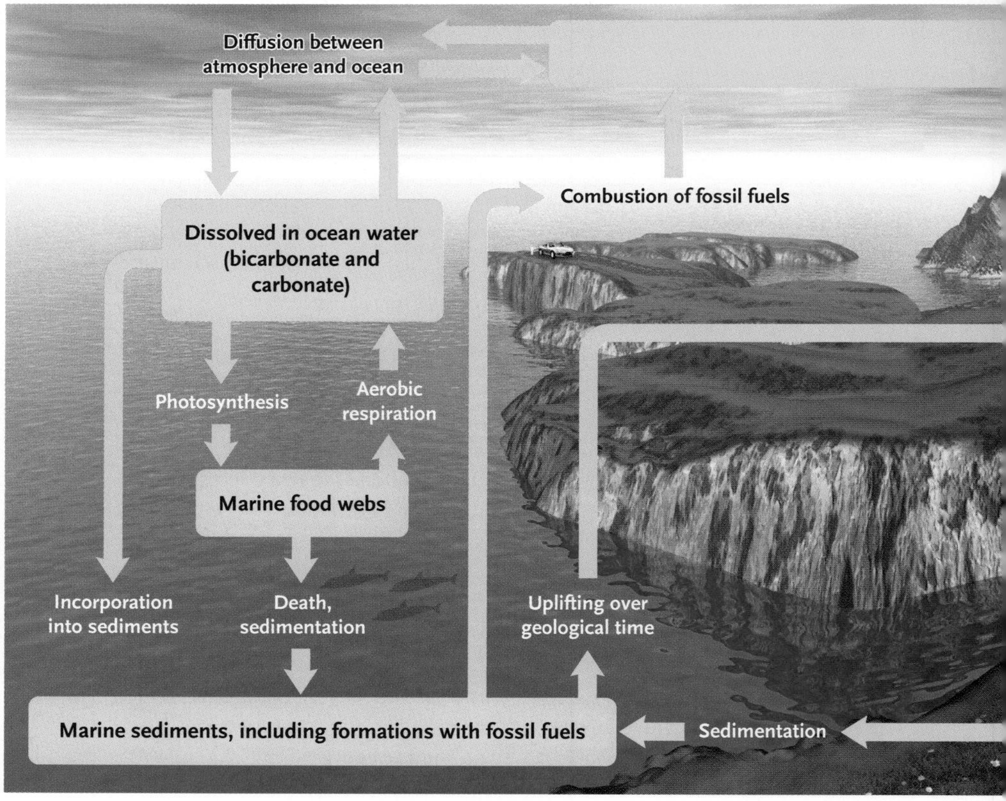

Figure 51.13

The carbon cycle. Marine and terrestrial components of the global carbon cycle are linked through an atmospheric reservoir of carbon dioxide. **(a)** By far, the largest amount of Earth's carbon is found in sediments and rocks. **(b)** Earth's atmosphere mediates most movements of carbon. **(c)** In this illustration of the carbon cycle, boxes identify major reservoirs, and labels on the arrows identify the processes that cause carbon to move between reservoirs.

Although carbon moves somewhat independently in the sea and on land, a common atmospheric pool of CO_2 creates a global **carbon cycle (Figure 51.13).**

The largest reservoir of carbon is sedimentary rock, such as limestone or marble. Rocks are in the unavailable-inorganic compartment, and they exchange carbon with living organisms at an exceedingly slow pace. Most *available* carbon is present as dissolved bicarbonate ions (HCO_3^-) in the ocean. Soil, the atmosphere, and plant biomass form other significant, but much smaller, reservoirs of available carbon. Atmospheric carbon is mostly in the form of molecular CO_2, a product of aerobic respiration. Volcanic eruptions also release CO_2 into the atmosphere.

Sometimes carbon atoms leave the organic compartments for long periods of time. Some organisms in marine food webs build shells and other hard parts by incorporating dissolved carbon into calcium carbonate ($CaCO_3$) and other insoluble salts. When shelled organisms die, they sink to the bottom and are buried in sediments. The insoluble carbon that accumulates as rock in deep sediments may remain buried for millions of years before tectonic uplifting brings it to the surface, where erosion and weathering dissolve sedimentary rocks and return carbon to an available form.

Carbon atoms were also transferred to the unavailable-organic compartment when soft-bodied organisms were buried in habitats where low oxygen concentration prevented decomposition. Under suitable geological conditions, these carbon-rich tissues were slowly converted to gas, petroleum, or coal, which humans now use as fossil fuels. Human activities, especially the burning of fossil fuels, are transferring carbon into the atmosphere at a high rate. The resulting change in the worldwide distribution of carbon is having profound consequences for Earth's atmosphere and climate, including a general warming of the climate and a rise in sea level, as described in *Focus on Applied Research.*

The Nitrogen Cycle Depends upon the Activity of Diverse Microorganisms

All organisms require nitrogen to construct nucleic acids, proteins, and other biological molecules. Earth's atmosphere had a high nitrogen concentration long

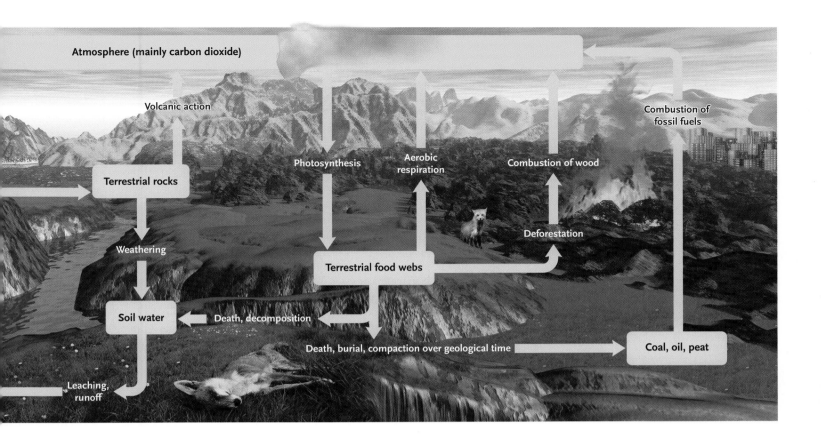

Labels in figure:
Atmosphere (mainly carbon dioxide)
Volcanic action
Combustion of fossil fuels
Photosynthesis
Aerobic respiration
Combustion of wood
Terrestrial rocks
Deforestation
Weathering
Terrestrial food webs
Soil water
Death, decomposition
Death, burial, compaction over geological time
Coal, oil, peat
Leaching, runoff

before life originated. Today, a global **nitrogen cycle** moves this element between the huge atmospheric pool of gaseous molecular nitrogen (N_2) and several much smaller pools of nitrogen-containing compounds in soils, marine and freshwater ecosystems, and living organisms **(Figure 51.14)**.

Nitrogen Cycling within Ecosystems. Molecular nitrogen is abundant in the atmosphere, but triple covalent bonds bind its two atoms so tightly that most organisms cannot use it. However, three biochemical processes—nitrogen fixation, ammonification, and nitrification **(Table 51.2)**—convert nitrogen into nitrogen compounds that primary producers can incorporate into biological molecules such as proteins and nucleic acids. Secondary consumers obtain their nitrogen by consuming primary producers, thereby initiating the movement of nitrogen through the food webs of an ecosystem.

In **nitrogen fixation** (see Section 33.3), molecular nitrogen (N_2) is converted into ammonia (NH_3) and ammonium ions (NH_4^+). Certain bacteria, including *Azotobacter* and *Rhizobium,* which collect molecular nitrogen from the air between soil particles, are the major nitrogen fixers in terrestrial ecosystems. The cyanobacteria partners in some lichens (see Section 28.3) also fix molecular nitrogen. Other cyanobacteria, such as *Anabaena* and *Nostoc,* are important nitrogen fixers in

aquatic ecosystems; the water fern (genus *Azolla*) plays that role in rice paddies. Collectively, these organisms fix an astounding 200 million metric tons of nitrogen each year; nitrogen fixation can also result from lightning and volcanic action. Plants and other primary producers assimilate and use this nitrogen in the biosynthesis of amino acids, proteins, and nucleic acids, which then circulate through food webs.

Some plants, including legumes (such as beans and clover), alders (*Alnus* species), and some members of the rose family (Rosaceae), are mutualists with nitrogen-fixing bacteria. These plants acquire nitrogen from soils much more readily than plants that lack such mutualists. Although these plants have the competitive edge in nitrogen-poor soil, nonmutualistic species often displace them in nitrogen-rich soil.

In addition to nitrogen fixation, several other biochemical processes make large quantities of nitrogen available to producers. **Ammonification** of detritus by bacteria and fungi converts organic nitrogen into ammonia (NH_3), which dissolves into ammonium ions (NH_4^+) that plants can assimilate; some ammonia escapes into the atmosphere as a gas. **Nitrification** by certain bacteria produces nitrites (NO_2^-) that are then converted by other bacteria to usable nitrates (NO_3^-). All of these compounds are water-soluble, and water rapidly leaches them from soil into streams, lakes, and oceans.

Applied Research: Disruption of the Carbon Cycle

Concentrations of gases in the lower atmosphere have a profound effect on global temperature, which in turn has enormous impact on global climate. Molecules of carbon dioxide (CO_2), water, ozone, methane, nitrous oxide, and other compounds collectively act like a pane of glass in a greenhouse (hence the term *greenhouse gases*). They allow the short wavelengths of visible light to reach Earth's surface; but they impede the escape of longer, infrared wavelengths back into space, trapping much of that energy as heat. In short, greenhouse gases foster the accumulation of heat in the lower atmosphere, a warming action known as the **greenhouse effect,** which prevents Earth from being a cold and lifeless planet.

Since the late 1950s, scientists have measured atmospheric concentrations of CO_2 and other greenhouse gases at remote sampling sites, which are free of local contamination and reflect the average concentrations of these gases in the atmosphere. Results indicate that concentrations of greenhouse gases have increased steadily for as long as they have been monitored.

The graph for atmospheric CO_2 concentration **(Figure a)** has a regular zigzag pattern that follows the annual cycle of plant growth in the northern hemisphere. Photosynthesis withdraws so much CO_2 from the atmospheric available-inorganic pool during the northern hemisphere summer that

its concentration falls. The concentration is higher during the northern hemisphere winter, when aerobic respiration continues, returning carbon to the atmospheric available-inorganic pool, and photosynthesis slows. The zigs and the zags in the data for CO_2 represent seasonal highs and lows, but the midpoint of the annual peaks and troughs has increased steadily for 40 years. Many scientists interpret these data as evidence of a rapid buildup of atmospheric CO_2, which represents a shift in the distribution of carbon in the major reservoirs on Earth. The best estimates suggest that CO_2 concentration has increased by 35% in the last 150 years and by more than 10% in the last 30 years.

What has caused the increase in the atmospheric concentration of CO_2? Burning of fossil fuels and wood is the largest contributor, because CO_2 is a combustion product of this process. Today, humans burn more wood and fossil fuels than ever before. Vast tracts of tropical forests are being cleared and burned (see Section 53.2). To make matters worse, deforestation reduces the world's biomass of plants, which assimilate CO_2 and help maintain the carbon cycle as it existed before human activities disrupted it.

Why is an increase in the atmospheric CO_2 concentration so alarming? Recent research suggests that plants with C_3 metabolism will respond to increased CO_2 concentrations with increased growth rates, but that C_4 plants will not (review Section 9.4 on C_3 and C_4 plants). Thus, rising atmospheric levels of CO_2 will probably alter the relative abundances of many plant species, changing the composition and dynamics of their communities.

Simulation models by scientists who study the global climate suggest that increasing concentrations of any greenhouse gas may also intensify the greenhouse effect, contributing to a trend of global warming. Should we be alarmed about the prospect of a warmer planet? Some models predict that the mean temperature of the lower atmosphere will rise by 4° C,

enough to increase ocean surface temperatures. Water expands when heated, and global sea level could rise as much as 0.6 m just from this expansion. In addition, atmospheric temperature is rising fastest near the poles. Thus, global warming may also foster melting of glaciers and the Antarctic ice sheet, which might raise sea level much more, inundating low coastal regions. Waterfronts in Vancouver, Los Angeles, San Diego, Galveston, New Orleans, Miami, New York, and Boston could be submerged. So might agricultural lands in India, China, and Bangladesh, where much of the world's rice is grown. Moreover, global warming could disturb regional patterns of precipitation and temperature. Areas that now produce much of the world's grains, including parts of Canada and the United States, would become arid scrub or deserts, and the now-forested areas to their north would become dry grasslands.

Many scientists believe that atmospheric levels of greenhouse gases will continue to increase at least until the middle of the twenty-first century and that global temperature may rise by several degrees. At the Earth Summit in 1992, leaders of the industrialized countries agreed to try to stabilize CO_2 emissions by the end of the twentieth century. We have already missed that target, and some countries, including the United States, which is the largest producer of greenhouse gases, have now abandoned that goal as too costly. Stabilizing emissions at current levels will not reverse the damage already done, nor will it stop the trend toward global warming. Many scientists agree that we should begin preparing for the consequences of global warming now. For example, we might increase reforestation efforts because a large tract of forest can withdraw significant amounts of CO_2 from the atmosphere. We might also step up genetic engineering studies to develop heat-resistant and drought-resistant crop plants, which may provide crucial food reserves in regions of climate change.

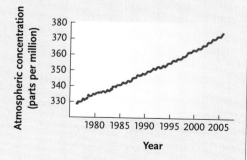

Figure a
Increases in atmospheric concentration of carbon dioxide, mid-1970s through 2004. The data were collected at a remote monitoring station in Australia (Cape Grim, Tasmania) and compiled by scientists at the Commonwealth Scientific and Industrial Research Organization, an agency of the Australian government.

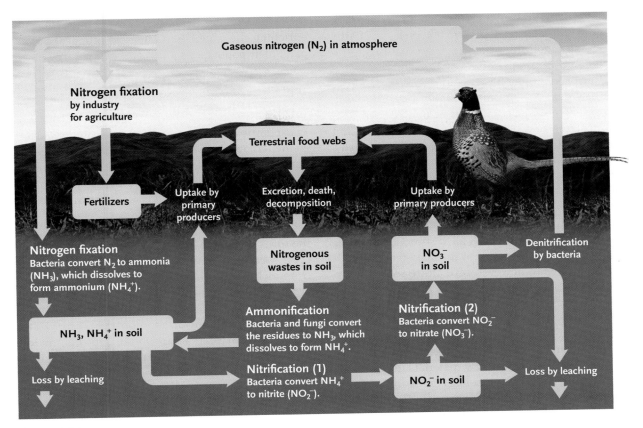

Figure 51.14

The nitrogen cycle in terrestrial ecosystems. Nitrogen cycles through terrestrial ecosystems when unavailable molecular nitrogen is made available through the action of nitrogen-fixing bacteria. Other bacteria recycle nitrogen within the available organic compartment through ammonification and two types of nitrification, converting organic wastes into ammonium ions and nitrates. Denitrification converts nitrate to molecular nitrogen, which returns to the atmosphere. Runoff carries various nitrogen compounds from terrestrial ecosystems into oceans, where it is recycled in marine food webs.

Under conditions of low oxygen availability, **denitrification** by still other bacteria converts nitrites or nitrates into nitrous oxide (N_2O) and then into molecular nitrogen (N_2), which enters the atmosphere, completing the cycle. This action can deplete supplies of soil nitrogen in waterlogged or otherwise poorly aerated environments, such as bogs and swamps. In an interesting twist on the usual predator–prey relationships, several species of flowering plants that live in nitrogen-poor soils, such as Venus' fly trap *(Dionaea muscipula)*, capture and digest small insects as their primary nitrogen source.

Table 51.2	**Biochemical Processes That Influence Nitrogen Cycling in Ecosystems**		
Process	**Organisms Responsible**	**Products**	**Outcome**
Nitrogen fixation	Bacteria: *Rhizobium, Azotobacter, Frankia* Cyanobacteria: *Anabaena, Nostoc*	Ammonia (NH_3), ammonium ions (NH_4^+)	Assimilated by primary producers
Ammonification of organic detritus	Soil bacteria and fungi	Ammonia (NH_3), ammonium ions (NH_4^+)	Assimilated by primary producers
Nitrification			
(1) Oxidation of NH_3	Bacteria: *Nitrosomonas, Nitrococcus*	Nitrite (NO_2^-)	Used by nitrifying bacteria
(2) Oxidation of NO_2^-	Bacteria: *Nitrobacter*	Nitrate (NO_3^-)	Assimilated by primary producers
Denitrification of NO_3^-	Soil bacteria	Nitrous oxide (N_2O), molecular nitrogen (N_2)	Released to atmosphere

Human Disruption of the Nitrogen Cycle. Human activities are altering the nitrogen cycle, primarily through the application of nitrogen-containing fertilizers. Of all nutrients required for primary production, nitrogen is often the least abundant. Agriculture routinely depletes soil nitrogen: with each harvest, nitrogen is removed from fields through the harvesting of plants that have accumulated nitrogen. Soil erosion and leaching remove more. Traditionally, farmers rotated their crops, alternately planting legumes and other crops in the same fields. In combination with other soil-conservation practices, crop rotation stabilized soils and kept them productive, sometimes for thousands of years.

Until 50 years ago, nearly all the nitrogen in living systems was made available by nitrogen-fixing microorganisms. Today, however, agriculture relies on the application of synthetic fertilizers. Some yields have quadrupled over the past 50 years. But 50 years is just an instant in the history of agriculture, and such high yields may not be sustainable for very long. Moreover, the production of synthetic fertilizers is expensive. It uses fossil fuels both as a raw material and as an energy source, so that fertilizer becomes increasingly costly as supplies of fossil fuels dwindle. Furthermore, rain and runoff leach excess fertilizer from agricultural fields and carry it into aquatic ecosystems. Like the phosphorus in Lake Erie, nitrogen has become a major pollutant of freshwater ecosystems, artificially enriching the waters and allowing producers to expand their populations.

The Phosphorus Cycle Includes a Large Sedimentary Reservoir

Phosphorus compounds lack a gaseous phase, and this element moves between terrestrial and marine ecosystems in a sedimentary cycle **(Figure 51.15).** Earth's crust is the main reservoir of phosphorus, as it is for other minerals such as calcium and potassium that undergo sedimentary cycles.

Phosphorus is present in terrestrial rocks in the form of phosphates (PO_4^{3-}). In the **phosphorus cycle,** weathering and erosion carry phosphate ions from rocks to soil and into streams and rivers, which eventually transport them to the ocean. Once there, some phosphorus enters marine food webs, but most of it precipitates out of solution and accumulates for millions of years as insoluble deposits, mainly on continental shelves. When parts of the seafloor are uplifted and exposed, weathering releases the phosphates.

Plants absorb and assimilate dissolved phosphates directly, and phosphorus moves easily to higher trophic levels. All heterotrophs excrete some phosphorus as a waste product in urine and feces, which are decom-

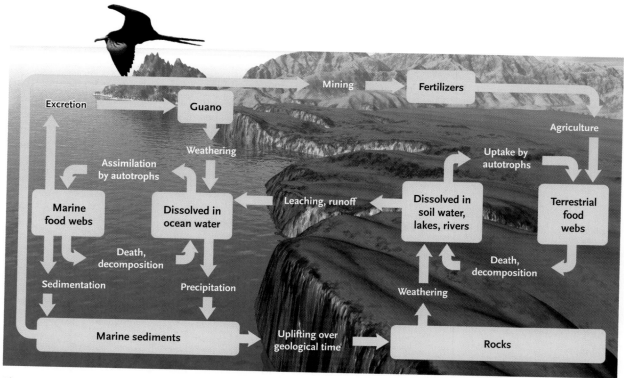

Figure 51.15

The phosphorus cycle. Phosphorus becomes available to biological systems when wind and rainfall dissolve phosphates in rocks and carry them into adjacent soil and freshwater ecosystems. Runoff carries dissolved phosphorus into marine ecosystems, where it precipitates out of solution and is incorporated into marine sediments.

posed, and producers readily absorb the phosphate ions that are released. Thus, phosphorus cycles rapidly *within* terrestrial communities.

Supplies of available phosphate are generally limited, however, and plants acquire it so efficiently that they reduce soil phosphate concentration to extremely low levels. Thus, like nitrogen, phosphorus is a common ingredient in agricultural fertilizers, and excess phosphates are pollutants of freshwater ecosystems. For many years, phosphate for fertilizers was obtained from *guano* (the droppings of seabirds that consume phosphorus-rich food), which was mined on small islands off the Pacific coast of South America. Most phosphate for fertilizer now comes from phosphate rock mined in Florida and other places with abundant marine deposits.

STUDY BREAK

1. In the generalized compartment model of biogeochemical cycling, how are the compartments where nutrients accumulate classified?
2. How does the global hydrologic cycle maintain its balance?
3. What process moves large quantities of carbon from an organic compartment to an inorganic compartment?
4. What microorganisms drive the global nitrogen cycle, and how do they do it?
5. What is Earth's main reservoir for phosphorus, and why is it recycled at such a slow rate from that reservoir?

51.3 Ecosystem Modeling

Ecologists Use Conceptual Models and Simulation Models to Understand Ecosystem Dynamics

To make predictions about how an ecosystem will respond to specific changes in physical factors, energy flow, or nutrient availability, ecologists turn to ecosystem modeling. Analyses of energy flow and nutrient cycling allow us to create a *conceptual model* of how ecosystems function **(Figure 51.16)**. Energy that enters ecosystems is gradually dissipated as it flows through a food web. By contrast, nutrients are conserved and recycled among the system's living and nonliving components. This very general model does not include processes that carry nutrients and energy out of one ecosystem and into another.

Note that the conceptual model ignores the nuts-and-bolts details of exactly how specific ecosystems function. Although it is a useful tool, a conceptual

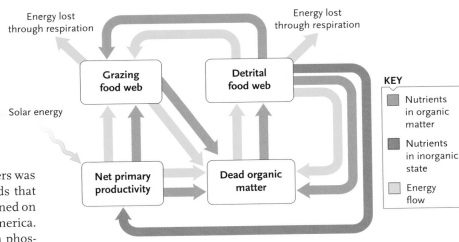

Figure 51.16
A conceptual ecosystem model. A simple conceptual model of an ecosystem illustrates how energy flows through the system and is lost from both detrital and grazing food webs. Nutrients are recycled and conserved.

model doesn't really help us predict what would happen, say, if we harvested 10 million tons of introduced salmon from Lake Erie every year. We could simply harvest the fishes and see what happens. But ecologists prefer less intrusive approaches to study the potential effects of disturbances.

One approach to predicting "what would happen if . . ." is **simulation modeling.** Using this method, researchers gather detailed information about a specific ecosystem. They then create a series of mathematical equations that define its most important relationships. For example, one set of equations might describe how nutrient availability limits productivity at various trophic levels. Another might relate population growth of zooplankton to the productivity of phytoplankton. Other equations would relate the population dynamics of primary carnivores to the availability of their food, and still others would describe how the densities of primary carnivores influence reproduction in populations at both lower and higher trophic levels. Thus, a complete simulation model is a set of interlocking equations that collectively predict how changes in one feature of an ecosystem might influence others.

Creating a simulation model is no easy task, because the relationships within every ecosystem are complex. First, you must identify the important species, estimate their population sizes, and measure the average energy and nutrient content of each. Next, you would describe the food webs in which they participate, measure the quantity of food each species consumes, and estimate the productivity of each population. And, for the sake of completeness, you would determine the ecosystem's energy and nutrient gains and losses caused by erosion, weathering, precipitation, and runoff. You would repeat these measurements seasonally to identify annual variation in these factors. Finally, you might repeat the measurements over several years to determine the effects of year-to-year variation in climate and chance events.

After collecting these data, you would write equations that quantify the relationships in the ecosystem, including information about how temperature and other abiotic factors influence the ecology of each species. Having completed that job, you could begin to predict—possibly in great detail—the effects of harvesting 10 million or even 50 million tons of salmon annually from Lake Erie. Of course, you would have to refine the model whenever new data became available.

Some ecologists devote their professional lives to the study of ecosystem processes. The long-term initiative at the Hubbard Brook Forest provides a good example. As we attempt to understand larger and more complex ecosystems—and as we create larger and more complex environmental problems—modeling becomes an increasingly important tool. If a model is based on well-defined ecological relationships and good empirical data, it can allow us to make accurate predictions about ecosystem changes without the need for costly and environmentally damaging experiments. But like all ideas in science, a model is only as good as its assumptions, and models must constantly be adjusted to incorporate new ideas and recently discovered facts.

In the next chapter we examine how interactions among ecosystems establish the global phenomena that characterize the biosphere.

STUDY BREAK

1. What are the advantages and disadvantages of relying on conceptual models that describe ecosystem function?
2. What data must ecologists collect before constructing a simulation model of an ecosystem?

Unanswered Questions

How does the carbon cycle of a forest respond to climate change and urbanization?

As you've read in this chapter, human influences on the environment can have dramatic unforeseen consequences for ecosystems, altering energy flow and nutrient cycling. Given the complexity of ecosystems—the myriad scales of influence and multiple interactions among the organisms, the physical environment, and climate change variables—the precise response of an ecosystem is difficult to predict, even with advanced ecosystem models. We do known with certainty, however, that the carbon, nitrogen, and water cycles of forested ecosystems in the northeastern United States are changing, and they are likely to continue to do so.

Carbon cycle research in forested ecosystems often entails building an ecosystem model from quantitative data on the various pools and fluxes of carbon in the ecosystem and how these change with time. Scientists then correlate these changes with the environmental conditions and derive a mechanistic understanding that they can use to make predictions about how the ecosystem will respond to future changes. In theory, *gross primary productivity* (GPP) should be predictable from a basic understanding of photosynthesis and a general description of the ambient environmental conditions. In practice, however, the complexity of canopy architecture and leaf positioning, the timing of recurring natural phenomena, and the effects of herbivory and leaf losses from abiotic factors all make accurate predictions more difficult. Furthermore, the problem is dynamic because age-related changes in stand structure, disturbance, invasion, drought, seasonality, and pests or pathogens all add spatial and temporal complexities. Scientists should be able to predict *net primary productivity* (NPP), a key parameter used by ecologists to classify the world's ecosystems, from measurements of the cellular respiration and the relative abundances of representative organisms from the ecosystem.

Quantifying GPP and NPP on a large spatial scale can be challenging, and discovering the underlying mechanisms that control ecosystem responses to changes in environmental conditions is difficult. For example, studies at Black Rock Forest, a deciduous-oak-dominated forest in New York State, revealed that temporal heterogeneity (seasonal variation in leaf and stem respiration) and spatial heterogeneity (variations in canopy and hill slope position) are important factors that must be included in models of canopy respiration. Nevertheless, some simplifications may be possible. For example, while the basal rate of respiration is quite variable and subject to acclimation, it may be predictable from basic plant properties such as their nitrogen concentrations. Furthermore, the temperature coefficient of respiration is relatively constant, greatly simplifying the construction of an ecosystem model. To consider the impact of tree respiration on ecosystem form and function fully, my research team experimented with models that explicitly consider physiological linkages between photosynthesis and respiration, as mediated by leaf carbohydrate pools. We found that when we included direct linkages to carbon gain in the analysis, the model correctly predicted a large (23%) decrease in the estimated nighttime canopy respiration during the growing season. This result emphasizes the need for a process-based modeling approach when estimating forest productivity.

Our research at Black Rock Forest has also demonstrated that human activities in New York City (60 miles to the south) may be influencing tree growth in both urban and rural areas, with significant changes in seedling size, biomass allocation, herbivory, stomatal densities, nutrient concentrations, efficiency of water use, and rates of key physiological processes such as photosynthesis and respiration. Urbanization has a clear effect on the land area developed, but current research is showing that human activities in urban areas also influence forested ecosystems in the surrounding rural areas. Understanding how human activity, climate change, and forest ecosystems interact is crucial if we are to make prudent and sustainable development decisions, preserving the health of the ecosystems and the services they provide.

Kevin Griffin is an associate professor at Columbia University's Lamont-Doherty Earth Observatory. His research centers on processes in plant and ecosystem ecology, the goal of which is to increase our understanding of both the role of vegetation in the global carbon cycle and the interactions between the carbon cycle and Earth's climate system. To learn more about Dr. Griffin's research, go to http://www.ldeo.columbia.edu/.

Review

Go to ThomsonNOW™ at www.thomsonedu.com/login to access quizzing, animations, exercises, articles, and personalized homework help.

51.1 Energy Flow and Ecosystem Energetics

- Ecosystems include biological communities and the abiotic environmental factors with which they interact (Figure 51.1).

- Food webs define the pathways along which energy and nutrients move through the biological components of an ecosystem. Ecosystems include both grazing and detrital food webs, which are closely interconnected (Figure 51.2).

- Only a small portion of the solar energy that reaches Earth is converted into chemical energy through the process of photosynthesis.

- An ecosystem's gross primary productivity is the rate at which producers convert solar energy into chemical energy. Producers use some energy for respiration; some is converted to heat; and some remains in the ecosystem as net primary productivity.

- Primary productivity is measured in units of energy captured or biomass produced per unit area per unit time. Net primary productivity indexes the energy available to support heterotrophs. Ecosystems vary in productivity and in their contributions to Earth's total productivity (Figure 51.3, Table 51.1).

- On land, primary productivity is limited by the availability of sunlight, water, and nutrients; temperature; and how much photosynthetic tissue is present. In marine and aquatic ecosystems, primary productivity is limited when sunlight and nutrients are not available in the same place (Figures 51.4 and 51.5).

- Only a fraction of the energy at any trophic level is converted into biomass at higher trophic levels. Ecological efficiencies generally range from 5% to 20%. As energy passes through a food web, an average of 90% is lost at each transfer between trophic levels, limiting the number of trophic levels that a food web can support (Figure 51.6).

- Ecological pyramids portray the effects of energy losses. For terrestrial ecosystems, pyramids of energy, biomass, and numbers generally have broad bases and narrow tops (Figures 51.7–51.9).

- The food preferences of consumers can influence primary productivity through a trophic cascade (Figure 51.10).

Animation: The role of organisms in an ecosystem

Animation: Food webs

Animation: Energy flow at Silver Springs

51.2 Nutrient Cycling in Ecosystems

- Earth is a closed system with respect to matter.

- Nutrients circulate in biogeochemical cycles between living organisms and nonliving reservoirs. Nutrients accumulate in four compartments, defined by whether the nutrients are available or unavailable and whether they are in organic or inorganic material (Figure 51.11). Nutrients move rapidly between available compartments. Exchange rates for the unavailable compartments are slow. Some biogeochemical cycles are atmospheric; others are sedimentary.

- Water circulates through the atmosphere, oceans, and terrestrial and freshwater ecosystems in a global hydrologic cycle. Water evaporates from the oceans and continents and falls as precipitation. Runoff and streamflow return excess precipitation from the land to the oceans (Figure 51.12).

- The carbon cycles in terrestrial and aquatic ecosystems are linked through an atmospheric pool of CO_2, which primary producers assimilate. Respiration returns carbon to the atmosphere as CO_2. Earth's largest reservoir of carbon is unavailable in sedimentary rock. Other large reservoirs include coal, oil, and peat as well as dissolved bicarbonate and carbonate ions in seawater (Figure 51.13).

- Nitrogen is cycled between living organisms and an atmospheric pool of nitrogen gas. Bacteria and cyanobacteria make nitrogen available to the food web through the processes of nitrogen fixation, ammonification, and nitrification. Denitrification converts nitrogen compounds to molecular nitrogen, which enters the atmosphere (Figure 51.14, Table 51.2). The use of synthetic fertilizers disrupts the nitrogen cycle.

- Phosphorus undergoes a sedimentary cycle. Weathering and erosion of rock make phosphorus available; it is leached from soil and carried to the ocean. Dissolved phosphates precipitate out of seawater, forming insoluble deposits, which are eventually uplifted by tectonic processes (Figure 51.15).

Animation: Hydrologic cycle

Animation: Hubbard Brook experiment

Animation: Carbon cycle

Animation: Greenhouse effect

Animation: Greenhouse gases

Animation: Carbon dioxide and temperature

Animation: Nitrogen cycle

Animation: Phosphorus cycle

51.3 Ecosystem Modeling

- Conceptual models describe energy flow and nutrient cycling in ecosystems (Figure 51.16).

- Simulation models are interlocking mathematical equations that define the relationships between populations and between populations and the physical environment. They allow users to predict the effects of changes in ecosystem structure and function.

Questions

Self-Test Questions

1. Which of the following events would move energy and material from a detrital food web into a grazing food web?
    a. A beetle eats the leaves of a living plant.
    b. An earthworm eats dead leaves on the forest floor.
    c. A robin catches and eats an earthworm.
    d. A crow eats a dead robin.
    e. A bacterium decomposes the feces of an earthworm.

2. The total dry weight of plant material in a forest is a measure of the forest's:
    a. gross primary productivity.
    b. net primary productivity.
    c. cellular respiration.

d. standing crop biomass.
e. ecological efficiency.

3. Which of the following ecosystems has the highest rate of net primary productivity?
   a. open ocean
   b. temperate deciduous forest
   c. tropical rain forest
   d. desert and thornwoods
   e. agricultural land

4. Endothermic animals exhibit a lower ecological efficiency than ectothermic animals because:
   a. endotherms are less successful hunters than ectotherms.
   b. endotherms eat more plant material than ectotherms.
   c. endotherms are larger than ectotherms.
   d. endotherms produce fewer offspring than ectotherms.
   e. endotherms use more of their energy to maintain body temperature than ectotherms.

5. The amount of energy available at the highest trophic level in an ecosystem is determined by:
   a. only the gross primary productivity of the ecosystem.
   b. only the net primary productivity of the ecosystem.
   c. the gross primary productivity and the standing crop biomass.
   d. the net primary productivity and the ecological efficiencies of herbivores.
   e. the net primary productivity and the ecological efficiencies at all lower trophic levels.

6. Some freshwater and marine ecosystems exhibit an inverted pyramid of:
   a. biomass.   d. turnover.
   b. energy.   e. ecological efficiency.
   c. numbers.

7. Which process moves nutrients from the available-organic compartment to the available-inorganic compartment?
   a. respiration   d. sedimentation
   b. erosion   e. photosynthesis
   c. assimilation

8. Which of the following materials has a sedimentary cycle?
   a. water   d. phosphorus
   b. oxygen   e. carbon
   c. nitrogen

9. Which of the following statements is supported by the results of studies at the Hubbard Brook Experimental Forest?
   a. Most of the energy captured by primary producers is lost before it reaches the highest trophic level in an ecosystem.
   b. Deforested watersheds experience significantly less run-off than undisturbed watersheds.
   c. Deforested watersheds lose more calcium and nitrogen in runoff than undisturbed watersheds.
   d. Nutrients generally move through biogeochemical cycles very quickly.
   e. Deforested watersheds generally receive more rainfall than undisturbed watersheds.

10. Nitrogen fixation converts:
    a. atmospheric molecular nitrogen to ammonia.
    b. nitrates to nitrites.
    c. ammonia to molecular nitrogen.
    d. ammonia to nitrates.
    e. nitrites to nitrates.

Questions for Discussion

1. A lake near your home became overgrown with algae and pondweeds a few months after a new housing development was built nearby. What data would you collect to determine whether the housing development might be responsible for the changes in the lake?

2. Some politicians question whether recent increases in atmospheric temperature result from our release of greenhouse gases into the atmosphere. They argue that atmospheric temperature has fluctuated widely over Earth's history, and the changing temperature is just part of an historical trend. What information would allow you to refute or confirm their hypothesis? In addition, describe the pros and cons of reducing greenhouse gases as soon as possible versus taking a "wait and see" approach to this question.

3. If you could design the ideal farm animal—one that was grown as food for humans—from scratch, what characteristics would it have?

4. If you were growing a vegetable garden, identify the factors that might affect its primary productivity. How would you increase productivity? Identify some of the possible consequences of your gardening activities to nearby ecosystems.

Experimental Analysis

Design an experiment to test the hypothesis that the top predator in an aquatic ecosystem regulates the ecosystem's productivity. Establish as many experimental ponds as you wish, and imagine stocking them with organisms at different trophic levels. If the hypothesis is correct, describe the results you would expect to record from each of your experimental treatments.

Evolution Link

In the discussion of trophic cascades, we described how herbivorous zooplankton of different sizes eat phytoplankton of different sizes and how different types of predators preferentially feed on different sizes of zooplankton. Develop hypotheses about how these feeding preferences might establish different patterns of natural selection on the phytoplankton and zooplankton. How could you test your hypotheses?

How Would You Vote?

Emissions from motor vehicles are a major source of greenhouse gases. Many people buy large vehicles that use more fuel but are viewed as safer and more useful. Should such vehicles be taxed extra to discourage sales and offset their environmental costs? Can we expect the emergence of better fuels as well as more of the fuel-efficient, larger vehicles that are becoming available? Go to www.thomsonedu.com/login to investigate both sides of the issue and then vote.

Stormy weather in the biosphere. Atmospheric disturbances like Hurricane Fran, seen here in a satellite photograph taken on September 4, 1996, often have a dramatic impact on living systems.

NOAA/Photo Researchers, Inc.

STUDY PLAN

52.1 Environmental Diversity of the Biosphere

Variations in incoming solar radiation create global climate patterns

Regional and local effects overlay global climate patterns

52.2 Organismal Responses to Environmental Variation

Organisms use homeostatic responses to cope with environmental variation

Global warming is changing the ecology of many organisms

52.3 Terrestrial Biomes

Environmental variation governs the distribution of terrestrial biomes

Tropical forests include the most species-rich communities on Earth

Savannas grow where moderate rainfall is highly seasonal

Deserts develop in places where little precipitation falls

Chaparral grows where winters are cool and wet and summers are hot and dry

Temperate grasslands are held in a disclimax state by periodic disturbance

Temperate deciduous forests experience seasonal dormancy

Evergreen coniferous forests predominate at high northern latitudes

Tundra comprises a vast, treeless plain in the northernmost habitats

52.4 Freshwater Biomes

Streams and rivers carry water downhill to a lake or the sea

Lakes are bodies of standing water that accumulates in basins

52.5 Marine Biomes

Estuaries form where rivers meet the sea

Rocky and sandy coasts experience cyclic periods of exposure and submergence

Light penetrates the shallow water over continental shelves and oceanic banks

In the open ocean, photosynthesis occurs only in the upper layers

The benthic province includes the rocks and sediments of the ocean bottom

52 The Biosphere

WHY IT MATTERS

The winter of 1997–1998 was one for the books. Record rainfall caused mudslides in California and flooding along the normally arid coast of Ecuador and Peru. But the annual rains never arrived in Asia and Australia, and fires consumed tropical rain forests in Indonesia and Malaysia. What caused these major climatic dislocations? Every 3 to 7 years, interactions between the upper layers of the Pacific Ocean and the atmosphere produce El Niño, a climatic event with global consequences **(Figure 52.1)**. The 1997–1998 El Niño altered weather patterns worldwide, killing more than 2000 people and causing at least $30 billion in property damage.

In most years, air flows from a high pressure system over the eastern Pacific toward a low pressure system over the western Pacific. These winds move surface water from east to west and bring heavy rains to parts of Asia and Australia. Winds also usually blow from the poles toward the equator along the western sides of continents, and Earth's rotation causes these winds to push ocean surface water westward, away from the coast. The displaced surface water is replaced by cold, deep, nutrient-rich water carried by vertical currents called *upwellings* (see Figure 52.1a). The nutrients support complex marine

1203

a. Usual pattern of Pacific Ocean currents

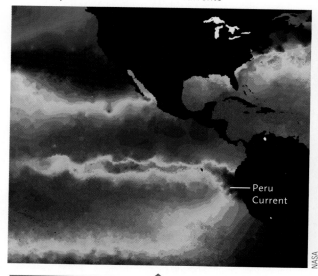

Peru Current

In most years, the powerful Peru Current carries cold water from the ocean bottom to the surface off the west coast of South America. The cold surface water then flows westward along the equator toward a large pool of warm water in the western Pacific Ocean. In this satellite photo taken on May 31, 1988, dark red indicates the warmest water and dark green the coldest upwelled water.

b. Pacific Ocean currents in an El Niño year

During an El Niño event, equatorial winds reverse directions and warm surface water flows eastward along the equator from the western Pacific Ocean toward South America. In this satellite photo taken on May 13, 1992, the warm water (red) spreads up and down the west coast of North and South America, suppressing the upwelling of cold water by the Peru Current.

Figure 52.1
El Niño and Pacific Ocean currents.

food webs in the shallow water above the continental shelf. For example, the Peru Current along the west coast of South America once supported a rich anchovy fishery.

Ocean currents vary seasonally, however, and in late December or early January, a warm, nutrient-poor current flows eastward along the equator and then north and south along the coastlines of Central and South America. Peruvian fishermen call this warm current El Niño (Spanish for "the child"), because it reaches their coast around Christmas. It usually persists for only a few weeks.

In strong El Niño years, atmospheric pressure systems change over the Pacific, altering the prevailing winds and ocean currents. Equatorial winds weaken; surface currents reverse direction, flowing from west to east; and a huge pool of warm ocean water accumulates in the eastern Pacific. During these shifts, the heavy rain that usually falls on Asia and Australia is instead delivered to the central and eastern Pacific. Thus, in El Niño years, Asia and Australia receive less rain than usual, and the west coasts of the Americas receive more. In the United States, winter temperatures are unusually high in the north central states and unusually low in the southern states.

El Niño episodes also alter sea surface temperature. When the warm current flowing from west to east reaches the continental shelf, it displaces the cold water of the Peru Current and prevents the usual upwelling (see Figure 52.1b). These changes in ocean currents have catastrophic effects on marine food webs. Lacking sufficient nutrients, phytoplankton die, followed by fishes that eat phytoplankton, and seabirds

that eat fishes. In combination with overfishing, the El Niño of 1972 drove the Peruvian anchovy population to the brink of extinction.

Some El Niño years are followed by a weather pattern called La Niña: the low pressure system over the western Pacific is accentuated, pulling air and ocean surface water from east to west. Low ocean surface temperatures extend from the coast of South America to Samoa. La Niña's effect on winter weather is opposite that of El Niño: parts of Asia and Australia are unusually wet; and the northern United States experiences periods of cold, wet weather, whereas the southern region is unusually warm and dry.

El Niño and La Niña are two extremes of a global climate cycle called the El Niño Southern Oscillation, or ENSO (the name refers to fluctuations in air pressure over the tropical Pacific). ENSO is a product of large-scale interactions between the ocean and atmosphere and has a major impact on the **biosphere**, all the places on Earth where organisms live. The biosphere has three abiotic components, which surround Earth's geological bulk like a skin. The **hydrosphere** encompasses all the water, including oceans and polar ice caps. The **lithosphere** includes the rocks, sediments, and soils of the crust. Finally, the **atmosphere** includes gases and airborne particles that envelop the planet.

In this chapter, we survey the biosphere with a wide-angle lens. First, we examine its environmental diversity and how organisms cope with it. We then consider how variations in the physical environment influence the large-scale distributions of ecosystems on land, in fresh water, and in the sea.

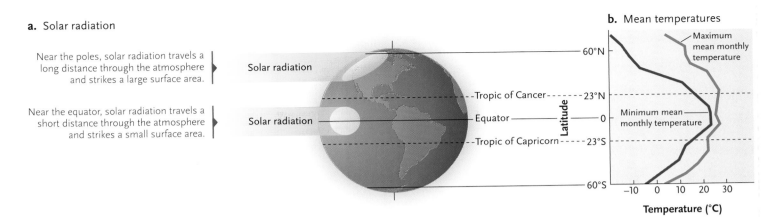

a. Solar radiation

Near the poles, solar radiation travels a long distance through the atmosphere and strikes a large surface area.

Solar radiation

Near the equator, solar radiation travels a short distance through the atmosphere and strikes a small surface area.

Solar radiation

b. Mean temperatures

Maximum mean monthly temperature

60°N

Tropic of Cancer — 23°N

Equator — 0

Minimum mean monthly temperature

Tropic of Capricorn — 23°S

60°S

Latitude

Temperature (°C)

Figure 52.2

Latitudinal variation in solar radiation and temperature. **(a)** Solar radiation is more intense near the equator than near the poles. **(b)** Minimum and maximum mean monthly temperatures as well as the range of mean monthly temperatures vary with latitude.

52.1 Environmental Diversity of the Biosphere

Numerous abiotic factors—sunlight, temperature, humidity, wind speed, cloud cover, and rainfall—contribute to a region's **climate,** the weather conditions prevailing over an extended period of time. Climates vary on global, regional, and local scales, and they undergo seasonal changes almost everywhere.

Variations in Incoming Solar Radiation Create Global Climate Patterns

A global pattern of environmental diversity results from latitudinal variation in incoming solar radiation, Earth's rotation on its axis, and its orbit around the sun.

Solar Radiation. Earth's spherical shape causes the intensity of incoming solar radiation to vary from the equator to the poles **(Figure 52.2).** When sunlight strikes Earth directly at a 90° angle, as it does near the equator, it travels the shortest possible distance through the radiation-absorbing atmosphere and falls on the smallest possible surface area (see Figure 52.2a). When sunlight arrives at an oblique angle, as it does near the poles, it travels a longer distance through the atmosphere and shines on a larger area. Thus, solar radiation is more concentrated near the equator than it is at higher latitudes, causing latitudinal variation in temperature (see Figure 52.2b).

Seasonality. Earth is tilted on its axis at a fixed position of 23.5° from the perpendicular to the plane on which it orbits the sun **(Figure 52.3).** This tilt produces seasonal variation in the duration and intensity of incoming solar radiation. The Northern Hemisphere receives its maximum illumination—and the Southern Hemisphere its minimum—on the June solstice (around

June 22), when the sun shines directly over the Tropic of Cancer (23.5° N latitude). The reverse is true on the December solstice (around December 22), when the sun shines directly over the Tropic of Capricorn (23.5° S latitude). Twice each year, on the vernal and autumnal equinoxes (around March 21 and September 23, respectively), the sun shines directly over the equator.

Earth's tilt is permanent, and only the **tropics**—the latitudes between the Tropics of Cancer and Capricorn—ever receive intense solar radiation from directly overhead. Moreover, the tropics experience only small seasonal changes in temperature and day length: environmental temperature is high and days last approximately 12 hours throughout the year. (Tropical seasonality is reflected in the alternation of wet and dry periods, rather than warm and cold seasons.) Seasonal variation in temperature and day length increases steadily toward the poles. Polar winters are long and cold with periods of continuous dark-

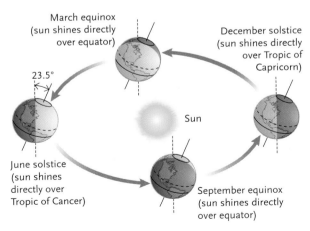

March equinox (sun shines directly over equator)

December solstice (sun shines directly over Tropic of Capricorn)

23.5°

Sun

June solstice (sun shines directly over Tropic of Cancer)

September equinox (sun shines directly over equator)

Figure 52.3

Seasonal variation in solar radiation. Earth's fixed tilt on its axis causes the Northern Hemisphere to receive more sunlight in June and the Southern Hemisphere to receive more in December. These differences are reflected in seasonal variations in day length and temperature, which are more pronounced at the poles than at the equator.

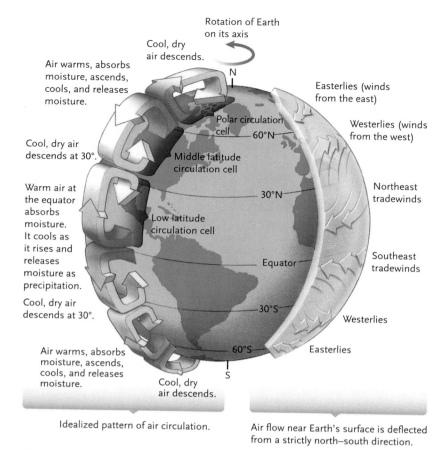

Rotation of Earth on its axis

Cool, dry air descends.

Air warms, absorbs moisture, ascends, cools, and releases moisture.

Easterlies (winds from the east)

Polar circulation cell — 60°N

Westerlies (winds from the west)

Middle latitude circulation cell

Cool, dry air descends at 30°.

Northeast tradewinds

30°N

Warm air at the equator absorbs moisture. It cools as it rises and releases moisture as precipitation.

Low latitude circulation cell

Equator

Southeast tradewinds

Cool, dry air descends at 30°.

30°S

Westerlies

Air warms, absorbs moisture, ascends, cools, and releases moisture.

60°S

Easterlies

S

Cool, dry air descends.

Idealized pattern of air circulation.

Air flow near Earth's surface is deflected from a strictly north–south direction.

Figure 52.4

Global air circulation. Latitudinal variations in the intensity of solar radiation cause equatorial air masses to warm and rise, initiating a global pattern of air movement in three circulation cells in each hemisphere. Air masses moving near Earth's surface create easterly and westerly winds, which are deflected from a strictly north–south flow by the planet's rotation.

ness, and polar summers are short with periods of continuous light.

Air Circulation. Sunlight warms air masses, causing them to expand, lose pressure, and rise in the atmosphere. The unequal heating of air at different latitudes initiates global air movements, producing three circulation cells in each hemisphere **(Figure 52.4)**. Warm equatorial air masses rise to high altitude before spreading north and south. They eventually sink back to Earth at about 30° N and S latitude. At low altitude, some air masses flow back toward the equator, completing low-latitude circulation cells. Others flow toward the poles, rise at 60° latitude, and divide at high altitude. Some air flows toward the equator, completing the pair of middle-latitude circulation cells. The rest moves toward the poles, where it descends and flows toward the equator, forming the polar circulation cells.

The flow of air masses at low altitude creates winds near the planet's surface. But the surface ro-

tates beneath the atmosphere, moving rapidly near the equator, where Earth's diameter is greatest, and slowly near the poles. Latitudinal variation in the speed of rotation deflects the movement of the rising and sinking air masses from a strictly north–south path into belts of easterly and westerly winds (see Figure 52.4); this deflection is called the Coriolis effect. Winds near the equator are called the trade winds; those further from the equator are the temperate westerlies and easterlies, named for the direction from which they blow.

Precipitation. Differences in solar radiation and global air circulation create latitudinal variations in rainfall **(Figure 52.5)**. Warm air holds more water vapor than cool air does. As air near the equator heats up, it absorbs water, primarily from the oceans. However, the warm air masses expand as they rise, and their heat energy is distributed over a larger volume, causing their temperature to drop. A decrease in temperature without the actual *loss* of heat energy is called **adiabatic cooling.** After cooling adiabatically, the rising air masses release moisture as rain. Torrential rainfall is characteristic of warm equatorial regions, where rising, moisture-laden air masses cool as they reach high altitude.

As cool, dry air masses descend at 30° latitude, increased air pressure at low altitude compresses them, concentrating their heat energy, raising their temperature, and increasing their capacity to hold moisture. The descending air masses absorb water from the land, so these latitudes are typically dry. Some air masses continue moving poleward in the lower atmosphere. When they rise at 60° latitude, they cool adiabatically and release precipitation (see Figure 52.4), creating moist habitats in the northern and southern temperate zones.

Ocean Currents. Latitudinal variations in solar radiation also warm the oceans' surface water unevenly. Because the volume of water increases as it warms, sea level is about 8 cm higher at the equator than at the poles. The volume of water associated with this "slope" is enough to cause surface water to move in response to gravity. The trade winds and temperate westerlies also contribute to the mass flow of water at the ocean surface. Thus, surface water flows in the direction of prevailing winds, forming major currents. Earth's rotation, the positions of landmasses, and the shapes of ocean basins also influence their movement.

Oceanic circulation is generally clockwise in the Northern Hemisphere and counterclockwise in the Southern **(Figure 52.6)**. The trade winds push surface water toward the equator and westward until it contacts the eastern edge of a continent. Swift, narrow, and deep currents of warm, nutrient-poor water run toward the poles, parallel to the east coasts of continents. For ex-

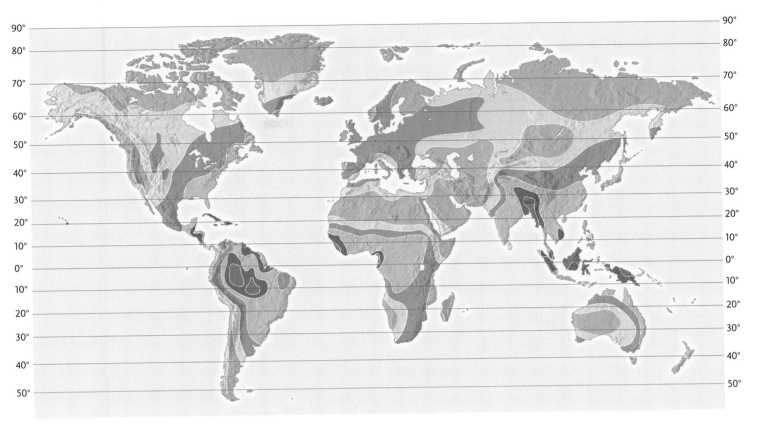

Precipitation (cm)

Under 25	50 to 100	200 to 250
25 to 50	100 to 200	Over 250

Figure 52.5
Variations in precipitation. The tropics receive high annual rainfall, whereas regions near 30° latitude are usually dry. Local topographic features and ocean currents also influence precipitation patterns.

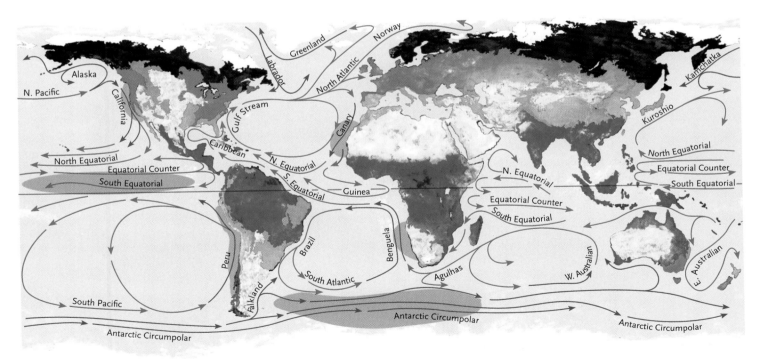

KEY

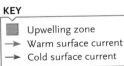

Upwelling zone
→ Warm surface current
→ Cold surface current

Figure 52.6
Ocean currents. Prevailing winds, Earth's rotation, gravity, the shape of ocean basins, and the positions of landmasses establish the direction and intensity of surface currents in the oceans. In general, warm currents flow away from the equator, and cold currents flow toward it.

ample, the Gulf Stream flows northward along the east coast of North America, carrying warm water toward northwestern Europe. Cold water returns from the poles toward the equator in slow, broad, and shallow currents, such as the California Current, that parallel the west coasts of continents.

Regional and Local Effects Overlay Global Climate Patterns

Although global and seasonal patterns determine a site's climate, regional and local effects also influence abiotic conditions.

Proximity to the Ocean. Currents running along seacoasts exchange heat with air masses flowing above them, moderating the temperature over nearby land. Breezes often blow from the sea toward the land during the day and in the opposite direction at night **(Figure 52.7)**. These local effects sometimes override latitudinal variations in temperature. For example, the climate in London is much milder than that in Minneapolis, even though Minneapolis is slightly further south. Minneapolis has a **continental climate** that is not moderated by the distant ocean, but London has a **maritime climate**, tempered by winds that cross the nearby North Atlantic Current.

a. Daytime: land warmer than sea

2 Cool air descends and replaces air over land through onshore flow.

1 Warm air ascends.

b. Nighttime: sea warmer than land

1 Warm air ascends.

2 Cool air descends and replaces air over sea through offshore flow.

Figure 52.7
Sea breezes and land breezes. On a summer afternoon **(a)**, warm air rises over the land, and a cool sea breeze blows inland from the ocean. At night **(b)**, when the ocean is warmer than the land, the pattern is reversed.

Ocean currents also affect moisture conditions in coastal habitats. For example, air masses absorb water as they move from west to east across the Pacific Ocean. They cool as they cross the cold California Current, and when they reach land in northern California and Oregon during winter, their water vapor condenses into heavy fog and rain. During summer, however, land is warmer than the adjacent ocean. The air masses heat up as they cross the land, and they accumulate water, creating dry conditions.

Some regions experience **monsoon cycles** caused by seasonal reversals of wind direction. In the North American southwest, for example, summer heat causes air masses over land to rise, creating a zone of low pressure. Moist air from the nearby Gulf of California flows inland, where it rises and cools adiabatically, releasing substantial precipitation. Summer monsoon rains deliver one-third to one-half of the annual rainfall in Arizona and New Mexico. During the winter, when land is cooler than the nearby ocean, low-pressure systems form over the ocean, and winds blow from the land to the sea; thus, winters in the southwest are generally dry. Seasonal monsoon cycles also deliver torrential rainfall to parts of Africa, Asia, and South America.

The Effects of Topography. Mountains, valleys, and other topographic features also influence regional climates. In the Northern Hemisphere, south-facing slopes are warmer and drier than north-facing slopes, because they receive more solar radiation. In addition, adiabatic cooling causes air temperature to decline 3° to 6°C for every 1000 m increase in altitude.

Mountains also establish regional and local rainfall patterns. For example, after a warm air mass picks up moisture from the Pacific Ocean, it moves inland and reaches the Sierra Nevada, which parallels the California coast. As it rises to cross the mountains, the air cools adiabatically and loses moisture, releasing heavy rainfall on the windward side **(Figure 52.8)**. After the now-dry air crosses the peaks, it descends and warms, absorbing moisture and forming a **rain shadow**. Habitats on the leeward side of mountains, such as the Great Basin Desert in western North America, are typically drier than those on the windward side.

Microclimate. Although climate influences the overall distributions of organisms, the abiotic conditions that immediately surround them—the **microclimate**—have the greatest effect on survival and reproduction. For example, a fallen log on the forest floor creates a microclimate in the underlying soil that is shadier, cooler, and moister than surrounding soil exposed to sun and wind. Many animals, including some insects, worms, salamanders, and snakes, occupy these sheltered sites and avoid the effects of prolonged exposure to the elements.

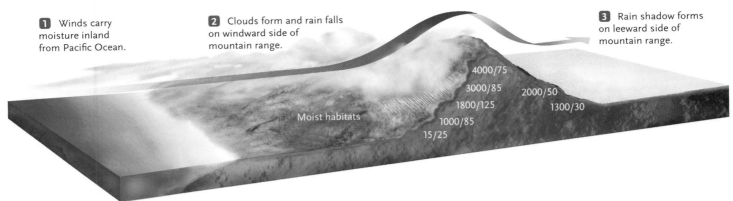

1 Winds carry moisture inland from Pacific Ocean.

2 Clouds form and rain falls on windward side of mountain range.

3 Rain shadow forms on leeward side of mountain range.

4000/75
3000/85
1800/125
1000/85
15/25
Moist habitats
2000/50
1300/30

Figure 52.8
Formation of a rain shadow. White numbers indicate altitude in meters followed by mean annual precipitation in centimeters for the Sierra Nevada of California.

52.2 Organismal Responses to Environmental Variation

Daily and seasonal variations in physical factors have profound effects on the biology of individual organisms. Moreover, large-scale variations in environmental conditions often influence the distributions of populations.

Organisms Use Homeostatic Responses to Cope with Environmental Variation

Animals in particular exhibit diverse homeostatic responses—biochemical, behavioral, physiological, and morphological—that enable them to maintain relatively constant conditions within their cells and tissues. Although the ability to use these responses almost certainly has a genetic basis, only some responses to environmental variation are *obligate* (that is, they must always be used). *Insights from the Molecular Revolution* describes one such evolutionary response at the biochemical level. Many behavioral and physiological responses are *facultative*. In other words, animals may use them or not, as their immediate conditions demand. Here we provide two brief examples of facultative behavioral and physiological responses to variations in environmental temperature.

Like many ectothermic animals, lizards often use behaviors to regulate body temperature (see Figure 1.15 and Section 46.6). They commonly *bask* in sunny spots to raise body temperature and seek shaded places to cool off. Many *Anolis* lizard species (see *Focus on Research Organisms* in Chapter 30) are distributed over broad altitudinal ranges, and populations living at high altitude encounter cooler environments than do those at low altitude. While they were graduate students at Harvard University, Paul E. Hertz, now of Barnard College, and Raymond B. Huey, now of the University of Washington, hypothesized that *Anolis* populations living at cool, high altitudes would bask more frequently than those living at warm, low altitudes. Hertz and Huey tested their hypothesis by observing *Anolis cybotes* and its close relative *Anolis shrevei* along an altitudinal gradient in the Dominican Republic **(Figure 52.9).** Their results indicate that basking frequency increases steadily with altitude. Moreover, the body temperatures of the lizards vary much less with altitude than do air temperatures at the same localities. The researchers therefore concluded that increased basking frequency by lizards at high altitude partially compensates for the lower environmental temperatures they encounter.

The state of extreme physiological sluggishness called *torpor* is a facultative response to daily variations in environmental temperature. Endothermic animals use the heat generated by the metabolic breakdown of food to maintain high body temperature (see Section 46.6). However, small endotherms, such as hummingbirds, have a large relative surface area through which they lose body heat. When environmental temperature is low, they may lose heat faster than they can generate it, risking the total depletion of their energy reserves and death by starvation. The problem is particularly acute at night, when hummingbirds cannot feed to replenish their energy stores. F. Reed Hainsworth and Larry Wolf of Syracuse University discovered that the purple-throated carib *(Eulampis jugularis),* a West Indian hummingbird, often becomes torpid at night, lowering its body

Fish Antifreeze Proteins

Polar-dwelling fishes, such as winter flounder, Alaskan plaice, and Arctic sculpin, have "antifreeze proteins" that prevent their bodies from freezing into solid ice at the extremely low environmental temperatures they encounter. As ice crystals begin to form within a fish's cells and tissues, the antifreeze proteins bind to the crystals and cover them with a protein coat that prevents further crystal growth and fusion. As a result, the fishes freeze only to an ultrafine slush that allows continued activity, including movement and feeding. The antifreeze proteins are small molecules containing between 30 and 50 amino acids.

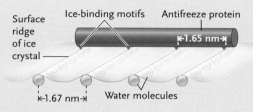

Figure a

How winter flounder antifreeze protein may bind to the surface ridges of an ice crystal. Only the water molecules forming the tips of the ridges are shown.

Researchers do not fully understand how the antifreeze proteins bind to ice crystals. Frank Sicheri and D. S. C. Yang at McMaster University in Hamilton, Ontario, and the Bio-Crystallography Laboratory of the VA Medical Center in Pittsburgh, Pennsylvania, sought a molecular solution to this problem. They used X-ray diffraction (see Section 14.2) to work out the molecular structure of the antifreeze protein from the winter flounder (*Pseudopleuronectes americanus*). They grew protein crystals in a solution at 4°C and then examined them by X-ray diffraction at 4°C and −180°C.

The X-ray diffraction data indicated that the 37 amino acids of the antifreeze protein wind into a single, linear alpha helix. Along one side of the helix, side groups of two polar amino acids, threonine and asparagine, extend from the surface at four evenly spaced locations in a flat plane, one at either end of the molecule and two within the helix. Sicheri and Yang propose that these locations, which are spaced 1.65 nm apart, are *ice-binding motifs*.

The four motifs would fit nicely to the tips of ridges formed by water molecules on the surface of an ice crystal, which are spaced at intervals of 1.67 nm **(Figure a).** Hydrogen bonds between the polar amino acid side groups and water molecules along the ridges would hold the proteins tightly to the surface of the ice crystal. The tight fit would prevent more water molecules from adding to the ice surface and thereby prevent further crystal growth.

Understanding how the fish antifreeze proteins work is not just a fascinating scientific issue. The description and characterization of the antifreeze proteins could lead to medical and industrial applications in situations where procedures and equipment must tolerate freezing conditions. In fact, the U.S. Food and Drug Administration recently approved the use of an antifreeze protein—originally discovered in an arctic fish, but now produced by genetically engineered yeast—as an ingredient in ice cream to enhance its creamy texture.

---

temperature from 40° to 20°C. Because torpor reduces the temperature difference between their bodies and the environment, torpid birds lose heat less rapidly. At the nighttime environmental temperatures they usually encounter, the torpid hummingbirds may use 80% less energy than they would if they had not entered a temporarily dormant state.

Global Warming Is Changing the Ecology of Many Organisms

As described in Chapter 51's *Focus on Applied Research,* most scientists agree that the atmosphere is getting warmer. What effect will global warming have on biological systems? Biologists hypothesize that, on the spatial scale of the biosphere, rising temperatures will affect the geographical distributions of populations, species, and communities. Models of climate change predict that the distributions of polar species will contract to even higher latitudes, and the ranges of temperate and tropical species will expand or shift toward the poles. The models also predict that global warming will change the timing of important biological events. For example, plants whose flowering is triggered by warm

springtime temperatures will flower earlier in the season; similarly, migratory animals will return from their wintering grounds and begin reproducing earlier in the year.

Camille Parmesan of the University of Texas at Austin and Gary Yohe of Wesleyan University tested these predictions with a massive literature review. They surveyed studies of changes in the geographical distributions and timing of springtime activities in a wide variety of herbaceous plants, trees, invertebrates, and vertebrates over roughly the past 100 years. Their analysis, published in 2003, suggests that the geographical ranges of 99 species of butterflies, birds, and alpine herbs in the Northern Hemisphere have shifted dramatically into habitats that had previously been too cold for them. Some species have expanded their distributions northward an average of 6.1 km per decade. Other species have shifted their distributions to higher altitude, an average of 6.1 m per decade. Their analysis also indicated that for 172 species of plants, butterflies, amphibians, and birds, springtime growth and reproduction has occurred on average 2.3 days earlier per decade. If these trends continue at the same rate, spring flowering and animal reproduction will occur

Figure 52.9 Observational Research

How Lizards Compensate for Altitudinal Variations in Environmental Temperature

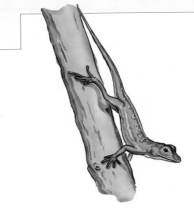

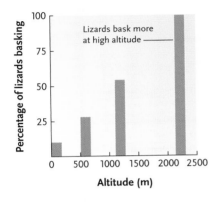
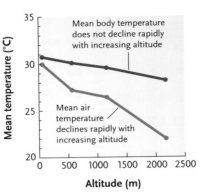

HYPOTHESIS: Lizards living at high altitude can use behaviors to compensate for the low environmental temperatures they encounter.

PREDICTION: The percentage of lizards observed basking in the sun will increase with altitude, and mean air temperatures will vary more than lizard body temperatures among study sites distributed along an altitudinal gradient.

METHOD: Hertz and Huey measured the basking behavior as well as air temperatures and body temperatures of two closely related species of *Anolis* lizards distributed along an altitudinal gradient in the Dominican Republic. They surveyed populations of lizards at sea level, 550 m, 1100 m, and 2200 m altitude. They then compared the percentages of lizards basking and the mean air and lizard body temperatures at the four study sites.

RESULT: The percentage of lizards basking increased steadily with altitude. Mean air temperature differed by as much as 8°C among study sites, but mean body temperature differed by only 2°C.

CONCLUSION: Lizards living at high altitude bask in patches of sun more frequently, partially compensating for the low environmental temperatures in their habitats.

---

one full month earlier in the year 2130 than it did in 2000.

A parallel analysis of less detailed data on 677 species of plants and animals suggests that 62% of the species surveyed showed trends toward earlier flowering, breeding, or growth. And for 434 species in which researchers documented a change in geographical distribution, 80% of the shifts were in the direction predicted by climate change models. Parmesan and Yohe noted that geographical distributions change rapidly, and that species respond to both cooling and warming trends. Marine species in Europe expanded their ranges northward during two warming periods in the twentieth century (1930–1945 and 1975–1999), but shifted their ranges southward during a cooling period (1950–1970).

Global warming is also changing species composition and relative abundance within ecological communities. For example, among invertebrates and fishes on the California coast, cold-adapted species have become less abundant and warm-adapted species more abundant. Comparable changes have been noted in communities from Antarctica to the Arctic.

The geographical distributions of species and communities have often changed with climate shifts over evolutionary time, but the rate of global warming has accelerated in your lifetime. As you know from preceding chapters, the factors that govern the structures of communities and ecosystems are complex, and scientists are far from being able to predict all of the consequences of these changes in detail. In the next section, we describe how today's climate affects species and community distributions on a biosphere-wide scale. You can be certain that biology texts in the twenty-second century will paint a very different portrait of these large-scale associations.

STUDY BREAK

1. How does the behavior of *Anolis* lizards in the Dominican Republic change over altitude?
2. What effect is global warming likely to have on the geographical distributions of organisms?

52.3 Terrestrial Biomes

In Section 22.2 we described how convergent evolution produces morphological and physiological similarities in species that occupy similar environments. Early

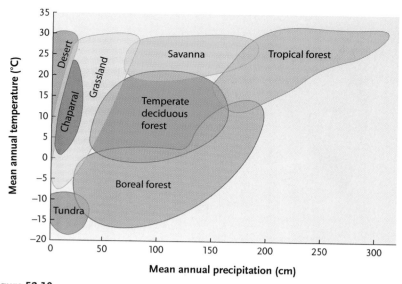

Figure 52.10
Climograph. Each of the major terrestrial biomes occupies a characteristic combination of temperature and moisture conditions.

example, in eastern North America, the temperate deciduous forest biome includes beech-maple forests in the north and oak-hickory forests in the south. Before surveying eight major terrestrial biomes—*tropical forests, savannas, deserts, chaparral, temperate grasslands, temperate deciduous forests, evergreen coniferous forests,* and *tundra*—we consider how environmental factors influence their overall distribution.

Environmental Variation Governs the Distribution of Terrestrial Biomes

Because organisms—and the communities they form—are sensitive to abiotic factors, climate is the main determinant of biome distribution. A **climograph** portrays the particular combination of temperature and rainfall conditions where each terrestrial biome occurs (**Figure 52.10**). For example, some deserts, grasslands, savannas, and tropical forests occur in areas that have comparable mean annual temperatures but vastly different rainfall. Conversely, some biomes, such as boreal forests, temperate deciduous forests, and savannas, are found under similar moisture conditions but different temperature regimes.

Although the climograph provides a general portrait of the temperature and moisture conditions where the different biomes occur, it does not address the de-

in the twentieth century, two American ecologists, Frederic Clements of the Carnegie Institution in Washington and Victor Shelford of the University of Illinois, generalized this observation within a larger perspective by defining the **biome** as a vegetation type plus its associated microorganisms, fungi, and animals. Although vegetation is superficially similar throughout a biome, its species composition varies from place to place. For

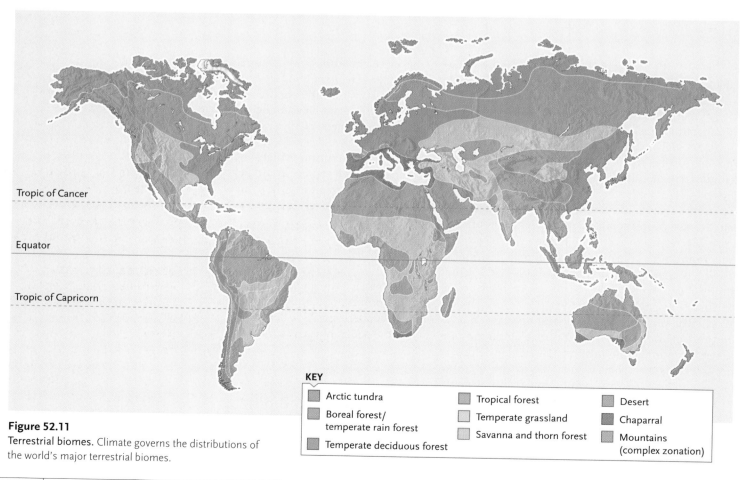

Figure 52.11
Terrestrial biomes. Climate governs the distributions of the world's major terrestrial biomes.

KEY

- Arctic tundra
- Boreal forest/ temperate rain forest
- Temperate deciduous forest
- Tropical forest
- Temperate grassland
- Savanna and thorn forest
- Desert
- Chaparral
- Mountains (complex zonation)

tails of environmental variation. For example, the climograph includes only mean annual temperature and rainfall, not seasonal variation in these factors. Two regions may have the same mean temperature even though one experiences blazingly hot summers and bitterly cold winters and the other has moderate temperature throughout the year; we would expect them to harbor different organisms. Moreover, the distributions of communities are also influenced by nonclimatic factors, such as regional variations in soil structure and mineral composition (see Section 33.2).

Because temperature and rainfall exhibit latitudinal patterns (displayed in Figures 52.2 and 52.5), the distributions of some terrestrial biomes appear as bands on a world map (Figure 52.11). But regional and local climatic variations influence these broad patterns. For example, chaparral is common in certain coastal habitats, whereas grasslands occur further inland at similar latitudes. Comparable bands of distinct vegetation form on mountainsides because temperature and moisture conditions also change with altitude.

Tropical Forests Include the Most Species-Rich Communities on Earth

Three types of **tropical forests**—rain forest, deciduous forest, and montane forest—sweep across the parts of Africa, Asia, Australia, and Central and South America that receive intense solar radiation and heavy rainfall.

Tropical rain forests grow where some rain falls every month, mean annual rainfall exceeds 250 cm, mean annual temperature is at least 25°C, and humidity is above 80%. Limited by neither temperature nor water, the productivity of a tropical rain forest is exceptionally high (see Section 51.1). Trees replace their leaves throughout the year, producing a continuous rain of detritus that ants, land crabs, and other detritivores quickly consume. Decomposers are also active in the hot, moist environment, and almost no litter accumulates on the ground. Because nutrients released by decomposition are promptly absorbed by vegetation or leached by rain, soil in tropical rain forests is nutrient-poor, with low humus content (see Section 33.2).

Tropical rain forests are usually layered (see Figure 50.19). The crowns of tall trees form a dense, tangled canopy that intercepts most incoming sunlight 40 to 45 m above the ground (Figure 52.12). Even the largest trees grow only shallow roots in the thin soil, but many have wide *buttresses*, woody lateral extensions of their trunks, that stabilize them in the ground. Shade-tolerant shrubs and small trees form understory layers below the canopy. The woody stems of lianas climb through both layers, and epiphytes, such as bromeliads and orchids, cover the trunks and branches of trees, especially in sunlit openings. In mature rain forests, the ground is surprisingly bare of leafy vegetation, because very little sunlight reaches the forest floor.

© 1991 Gary Braasch

Figure 52.12
Tropical forest. Many tropical rain forest trees are covered with lianas and epiphytes.

Tropical rain forests probably harbor more plant and animal species than all other terrestrial biomes combined. Ecologists have proposed numerous hypotheses to explain both the evolution and the maintenance of high species richness in these communities (see Section 50.7), but no single hypothesis explains the pattern adequately. In fact, we do not even have a complete species list for any rain forest community, largely because most animals live in the highly productive canopy, which ecologists have only recently begun to study in detail (see *Focus on Research*). The most extensive tracts of tropical rain forest occur in South America, central and western Africa, and Southeast Asia. Unfortunately, they are being cleared at an alarming rate (see Chapter 53); some experts predict that this biome will all but disappear before the middle of the twenty-first century.

Habitats centered at 20° north and south of the equator experience a pronounced summer rainy season and winter dry season. **Tropical deciduous forests** occur where winter drought reduces photosynthesis, and most trees drop their leaves. For example, the monsoon forests of Southeast Asia, which harbor teak and other tropical hardwoods, are as lush as tropical rain forests in the rainy season; but many trees are bare in the dry season.

High altitudes in the tropics support distinctive **tropical montane forests,** or "cloud forests," which are frequently enveloped in mist. The trees, often no more

FOCUS ON RESEARCH

Basic Research: Exploring the Rain Forest Canopy

Biological diversity in tropical rain forests has fascinated naturalists for centuries. Sadly, most of its organisms live beyond our reach. The forest canopy extends from 9 or 10 m above the ground to heights as great as 45 m, making the canopy inaccessible and largely unexplored. Early ecologists were able to study canopy-dwelling species only when they found a fallen tree or followed loggers into the forest. In the 1930s, a clever botanist trained monkeys to retrieve plants from the canopy, but these efforts provided little

A platform in the canopy of a tropical rain forest in Costa Rica provides a comfortable perch for Donald Perry to survey the pollinating activity of birds and bees.

data about the ecological interactions that govern life in the treetops.

Many ecologists still study canopy-dwelling organisms from the safety of the ground. Binoculars provide a good view of fairly large vertebrates. And a hike along a ridge top can provide a canopy-level view of trees growing in an adjacent valley or ravine. Some researchers use ropes to hoist nets or traps into the canopy, lowering them periodically to see what they have caught. Others spray a fog of insecticide into the canopy to kill small invertebrates, which then rain down onto plastic sheets spread below the trees. These ground-based techniques have led to the discovery of hundreds—perhaps thousands—of new arthropod species. Ecologists now collect huge samples of arthropods to study the species composition and structure of communities and to monitor changes in these communities over time. But distant observations and mass sampling techniques don't provide detailed data about which insects are feeding on a tree, how often hummingbirds pollinate a flower, or when a tiny lizard hunts its prey.

Today many ecologists routinely risk life and limb to collect detailed ecological data in the rain forest canopy. They climb trees and crawl along stout branches. Many build stable observation decks with walkways, allowing study on either side of the "trail."

What does this newfound access to the rain forest canopy add to our knowledge of organisms that live there? Researchers can measure the

physical environment of the canopy and observe the physiological and behavioral adaptations of its plants and animals. For example, researchers are gathering data on the feeding habits and behavior of small animals that never venture to the ground, such as fruit-eating bats and birds. When coupled with information about the movement patterns of these animals, the data provide insight into the dispersal of seeds in the fruits. And an understanding of seed dispersal provides information for studies of the population ecology of rain forest trees.

Canopy ecologists have also discovered fascinating relationships between plants and their animal pollinators. For example, Donald Perry, a freelance biologist, discovered that birds are attracted to the sweet nectar of the vine *Norantea sessilis*. Feeding birds step on the vine's sturdy flowers; their feet become covered with the plant's pollen, which is embedded in a gummy substance. When the birds visit another vine of the same species, they transfer the pollen to that plant's flowers, providing cross-pollination, which appears to be necessary for the vine's reproduction.

Research in the tropical rain forest canopy promises exciting discoveries about ecological relationships in this unique biome, which is the most threatened on Earth (see Chapter 53). Such research is essential for developing a public appreciation of tropical forests and for creating conservation plans to preserve them.

than 3 m tall, are densely covered with epiphytes, which thrive in the moisture-laden air. Cloud-forest plants grow slowly because productivity is limited by low temperatures, high humidity (making transpiration difficult), and sunlight-blocking clouds.

Savannas Grow Where Moderate Rainfall Is Highly Seasonal

Grasslands with few trees, the biome called **savanna**, grow in areas adjacent to tropical deciduous forests **(Figure 52.13)**. Seasonality in tropical and subtropical savannas is determined by the availability of water; al-

though annual rainfall averages 90 to 150 cm, droughts typically last for months. Grasses are successful in semiarid conditions because their shallow roots harvest water efficiently. With the onset of seasonal rains, they grow quickly, reaching a height of 2 to 3 m. During the dry season, grasses die back and frequently burn, but their underground parts remain alive and resprout when water again becomes available. Shrubby trees outcompete grasses in moist, low-lying areas or on rocky ground, but periodic fires and grazing mammals eliminate most trees as seedlings.

The largest savannas stretch across eastern and southern Africa; smaller patches occur in India, Aus-

Figure 52.13

Savanna. The African savanna, a warm grassland with scattered stands of shrubby trees, has an enormous concentration of large ungulates (hoofed, herbivorous mammals), such as these wildebeests (*Connochaetes taurinus*).

tralia, and South America. African savannas are home to large herbivorous mammals, including antelopes, zebras, giraffes, and elephants, some of which fall prey to savanna predators, such as lions, leopards, cheetahs, and wild dogs. Grazing mammals follow the seasonal cycle of grasses, migrating away from dry areas to greener pastures.

Thorn forests grow at the arid borders of true savanna, where large mammals are less abundant. Grasses and other plants that store energy in large underground root systems grow among scrubby trees. Thorn forests are also highly seasonal, growing dramatically in the rainy season and dying back during the annual dry season, which may last for 8 to 9 months.

Deserts Develop in Places Where Little Precipitation Falls

Deserts form where rainfall averages less than 25 cm per year. The hot deserts of the American Southwest, northern Chile, Australia, northern and southern Africa, and Arabia occur near 30° latitude, where descending air masses create very dry conditions. Cool deserts, such as the Gobi and Kyzyl-Kum of Asia and the Great Basin of North America, form in massive rain shadows at higher latitudes.

Desert conditions are often extreme. Rainfall arrives infrequently in heavy, brief pulses; and sudden runoff erodes topsoil, which often has high mineral content but little organic matter. Dry air and scant cloud cover allow most sunlight to reach the ground, raising daytime air and ground temperatures as high as 45°C and 70°C, respectively. At night, the surface loses heat quickly; in some deserts, temperatures drop below freezing in winter.

Desert vegetation is always sparse because arid environments do not favor large, leafy plants. Some deserts, such as the Namib of Africa and the Atacama-Sechura of South America, receive so little rainfall that large areas are practically devoid of vegetation. By contrast, the hot Sonoran Desert in northern Mexico, southeastern California, and southern Arizona harbors a diverse flora, including deep-rooted shrubs and shallow-rooted cacti **(Figure 52.14)**. Mesquite and cottonwood trees grow deep taproots into the permanent water supply below streambeds. Perennial plants often protect their tissues from herbivores with spines or toxic chemicals, and many use CAM photosynthesis to conserve water (see Section 9.4). After seasonal rains, annual plants germinate, mature, flower, and produce seeds before brutally dry conditions resume.

Deserts also support abundant animals, most of them fairly small. Ants, birds, and rodents often subsist on seeds. Some seed-eating mammals survive on the water they extract from food. Insects, some lizards, and mammals consume the sparse vegetation. Scorpions, lizards, and birds feed primarily on insects; snakes, owls, and foxes prey on other animals. Most

Figure 52.14

Desert. The warm Sonoran Desert near Tucson, Arizona, is home to columnar saguaro cacti (*Carnegiea gigantea*) and other drought-adapted plants.

Figure 52.15
Chaparral. Chaparral covers broad expanses of hills in coastal areas of central and southern California.

central and southern California, central Chile, southwestern Australia, southern Africa, and the Mediterranean region.

Chaparral shrubs are dense, with hard, tough, evergreen leaves **(Figure 52.15)**. They build woody stems above ground and large root systems in the soil. Many species, such as sages (genus *Salvia*), produce toxic, aromatic compounds that inhibit the germination and growth of potential competitors. Just after the winter rains, the shrubs are covered with new leaves and flowers, and the vegetation teems with insects and breeding birds. During the hot, dry summers, however, most plants are dormant, and lightning sparks frequent fires. The aromatic oils and resins of many species, such as eucalyptus, make them highly flammable. Their aboveground parts burn swiftly, but they quickly resprout from large root crowns. Other species release seeds from fire-resistant cones or pods, and their seedlings grow in ash-enriched soil.

Temperate Grasslands Are Held in a Disclimax State by Periodic Disturbance

Temperate grasslands include the prairies of North America, the steppes of central Asia, the pampas of South America, and the veldt of southern Africa **(Figure 52.16)**. They stretch across the interiors of continents, where winters are cold and snowy and summers are warm and fairly dry. Only 25 to 100 cm of rain falls unevenly through the year. Temperate grasslands are disclimax communities: seasonal drought, periodic fires, and grazing by mammals inhibit succession, preventing shrubs and trees from displacing perennial grasses and herbaceous plants (see Section 50.6). Grassland soil is rich in organic matter because the aboveground parts of most plants die and decompose annually.

desert animals avoid the midday heat and dehydrating conditions; many retreat into underground burrows, where water vapor from their respiration cools and moistens the air. Many species are nocturnal or active only in the early morning and late afternoon.

Chaparral Grows Where Winters Are Cool and Wet and Summers Are Hot and Dry

A scrubby mix of short trees and low shrubs called **chaparral** dominates narrow sections of coastal land between 30° and 40° latitude, where winters are cool and wet and summers hot and dry. Seasonal rainfall averages only 25 to 60 cm per year. Chaparral occurs in

a. Shortgrass prairie

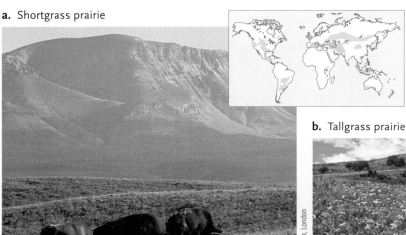

b. Tallgrass prairie

Figure 52.16
Temperate grassland. **(a)** The western plains of North America were once covered with shortgrass prairie, as shown here east of the Rocky Mountains. Bison were the dominant large herbivores. **(b)** Tallgrass prairie, like this lush patch in eastern Kansas, once covered the eastern plains.

In North America (see Figure 52.16a) shortgrass prairie covers much of the west, where winds are strong, rainfall light and infrequent, and evaporation rapid. Drought-tolerant perennials have deep roots, and their underground rhizomes, which store energy, resprout quickly after a fire. Tallgrass prairie (see Figure 52.16b) once occupied moister regions to the east of the shortgrass prairie. It boasted an abundance of legumes and sunflowers, often 3 m tall, but most of it was converted to farmland long ago; small patches still exist in nature preserves and in glades within eastern deciduous forests.

North American grasslands are still occupied by large grazing mammals, including pronghorns and bison, which once numbered in the millions. The most familiar burrowing mammal is the prairie dog, a rodent, but pocket gophers, ground squirrels, and jackrabbits are also common. Wolves were the primary large predators until they were hunted nearly to extinction. Coyotes, foxes, ferrets, hawks, and owls still take small prey today.

Temperate Deciduous Forests Experience Seasonal Dormancy

At temperate latitudes, with warm summers, cold winters, and annual precipitation between 75 and 250 cm, **temperate deciduous forests** grow at low to middle altitudes. In winter, low temperatures reduce photosynthetic rates, and snow and ice can damage leaves. Thus, most plants shed their leaves and grow new ones in spring **(Figure 52.17)**. The thick layer of leaf litter, which releases mineral nutrients as it decomposes, enriches the soil. Decomposition is slow, however, because the growing season is only about 7 months long.

Temperate deciduous forests have much lower species richness than tropical forests. Trees form a canopy 10 to 35 m high, and woody shrubs form an understory below it. Herbaceous plants and a ground layer of mosses or liverworts grow below the shrubs. Many herbaceous plants, including some terrestrial orchids, flower early in spring, before trees produce sunlight-blocking leaves; others flower near the end of the growing season.

Forests of ash, beech, birch, chestnut, elm, and oak stretched unbroken across eastern North America, Europe, and eastern Asia before farmers cleared the land. In North America, introduced diseases and insects have nearly eliminated the once dominant species, such as American chestnut and American elm. Today, beech, birch, and maple predominate in the Northeast; oak–hickory forests dominate farther south and west; and oak woodlands merge into tallgrass prairie to the west. Before the arrival of Europeans, deer, bison, bears, and pumas roamed the forests with many smaller species of animals. Today, small mammals such as voles, mice, chipmunks, squirrels, rabbits,

Summer

Winter

Thomas E. Hemmerly

Thomas E. Hemmerly

Figure 52.17
Temperate deciduous forest. Seasonal variations in temperature and water change the character of this forest south of Nashville, Tennessee.

opossums, and raccoons predominate, although deer and bears have recently surged in abundance.

Evergreen Coniferous Forests Predominate at High Northern Latitudes

The **boreal forest**, or **taiga** (Russian for "swamp forest"), is a circumpolar expanse of evergreen coniferous trees in Europe, Asia, and North America **(Figure 52.18)**. Snow blankets the ground during long and extremely cold winters, and most precipitation falls during the short summer. In the northernmost taiga, plants grow quickly during long (18-hour) summer days.

Stands of white spruce and balsam fir dominate North America's boreal forest. Their needle-shaped leaves have a thick cuticle and recessed stomata that conserve water during winter, when ground water is frozen. Fallen needles acidify the thin soil, which speeds the leaching of most nutrients, and few shrubs and herbaceous plants grow beneath the conifers. Lightning-sparked fires are common; some deciduous trees grow in areas opened by fire, but conifers eventually replace them. Cold streams, marshes, ponds, and lakes often dot the landscape; at flat, poorly drained sites, peat mosses, shrubs, and stunted trees dominate acidic bogs, called muskegs.

Most taiga is relatively undisturbed by humans, and it still harbors its native animals. Moose, elk, and deer are the dominant large herbivores. Hare as well as squirrels, porcupines, and other rodents also feed on plants. Some small animals are active all winter in runways they dig beneath the snow. Wolves, lynx, and

Figure 52.18

Boreal forest. Single-species stands of spruce dominate this boreal forest, the predominant forest at high latitudes in the Northern Hemisphere.

on the Olympic Peninsula receives 500 cm of rainfall per year, as much as some tropical forests. This temperate rain forest harbors some of the world's tallest trees, including Douglas fir and Sitka spruce to the north and coast redwoods to the south.

Tundra Comprises a Vast, Treeless Plain in the Northernmost Habitats

The treeless **arctic tundra** stretches from the boreal forests to the polar ice cap in Europe, Asia, and North America. Covering almost 5% of the land, this biome is windswept and wet. Winter temperatures are consistently below freezing. The 2-month summer is so cool that only the topmost layer of soil ever thaws, leaving the ground below perpetually frozen; in some areas, this **permafrost** is more than 500 m thick. Although less than 25 cm of precipitation falls each year, evaporation is slow, and permafrost is impermeable; thus, low-lying soil remains permanently waterlogged, forming bogs **(Figure 52.19a)**. Anaerobic conditions and low temperatures retard decomposition, and soggy masses of detritus accumulate.

Plants in the tundra are short because the weak sunlight and minimal growing season provide barely enough energy and warmth for net primary productivity; moreover, strong winter winds shred any plants with a high profile. The vegetation consists of low-growing lichens, mosses, grasses, perennial herbs, dwarf shrubs, and a few stunted trees, usually less than 1 m tall. During summer's nearly continuous sunlight, plants flower profusely, and their fruits ripen fast.

Some animals, including herbivorous arctic hares, lemmings, and willow ptarmigans as well as predatory

wolverines prey on herbivores. Grizzly bears and black bears roam the forest, devouring seeds, berries, fishes, and small animals. Mosquitoes, black flies, and gnats are superabundant near bogs and lakes in summer.

Other types of coniferous forest grow in more southerly coastal lowlands where winters are mild and wet and the summers are cool. For example, a **temperate rain forest,** supported by heavy rain and fog, parallels the coast from Alaska into northern California. In western Washington State, the rain forest

a. Arctic tundra

b. Alpine tundra

Figure 52.19

Tundra. **(a)** Rain and snowmelt cannot percolate through the arctic tundra's permafrost. In summer water accumulates in ponds and bogs as shown in this aerial photograph of the tundra in northern Russia. **(b)** Compact, short plants form the alpine tundra, which grows on mountaintops at temperate latitudes, such as the Cascade Range of Washington state.

snowy owls, wolves, foxes, and lynx, are permanent tundra residents. In summer, herds of herbivorous musk oxen, caribou, and reindeer migrate there from boreal forests, and migratory shorebirds and waterfowl arrive to breed. Flying insects abound in summer, especially mosquitoes and black flies, which reproduce in boggy habitats.

A similar biome, called **alpine tundra,** occurs on high mountaintops throughout the world **(Figure 52.19b).** Dominant plants form cushions and mats that withstand the buffeting of strong winds. Winter temperatures are well below freezing, and shaded patches of snow persist even in summer. The thin, fast-draining soil is nutrient-poor, and primary productivity is low.

STUDY BREAK

1. Which terrestrial biomes occur in habitats that receive the greatest amount of rainfall?
2. Which terrestrial biomes are renewed by periodic fires?
3. Which terrestrial biomes have the tallest vegetation? Which ones have the shortest?
4. In which terrestrial biomes are the trees usually evergreen?

52.4 Freshwater Biomes

Aquatic biomes comprise several distinctive habitats in either freshwater or marine environments. Freshwater biomes occur where water with a salt concentration below 0.5% accumulates or moves through a landscape. Ecologists distinguish between *lotic* biomes, where water flows through channels, and *lentic* biomes, where water stands in an open basin. All freshwater biomes interact with surrounding land, because runoff carries a nearly constant input of nutrients. Highly productive ecotones, called **wetlands,** often define the borders of freshwater biomes. These marshes and swamps may harbor an astounding array of microorganisms, algae, plants, invertebrates, and vertebrates.

Streams and Rivers Carry Water Downhill to a Lake or the Sea

The flowing-water biomes start as seeps on high ground. As they flow downhill, they grow into narrow streams, which merge to form wide rivers **(Figure 52.20).** Streams and rivers include three habitats. *Riffles* are shallow, fast-moving, turbulent stretches over a rough bottom of pebbles or rocks. *Pools* are deep, slow-moving areas with a smooth sand or mud bottom. *Runs* are deep, fast-moving stretches over smooth bedrock or sand. Streams generally have high flow rate, low volume, and lots of riffles and pools. As they merge into rivers, flow rate declines, but flow volume increases, and runs and pools predominate. Flow rate and volume also vary seasonally with the rate of water input from rainfall and snowmelt and geographically with altitude and topography.

Physical factors change over the length of a flowing-water system. The concentration of suspended particulate material is low in streams, but high in rivers, which are often turbid with silt. Temperature also increases as water flows downstream to warmer lowland habitats. Because oxygen is more soluble in cold water than in warm water, dissolved oxygen is usually higher in streams than in rivers. Erosion of the streambed and surrounding land provides the solute content of flowing water. Today, agricultural runoff and industrial and municipal wastes provide major input. In unpolluted streams, organic detritus provides more than 95% of the nutrients and energy entering aquatic food webs. This input is particularly important in streams flowing through dense forests, where vegetation blocks the sunlight necessary for primary productivity.

The flow of water affects every aspect of life in streams and rivers. In swift-moving riffles, primary producers cling permanently to fixed substrates, because phytoplankton are swept away by the current. Insect larvae and other invertebrates attach to the un-

a. A stream

E. F. Benfield, Virginia Tech

b. A river

© Dr. Morley Read/Photo Researchers, Inc.

Figure 52.20
Stream and river habitats. **(a)** In streams, such as this one in Virginia, water flows quickly through narrow channels, often with a rocky bottom. **(b)** In rivers, like the Rio Napo in Ecuador, water flows more slowly through broad channels, and suspended sediments often make the water murky.

Figure 52.21
Lakes. A lake in Torres del Paine National Park, a biosphere reserve in Chile.

dersides of rocks, and many species are flattened, maintaining a low profile in the current. By contrast, large rivers have dense populations of algae and cyanobacteria, which attach to rocks and other substrates, and rooted aquatic plants at the river's edge.

Lakes Are Bodies of Standing Water That Accumulates in Basins

Lakes and other standing-water biomes are generally fed by rainfall and by streams and rivers that drain surrounding watersheds **(Figure 52.21).** Because the availability of light affects a lake's primary productivity, ecologists often distinguish between the **photic zone** of a lake, the surface water that sunlight penetrates, and the deeper **aphotic zone,** which is always dark.

Lake Zonation. Every lake includes zones, defined by depth and distance from the shore, that provide distinctive environments **(Figure 52.22).** In the **littoral zone,** the shallow water near the shore, sunlight penetrates to the bottom. Enriched by nutrients made available by decomposers and runoff, the littoral zone has high productivity and species richness. Rooted aquatic plants, such as cattails and water lilies, grow above the surface, and "floating aquatics," such as duckweed, are common. Submerged vegetation harbors a rich community of microorganisms, epiphytes, and invertebrates. Numerous animals—insects, worms, snails, crayfish, fishes, frogs, turtles, and water birds—use the littoral zone to feed and reproduce.

The **limnetic zone,** the sunlit water beyond the littoral, supports plankton communities: the primary producers are phytoplankton—cyanobacteria, diatoms, and green algae; the primary consumers are zooplankton—rotifers, copepods, and other tiny heterotrophs. Small fishes, which feed on plankton, are themselves consumed by larger fishes, such as bass.

Photosynthesis is impossible, however, in the **profundal zone,** the perpetually dark water below the limnetic zone. Nevertheless, a constant rain of detritus from the limnetic zone supports a community of bacterial decomposers and animal detritivores, including worms, clams, insect larvae, and catfish.

Seasonal Changes in Temperate Lakes. In temperate areas, seasonal temperature variations induce changes in the vertical zonation of lakes **(Figure 52.23).** Like other liquids, water gets denser as it cools. But water has a unique property: it reaches maximum density at 4°C, with the density declining as it gets colder. Thus, water at 4°C sinks below water that is either warmer or colder; ice floats because it is less dense than very cold water.

During winter, ice forms on the surface of temperate zone lakes. Water temperature varies from near freezing just below the ice to 4°C at the bottom. Differences in the density of water at 0° and 4°C maintain this thermal stratification. In spring, as the ice melts, the warmer, denser water sinks; and the surface temperature gradually rises to 4°C. For a brief time, the temperature is uniform at all depths. Winds blowing across the lake create vertical currents that cause a **spring overturn,** mixing surface water with deep water. Oxygen at the surface moves to the bottom, and nutrients from the bottom move to the surface.

By midsummer, sunlight heats the top layer of the limnetic zone, called the **epilimnion,** to temperatures above 4°C. In large lakes, the epilimnion may be more than 10 m deep. In the deep water of the lake's profundal zone, called the **hypolimnion,** the temperature remains near 4°C. However, at the boundary between the epilimnion and the hypolimnion, water temperature changes abruptly over a narrow depth range, called the **thermocline.** The thermocline prevents vertical mixing because warm surface water floats above the thermocline, and cool deep water stays below it. During summer, nutrient-rich detritus sinks to the bottom of the lake, where decomposition depletes the oxygen dissolved in the hypolimnion. In autumn, declining sunlight and winds cause the epilimnion to cool, and as the water becomes denser, it sinks, eliminating the thermocline. Winds then mix the water vertically once again during an **autumn overturn,** and dissolved gases and nutrients are equalized at all depths.

Primary productivity in the limnetic zone varies with the seasonal overturns. In spring, increased sunlight, warm temperatures, and the sudden

Figure 52.22
Lake zonation. The zonation in a lake is based upon the water's depth and its distance from shore.

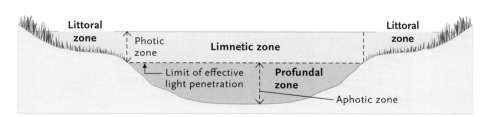

availability of nutrients induce a bloom of productivity. As the season progresses and the thermocline prevents vertical mixing, nutrient levels dwindle in the epilimnion, and primary productivity declines. By late summer, nutrient shortages limit photosynthesis. After the autumn overturn, nutrient cycling drives a short burst of primary productivity. But as days get shorter and temperature declines, primary productivity remains low until spring.

Trophic Nature of Lakes. Ecologists classify lakes by their nutrient content and rates of productivity. **Oligotrophic lakes** are poor in nutrients and organic matter, but rich in oxygen. Their low primary productivity keeps the water crystal clear, making them popular recreational sites. By contrast, **eutrophic lakes** are rich in nutrients and organic matter. The decomposition of organic matter depletes oxygen in the hypolimnion when the lake is stratified, and high primary productivity in the epilimnion often chokes the water with seasonal blooms of cyanobacteria and filamentous algae. Eutrophic lakes are often thick and "soupy," making them unattractive for recreation. Over long periods of time, as sediments accumulate, lakes naturally change from oligotrophic to eutrophic; their basins eventually fill with sediments, and terrestrial plants invade.

As you learned in the description of changes in Lake Erie at the beginning of Chapter 51, the addition of nutrients to a lake often disrupts its trophic condition. In a classic experiment conducted in the late 1960s, David Schindler and his colleagues at The Experimental Lakes Project in Ontario, Canada, experimentally separated the two basins of a lake with a plastic curtain. The researchers added phosphates to one basin and used the other basin as a control. Within 2 months, the artificially enriched basin sported a bloom of cyanobacteria, a sign of eutrophication; the control basin remained oligotrophic and crystal clear **(Figure 52.24)**.

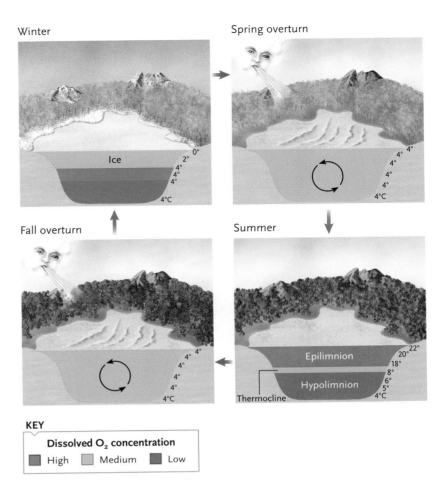

Figure 52.23
Seasonal overturns in lakes. The waters of shallow temperate-zone lakes mix twice each year. During the spring and autumn overturns, temperature is equalized at all depths; nutrients are carried upward from the bottom; and oxygen is carried downward from the surface.

52.5 Marine Biomes

Marine biomes, in which salinity (salt concentration) averages about 3%, cover nearly three-fourths of Earth's surface and account for a large fraction of its primary productivity. They also mediate important global processes: marine phytoplankton process large amounts of carbon dioxide, generating oxygen and moderating the greenhouse effect.

As with standing freshwater biomes, depth and distance from shore govern the physical characteristics of marine habitats. Ecologists describe ocean zonation in several ways **(Figure 52.25)**, including the distinction between the photic and aphotic zones. Another major distinction is between the **pelagic province**, the water, and the **benthic province**, the bottom sediments. The pelagic province includes the **neritic zone**, the shallow water above the continental shelves, and the **oceanic zone**, the deep water beyond them. The benthic province is divided into the **intertidal zone**, the shoreline that is alternately submerged and exposed by tides, and the **abyssal zone**, the bottom sediments that lie permanently below deeper water. Here we describe five marine biomes—*estuaries, rocky and sandy coasts, continental shelves and oceanic banks, open ocean,* and *benthic regions*—that represent particular associations of organisms occupying different marine zones and provinces.

Figure 52.24 Experimental Research

Artificial Eutrophication of a Lake

QUESTION: Does the addition of excess phosphorus to a lake encourage the growth of primary producers, such as cyanobacteria?

EXPERIMENT: Schindler and his colleagues experimentally separated the two basins of a lake in Ontario, Canada, with a plastic curtain. The researchers added phosphates to one basin and used the other basin as a control.

RESULTS: Within 2 months, the artificially enriched basin (in the upper left of the photo) sported a pale green bloom of cyanobacteria, a sure sign of eutrophication; the control basin remained oligotrophic and crystal clear.

D. W. Schindler, *Science*, 897–899

CONCLUSION: The addition of excess phosphorus to a lake encourages blooms of cyanobacteria, causing the lake to change from oligotrophic to eutrophic.

Estuaries Form Where Rivers Meet the Sea

Estuaries are coastal regions where seawater mixes with fresh water from rivers, streams, and runoff **(Figure 52.26).** Salinity is low where fresh water enters the estuary and high on the tidal side. After heavy rainfall, fresh water floods into the habitat, reducing salinity and raising water temperature. At high tide, cold, salty water flows in from the sea. All estuarine organisms must tolerate these variable conditions.

Variations in local topography influence an estuary's physical features. Chesapeake Bay in Maryland, Mobile Bay in Alabama, and San Francisco Bay in California are broad, shallow estuaries. The estuaries in Alaska and British Columbia are narrow and deep, as are Norway's fjords. Many estuaries are bordered by **salt marshes,** tidal wetlands dominated by emergent grasses and reeds (see Figure 52.26a). In tropical estuaries, the roots of densely packed mangrove trees penetrate the muddy bottom, accumulating sediments and slowly adding land to the shoreline (see Figure 52.26b).

The constant input of nutrients and removal of wastes by the tides contribute to exceptionally high productivity in estuaries. Primary producers include phytoplankton, salt-tolerant grasses and reeds that can withstand submergence at high tide, and algae that grow in mud and on plant surfaces. Roots and stems trap organic matter, which enters detrital food webs. The detritus (and bacteria clinging to it) supports nematodes, snails, crabs, and fishes; suspension-feeding mollusks and arthropods capture edible particles in the slowly moving water. Many marine arthropods and fishes breed in calm, shallow estuaries, where their young find abundant food and refuge from predators in the complex vegetation. Migratory birds use estuaries as rest stops, and shore birds and waterfowl use their muddy bottoms as rich feeding grounds, particularly at low tide.

Rocky and Sandy Coasts Experience Cyclic Periods of Exposure and Submergence

The intertidal zone, the area between low and high tide marks, is one of the most stressful habitats on Earth. On rocky shores, residents are battered by waves and floating debris. Sessile species, such as mussels and barnacles, attach to substrates with special structures or cement. Motile species, such as limpets and sea stars, simply hang onto rocks. Organisms that live high on the shore dry out at low tide, freeze in winter, and bake in summer. Exposed animals often seal themselves inside shells, and intertidal algae have thick polysaccharide coats that adsorb water and prevent dehydration.

Biotic interactions also take their toll. Organisms throughout the intertidal zone compete for attachment sites to avoid being washed away (see Figure 50.12). At low tide, predatory birds and mammals attack from above; at high tide, predatory fishes move in from the sea. Because the tides often scour detritus from the rocky intertidal, grazing food webs predominate.

Rocky shores often have three zones **(Figure 52.27).** The *upper intertidal* is submerged only during the highest tide of the lunar cycle. It is sparsely populated by barnacles, sturdy algae, and grazing and predatory snails. The *middle intertidal* is submerged daily during the highest regular tide and exposed during the lowest. Its tide pools are occupied by red, brown, and green algae, grazing and predatory mollusks, sponges, sea anemones, worms from several phyla, hermit crabs, echinoderms, and small fishes. Biodiversity is greatest in the *lower intertidal,* which is exposed only during the lowest tide of the lunar cycle. It is occupied by dense beds of algae, tunicates, echinoderms, other invertebrates, and fishes.

Sandy shores are composed of loose sediments that waves and currents constantly rearrange. Large plants cannot grow on such unstable substrates, so grazing food webs are rare. Organic debris imported from offshore or from nearby land supports detrital food webs. Animals live in burrows, which they must frequently repair as the substrate shifts. Crabs and shorebirds live as scavengers or predators above the high tide mark. At night, beach hoppers and ghost crabs leave their burrows, seeking food. Marine worms, clams, crabs, and other invertebrates live in the sand between the high and low tide marks.

Light Penetrates the Shallow Water over Continental Shelves and Oceanic Banks

The neritic zone includes the shallow water over continental shelves and oceanic banks, underwater landmasses that rise to within 300 m of the surface. Although small in area, the neritic zone is highly productive and species-rich **(Figure 52.28).** Runoff from the land brings a steady inflow of nutrients; and upwelling and waves circulate nutrients from the bottom to the photic zone.

In temperate regions, giant kelp forests, which are among Earth's most productive ecosystems, occupy some continental shelves and banks (see Figure 52.28a). Kelp are enormous algae that attach to the bottom with giant holdfasts; their stipes ("stems") reach upward with fronds fanning out into the water. Sea anemones, snails, echinoderms, lobsters, and other invertebrates live in the kelp, where fishes and other predators consume them. Even where kelp does not grow, continental shelves and banks teem with life. Most of the important fisheries in the temperate zone occur there.

In the tropics, the warm but nutrient-poor water above continental shelves is often occupied by **coral reefs** (see Figure 52.28b). Sunlight penetrates the clear water all the way to the bottom. Photosynthetic dinoflagellates, living as endosymbionts of the coral animals (see Section 26.2), and coralline algae are largely responsible for primary productivity. Coral animals also feed on microscopic organisms and suspended particles. The reefs are the remains of corals, algae, and

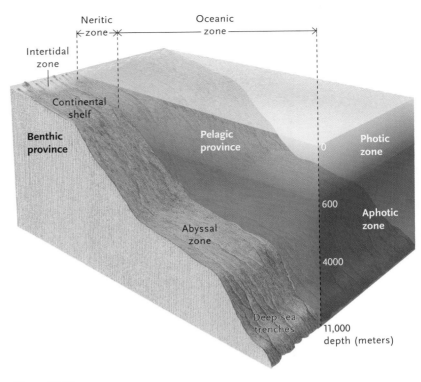

Figure 52.25
Oceanic zonation. Ecologists divide the ocean into the pelagic province (the water) and the benthic province (the ocean bottom). Zones are defined according to the depth of water (photic versus aphotic zones) and distance from shore (neritic versus oceanic zones in the pelagic province, intertidal versus abyssal zones in the benthic province). The different zones are not drawn to scale.

other organisms, and their structural complexity rivals that of tropical rain forests. Tides and currents carve ledges and caverns; and storms frequently disturb the reefs, creating openings in which new coral colonies can grow (see Section 50.5). A reef may be festooned with as many as 750 species of corals and a dizzying variety of algae. The diversity of coral skeletons provides a complex structure that is used by invertebrates from nearly every phylum and by a host of herbivorous and carnivorous fishes.

a. Salt marsh grasses

© Annie Griffiths Belt/Corbis

b. Mangroves

© Doug Peebles/Corbis

Figure 52.26
Estuaries. **(a)** The salt marsh grass (*Spartina* species) is the major producer in a South Carolina estuary. **(b)** Red mangroves (*Rhizophora mangle*) are abundant in Florida's Everglades.

Courtesy J. L. Sumich, Biology of Marine Life, fifth edition, W. C. Brown, 1992

Upper intertidal

Middle intertidal

Lower intertidal

Figure 52.27
Vertical zonation in the intertidal. A rocky shore in the Pacific Northwest clearly exhibits the characteristic vertical zonation. The distance between low and high tide marks on this rocky shore is about 3 m.

In the Open Ocean, Photosynthesis Occurs Only in the Upper Layers

The oceanic zone lies beyond the continental shelves. Though generally low in nutrients, it is locally enriched by runoff from land and by upwelling bottom waters. The open ocean is typically cold, except in the tropics. The surface water is illuminated by sunlight, which warms it somewhat and allows photosynthesis. Most primary productivity is restricted to a depth of about 50 m, however, because seawater filters light. Photosynthetic activity varies seasonally, as it does on land.

"Pastures" of phytoplankton are eaten by zooplankton, including copepods, shrimplike krill, small worms, cnidarians, and the larvae of invertebrates and fishes. Consumers that can actively swim against the currents, such as squids, fishes, marine turtles, and whales, are called **nekton (Figure 52.29).** Some consumers feed on plankton, and some prey on other nekton. Low light levels in water between about 50 and 600 m allow little photosynthesis, but many fishes and some mobile invertebrates are active at these depths, traveling into the sunlit zone to feed on organisms near the surface.

No sunlight ever penetrates the deepest part of the oceanic zone below 600 m. Some of these abyssal regions, such as the Marianas Trench, are more than 9 km below the surface. Scientists have explored the deepest water in the ocean only during the past few decades, but we know that it is a cold (2° to 3°C), dark environment, where organisms live under tremendous pressure from the ocean above. Abyssal communities are surprisingly diverse, although population densities tend to be low. The denizens of the abyssal zone include invertebrates, bony fishes, and sharks. Some fishes and invertebrates are bioluminescent, producing spots of light that may serve for communication or as lures to entice prey within reach of their large jaws **(Figure 52.30).**

The Benthic Province Includes the Rocks and Sediments of the Ocean Bottom

The benthic province extends from the intertidal zone to the deep-sea trenches. In the oceanic zone, bottom sediments are composed of soft mud, fine particles of silt, detritus, and the shells of dead microscopic organ-

a. Kelp forest   **b.** Coral reef

© Georgette Douwma/Photo Researchers, Inc.

© Norbert Wu

Figure 52.28
Neritic zone. (a) Kelp forests, such as this one off the coast of California, often grow in the neritic zone along the coast at temperate latitudes. **(b)** This coral reef in the Raja Ampat islands of Indonesia illustrates the structural complexity and biological diversity found in reef communities.

isms. Species living in and on the bottom are collectively called **benthos.** Sunlight never strikes the benthic province of the open ocean, which is inhabited by bacteria, fungi, and a variety of animals. Sessile invertebrates, such as sponges, sea anemones, and clams, live amidst the sediments, and many motile animals, including worms, mollusks, crustaceans, echinoderms, and fishes, form detrital food webs supported by organic remains that sink from pelagic communities.

In 1976, researchers found communities thriving near hydrothermal vents at a depth of 3000 m near the Galápagos Rift, a volcanically active boundary between two crustal plates. Near-freezing water seeps into fissures where it is heated to temperatures of 350°C or higher. Pressure forces the heated water upward, and minerals are leached from porous rocks as the water spews out through vents in the seafloor. This hydrothermal outpouring releases hydrogen sulfide, which serves as an energy source for chemoautotrophic bacteria, the primary producers in hydrothermal vent communities. Some of these bacteria live as endosymbionts of giant clams and tube-dwelling worms **(Figure 52.31).** Deep-sea food webs also include sea anemones, crustaceans, and fishes. Researchers have located hydrothermal vent ecosystems in the South Pacific, near Easter Island; in the North Pacific, off the coast of British Columbia; the Gulf of California, about 150 miles south of the tip of Baja California, Mexico; and the Atlantic.

Recent research in the deepest reaches of the ocean reveals that communities also exist in areas far from hydrothermal vents. These "cold seep" communities thrive on broad expanses of the seafloor, where extremely salty water percolates upward from the underlying rocks and sediment, carrying abundant minerals, hydrogen sulfide, and methane to areas that are accessible to organisms. Chemosynthetic bacteria, which grow in large mats, can metabolize these molecules, forming the base of food webs that also include sponges, worms, and bivalve mollusks.

Figure 52.29

Open ocean. A humpback whale *(Megaptera novaeangliae)* breaches (leaps out of the water) near a British Columbia coastline.

Figure 52.30

Deep sea. A deep-sea anglerfish *(Himantolophus* species) uses a bioluminescent lure to attract prey to its formidable jaws.

Figure 52.31

Deep benthos. Giant tube-dwelling worms are common in the hydrothermal vent communities on the deep ocean floor.

STUDY BREAK

1. What is the difference between the benthic and pelagic provinces of the ocean?
2. Which marine biomes experience the largest fluctuations in salinity (salt concentration) over time?
3. Which marine biomes or regions within marine biomes receive abundant energy input from sunlight?
4. What is the source of nutrients and energy for the benthos of the oceanic zone?
5. What organisms are the primary producers in hydrothermal vent and cold seep communities, and how do they differ from the primary producers in the photic zone?

How will biomes change in response to anthropogenic (human-induced) global warming?

Will biomes remain largely intact and simply shift their geographical distributions northward or upward (to higher altitudes)? Or will the biomes we recognize today become disrupted, and new biomes arise as species associate in different combinations? Parmesan and Yohe estimated that 59% of wild species around the world have already shown some change in their geographical distributions in response to the relatively small level of global warming—a 0.7°C rise in average temperature—over the past 100 years. Documented responses to global warming vary from species to species, however. For example, only 20% of butterfly species in Spain, France, and North Africa have shifted their southern range boundaries northward, but 70% of butterfly species in the United Kingdom and Scandinavia have expanded their northern range borders further northward, sometimes by as much as 300 km over the past 30 years.

Thus, although the distributions of some species appear stable, the ranges of others are showing strong responses to global warming. At least two hypotheses may explain these patterns. First, some species may be stressed by rising temperatures but have not yet shown a measurable response. Second, the geographical distributions of some species may not be governed primarily by climate. Whatever the reason, the fact that we observe large variation in the response of different species suggests that not all species in a community are moving together. Thus, the existing communities of birds, butterflies, and trees are being disrupted—with some species moving and others not. Biologists have also noted differences in the response of different taxonomic groups. Butterflies in Europe and North America seem to be shifting their distributions northward and upward at about the same rate that temperatures are changing, but plants appear to lag behind. Alpine herbs in Switzerland, for example, have shifted their distributions upward at about half the rate that one might expect from the rate of regional warming, and it was 30 years after warming began before tree seedlings in Sweden started to colonize alpine habitats at higher elevations, shifting the treeline upward.

Even bigger questions remain. Will the vegetation that currently lives in the tropics expand into what is now the temperate zone and cover more of the planet? Some studies suggest that tropical lowland trees are already at their physiological limit—already showing signs of stress by shutting down photosynthesis on the hottest and driest days. Furthermore, climate model projections from a 2007 report of the Intergovernmental Panel on Climate Change consistently show substantial drying as well as warming in midlatitudes. If this projection is correct, many plants and animals now living in the wet tropics will be unable to shift northward into what may become an extreme desert climate. And what will happen to the arctic tundra? Researchers have already collected strong evidence that shrubs and trees are encroaching northward into the tundra of Alaska and Canada. The permafrost is melting, and the soil is drying. How do these observed changes in plant and animal distributions relate to the future? Human activity has already caused Earth's mean annual temperature to rise by 0.7°C in the past 100 years. Climate model projections suggest that further increases between 1.8°C and 4.0°C are likely; some models suggest the rise will be over 6.0°C. Can the tundra biome survive even the lowest projections—more than twice the warming it has already experienced?

How will evolution shape the ways that wild species respond to climate change? Which groups of organisms are likely to adapt, and which are likely to become extinct?

Populations are evolving all the time in response to changing selection regimes. Global warming is one of many human-driven environmental changes that could foster genetic change. Biologists have known for decades that organisms are locally adapted to the climatic conditions under which they routinely live. Scientists have documented local genetic changes toward more warm-adapted genotypes in fruit flies, mosquitoes, and the algal symbionts of corals. Do the observed genetic changes suggest that these species are adapting to anthropogenic global warming? Will other species follow suit? The fossil record suggests that during the Pleistocene glaciations, when Earth's temperature shifted between glacial periods (4°–8°C colder than now) and interglacial periods (today's temperatures), very few species became extinct and few experienced substantial morphological evolution. But, before the Pleistocene, Earth was much hotter than it is today, and the atmosphere had higher levels of CO_2. During the transition from these very warm, high CO_2 conditions to the colder, low CO_2 conditions of the Pleistocene, a large proportion of species became extinct. To how much climate change can organisms adapt? At what point is climate change extreme enough that species come to the limit of their genetic variation, can no longer adapt, and become extinct?

Camille Parmesan is an associate professor in the Section of Integrative Biology at the University of Texas at Austin. Her recent work has focused on current impacts of climate change on wildlife, and especially on butterfly range shifts. To learn more about her research, go to http://cluster3 .biosci.utexas.edu/research/parmesanLab.

Review

52.1 Environmental Diversity of the Biosphere

- The biosphere encompasses all the regions on Earth where organisms live, including the atmosphere, hydrosphere, and lithosphere.

- Latitudinal variations in solar radiation establish global climate patterns (Figure 52.2). Earth's tilt on its axis causes seasonal variation in solar radiation and climate (Figure 52.3). Seasonal variations in day length and temperature increase steadily from tropical latitudes toward the poles.

- Unequal heating of the atmosphere causes air masses to flow in circulation cells that create worldwide wind and precipitation patterns (Figures 52.4 and 52.5). Ocean currents generally flow clockwise in the Northern Hemisphere and counterclockwise in the Southern Hemisphere (Figure 52.6).

- The oceans and local topographical features influence regional and local climates. Proximity to the ocean has a moderating effect on terrestrial climates (Figure 52.7). Habitats are generally wetter on the windward sides of mountains than on the leeward sides (Figure 52.8).

Animation: Global air circulation patterns

Animation: Air circulation and climate

Animation: Major climate zones and ocean currents

Animation: Rain shadow effect

Animation: El Niño Southern Oscillation

52.2 Organismal Responses to Environmental Variation

- Organisms use homeostatic responses to cope with environmental variation. Animals often use facultative behavioral and physiological mechanisms to respond to environmental temperature (Figure 52.9).

- Global warming is affecting the ecology of many organisms. Many species are experiencing changes in their geographical ranges or in the timing of their reproduction.

52.3 Terrestrial Biomes

- Biomes are general types of vegetation and other associated organisms. Climate is the major determinant of terrestrial biome distributions (Figures 52.10 and 52.11).

- Tropical forest occurs at low latitudes where seasonality is determined by variations in rainfall rather than by day length and temperature (Figure 52.12). Tropical rain forests are the most species-rich terrestrial biome, but they grow on nutrient-poor soils.

- Savanna is tropical and subtropical grassland with scattered trees (Figure 52.13). Long dry seasons, fires, and grazing by large mammals prevent trees from replacing perennial grasses.

- Deserts form in arid regions where precipitation is low and temperature varies widely on a daily and seasonal basis (Figure 52.14).

- Chaparral is a coastal biome dominated by dense, woody shrubs and trees that resprout after periodic fires (Figure 52.15). Chaparral occurs where winters are mild and wet and summers hot and dry.

- Temperate grassland grows where winters are cold, summers are warm, and rainfall is moderate (Figure 52.16). Tree seedlings are eliminated from grasslands by droughts, periodic fires, and grazing by mammals. Grassland soils are rich and deep.

- Temperate deciduous forest flourishes at middle latitudes with abundant rainfall. The seasonality of the climate is reflected in the annual loss and regrowth of leaves (Figure 52.17).

- The boreal forest, or taiga, includes dense stands of coniferous trees at high latitudes, where winters are long and cold (Figure 52.18).

- Tundra is the northernmost biome, where plants grow in shallow topsoil over a layer of permafrost (Figure 52.19). The brief growing season and winter winds cause tundra plants to be very short.

Animation: Terrestrial biomes

Animation: Soil profiles

52.4 Freshwater Biomes

- Freshwater biomes include both flowing-water and standing-water systems.

- The physical characteristics of flowing-water biomes change from the headwaters of a stream to the mouth of a river (Figure 52.20), and their food webs are largely detrital.

- The physical characteristics of standing-water biomes change with the depth of water and distance from shore (Figures 52.21 and 52.22). Lakes exhibit marked vertical zonation and, in the temperate zone, undergo a seasonal mixing of their waters (Figure 52.23). Lakes are generally classified by their nutrient status and productivity (Figure 52.24).

Animation: Lake zonation

Animation: Lake turnover

Animation: Trophic nature of lakes

52.5 Marine Biomes

- The world's oceans exhibit marked zonation based on water depth and distance from shore (Figure 52.25).

- Estuaries are highly productive tidal biomes where rivers provide a constant input of nutrients and freshwater, and the tides carry away wastes (Figure 52.26).

- The intertidal zone is a stressful environment that is alternately submerged and exposed (Figure 52.27).

- Highly productive and diverse shallow-water biomes grow on continental shelves and oceanic banks. Kelp forests predominate at high latitudes, whereas coral reefs occur in the tropics (Figure 52.28).

- The open ocean is highly stratified because photosynthesis is possible only in the uppermost 50 m of water. Plankton are the primary producers in the uppermost layers; they support grazing food webs (Figure 52.29). The deep sea includes many predatory species (Figure 52.30).

- Organisms of the sea floor occupy the benthic province. Falling detritus supports most benthic communities, but chemoautotrophic bacteria support communities near deep-sea hydrothermal vents (Figure 52.31).

Animation: Rocky intertidal zones

Animation: Three types of reefs

Animation: Oceanic zones

Animation: Coastal upwelling

Animation: Hydrothermal vent community

Questions

Self-Test Questions

1. The lithosphere includes all:
   a. oceans.
   b. ice caps.
   c. rocks, soils, and sediments.
   d. gases and airborne particles.
   e. places where organisms live.

2. Earth's 23.5° tilt on its axis directly causes:
   a. latitudinal variation in average annual rainfall.
   b. ocean currents to rotate clockwise in the Northern Hemisphere.
   c. microclimates to vary dramatically over short distances.
   d. low rainfall on the leeward side of mountain ranges.
   e. seasonal variation in the amount of solar radiation.

3. Adiabatic cooling causes rising air masses to:
   a. absorb moisture from Earth's surface.
   b. release precipitation.
   c. change the direction of the El Niño current.
   d. flow toward the equator from the poles.
   e. be deflected from a strictly northward or southward flow.

4. The term "rain shadow" describes the:
   a. low rainfall that is typical on the leeward side of mountains.
   b. low rainfall that is typical at 30° latitude.
   c. high rainfall that is typical on the windward side of mountains.
   d. blocking of rain by vegetation in dense tropical forests.
   e. low rainfall that is typical in the interior of continents.

5. The major climatic factors that govern the distributions of terrestrial biomes are:
   a. temperature only.
   b. rainfall only.
   c. wind speed only.
   d. temperature and rainfall.
   e. temperature, rainfall, and wind speed.

6. Which biome experiences the highest annual rainfall?
   a. tropical rain forest
   b. tropical savanna
   c. chaparral
   d. temperature grassland
   e. arctic tundra

7. From which biome are trees excluded by periodic fires and grazing herbivores?
   a. tropical rain forest
   b. thorn forest
   c. chaparral
   d. temperate grassland
   e. arctic tundra

8. The major source of nutrients in the headwaters of a small stream is from:
   a. dead leaves and other organic matter from adjacent land.
   b. photosynthesis by phytoplankton.
   c. photosynthesis by floating aquatic plants.
   d. the activity of chemoautotrophic bacteria.
   e. minerals from the underlying bedrock.

9. During the spring overturn in a temperate zone lake:
   a. oxygen is carried from the surface to the bottom, and nutrients are carried from the bottom to the surface.
   b. nutrients are carried from the surface to the bottom, and oxygen is carried from the bottom to the surface.
   c. nutrients and oxygen are carried from the bottom waters to the surface waters.
   d. nutrients and oxygen are carried from the surface waters to the bottom waters.
   e. oxygen concentration remains constant at all depths, and nutrients sink to the bottom.

10. In which habitat must organisms adjust regularly to changing salinity?
    a. salt marsh
    b. coral reef
    c. benthic province
    d. estuary
    e. riffle

Questions for Discussion

1. Temperate grassland and chaparral often burn in lightning-induced fires, which stimulate the germination of seeds and regrowth of existing vegetation. Do you think that companies or the government should sell fire insurance to people who build expensive homes in places where periodic fires are virtually inevitable?

2. Boreal forests generally harbor many fewer species of trees than tropical forests do. Develop three hypotheses to explain this pattern. What data would you collect to test your hypotheses?

3. Describe the biome in which you live, noting the prevailing climate and microclimates and any other factors that govern the characteristics of the organisms that naturally occur there.

4. Many regions on Earth have been developed for agriculture, industry, and human habitation. Have our activities created new biomes? What physical environments are created by development, and what plants and animals occupy developed areas?

Experimental Analysis

Design an experiment to test the hypothesis that streams receive much of their nutrients and energy from material that falls into them from overhanging vegetation.

Evolution Link

If the geographical ranges of species change in response to global warming, what new selection pressures will organisms face as they move into ecological communities where they have not previously occurred? Your answer should address the effects of novel species interactions as well as the effects of encountering different physical environments.

How Would You Vote?

We cannot stop an El Niño from happening, but we might be able to minimize its environmental, social, and economic impacts. Would you support the use of taxpayer dollars to fund research into the causes and effects of El Niño? Go to www.thomsonedu.com/login to investigate both sides of the issue and then vote.

Florida panther *(Puma concolor coryi)*. Fewer than 100 individuals of this endangered subspecies survive.

JH Pete Carmichael/Getty Images Inc.

STUDY PLAN

53.1 The Benefits of Biodiversity

Biodiversity benefits humans directly

Ecosystem services benefit all forms of life

Biodiversity has intrinsic worth beyond its utility to humans

53.2 The Biodiversity Crisis

Human activities disturb and fragment habitats

Deforestation may lead to desertification

Many forms of pollution overwhelm species and ecosystems

Exotic species often eliminate native species

Overexploitation greatly reduces population sizes

Human activities are causing a dramatic increase in extinction rates

53.3 Biodiversity Hotspots

Conservation biologists focus their efforts in areas where biodiversity is both concentrated and endangered

53.4 Conservation Biology: Principles and Theory

Systematics organizes our knowledge of the biological world

Population genetics informs strategies for species preservation

Studies of population ecology and behavior are essential elements of conservation plans

Community and landscape ecology help large-scale preservation projects

53.5 Conservation Biology: Practical Strategies and Economic Tools

Conservation efforts aim to preserve, conserve, and restore habitats

Successful conservation plans must incorporate economic factors

53 Biodiversity and Conservation Biology

WHY IT MATTERS

Someone seems to be missing. Investigators thoroughly checked the subject's known haunts, but found no trace. They questioned others in the neighborhood, but came up with few leads. The case is especially difficult because the subject was last seen alive in 1978. With so cold a trail to follow, investigators reluctantly marked the case file "Missing and Presumed Extinct."

The subject in this case was Miss Waldron's red colobus monkey, *Procolobus badius waldroni* **(Figure 53.1).** Named for a traveling companion of the taxonomist who first described it in 1933, this distinctively colored subspecies lived in large and noisy social groups in a remote forest on the border between Ivory Coast and Ghana in West Africa.

John Oates of the City University of New York recently led a research team that tried to locate Miss Waldron's red colobus. They used every imaginable method, including visual and auditory censuses, searching for scat (dung) in natural habitats, interviewing local people, and looking in marketplaces where monkey meat is commonly traded. In 2000, more than 20 years after the last confirmed sighting, the researchers concluded that this monkey is probably extinct. A later

1229

Figure 53.1
Miss Waldron's red colobus. *Procolobus badius waldroni*, which weighed about 10 kg, may be the first primate subspecies to become extinct in more than 100 years.

search by a member of the team, William S. McGraw of Ohio State University, did find the skin of one monkey that a hunter had shot 6 months before. But McGraw searched in vain for a living monkey, and he concluded that even if a few are still alive, the population is so small that continued hunting will surely eliminate it.

Procolobus badius waldroni may be the first primate subspecies to become extinct in more than 100 years—and only the second in the last 500 years. Monkeys and other primates are among the most closely monitored and protected species on Earth. Nonetheless, Oates and his colleagues concluded, these monkeys probably became extinct because they were hunted locally for food by a growing human population and because humans have destroyed their natural habitats.

Miss Waldron's red colobus is just one of many species driven to extinction every year. Current threats to biodiversity, all of which ultimately result from human activities, are massive. The likely loss of this monkey should warn us that many taxa are at risk, even those that are most rigorously protected.

When ecologists speak of **biodiversity**, they are referring to the richness of living systems. At the most fundamental level of biological organization, biodiversity encompasses the *genetic variation* that is raw material for adaptation, speciation, and evolutionary diversification (see Chapters 20 and 21). At a higher level of organization, biodiversity includes *species richness* within communities (see Section 50.3). The number and variety of species within a community influences its overall characteristics, population interactions, and trophic structure. Finally, biodiversity exists at the *ecosystem level*. Complex networks of interactions bind species in an ecosystem together, and because different ecosystems interact within the biosphere, damage to one ecosystem can reverberate through others.

In this chapter we reflect on the importance of biodiversity and describe how human activities threaten it. We also consider theoretical and practical approaches to conservation biology, the scientific discipline that focuses on preserving Earth's biological resources.

53.1 The Benefits of Biodiversity

What is the value of biodiversity, and why should humans preserve it? Arguments for conserving biodiversity fall into three general groups: its direct benefit to humans, its indirect benefit to all living systems, and its intrinsic worth.

Biodiversity Benefits Humans Directly

Scientists constantly search for natural products that might provide humans with better food, clothing, or medicine. The development of a new medicine often begins when a scientist analyzes a traditional folk remedy or screens naturally occurring compounds for curative properties. Chemists then isolate and purify the active ingredient and devise a way to synthesize it in the laboratory. More than half of the 150 most commonly prescribed drugs were developed from natural products in this manner.

For example, *Taxol,* a drug treatment for breast and ovarian cancer, was isolated from the narrow strip of vascular cambium beneath the bark of the Pacific yew tree, *Taxus brevifolia* **(Figure 53.2).** Unfortunately, a

Figure 53.2
The Pacific yew tree. The slow-growing Pacific yew *(Taxus brevifolia)* is the original source of Taxol, a compound that effectively fights several cancers.

fully grown, 100-year-old tree produces only a tiny amount of Taxol, and six trees must be destroyed to extract enough to treat one patient. Pacific yew trees are not abundant, and they grow slowly. Harvesting them for Taxol extraction could quickly lead to their extinction—and an end to the natural source of this life-saving compound. However, after much research, scientists can now synthesize this widely used drug in the laboratory.

Wild plants and animals also serve as sources of genetic traits that may improve agricultural crops and domesticated livestock. For example, corn *(Zea mays)* is an annual plant. Its cultivation requires yearly tilling of the soil, a labor-intensive activity that leads to erosion and loss of topsoil. Farmers have yearned for a perennial strain of corn, one that would produce grain for years after a single planting. In 1978, botanists discovered teosinte *(Zea diploperennis)* a perennial plant closely related to corn, in the mountains of western Mexico. Researchers crossed the two species, producing a *perennial* corn. If they can increase the yield of this hybrid, it may prove to be an economically valuable crop **(Figure 53.3)**.

Today, many agricultural researchers use genetic engineering, the transfer of selected genes from one species into another (see Section 18.2), to alter crop plants more precisely than they can using hybridization. The transferred genes may be chosen to increase resistance to pests or environmental stress, promote faster growth, or increase shelf life after harvesting. However, many scientists and environmentalists fear that genetically modified crops may create environmental hazards that will inadvertently endanger biodiversity. For example, a genetically modified plant or animal that escaped into a natural habitat might compete with naturally occurring species. Or a genetically modified plant might poison harmless animals as well as insect pests.

Ecosystem Services Benefit All Forms of Life

Humans and other species derive indirect benefits when ecosystems perform the ecological processes on which all life depends. These **ecosystem services,** as they are called, include the decomposition of wastes, nutrient recycling, oxygen production, maintenance of fertile topsoil, and air and water purification.

Some ecosystem services can even mitigate environmental damage caused by humans. As you may recall from *Focus on Applied Research* in Chapter 51, the combustion of fossil fuels produces CO_2 and other waste products that accumulate in the atmosphere, increasing the greenhouse effect and fostering global warming. Photosynthetic organisms use CO_2 for essential metabolic processes; thus, forests and, even more importantly, communities of marine phytoplankton withdraw CO_2 from the atmosphere and

Figure 53.3

Teosinte and domesticated corn. Ears of domesticated corn (*Zea mays*, right) are much larger than those of its wild relative teosinte (*Zea diploperennis*, left). Scientists crossed the two species in the hope of producing a perennial corn; the hybrids produce ears of an intermediate size (middle).

incorporate it into living organisms (see Figure 51.13), a phenomenon called *carbon sequestration*. Recent research indicates that these organisms are essential for limiting the damage caused by the burning of fossil fuels. In the long run, biodiversity's indirect benefits, provided in the form of ecosystem services, may be even more valuable to humans than the direct benefits.

Biodiversity Has Intrinsic Worth beyond Its Utility to Humans

Some ethicists argue that we should preserve biodiversity because it has intrinsic worth, independent of its direct or indirect value to humans. They note that humans are just one species among millions in the remarkable network of life. Countering this position is the view that our immediate needs should always rank above those of other species and that we should use them to maximize our own welfare. The latter view inevitably leads to the disruption of natural environments and the loss of biodiversity. Framed in this way, the debate lies more within the realms of philosophy and public policy than biology. Nevertheless, many people feel an emotional or spiritual connection to natural landscapes and the plants and animals they harbor. Thus, biodiversity enhances human existence in intangible ways.

Figure 53.4 Experimental Research

Predation on Songbird Nests in Forests and Forest Fragments

QUESTION: Are songbird nests in small forest fragments more likely to be found by predators than nests in large forest patches?

EXPERIMENT: Wilcove placed between 13 and 50 artificial bird nests, each containing three quail eggs, in three habitat types: large areas of intact forest, rural forest fragments, and suburban forest fragments. He placed about half the nests at each study site on the ground at the base of a tree or shrub and half the nests 1 to 2 m above the ground in a sapling or shrub. He checked the nests after 7 days to determine what proportion of the nests had been subjected to predation.

RESULT: Predators generally found a larger proportion of the artificial bird nests in small forest fragments than they did in large forest patches.

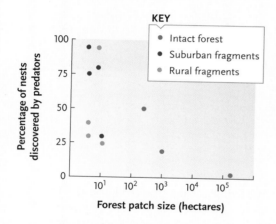

CONCLUSION: Songbirds' nests are much more likely to suffer from predation in small forest fragments than they are in large patches of intact forest.

53.2 The Biodiversity Crisis

Earth's biodiversity is currently declining dramatically. Although the proximate causes of the decline may vary from one group of organisms to another, the ultimate cause is always the same: human disruption of natural communities and ecosystems.

Human Activities Disturb and Fragment Habitats

When humans first enter undisturbed habitats, they typically build roads to gain access to resources, such as oil, wood, or game animals, or to begin agricultural development. The roads bring in settlers, who clear isolated areas for specific uses. Nonnative organisms are often introduced by humans or migrate into the now-disturbed area under their own power. These invaders then consume, parasitize, or compete with the native plants and animals. As the land is further changed and degraded, the habitat is altered dramatically, possibly forever. Although this pattern of development initially affects only locally distributed species, the negative effects spread rapidly to a regional scale. The remaining areas of *intact* habitat are inevitably reduced to small, isolated patches, a phenomenon that ecologists describe as **habitat fragmentation.**

Habitat fragmentation is a threat to biodiversity because small habitat patches can sustain only small populations. As you learned in Section 49.5, a habitat's *carrying capacity,* the maximum population size that it can support, varies with available resources. Populations that occupy small habitat patches inevitably experience low carrying capacities, a problem that is especially acute for species at the higher trophic levels (see Section 51.1). Furthermore, fragmented habitat patches are often separated by unsuitable habitat that organisms may be unable or unwilling to cross. As a result, individuals from one isolated population are unlikely to migrate into another, reducing gene flow between them. The combination of small population size and genetic isolation fosters genetic drift, which reduces genetic variability and fosters extinction (see Section 20.3).

Habitat fragmentation not only reduces the amount of undisturbed habitat; it also jeopardizes the quality of the habitat that remains. Human activities create noise and pollution that spread into nearby areas. The removal of natural vegetation disrupts the local physical environment, exposing the borders of the remaining habitat to additional sunlight, wind, and rainfall. Increased runoff compacts the soil and makes it waterlogged. These phenomena are collectively described as **edge effects.**

The effects of habitat fragmentation are often profound. For example, populations of forest-dwelling, migratory songbirds have declined markedly in eastern North America since the late 1940s, largely because of habitat fragmentation in their North American breeding grounds and in their Caribbean and South American wintering grounds.

In 1994, Scott K. Robinson of the Illinois Natural History Survey and David S. Wilcove of the Environmental Defense Fund identified three factors that decrease populations of migratory songbirds in fragmented breeding habitats. First, small forest patches often lack specific habitat types—such as streams, cool ravines, or dense ground cover—that many songbird species require.

Second, songbirds breeding in forest patches are more likely to suffer from brood parasitism (described in the opening of Chapter 50) by brown-headed cowbirds (*Molothrus ater*) than are those breeding in intact forests. Brown-headed cowbirds, which prefer open habitats, were rare in eastern North America before European settlers converted forests to farmland. Today, cowbirds are abundant in open agricultural fields and suburban gardens, and they locate the nests of unwitting "foster parents" in nearby forest fragments where the host species breed. Parasitized songbirds rear fewer than half as many young as they might otherwise raise, and their populations decline accordingly.

The third factor that reduces songbird numbers in forest fragments is increased nest predation by blue jays (*Cyanocitta cristata*), American crows (*Corvus brachyrhynchus*), common grackles (*Quiscalus quiscula*), squirrels (genus *Sciurus*), raccoons (*Procyon lotor*), and domestic dogs and cats. These predators, which feed on songbird eggs and young, are now superabundant in rural and suburban areas, and they enter adjacent forest fragments in search of an easy meal. Wilcove tested the predation hypothesis experimentally by placing artificial nests with quail eggs in intact forests and in forest fragments. Although he did not observe predation directly, he found that predators discovered only 2% of the nests in the largest intact forest, but they often found 50% or more of the nests placed in small, suburban forest fragments **(Figure 53.4)**.

Deforestation May Lead to Desertification

Forests are among the habitats that humans most frequently clear and convert. According to the United Nations Forest Resources Assessment released in 2005, global deforestation is occurring at a rate of about 13 million hectares per year, or 25 hectares per minute. In other words, an area of forest equivalent to 42 football fields is cleared of all trees every minute of every day.

Deforestation does not occur uniformly across the globe. Today, more than 90% of the deforestation occurs in tropical regions, mostly to clear land for grazing. Brazil has experienced the most extensive recent damage, accounting for 25% of all deforestation during the late twentieth century **(Figure 53.5)**. This assessment is particularly troubling because Brazil contains approximately 27% of the planet's total aboveground woody biomass. Compounding the environmental damage, most tropical forests are burned as they are cleared, a process that adds CO_2 to the atmosphere, enhancing the greenhouse effect and increasing the rate of global warming (see *Focus on Applied Research* in Chapter 51).

Once a forest has been cut, heavy grazing or farming drains nutrients from the soil. To remain productive, even the best agricultural or grazing lands require either the application of fertilizers or long periods during which the land is fallow, allowing plants to replenish the soil naturally. Unfortunately, the soil where tropical forests grow is often of marginal value right from the start (for reasons described in Chapter 52), and it is rapidly degraded; it becomes hard, even more nutrient-poor, unable to retain water, and likely to wash away.

When large tracts of subtropical forest are cleared and overused, the land often undergoes **desertification:** the groundwater table recedes to deeper levels; less surface water is available for plants; soil accumulates high concentrations of salts (a process called *salinization*); and topsoil is eroded by wind and water. In other words, the habitat is converted to desert.

Desertification speeds the loss of biodiversity locally and can eliminate entire ecosystems. For example, desertification has decimated habitats in the Sahel re-

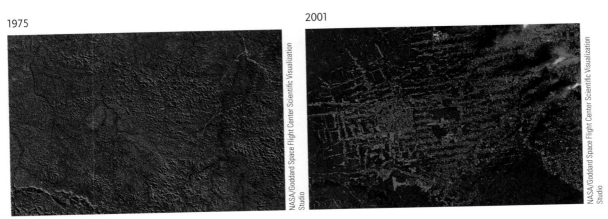

1975

2001

NASA/Goddard Space Flight Center Scientific Visualization Studio

Figure 53.5
Deforestation in the Amazon Basin. Satellite photos of Rondonia, in the Brazilian Amazon, show how much of the Amazon forest was cut (light green) between 1975 and 2001. Each photo illustrates an area approximately 60 by 85 km.

a. The Sahel region of Africa

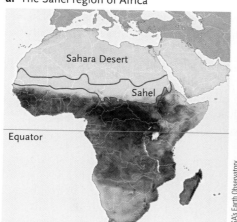

b. Women preparing millet, a grain, in the Sahel

NASA's Earth Observatory

Romano Cagnoni/Peter Arnold, Inc.

Figure 53.6

Desertification in the Sahel. **(a)** A satellite photo taken near the end of the dry season in June 2005 illustrates the severe desertification in parts of the Sahel region of Africa. Dark green areas are densely vegetated; light green areas are sparsely vegetated, and sand-colored areas are barren. **(b)** People who live in this region can barely eke out a living on the land.

gion of Africa, just south of the Sahara Desert **(Figure 53.6)**. Excessive grazing of cattle and goats by an ever-expanding human population is the main reason for the Sahara's southward expansion at a rate of 5.5 to 8 km per year. Because the sand dunes of the expanding desert shift constantly, agriculture and grazing are nearly impossible, resulting in frequent famines among the people of the Sahel.

Desertification and salinization have also begun in the Everglades, a unique, shallow "river of grass" that covers much of southern Florida. The amount of fresh water flowing through South Florida to the Everglades has decreased approximately 70% since 1948, when an extensive network of canals and levees was built to reduce flooding. The rapidly growing human population in South Florida contributes directly to desertification, as groundwater is tapped for domestic use and to irrigate lawns, golf courses, and agricultural fields. Salt water from the Gulf of Mexico now intrudes into the water table, causing salinization of the soil. The Comprehensive Everglades Restoration Plan (CERP), approved by the U.S. Congress in 2000, seeks to restore the natural flow of the Everglades over the next 30 years. This project may halt or reverse the desertification process.

Sadly, deforestation, desertification, and global warming reinforce each other in a positive feedback cycle (see *Focus on Applied Research* in Chapter 51). If scientists' projections are correct, desertification will lead to an increase in the average global temperature, speeding evaporation and the retreat of forests, which, in turn, will increase rates of desertification. If deforestation and desertification continue, we will soon lose a large proportion of Earth's forests and face a decrease in the area of habitable land.

Many Forms of Pollution Overwhelm Species and Ecosystems

The release of **pollutants**—materials or energy in forms or quantities that organisms do not usually encounter—poses another major threat to biodiversity.

Although chemical pollutants, the by-products or waste products of agriculture and industry, are released locally, many spread in water or air, sometimes on a continental or global scale. Within North America, for example, winds carry airborne pollutants from coal-burning power plants to the Northeast **(Figure 53.7)**. Sulfur dioxide (SO_2), which dissolves in water vapor in the air and forms sulfuric acid, falls as **acid precipitation**, acidifying soil and bodies of water. Many lakes in northeastern North America have experienced a precipitous drop in pH from historical readings near 6 to values that are now well below 5—a 10-fold increase in acidity. Although the lakes once harbored lush aquatic vegetation and teemed with fishes, they are now crystal clear and nearly devoid of life.

As residents of major cities and industrial areas know all too well, carbon wastes from factories and automobile engines cause terrible local pollution, increasing rates of asthma and other respiratory ailments. Some airborne pollutants, notably CO_2, also join the general atmospheric circulation, where they contribute to the greenhouse effect and global warming.

Like air pollution, water pollution originates locally but has a much broader impact. Oil spills, for example, disrupt local ecosystems, killing most organisms near the spill. Because oil floats on water, it spreads rapidly to nearby areas. The wreck of an oil tanker off the coast of Spain in 2002 destroyed many fertile fishing grounds within a few weeks. Scientists

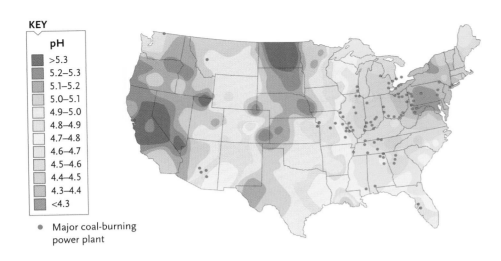

pH
>5.3
5.2–5.3
5.1–5.2
5.0–5.1
4.9–5.0
4.8–4.9
4.7–4.8
4.6–4.7
4.5–4.6
4.4–4.5
4.3–4.4
<4.3

● Major coal-burning power plant

Figure 53.7
Acid precipitation. Coal-burning power plants (indicated by red dots) release air pollution that is carried northeast, where it falls as acid precipitation. The map shows the average pH of rainfall.

expect oil to continue leaking from the sunken ship for another 50 years; the effects of the long-term leakage may linger for centuries.

Pollution can also have serious effects on terrestrial ecosystems. As a recent disaster in India, Nepal, and Pakistan illustrates, the application of synthetic compounds to agricultural fields or livestock can have dire and far-reaching consequences. For thousands of years, gigantic populations of vultures (several *Gyps* species)—estimated at more than 40 million birds—performed an important ecosystem service by consuming the abandoned carcasses of farm animals across South Asia. In the early 1990s, however, farmers began to administer diclofenac, a new and inexpensive anti-inflammatory drug, to injured livestock. Within a few years, vultures began to disappear; in 2006, scientists estimated that their populations had declined by more than 97%. Recent research revealed that diclofenac, which causes fatal kidney failure in birds, was responsible for the deaths: vultures were ingesting substantial doses of the drug from the livestock carcasses they ate. All vulture species in South Asia are now on the verge of extinction, and although governments in the region have banned the sale of diclofenac, wildlife experts say that the vulture populations are unlikely to recover soon, if ever.

The decline in vulture populations has had a disastrous impact on urban and rural communities in South Asia. Livestock carcasses are now consumed by growing populations of wild dogs, many of which carry rabies. India has the world's highest human death toll from rabies—30,000 per year—and two-thirds of the cases are caused by dog bites. Populations of rats and flies also appear to be increasing. *Focus on*

Research (p. 1236) describes another example of how pesticides and other chemicals accumulate at lethal concentrations in organisms living at higher trophic levels.

Some forms of pollution have more subtle effects. Light and noise pollution disrupt the activities of nocturnal animals or those that rely on vision or hearing for orientation. For example, light pollution in beachfront communities disrupts the reproduction of marine turtles, all species of which are declining in numbers **(Figure 53.8)**. Female turtles crawl up on beaches at night to lay their eggs in the sand; after the eggs hatch, the young dig their way out of the nest and head for the ocean. But female turtles are reluctant to come ashore on beaches with artificial light. And lights may later confuse and misdirect their hatchlings, making them even easier prey for predators, or cause them to stay too long on shore, where they dehydrate and die.

Exotic Species Often Eliminate Native Species

As humans travel from one habitat to another, we inevitably carry other species with us. Seeds cling to our legs, insects accompany us in our food and possessions, and some organisms hitch a ride on boats or cars. The introduction of nonnative organisms, called

Figure 53.8
Light pollution disrupts green turtle reproduction. **(a)** Female green turtles (*Chelonia mydas*) are reluctant to nest on beaches affected by light and noise pollution. **(b)** Artificial light confuses hatchling turtles, hindering their escape from eager predators like this great blue heron (*Ardea herodias*).

a. Female green turtle digging a nest

Science Photo Library/Photo Researchers, Inc.

b. Heron eating hatchling green turtle

Bertram G. Murray/Animals, Animals—Earth Scenes

Applied Research: Biological Magnification

The synthetic organic pesticide DDT (dichloro-diphenyl-trichloroethane) was first used widely during World War II. In the tropical Pacific, it killed the mosquitoes that transmitted malarial parasites (*Plasmodium* species) to soldiers. In war-ravaged European cities, it controlled body lice that carried the bacteria causing typhus *(Rickettsia rickettsii)*. After the war, people started using DDT to kill agricultural pests, disease vectors, and insects in homes and gardens.

Although DDT is a stable hydrocarbon compound that is nearly insoluble in water, it is more mobile than its users expected. Winds carry it as a vapor, and water transports it as fine particles. DDT is also highly soluble in fats, accumulating in animal tissues—and it travels with animals wherever they go.

Unfortunately, consumers accumulate the DDT from all of the organisms they eat in their lifetimes. Primary consumers, like herbivorous insects, may ingest relatively small quantities. But a songbird that eats many insects will accumulate a moderate amount, and a predator that feeds on songbirds will accumulate even more. Thus, DDT and other nondegradable poisons become concentrated in organisms at higher trophic levels, a phenomenon called **biological magnification** (see **figure**). Although many organisms can partially metabolize DDT to other compounds, these products are also toxic or physiologically disruptive.

After the war, DDT moved rapidly through ecosystems, affecting organisms in ways that no one had predicted. In cities where DDT controlled Dutch elm disease, songbirds died after eating contaminated insects and seeds. In streams flowing through forests where DDT killed spruce budworms, salmon died because runoff carried the pesticide into their habitat. And in croplands around the world, new pests flourished because DDT indiscriminately killed the natural predators that had kept their populations in check.

Eventually, the effects of biological magnification began to show up in places far removed from the sites of DDT application. Top carnivores in some food webs were pushed to the brink of extinction. The reproduction of bald eagles, peregrine falcons, ospreys, and brown pelicans was disrupted because one DDT breakdown product interferes with the deposition of calcium in their eggshells. When birds tried to incubate their eggs, the shells cracked beneath the parents' weight. Even today, traces of DDT are found in the bodies of nearly all species, including in human fat and breast milk.

Since the 1970s, DDT has been banned in the United States, except for restricted applications to protect public health. Many hard-hit species have partially recovered, but some birds still lay thin-shelled eggs because they pick up DDT at their winter ranges in Latin America. As recently as 1990, the California State Department of Health recommended that a fishery off the coast of California be closed; DDT from industrial waste discharged 20 years earlier was still moving through that ecosystem. Moreover, DDT is still used in other countries, and some enters the United States on imported fruit and vegetables.

Biological magnification is a problem that applies to many compounds that humans release into the environment. For example, polychlorinated biphenyls (PCBs), commonly used in the manufacture of plastics and electrical insulation, enters aquatic ecosystems in factory wastes. Their use has been banned in the United States since the 1970s. But these compounds break down very slowly, and vast deposits have accumulated in the bottom sediments of rivers and lakes. Once bottom-feeding organisms ingest them, the toxins work their way up food webs, accumulating at higher and higher concentrations in consumers. The effects on humans can be severe; pregnant women who regularly eat fish from the Great Lakes often give birth to children with below-average weight and neonatal behavioral problems. The pollution in some areas of New York State was so severe that the Department of Health advised people to avoid eating freshwater fish more than once a month. PCBs can be removed from aquatic ecosystems by dredging, but the dredging activity itself stirs up the polluted sediments, releasing the toxins into the water that flows above.

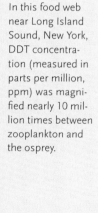

In this food web near Long Island Sound, New York, DDT concentration (measured in parts per million, ppm) was magnified nearly 10 million times between zooplankton and the osprey.

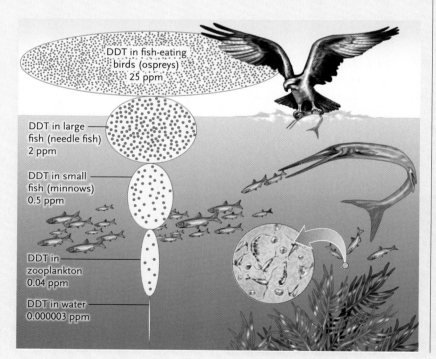

DDT in fish-eating birds (ospreys) 25 ppm

DDT in large fish (needle fish) 2 ppm

DDT in small fish (minnows) 0.5 ppm

DDT in zooplankton 0.04 ppm

DDT in water 0.000003 ppm

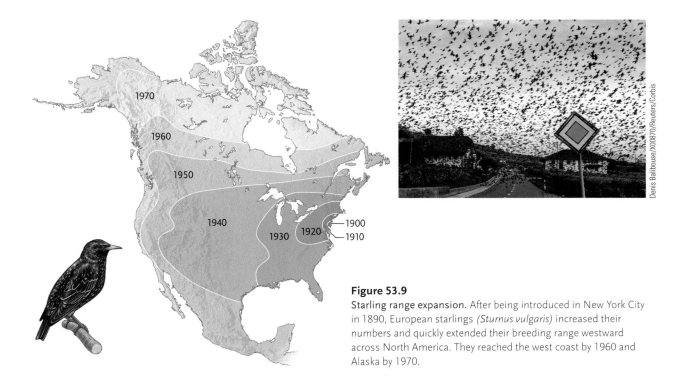

Figure 53.9

Starling range expansion. After being introduced in New York City in 1890, European starlings *(Sturnus vulgaris)* increased their numbers and quickly extended their breeding range westward across North America. They reached the west coast by 1960 and Alaska by 1970.

exotic species, into new habitats poses one of the most serious threats to biodiversity.

Exotic species often prey upon, parasitize, or out-compete native species, leading to their extinction. Many have *r*-selected life histories (see Section 49.6); they mature quickly and reproduce prodigiously, and they thrive in the degraded habitats that humans so frequently create. In the absence of natural checks on population growth—such as competitors, predators, and parasites—exotics often experience exponential population growth (see Section 49.5).

The European starling *(Sturnus vulgaris)* provides an example of the explosive population growth and range expansion of an exotic species. These birds were released in North America in 1890 when a misguided individual, who wanted to introduce all of the bird species mentioned by Shakespeare into North America, imported them into Brooklyn, New York. Within 70 years, they had spread across the continent **(Figure 53.9);** their population size is now estimated at 200 million. Starlings pose a serious threat to native birds, including several woodpecker species, because they successfully compete with them for nesting sites in natural cavities in trees.

Introduced plants often transform entire ecosystems. One of the best-known examples is kudzu *(Pueraria lobata),* a fast-growing species from Asia. In the early 1900s, it was widely planted in the southeastern United States as a source of animal feed. Later, a government agency promoted it as a plant that could stabilize soils and decrease erosion on deforested hillsides; we now know that it does not perform those functions effectively. But when kudzu has access to abundant nutrients and water, its branches can grow up to 30 cm per day. It spread quickly across the South, literally overgrowing almost all native plants **(Figure 53.10).**

Exotic insects often become pests of agricultural crops and native plants. The hemlock woolly adelgid *(Adelges tsugae)* was accidentally introduced into North America from Asia. The adelgid kills eastern hemlocks *(Tsuga canadensis)* by feeding on their sap. It now threatens the trees from North Carolina to Massachusetts **(Figure 53.11).** But adelgids endanger far more than these evergreen trees. Hemlocks buffer the physical

Figure 53.10
Kudzu, the vine that ate the South. Kudzu *(Pueraria lobata),* an introduced vine, grows so quickly that it often covers living trees or even abandoned buildings.

a. Woolly adelgids

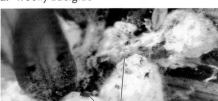

Adelgids

b. Hemlocks killed by woolly adelgids

c. Eastern hemlock and woolly adelgid ranges

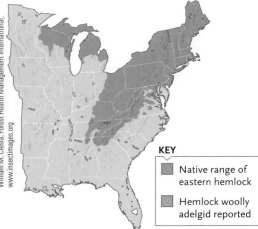

KEY

■ Native range of eastern hemlock

■ Hemlock woolly adelgid reported

Figure 53.11

Hemlock woolly adelgid. **(a)** The aphidlike woolly adelgid *(Adelges tsugae)* feeds on the sap of **(b)** eastern hemlock *(Tsuga canadensis)*, often killing the tree. This insect pest is spreading northward **(c)** and may someday endanger hemlocks throughout their geographical range.

conditions below them: hemlock stands are cool in summer and warm in winter, sustaining a unique community of organisms that includes ruffed grouse *(Bonasa umbellus)*, turkey *(Meleagris gallopavo)*, white-tailed deer *(Odocoileus virginianus)*, and snowshoe hare *(Lepus americanus)*. Infested stands rarely survive more than a few years, and the communities established under pure stands of eastern hemlock will likely become extinct because of feeding by the adelgid.

Overexploitation Greatly Reduces Population Sizes

Many local extinctions result from **overexploitation**, the excessive harvesting of an animal or plant species. At a minimum, overexploitation leads to declining population sizes in the harvested species. In the most extreme cases, a species may be wiped out completely. Overexploitation also can foster evolutionary changes in the exploited population, much the way guppies respond to natural predators in the streams of Trinidad (described in *Focus on Research* in Chapter 49).

The fishery on the Grand Banks off the coast of Newfoundland, Canada, provides a sad example of overexploitation **(Figure 53.12)**. For hundreds of years, fishermen used traditional line and small-net fishing to harvest a large but sustainable catch. During the twentieth century, however, new technology allowed them to locate and exploit schools of fishes more efficiently. As a result, 45% of the fish species harvested there are now overfished. Haddock *(Melanogrammus aegelfinus)* and yellowtail flounder *(Limanda ferruginea)*

a. The Grand Banks

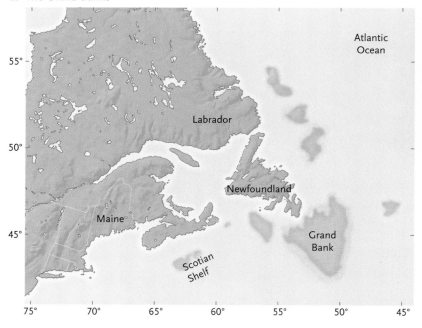

b. Atlantic cod

Figure 53.12

Overexploitation of North Atlantic fisheries. **(a)** The Grand Banks (sand-colored shading) were severely overfished in the late twentieth century, leading to the near extinction of many species, including the **(b)** Atlantic cod *(Gadus morhua)*.

have been essentially eliminated from the Grand Banks, and their populations will probably never recover. And because fishermen preferentially harvest the oldest and largest individuals, which fetch a higher market price, Atlantic cod *(Gadus morhua)* now mature at a younger age (3 years compared with 5 or 6 years) and smaller size.

As a consequence of overfishing, the average yield of the Grand Banks has declined to less than 10% of the highest historic levels. In the mid-1960s, Atlantic cod yielded a minimum of 350,000 tons per year. By the mid-1970s, the catch dropped to 50,000 tons per year. The Canadian government finally closed the fishery in 1993, after the cod catch fell below 20,000 tons for several consecutive years. But the damage had already been done: the most heavily exploited species are less marketable because of their smaller size, fish populations have decreased to dangerously low levels, and the fishing industry is itself imperiled. This sequence of events has been replicated in fisheries around the world. Indeed, in a report published in 2003, Ransom A. Myers and Boris Worm of Dalhousie University in Nova Scotia estimated that modern fishing techniques have reduced the biomass of large predatory fishes by about 90% in marine ecosystems.

Overexploitation is not inevitable; careful management of fisheries can achieve sustainable harvests. Many approaches are possible, such as providing supplemental food or shelter; maintaining captive breeding populations, from which individuals are introduced into the wild; limiting the times of harvest to avoid disrupting reproductive cycles; and limiting the size and character of the catch. Similar strategies can be devised for other resource populations.

Human Activities Are Causing a Dramatic Increase in Extinction Rates

As you may remember from Section 22.5, extinction has been common in the history of life: roughly 10% of the species alive at any time in the past became extinct within 1 million years. These *background extinction rates* eliminated perhaps seven or eight species per year. Paleobiologists have also documented at least five *mass extinctions*, during which extinction rates increased greatly above the background rate for short periods of geological time (see Figure 22.18).

At present, Earth appears to be experiencing the greatest mass extinction of all time. According to Edward O. Wilson of Harvard University, extinction rates today may be 1000 times the historical background rate, meaning that thousands of species are being driven to extinction each year. The vast majority of extinctions are a direct result of habitat fragmentation, desertification, rising levels of pollution, the introduction of exotic species, and the overexploitation of natural populations.

If humans are the cause of the current mass extinction, why has it taken so long to occur? Why didn't the mass extinction begin long ago? The answer lies in our increased rate of population growth (see Section 49.7). During the nineteenth and twentieth centuries, improvements in food production, sanitation, and health care increased human life expectancy. Our ever-increasing population consumes resources and produces wastes at an escalating rate. As global population continues to increase, so will the habitat destruction that inevitably accompanies population growth.

STUDY BREAK

1. How has habitat fragmentation affected breeding songbird populations in eastern North America?
2. What factors have increased the likelihood of desertification in southern Florida?
3. What are the consequences of the overexploitation of fish populations?
4. How do extinction rates today compare with the background extinction rate evident in the fossil record?

53.3 Biodiversity Hotspots

Given the detrimental effects of human activities on biodiversity and natural environments, conservation biologists are constantly seeking ways to minimize or reverse the damage.

Conservation Biologists Focus Their Efforts in Areas Where Biodiversity Is both Concentrated and Endangered

If we are to limit the effects of human activities and preserve biodiversity, we must know how and where biodiversity is distributed. Although species richness generally increases from the poles to the tropics within many communities (see Section 50.7), these large global patterns do not help biologists pinpoint those areas where conservation efforts will have the greatest impact.

In a survey published in 2000, Norman Myers of Oxford University and his colleagues in England and the United States pinpointed 25 **biodiversity hotspots,** areas where biodiversity is both concentrated and endangered **(Figure 53.13).** As defined by the Endangered Species Act, adopted by the U.S. Congress in 1973, an **endangered species** is one that is "in danger of extinction throughout all or a significant portion of its range." (Species that are likely to become endangered in the near future are designated as *threatened.*) Thus, to qualify as a biodiversity hotspot, an

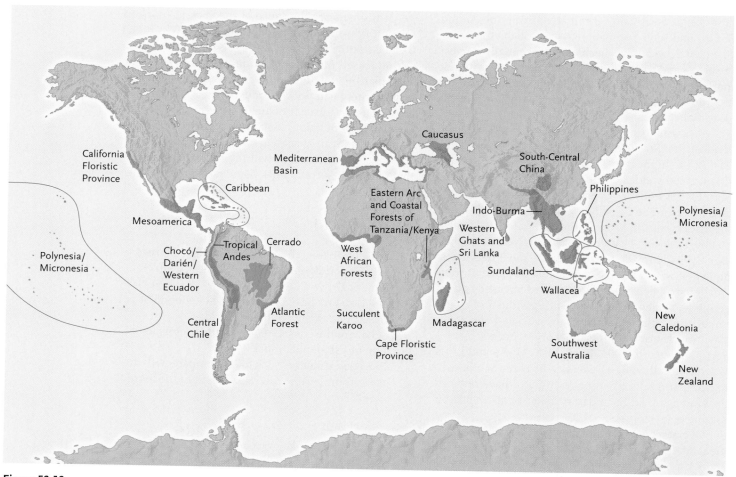

Figure 53.13

Biodiversity hotspots. Norman Myers and his colleagues identified 25 places that harbor many endemic species that are threatened by human encroachment.

area must harbor a large number of **endemic species** (those that are found nowhere else), and it must have already lost much of its natural vegetation to human encroachment.

Endemic species tend to have highly specific habitat or dietary requirements, low dispersal ability, and restricted geographical distributions. Myers used the number of endemic species as a criterion for identifying hotspots because locally distributed species account for much of Earth's biodiversity; and if the local habitats where these species occur are at risk of development, the species are also at risk. Although the 25 hotspots occupy only 1.4% of Earth's land surface, they include the only remaining habitat for approximately 45% of all terrestrial plant species and 35% of all terrestrial vertebrate species.

Sixteen of the 25 hotspots are in the tropics, where humans have already cut much of the natural vegetation. For example, until the mid-1900s, the Brazilian Atlantic Forest stretched undisturbed along the southern coast of Brazil and parts of Paraguay and Argen-

tina. Since then, 93% of the forest has been cleared for agriculture and grazing, making it one of the most endangered ecosystems on Earth. Today more than 70% of Brazil's population lives within the historical distribution of the Atlantic Forest, and most of its endemic species are threatened. Yet the Atlantic Forest still harbors more than 5% of Earth's butterfly species, 7% of the primate species, and more than 430 tree species per hectare.

Nearly all tropical islands fall within one of the designated hotspots, and 9 of the 25 hotspots are mostly or completely made up of tropical islands. As you may recall from Chapter 21, island clusters often harbor many species because their geography fosters adaptive radiations. By definition, because island-dwelling species have limited geographical ranges, their population sizes tend to be small; and small populations always face a high likelihood of extinction. Because most tropical islands also house dense human populations, it is not surprising that they are well represented on the hotspot list.

53.4 Conservation Biology: Principles and Theory

Conservation biology is an interdisciplinary science that focuses on the maintenance and preservation of biodiversity. Conservation biologists use theoretical concepts from systematics, population genetics, behavior, and ecology to develop ways to protect threatened wildlife. We introduce theoretical aspects of conservation biology in this section and practical applications in the next.

Systematics Organizes Our Knowledge of the Biological World

To develop a conservation plan for any habitat, scientists must start with an inventory of its species. Their primary tool is systematics, the branch of biology that discovers, describes, and organizes our knowledge of biodiversity (see Chapter 23).

Cataloguing the diversity of life may be the most daunting task that biologists face. After more than 200 years of work, systematists have described and named approximately 1.6 million species. However, they realize that this number represents only a fraction of existing species.

In 1982, Terry Erwin of the Smithsonian Institution studied beetle biodiversity at the Tambopata National Reserve in Southern Peru. He sprayed biodegradable insecticide into the canopy of one large tree and collected 15,869 individual beetles, which he sorted into 3429 species. More than 90% of the individual beetles he collected belonged to species that had not yet been described. Erwin used this astounding result and a complex mathematical model to predict that approximately 30 million species currently exist.

Nigel Stork of the Natural History Museum in London later questioned Erwin's conclusions. Using additional data and a modified set of assumptions, he estimated that the actual number of living species was closer to 100 million. If his figure is correct, more than 98% of species—most of them arthropods, nematodes, bacteria, and archaeans—are still unknown to science. Regardless of whether biodiversity encompasses 30 million species or 100 million, systematists clearly have much work to do.

Recently, conservation biologists and systematists have begun to develop a new technology that will simplify the identification of species in the field, thereby facilitating the creation of a catalog of biodiversity. *Insights from the Molecular Revolution* describes the effort to develop a "DNA barcode scanner."

Population Genetics Informs Strategies for Species Preservation

When populations are reduced to small size, genetic drift inevitably reduces their genetic variability (see Section 20.3) and the evolutionary potential to adapt to changing environments. Thus, the loss of even a small fraction of a species' genetic diversity reduces its survival potential. To avoid this problem, conservationists strive not only to increase the population sizes of threatened or endangered species but to maintain or increase their genetic variation, both within and between populations.

For example, the whooping crane *(Grus americana)* was once an abundant bird in wet grassland environments through much of central North America **(Figure 53.14).** By the early 1940s, excessive hunting and habitat destruction had caused their numbers to decline to just 21 individuals in two isolated populations. This population bottleneck and the resultant loss of genetic variability apparently contributed to developmental deformities of the spine and trachea that had not been seen previously.

During the 1970s, biologists began an aggressive conservation program. In addition to preserving habitats in the crane's summer and winter ranges, they initiated a carefully controlled captive breeding program designed to minimize the effects of inbreeding. Although more than 300 whooping cranes now survive in several wild and captive populations, recent research reveals that they still have a remarkably low level of genetic variability. As expected, the genetic effects of a severe population bottleneck may persist long after a population begins to increase in size.

Figure 53.14
Whooping cranes. Endangered whooping cranes *(Grus americana)* winter in the Aransas National Wildlife Refuge in Corpus Christi, Texas.

Brian K. Miller/Animals, Animals—Earth Scenes

Developing a DNA Barcode System

Everyone is familiar with the checkout scanners at supermarkets and other stores. The cashier quickly passes an item's barcode over the scanner, and the register identifies it and records its price. The system works because the barcode on every item contains unique identifying information. Some biologists have proposed an analogous method, called DNA barcoding, for identifying animal and plant species quickly and accurately. The researchers envision using a handheld device to rapidly analyze DNA in the field; the resulting data would be sent to a database by cell phone, and minutes later an identification and a description of the species would appear on the instrument's screen.

While the analytical device is not yet ready for use in the field, DNA barcoding is now being tested. This technique is the brainchild of Paul Hebert, a population geneticist at the University of Guelph, Ontario, Canada. His idea has caught on, and in 2004 a consortium of major natural history museums and herbariums started the Barcode of Life Initiative, with the goal of creating a database of DNA barcodes linked to specimens already identified in their collections. The approach potentially could replace the traditional methods of systematic analysis using organismal and genetic characters to identify species.

Hebert proposed using the first part of the COI (cytochrome oxidase 1) gene—a sequence of about 500 nucleotides—as the DNA barcode to distinguish animal species. This mitochondrial gene tends to vary greatly between species. Moreover, it appears to have no inserted or deleted DNA

segments in most animal species, making the alignment and comparison of sequences straightforward. Hebert's hope is that any COI gene sequence obtained in the field will provide a unique identifier for the species from which the DNA sample was obtained.

Early tests of Hebert's barcode approach have been promising. He and his collaborators first analyzed the COI gene sequence in the skipper butterflies of Costa Rica. Although adult skippers look pretty much alike, their caterpillars vary in appearance and in their food plant preferences, leading researchers to wonder if butterflies that had been assigned to one species (Astraptes fulgerator) might actually represent several. Analyses of the COI gene sequence sampled from 484 adults allowed Hebert and his colleagues to identify 10 distinctive DNA barcodes, suggesting that there are at least 10 species of skipper in Costa Rica rather than just one.

In early 2007, Hebert and his colleagues reported that they had used the DNA barcode to analyze 2500 specimens of 643 North American bird species. The results were impressive: barcode differences between species were an order of magnitude greater than the differences within species, allowing the unambiguous identification of species from a short DNA sequence. Interestingly, the barcode analysis identified 15 probable new species that had not been previously identified and revealed that 8 supposed species of gull may be variants of just one species.

Taken together, the results of the two research studies provide support for the use of DNA barcodes, and for

using the COI gene sequence specifically for the barcode analysis, as a means of identifying animal species.

Can DNA barcodes be used to identify plant species? The mitochondrial COI gene sequence used for barcoding animals is not suitable for barcoding plants because the gene has evolved much more slowly in plants and therefore exhibits less variability among species. However, in 2005, researchers reported on a study that used two different DNA sequences, one from the nuclear genome and the other from the chloroplast genome, to barcode flowering plants. Trials involving 53 plant families, with a total of 99 species from 80 genera, suggested that the two sequences could distinguish a large number of flowering plant species, making barcoding of flowering plants a feasible proposition.

Despite these early successes, many skeptics believe that DNA barcoding will prove to be inaccurate and may, in fact, produce false conclusions about species designations and incorrect counts of biodiversity. The skeptics argue that the approach has not been tested sufficiently in closely related species, which may exhibit only small differences in the barcode sequence. They also point out that the assumption that organisms have a fixed genetic characteristic—like the barcodes on items at the supermarket—contradicts fundamental ideas about genetic variability that are at the core of contemporary evolutionary theory. The DNA barcoding efforts continue nonetheless, and new data will continue to fuel the debate between the supporters of this approach and the naysayers.

Studies of Population Ecology and Behavior Are Essential Elements of Conservation Plans

Conservation programs also require data about target species' ecology and behavior, including their feeding habits, movement patterns, and rates of reproduction.

Sea otters (Enhydra lutris) are predatory marine mammals that live along the coastline of the North Pacific Ocean. In the early 1700s, they numbered ap-

proximately 300,000 individuals (Figure 53.15), but commercial hunting reduced their numbers to about 3000 individuals by the start of the twentieth century. Sea otters are keystone predators (see Section 50.4), and the destruction of sea otter populations had profound effects on the communities in which they lived. As the numbers of sea otters plummeted, populations of sea urchins, one of their favored prey, exploded; burgeoning sea urchin populations decimated local kelp

a. Sea otter

b. Geographical range of sea otters

Malcolm Schuyl/Peter Arnold, Inc.

KEY

▮ Absent from historic range

▯ Present range

Figure 53.15
Sea otters. After being hunted nearly to extinction, **(a)** sea otters (*Enhydra lutris*) have been reintroduced in many parts of their historical range **(b)**.

beds, disrupting the communities of animals that live among these giant algae.

International treaties ended nearly all hunting of sea otters in 1911, and the populations subsequently recovered to about one-third of their original levels. Conservation biologists facilitated the recovery by reintroducing otters into southeastern Alaska, British Columbia, Washington, and California. Before deciding where otters should be reintroduced, scientists had to assess the resources available at different sites and determine how far individual otters would move, how rapidly they would reproduce, and how quickly their populations would spread. The reintroduction effort was successful at first. However, populations in California have experienced high mortality since the mid-1990s, and nearly half of those dying have been adults in their reproductive prime. Researchers have identified parasitic infections and heart disease as leading causes of death, suggesting that some coastal environments are so badly degraded that they may no longer support populations of this species.

Given the complexities of the ecological relationships in natural communities and ecosystems, conservation biologists have developed two sophisticated types of population analysis, *population viability analysis* and *metapopulation dynamics,* to design effective conservation plans.

Population Viability Analysis. Using complex mathematical models, conservation biologists can conduct a **population viability analysis** (PVA) to determine how large a population must be to ensure its long-term survival. PVAs evaluate phenomena that may influence the longevity of the population or species: habitat suitability, the likelihood of catastrophic events, and other factors that may cause fluctuations in demo-

graphics, population size, or genetic variability. When conducting a PVA, researchers must decide what level of risk is acceptable for a given survival time. For example, should a conservation plan attempt to ensure a 95% probability that the species will survive for 100 years, or should it specify a 99% survival probability? An increase in either the survival probability or the survival time requires an increase in the size of the population that must be conserved. The **minimum viable population size** identifies the smallest population that fits the desired specifications of the conservation plan. *Focus on Research* describes how biologists used PVA in the conservation of an Australian marsupial, the yellow-bellied glider.

Metapopulation Dynamics. In many species, individuals move frequently from one local population to another. To describe the dynamics of such movements, ecologists define a **metapopulation** as a group of neighboring populations that exchange individuals. Local populations within a metapopulation are not all equal: they often differ in size, population growth rates, the suitability of their habitats, their exposure to predators, and other factors. Moreover, some may decline steadily in size, while others may increase.

Under favorable circumstances, a population may produce numerous offspring, some of which emigrate and join nearby populations, where they breed, providing a genetic connection between local populations (see the discussion of gene flow in Section 20.3). Thus, dispersal and gene flow between local populations maintain the metapopulation.

Populations that are either stable or increasing in size are described as **source populations** because they are a possible source of immigrants to other populations. Those that decline in size are called **sink**

Applied Research: Preserving the Yellow-Bellied Glider

Predicting the future is never easy, especially the future of a threatened species. But population viability analysis (PVA) allows conservation biologists to predict how a species will fare under a range of possible scenarios. An effective PVA for an animal species requires detailed information about its diet, predators, mating habits, habitat preferences, space requirements, demography, geographical distribution, responses to climatic fluctuations and human disturbances, and a host of other aspects of its biology.

The Australian yellow-bellied gliding marsupial, *Petaurus australis*, better known as the yellow-bellied glider, provides an example of how PVA is essential for a conservation effort. This mammal, about the size of a squirrel, lives in small family groups in undisturbed *Eucalyptus* forests along Australia's eastern coast. Each glider family maintains a home range (the area it uses for feeding and other activities) of 25 to 85 hectares; the home ranges of neighboring families do not overlap. As a result, the population density of gliders has never been high. But glider populations have declined precipitously as forests have been cleared, and the species is now considered threatened.

Using data from nearly 20 published papers, two Australian conservation biologists, Russ Goldingay of the University of Wollongong and Hugh Possingham of the University of Adelaide, conducted a PVA for this species. They estimated age distributions in glider populations as well as survival probabilities, litter sizes, sex ratios, lifespan, and home range sizes. They analyzed these data using a mathematical model that predicts the viability for populations of various sizes. In most PVAs, a population is considered viable if it has a 95% probability of surviving for 100 years. Goldingay and Possingham introduced additional complexity to their analysis by assessing the effects of unpredictable environmental events, such as drought, on breeding success. They also conducted sensitivity analyses to examine how changing the values of specific parameters—such as litter size, mortality rates of the different age classes, or the frequency and severity of droughts—might influence the general predictions of the viability model.

Once Goldingay and Possingham had completed many thousands of these calculations, they concluded that a viable population of gliders would require at least 150 family groups. They also suggested that a population of that size would need approximately 18,000 hectares (roughly 70 square miles). Currently, only 1 of the 15 existing conservation reserves is that large.

Goldingay and Possingham did not factor some common environmental disturbances—fire, disease, or predation by introduced species—into their analyses. Such disturbances could decimate a small glider population in short order. Thus, the outlook for gliders may be bleaker than the researchers suggest, because their estimates of minimum viable population size and minimum necessary habitat size are almost certainly too low. Given only this information, we might predict that the glider will inevitably become extinct.

However, there is some hope for the yellow-bellied glider. Goldingay and Possingham assumed that gliders don't move between populations, a behavior that promotes gene flow. They ignored this aspect of metapopulation dynamics because they had no data on gene flow in this species. The movement of individuals between populations could reduce the required minimum viable population size by decreasing the likelihood of genetic drift and the extinction of local populations. Biologists may even be able to transplant gliders from one population to another, effectively creating source and sink populations. This procedure might increase population size and genetic diversity in the most endangered populations. If successful, such an approach could stave off extinction.

As a result of this PVA, conservation biologists can determine which of the remaining forest tracts are large enough to sustain a yellow-bellied glider population. Thus, they now know where to concentrate their limited resources to secure the future survival of this species. Although predicting the future is difficult, PVAs allow conservation biologists to make accurate and reliable recommendations for selective transplants that will contribute to the conservation of threatened species.

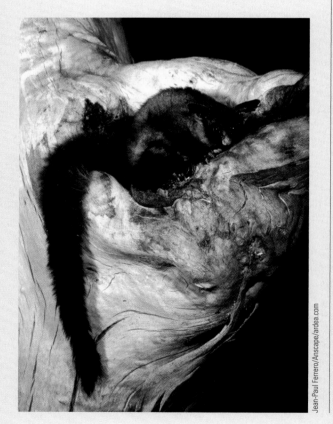

Jean-Paul Ferrero/Ariscape/ardea.com

Figure 53.16 Observational Research

Metapopulation Structure of the Bay Checkerspot Butterfly

Paul Ehrlich

Map of serpentine habitat patches near Morgan Hill

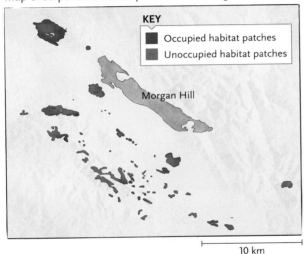

KEY

■ Occupied habitat patches

■ Unoccupied habitat patches

Morgan Hill

10 km

HYPOTHESIS: Populations of the bay checkerspot butterfly *(Euphydryas editha bayensis)* living on small patches of suitable habitat are "sink" populations that frequently become extinct. Populations in large habitat patches can serve as a "source" of individuals to recolonize small habitat patches nearby.

PREDICTION: Because the bay checkerspot butterfly is a weak flyer, small patches of suitable habitat that are close to a large source population will be recolonized frequently. Patches of suitable habitat that are far from a large source population will be recolonized only rarely.

METHOD: Susan Harrison, Dennis D. Murphy, and Paul R. Ehrlich of Stanford University surveyed 59 small patches of serpentine grassland near San Jose, California, in 1986 and 1987. They estimated each patch's "quality" based on the presence or absence of food plants on which bay checkerspots depend and on aspects of the physical environment that are important to these butterflies. They also measured each patch's distance from Morgan Hill, a very large patch of suitable habitat that had sustained a bay checkerspot population for years. In patches where they found butterflies, they estimated bay checkerspot population sizes.

RESULTS: A complex statistical analysis revealed that both distance from the Morgan Hill population and habitat patch quality were important factors in determining whether bay checkerspots would be present or absent in a small habitat patch. The authors noted that only the nine high-quality habitat patches near Morgan Hill (red on the map) were occupied by bay checkerspots. Of 50 unoccupied habitat patches, 6 were near Morgan Hill but of low quality; 18 were of high quality but far from Morgan Hill; and 26 were too far from Morgan Hill and of too low quality to support a population of bay checkerspots.

CONCLUSION: Populations of bay checkerspot butterflies that occupy large patches of suitable habitat serve as source populations for individuals that recolonize small patches of suitable habitat where butterfly populations frequently become extinct. However, because the bay checkerspot is a weak flyer, it recolonizes small patches of suitable habitat only if they are close to a source population.

populations because they represent a drain on the supply of available immigrants. Individuals usually move from source populations to sink populations, and sink populations persist because they receive immigrants from source populations in the metapopulation.

The bay checkerspot butterfly, *Euphydryas editha bayensis* **(Figure 53.16)**, provides an example of metapopulation dynamics. This species is restricted to serpentine grassland in the San Francisco Bay area (see Figure 50.18) because its larvae eat plants that grow only in that community. Human disturbance has fragmented much of the butterfly's natural habitat into patches of varying size, each of which may support a local butterfly population. The life cycle of these butterflies is always a race against time, because the larvae must feed and mature before dry summer weather kills their food plants. Populations in small patches often become extinct, but those occupying larger patches, where food plants stay alive longer, generally survive the seasonal drought. Butterfly populations in larger

habitat patches therefore serve as source populations for emigrants that repopulate small habitat patches the following year. But the bay checkerspot is a poor flyer, and it cannot disperse long distances. Thus, small patches of suitable habitat harbor bay checkerspots only if they are close to a larger patch that serves as a source. A conservation plan for this butterfly would therefore aim to preserve habitat patches of sufficient size to serve as sources for nearby smaller patches.

Community and Landscape Ecology Help Large-Scale Preservation Projects

Many conservation efforts focus on the preservation of entire communities or ecosystems. These projects often depend on the work of community and landscape ecologists.

Species/Area Relationships. As you know from Chapter 50, community composition is dynamic: some species become extinct and others join the community

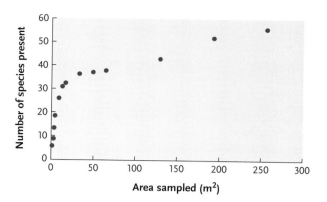

Figure 53.17
The species/area relationship. Data on plant distributions in Quarry Meadow in Austin, Texas, illustrate the relationship between habitat area and number of species present.

Figure 53.18
Edge effects and patch size. This hypothetical example illustrates how a 20 m wide edge disrupts a larger fraction of a small habitat patch than a large habitat patch.

tains 25,600 m² of intact habitat. Although the large patch is only four times larger than the small patch, the large patch contains more than seven times as much *intact* habitat.

Landscape Ecology. Researchers in the field of **landscape ecology** determine how large-scale ecological factors—such as the distribution of plants, topography, and human activity—influence local populations and communities. Knowing that larger protected areas will preserve more species, conservation biologists have debated whether nature preserves should comprise one large habitat patch or several smaller patches. Ecologists identify this debate with the acronym **SLOSS** (*Single Large Or Several Small*). Jared Diamond of the University of California, Los Angeles, initiated the SLOSS debate in 1975. Applying the lessons of island biogeography, Diamond concluded that a single large preserve was preferable to several smaller ones, even if they encompassed an equivalent area.

Conservation biologists have since concluded that no single design is best for all organisms. For large animals, such as predatory cats, one large preserve may be best, because individuals must patrol large areas to search for food. For smaller animals, such as insects, several small preserves, each providing a slightly different environment that supports one population, is preferable; if a population in one preserve becomes extinct, individuals from elsewhere in the metapopulation can recolonize the area.

Diamond also suggested that small preserves would function better if corridors of intact habitat connected them. Individuals could move between preserves, reviving any local populations that experienced a decline. These landscape corridors might effectively join the smaller constituent populations into one larger population, which would avoid some of the genetic difficulties encountered by small populations.

However, some conservation biologists argued that landscape corridors connecting small preserves may actually threaten biodiversity. Corridors are usually narrow and thus subject to strong edge effects. In some environments, they are drier and more susceptible to fires that could spread into the preserves they connect. Corridors might also provide entry points for exotic species and disease-causing organisms. Finally, species that don't enter habitat edges would be unlikely to use the corridors at all.

Ellen I. Damschen of North Carolina State University and several colleagues conducted an ambitious long-term field experiment on the effect of landscape corridors on plant species richness **(Figure 53.19)**. Their results, published in late 2006, suggest that habitat patches connected by corridors retain more native plant species than isolated patches and that corridors did not promote the entry of exotic species. Thus, based on limited experimental evidence, corridors appear to be a useful feature in the design of nature preserves.

through immigration. If we view fragmented patches of intact habitat as islands in a sea of unsuitable terrain, we can apply the predictions of the theory of island biogeography (see Section 50.7) to the design of protected areas. For example, we might expect that the number of species a patch will support depends on its size and proximity to larger patches.

Indeed, ecologists recognized long ago that large habitat patches sustain more species than small patches do **(Figure 53.17)**. When plotted on an arithmetic scale, the relationship between species richness and habitat area increases sharply at first and then flattens. In other words, for relatively small habitat patches, even minor increases in area allow a large increase in the number of resident species; but as habitat patches get larger, the number of species present eventually levels off. You encountered an example of this relationship in our discussion of bird species richness on islands of different sizes (see Figure 50.30b).

As habitats become increasingly fragmented, edge effects exaggerate the species/area relationship in mainland habitat patches **(Figure 53.18)**. Consider two hypothetical patches of habitat: one is 100 m on a side, with a total area of 10,000 m²; the other is 200 m on a side, with a total area of 40,000 m². Now, imagine that edge-effect disturbances penetrate 20 m into each patch from all directions. The small patch contains only 3600 m² of intact habitat, but the large patch con-

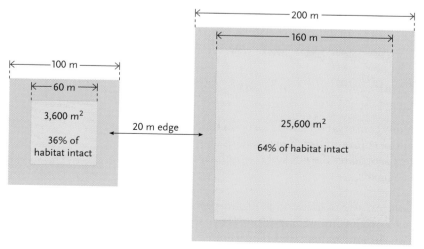

Landscape corridors may also allow large animals to move freely between patches of suitable habitat. For example, the Florida panther (*Puma concolor coryi*), shown on page 1229, is critically endangered: only 70 to 100 individuals of this subspecies remain from a population that once ranged throughout the southeastern United States; other panther subspecies still inhabit the western states. Panthers are large predators, and each female requires nearly 20,000 hectares (more than 75 square miles) for hunting and breeding; males each require more than twice as much space.

Although the state and federal governments have set aside several panther conservation areas in Florida, 52% of the habitat panthers occupy is privately owned, and most of it is highly fragmented. Panthers frequently cross roads, and most panther deaths in Florida are caused by accidents with motor vehicles. Protected landscape corridors might enable panthers to move more safely between conservation areas. A preliminary study found that panthers already use such corridors, typically along wooded riverbanks, when they are available. The Florida Fish and Wildlife Service has proposed the creation of an ambitious 6100-hectare network of such corridors alongside the Caloosahatchee River to link several significant habitat fragments in neighboring counties.

STUDY BREAK

1. How does a population bottleneck change the likelihood that a species will become extinct?
2. How does a population viability analysis assist in the development of a conservation plan for a species?
3. Would a single large nature preserve or several small preserves experience greater edge effects?

53.5 Conservation Biology: Practical Strategies and Economic Tools

Conservation biology seeks to protect native species, communities, and ecosystems from the effects of human activity. Meeting that goal and reversing some of the existing damage requires the integration of biological research with economic and social realities.

Conservation Efforts Aim to Preserve, Conserve, and Restore Habitats

Conservation groups often highlight efforts to preserve individual animal species, such as the giant panda (*Ailuropoda melanoleuca*) or California condor (*Gymnogyps californianus*). The preservation of "charismatic megavertebrates," as these large animals are sometimes described, attracts substantial public sup-

Figure 53.19 Experimental Research

Effect of Landscape Corridors on Plant Species Richness in Habitat Fragments

QUESTION: Do landscape corridors connecting habitat patches influence the species richness of native and exotic plants within the habitat patches?

EXPERIMENT: Damschen and her colleagues studied changes in the community composition and species richness of the plants in open habitat patches within a longleaf pine (*Pinus palustris*) forest in South Carolina. Their experimental design included both isolated patches and patches that were connected to one another by a landscape corridor. All patches included the same land area, and their large size (1.375 ha each, including the landscape corridors) allowed the researchers to make a realistic assessment of the effects of landscape corridors. After creating the patches of open habitat within the forest in 2000, the researchers catalogued all plant species occurring in the patches through 2005, although they were unable to collect data in 2004.

RESULTS: Over the course of the study, habitat patches that were connected by landscape corridors harbored increasingly more plant species than did isolated habitat patches. The researchers also noted that the difference in species richness between the two experimental treatments was caused by a difference in the number of native plant species present. The number of exotic species in connected and isolated habitat patches was similar.

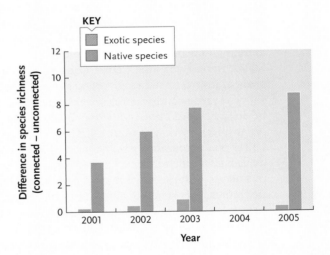

CONCLUSION: Landscape corridors between patches of open habitat in longleaf pine forests increase the species richness of native species in open habitat patches, but they do not foster the entry of exotic species.

port. Nonetheless, there is little point in trying to preserve natural populations of individual species if their habitats are in jeopardy. An alternative to species-based conservation focuses on the preservation of intact habitats; individual species are conserved as a consequence of preserving the habitats on which they depend. Conservation biologists approach this goal with a continuum of approaches, which fall into three general categories: *preservation, mixed-use conservation,* and *restoration.*

a. Albany Pine Bush

b. Karner Blue butterfly

Figure 53.20

The Albany Pine Bush habitat. **(a)** The Pine Bush lies entirely within the city limits of Albany, New York. It is home to about 50 threatened or endangered plant and animal species, including **(b)** the Karner Blue butterfly *(Lycaeides melissa samuelis)*.

Conservation through Preservation. In many countries, habitats are preserved when an individual or organization purchases them and enforces strict standards of land use. In sensitive habitats, people may be excluded altogether; in other cases, access is restricted and the exploitation of resources is controlled. This approach works well in countries with efficient law enforcement and a tradition of private land ownership. In the United States, for example, the Nature Conservancy has purchased large tracts of land to preserve native species.

The preservation approach has been successful in preserving portions of the Pine Bush habitat near Albany, New York **(Figure 53.20).** This unique ecosystem arose approximately 11,000 years ago at the end of the last glacial period, when a massive deposit of sand was left near the western margin of Albany's current city limits. This sandy region formed an inland pine-barrens habitat in which pitch pine *(Pinus rigida)*, scrub oak *(Quercus ilicifolia)*, and dwarf chestnut oak *(Quercus prinoides)* are now the dominant vegetation. The Pine Bush is home to more than 50 plant and animal species that the state and federal government list as threatened or endangered. The habitat itself was once vulnerable because it lies within Albany's city limits; however, since 1988, the Pine Bush has been jointly owned and protected by New York state, local municipalities, and the Nature Conservancy.

Mixed-Use Conservation. The preservation approach does not work under all circumstances. Where outright preservation is impractical, conservation biologists advocate mixed-use conservation, which combines the protection of some land parcels with the controlled development of others.

The Ngorongoro Conservation Area (NCA) in Tanzania provides an example of mixed-use conservation. The NCA covers 829,000 hectares of grassland and borders the Serengeti National Park. Because it houses a high concentration of wildlife, the NCA is one of the most heavily visited tourist destinations in eastern Africa. For the past several hundred years, the Maasai people have herded cattle, goats, and sheep in the Serengeti and Ngorongoro **(Figure 53.21).** The Maasai are nomadic pastoralists who frequently move their relatively small herds to new grazing areas in the region. As a result, their activities do not degrade the land or exclude native wildlife. In 1959 the Maasai agreed to vacate the Serengeti, which was converted into a national park, in return for retaining the rights to live and herd livestock within the NCA. The government of Tanzania helped create the necessary infrastructure within the NCA, including a constant water supply as well as social services. Under this agreement, 40,000 indigenous residents, most of them Maasai, live in this large and valuable conservation area.

Conservation through Restoration. Conservation biologists sometimes create restoration plans to reestablish the vitality of a previously disrupted community or ecosystem. This effort requires the removal of contaminants, impediments to the natural flow of water, and barriers to animal movement as well as the restoration of natural processes, such as periodic fires or floods. Most restoration projects also require replanting key plant communities and long-term management once restoration is complete.

Not all degraded habitats can be restored, and not all potential restoration projects are equally feasible. When making project decisions, restoration ecologists

Figure 53.21
Mixed-use conservation. The Maasai use the Ngorongoro Conservation Area to graze cattle and goats.

consider a number of factors: Will the restored habitat be suitable for rare or endangered species, and will its creation increase endemic biodiversity? Would the restoration reunite previously fragmented land parcels? Will the restored habitat experience the periodic disturbances, such as fires or floods, that are essential for its continued existence? What are the costs of implementing the plan and maintaining the area? Finally, would the restored land be valued by local residents, and will they support and maintain it?

A successful restoration project is currently underway in the Brazilian Atlantic Forest, sponsored by the Instituto de Pesquisas Ecológicas (IPÊ), a Brazilian nongovernmental organization. In western São Paulo state, near the Morro do Diabo state park, IPÊ is trying to recreate the natural Brazilian Atlantic Forest ecosystem by planting native trees in habitat corridors between remaining forest fragments. These corridors of native tree species should facilitate the preservation of species in those forest patches and supply valuable botanical resources for endemic wildlife and local residents.

Successful Conservation Plans Must Incorporate Economic Factors

Biologists can almost always develop a plan to conserve a species, community, or ecosystem. But to be successful, a plan must be economically feasible, and it must provide direct benefits to local residents whose lives it will affect.

Local Involvement. Early conservation efforts simply set aside protected areas in which most human activities were banned. Local people were denied access to resources within the preserve—resources that were sometimes essential for their survival. Not surprisingly, these plans generated antipathy towards conservationists and the organisms they were trying to preserve.

For example, the northern spotted owl (*Strix occidentalis caurina*) lives only in old growth coniferous forests of the Pacific Northwest, where many local residents worked in forestry or supporting industries. The suggestion that the owl be listed as an endangered species triggered a bitter political battle between conservationists and local residents because the conservation plan for the owls required closing large tracts of forest to logging. Washington State listed the owl as an endangered species in 1988, but local residents, who lost jobs when logging was reduced, remain hostile to these conservation efforts.

Conservation plans are more successful if they provide local residents with benefits that depend on the existence of a preserve. Royal Chitwan National Park provides an excellent example. For more than 100 years, this area, located in south central Nepal near the northern border of India, was a privately owned "preserve" used for big game hunting by royalty. These activities decimated local populations of large mammals, especially the Bengal tiger (*Panthera tigris*) and one-horned or Indian rhinoceros (*Rhinoceros unicornis*). Populations of both species dwindled to approximately 100 individuals by the mid-1960s. The area was subsequently opened for settlement, and, as immigrants swarmed into the fragile grassland, its human population exploded.

The area was converted into Royal Chitwan National Park in 1973. Today, humans are excluded from the park for most of the year. But each January, after the monsoon rains end and the grasses have dried, local residents are welcomed into the park for their annual harvest festival. They cut the grass and carry it away to thatch roofs, make mats, and feed domestic animals **(Figure 53.22)**. The local people value Chitwan and argue for its preservation. Today, more than 600 one-horned rhinos survive in Nepal, most of them in Royal Chitwan National Park. And the Bengal tiger population of Chitwan has increased to approximately 250 individuals.

Ecotourism. In some preserves, governments enlist local residents in park development and operations, providing them with a viable livelihood. The most successful approach has been the development of **ecotourism,** in which visitors, often from wealthier countries, pay a fee to visit a nature preserve. Local people work as guides, cooks, and logistical and support staff.

Not everyone agrees that ecotourism is helpful. Critics note that increased human traffic may degrade habitats, and unregulated ecotourism can eventually lead to overdevelopment. For example, several million people visit national parks in the western United States annually. Traffic jams, automobile accidents, and long lines routinely plague visitors at the most popular sites. Cranky ecotourists call for the construction of more roads and parking lots, which are inconsistent with the purpose of a national park because cars increase local air pollution and occasionally kill wildlife. In 2006, the government began charging a $20 fee for each automobile entering Yosemite Na-

Figure 53.22
Conservation and the local economy. Local residents support conservation efforts at the Royal Chitwan National Park, Nepal, because officials open the park for a grass harvest each year.

Ed Degginger/Color-Pic

tional Park in California, hoping to limit the number of visitors arriving in private vehicles and to increase reliance on public transportation.

Countrywide Economic Approaches. In the mid-1990s, conservation biologists and economists developed the concept of **ecosystem valuation,** in which ecosystem services—such as carbon dioxide processing or water retention and purification, which are best provided by intact ecosystems—are assigned an economic value. These estimated values are used to negotiate contracts in which a private company or conservation organization pays a community, state, or country to maintain intact ecosystems. By one 1997 estimate, the gross global ecosystem valuation is roughly 18 trillion U.S. dollars per year. If less obvious benefits provided by nature are tallied—soil formation, crop pollination, and nutrient cycling—the total value of ecosystem ser-

vices rises to 33 trillion U.S. dollars—almost twice the value of all goods produced by all humans on the planet!

The implementation of ecosystem valuation exchanges is determined on a case-by-case basis, depending on what ecosystem services the paying organization wants to preserve. Costa Rica is leading the way in this effort by creating valuation contracts with several corporations. For example, in 1998, the Monteverde Conservation League signed a contract with a local electrical company to ensure the continued flow of water from the Bosque Eternal de los Niños, a forest preserve. The company had plans to build a hydroelectric dam on the Rio Esperanza, and feared that deforestation upstream would disrupt water flow through the dam. The contract specifies that the electrical company will pay the people who live upstream to preserve their forests rather than cutting them.

Unanswered Questions

Are there general patterns in networks of interacting species? What causes those patterns, and what are their consequences for biodiversity conservation?

Biodiversity is a buzzword referring to the diversity of life at its different levels of organization—from molecules and genes to organisms and ecosystems. The term describes not only the component parts but also the way the parts are assembled and how they function. In any ecosystem, organisms interact with each other through a variety of beneficial and detrimental interactions. Ecologists usually describe and summarize these complex interactions as *networks,* in which species are represented as nodes and interspecific interactions as links. Ecologists have long sought to uncover the emergent properties of interaction networks, their causes, and their consequences.

Some early theoretical models of food webs (networks depicting who eats whom in an ecosystem) suggested that species-rich, highly connected networks would be less stable than smaller, simpler networks. This finding astounded many ecologists because it contradicted a widely accepted hypothesis that larger, more complex food webs were more stable and thus more resistant to disturbance, such as invasions by exotic species. Supporters of the "complex food webs are more stable" hypothesis noted that many large, complex food webs—such as those in tropical forests—exist in nature, and their persistence over time had to be explained.

Later theoretical research suggested that the solution to the contradiction might lie in the strength of species interactions (that is, the impact that one species has on others). Food webs with a few very strong interactions and many weaker ones tend to be more stable than food webs with many interactions of medium strength. In other words, food webs that include one or two keystone species tend to be more stable than those without keystone species. Further research has revealed that many real food webs include a few strong interactions and many weaker ones, but ecologists still do not know what factors determine interaction strength and why we often observe only a few strong interactions and many weak ones.

For many years, research on ecological interaction networks has focused almost exclusively on food webs—that is, on predator–prey interactions. More recently, however, other types of interactions (including those between plants and their mutualistic pollinators and seed dispersers, or between hosts and their parasites) have started to receive more attention. This new research has uncovered some intriguing general patterns. For example, mutualistic networks include asymmetric interactions of two general types: (1) species that have few links to other species ("specialists") tend to interact with species that have many links to other species ("generalists"); and (2) species that exhibit weak interactions tend to be associated with species that affect them strongly. These two types of asymmetry in mutualistic networks seem to result from the fact that only a few species have many links and only a few interactions are strong. As was the case for the analysis of food webs, the unresolved issue is *why* only a few species have many links or have strong interactions.

Answering these questions is important not only because they improve our understanding of the complexity of ecological systems, but also because they may have important implications for biodiversity conservation. For example, the widespread existence of asymmetric interactions makes interaction networks highly resistant to perturbations (such as habitat modification and species invasions) that could result in the local extinction of some species. The removal of highly linked species with strong interactions will affect many other species in the community, but because most species have few links and weak interactions, most extinctions will have only minor effects on the overall structure and functioning of the network.

Diego Vázquez is a researcher with the National Research Council of Argentina at the Argentine Institute of Dryland Research in Mendoza, Argentina. He received his *licenciatura* (undergraduate degree) in biology from the University of Buenos Aires and his Ph.D. in ecology and evolutionary biology from the University of Tennessee. His interests include community ecology, plant-animal interactions, mutualism, biological invasions, and conservation biology. You can learn more about his research at http://www.cricyt.edu.ar/interactio.

Thus, both the forests and water flow are preserved, maintaining the forest ecosystem and generating badly needed electricity.

Biodiversity is a precious resource that is disappearing rapidly throughout the world. It can still be conserved through a monumental effort to catalog the diversity of living organisms and develop an understanding of their ecological relationships. Perhaps the major challenge for conservation biologists is the education of the human population about the value of biodiversity and the development of conservation plans that will enlist the support of people who live among the threatened species.

STUDY BREAK

1. Is the Pine Bush habitat in New York State an example of preservation, mixed-use conservation, or restoration?
2. How has the establishment of the Royal Chitwan National Park in Nepal been a successful conservation effort? How do conservation biologists measure its success?
3. How can the concept of ecosystem services be used to foster conservation of threatened habitats and species?

Review

Go to ThomsonNOW™ at www.thomsonedu.com/login to access quizzing, animations, exercises, articles, and personalized homework help.

53.1 The Benefits of Biodiversity

- Biodiversity provides direct benefits to humans because natural populations of organisms can be sources of useful natural products as well as genetic resources that can improve domesticated crops and animals (Figures 53.2 and 53.3).
- Biodiversity provides indirect benefits to humans by maintaining normal ecosystem processes, some of which help to counteract the harmful effects of human activities.
- Ethicists and environmentalists argue that biodiversity should be preserved simply because of its intrinsic worth.

53.2 The Biodiversity Crisis

- Human disruption of a habitat usually begins with the construction of a road that provides access to resources; the disruption spreads rapidly. Habitat fragmentation reduces the size of intact habitat patches, and edge effects diminish the quality of remaining habitat (Figure 53.4). Only small populations, which are subject to genetic drift and an increased likelihood of extinction, can inhabit small habitat patches.
- Deforestation is occurring at an alarming rate, especially in tropical regions (Figure 53.5). Excessive deforestation may lead to desertification and the loss of entire ecosystems (Figure 53.6). Deforestation, desertification, and global warming reinforce each other in a positive feedback cycle.
- Although pollution is released locally, it often spreads to regional and global scales, especially in bodies of water and the atmosphere (Figure 53.7). Pollution can take many forms (Figure 53.8).
- Exotic species often contribute to the extinction of native species through competition, predation, or parasitism (Figures 53.9–53.11). Humans frequently introduce exotics into communities either intentionally or inadvertently.
- Overexploitation of natural populations reduces their sizes and may induce evolutionary responses in the exploited populations (Figure 53.12).
- Although extinction has been common in the history of life, human activities have recently initiated what may be the greatest mass extinction of all time. Some biologists estimate that extinction rates today may be 1000 times the background extinction rate.

Animation: Five major extinctions

Animation: Effects of deforestation

Animation: Effect of air pollution in forests

53.3 Biodiversity Hotspots

- Biodiversity hotspots harbor large numbers of endemic species and are threatened by human activities (Figure 53.13). Although hotspots encompass only 1.4% of the land, a much larger proportion of biodiversity inhabits these areas.
- More than half the identified hotspots are in the tropics, and nearly all tropical islands are included within the hotspot designation.
- Preserving the hotspots will conserve a substantial part of Earth's biodiversity.

Animation: Global crises by region and habitat

Animation: Three types of reefs

53.4 Conservation Biology: Principles and Theory

- Conservation biology draws its theoretical foundation from systematics, population genetics, population ecology, behavior, community ecology, and landscape ecology.
- Systematists provide taxonomic inventories of biodiversity that are helpful for establishing conservation priorities.
- Conservation biologists design breeding programs to maintain or increase the genetic variability of species being preserved (Figure 53.14).
- Besides studying the population ecology and behavior of targeted species (Figure 53.15), conservation biologists use population viability analyses to determine the minimum viable population size necessary to conserve threatened species. Analyses of metapopulation dynamics can help conservation biologists understand the interactions among small populations of threatened species (Figure 53.16).
- Studies in community ecology have established the generality of the species/area effect: large habitat patches harbor more species than small habitat patches do (Figure 53.17).
- From the perspective of landscape ecology, biologists have debated the advantages and disadvantages of establishing one large reserve versus several smaller ones that are connected by habitat corridors (Figures 53.18 and 53.19).

53.5 Conservation Biology: Practical Strategies and Economic Tools

- Efforts to conserve communities or ecosystems follow one of three general strategies. *Preservation* requires the restriction or prohibition of human access to the area (Figure 53.20). *Mixed-use conservation,* an approach that balances the conflicting demands of habitat preservation and development, allows local residents to use the protected area in limited ways (Figure

53.21). *Restoration* attempts to recreate natural communities and ecosystems in places that have already been degraded by human activities.

- Conservation plans must also incorporate economic and social factors to win local support. Most conservation plans now include the involvement of local residents to generate revenue for their communities (Figure 53.22). Ecosystem valuation also encourages the preservation of ecosystems by assigning them a significant economic value.

Animation: Sustainable resource management

Questions

Self-Test Questions

1. The greatest extinction in the history of life on Earth:
   a. occurred at the end of the Permian period.
   b. occurred at the end of the Cretaceous period.
   c. occurred at the end of the Ordovician period.
   d. occurred at the end of the Cambrian era.
   e. may be occurring now.

2. Which of the following is usually the first step in the disruption of a natural habitat by humans?
   a. establishment of small villages
   b. planting of crops
   c. building of a road
   d. invasion by exotic species
   e. overexploitation of resources

3. Habitat fragmentation has damaged populations of breeding birds in North America because:
   a. the remaining habitat patches rarely contain enough food for birds to rear their offspring.
   b. the nests of birds in small habitat patches are frequently attacked by predators.
   c. pairs of breeding birds cannot easily move from one habitat patch to another.
   d. female birds cannot locate potential mates in small habitat patches.
   e. small habitat patches do not have enough edges to provide adequate hiding places.

4. Deforestation:
   a. is a problem only in the tropics.
   b. may speed desertification.
   c. is slowed by grazing and farming.
   d. permanently enriches the soil.
   e. leads to the formation of lush grasslands.

5. Chemical pollutants:
   a. can spread rapidly from the places they are released.
   b. do not appear to influence global climate change.
   c. have contributed to global mass extinctions.
   d. rarely affect natural bodies of water.
   e. rarely influence animals feeding at higher trophic levels.

6. Which of the following is most likely to be a biodiversity hotspot?
   a. a patch of forest in the middle of North America that is 500 km from the nearest big city
   b. a series of uninhabitable sand dunes in the Sahara Desert
   c. a botanical garden that houses representatives of 25,000 plant species
   d. a tropical island with many endemic species and a growing human population
   e. a suburban neighborhood where fields have been converted to backyards and playgrounds

7. Population viability analyses allow conservation biologists to:
   a. identify the source population from which an individual dispersed to a sink population.
   b. determine how large an area must be preserved for the protection of a threatened species.
   c. identify whether individuals of a threatened species are reproductively mature.
   d. predict the minimum population size of a threatened species that is likely to survive.
   e. predict whether a threatened species will use habitat corridors.

8. Metapopulations are defined as:
   a. neighboring populations that exchange individuals.
   b. populations that steadily decrease in size.
   c. populations that steadily increase in size.
   d. populations that produce numerous fertile offspring.
   e. populations that never receive immigrants.

9. For which of the following species has the use of habitat corridors been proposed as an important conservation tool?
   a. sea otters
   b. bay checkerspot butterflies
   c. Florida panthers
   d. whooping cranes
   e. Eastern hemlocks

10. The main goal of restoration ecology is the reestablishment of:
    a. natural patterns of water flow.
    b. the vitality of a degraded ecosystem.
    c. the historical corridors linking forest fragments.
    d. the natural barriers to animal movement.
    e. ecotourism.

Questions for Discussion

1. National parks are often established in ecologically sensitive areas. In many places they have become so popular that visitors endanger the ecosystems the parks were originally designed to preserve. How can the goals of conservationists, who work to maintain intact ecosystems, be balanced with those of citizens who wish to visit intact ecosystems? In other words, how would you regulate domestic ecotourism?

2. How do the principles of population genetics and the principles of metapopulation dynamics apply to the SLOSS debate? Do they suggest different ideal designs for nature preserves?

3. Imagine that you are a conservation biologist who has been asked to develop a conservation plan for a species of lizard that lives in the deserts of the American Southwest. What sorts of data would you collect before developing a final plan?

Experimental Analysis

Devise a field study to determine whether the species/area relationship applies to aquatic ecosystems, such as ponds and lakes, as it does to terrestrial habitats.

Evolution Link

Overexploitation of marine fish stocks has depleted natural populations and caused a reduction in the age and size at which many fish species become reproductively mature. What sort of government regulations of fishing might reverse the current trend toward smaller adult size? Explain your answer in terms of the selection pressures that fishing places on targeted species.

How Would You Vote?

Material goods can be manufactured in ways that protect biodiversity but often are more expensive than comparable goods produced without regard for the environment. As a consumer, are you willing to pay extra for the first kind? Go to www.thomsonedu.com/login to investigate both sides of the issue and then vote.

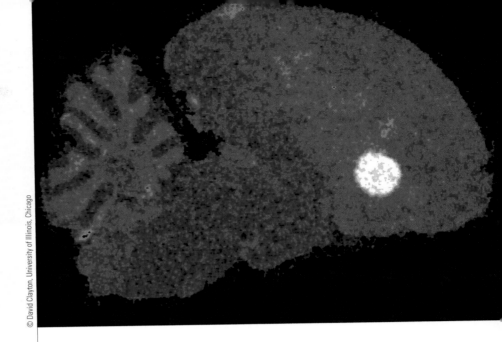

A section of zebra finch *(Taeniopygia guttata)* brain, stained to illuminate expression of the *zenk* gene, which helps a male bird reproduce his species' song.

© David Clayton, University of Illinois, Chicago

STUDY PLAN

54.1 Genetic and Environmental Contributions to Behavior

Most behaviors have both instinctive and learned components

54.2 Instinctive Behaviors

Many instinctive behaviors are highly stereotyped

Behavioral differences between individuals may reflect underlying genetic differences

54.3 Learned Behaviors

Learned behaviors are modified by an animal's prior experiences

54.4 The Neurophysiological Control of Behavior

Discrete neural circuits in specific brain regions control singing behavior in songbirds

The activation of specific genes fosters the development of nuclei that regulate a bird's song

54.5 Hormones and Behavior

Hormones regulate the development of cells and networks that form the neural basis of behavior

Changing hormone concentrations alter the behavior of animals as they mature

Hormone levels affect reproductive activity in many animals

54.6 Nervous System Anatomy and Behavior

Hard-wired connections between sensory and motor systems provide rapid behavioral responses to life-threatening stimuli

The structure of sensory systems allows animals to respond appropriately to different stimuli

The amount of brain tissue devoted to analyzing sensory information varies from one sensory system to another

54 The Physiology and Genetics of Animal Behavior

WHY IT MATTERS

Male white-crowned sparrows *(Zonotrichia leucophrys)* are handsome birds with a song that birdwatchers describe as a "plaintive whistle" followed by a "husky trilled whistle." This distinctive song is a critical part of a male white-crown's **behavioral repertoire**, the set of actions that it can perform in response to stimuli in its environment. An adult male sparrow's song is one of the ways he struts his stuff. The song not only announces his presence to rival males, but it also signals to females that he is available as a potential mate. Experienced birders easily recognize this song, which differs from that of song sparrows *(Melospiza melodia)* and swamp sparrows *(Melospiza georgiana)*, as sound spectrograms illustrate **(Figure 54.1)**. In fact, every songbird species produces vocal signals that are characteristic of its species and its species alone.

The study of **animal behavior** involves discovering how animals respond to specific stimuli and why they respond in predictable and characteristic ways. A comprehensive approach to animal behavior studies first crystallized in the 1930s, when European researchers—notably Konrad Lorenz, Niko Tinbergen, and Karl von Frisch, who shared a Nobel Prize for their work in 1973—developed the discipline

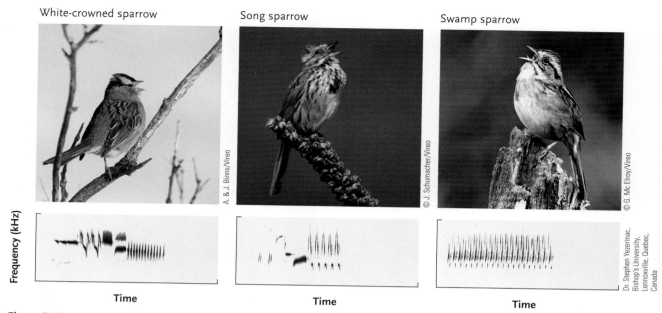

Figure 54.1

Songbirds and their songs. Sound spectrograms (visual representations of sound graphed as frequency versus time) illustrate differences in the songs of the white-crowned sparrow (*Zonotrichia leucophrys*), the song sparrow (*Melospiza melodia*), and the swamp sparrow (*Melospiza georgiana*).

of **ethology**, which focuses on how animals behave in their natural environments. They analyzed how evolutionary processes shape inherited behaviors and the ways that animals respond to specific stimuli. Tinbergen identified four basic questions that any broad study of animal behavior should address: (1) What mechanisms trigger a specific behavioral response? (2) How does the expression of a behavior develop as an animal matures? (3) What is the behavior's function and how does it increase an animal's chances of surviving and reproducing? (4) How did the behavior evolve?

Advances in **neuroscience**—the integrated study of the structure, function, and development of the nervous system—now allow researchers to explore the first and second questions in detail. Comparable advances in genetic analysis and evolutionary theory enable scientists to address the third and fourth questions. In this chapter, we examine the *proximate causes* of behavior—the genetic, cellular, physiological, and anatomical mechanisms that underlie an animal's ability to detect internal stimuli and environmental cues and react to them in species-specific ways. In Chapter 55, we consider the *ultimate causes* of animal behavior—its adaptive value and evolution.

54.1 Genetic and Environmental Contributions to Behavior

For many years, animal behaviorists debated whether animals are born with the ability to perform most behaviors completely or whether experience is necessary to shape their actions. However, extensive research in neuroscience has demonstrated that no behavior is determined entirely by genetics or entirely by environmental factors. Instead, behaviors develop through complex gene–environment interactions. We illustrate such an interaction below with a detailed description of the process through which male white-crowned sparrows learn their adult song.

Most Behaviors Have both Instinctive and Learned Components

Why do adult male white-crowned sparrows sing a song that no other species sings? One possible explanation is that they possess an innate (inborn) ability to produce their particular song, an ability so reliable that young males sing the "right" song the first time they try. According to this hypothesis, their distinctive song would be an example of an **instinctive behavior**, a genetically "programmed" response that appears in complete and functional form the first time it is used. An alternative hypothesis is that they acquire the song as a result of certain experiences, such as hearing the songs of adult male white-crowns that live nearby. In other words, this species' distinctive song might be an example of a **learned behavior**, one that is dependent upon having a particular kind of experience during development.

How can we determine which of these two hypotheses is correct? If the white-crowned sparrow's song is instinctive, isolated male nestlings that have never heard other members of their species should be able to sing their species' song when they mature. But if the learning hypothesis is correct, young birds deprived of certain essential experiences should not sing "properly" when they become adults.

In a set of pioneering experiments conducted at Rockefeller University, Peter Marler tested these alternative hypotheses. He took newly hatched white-crowns from nests in the wild and reared them individually in soundproof cages in his laboratory. Some of the chicks listened to recordings of a male white-crowned sparrow's song when they were 10 to 50 days old; others did not. The juvenile males in both groups first started to vocalize when they were about 150 days old. For many days, they produced whistles and twitters that only vaguely resembled the songs of adults. But gradually the young males that had listened to tapes of their species' song began to sing better and better approximations of that song. At about 200 days of age, they were right on target, producing a song that was nearly indistinguishable from the one they had heard months before. By contrast, males in the group that had not heard tape-recorded white-crown songs never came close to singing the way wild males do.

These results revealed that learning is essential for a young male white-crowned sparrow to acquire the full song of its species. Although birds isolated as nestlings did sing instinctively, they needed the acoustical experience of listening to their species' song early in life if they were to reproduce it months later. We can therefore reject the hypothesis that white-crowned sparrows hatch from their eggs with the ability to produce the "right" song. Their species-specific song—and presumably those of other songbirds—has both instinctive and learned components.

Although early researchers generally classified behaviors as *either* instinctive *or* learned, we now know that most behaviors include both instinctive and learned components. Nevertheless, some behaviors have a strong instinctive component, whereas others are mostly learned.

STUDY BREAK

1. What is the difference between an instinctive behavior and a learned behavior?
2. How did the isolation of young male sparrows in soundproof cages allow Marler to conclude that learning was important to song acquisition?

54.2 Instinctive Behaviors

Instinctive behaviors—which are often grouped into functional categories, such as feeding behaviors, defensive responses, mating behaviors, and parental care activities—can be performed without the benefit of prior experience. We therefore assume that they have a strong genetic basis and that natural selection has preserved them as adaptive behaviors.

Many Instinctive Behaviors Are Highly Stereotyped

Many instinctive behaviors are highly stereotyped; in other words, when triggered by a specific cue, they are performed over and over in almost exactly the same way. Such behaviors are called **fixed action patterns**, and the simple cues that trigger them are called **sign stimuli.** For example, sign stimuli and fixed action patterns govern the transfer of food from herring gull *(Larus argentatus)* parents to their offspring. Researchers found that very young chicks secure food from their parents through a begging response (the fixed action pattern), which is triggered when they see a red spot on the lower bill of an adult (the sign stimulus). This cue "releases" the begging behavior of hungry baby gulls, which peck at the spot on the parent's bill. In turn, the tactile stimulus delivered by the pecking chick serves as a sign stimulus that induces the adult bird to regurgitate food stored in its crop. The baby gulls then feed on the chunks of fish, clams, or other food that lie before them. We know that the spot on the parent's bill releases the begging response of the young gull because the same response is triggered by an artificial bill that looks only vaguely like a herring gull's bill, provided it has a dark contrasting spot near the tip **(Figure 54.2).** Thus, even very simple cues can activate fixed action patterns.

Human infants often respond innately to the facial expressions of adults **(Figure 54.3).** For example, researchers can trigger smiling in even very young babies simply by moving a mask toward the infant, as long as the mask possesses two simple, diagrammatic eyes. Clearly the infant, like a nestling herring gull, is not reacting to every feature of a face; instead it focuses on simple cues, which function as sign stimuli that release a fixed behavioral response.

Natural selection has molded the behavior of some parasitic species to exploit the relationship between sign stimuli and fixed action patterns for their own benefit. For example, birds that are brood parasites lay their eggs in the nests of other species (see *Why It Matters* at the beginning of Chapter 50). When the brood parasite's egg hatches, the alien nestling mimics and even exaggerates sign stimuli that are ordinarily exhibited by its hosts' own chicks: opening its mouth, bobbing its head, and calling vigorously. These exaggerated behaviors elicit feeding by the foster parents, and the young brood parasite often receives more food than the hosts' own young **(Figure 54.4).**

Although instinctive behaviors are often performed completely the first time an animal responds to a stimulus, they can by modified by an individual's experiences. For example, the fixed action patterns of a young herring gull change through time. Although

Figure 54.2 Experimental Research

The Role of Sign Stimuli in Parent-Offspring Interactions

QUESTION: What feature of the parent's head triggers pecking behavior in young herring gulls?

EXPERIMENT: Niko Tinbergen and A. C. Perdeck tested the responses of young herring gull *(Larus argentatus)* chicks to cardboard cutouts of an adult herring gull's head and bill. They waved these models in front of the chicks and recorded how often a particular model elicited a pecking response from the chicks. One cutout included an entire gull's head with a red spot near the tip of the bill; another cutout included just the bill with the red spot; the third cutout included the entire head but lacked the red spot.

RESULT: Young herring gulls pecked at the model of the bill with a red spot almost as often as they pecked at the model of an entire head with a red spot, but they pecked much less frequently at the model of an entire head that lacked a red spot.

Herring gulls *(Larus argentatus)*

© Marie Read Natural History Photography

CONCLUSION: Begging behavior by young herring gulls is triggered by a simple sign stimulus, the red spot on the parent's bill. Experimental tests revealed that herring gull chicks respond more to the presence of the contrasting spot than they do to the outline of an adult's head.

Figure 54.3 Instinctive responses in humans. The smiling face of an adult is a sign stimulus that triggers smiling behavior in very young infants.

Evan Cerasoli

the youngster initially begs by pecking at almost anything remotely similar to an adult gull's bill, it eventually learns to recognize the distinctive visual and vocal features associated with its parents. The chick uses this information to become increasingly selective about the stimuli that will elicit its begging behavior. Thus, instinctive behaviors can be modified in response to particular experiences during their early performances.

Behavioral Differences between Individuals May Reflect Underlying Genetic Differences

Because the performance of instinctive behaviors does not depend on prior experience, behavioral differences between individuals may reflect genetic differences between them. Stevan Arnold, then at the University of Chicago, tested that hypothesis by studying the innate responses of captive newborn garter snakes *(Thamnophis elegans)* to the olfactory stimuli provided by potential food items that they had never before encountered. Arnold measured the snakes' responses to cotton swabs that had been dipped in a smelly extract of banana slug *(Ariolimax columbianus)*, a shell-less mollusk. A snake "smells" by tongue-flicking, which draws volatile chemicals into a special sensory organ in the roof of its mouth. If the young snake had been born to a mother captured in coastal California, where adult garter snakes regularly eat banana slugs, it almost always began tongue-flicking at the slug-scented cotton swab **(Figure 54.5)**. By contrast, newborn snakes whose parents came from inland California, where banana slugs do not occur, rarely tongue-flicked at the swabs. Thus, although the coastal and inland snakes belong to the same species, their instinctive responses to the volatile chemicals associated with banana slugs were markedly different.

In another experiment, Arnold tested whether newborn snakes would feed on bite-sized chunks of slug. After a brief flick of the tongue, 85% of the newborn snakes from a coastal population routinely struck at the slug and swallowed it, despite having had no prior experience with this prey. By contrast, only 17% of newborn snakes from the inland population ate slugs consistently, even when no other food was available. Arnold hypothesized that coastal and inland garter snakes possess different alleles at one or more gene loci controlling their odor-detection mechanisms, leading to differences in their behavior. To test this hypothesis, Arnold crossbred coastal and inland snakes. If genetic differences contribute to the different food preferences of the two snake populations, then hybrid offspring, which receive genetic information from each parent, should behave in an intermediate fashion. Results of the experiment confirmed his prediction: when presented with bite-sized chunks of slug, 29% of the newborn snakes of mixed parentage consumed them every time.

Many additional experiments have confirmed that genetic differences between individuals can translate into behavioral differences between them. *Insights from the Molecular Revolution* describes a striking example of a single gene that influences the grooming behavior of mice. Bear in mind that single genes do not control complex behavior patterns directly. Instead, the alleles present affect the kinds of enzymes that cells can produce, influencing the biochemical pathways involved in the development of an animal's nervous system. The resulting neurological differences can translate into a behavioral difference between individuals that have certain alleles and those that do not.

Figure 54.4

Exploitation of a releaser. This young European cuckoo *(Cuculus canorus)*, a brood parasite, stimulates feeding behavior by its foster parent, a hedge sparrow *(Prunella modularis)*. It secures food by displaying exaggerated versions of the sign stimuli used by the host offspring to release feeding behavior by the parents.

a. Banana slug

b. Adult coastal garter snake eating a banana slug

c. Newborn coastal garter snake "smelling" slug extract

Figure 54.5

Genetic control of food preference. (a) Banana slugs *(Ariolimax columbianus)* are a preferred food of **(b)** an adult garter snake *(Thamnophis elegans)* from coastal California. **(c)** A newborn garter snake from a coastal population flicks its tongue at a cotton swab drenched with tissue fluids from a banana slug.

STUDY BREAK

1. How do the chicks of brood parasites stimulate unwitting foster parents to feed them?
2. How did Arnold demonstrate that the receptiveness of garter snakes to a meal of banana slugs had a genetic basis?

54.3 Learned Behaviors

Unlike instinctive behaviors, learned behaviors are not performed completely the first time an animal responds to a specific stimulus. Instead, they change in response to environmental stimuli that an individual experiences as it develops.

Learned Behaviors Are Modified by an Animal's Prior Experiences

Behavioral scientists generally define **learning** as a process in which experiences change an animal's behavioral responses. Different types of learning occur un-

A Knockout by a Whisker

Almost all eukaryotic organisms share a series of developmental interactions called the *wingless/Wnt* pathway. The name comes from the original discovery of the pathway in the fruit fly *Drosophila melanogaster*, in which mutant genes of the pathway cause alterations in the wings and other segmental structures. Recently, three genes closely related to *disheveled (dsh)*, one of the genes of the *Drosophila wingless/Wnt* pathway, were isolated and identified in mice. No functions have yet been identified for the proteins encoded in the three mouse disheveled genes, but tests show that they are highly active in both embryos and adults. Their function must be important, but what could it be?

Nardos Lijam and his coworkers in several laboratories, including Case Western University, the Universities of Colorado and Maryland, and the National Institutes of Health in Bethesda, Maryland, decided to seek an answer to this question by developing a line of mice that totally lacked one of the disheveled genes, called *Dvl-1* in genetic shorthand. First they constructed an artificial copy of the *Dvl-1* gene with the central section scrambled so that no functional proteins could be made from its encoded directions. Next they introduced the artificial gene into em-

bryonic mouse cells. Cells that successfully incorporated the gene were then injected into very early mouse embryos. Some of the mice grown from these embryos were heterozygotes, with one normal copy of the *Dvl-1* gene and one nonfunctional copy. Interbreeding of the heterozygotes produced some individuals that carried two copies of the altered *Dvl-1* gene and no normal copies. Such individuals, in which the normal gene is eliminated, are called knockout mice for the missing gene. (Making knockout mice is described in Section 18.2.)

Surprisingly, the knockout mice grew to maturity with no apparent morphological defects in any tissue examined, including the brain. Their motor skills, sensitivity to pain, cognition, and memory all appeared to be normal. Their social behavior was a different story, however. When housed with normal mice, the knockouts failed to take part in the common activities of mouse social groups: social grooming, tail pulling, mounting, and sniffing. Rather than building nests and sleeping in huddled groups, as normal mice do in the cages, the knockouts tended to sleep alone, without constructing full nests from cage materials. Mice heterozygous for the *Dvl-1* gene—that is, with one normal and one altered

copy of the gene—behaved normally in all these social activities.

The knockout mice also jumped around wildly in response to an abrupt, startling sound while the response of normal mice was less extreme. Since it is known that a neural circuit of the brain inhibits the startle response of normal mice, the reaction of the knockout mice suggested that this inhibitory circuit was probably altered. Humans with schizophrenia, obsessive-compulsive disorders, Huntington disease, and some other brain dysfunctions also show an intensified startle reflex similar to that of the *Dvl-1* knockout mice.

The researchers' analysis revealed that the *Dvl-1* gene modifies developmental pathways affecting complex social behavior in mice, and probably in other mammals. It is one of the first genes affecting mammalian behavior to be identified. The similarity in startle-reflex intensity between the knockout mice and humans with neurological or psychiatric disorders also suggests that mutations in the *Dvl* genes and the *wingless* developmental pathway may underlie some human mental diseases. If so, further studies of the *Dvl* genes may give us clues to the molecular basis of these diseases, and a possible means to their cure.

© Nina Leen/Time and Life Pictures/Getty Images

Figure 54.6

Imprinting. Having imprinted on him shortly after hatching, young greylag geese *(Anser anser)* frequently joined Konrad Lorenz for a swim.

der different environmental circumstances. In this section we consider *imprinting, classical conditioning, operant conditioning, insight learning,* and *habituation.*

Some animals learn the identity of a caretaker or the key features of a suitable mate during a **critical period,** a restricted stage of development early in life. This type of learning is called **imprinting.** For example, newly hatched geese imprint on their mother's appearance and identity, staying near her for months. And

when they reach sexual maturity, they try to mate with other geese, which exhibit the visual and behavioral stimuli on which they had imprinted as youngsters. When Konrad Lorenz, one of the founders of ethology, tended a group of newly hatched greylag geese *(Anser anser),* they imprinted on him instead of an adult of their own species **(Figure 54.6).** The male geese not only followed Lorenz about, but they also courted humans when they achieved sexual maturity.

Other forms of learning can occur throughout an animal's lifetime. Russian physiologist Ivan Pavlov's classic experiments with dogs explored **classical conditioning,** a type of learning in which animals develop a mental association between two phenomena that are usually unrelated. Dogs generally salivate when they eat. The food is called an *unconditioned stimulus* because the dogs respond to it instinctively; no learning is required for the stimulus (food) to elicit the response (salivation). In his experiment, Pavlov rang a bell just before offering food to dogs. After about 30 trials in which dogs received food immediately after the bell rang, the dogs associated the bell with feeding time, and they drooled profusely whenever it rang—even when no food was forthcoming. Thus, the bell became a *conditioned stimulus,* one that elicited a particular learned response. In classical conditioning, an animal learns to respond to a conditioned stimulus when it precedes an unconditioned stimulus that normally triggers the response. For example, your cat may become exceptionally friendly whenever she hears the sound of a can opener, another example of classical conditioning.

In another form of associative learning, called trial-and-error learning or **operant conditioning,** animals learn to link a voluntary activity, called an *operant,* with its favorable consequences, called a *reinforcement.* For example, a laboratory rat will explore a new cage randomly. If the cage is equipped with a bar that releases food when it is pressed, the rat will eventually lean on the bar by accident (the operant) and immediately receive a morsel of food (the reinforcement). After just a few such experiences, a hungry rat will learn to press the bar in its cage more frequently—as long as bar-pressing behavior is followed by access to food. Laboratory rats have also learned to press bars to turn off disturbing stimuli, such as bright lights.

A few animal species can abruptly solve problems without apparent trial-and-error attempts at the solution; researchers call this **insight learning.** For example, captive chimpanzees (*Pan troglodytes*) were able to solve a novel problem that their keepers devised: how to get bananas hung far out of reach. The chimps studied the situation, then stacked and stood on several boxes, and used a stick to knock the fruit to the floor.

Animals typically lose their responsiveness to frequent stimuli that are not quickly followed by the usual reinforcement. This learned loss of responsiveness, called **habituation,** saves the animal the time and energy of responding to stimuli that are no longer important. For example, the sea hare *Aplysia,* a shell-less mollusk, typically responds to a touch on the side of its body by retracting its delicate gills, a response that helps protect it from approaching predators. But if an *Aplysia* is touched repeatedly over a short period of time with no harmful consequences, it stops retracting its gills.

STUDY BREAK

1. Dogs typically wag their tails when they see their owners pick up a leash. What kind of learning does this demonstrate?
2. What type of learning allows you to sleep through your alarm clock when it rings to awaken you for biology class?

54.4 The Neurophysiological Control of Behavior

Research in neuroscience has shown that all behavioral responses, even those that are either mostly instinctive or mostly learned, depend on an elaborate physiological foundation provided by the biochemistry and structure of neurons (nerve cells). The neurons that regulate an innate response as well as those that make it possible for an animal to learn something are products of a complex developmental process in which genetic information and environmental contributions are intertwined. Although the anatomical and physiological basis for some behaviors is present at birth, an individual's experiences alter cells of its nervous system in ways that produce particular patterns of behavior. In this section we use examples from research on the singing behavior of songbirds to explore general principles about the physiological basis of behavior that apply to many other kinds of animals.

Discrete Neural Circuits in Specific Brain Regions Control Singing Behavior in Songbirds

Marler's experiments (see Section 54.1) help explain the physiological underpinnings of singing behavior in male white-crowned sparrows. If acoustical experience shapes this behavior, a sparrow chick's brain must be able to acquire and store information present in the songs of other males. Then, months later, when the young male starts to sing, its nervous system must have special features that enable the bird to match its vocal output to the stored memory of the song that it had heard earlier. Eventually, when it achieves a good match, the sparrow's brain must "lock" on the now complete song and continue to produce it when the bird is singing.

Additional experiments have provided detailed information about the nature of the sparrow's nervous system. Young birds that did not hear taped song during their critical period, between 10 and 50 days old, never produced the full song of their species, even if they heard it later in life. In addition, young birds that heard recordings of *other* bird species' songs during the critical period never generated replicas of those songs as they matured. These and other findings suggested that certain neurons in the young male's brain are influenced only by appropriate stimuli, namely the acoustical signals from individuals of its own species, and only during the critical period. Neuroscientists have identified the neurons clusters, called *nuclei* (singular, *nucleus*), that make song learning and song production possible.

Moreover, every behavioral trait appears to have its own neural basis. For example, a male zebra finch, *Taeniopygia guttata* (**Figure 54.7**), another songbird, can

Figure 54.7
Zebra finches.
Native to Indonesia, zebra finches *(Taeniopygia guttata)*, have played an important role in studies of the physiological basis of song learning. The male has a striped throat.

truders of its own species, which represent a real threat to its continued control of its territory.

STUDY BREAK

1. What research results suggest that certain neurons in the young male bird's brain are influenced only by acoustical signals from members of its own species and only during a critical period?
2. What happens to cells in the nucleus in the forebrain of a zebra finch after it hears a neighboring bird's song many times?
3. What is the role of the ZENK enzyme in song learning?

discriminate between the songs of strangers and the songs of established neighbors on adjacent **territories.** (In many bird species, territories are plots of land, defended by individual males or breeding pairs, within which the territory holders have exclusive access to food and other necessary resources. Territories are discussed further in Chapter 55.) The ability to discriminate between the songs of neighbors and those of strangers also involves a nucleus in the forebrain. Cells in this nucleus fire frequently the first time that the song of a new zebra finch is played to a test subject. But as the song is played again and again, these cells cease to respond, indicating that the bird becomes habituated to a now familiar song, although it still reacts to the songs of strangers. The neurophysiological networks that make this selective learning possible enable male zebra finches to behave differently toward familiar neighbors, which they largely ignore, and unfamiliar singers, which they attack and drive away.

The Activation of Specific Genes Fosters the Development of Nuclei That Regulate a Bird's Song

The role of genes in learning has been identified by research using new molecular and cellular techniques that reveal when a specific gene is active in neurons. When a bird is exposed to relevant acoustical stimuli, such as the songs of potential rivals of its own species, certain genes are "turned on" within neurons in the song-controlling nuclei of the bird's brain. For example, when a zebra finch hears the elements of its species' song, a gene called *zenk* becomes active in the brain, producing an enzyme that changes the structure and function of the neurons (see photo on p. 1253). In effect, the ZENK enzyme programs the neurons of the bird's brain to "anticipate" key acoustical events of potential biological importance. When these events occur, they trigger additional changes in the bird's brain that affect its actions. As a result, a territory owner habituates to (that is, learns to ignore) a singing neighbor with which it has already adjusted territorial boundaries; but it retains the ability to detect and repel new in-

54.5 Hormones and Behavior

Research on many animal species has revealed that hormones are the chemical signals triggering the performance of specific behaviors. They often accomplish this function by regulating the development of neurons and neural networks or by stimulating the cells within endocrine glands to release chemical signals.

Hormones Regulate the Development of Cells and Networks That Form the Neural Basis of Behavior

How did the neurons in an adult zebra finch acquire the remarkable capacity to change in response to specific stimuli? In zebra finches, only males produce courtship songs. Very early in a male songbird's life, certain cells in its brain produce the hormone estrogen, which affects target neurons in an area of the developing brain called the *higher vocal center.* The presence of this hormone leads to a complex series of biochemical changes that result in the production of more neurons in parts of the brain that regulate singing. By contrast, the brains of developing females do not produce estrogen, and in the absence of this hormone, the number of neurons in the higher vocal center of females *declines* over time **(Figure 54.8).** Experiments have shown that when young female zebra finches are given estrogen, they produce more neurons in the higher vocal center; but the treated females do not sing later in life unless they are also treated with androgens (male hormones).

Thus, genetically induced hormone production contributes to song learning and singing behavior in male zebra finches by regulating the numbers and types of neurons in the brain centers that produce those behaviors. The development of these neurons primes them for additional changes in response to specific acoustical experiences during the bird's development. Moreover, specific stimuli, such as the songs of either familiar or unfamiliar males, can alter the genetic activity of the neurons that control the behavior of adult birds.

Changing Hormone Concentrations Alter the Behavior of Animals as They Mature

Just as estrogen influences the development of singing ability in zebra finches, other hormones mediate the development of the nervous system in other species. Indeed, a change in the concentration of a certain hormone is often the physiological trigger that induces important changes in an animal's behavior as it matures.

In honeybees *(Apis mellifera)*, worker bees perform different tasks for the colony's welfare as they grow older: bees that are less than 15 days old tend to care for larvae and maintain the hive, whereas those that are more than 15 days old often make foraging excursions from the hive to collect the nectar and pollen that bees eat **(Figure 54.9).** These behavioral changes are induced by rising concentrations of juvenile hormone (see Section 40.5), which is released by a gland near the bee's brain. Despite its name, circulating levels of juvenile hormone actually increase as a honeybee gets older.

Juvenile hormone may exert its effect on the bee's behavior by stimulating genes in certain brain cells to produce proteins that affect nervous system function. One such chemical, *octopamine,* stimulates neural transmissions and reinforces memories. It is concentrated in the antennal lobes, a part of the bee's brain that contributes to the analysis of chemical scents in the bee's external environment. Octopamine is found at higher concentrations in the older, foraging bees that have higher levels of juvenile hormone. And when extra juvenile hormone is administered to bees experimentally, their production of octopamine increases. Thus, increased octopamine levels in the antennal lobes may help a foraging bee home in on the odors of flowers where it can collect nectar and pollen.

The honeybee example illustrates how genes and hormones interact in the development of behavior. Genes code for the production of hormones, which change the intracellular environment of assorted target cells. The hormones then directly or indirectly change the genetic activity and enzymatic biochemistry in their targets. If the cells in question are neurons, the changes in their biochemistry translate into changes in the animal's behavior.

Hormone Levels Affect Reproductive Activity in Many Animals

The African cichlid fish *(Haplochromis burtoni)* provides an example of how hormones regulate reproductive behavior. Some adult males maintain nesting ter-

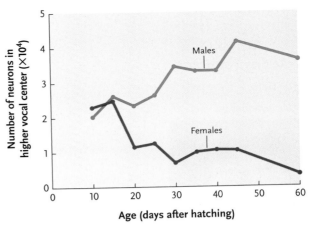

Figure 54.8
Hormonally induced changes in brain structure. The brains of young male zebra finches secrete estrogen, which stimulates the production of additional neurons in the higher vocal center. Lacking this hormone, the brains of young female zebra finches lose neurons in this brain region.

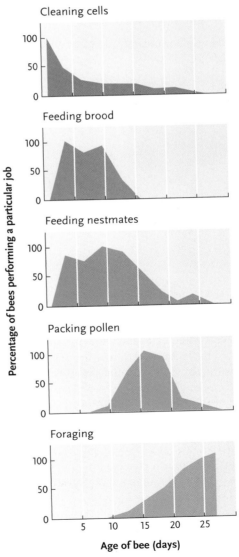

Figure 54.9
Age and task specialization in honeybee *(Apis mellifera)* workers. Young bees typically clean cells and feed the brood; older workers leave the hive to forage for food.

Figure 54.10 Experimental Research

Effects of the Social Environment on Brain Anatomy and Chemistry

a. African cichlid fish *(Haplochromis burtoni)*

Nonterritorial male

Territorial male

Russell Fernald, Stanford University

QUESTION: How does the acquisition or the loss of a territory affect the brain anatomy and chemistry of an African cichlid fish *(Haplochromis burtoni)*?

EXPERIMENT: Fernald and his students housed groups of male cichlids in aquariums in their laboratory. Males that established and maintained territories in the aquariums were brightly colored, whereas those that could not hold territories were pale and drab. The researchers then moved some small territorial males into tanks where larger males had already established territories. The newly introduced males could not establish and maintain territories under these experimental conditions, and therefore changed status from territorial to nonterritorial. The researchers also moved some large nonterritorial males into tanks with smaller territorial males. Under these experimental conditions, the newly introduced males quickly established and maintained territories, changing their status from nonterritorial to territorial. Other males, left in their original tanks so that their territorial status did not change, served as controls. Four weeks later, the researchers examined the brains of the experimental and control fish and measured the size of the neurons that produce GnRH, a hormone that stimulates bright coloration as well as aggressive behavior and mating behavior in males.

RESULT: The GnRH-producing cells in the brains of experimental males that had lost their territories were much smaller than those in the brains of control males that had maintained their territories. By contrast, the GnRH-producing cells in the brains of experimental males that had gained territories were much larger than those of control males that had never held territories.

b. GnRH-secreting cells

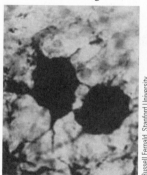

Territorial control

Russell Fernald, Stanford University

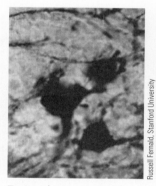

Territorial to nonterritorial experimentals

Russell Fernald, Stanford University

Nonterritorial control

Russell Fernald, Stanford University

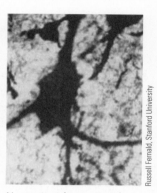

Nonterritorial to territorial experimentals

Russell Fernald, Stanford University

CONCLUSION: Changes in social status influence the size of brain cells producing hormones that influence the color and behavior of males.

---

ritories on the bottom of Lake Tanganyika in East Africa. Territory holders are brightly colored, and they exhibit elaborate behavioral displays that attract egg-laden females to their territories. These males defend their real estate aggressively against neighboring territory holders and against incursions by males that have no territories of their own. By contrast, nonterritorial males are much less colorful and aggressive; they do not control a patch of suitable nesting habitat, and they make no effort to court females.

The behavioral differences between the two types of males are caused by differences in their levels of circulating sex hormones. Recall from Section 47.3 that gonadotropin-releasing hormone (GnRH) stimulates the testes to produce testosterone and sperm. When the circulating testosterone is carried to the brain, it modulates the activity of neurons that regulate sexual and aggressive behavior. In territorial fish the GnRH-producing neurons in the hypothalamus are large and biochemically active, but in nonterritorial fish they are small and inactive. In the absence of GnRH, the testes do not produce testosterone; the testosterone-deficient fish do not court females with sexual displays, nor do they usually attack other males.

What causes the differences in the neuronal and hormonal physiology of the two types of male fish? Russell Fernald and his students at Stanford University conducted laboratory experiments in which they manipulated the territorial status of males: some territorial males were changed into nonterritorial males; some nonterritorial males were changed into territorial males; and the territorial status of other males was left unchanged as a control **(Figure 54.10)**. Four weeks later, they compared the coloration and behavior as well as the size of the GnRH-producing cells in the brains of the experimental fishes with those of the control males that had retained their original status. Males that had held territories in the past, but had then been defeated by another male, quickly lost their bright colors and stopped being combative. Moreover, their GnRH-producing cells were smaller than those of the successful territory-holding controls. Conversely, males that gained a territory in the experiment quickly developed bright colors and displayed aggressive behaviors towards other males. And the GnRH-producing cells in their brains were larger than those of fishes that had maintained their status as non-territory-holding controls.

The neuronal, hormonal, and behavioral differences between the two experimental groups of males are therefore correlated with a key environmental variable: success or failure in the acquisition and maintenance of a territory. The fish can detect and store information about their aggressive interactions. The neurons that process this information transmit their input to the hypothalamus where it affects the size of the GnRH cells, which in turn dictates the hormonal state of the male. A decrease in GnRH production can turn a feisty territorial male into a subdued drifter, biding his time and building his energy reserves for a future attempt at defeating a weaker male and taking over his territory. If successful in regaining territorial status, the male's GnRH levels will increase again, and the once-peaceful male will revert to vigorous sexual and aggressive behavior.

Note the general similarity of these processes to those described for the white-crowned sparrow's song learning: the fish's brain possesses cells that can change their biochemistry, structure, and function in response to well-defined social stimuli. These physiological changes make it possible for the fish to modify its behavior, depending on its social circumstances. In the next section, we examine how the structure of the nervous system allows animals to respond to important environmental stimuli.

STUDY BREAK

1. What is the effect of estrogen on the development of neurons in the higher vocal center of young zebra finches?
2. How might juvenile hormone production influence a bee's ability to recognize and locate appropriate food sources?
3. How does the loss of its territory change the brain chemistry of an African cichlid fish?

54.6 Nervous System Anatomy and Behavior

Although many behaviors result from gene–environment interactions and changes in hormone concentrations, some specific behaviors are produced by the anatomical structure of an animal's nervous system. Studies on a wide range of animal species demonstrate that the nervous systems of many animals provide rapid responses to key stimuli. In other species, sensory systems are structured to acquire a disproportionately large amount of information about those stimuli that are most important to survival and reproductive success.

Hard-Wired Connections between Sensory and Motor Systems Provide Rapid Behavioral Responses to Life-Threatening Stimuli

In some animals, important information acquired by the senses is relayed directly to motor neurons. Such a system provides crickets with a potentially lifesaving predator avoidance behavior. Crickets and some other insects fly mainly at night, a behavior that allows them to avoid day-flying predatory birds. But flying crickets aren't safe even at night, because insect-eating bats can detect them in pitch darkness.

Bats detect potential prey by echolocation (see *Why It Matters* at the beginning of Chapter 39). They call almost continuously while flying at night, and the sound waves they produce bounce off items in their path, creating echoes that the bats hear and use to track their prey. Their vocalizations are of such high frequency (up to 100,000 hertz) that they lie outside the upper limit (20,000 hertz) of unaided human hearing. However, a bat's auditory apparatus and brain not only can hear ultrasound, as these high frequency sounds are called, but also can analyze ultrasonic echoes in a way that permits the bat to identify, approach, and capture flying insect prey. With enemies of this sort in its environment, a cricket flying at night is in real danger of being intercepted and eaten.

Crickets are not defenseless, however. Black field crickets *(Teleogryllus oceanicus)*, for example, hear ultrasound through ears in their front legs (see Figure 39.7). The approach of a calling bat causes sensory neurons connected to the ears to fire. However, to be of any use to the cricket, this information must be translated immediately into evasive action—and crickets have the anatomical and physiological equipment to do exactly that.

Imagine that a bat is zeroing in for the kill, rushing toward the left side of a flying cricket. The cricket's left ear will be bombarded with more intense ultrasound than the right ear, and the neurons that receive input from the ears will also be stimulated unequally. The cricket's nervous system is structured to relay

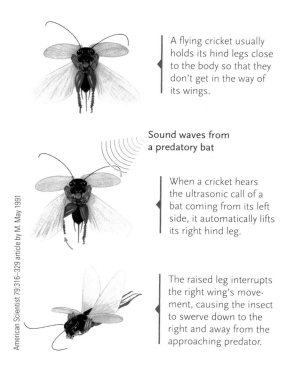

A flying cricket usually holds its hind legs close to the body so that they don't get in the way of its wings.

Sound waves from a predatory bat

When a cricket hears the ultrasonic call of a bat coming from its left side, it automatically lifts its right hind leg.

The raised leg interrupts the right wing's movement, causing the insect to swerve down to the right and away from the approaching predator.

American Scientist 79:316–329 article by M. May 1991

Figure 54.11
A neural mechanism for escape behavior in the black field cricket (*Teleogryllus oceanicus*).

incoming messages from the *left* ear to the motor neurons (muscle-regulating nerve cells) that control the *right* hind leg. Sufficient ultrasonic stimulation on the left side of the body will induce the motor neurons for the right hind leg to fire, causing muscle contractions in that leg. As the right hind leg jerks up, it blocks movement of the right hind wing, reducing the flight power generated on the right side of the cricket's body. The flying cricket then swerves sharply to the right and loses altitude, diving down and away from the approaching bat **(Figure 54.11).** Thus, the anatomical structure of the cricket's nervous system produces a behavioral response that takes the cricket out of harm's way.

If all goes well, the cricket will gain the safety of foliage or leaf litter on the ground before the bat can reach it. Once there, echoes bouncing off the materials all around it will mask any ultrasonic echoes coming from its body. The thwarted bat will be forced to look elsewhere for prey that responds less rapidly.

The Structure of Sensory Systems Allows Animals to Respond Appropriately to Different Stimuli

In some animals, the structure and neural connections of sensory systems allow them to distinguish potentially life-threatening stimuli from those that are more mundane. For example, fiddler crabs *(Uca pugilator)* live and feed on mud flats where they build burrows that provide safe refuge from predators, including crab-hunting shorebirds. But to use its burrow wisely,

a crab must be able to distinguish between predatory gulls and its fellow fiddler crabs. Otherwise, it would dash for cover whenever anything moved in its field of vision.

Fiddler crabs possess long-stalked eyes that they hold above their carapace perpendicularly to the ground. John Layne, a neurophysiologist at Duke University, wondered whether a crab might distinguish between dangerous predators and fellow crabs by having a divided field of vision. A large predatory gull sailing in for the kill would stimulate receptors on the upper part of the eye, whereas a fellow crab, whose movements would be slightly below the midpoint of the eyes, would stimulate a lower set of visual receptors. If the receptors above and below the retinal equator relayed their signals to different groups of neurons, the crab's nervous system could be "wired" to provide different responses to the different stimuli.

Layne hypothesized that receptors above the midline of the eye activate neurons that control an escape response, so that stimulation from above would reliably trigger a dash for the burrow. By contrast, a moving stimulus at or below eye level, as when one crab approached another, would provide input to the neurons that allow a crab to behave appropriately to a male or female of its species.

Layne tested this hypothesis by placing crabs on an elevated platform in a glass jar. He then presented the same moving stimulus, a black square, to each crab at two heights; sometimes the stimulus circled the jar above the crab's eyes and sometimes below. Stimuli that activated the upper part of the retina did indeed induce escape behavior, but if the stimuli were below the retinal equator, the animal generally ignored the moving objects altogether **(Figure 54.12).** Thus, specific nervous system connections between a fiddler crab's eyes and brain provide appropriate responses to different specific stimuli.

The Amount of Brain Tissue Devoted to Analyzing Sensory Information Varies from One Sensory System to Another

The match between the structure of an animal's nervous system and the real-world challenges it faces extends beyond the ability to avoid predators. For example, the star-nosed mole *(Condylura cristata)*, which lives in wet tunnels in North American marshlands, spends almost all of its life in complete darkness. Like nocturnal insect-eating bats, the mole must find food without benefit of visual cues; and, like the bats, it has a receptor-perceptual system that enables it to feed effectively. The star-nosed mole subsists largely on earthworms, and it uses its nose to locate them—but not by smell. Instead, as the mole proceeds down a tunnel, 22 fingerlike tentacles from its nose sweep the area di-

Figure 54.12 Experimental Research

Nervous System Structure and Appropriate Behavioral Responses

QUESTION: Do fiddler crabs respond differently to stimuli that are presented above the midline of their visual field than they do to stimuli presented below it?

EXPERIMENT: Layne investigated this question by placing crabs on an elevated platform in a glass jar. He then presented the same moving stimulus, a black square, to each crab at varying heights; sometimes the stimulus circled the jar above the crab's eyes and sometimes below.

RESULTS: Stimuli that activated the upper part of the retina did indeed induce escape behavior, but when the stimuli were below the retinal equator, the animal generally ignored the moving objects altogether.

Fiddler crab (*Uca pugilator*)

© Jeff Foott/Dcom/DRK Photo

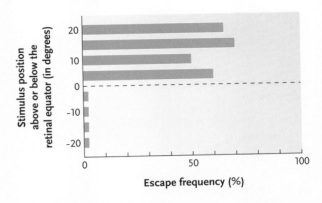

CONCLUSION: Specific nervous system connections between a fiddler crab's eyes and brain provide appropriate responses to different specific stimuli.

UNANSWERED QUESTIONS

How does an animal choose which behavior to perform?

Animals must perform many types of behaviors in their lifetimes. But what determines which behavior is most appropriate in any particular situation? In some cases, it appears that an animal performs whichever behavior it is most provoked to do at the time. For example, when the Eimer's organs of a star-nosed mole tell it that its tentacles have contacted an earthworm, the mole responds by biting into whatever is in front of its mouth. In other cases, an animal may perform whichever behavior will bring it the most reward. Researchers are studying how animals choose behaviors, using a combination of methods including theoretical modeling, neuroscience, and molecular biology. Male fruit flies often must choose between aggression and courtship, and molecular manipulations that result in the inability of a fly to discriminate between males and females often result in the fly making the wrong choice. Understanding the behavioral choices that animals make will help us understand both the mechanisms and evolution of behavior.

How does experience change the brain to affect behavior?

Some behaviors develop only after an animal has had certain experiences. How does experience translate into a new or improved behavior? Researchers are tracing the paths from stimulus perception, to activation of specific neural circuits and molecular pathways in the brain, to changes in brain structure and chemistry that then lead to behavioral change. This research is helped enormously by the burgeoning field of molecular neuroscience. Knowing how experience translates into a new or improved behavior has broad implications for our understanding of learning and memory, brain plasticity, brain disease, and recovery from stroke.

How do genes influence behavior?

Genetic differences between individuals can translate into behavioral differences between them. But genes encode proteins, and the road between the transcription of a gene and the performance of a specific behavior is a "long and winding" one. Researchers are studying how genetic differences between individuals influence the biochemical pathways that shape the development of the nervous system or later modify its function. Insights into how genes influence behavior have profound implications for our understanding of brain function and perhaps even policy decisions about screening people for genes that may influence their behavior.

John Dixon The Champaign–Urbana News–Gazette

Gene E. Robinson is the G. William Arends Professor of Integrative Biology at the University of Illinois at Urbana-Champaign, where he studies social behavior and genomics using the honeybee. To learn more about his research, go to http://www.life.uiuc.edu/robinson.

a. Sensory organs on the tentacle of a star-nosed mole

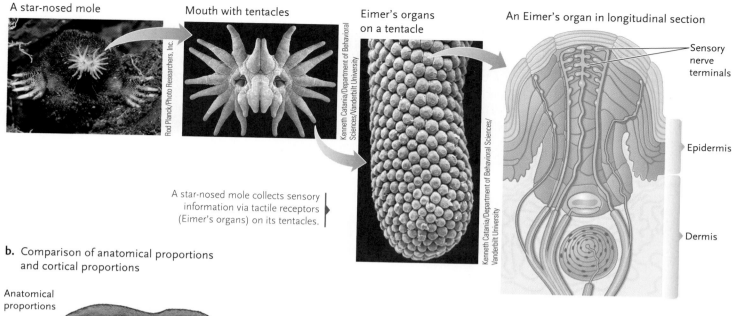

A star-nosed mole

Mouth with tentacles

Eimer's organs on a tentacle

An Eimer's organ in longitudinal section

Sensory nerve terminals

Epidermis

Dermis

A star-nosed mole collects sensory information via tactile receptors (Eimer's organs) on its tentacles.

Rod Planck/Photo Researchers, Inc.

Kenneth Catania/Department of Behavioral Sciences/Vanderbilt University

Kenneth Catania/Department of Behavioral Sciences/Vanderbilt University

b. Comparison of anatomical proportions and cortical proportions

Anatomical proportions

Cortical proportions

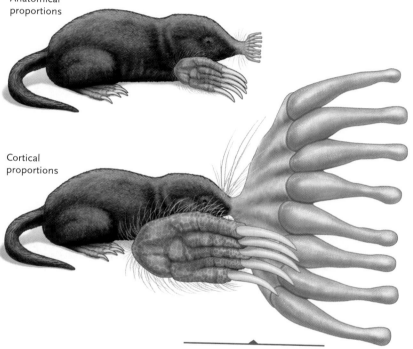

Most of the mole's cereberal cortex is devoted to the tentacles and front, digging feet.

Figure 54.13

The collection and analysis of sensory information by the star-nosed mole *(Condylura cristata)*. **(a)** The mole's nose has 22 fleshy tentacles covered with cylindrical tactile receptors called Eimer's organs, each containing sensory nerve terminals. **(b)** The star-nosed mole's cerebral cortex devotes far more space and neurons to the analysis of tactile inputs from the tentacles than from elsewhere on the body. These drawings compare the mole's actual body proportions with the relative amount of cortical tissue that processes sensory information from the various parts of the body.

rectly ahead. These tentacles are covered with thousands of tactile (touch) receptors called Eimer's organs **(Figure 54.13a)**. Sensory nerve terminals in the Eimer's organs generate complex and detailed patterns of signals about the objects they contact. These messages are relayed by neurons to the cortex of the mole's brain, much of which is devoted to the analysis of information from the nose's tactile receptors.

The structural basis of the mole's sensory analysis is apparent when we consider that the amount of brain tissue responding to signals from the mole's nose contains many more cells than do the tissues

that decode tactile signals from all other parts of the animal's body combined **(Figure 54.13b)**. Moreover, the brain does not treat inputs from all 22 of the mole's "fingers" equally. Instead, the brain devotes more cells to the tentacles closest to the mole's mouth, and fewer cells to analyzing messages from tentacles that are farther away.

The processing of tactile information in this species is clearly related to the importance of finding food in totally dark underground tunnels. Moreover, the extra attention given to signals from certain tentacles almost certainly helps the mole locate prey that are close to its mouth, allowing it to bite worms before they can move away after being touched.

As these examples illustrate, animal nervous systems do not offer neutral and complete pictures of the environment. Instead, distorted and unbalanced perceptions of the world are advantageous because certain types of information are far more important for survival and reproductive success than others. In the next chapter, we examine the ecological circumstances and selective forces that have promoted the evolution of specific behaviors.

STUDY BREAK

1. How does the anatomy of the cricket's nervous system helps it avoid an approaching bat?
2. What behavioral response is elicited in a fiddler crab when its eyes detect movement above the midline of its visual field?
3. Explain the sensory mechanism that allows a star-nosed mole to locate earthworms in its tunnels.

Review

Go to **Thomson**NOW™ at www.thomsonedu.com/login to access quizzing, animations, exercises, articles, and personalized homework help.

54.1 Genetic and Environmental Contributions to Behavior

- Most behaviors have both instinctive and learned components. Some behaviors can be produced only if the animal's nervous system acquires inputs from specific experiences during a critical stage of its development (Figure 54.1).

54.2 Instinctive Behaviors

- Instinctive behaviors are those that an animal performs completely the first time it is presented with a stimulus.
- Fixed action patterns are highly stereotyped behaviors that animals exhibit in response to simple cues called sign stimuli (Figures 54.2–54.4). Fixed action patterns often change through time in response to an animal's experiences.
- Behavioral differences between individuals often reflect underlying genetic differences. Research on garter snakes suggests that certain food preferences are genetically based (Figure 54.5).

    Animation: Instinctive behavior in infants

    Animation: Adaptive behavior in starlings

    Animation: Snake taste preference

    Animation: Cuckoo and foster parent

54.3 Learned Behaviors

- Learned behaviors develop only after an animal has had certain experiences in its environment. The different forms of learning include imprinting (Figure 54.6), classical conditioning, operant conditioning, and insight learning. Habituation is a learned loss of responsiveness to specific stimuli.

54.4 The Neurophysiological Control of Behavior

- Animal behavior requires an anatomical, physiological, and biochemical foundation based in the nervous system. An individual's experience alters cells of the nervous system in ways that produce particular patterns of behavior.

- The physiological basis of bird singing behavior resides in specific neuron clusters, called nuclei, that communicate with each other in the bird's brain.
- Bird song and some other behaviors develop only after specific genes are activated within the neurons that produce the behavior.

    Video: Development and elicitation of bird song

54.5 Hormones and Behavior

- Hormones can mediate the expression of specific behaviors by activating genes that change the biochemistry, morphology, and number of neurons in specific nuclei. Estrogen stimulates the production of neurons in the higher vocal center of male zebra finches (Figures 54.7 and 54.8).
- Age-related changes in hormone levels can alter the behavior of animals over the course of their lives. Changes in juvenile hormone concentration are correlated with changes in task specialization in honeybees (Figure 54.9).
- Behavioral interactions with other individuals can alter an animal's hormone levels, inducing changes in its behavior. Research on male cichlid fishes suggests that variations in coloration and aggressive behavior associated with territorial status and social interactions are mediated by the production of certain hormones (Figure 54.10).

    Animation: Hormonal control of behavior

54.6 Nervous System Anatomy and Behavior

- Sensory information enables animals to make adaptive behavioral responses to their environments. In crickets and some other animals, sensory systems relay information to motor systems, inducing an almost instantaneous response (Figure 54.11).
- Nervous systems can generate prompt and effective responses to those environmental stimuli that may have a large impact on an animal's survival and reproductive success. Fiddler crabs respond differently to movements that occur either above or below the midline of their visual field (Figure 54.12).
- Sensory systems are often structured to provide more information about important environmental factors. A large fraction of a star-nosed mole's brain is devoted to analyzing input from tactile receptors on its nose (Figure 54.13).

Questions

Self-Test Questions

1. Marler concluded that white-crowned sparrows can learn their species' song only:
   a. after receiving hormone treatments.
   b. during a critical period of their development.
   c. under natural conditions.
   d. from their genetic father.
   e. if they are reared in isolation cages.

2. Instinctive behaviors are:
   a. performed completely the first time they are used.
   b. always modified by an animal's experiences.
   c. performed only by very young animals.
   d. often the product of habituation.
   e. sometimes called trial-and-error responses.

3. A stimulus that always causes an animal to behave in a highly stereotyped way is called:
   a. a fixed action pattern.
   b. an instinct.
   c. habituation.
   d. a sign stimulus.
   e. a reinforcement.

4. Arnold's experiments on the feeding preferences of garter snakes demonstrated that food choice is largely governed by a snake's:
   a. early experiences.
   b. genetics.
   c. size and color.
   d. diet while it was developing inside its mother.
   e. trial-and-error learning.

5. Learning in which an animal associates two phenomena that it experiences at approximately the same time is called:
   a. imprinting.
   b. operant conditioning.
   c. classical conditioning.
   d. insight learning.
   e. habituation.

6. The development of the song system in male songbirds depends on:
   a. direct connections between sensory neurons and motor neurons.
   b. a decrease in the number of neurons in the song system.
   c. the behaviors of females, which stimulate hormone production.
   d. the successful defense of a territory.
   e. the production of estrogen early in life.

7. One of the functions of octopamine in foraging honeybees is to:
   a. increase the production of juvenile hormone.
   b. decrease the production of juvenile hormone.
   c. make the bees defend their territory more aggressively.
   d. stimulate neural transmissions and reinforce memories.
   e. increase the time they spend caring for larvae.

8. In cichlid fishes, high levels of the hormone GnRH:
   a. make females more receptive to male attention.
   b. cause males to be sexually aggressive but not territorial.
   c. stimulate a male to defend its territory.
   d. cause males to abandon their territories.
   e. cause males to lose their bright colors.

9. Sensory bias in the nervous system of a cricket ensures that ultrasound perceived on one side of the body will cause:
   a. a movement in a leg on the same side of the body.
   b. a movement in a leg on the opposite side of the body.
   c. the cricket to respond with a vocalization.
   d. the cricket to stop vocalizing.
   e. the cricket to fly toward the sound.

10. In the brain of a star-nosed mole, more cells decode:
    a. tactile information from its feet than from all other parts of its body.
    b. tactile information from the tentacles on its nose than from all other parts of its body.
    c. tactile information from its mouth than from all other parts of its body.
    d. visual information from the top part of its visual field than the bottom part.
    e. visual information from the bottom part of its visual field than the top part.

Questions for Discussion

1. One day, while walking in the country, you see a rooster wade into a pond and begin to court a female mallard duck. What probably happened to the rooster early in life?

2. Using an example from your own experience, explain why habituation to a frequent stimulus might be beneficial. Also describe an example in which habituation might be harmful or even dangerous.

3. Is learning always superior to instinctive behavior? If you think so, why do so many animals react instinctively to certain stimuli? Are there some environmental circumstances in which being able to respond "correctly" the first time would have a big payoff?

4. Cockroaches have two small projections called *cerci* at the tip of the abdomen. You suspect that the cerci might be responsible for the insects' ability to detect predators, such as lizards, rushing toward them from behind. Under the microscope you see that each cercus is covered with fine hairs. What properties should these hairs have if they are part of a system that detects moving air pushed ahead by an approaching predator? How might the roach determine whether the danger was coming from the right or left side? How quickly should cercal information be processed compared with information about the chemicals in a food item?

Experimental Analysis

You find that some fruit flies in your lab are quick to come to a dish containing citrus oils, but others are not as responsive. How could you test whether these behavioral differences are caused by genetic differences among the flies? What should happen if you performed an artificial selection experiment (see Section 19.2) in which you tried to select for quick versus slow responses to citrus oils?

Evolution Link

Some birds that are frequently kept as pets, such as parrots and myna birds, have the uncanny ability to imitate human speech. Develop a hypothesis that explains why the ability to be a good mimic might have evolved in these species. What features of the birds' brains might be involved in this behavior?

Musk oxen *(Ovibos moschatus)*. The social behavior of a herd of musk oxen includes encircling their young to protect them from predators.

© Paul Nicklen/National Geographic/Getty Images

STUDY PLAN

55.1 Migration and Wayfinding

Migrating animals make long round-trips on a seasonal cycle

Animals use wayfinding mechanisms to guide their movements

Environmental cues trigger hormonal changes that induce seasonal migration

Seasonal variation in food supply may explain the evolution of migratory behavior

55.2 Habitat Selection and Territoriality

Animals use multiple criteria for selecting habitats

Genetics and learning influence habitat selection

Animals sometimes defend patches of habitat for their exclusive use

55.3 The Evolution of Communication

Animal signals can activate different sensory receptors in receivers

Honeybees use several communication channels to transmit complex messages

Biologists use evolutionary hypotheses to analyze communication systems

55.4 The Evolution of Reproductive Behavior and Mating Systems

Males and females use different reproductive strategies

Male competition for females and female mate choice foster sexual selection

Patterns of parental care and territoriality influence mating systems

55.5 The Evolution of Social Behavior

Group living carries both benefits and costs

Fitness varies among the members of a dominance hierarchy

In some animal societies, individuals exhibit altruistic behavior

An unusual genetic system may explain altruism in eusocial insects

55.6 An Evolutionary View of Human Social Behavior

Evolutionary analyses may help to explain human social behavior

55 The Ecology and Evolution of Animal Behavior

WHY IT MATTERS

In early spring, male white-crowned sparrows leave their wintering grounds in Mexico and fly thousands of kilometers to their northern breeding range. There, they select patches of habitat that contain the resources necessary for breeding—suitable cover, potential nesting sites, and abundant food. Then, they start to sing and sing, repeating their song thousands of times every day. The songs are a form of communication through which males announce their presence to rival males and to females. Males also perform elaborate courtship behaviors. And once the young hatch, they communicate with their parents, eliciting the care they need before leaving the nest.

All of these behaviors carry significant costs and risks. For example, migration requires enormous energy expenditure, and many migrating birds die before completing their trip. Moreover, singing males are conspicuous, and they may attract the attention of a hawk or some other predator. Given the costs and dangers associated with these behaviors, what benefits do the birds gain from performing them? The ultimate evolutionary benefit is obvious: with luck, individuals performing these complex and diverse behaviors may leave surviving offspring **(Figure 55.1)**.

1269

Figure 55.1

Reproductive success. Parental care is just one of many behaviors required for successful reproduction in white-crowned sparrows *(Zonotrichia leucophrys)* and many other animal species. The number of surviving nestlings will determine the reproductive success of their parents.

Questions about ultimate benefits are fundamentally different from the questions we considered in the previous chapter, where we focused on *how* underlying physiological and genetic mechanisms enable animals to behave. In this chapter, we try to explain *why* animals behave as they do. Why do sparrows migrate to their breeding grounds, breed in certain habitats but not others, and expose themselves to predation by singing? The behavior of animals is closely tied to ecological circumstances, and evolutionary biologists view most behaviors as an individual's responses to its environment. Moreover, like morphological traits, behaviors are subject to microevolutionary change (see Chapter 20). If particular alleles contribute even slightly to the development of a behavior that enhances an animal's fitness, natural selection will cause the frequency of those alleles to increase in the next generation.

Behavioral biologists apply ecological and evolutionary analyses to all forms of animal behavior, including those described above. In this chapter, we examine the ecology and evolution of several categories of animal behavior: orientation, navigation, and migration; habitat selection and territoriality; communication; reproductive behavior and mating systems; and social behavior, including behaviors described as altruistic. We close the chapter with a brief look at human behavior.

55.1 Migration and Wayfinding

Most animals move through their environments at some stage of their life cycles. Although some species move only short distances to find suitable environmen-

tal conditions, many others undertake large-scale movements on a seasonal schedule.

Migrating Animals Make Long Round-Trips on a Seasonal Cycle

Many animal species undertake a seasonal **migration**, traveling from the area where they were born to a distant and initially unfamiliar destination, and returning to their birth site later. The Arctic tern *(Sterna paradisaea)*, a seabird, makes an annual round-trip migration of 40,000 km **(Figure 55.2)**. Many other vertebrate species, including gray whales and salmon, also undertake long and predictable journeys. Even some arthropods migrate long distances. For example, spiny lobsters *(Panulirus* species) form long conga lines as they move between coral reefs and the open ocean floor on a seasonal cycle **(Figure 55.3)**.

Animals Use Wayfinding Mechanisms to Guide Their Movements

Moving animals use various wayfinding mechanisms to arrive at their destination. Biologists group these mechanisms into three general categories: *piloting, compass orientation,* and *navigation.* Many species probably use a combination of these mechanisms to guide their movements.

The simplest wayfinding mechanism is **piloting**, in which animals use familiar landmarks to guide their journey. For example, gray whales *(Eschrichtius robustus)* migrate from Alaska to Baja California and back using visual cues provided by the Pacific coastline of North America. When it is time to breed and lay eggs, Pacific salmon (genus *Oncorhynchus*) use olfactory cues to pilot their way from the ocean back to the stream where they themselves hatched.

Animals that do not undertake long migrations also use specific landmarks to identify their nest site or places where they have stashed food. In a famous experiment published in 1938, Niko Tinbergen showed that female digger wasps *(Philanthus triangulum)*, which nest in soil, use visual landmarks to find their nests after flying off in search of food **(Figure 55.4)**. Tinbergen arranged pinecones in a circle around one nest while the female was still inside. As she left, she flew around the area, apparently noting nearby landmarks. Tinbergen then moved the circle of pinecones a short distance away. Each time a female returned, she searched for her nest within the pinecone circle—and never once found it unless the pinecones were returned to their original position. In a follow-up study, Tinbergen rearranged the circle of pinecones into a triangle after females left their nests and added a ring of stones nearby. The returning females looked for their nest in the stone circle. Tinbergen concluded that digger wasps respond to the general outline or geometry of landmarks around their nests and not to the specific objects that create those landmarks.

A more sophisticated wayfinding mechanism, **compass orientation**, allows animals to move in a particular direction, often over a specific distance or for a prescribed length of time. Some day-flying migratory birds, for example, orient themselves using the sun's position in the sky in conjunction with an internal biological clock (see Section 40.4). The internal clock allows the bird to use the sun as a compass, compensating for changes in its position through the day; the clock may also allow some birds to estimate how far they have traveled since beginning their journey. Other migratory animals use polarized light or Earth's magnetic field as a compass.

Some birds that migrate at night use the positions of stars to determine their direction. The indigo bunting *(Passerina cyanea),* for example, flies about 3500 km from the northeastern United States to the Caribbean or Central America each fall and makes the return journey each spring. Stephen Emlen of Cornell University demonstrated that these birds use celestial cues to direct their migration **(Figure 55.5).** Emlen confined individual buntings in cone-shaped test cages. He lined the sides of the cages with blotting paper, placed inkpads on the bottom, and kept the cages in an outdoor enclosure so that the birds had a full view of the sky. Whenever a bird made a directed movement, its inky footprints indicated the direction in which it was trying to fly. Emlen found that on clear nights in fall, the footprints pointed to the south; on clear nights in spring, they pointed north. On cloudy nights, when the buntings could not see the stars, their footprints were evenly distributed in all directions, indicating that their compass required a view of the stars.

The most complex wayfinding mechanism is **navigation**, in which an animal moves toward a specific destination, using both a compass and a "mental map" of where it is in relation to the destination. Human hikers in unfamiliar surroundings routinely use navigation to find their way home: they use a map to determine their current position and the necessary direction of movement and a compass to orient themselves in that direction. Scientists have documented true navigation in only a few animal species. Perhaps the most notable is the homing pigeon *(Columba livia),* which can navigate to its home coop from any direction. Recent research suggests that homing pigeons probably use the sun's position as their compass and olfactory cues as their map.

Environmental Cues Trigger Hormonal Changes That Induce Seasonal Migration

For white-crowned sparrows and many other species, researchers have shown that decreasing (or increasing) day length, a correlate of the approaching autumn (or

Figure 55.2

Long-distance migration. Arctic terns *(Sterna paradisaea)* migrate from the high Arctic to Antarctica each year, a round-trip journey of 40,000 km. This species' summer breeding range is shaded on the map.

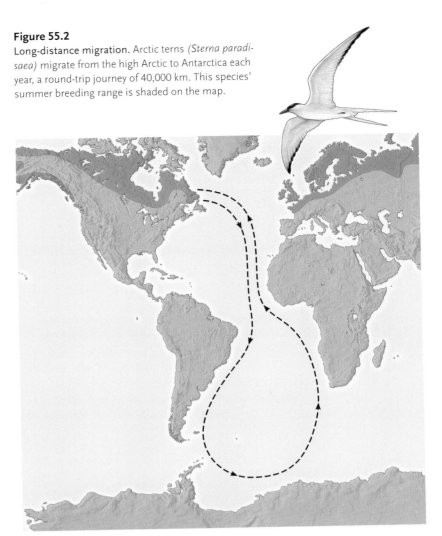

Figure 55.3

Migrating arthropods. Spiny lobsters *(Panulirus argus)* make seasonal migrations between coral reefs and the open ocean floor. As many as 50 individuals march in single file for several days.

Howard Hall/Oxford Scientific/Index Stock

Figure 55.4 Experimental Research

Using Landmarks to Find the Way Home

QUESTION: How do female digger wasps *(Philanthus triangulum)* relocate their nests after flying off to search for food?

EXPERIMENT: Tinbergen arranged pinecones in a circle around the nest of a female digger wasp while she was still inside. After leaving the nest, she circled the area a few times, apparently noting nearby landmarks. Tinbergen then moved the circle of pinecones a short distance away.

RESULT: Each time the female returned, she searched for her nest within the pinecone circle. She was unable to find the nest unless Tinbergen replaced the pinecones in their original position.

Wasp's flight pattern on leaving nest

Wasp's return, looking for nest

CONCLUSION: Female digger wasps use the location of local landmarks to find the entrances to their underground nests.

---

spring), stimulates the anterior pituitary of the bird's brain to generate a series of hormonal changes. The birds then feed heavily and accumulate the fat reserves necessary to fuel the long journey. Sparrows also become increasingly restless at night, until one evening they launch themselves into their nocturnal migration. Their ability to adopt and maintain a southerly orientation in autumn (and a northerly one in spring) rests in part on their capacity to use the positions of stars to provide them with directional information.

Seasonal Variation in Food Supply May Explain the Evolution of Migratory Behavior

Migratory behavior entails obvious costs, such as the time and energy devoted to the journey and the risk of death from exhaustion or predator attack. Why then do some species migrate? What benefits accrue to an individual that undertakes a costly migration?

For migratory birds, the most widely accepted hypothesis focuses on seasonal changes in food supplies. The amount of insect food available in northern forests increases explosively during the warm spring and summer, providing abundant resources to produce eggs and rear offspring. Then, during the late fall and winter, insects all but disappear. A few bird species that forage on seeds and dormant insects do not head south. However, energy supplies are more predictably available in tropical overwintering grounds, and migratory birds may have a better chance of surviving there. The following spring they return north to exploit the food bonanza on their summer breeding grounds.

The two-way migratory journeys may provide other benefits as well. Avoiding the northern winter is probably adaptive because endotherms must increase their metabolic rates just to stay warm in cold climates (see Section 46.8). But in summer the days are longer at high latitudes than they are in the tropics (see Section 52.1), giving adult birds more time to collect enough food to rear a brood.

Seasonal changes in food supply also underlie the migration of monarch butterflies *(Danaus plexippus),* which eat milkweed *(Asclepias* species) leaves as larvae and the nectar of milkweeds and other plants as adults.

Figure 55.5 Experimental Research

Experimental Analysis of the Indigo Bunting's Star Compass

QUESTION: Do indigo buntings *(Passerina cyanea)* use the positions of stars in the night sky to orient their migrations?

EXPERIMENT: Emlen placed individual buntings in cone-shaped test cages. He lined the sides of the cages with blotting paper, placed inkpads on the bottom, and kept the cages in an outdoor enclosure so that the birds had a full view of the sky. Whenever a bird made a directed movement, its inky footprints indicated the direction in which it was trying to fly. Emlen predicted that the footprints would show the buntings' inclination to migrate south in autumn and north in spring.

Indigo bunting

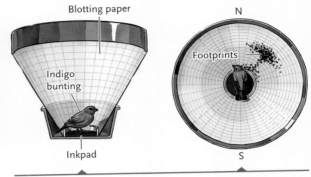

Side (left) and overhead (right) views of the test cage with blotting paper on the sides and an inkpad on the bottom

RESULTS: On clear nights in autumn, the footprints pointed to the south; on clear nights in spring, they pointed north. On cloudy nights, when buntings could not see the stars, their footprints were evenly distributed in all directions.

In autumn, the bunting footprints indicated that they were trying to fly south.

In spring, the bunting footprints indicated that they were trying to fly north.

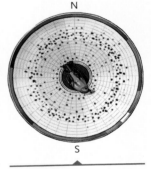

On cloudy nights, when buntings could not see the stars, their footprints indicated a random pattern of movement.

CONCLUSION: Indigo buntings use the positions of the stars to direct their seasonal migrations. When they could see the stars above their test cages, they moved in the predicted direction; but when clouds obscured their view of the stars, they moved in random directions.

In eastern North America, milkweed plants grow only during spring and summer. Many adult monarchs head south in late summer, when milkweeds are beginning to die, migrating as much as 4000 km from eastern and central North America to central Mexico, where they cluster in spectacular numbers **(Figure 55.6).** Unlike migrant birds, these insects do not feed on their overwintering grounds. Instead, their metabolic rate decreases in the cool mountain air, and the butterflies become inactive for months, thereby conserving precious energy reserves. When spring arrives, the butterflies become active again and begin the return migration to northern breeding habitats. The northward migration is slow, however, and many individuals stop along the way to feed and lay eggs. But their offspring, and their offspring's offspring, continue the northward

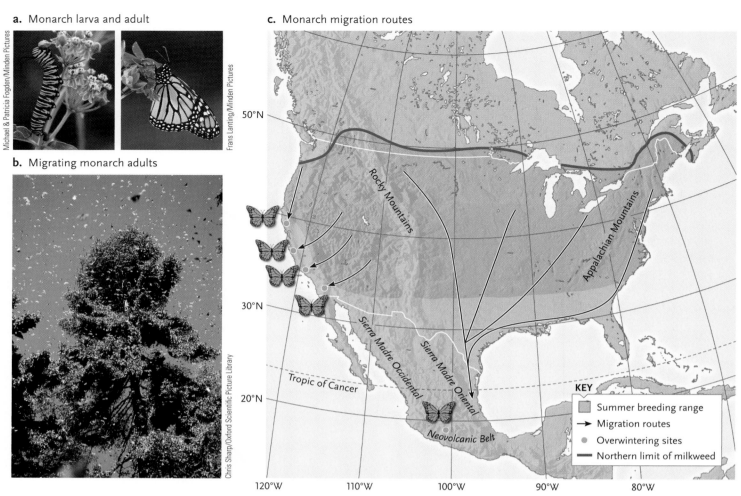

Figure 55.6
Monarch butterfly migrations. **(a)** Monarch butterflies *(Danaus plexippus)* feed primarily on milkweed plants. **(b)** When milkweed plants in their breeding range die back at the end of summer, millions of monarchs begin a southward migration. **(c)** Butterflies that live and breed east of the Rocky Mountains migrate to Mexico. After passing the winter in a semidormant state, they migrate northward the following spring. Monarchs living west of the Rocky Mountains winter in coastal California.

migration through the summer; some descendants eventually reach Canada for a final round of breeding. The summer's last generation then returns south to the spot where their ancestors, two to five generations removed, spent the previous winter.

For other animals, the migration to breeding grounds may provide special conditions necessary for reproduction. For example, gray whales migrate south to breeding grounds in quiet, shallow lagoons where predators are rare and warm water temperatures will not stress their calves.

STUDY BREAK

1. What is the difference between piloting, compass orientation, and navigation?
2. What is the most probable selection pressure that has fostered seasonal migrations in birds?

55.2 Habitat Selection and Territoriality

The geographical range of nearly every animal species includes a mosaic of habitat types. The breeding range of white-crowned sparrows, for example, encompasses forests, meadows, housing developments, and city dumps. An animal's choice of habitat is critically important because the habitat provides food, shelter, nesting sites, and the other organisms with which it interacts. If an animal chooses a habitat that does not provide appropriate resources, it will not survive and reproduce.

Animals Use Multiple Criteria for Selecting Habitats

On a large spatial scale, animals almost certainly use multiple criteria to select the habitats they occupy, but no research has yet established any general principles about how animals make these choices. When a migrating bird arrives at its breeding range, for example, it probably cues on large-scale geographical features, such as a pond or a patch of large trees. If it does not find the food or nesting resources it needs—or if other individuals have already depleted those resources—it may move to another habitat patch.

On a very fine spatial scale, basic responses to physical factors enable some animals to find suitable habitats. The simplest such mechanism is called a **kinesis** (*kinesis* = movement), a change in the rate of movement or the frequency of turning movements in response to environmental stimuli. For example, the terrestrial crustaceans known as wood lice (Isopoda) typically live under rocks and logs or in other damp places. Although these arthropods are not attracted to moisture per se, laboratory experiments have shown that when a wood louse encounters dry soil, it exhibits a kinesis, scrambling around and turning frequently; when it reaches a patch of moist soil, it moves much less. As a result, these animals accumulate in moist habitats. Biologists infer that this behavior is adaptive because wood lice exposed to dry soil quickly dehydrate and die. Other animals may exhibit a **taxis** (*taxis* = ordered movement), a response that is directed either toward or away from a specific stimulus. For example, cockroaches (order Blattodea) exhibit negative phototaxis: they actively avoid light and seek darkness, a behavior that makes them harder for visually oriented predators to detect.

Genetics and Learning Influence Habitat Selection

Biologists generally assume that habitat selection is adaptive and has been shaped by natural selection in most animal species. For example, some animals instinctively select habitats where they are well camouflaged, a means of avoiding detection by predators (see Figure 50.3); predators would discover and eliminate any individual that fails to select a matching background—along with any alleles responsible for the mismatch. Many insects have a genetically determined preference for the plants that they eat during their larval stage. Adults often restrict their mating and egg-laying activities to these food plants, effectively selecting the habitats where their offspring will live and feed, as described in the discussion of sympatric speciation (see Section 21.3).

Even vertebrates sometimes exhibit such innate preferences, as demonstrated by two closely related European bird species, blue tits (*Parus caeruleus*) and coal tits (*Parus ater*). Adult blue tits feed mostly in oak trees, whereas coal tits prefer to feed in pines. When researchers reared the young of both species in cages without any vegetation at all and then offered them a choice between oak branches and pine branches, coal tits immediately gravitated toward pines and blue tits toward oaks, strongly suggesting that the preference is

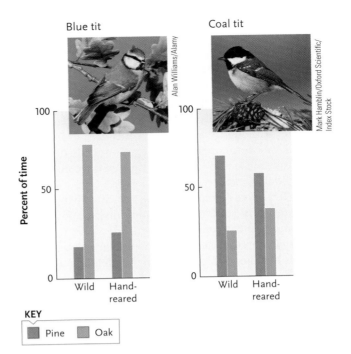

Figure 55.7

Habitat selection by birds. Wild blue tits *(Parus caeruleus)* show a strong preference for oak trees, and coal tits *(Parus ater)* show a strong preference for pine trees. Hand-reared birds that were raised in a vegetation-free environment showed identical, though slightly weaker, preferences.

innate **(Figure 55.7).** Further research demonstrated that each species feeds most successfully in the tree species it prefers. Thus, natural selection probably fostered these preferences.

Habitat preferences can be molded by experiences early in life, however. For example, the tadpoles of red-legged frogs (*Rana aurora*) usually live in aquatic habitats cluttered with sticks, strands of algae, and plant stems; when given a choice in the laboratory, they prefer striped backgrounds to plain ones. By contrast, tadpoles of the closely related cascade frog *(Rana cascadae)* live over gravel bottoms, and they prefer plain substrates over striped ones. However, when red-legged frogs are reared over plain substrates and cascade frogs over striped substrates, they no longer exhibit preferences for their usual substrates.

Animals Sometimes Defend Patches of Habitat for Their Exclusive Use

Under some circumstances, animals may defend a **territory** from other members of their species, retaining more or less exclusive use of the resources it contains. Territorial behavior occurs in all major groups of vertebrates, many insects, and some other invertebrates, but it is by no means universal. In many organisms, territorial behavior occurs only during the breeding season.

Animals establish and defend territories only when some critical resource is in short supply. Moreover, the resource must be fixed in space so that the area around it can be defended. For example, during the breeding season, most songbirds defend a territory within which they build a nest and collect food for their young. By contrast, many sea bird species, such as terns and penguins, do not defend a feeding territory. They catch fish in the ocean and build nests on the shore. Although they defend a tiny area around the nest, they never attempt to defend a section of ocean; fishes come and go at will and thus do not constitute a defendable resource.

Territorial defense is always a costly activity. Patrolling territory borders, performing displays hundreds of times per day, and chasing intruders take time and energy. Moreover, territorial displays increase an animal's likelihood of being injured or detected and captured by a predator.

Experiments conducted by Catherine Marler and Michael Moore of Arizona State University illustrate the cost of territorial behavior in Jarrow's spiny lizard *(Sceloporus jarrovi)*. Male lizards ordinarily exhibit strong territoriality only during the autumn mating season, when elevated blood levels of testosterone stimulate their aggressive behavior. The researchers implanted small doses of testosterone under the skin of experimental animals in June and July, during the nonmating season; controls received a placebo treatment. Testosterone-enhanced males were more active and displayed more often than control males. But experimental males spent less time feeding, even though they used about 30% more energy per day than control males. Over the course of about 7 weeks, a significantly higher percentage of experimental males died—a clear sign that engaging in territorial behavior is costly.

On the other hand, the benefits of maintaining a territory include having access to nesting sites, food supplies, and refuges from predators. For example, the surgeonfish *(Acanthurus lineatus)*, which lives in the coral reefs around American Samoa, may engage in as many as 1900 chases per day to defend a small territory from other algae-eating fish species. But territory holders may consume up to five times as much food as nonterritory holders.

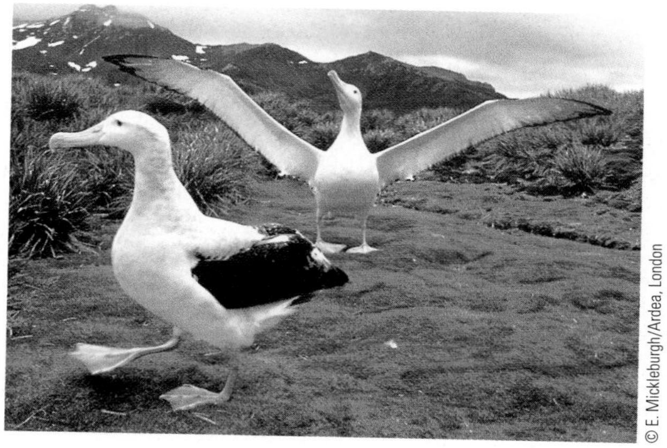

© E. Mickleburgh/Ardea, London

Figure 55.8 Visual displays. The courtship display of a male wandering albatross *(Diomedea exulans)* includes ritualized postures and movements of the wings and body.

Study Break

1. Why do wood lice tend to occur in moist parts of their habitat?
2. What are the costs of maintaining a territory, and what are the benefits?

55.3 The Evolution of Communication

When resident animals advertise their presence in their territories, they are communicating information to nearby animals. In the formal language of animal behavior studies, all communication systems involve an interaction between a *signaler,* the animal that transmits information, and a *signal receiver,* the animal that intercepts the information and makes a behavioral response. Natural selection has adjusted the ability of signalers to transmit information and the ability of receivers to get the message.

Animal Signals Can Activate Different Sensory Receptors in Receivers

Biologists categorize animal signals according to the sensory receptors, or "channels," through which the signal acts: *acoustical, visual, chemical, tactile,* or *electrical.* Each channel has specific advantages.

Bird songs are examples of **acoustical signals;** a signaler produces a sound that is heard by a signal receiver. Many animals use the acoustical channel, including a host of nocturnal and burrow-dwelling insects and amphibians. These signals reach distant receivers, even at night and in cluttered environments where visual signals are less effective.

Because humans frequently use facial expressions and body language to send messages, **visual signals** are a familiar form of communication. In many animals, visual signals are *ritualized;* in other words, they have become exaggerated and stereotyped over evolutionary time, forming an easily recognized visual display **(Figure 55.8).** Visual displays can even be useful at night or in the darkness of the deep sea; some animals, such as fireflies and certain fishes, send bioluminescent signals to distant receivers.

Many species release **chemical signals,** which carry messages to signal receivers through the olfactory channel. Scent marking (spraying) by male cats is an example. In particular, mammals and insects often communicate through **pheromones,** distinctive volatile chemicals released in minute amounts to influence the behavior of members of the same species. For example, a worker ant's body contains a battery of glands, each releasing a different pheromone **(Figure 55.9).** One set of pheromones recruits fellow workers to battle colony invaders; another set stimulates workers to collect food that has been discovered outside the colony. Other animals release pheromones to attract mates. Female silkworm moths *(Bombyx mori)* produce bombykol, a single molecule of which can generate a message in specialized receptors on the antennae of any male silkworm moth that is downwind (see Figure 39.19).

In many species, touch conveys important messages from a signaler to a receiver. **Tactile signals** can operate only over very short distances, but for social animals living in close company, they play a significant role in the development of friendly bonds between individuals **(Figure 55.10)**.

Some freshwater fish species, especially those that occupy murky tropical rivers where visual signals could not be seen, use weak **electrical signals** to communicate. These fishes have electric organs that can release charges of variable intensity, duration, and frequency, allowing substantial modulation of the message that a signaler sends. Among the New World knifefishes (order Gymnotiformes), including the electric eel *(Electrophorus electricus)*, electrical discharges can signal threats, submission, or a readiness to breed.

Honeybees Use Several Communication Channels to Transmit Complex Messages

When animals need to convey a complex message, they may use several channels of communication simultaneously. For example, as Karl von Frisch demonstrated, the famous dance of the honeybee *(Apis mellifera)* involves tactile, acoustical, and chemical communication **(Figure 55.11)**. When a foraging honeybee discovers pollen or nectar, it returns to its colony and performs a complex dance on the vertical surface of the honeycomb in the complete darkness of the hive. The dancer moves in a circle, attracting a crowd of workers, some of which follow and maintain physical contact with the dancer. From the dance, they acquire information about the distance and direction they will need to fly to locate the food source.

When the food source is less than about 75 m from the hive, the bee performs a "round dance" (see Figure 55.11a). It moves in tight circles, swinging its abdomen back and forth. Bees surrounding the dancer produce a brief acoustical signal, which stimulates the dancer to regurgitate a sample of the food it discovered. The regurgitated sample serves as a chemical cue to other workers, which then leave the hive to search for that type of food.

If the food source is more distant, the forager performs what von Frisch described as the "waggle dance." The bee dances a half circle in one direction and then dances in a straight line while waggling its abdomen before dancing a half circle in the other direction (see Figure 55.11b). With each waggle, the dancer produces a brief buzzing sound. Von Frisch determined that the angle of the straight run relative

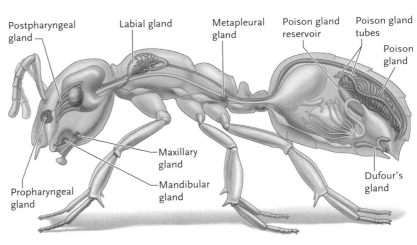

Postpharyngeal gland · Labial gland · Metapleural gland · Poison gland reservoir · Poison gland tubes · Poison gland · Maxillary gland · Mandibular gland · Dufour's gland · Propharyngeal gland

Figure 55.9
Chemical signals. An ant's body contains a host of pheromone-producing glands, each of which manufactures and releases its own volatile chemical or chemicals.

to the vertical honeycomb indicates the direction of the food source relative to the position of the sun (see Figure 55.11c). The duration of the waggles and buzzes that the bee makes on the straight run carries information about the distance to the food: the longer the time spent waggling and buzzing, the further the food is from the hive.

© Kenneth W. Fink/Photo Researchers, Inc.

Figure 55.10
Tactile signals. Grooming by hyacinth macaws *(Anodorhynchus hyacinthinus)* removes parasites and dirt from feathers. The close physical contact also promotes friendly relations between groomer and groomee.

a. Round dance

b. Waggle dance

c. Coding direction in the waggle dance

When the bee moves straight down the comb, other bees fly to the source directly away from the sun.

When the bee moves 45° to the right of vertical, other bees fly at a 45° angle to the right of the sun.

When the bee moves straight up the comb, other bees fly straight toward the sun.

Figure 55.11

Dance communication by honeybees. Foraging honey bees *(Apis mellifera)* transmit information about the location of a food source by dancing on the vertical honeycomb. **(a)** If the food source is close to the hive, the forager performs a "round dance." **(b)** If the food source is more than about 75 m from the hive, the forager performs a "waggle dance," which indicates the distance to the food source. **(c)** The dancing bee indicates the direction to a distant food source by the angle of the waggle run.

Figure 55.12

Threat displays. The threat display of a dominant male mandrill *(Mandrillus sphinx)*, used to drive away rival males, features exposed canines.

Tom and Pat Leeson

Biologists Use Evolutionary Hypotheses to Analyze Communication Systems

Signal receivers often respond to communication from signalers in predictable ways. For example, a male white-crowned sparrow generally avoids entering a neighboring territory simply because it hears the song of the resident male. Similarly, young male baboons often retreat without a fight when they see an older male's visual threat display **(Figure 55.12)**, even though they may lose the chance to mate with a female. Why do these receivers behave in ways that appear to be beneficial to their rivals, but not to themselves?

When biologists try to explain behavioral interactions, their hypotheses focus on how an animal's actions may allow it to contribute more offspring to the next generation. In our first example, the retreating male sparrow avoids wasting time and energy on a battle he

is likely to lose—as well as the possibility of being injured or killed by another male. Moreover, ousting the current resident might be more tiring and risky than finding a suitable unoccupied breeding site. This hypothesis predicts that resident males should almost always win physical contests. In cases when an intruder wins a territory from a resident, it may do so only after a prolonged series of exhausting clashes. Observations of territorial species—whether birds, lizards, frogs, fish, or insects—generally support these predictions.

Applying a similar argument to competition among male baboons, we can predict that smaller or younger males will concede females to threatening older rivals without fighting. The signal receiver retreats after receiving the threat because he judges that he would be demolished in real combat—after all, a male baboon's canine teeth are not just for show. Evolutionary analyses therefore suggest that both the signaler and the signal receiver benefit from the transfer of information in their communication system.

An evolutionary analysis also helps to explain the strange yell of ravens *(Corvus corax),* which scavenge carcasses of deer, elk, or moose in northern forests during winter. When one of these large birds comes across a food bonanza, it may call loudly, attracting a crowd of hungry ravens. The calling behavior puzzled Bernd Heinrich of the University of Vermont. Wouldn't a quiet raven eat more, survive longer, and produce more offspring than a noisy bird? If natural selection favored the raven's calling behavior, we might expect that the cost of calling (in terms of lost food) would be offset by a reproductive benefit for the individual caller. Heinrich noticed that paired, territory-owning adults did not yell loudly when they found goat carcasses that he had hauled into the Maine woods; instead, they fed quietly. Only young, wandering ravens that happened upon a carcass in another bird's territory advertised their discovery. The signals of these birds attracted other nonterritorial ravens, which collectively overwhelmed the resident pair's attempts to defend their territory. Only then was a wanderer likely to have a chance to feed in an area that would otherwise be off-limits.

STUDY BREAK

1. Which channels do humans consciously use to communicate with each other?
2. How does a honeybee tell its hive-mates that it has discovered a distant food source?
3. Why do ravens sometimes announce their discovery of food?

55.4 The Evolution of Reproductive Behavior and Mating Systems

In many animal species, communication coordinates the reproductive activities of males and females and governs the interactions between parents and offspring. In this section, we examine how several elements of behavior contribute to the reproductive success of individuals.

Males and Females Use Different Reproductive Strategies

In sexually reproducing species, males and females often differ in their overall **reproductive strategies,** the set of behaviors that lead to reproductive success. This difference arises in part from a fundamental difference in the amount of **parental investment,** the time and energy devoted to the production and rearing of offspring, provided by the two sexes. Because eggs are much larger than sperm, females almost always contribute more energy than males to the production of a gamete.

A male might increase the number of offspring that carry his alleles simply by mating with multiple females, especially if he does not spend time and energy providing parental care to his offspring. Thus, in many animal species, males compete intensely for access to females, and any trait that increases a male's access or attractiveness to females has a big reproductive payoff.

Entirely different selection pressures operate on females, whose reproductive output is generally limited by the number of eggs they can produce. Mating with multiple males will not increase that number. But the success of her offspring may depend on the attributes of their father or the territory he holds. Thus, the females of many species choose their mates carefully. In some cases, females mate with males whose territories include abundant resources, ensuring an ample food supply for their young. In other cases, females choose robust males that will contribute "good genes" (that is, alleles that confer a high likelihood of surviving and reproducing) to her offspring, increasing their chances of long-term success.

Male Competition for Females and Female Mate Choice Foster Sexual Selection

Male competition for access to females coupled with the females' choice of mates establishes a form of natural selection called **sexual selection,** that is, selection for mating success (see Section 20.3). As a result of sexual selection, males are larger than females in many species, and males have ornaments and weapons, such as horns and antlers, that are useful for attracting females as well as for butting, stabbing, or intimidating rival males. Males typically show off these elaborate structures in complex **courtship displays** to attract the attention of females. For example, male peafowl (*Pavo cristatus*) strut in front of females while spreading a gigantic fan of tail feathers, which they shake, rattle, and roll.

Why should females choose males with exaggerated structures that they display conspicuously? Biologists have developed several hypotheses to explain the attraction. First, a male's large size, bright feathers, or large horns might indicate that he is particularly healthy, that he can harvest resources efficiently, or simply that he has managed to survive to an advanced age. These traits are, in effect, signals of male quality; and if they reflect a male's genetic makeup, he is likely to fertilize a female's eggs with sperm containing successful alleles. In some cases, big, showy males hold large, rich territories, and females that choose them gain access to the resources their territories contain.

The degree to which females *actively* choose genetically superior mates varies among species. In the northern elephant seal (*Mirounga angustirostris*), for example, female choice is more or less passive. Large numbers of females gather on beaches to give birth to their pups before becoming sexually receptive again. Males locate these clusters of females and fight to keep other males away (see Figure 20.8). Males that win have exceptional reproductive success, but only after engaging in violent and relentless combat with rival males. In this kind of mating system, females are practically guaranteed to receive sperm from large and powerful males in superb physiological condition, attributes that may well be associated with alleles that will increase their offspring's chances of living long enough to reproduce.

In other species, females exercise more active mate choice, copulating only after inspecting a group of potential partners. Among birds, active female mate choice is most apparent at **leks,** display grounds where each male possesses a small territory from which it courts attentive females. Male sage grouse (*Centrocercus urophasianus*), a lekking bird of western North America, gather in open areas among stands of sagebrush. Each male defends just a few square meters, where it struts in circles while emitting booming calls and showing off its elegant tail feathers and big neck pouches (**Figure 55.13**). Females wander among the displaying males, presumably analyzing the males' visual and acoustical displays. Eventually, each female selects a mate from among the dozens of males present. Females repeatedly favor males that come to the lek daily, defend their small area vigorously, and display more frequently than the average lek participant. In other words, favored males can sustain their territorial defense and high display rate over long periods, an ability that may correlate with useful genetic traits.

Figure 55.13

Lekking behavior. Male sage grouse *(Centrocercus urophasianus)* use their ornamental feathers in visual courtship displays performed at a lek, where each male has his own small territory. The smaller brown females observe the prancing males before choosing a mate.

Experimental studies of peafowl suggest that the top peacocks at a lek may indeed supply advantageous alleles to their offspring. In nature, peahens prefer males whose tails have many ornamental eyespots **(Figure 55.14)**. In an experiment on captive birds, some females were mated to males with highly attractive tails, but others were paired with males whose tails were less impressive. The offspring of both groups were reared under uniform conditions for several months and then released into an English woodland. After 3 months on their own, the offspring of fathers with impressive tails survived better and weighed significantly more than did those whose fathers had less

Figure 55.14

Sexual selection for ornamentation. The attractiveness of a peacock *(Pavo cristatus)* to females depends in part on the number of eyespots on his extraordinary tail. The offspring of males with elaborate tails are more successful than the offspring of males with plainer tails.

attractive tails. Apparently, a peahen's mate choice does provide her offspring with a survival advantage.

Another hypothesis argues that females select showy males even though their ornate structures may impede their locomotion or their elaborate displays may attract the attention of a predator. According to this hypothesis, any male that survives *despite* carrying such a handicap must have a very strong constitution indeed, and he will pass those successful alleles—as well as the alleles responsible for the ornamental handicap—to the female's offspring.

Patterns of Parental Care and Territoriality Influence Mating Systems

In the examples of mate choice just described, successful males inseminate many females, increasing their reproductive success dramatically. But one male mating with many females is only one of several **mating systems**, the ways in which males and females pair up. Some species are **promiscuous**: individuals do not form close pair bonds, and both males and females mate with multiple partners. Other species are **monogamous**: one male and one female form a long-term association. Finally, some species are **polygamous**: *either* males *or* females may have many mating partners. If one male mates with many females, the relationship is called **polygyny**; if one female mates with multiple males, it is described as **polyandry**.

Mating systems appear to have evolved to maximize reproductive success, partly in response to the amount of parental care that offspring require and partly in response to other aspects of a species' ecology. For example, the young of most songbird species, like the white-crowned sparrow, are helpless upon hatching; all they can do is open their mouths and peep, signaling to their parents that they are ready to be fed. These young require lots of parental care, and they are more likely to flourish if both parents bring food to the nest. As you might expect, nearly all songbirds are monogamous, and males and females team up to provide parental care to their offspring.

In some other bird species, such as red-winged blackbirds *(Agelaius phoenecius)*, males establish large, resource-filled territories, and females select mates largely by the quality of the real estate a male holds. Any male with an exceptionally fine territory will be desirable, even if another female has already established herself there. A second female may judge that more resources are available in his territory than in a neighboring one, despite competition with the other female. However, if many females have already settled in a male's territory, intense competition from them may make it less attractive. Given this pattern of habitat and mate choice by females, red-winged blackbirds have a polygynous mating system; males may fertilize the eggs of multiple females and provide little if any direct care to their offspring.

1. For monogamous species, what characteristics of males should increase their attractiveness to females?
2. What activities do male and female sage grouse perform at a lek?
3. Why might a female red-winged blackbird settle on a territory that was already occupied by another female?

55.5 The Evolution of Social Behavior

Social behavior, the interactions that animals have with other members of their species, has profound effects on an individual's reproductive success. Some animals are solitary, getting together only briefly to mate (rhinoceroses and leopards); others spend most of their lives in small family groups (gorillas); still others live in groups with thousands of relatives (termites and honey bees). Some species, such as some African antelopes and humans, live in large social units composed primarily of nonrelatives.

Group Living Carries both Benefits and Costs

Ecological factors have a large impact on the reproductive benefits and costs of social living. Groups of cooperating predators frequently capture prey more effectively than they would on their own. For example, white pelicans *(Pelecanus erythrorhynchos)* often encircle a school of fish before moving in for the kill. Conversely, prey that are subject to intense predation often gain safety in numbers. Those living in groups have more watchful eyes to detect an approaching predator. In ad-

dition, a predator may be confused when multiple prey scatter in many directions. Finally, few predators have the capacity to capture every individual in a prey cluster, so that some prey escape while the predator pursues others.

Some prey species, such as musk oxen *(Ovibos moschatus),* join forces to defend themselves actively (see the photo that opens this chapter). Even some insects, such as Australian sawfly caterpillars *(Perga dorsalis),* exhibit cooperative defensive behavior **(Figure 55.15).** When predators disturb the caterpillars, all members of the group rear up and writhe about, regurgitating sticky, pungent oils that they have collected from the eucalyptus leaves they eat. Although the caterpillars can store these oils safely, they are toxic and repellent to bird predators.

A group of sawflies regurgitates more repellent eucalyptus oils than a single individual, which may explain why these insects form their simple societies. If this hypothesis is correct, solitary individuals should be at greater risk of being eaten than those that live communally. Birgitta Sillén-Tullberg of the University of Stockholm, Sweden, tested this prediction by offering sawfly caterpillars to young great tits *(Parus major),* a songbird species. Birds that received caterpillars one at a time consumed an average of 5.6, but those that received them in groups of 20 ate an average of only 4.1 caterpillars. As Sillén-Tullberg had predicted, the caterpillars were somewhat safer in a group than on their own.

In some environments, the costs of social clumping can be significant. These costs may include increased competition for food. For example, when thousands of royal penguins *(Eudyptes schlegeli)* crowd together in huge colonies **(Figure 55.16),** the pressure

Figure 55.15
Social defensive behavior. Australian sawfly *(Perga dorsalis)* caterpillars clump together on tree branches. These larvae each regurgitate yellow blobs of sticky, aromatic fluid. The accumulation of fluid from a large group of caterpillars successfully deters some predators.

John Alcock/Arizona State University

Figure 55.16
Colonial living. Royal penguins *(Eudyptes schlegeli)* on Macquarie Island, between New Zealand and Antarctica, experience both benefits and costs from living together in huge colonies.

A. E. Zuckerman/Tom Stack & Associates

on the local food supplies is great, increasing the risk of starvation. Communal living also facilitates the spread of contagious diseases and parasites. Nestlings in large colonies of cliff swallows (*Petrochelidon pyrrhonota*) are often stunted in growth because their nests are swarming with blood-feeding parasites, which move easily from nest to nest under crowded conditions. Such costs are probably why the vast majority of animals do not live in large, complex societies.

Fitness Varies among the Members of a Dominance Hierarchy

Recognizing the costs as well as the benefits of social living, biologists have examined features of social living that appear to reduce the fitness of some individuals. For example, some animal species form **dominance hierarchies,** social systems in which each individual's behavior is governed by its place in a highly structured social ranking. In a typical dominance hierarchy, the dominant or *alpha* individual rules the roost; subordinate individuals typically concede valuable resources to more dominant animals without so much as a peep of protest.

Although dominant individuals gain first access to resources, they also incur costs. Frequent challenges from lower ranking individuals may induce a stress response in dominant animals, which must constantly defend their status. For example, in some primates, wild dogs, and other mammals, dominant males have higher blood levels of cortisol and other stress-related hormones (see Section 40.4) than do subordinates. Elevated cortisol levels may induce high blood pressure, the disruption of sugar metabolism, and other pathological conditions.

Why does a subordinate remain in the group when dominant companions reduce its chances for reproductive success? A possible explanation is that survival rates and reproductive success may be even lower for animals that live by themselves: a solitary baboon surely quickens the pulse of a passing leopard **(Figure 55.17).** A subordinate member of a group gains the benefits, such as protection against predators, that come from being part of the group. Low-ranking males may even have the chance to copulate with one of the group's females when dominant males are not watching, thus ensuring some representation of their alleles in the next generation. And if a low-ranking individual can live long enough, its social superiors may be toppled by predation, accidents, or old age, and a one-time subordinate may find itself high on the social register with food and mates galore.

In Some Animal Societies, Individuals Exhibit Altruistic Behavior

In some species, group members appear to sacrifice their own reproductive success to help individuals that are not their direct descendants; such behaviors are collectively called **altruism.** For example, subordinate members of a wolf pack do not reproduce, but they share captured prey with the dominant pair and that pair's offspring. Altruistic behavior, by its very definition, appears to contradict a basic premise of Darwinian evolutionary theory, namely that natural selection favors traits that increase an *individual's* relative fitness. Why don't subordinate wolves simply save the energy spent on helping, bide their time until they can become dominant, and then produce their own offspring?

Behavioral ecologist William D. Hamilton of University College, London, provided a solution to this puzzle. He recognized that alleles favoring altruism could be propagated indirectly if altruistic individuals sacrificed personal reproduction to help their relatives reproduce. Helping relatives in this way can propagate the helper's own genes because the family shares alleles inherited from their ancestors.

We can quantify the average percentage of alleles that relatives are likely share by calculating their degree of relatedness **(Figure 55.18).** We start by considering half siblings who, by definition, share only one genetic parent. Half siblings share on average 25% of their alleles by inheritance from their shared parent, making their degree of relatedness 0.25. By contrast, full siblings, who share the same genetic mother *and* father, share 25% of their alleles through the mother and 25% of their alleles through the father, for a total, on average, of 25% + 25% = 50% of their alleles. In other words, the degree of relatedness for full siblings is 0.50. The degree of relatedness between a nephew or niece and an aunt or uncle is 0.25, and the degree of relatedness between first cousins is 0.125. Thus, individuals should be more likely to help close relatives because, by increasing a close relative's fitness, the individual is helping to propagate some of its own alleles.

If altruistic behavior reduces the reproductive success of an individual exhibiting that behavior, how could an allele that promotes altruistic behavior persist

Figure 55.17
The cost of living alone. A solitary olive baboon (*Papio anubis*) confronts a leopard (*Felis pardalis*) bravely but without much chance of survival.

John Dominis/Time & Life Pictures/Getty Images

Figure 55.18 Research Method

Calculating Degrees of Relatedness

PURPOSE: The kin-selection hypothesis suggests that the extent of altruistic behavior exhibited by one individual to another is directly proportional to the percentage of alleles that they share. The hypothesis therefore predicts that individuals are more likely to help close relatives because, by increasing a close relative's fitness, the individual is helping to propagate some of its own alleles. Researchers calculate the degree of relatedness between individuals to test this prediction.

PROTOCOL: To calculate the degree of relatedness between any two individuals, we first draw a family tree that shows all of the genetic links between them. The alleles of a parent are shuffled by recombination and independent assortment in the gametes they produce, so we can calculate only the average percentage of a parent's alleles that offspring are likely to share.

We start by considering *half* siblings, those who share only one genetic parent. Each sibling receives half of its alleles from its mother. Because a parent has only two alleles at each gene locus, the probability of sibling A getting a particular allele from its mother is 0.5 (decimal notation for 50%). Similarly, the probability of sibling B getting the same allele from its mother is also 0.5. Statistically, the probability that two independent events—in this case, the transfer of an allele to sibling A and the transfer of *the same* allele to sibling B—will both occur is the product of their separate probabilities. Thus, the likelihood that both siblings receive the same allele from their mother is $0.5 \times 0.5 = 0.25$.

Now consider two *full* siblings, who share the same genetic mother *and* father. They share 25% of their alleles through the mother *plus* 25% of their alleles through the father, for a total of 50% (half their alleles). In other words, the degree of relatedness for full siblings is 0.50.

INTERPRETING THE RESULTS: Each link drawn between a parent and an offspring or between full siblings indicates that those two individuals share, on average, 50% of their alleles. We can calculate the total relatedness between any two individuals by multiplying out the probabilities across all of the links between them. Thus, the degree of relatedness between a niece and an uncle is 0.25, and the degree of relatedness between first cousins is 0.125.

Half Siblings

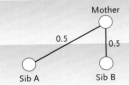

Relatedness = (0.5)(0.5) = 0.25

Full Siblings

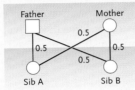

Relatedness
Through mother = (0.5)(0.5) = 0.25
Through father = (0.5)(0.5) = 0.25
Total relatedness = 0.25 + 0.25 = 0.5

First Cousins

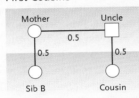

Relatedness = (0.5)(0.5)(0.5) = 0.125

or even increase in frequency in a population? The answer depends on the overall number of offspring that carry the allele in the next generation. Altruistic behavior may increase the survival of the altruist's relatives—and they may share the allele in question. If the altruistic behavior allows the assisted relatives to produce proportionately more offspring than the altruist might have produced without helping them, the allele for altruism can increase in frequency in the population. This form of natural selection is aptly called **kin selection.**

For example, suppose a male wolf helps his parents rear four pups to adulthood, pups that would have died without the extra assistance provided by the altruist. Because the pups are his siblings, they share 50% of his genes; thus, on average the helper wolf has created "by proxy" two ($0.50 \times 4 = 2$) copies of any allele that contributed to his altruistic behavior. The costs of his altruism must be measured against this indirect reproductive success. If he had abstained from altruism, the helper wolf might have raised, say, two surviving offspring of his own. Each of his offspring would

carry half of his alleles, preserving just one ($0.50 \times 2 = 1$) copy of a given allele. Under these hypothetical circumstances, reproducing on his own would have produced fewer copies of his alleles in the next generation than helping to raise his siblings.

Although our example of the altruistic wolf is hypothetical, biologists have observed sibling helpers in many bird and mammal species. The phenomenon is especially common among animals in which inexperienced young adults are unable to control sufficient resources to reproduce successfully on their own. Their altruistic behavior not only assists reproduction by their close relatives, but it may also provide useful practice for rearing their own future offspring.

Hamilton's kin-selection hypothesis explains altruistic behavior between closely related individuals, but behavioral biologists have also observed examples of altruism between nonrelatives. For example, the common vampire bat *(Desmodus rotundus),* which feeds on the blood of sleeping mammals, must consume a meal every 2 days to avoid starving to death. Bats that have consumed a large meal often share

a. Queen with sterile workers

b. Workers sharing food and passing pheromones

Figure 55.19
Life in a honeybee *(Apis mellifera)* colony. **(a)** A court of sterile worker daughters surrounds a queen bee, the only female of the colony that reproduces. **(b)** Worker bees routinely share food and transfer pheromones to one another.

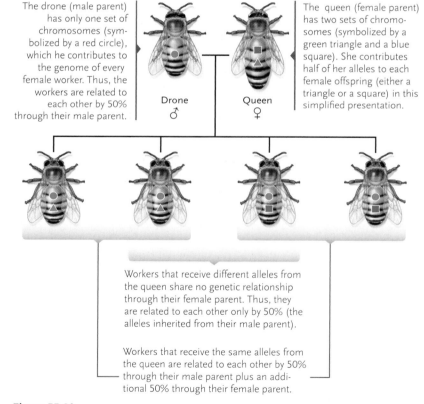

The drone (male parent) has only one set of chromosomes (symbolized by a red circle), which he contributes to the genome of every female worker. Thus, the workers are related to each other by 50% through their male parent.

The queen (female parent) has two sets of chromosomes (symbolized by a green triangle and a blue square). She contributes half of her alleles to each female offspring (either a triangle or a square) in this simplified presentation.

Drone ♂

Queen ♀

Workers that receive different alleles from the queen share no genetic relationship through their female parent. Thus, they are related to each other only by 50% (the alleles inherited from their male parent).

Workers that receive the same alleles from the queen are related to each other by 50% through their male parent plus an additional 50% through their female parent.

Figure 55.20
Haplodiploidy. The genetic system of eusocial insects produces full siblings that have an exceptionally high degree of relatedness. Although this simplified model ignores recombination between the queen's two sets of chromosomes, it demonstrates how half the workers are related to each other by 50% and half are related to each other by 100%. Thus, the average degree of relatedness among workers is 75%. Including recombination would complicate the illustration, but the conclusion would be the same.

their bounty with unrelated members of their group. Why would one bat share its resources with a non-relative? Robert Trivers, then of Harvard University, proposed that individuals will help nonrelatives if they are likely to return the favor in the future. Trivers called this form of altruistic behavior **reciprocal altruism**, because each member of the partnership can potentially benefit from the relationship. Trivers hypothesized that reciprocal altruism would be favored by natural selection as long as individuals that do *not* reciprocate—called "cheaters" by behavioral biologists—are denied future aid. Observations of vampire bats and some other animals have confirmed Trivers' hypothesis: when a vampire bat accepts a "blood donation" from another bat, but then refuses to share food that it has collected, the other bats refuse to share their food with it in the future.

An Unusual Genetic System May Explain Altruism in Eusocial Insects

Hamilton's insights lead to a critical prediction about the occurrence of self-sacrificing behavior: altruism should usually be directed to close relatives. The evidence from many animal species overwhelmingly supports this prediction, but some species of ants, bees, wasps, and termites, those known as eusocial insects, provide a truly remarkable example. In **eusocial** insects, thousands of related individuals—a large percentage of them sterile female workers—live and work together in a colony for the reproductive benefit of a single queen and her mate(s). The workers may even die in defense of their colonies. How did this self-sacrificing social behavior evolve, and why does it persist over time? The failure of altruistic workers to reproduce should doom any alleles that promote altruism to early extinction.

For example, in a honeybee *(Apis mellifera)* colony, which may contain 30,000 to 50,000 related individuals, the only fertile female is the queen bee; all of the workers are her daughters **(Figure 55.19)**. The queen's role in the colony is to reproduce. The workers perform all other tasks in maintaining the hive, from feeding the queen and her larvae to constructing new honeycomb and foraging for nectar and pollen. They also transfer food to one another and sometimes guard the entrance to the hive. Some pay the ultimate sacrifice when they sting intruders: this act of defense tears open the bee's abdomen, leaving the stinger and the poison sac behind in the intruder's skin, but killing the bee.

Why do bees and other eusocial insects devote their entire lives to helping their mother produce hundreds of thousands of eggs? One answer may lie in a genetic phenomenon called **haplodiploidy**, an unusual pattern of sex determination in these insects **(Figure 55.20)**. Like many other organisms, female bees are diploid, receiving one set of chromosomes from each parent. But male bees are haploid: they hatch from unfertilized eggs. When a queen bee mates with one drone (a male), all of the sperm he delivers are genetically identical because males have only one set of chromosomes. Thus, all workers inherit exactly the same set of alleles from their male parent, producing a 50% degree of relatedness among them. Like other diploid organisms, the workers are also related to each other by an average of 25% through their female parent. Adding these two components of relatedness, we see that workers are related to each other by an average of 75%, a higher degree of relatedness than they would have to any offspring they produced if they were fertile.

This extremely high degree of relatedness among the workers in some eusocial insect colonies may explain their exceptional level of cooperation. When Hamilton first worked out this explanation of social behavior in these insects, he suggested that the workers devote their lives to caring for their siblings—the queen's other offspring—because a few of those siblings, which carry 75% of the workers' alleles, may become future queens producing enormous numbers of offspring themselves.

Nonbreeding workers also exist in a mammalian species, the naked mole-rat *(Heterocephalus glaber),* a small, almost hairless animal that lives in underground colonies of 70 to 80 individuals in eastern Africa. As described in *Insights from the Molecular Revolution,* recent studies have shown that the highly cooperative individuals occupying a colony share an exceptionally high proportion of their alleles.

STUDY BREAK

1. What do the social behaviors of musk oxen and sawfly larvae have in common?
2. Which animals in a dominance hierarchy are most likely to reproduce?
3. Why might the genetic system of many eusocial insects promote altruistic behavior?

55.6 An Evolutionary View of Human Social Behavior

Evolutionary Analyses May Help to Explain Human Social Behavior

If we can analyze the evolutionary basis of the behavior of honeybees, naked mole-rats, and other animals, perhaps we can do the same for human behavior. According to Hamilton's kin selection hypothesis, we would expect human altruism toward nonrelatives to be rare. And it is true that *most* acts of human altruism are directed toward family members; huge sacrifices to help nonrelatives are relatively uncommon. But why, from an evolutionary perspective, do such charitable acts toward strangers occur at all?

Many behavioral biologists believe that reciprocal altruism can explain why humans have an evolved willingness to engage in low-cost acts of charity. Such behavior demonstrates their capacity for cooperation, and generosity is a socially approved trait that may confer benefits on those who exhibit it. This hypothesis yields the prediction that people who engage in charity will usually let others know about it. That prediction is supported by data showing that when organizers of blood drives offer small participation pins to donors, more people sign up to give blood.

Sometimes researchers employ an evolutionary perspective to study difficult or painful societal issues, such as the occurrence of child abuse within families. Evolutionary theory leads us to predict that family members should generally help, not harm, one another. Margo Wilson and Martin Daly of McMaster University wondered if child abuse might be more common in reconstituted families, those with stepparents who are *not* genetically related to all the children in their care. To test this hypothesis, they examined data on criminal child abuse within families, made available by the police department of a Canadian city. In this city, the chance that a young child would be subject to criminal abuse was 40 times higher for children living with one stepparent and one genetic parent than for children who lived with both genetic parents **(Figure 55.21).**

This example illustrates the sort of insights that an evolutionary analysis of human behavior can provide. Wilson and Daly are not justifying or excusing child abusers. Neither are they claiming that abusive stepparenting is evolutionarily adaptive. Instead, their point is that humans may have some genetic characteristic that makes it more difficult to invest in children that they know are not their own, particularly if they also care for their own genetic children. These results are not just academic. Although a large majority of stepparents cope well with the difficulties of their role, a few do not. Knowledge of familial circumstances under which child abuse is more likely to occur may allow us to provide social assistance that would prevent some children from being abused in the future.

In recent years, the application of evolutionary thinking to human behavior has produced research on all sorts of questions. Sometimes the questions are interesting or even profound. Why do some tightly knit ethnic groups discourage intermarriage with members of other groups? At other times the issues may seem frivolous. Why do men often find women with certain physical characteristics attractive? Although evolution-

Unadorned Truths about Naked Mole-Rat Workers

Naked mole-rats (*Heterocephalus glaber*) are sightless and essentially hairless burrowing mammals (see the accompanying photo) that live in mazes of subterranean tunnels in parts of Ethiopia, Somalia, and Kenya. Mole-rat colonies, which may include from 25 to several hundred individuals, contain a single "queen" and one to three males as the only breeding individuals. The remaining males and females are nonbreeding workers that, like the worker bees, ants, and termites of insect colonies, do all the labor: digging and defending the tunnels and caring for the queen and her mates.

One of the many unanswered questions about these colonial mammals is the genetic structure of a colony. Is close kinship one of the relationships underlying the altruistic behavior of the workers? In other words, do they cooperate because they are all brothers and sisters?

H. Kern Reeve and his colleagues at Cornell University investigated this question using molecular techniques resembling the DNA fingerprinting analysis often used to determine human kinship. The technique (see Section 18.2) depends on a group of repeated DNA sequences that vary to a greater or lesser extent among individuals; that is, they are *polymorphic*. No two individuals (except identical twins) are likely to have exactly the same combination of sequences. Brothers and sisters with the same parents have the most closely related sequences; as relationships become more distant, the differences in the sequences increase.

The researchers began their work by capturing mole-rats living in four colonies in Kenya. Individuals from the same colony were placed together in a system of artificial tunnels. Samples of DNA, taken from individuals that died naturally in the artificial colonies, were subjected to DNA fingerprinting analysis (see Section 18.2). First, the extracted DNA was fragmented by treating with a restriction endonuclease. The DNA fragments were separated by agarose gel electrophoresis, then transferred to a membrane filter by the Southern blot technique (see Figure 18.9). Next, the DNA fragments on the filter were hybridized independently with three radioactively labeled probes that identify three distinct groups of polymorphic sequences in the mole-rat DNA. The hybridization patterns were visualized by autoradiography. The pattern of bands, different for each individual (other than twins), is the DNA fingerprint.

The fingerprint of each mole-rat was compared with the fingerprints of other members of the same and other colonies. In the comparisons, bands that were the same in two individuals were scored as "hits." The number of hits was then analyzed to assign relatedness by noting which individuals shared the greatest number of bands.

The comparisons revealed that individuals in the same mole-rat colony were indeed closely related—they shared an unusually high number of bands, higher than human siblings and approaching the band similarity of identical twins. The number of bands shared between individuals of different colonies was significantly lower, but still higher than that noted between unrelated individuals of other vertebrate species. The close relatedness of even separate colonies, as the investigators point out, may be due to similar selection pressures or to recent common ancestry among colonies in the same geographical region.

On the basis of their results, the researchers propose that the close genetic relatedness among individuals in a colony, which is assumed to increase the degree of altruistic behavior, is one of two major factors underlying the evolution and maintenance of the nonbreeding worker caste in the colonies. The second major factor, they propose, is the chance of survival, which is greater for mole-rats remaining in colonies than for those that attempt to live and breed on their own.

Naked mole-rats (*Heterocephalus glaber*) live in colonies containing many workers that are effectively sterile.

ary hypotheses about the adaptive value of behavior can be tested, helping us to understand why we behave as we do, they should never be used to *justify* behavior that is harmful to other individuals. Understanding why we get along or fail to get along with each other and the ability to make moral judgments about our behavior are uniquely human characteristics that set us apart from other animals.

STUDY BREAK

1. How might evolutionary biologists explain altruistic behavior that people exhibit to non-relatives?
2. Why might stepparents provide fewer resources to their children than birth parents do?

Figure 55.21 Observational Research

An Evolutionary Analysis of Human Cruelty

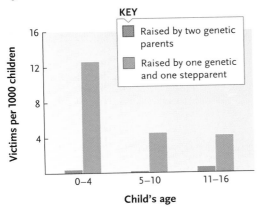

KEY
- Raised by two genetic parents
- Raised by one genetic and one stepparent

Victims per 1000 children (y-axis: 4, 8, 12, 16)

Child's age (x-axis: 0–4, 5–10, 11–16)

HYPOTHESIS: Wilson and Daly hypothesized that child abuse would be more common in families in which parents and children were not genetically related than in families in which parents raise their biological offspring.

PREDICTION: Stepparents abuse their stepchildren more frequently than birth parents abuse their biological children.

METHOD: The researchers analyzed data on criminal child abuse within families that had been collected by the police department of a large Canadian city.

RESULT: The data indicated that children living with one stepparent and one genetic parent were 40 times more likely to suffer criminal abuse than children living with two genetic parents.

CONCLUSION: Children raised by one genetic parent and one stepparent are significantly more likely to suffer abuse than those raised by two genetic parents.

UNANSWERED QUESTIONS

Who else is watching and listening?

Studies of communication have typically concentrated on the signaler and the intended receiver. But others are lurking in the background—eavesdroppers who are also attending to these signals. These third parties often use the signal to the signaler's detriment. Some flies, for example, listen for calling male crickets and deposit their larvae on the caller; the larvae eventually kill the cricket as they use him for a food source. Sometimes the signal that makes a male more attractive to females has the same effect on eavesdroppers. Male túngara frogs, for example, can add a "chuck" to the "whine" component of their call. These more complex calls make them more attractive to female frogs but also increase their risk of being captured by frog-eating bats or found by blood-sucking flies. Other animals circumvent this cruel bind by evolving signals in "private channels" to which intended receivers, but not predators, are privy. Ultraviolet signals in swordtail fishes and electrical signals in weakly electric fish are two examples. Increasingly, communication systems are being analyzed from the perspective of a complex communication network rather than the more simple two-way interactions.

Why do females prefer attractive mates?

One reason that females prefer attractive mates is that females can use the ornaments of males to judge their quality in terms of performance and survivorship. A superior male might provide better resources for the female and her offspring, or even pass on better genes to their young. Alternatively, male courtship traits must stand out against a sometimes chaotic background of environmental noise and signals from competing males. To stand out more than others, males might evolve signals that are better at stimulating the female's sensory, neural, and cognitive systems. If a species has evolved visual pigments that allow individuals to locate orange fruit, for example, males should evolve orange colors, because they will better stimulate the female's visual system. Such a scenario has been suggested for guppies. Researchers who study *sensory drive* try to understand how selection on sensory systems in one context, such as feeding, can influence functions in other contexts, such as mate choice.

To what degree is human behavior influenced by natural selection?

Cooperative and altruistic behavior sometimes evolve in animal societies. Since all societies face similar basic challenges, it is not surprising that cooperation can also be important in human societies. Strong evidence suggests that cooperation has evolved under selection in animal societies, but does logic demand that it has evolved under selection in humans as well? If not, what data would provide strong evidence for the evolution of human social behavior?

The field of evolutionary psychology, which poses these questions, is constrained because the invasive experimental approaches that have been so successful in animal biology cannot be applied to humans. Here the cross-cultural, comparative approach has made some important advances. These questions are being pursued by researchers in biology and psychology as well as anthropology and sociology. This mix of researchers with different approaches guarantees excitement and controversy about these very basic questions that ask why we are who we are.

Michael J. Ryan is the Clark Hubbs Regents Professor in Zoology at the University of Texas at Austin, where he studies the evolution and function of animal behavior. Most of his work has addressed sexual selection and communication in frogs and fish. To learn more about his research, go to http://www.sbs.utexas.edu/ryan/.

Review

Go to **Thomson**NOW™ at www.thomsonedu.com/login to access quizzing, animations, exercises, articles, and personalized homework help.

55.1 Migration and Wayfinding

- Some animals—including some arthropods, fishes, birds, and mammals—migrate seasonally, traveling from their birthplace to a distant locality and back again (Figures 55.2 and 55.3).

- Migrating animals may use piloting, compass orientation, or navigation to find their way. In piloting, animals use familiar landmarks to guide their journey (Figure 55.4). In compass orientation, animals use the position of the sun or stars, polarized light, or the Earth's magnetic field as a guide (Figure 55.5). In navigation, animals use mental maps of their position to find their destination.

- Biologists frequently interpret migratory behavior as an adaptive response to changing food supplies (Figure 55.6). Some animals occupy northern habitats when food is plentiful during the spring and summer breeding season. They generally head south to seasonally more productive habitats before the onset of winter.

55.2 Habitat Selection and Territoriality

- Animals use multiple criteria when selecting their habitats.

- Kineses and taxes help animals orient to appropriate portions of the habitats they occupy.

- Habitat selection often has a largely genetic basis, but learning and prior experience influence habitat selection in some species (Figure 55.7).

- Animals may establish and defend territories to gain exclusive use of defendable resources that are in short supply. The costs of territoriality include the time and energy devoted to territory defense and the risk of injury from fights or exposure to predators.

55.3 The Evolution of Communication

- Animal communication occurs between a signaler, which sends a message, and a signal receiver, which receives and interprets the message.

- Animals communicate using acoustical, visual, chemical, tactile, or electrical signals (Figures 55.8–55.10). Each sensory channel provides specific advantages. Animals may use more than one channel simultaneously.

- Honeybees use a combination of tactile, acoustical, and chemical channels to share information about the location of food sources (Figure 55.11).

Animation: Honeybee dances

55.4 The Evolution of Reproductive Behavior and Mating Systems

- Males and females exhibit different reproductive strategies. Males can increase their reproductive success by inseminating the eggs of many females. Females generally seek mates that provide successful alleles to offspring, have access to abundant resources, or help care for young.

- Males often compete for access to females (Figure 55.12). Sexual selection has produced elaborate structures that males use for displays and for aggressive interactions with other males (Figures 55.13 and 55.14). Females may choose to mate with males that have showy structures and great stamina, which may function as signs that they possess successful alleles.

- The type of mating system a species uses is tied to its pattern of territoriality and the amount of parental care the male parent provides.

55.5 The Evolution of Social Behavior

- Social interactions between individuals of the same species provide both benefits and costs. Group living may provide better protection from predators, more efficient feeding, and communal care of young (Figures 55.15 and 55.17). The costs of living in a group include increased competition for scarce resources and an increase in the spread of contagious diseases (Figure 55.16).

- Dominance hierarchies are highly structured societies in which some individuals have high status and first access to resources.

- Altruistic behavior appears to contradict a basic premise of Darwinian evolutionary theory, because altruistic individuals sacrifice their own fitness for the benefit of others. However, individuals generally display altruistic behavior to close relatives that share some of their alleles (Figures 55.18).

- An unusual mechanism of sex determination, haplodiploidy, makes the workers in some eusocial insect colonies more closely related to each other than siblings are in most species (Figures 55.19 and 55.20). Haplodiploidy may have fostered the evolution of highly altruistic behavior.

Animation: Sawfly defense

55.6 An Evolutionary View of Human Social Behavior

- Although humans are more likely to provide assistance to close relatives than to nonrelatives, acts of charity to strangers are common, especially if the altruist can advertise the generosity.

- An analysis of child abuse suggests that humans are more likely to abuse children to whom they are not genetically related than they are to abuse their biological children (Figure 55.21).

Questions

Self-Test Questions

1. Which of the following statements about animal migration is true?
   a. Piloting animals use the position of the sun to acquire information about their direction of travel.
   b. Animals migrating by compass orientation use mental maps of their position in space.
   c. Navigating animals use familiar landmarks to guide their journey.
   d. Navigating animals use a compass and a mental map of their position to reach a destination.
   e. Most migrating birds use olfactory cues to return to the place where they hatched from eggs.

2. In Marler and Moore's experiment with Jarrow's spiny lizard, what evidence from males that had received testosterone implants suggested that engaging in territorial behavior carries a heavy cost?
   a. They had to consume more water than control males.
   b. They mated with fewer females than control males.

c. They ate more frequently than control males.
d. They had higher death rates than control males.
e. They weighed more than control males.

3. Which signal type would provide the fastest communication between bats flying in a dark forest?
   a. chemical signals
   b. acoustical signals
   c. visual signals
   d. tactile signals
   e. electrical signals

4. Squashing an ant on a picnic blanket often attracts many other ants to its "funeral." What kind of signal did squashing the ant likely produce?
   a. an electrical signal
   b. a visual signal
   c. an acoustical signal
   d. a chemical signal
   e. a tactile signal

5. Which of the following behaviors might have been produced by sexual selection?
   a. A male frog calls loudly and clearly from a pond during the breeding season.
   b. A young male goat bleats plaintively when left by its mother.
   c. A hen clucks to call its chicks closer when a predator approaches.
   d. A female lion ignores the sexual advances of a young male.
   e. A male dog is attracted to the odor of a female dog.

6. In comparison to males, the females of many animal species:
   a. compete for mates.
   b. choose mates that are well camouflaged in their habitats.
   c. choose to mate with many partners.
   d. are always monogamous.
   e. choose their mates carefully.

7. Social behavior:
   a. is exhibited *only* by animals that live in groups with close relatives.
   b. cannot evolve in animals that maintain territories.
   c. evolved because group living provides benefits to individuals in the group.
   d. is never observed in insects and other invertebrate animals.
   e. can be explained only by the hypothesis of kin selection.

8. Altruism is a behavior that:
   a. cannot evolve.
   b. has been observed only in insects.
   c. increases the number of offspring an individual produces.
   d. can indirectly spread the altruist's alleles.
   e. can evolve only in animals with a haplodiploid genetic system.

9. The degree of relatedness between a parent and its biological offspring:
   a. is the same as that between brother and sister.
   b. is less than that between brother and sister.
   c. depends on how many siblings the parent has.
   d. promotes an individual's reproductive success.
   e. is the same as between first cousins.

10. The tendency for humans to be charitable to perfect strangers can be explained by the hypothesis of:
    a. sexual selection.
    b. kin selection.
    c. reciprocal altruism.
    d. polyandry.
    e. navigation.

Questions for Discussion

1. In Chapter 53, you learned about some of the environmental changes associated with global warming. What effects might global warming have on animal species that undertake seasonal migrations?

2. The yellow-rumped whippersnapper, an imaginary species of songbird, always established breeding territories in forests where trees are interspersed among many small ponds. Design an experiment to determine what features of the environment this species uses to select its habitat.

3. Although females provide parental care far more often than males in the animal kingdom as a whole, exceptions exist, especially among birds and fishes. Develop three evolutionary hypotheses to explain why male birds are so likely to involve themselves in caring for their broods.

Experimental Analysis

You discover that a particular butterfly species almost always lives in open meadows and almost never lives in nearby shaded forests. Design an experiment to test whether or not habitat selection by this species is adaptive.

Evolution Link

African honeyguides (family Indicatoridae) are birds that call to humans and other mammals, leading them to honeybee colonies in woodlands. The mammals then open the hives to extract the honey, and the honeyguide feeds on the beeswax. How could a communication system between two species evolve?

How Would You Vote?

Africanized bees are slowly expanding their range in North America. Some researchers think the more we know about them, the better we will be able to protect ourselves. Should we fund more research into the genetic basis of their behavior? Go to www.thomsonedu.com/login to investigate both sides of the issue and then vote.

Appendix A
Answers

Chapter 19

Study Break 19.1
1. Buffon did not understand how "anatomically perfect" animals could have useless structures.
2. Lamarck proposed that all species change through time, the changes are inherited by the next generation, the changes arise in response to environmental conditions, and specific mechanisms caused the changes.
3. If the geological features that are seen on Earth today were produced by the very slow processes seen today, acting over long periods of time, they must have taken more than 6000 years.

Study Break 19.2
1. Darwin observed that living organisms often resemble fossils found in the same area; that organisms found on South America resembled one another, even if they occupied different environments; and that many species found on the Galápagos Islands resembled species from the South American mainland.
2. Darwin realized that the effects of competition for resources in nature were similar to the action of a plant or animal breeder who used only certain individuals as parents for the next generation.
3. Darwin's theory relied on physical explanations for the origin of biodiversity; he recognized that evolutionary change takes place within a population rather than in individuals, he recognized that natural selection is a multistage process, and he emphasized the importance of environmental conditions to the process of natural selection.

Study Break 19.3
1. Two problems that slowed the acceptance of Darwin's theory are that Mendel's genetic studies focused on simple traits and that these traits often changed in just a few generations.
2. Microevolution refers to small, genetically based changes within populations; macroevolution refers to large-scale changes in patterns of biodiversity.
3. Evidence for evolution comes from studies of adaptation, the fossil record, historical biogeography, comparative morphology, comparative embryology, and comparative molecular biology.

Self-Test Questions
1. c 2. e 3. d 4. b 5. c 6. b 7. a 8. b
9. d 10. d

Chapter 20

Study Break 20.1
1. The variation in skunks is qualitative.
2. The researchers used artificial selection to change the activity levels of the mice.
3. Genetic variation, differing environmental effects on individuals, and interactions between genes and the environment affect phenotypic variation in a population.

Study Break 20.2
1. Genotype frequencies specify how alleles are combined in individuals, and allele frequencies specify how common the alleles are.
2. The Hardy-Weinberg model identifies the conditions under which evolution will *not* occur.
3. If genotype frequencies are not already in equilibrium, they will stop changing after one generation of random mating.

Study Break 20.3
1. Mutation and gene flow tend to increase genetic variation within populations, and natural selection and genetic drift tend to decrease it.
2. Stabilizing selection increases the representation of the average phenotype in a population.
3. Sexual selection, like directional selection, favors extreme phenotypes.

Study Break 20.4
1. Diploidy protects harmful recessive alleles because dominant alleles mask their effects in heterozygotes.
2. A balanced polymorphism is one in which two or more phenotypes are maintained in fairly stable proportions over many generations.
3. The sickle-cell allele is rare in Northern Europe because, in the absence of the malarial parasite, it confers no advantage on individuals that carry it.

Study Break 20.5
1. The adaptive value of a trait can be evaluated by comparing closely related species that live in different environments.
2. Natural selection preserves traits that were useful when the organisms subject to selection were alive and reproducing.

Self-Test Questions
1. c 2. b 3. c 4. d 5. b 6. e 7. a 8. b
9. c 10. d

Chapter 21

Study Break 21.1
1. The morphological species concept defines species based on morphological differences between them. The biological species concept defines species as populations that can successfully interbreed under natural conditions.
2. Clinal variation is a pattern of smooth variation along a geographical gradient.

Study Break 21.2
1. Prezygotic isolating mechanisms either prevent individuals of different species from mating or prevent sperm from one species from fertilizing the eggs of another. Postzygotic isolating mechanisms limit the survivorship or reproductive capability of hybrid individuals.
2. The scenario illustrates a behavioral isolating mechanism.

Study Break 21.3
1. In the first stage of allopatric speciation, populations become geographically separated. In the second stage, they become reproductively isolated.
2. Some populations of bent grass survive better on unpolluted soil, whereas others survive better on polluted soil.
3. Insects from different host races spend most of their time on different host plant species. Thus, they rarely encounter each other and would be unlikely to mate under natural conditions.

Study Break 21.4
1. Natural selection cannot promote reproduction isolation in allopatric populations directly, but it can lead to genetic divergence that results in reproductive isolation.
2. Polyploidy has frequently led to speciation in flowering plants.

Self-Test Questions
1. a 2. e 3. c 4. d 5. b 6. e 7. b 8. c
9. a 10. b

Chapter 22

Study Break 22.1
1. Hard parts, such as the shells or bones of animals, are the materials most likely to fossilize.
2. The fossil record provides an incomplete portrait of life in the past because not all organisms are equally likely to form fossils; fossils do not form in all

types of habitats; and fossils are often destroyed by geological processes and erosion.

3. The fossil record provides information about the morphology of ancient organisms; how structures changed over time; and the proliferation and extinction of evolutionary lineages. It also offers indirect evidence about the behavior, ecology, and physiology of organisms that lived in the past.

Study Break 22.2
1. Continental drift caused large-scale geographical separation of populations and lineages that subsequently evolved in isolation.
2. Distantly related species that live in widely separated parts of the world may resemble each other because convergent evolution fosters similar adaptations to the environments they occupy.

Study Break 22.3
1. The horse lineage was highly branched, and it included some species that were larger and some that were smaller than their ancestors.
2. The gradualist hypothesis predicts slow, continuous morphological changes in a lineage. The punctuated equilibrium hypothesis predicts that morphological changes occur rapidly as new species form and that most species experience little morphological change for long periods of time.

Study Break 22.4
1. In general, the average size of organisms has increased since life first appeared.
2. The processes that can produce evolutionary novelties include preadaptation, the differential growth of body parts (allometry), and changes in the timing of developmental events (heterochrony).

Study Break 22.5
1. A population of organisms may occupy a new adaptive zone after the evolution of a key morphological innovation that allows it to use the environment in a unique way or after a once-successful group of organisms declines.
2. The mass extinction at the end of the Cretaceous period took place over tens of thousands of years.
3. The first major adaptive radiation of animals took place in the Cambrian era.

Study Break 22.6
1. Similar developmental control genes are present in a wide variety of animals, plants, fungi, and prokaryotes. Their widespread distribution suggests that they were present in the ancestor of all these organisms and have been conserved through countless generations.
2. The *Pitx1* gene is expressed in fin buds that later produce spines, and it is not expressed in those that fail to produce spines.

Chapter 23

Study Break 23.1
1. A phylogenetic tree is a formal hypothesis about the evolutionary relationships among species. A classification is an arrangement of organisms into hierarchical groups that reflect their relatedness.
2. A phylogenetic tree and classification allow biologists to conduct research on just one species or a group of related species that share a genetic history.

Study Break 23.2
1. The system of binomial nomenclature avoids ambiguity in the naming of species because it assigns a unique two-part name to each species.
2. The taxonomic category immediately above the family is the order. The category immediately below family is the genus.

Study Break 23.3
1. Morphological traits are often useful for tracing the evolutionary relationships within a group of organisms because they can be observed and measured in fossils as well as in living organisms.
2. Prezygotic isolating mechanisms are often useful characters in systematic studies of animals because they are the characteristics that the animals use to identify the species of a potential mate.

Study Break 23.4
1. Systematists use homologous characters in their analyses because similarities in homologous characters indicate genetic relatedness and shared ancestry.
2. Outgroup comparison is a technique that compares the group under study to more distantly related organisms to identify the ancestral and derived versions of characters.

Study Break 23.5
1. A monophyletic taxon contains an ancestor and all of its descendants. A polyphyletic taxon includes species from different evolutionary lineages.
2. The traditional definition of Reptilia (including turtles, lizards and snakes, and crocodilians) is paraphyletic because it includes only some descendants of ancestral Sauropsida. It does not include birds, which are part of the sauropsid lineage.
3. In a cladistic analysis, organisms that share derived homologous characters are grouped together.

Study Break 23.6
1. Molecular characters provide several advantages in systematic analyses: (1) they provide abundant data; (2) molecular sequences can be compared between distantly related organisms that share no organismal characteristics or between closely related species with only minor morphological differences; and (3) proteins and nucleic acids are not directly affected by developmental or environmental factors that cause nongenetic morphological variation.
2. Mutations in some types of DNA appear to arise at a relatively constant rate. Thus, differences in the DNA sequences of two species can index their time of divergence. Large differences imply divergence in the distant past, whereas small differences suggest a more recent common ancestor.
3. Phylogenetic analyses of prokaryotes based on morphological data were not very successful because prokaryotes do not have many morphological features. Analyses based on molecular sequence data have been more successful because researchers can identify many molecular differences among the lineages of prokaryote.

Chapter 24

Study Break 24.1
1. Only in a reducing atmosphere can amino acids be produced from simpler chemicals and energy. Without amino acids there can be no life. Thus, any theory for the origin of life must consider the necessity of reducing conditions.
2. Current thinking is that early Earth's atmosphere per se was not reducing; hence, scientists have looked for localized regions where conditions were of a reducing nature, such as ocean floor hydrothermal vents. Another hypothesis is that some amino acids had an extraterrestrial origin.

Study Break 24.2
Lipid bilayers in the form of membranes are present in all present-day cellular organisms. Therefore, it is more intuitive to conceive of a path from a simple lipid bilayer system to a membrane-bound cell than it is to conceive of a transition of a clay-based system to a cellular system.

Study Break 24.3
The basic tenet of the theory of endosymbiont origins for mitochondria and chloroplasts is that organelles such as mitochondria and chloroplasts originated from symbiotic relationships between two prokaryotic organisms. An anaerobic prokaryotic is proposed to have ingested an aerobic prokaryote, which persisted in the cytoplasm, continuing to respire aerobically. A gradual process of mutual

adaptation transformed the cytoplasmic aerobes into mitochondria.

The same basic mechanism is believed to have led to the appearance of membrane-bound plastids (including chloroplasts) at a later time. In this case, nonphotosynthetic aerobic cells with mitochondria are proposed to have ingested photosynthetic prokaryotes that were perhaps similar to present-day cyanobacteria. Again, through mutual adaptation, the photosynthetic prokaryotes changed into plastids.

Self-Test Questions
1. d 2. e 3. d 4. a 5. b 6. b 7. c 8. e
9. c 10. e

Chapter 25

Study Break 25.1

1. Prokaryotes have no major cytoplasmic organelles equivalent to the endoplasmic reticulum, Golgi complex, mitochondria, or chloroplasts of eukaryotes, nor does it have lysosomes.

   The genetic material of a prokaryote generally is a single, circular DNA molecule localized in a non-membrane-bound central region of the cell called a nucleoid. By contrast, the genetic material of a eukaryote is distributed among a number of linear chromosomes, which consist of DNA complexed with basic proteins known as histones.

   Most prokaryotes are surrounded by a cell wall located outside the plasma membrane. Animal cells do not have a cell wall, but plants, fungi, and some other eukaryotes do. The compositions of the eukaryotic cell walls are chemically different from bacterial cell walls.

2. A chemoheterotroph oxidizes organic molecules as its energy source and obtains carbon in organic form. A photoautotroph uses light as its energy source and carbon dioxide as its carbon source.

3. Obligate anaerobes are poisoned by oxygen. They survive either by fermentation or by a form of respiration in which inorganic molecules are used as final electron acceptors. Facultative anaerobes use oxygen when it is present but live by fermentation when conditions are anaerobe.

4. Nitrogen fixation is the reduction of atmospheric nitrogen (N_2) to ammonia (NH_3). Nitrification is the conversion of ammonium (NH_4^+) to nitrate (NO_3^-).

   Nitrogen fixation, an exclusively prokaryotic process, is the only means of replenishing the nitrogen sources used by most microorganisms, and by all animals and plants.

5. A biofilm is a complex aggregation of microorganisms, many or all of them prokaryotes, attached to a surface. Biofilms are used in a variety of beneficial applications, including bioremediation of toxic organic chemicals contaminating the groundwater. On the other hand, biofilms can have adverse effects on human health. For instance, biofilms can result in antibiotic-resistant infections if they adhere to surgical materials, such as catheters and implants. Other beneficial and detrimental examples are in the chapter text.

Study Break 25.2

1. Comparing DNA sequences, RNA sequences, and protein sequences are the main methods used to classify prokaryotes. As they become available, genomic sequences are also being used for evolutionary comparisons.

   Ribosomal RNA sequences have been used extensively for the classification of prokaryotes. Of particular value to evolutionary studies are rRNA sequences. The amount of sequence divergence gives researchers an estimate of how much time has passed since any two species shared the same ancestor. Turning this around, an investigator uses the differences between rRNA sequences as an indicator of evolutionary relatedness.

2. The hypothesized evolutionary ancestor of present-day Proteobacteria is a purple, photosynthesizing bacterium.

3. Photosynthetic Proteobacteria carry out photosynthesis as either photoautotrophs (the purple sulfur bacteria) or photoheterotrophs (the purple nonsulfur bacteria). The photosynthesis process used does not use water as an electron donor and does not release oxygen as a by-product of photosynthesis. Their photosynthetic pigment is a type of chlorophyll distinct from that of plants.

   Cyanobacteria are photoautotrophs that carry out photosynthesis by the same pathways as eukaryotic algae and plants. They use the same chlorophyll as plants are their main photosynthetic pigment, and release oxygen as a by-product of photosynthesis.

4. An exotoxin is a toxic protein that leaks from or is secreted from the bacterium that makes it. An endotoxin is a normal lipopolysaccharide component of the outer membrane of Gram-negative bacteria; it is released when bacteria die and lyse. An exoenzyme is an enzymatic protein that is released from cells.

   An exotoxin interferes with biochemical processes of body cells. An endotoxin overstimulates the immune system, often causing inflammation. Depending on the bacterium, the endotoxin release has different effects, that may include organ failure and death. An exoenzyme digests plasma membranes, causing cells of the infected host to rupture and die. Exoenzymes may also digest extracellular materials and red and white blood cells.

Study Break 25.3

1. See Table 25.1 for a comparison of the properties of organisms in each of the three domains. The classification of Archaea as a distinct domain was based on comparisons of DNA and rRNA sequences. Unique to Archaea are chemical features of the plasma membrane and cell wall.

2. A methanogen lives in reducing environments, generating energy by converting substrates such as carbon dioxide, hydrogen gas, methanol or acetate into methane gas. All known methogens belong to the Euryarchaeota.

3. Extreme halophilic Archaea live in high-salt environments, requiring at least 1.5 M NaCl in order to live. Most of these organisms are aerobic chemoheterotrophs, obtaining energy from sugars, alcohols, and amino acids using pathways similar to those of bacteria. All known extreme halophilic Archaea belong to the Euryarchaeota.

4. Extreme thermophiles live in extremely hot environments such as thermal hot springs and hydrothermal ocean floor vents. Psychrophiles grow optimally in temperatures in the range -10 to $-20°C$, such as in the Antarctic and Arctic oceans.

Study Break 25.4

A virus is a nonliving entity that consists of genetic material (DNA or RNA) surrounded by a layer of protein (the coat or capsid). When a virus enters a cell, the virus's genetic material directs the replication of its genetic material and the production of progeny viruses. In other words, a virus depends upon a host cell for its life cycle, while directing that life cycle through the action of its own genes.

Viroids are infective plant pathogens consisting only of RNA.

A prion is an infectious agent capable of replication, but containing no nucleic acid molecule. Consisting only of a protein molecule, prions invade nerve cells of a variety of animals, causing fatal degeneration of the nervous system. The prion protein is an misfolded form of a normal cellular protein that is encoded by a nuclear gene. The misfolded prion protein typically forms aggregates, which in part are responsible for their adverse effects. The prion protein "replicates" by converting normal proteins to misfolded prion proteins.

Self-Test Questions
1. c 2. d 3. a 4. b 5. e 6. a 7. d 8. b
9. e 10. a

Chapter 26

Study Break 26.1

A protist is distinguished from a prokaryote by having typical eukaryotic cell features like a nuclear envelope surrounding its genetic material, and cell organelles such as mitochondria (in most protists), chloroplasts (in some protists), endoplasmic reticulum, Golgi complex, and so on.

Distinguishing protists from fungi, animals, and plants is more blurry. Fungi are nonmotile at all stages of their life cycles, while most protists are motile or have motile stages in their life cycles. Cell wall structure is also different from fungi, and from plants. Protists differ from both animals and plants by lacking highly differentiated structures and by not having complex developmental stages. Collagen, the extracellular support protein of animals, is absent in protists.

Study Break 26.2

1. Excavates' nuclear genomes contain genes that are of mitochondrial origin, arguing that they once had mitochondria.
2. The chloroplast will have two membranes: one derived from the plasma membrane of the engulfing eukaryote and the other from the plasma membrane of the cyanobacterium.

Self-Test Questions

1. e 2. d 3. d 4. b 5. b 6. c 7. b 8. a
9. b 10. c

Chapter 27

Study Break 27.1

1. Evolution of a root system gave land plants access to minerals and water in soil and provided physical support for aerial parts. The evolving shoot system of land plants, including lignified tissues in stems, allowed vascular plants to grow taller and stay erect, thereby gaining better access to sunlight for photosynthesis. Reproductive structures borne on aerial stems (such as flowers) might serve as platforms for more efficient dispersal of spores from the parent plant. Vascular tissues were innovations for distributing water (xylem) and sugars (phloem) through the plant body.
2. Homosporous plants produce a single type of sexual spore and are in effect bisexual, with each gametophyte capable of producing both sperm and eggs. Heterosporous species, including angiosperms and gymnosperms, produce two types of spores, which develop into sexually different gametophytes that produce either sperm or eggs. Plant scientists associate the evolution of heterospory with several key reproductive innovations in land plant evolution, including the protection of male gametes inside pollen grains and the protection of plant embryos inside seeds.

Study Break 27.2

1. Like aquatic plants, bryophytes produce flagellated sperm that must swim through water to reach eggs, and they lack a complex vascular system (although some have a primitive type of conducting tissue). Bryophytes do have parts that are rootlike, stemlike, and leaflike, although the "roots" are rhizoids, and bryophyte "stems" and "leaves" did not evolve from the same structures that vascular plant stems and leaves did. Sporophytes of some species have a water-conserving cuticle and stomata. Like most plants, bryophytes also have both sexual and asexual reproductive modes.
2. In general, mosses are the bryophytes that most closely resemble vascular plants. Some species produce structurally complex gametophytes that have a central strand of primitive water-conducting tissue that resembles the xylem of vascular plants, and in a few species the water-conducting cells are surrounded by sugar-conducting tissue resembling the phloem of vascular plants.

Study Break 27.3

1. In bryophytes, the gametophyte is much larger than the sporophyte and obtains its nutrition independently. The comparatively tiny sporophyte remains attached to the gametophyte and depends on the gametophyte for much of its nutrition. In modern lycophytes (club mosses and their close relatives), the gametophyte is free-living—though it is nourished by mycorrhizae instead of carrying out photosynthesis—and it is smaller than the sporophyte, which is a photosynthetic autotroph.
2. Ferns leaves often take the form of feathery fronds, and roots extend from underground stems called rhizomes. Whisk ferns lack true leaves and roots; instead, small leaflike scales dot an upright, green, branching stem, which arises from a horizontal rhizome system anchored by rhizoids. Horsetail sporophytes typically have underground rhizomes and roots that anchor the rhizome to the soil. The scalelike leaves are arranged in whorls about a photosynthetic stem.
3. In horsetails, the sporangia that produce spores are borne in strobili, and spores are carried away from the plant by air currents. In ferns, sporangia are produced on the lower surface or margin of leaves and spores are forcefully dispersed from the parent plant when contraction of a beltlike annulus rips open the sporangium and ejects the spores.

Study Break 27.4

1. The four major reproductive adaptations that evolved in gymnosperms include pollen and the ovule, both of which shelter spores; pollination rather than dispersal of swimming sperm; and the seed as a "package" that protects and often nourishes the embryo.
2. The three basic parts of a seed are the embryo sporophyte, endosperm, and outer seed coat. The endosperm nourishes the embryo sporophyte, and the seed coat protects it.
3. Features that make conifer sporophytes structurally more complex than other gymnosperms include anatomically complex needlelike or scalelike leaves that are adapted to aridity and the production of resins as metabolic by-products.

Study Break 27.5

1. The lack of fossil early angiosperms, including obvious transitional forms, has made it difficult to trace a clear evolutionary path for flowering plants. As result plant scientists have proposed several, often conflicting, classification schemes for angiosperms.
2. Seeds leaves (cotyledons) and pollen morphology are two major features used to distinguish monocots and eudicots. Monocots have a single seed leaf and pollen grain with a single groove. Eudicots have two seed leaves and pollen grains with three grooves. Monocots (such as grasses, lilies, and palms) also generally have fibrous root systems, leaves with parallel veins, flower parts in multiples of three, and scattered vascular bundles in stems. Eudicots (most flowering trees and shrubs, roses, sunflowers, beans) usually have netlike leaf venation, a primary taproot, flower parts in fours or fives, and vascular tissues arranged in a ring.
3. Adaptations that have contributed to the evolutionary success of angiosperms include vascular tissue modifications that make transport of water and nutrients more efficient; double fertilization, which results in enhanced nutrition (endosperm) for embryos; physical protection of embryos within ovaries and seeds; and coevolution with animal pollinators, which increases the likelihood that pollination will occur.

Self-Test Questions

1. c 2. b 3. a 4. d 5. d 6. e 7. c 8. c
9. e 10. d

Chapter 28

Study Break 28.1

1. Some fungi are multicellular while others, the yeasts, are single cells. (Some species alternate between these two forms at different life cycle stages.) The cells of all fungi are surrounded by a hardened wall; in most cases the hardener is the polysaccharide chitin. The body of a multicellular fungus consists of a dense mesh of filaments called hyphae, which in some groups are separated into cell-like compartments by cross walls (septa). Aggregations of hyphae are the structural foundation for all other parts that develop as part of a multicellular fungus. For example, in some species modified hyphae form rhizoids that anchor the fungus to its substrate.
2. Fungal spores are microscopic, usually nonmotile reproductive cells in which haploid nuclei are surrounded by a tough outer wall. They are produced sexually or asexually. Sexual spores are produced by

genetically different parent fungi and may unite in a sexual process that gives rise to a diploid life stage; asexual spores are genetically identical to the parent fungus and may give rise to a new, haploid individual.

3. Many fungal species have a life cycle stage called a dikaryon, which contains two haploid nuclei (a condition expressed as $n + n$). A dikaryon forms as the result of plasmogamy, a sexual stage in which the cytoplasms of two genetically different partners fuse. This fusion ensures genetic diversity in new individuals. At some point after a dikaryon forms, the nuclei fuse (karyogamy) to form a short-lived zygote. Meiosis in the zygote produces haploid nuclei that become packaged into sexual spores.

Study Break 28.2

1. The main phyla of fungi are the Chytridiomycota, Zygomycota, Glomeromycota, Ascomycota, and Basidiomycota. Chytrids are the only fungi that produce motile, flagellated spores. Zygomycetes often reproduce asexually, but sometimes reproduce sexually by way of hyphae that occur in + and − mating strains; haploid nuclei in the hyphae function as gametes. Following plasmogamy, further development produces zygospores in which karyogamy gives rise to diploid zygotes ($2n$ nuclei), which then undergo meiosis as sexual spores form. Glomeromycetes reproduce asexually, by way of spores that form at the tips of hyphae. In ascomycetes, chains of asexual spores called conidia, each containing a haploid nucleus, develop during asexual reproduction. Ascomycetes produce haploid sexual spores in pouchlike cells called asci. Most basidiomycetes reproduce only sexually: Club-shaped basidia develop on a basidiocarp (for example, a "mushroom") and bear sexual spores on their outer surface. When dispersed, the spores may germinate and give rise to a haploid mycelium. Cytoplasmic fusion may occur between hyphae of two compatible mating strains, producing a dikaryotic mycelium from which basidiocarps may grow. Microsporidia, single-celled parasites that may be related to zygomycetes, resemble spores but lack mitochondria. Fungi for which no sexual life stage has been identified are placed in a convenience grouping called "conidial fungi."

2. Anatomically, the simplest fungi are chytrids, which are microscopic, and zygomycetes, which have aseptate hyphae. Ascomycetes, basidiomycetes, and glomeromycetes all form septate (walled) hyphae.

3. Most chytrids are aquatic, but some are parasites on insects, plants, and some animals. Other are symbiotic partners in the gut of cattle and some other herbivores. Many zygomycetes are saprobes in soil, feeding on plant detritus. Their metabolic activities release mineral nutrients that plant roots can take up. Some zygomycetes are parasites of insects or spoil stored grains, bread, fruits, and vegetables such as sweet potatoes. Others are used in manufacturing products such as industrial pigments and pharmaceuticals. Many more are highly destructive plant pathogens and several can be human pathogens, causing athlete's foot, ringworm infections, and more serious illnesses. The pink bread mold *Neurospora crassa* has been crucial in genetic research.

Some basidiomycetes participate in vital mutualistic associations (mycorrhizae) with the roots of forest trees. Others produce prized edible mushrooms. Rusts and smuts are parasites that cause serious diseases in wheat, rice, and other plants.

Glomeromycetes are all specialized to form mycorrhizae with plant roots.

Study Break 28.3

1. A lichen is a communal life form representing a symbiosis between a photosynthetic green alga or species of cyanobacteria (the photobiont) and a nonphotosynthetic fungus (the mycobiont). The algal cells supply the lichen's carbohydrates, most of which are absorbed by the fungus. In some cases, the alga is protected from desiccation or some other environmental threat.

2. A mycorrhiza is a symbiotic association between a fungus and plant roots. The fungal hyphae make mineral ions and sometimes water available to the plant's roots, and in exchange the fungus absorbs carbohydrates, amino acids, and possibly other growth-enhancing substances provided by the plant. Mycorrhizae greatly enhance the plant's ability to extract various nutrients, especially phosphorus and nitrogen, from soil, and they are crucial to the survival of many plant species.

3. In endomycorrhizae the hyphae of a fungus (typically a glomeromycete) enter plant roots, where their tips branch into clusters called arbuscules where exchanges of nutrients take place. In ectomycorrhizae the fungal partners are basidiomycetes. Their hyphae surround plant roots but do not penetrate them.

Self-Test Questions

1. b 2. d 3. e 4. b 5. a 6. d 7. a 8. e
9. a 10. d

Chapter 29

Study Break 29.1

1. Several characteristics distinguish animals from plants: animal cells lack cell walls; animals are heterotrophic and, at some stage of the life cycle, motile.

2. The ability of animals to move through the environment allows them to search for and pursue the food items that supply them with nutrients and energy.

Study Break 29.2

1. A tissue is a group of cells that share a common structure and function. The three primary tissue layers that contribute to the bodies of most animals are endoderm, mesoderm, and ectoderm.

2. Humans are bilaterally symmetrical.

3. The coelom is a space within which internal organs can move independently of the body wall muscles. The fluid within it provides protection for internal organs. In some animals the coelom functions as a hydrostatic skeleton.

4. The advantages of having a segmented body include the redundancy of vital organ systems in different segments, allowing an animal to survive damage to some segments, and improved control over body movements.

Study Break 29.3

1. Molecular sequence studies have confirmed the distinctions between the Parazoa and the Eumetazoa, between the Radiata and the Bilateria, and between the Protostomia and the Deuterostomia.

2. A schizocoelom appears to be the ancestral body cavity among the protostomes.

Study Break 29.4

1. Sponges do not exhibit any kind of body symmetry.

2. A sponge gathers food from its environment by drawing water into its body through numerous small pores and harvesting particulate matter from the water with its choanocytes, or collar cells.

Study Break 29.5

1. Cnidarians capture animal prey by stinging it with their nematocysts and using their tentacles to pull it into their mouths.

2. The anthozoans, including sea anemones and corals, have only a polyp stage in their life cycle.

3. Ctenophores capture microscopic plankton in sticky filaments on their two tentacles, which are then drawn across the mouth.

Study Break 29.6

1. Free-living flatworms have digestive, excretory, nervous, and reproductive systems. Tapeworms lack a digestive system.

2. Ectoprocts, brachiopods, and phoronid worms all have a circular or U-shaped feeding structure called a lophophore, a characteristic that reveals their close evolutionary relationship.

3. The anatomical and physiological systems that allow squids and other cephalopods to be more active than other types of mollusks include a closed circulatory system with accessory hearts and a complex nervous system with giant nerve fibers. Many cephalopods use their excur-

rent siphon to expel jet of water, allowing them to move rapidly through the environment.

4. The organ systems that exhibit segmentation in most annelid worms include respiratory surfaces; parts of the nervous, circulatory, and excretory systems; and the body wall and coelom.

Study Break 29.7

1. The cuticle protects a nematode from the digestive enzymes of its host.
2. Although the rigid exoskeletons of arthropods do not expand, these animals grow a new, soft exoskeleton inside the existing one. After shedding the old exoskeleton, they grow to a larger size by expanding the new exoskeleton with either water or air before it hardens.
3. The body regions of the four living subphyla of arthropods differ in how they have become fused. Chelicerates have a fused cephalothorax and an abdomen. Crustaceans show variable patterns, but many have a fused cephalothorax and an abdomen. Myriapods have a head and a trunk. Hexapods have a separate head, thorax, and abdomen.
4. Insects with incomplete metamorphosis hatch from their eggs as wingless nymphs, which vary in how closely they resemble adults; nymphs then undergo metamorphosis into the adult form. Insects with complete metamorphosis hatch from eggs as larvae, which are always very different from adults. After becoming a pupa, their cells and tissues are reorganized into the adult form.

Self-Test Questions

1. b 2. d 3. c 4. a 5. c 6. e 7. e 8. a
9. d 10. b

Chapter 30

Study Break 30.1

1. Echinoderms have a water vascular system that operates their tube feet.
2. Water enters the pharynx of a hemichordate through its mouth, and it exits the pharynx through the gill slits. The animal extracts oxygen and particulate food from the water as it passes through the pharynx.

Study Break 30.2

1. The animal is not a chordate, because chordates have a dorsal nerve cord.
2. Vertebrates have an internal bony skeleton, including a cranium and vertebral column in most groups, as well as structures derived from neural crest cells.

Study Break 30.3

1. Vertebrates have multiple *Hox* gene complexes, which provide them with several copies of each *Hox* gene. Cephalochordates have just one *Hox* gene complex.

2. Of the three groups listed, Gnathostomata has the most species, and Amniota has the fewest.

Study Break 30.4

1. Hagfishes lack bone, paired fins, and scales in their skin. Hagfishes have neither a cranium nor a vertebral column, and lampreys have only rudimentary traces of vertebrae. These observations suggest that their lineages arose before these structures appeared in the vertebrates.
2. The derived traits possessed by conodonts and ostracoderms include structures made of bone or a bonelike material and, in some ostracoderms, a brain divided into three regions.

Study Break 30.5

1. Sharks are more efficient predators than acanthodians or placoderms were because they have well-developed sensory systems to detect prey; their lightweight skeletons and absence of heavy body armor allow them to pursue prey rapidly; and they have numerous teeth that are replaced when damaged or worn and a loosely attached upper jaw that permits them to suck in large chunks of food.
2. The air bladders of ray-finned bony fishes increase their locomotor abilities by allowing them to rise or sink easily in the water. Their fin rays allow them to engage in precise movements during locomotion.
3. The lungs of lungfish allow them to survive in environments with low oxygen content because they can acquire oxygen from the air.

Study Break 30.6

1. For the first tetrapods, the advantages of moving onto land included abundant food resources, an absence of predators, and readily available oxygen. The disadvantages included the need for more skeletal support against gravity, mechanisms to prevent dehydration in air, and modifications of sensory systems so that they would function in air.
2. The parts of the amphibian life cycle that are most dependent on water are the egg and larval stages.

Study Break 30.7

1. The amniote egg freed amniotes from a dependence on standing water because it can survive on land. The shells of amniote eggs mediate gas exchange and water exchange with the environment.
2. Among the three amniote lineages, the Synapsida includes the mammals, the Anapsida includes turtles, and the Diapsida includes lizards, snakes, crocodilians, and birds.
3. Because lizards are lepidosaurs and both birds and crocodilians are archosaurs, crocodilians are more closely related to birds than they are to lizards.

Study Break 30.8

The overall structure of turtles differs from other amniotes in that their bodies are enclosed within a bony, keratin-covered shell.

Study Break 30.9

1. Besides their loss of legs, snakes differ from their lizard ancestors in having smaller skull bones, and the connections between them are more elastic.
2. Several characteristics reveal the close evolutionary relationship of crocodilians and birds, including a four-chambered heart and maternal care of offspring.

Study Break 30.10

1. The specific adaptations that allow birds to fly either reduce their weight or increase their muscle power. Weight-reducing adaptations include a lightweight skeleton, the absence of teeth and a urinary bladder, and the habit of laying an egg as soon as it has a shell. Power-promoting adaptations include large wing muscles; efficient digestive, respiratory, and circulatory systems; and a high metabolic rate.
2. The structure of a bird's bill reflects its diet. For example, hummingbirds, which drink nectar, have long thin bills, and parrots, which eat hard nuts, have stout sharp bills. Wings and feet are adapted to birds' flying habits and habitats. For example, ducks have webbed feet that allow them to paddle in water, and albatrosses have long thin wings that work efficiently for long distance flight.

Study Break 30.11

1. Most mammals were probably active at night during the Mesozoic era to avoid competition with and predation by dinosaurs, which were active during the day.
2. The key adaptations that allow mammals to be active under many types of environmental conditions include insulating fur and fat and a high metabolic rate that generates lots of body heat.
3. The major groups of living mammals are distinguished on the basis of their reproductive habits. Monotremes lay eggs. Marsupials give birth to relatively undeveloped young after a short period of gestation. Placentals give birth to more developed young after a long period of gestation.

Study Break 30.12

1. The characteristics that allow many species of primates to spend a lot of time in trees include flexible shoulder and hip joints, grasping hands, and excellent depth perception.
2. The lowest taxonomic group that includes monkeys, apes, and humans is the Anthropoidea. The lowest taxonomic group that includes only apes and humans is the Hominoidea.
3. Gorillas, chimpanzees, and bonobos are the apes that spend the most time on the ground.

Study Break 30.13

1. Researchers usually use the criterion of bipedality to distinguish between humans and apes. Humans (that is, hominids) are bipedal, and apes are not.

2. The strongest evidence suggesting that Neanderthals and modern humans belong to different species comes from mtDNA sequence data: differences between gene sequences of Neanderthals and humans are three times greater than the differences between pairs of modern humans.

3. The African Emergence Hypothesis proposes that modern humans arose in Africa and then migrated to various other regions; all modern humans are descended from that wave of immigrants. The Multiregional Hypothesis argues that modern humans evolved simultaneously in many different regions from archaic human ancestors that had migrated from Africa.

Self-Test Questions

1. a 2. c 3. d 4. b 5. e 6. c 7. e 8. d
9. b 10. a

Chapter 49

Study Break 49.1

1. Studies of ecosystems are more "inclusive" than studies of populations because ecosystems include the populations of many different species.

2. Mathematical models are useful in ecological research because they help scientists formalize hypotheses about the relationships between variables and because they allow researchers to simulate the effects of changing variables before investing time and resources in experiments or observational studies.

Study Break 49.2

1. A population's size is simply the number of individual it contains. Its density is the number of individuals per area or volume of habitat occupied,

2. A clumped pattern of dispersion implies that individuals in the population help each other or that some vital resource also has a clumped distribution in the environment. A uniform pattern of dispersion implies that individuals in the population repel each other. A random pattern of dispersion does not imply either positive or negative interactions among individuals in the population.

Study Break 49.3

1. A life table usually summarizes statistics about the age-specific survival rates, age-specific mortality rates, and age-specific fecundity of a population.

2. Humans in the industrialized countries exhibit Type I survivorship curves because they provide lots of care to their offspring, thus reducing infant and childhood mortality to low levels.

Study Break 49.4

1. Children spend most of their energy on growth and maintenance, and devote energy to reproduction later in life.

2. Fecundity and the amount of parental care devoted to each offspring exhibit an inverse relationship because organisms that produce few offspring can devote substantial time and energy to each, whereas those that produce many offspring can devote only minimal time and energy to each.

Study Break 49.5

1. The model of exponential population growth predicts unlimited population growth through time, generating a J-shaped curve of population size versus time. The logistic model predicts that population growth slows down as the population approaches its carrying capacity, generating an S-shaped curve of population size versus time.

2. Carrying capacity is the maximum number of individuals in a population that an environment can support. The carrying capacity is thus a property of the environment with reference to a particular population.

3. A time lag is a delay in a population's response to a changing environment. It may cause a population's size to oscillate around its carrying capacity.

Study Break 49.6

1. The effects of density-dependent factors get stronger (that is, they affect a larger percentage of the individuals in the population) as the population's density increases. The effects of density-independent factors do not change (that is, they affect the same percentage of the individuals in a population) as the population's density changes.

2. The affects of infectious diseases are usually density-dependent because disease-causing pathogens spread more quickly through dense populations of the organisms they infect.

Study Break 49.7

1. Humans have sidestepped the controls that regulate the populations of other organisms by expanding their geographical range to include a wide variety of habitats, by increasing the carrying capacity through agricultural production, and by decreasing death rates through the introduction of medical care and sanitation.

2. The age structure of a population influences its future population growth by determining how many individuals will reach reproductive age in the future. Populations with a bottom-heavy age structure (that is, with many young children) will experience a growth spurt when children alive today reach sexual maturity.

Populations with a more even age structure will not experience a dramatic future increase in population size.

Self-Test Questions

1. b 2. c 3. b 4. d 5. a 6. e 7. d 8. e
9. d 10. b

Chapter 50

Study Break 50.1

1. Some carnivores will spend more time and energy capturing large prey than small prey because large prey provide a larger return on their investment of time and energy in the hunt.

2. Cryptic coloration makes an organism inconspicuous, allowing it to blend in with its surroundings. Aposematic coloration makes an organism highly conspicuous, advertising its unpalatability. Mimicry allows one organism, the mimic, to resemble another species, the model; models are usually unpalatable or poisonous. A mimic will have aposematic coloration if it resembles an aposematic model.

3. Field experiments can demonstrate that two species are competing for limiting resources if the removal of one species increases population size or density in the other or if the addition of a potential competitor decreases the population size or density of the other.

Study Break 50.2

1. Gleason's individualistic view of communities suggests that they are just chance assemblages of species that happen to be adapted to similar abiotic environmental conditions.

2. Ecologists find more species living in an ecotone than in the communities on either side of it because ecotones contain species from both neighboring communities as well as species that are adapted to transitional environmental conditions.

Study Break 50.3

1. The plant growth forms found in tropical forests include a canopy of tall trees, an understory of shorter trees and shrubs, an herb layer, vinelike lianas, and epiphytes.

2. The species richness of a community is the number of species it contains. Relative abundance refers to the commonness or rarity of species in the community.

3. Pigeons, which eat grain and other vegetable matter, are included in the second trophic level, primary consumers. Peregrine falcons, which feed on pigeons and other birds, are in the third trophic level, secondary consumers.

Study Break 50.4

1. On the one hand, the ecological literature on competition may overestimate

the importance of competition because ecologists are more likely to study and publish papers on interactions in which competition is important than on interactions in which it is not. On the other hand, the literature may underestimate the importance of competition because, if strong competition between species can not persist for long periods of time, we are unlikely to find populations competing strongly in nature.

2. Keystone species are those that have a substantial effect on community structure even if their populations are not very dense. Keystone species may either increase or decrease species richness in the communities they occupy.

Study Break 50.5

1. Strong storms allow coral communities to be rejuvenated through the recruitment of new individuals because they scour the seafloor, removing existing coral colonies from the community. These openings provide spaces where coral larvae may settle and initiate the growth of new colonies.

2. Moderately severe and moderately frequent disturbances increase a community's species richness by creating opportunities for *r*-selected species to colonize the habitat while allowing populations of *K*-selected species to persist.

Study Break 50.6

1. Primary succession occurs in places without soil; secondary succession occurs after a disturbance has destroyed vegetation.

2. A climax community differs from earlier successional stages in having taller, longer-lived vegetation, generally higher species richness, and a buffered physical environment under the vegetation.

3. Three hypotheses about the underlying causes of succession differ in how they view the role of population interactions. The facilitation hypothesis specifies no particular role for population interactions. The inhibition hypothesis suggests that species that are already present prevent other species from joining a community. The tolerance hypothesis suggests that as environmental conditions within the community change during succession, only species that can compete strongly under the changing conditions will persist.

Study Break 50.7

1. Some explanations of the high species richness in the tropics suggest that the benign climate and historically low levels of severe disturbance have fostered more rapid rates of speciation in tropical regions. Other explanations suggest that the year-round availability of food resources and complex food webs allow more species to coexist in tropical regions.

2. According to the equilibrium theory of island biogeography, large islands will harbor more species than small islands, and islands that are close to the mainland will harbor more species than those that are further away.

Self-Test Questions
1. c 2. c 3. a 4. b 5. b 6. e 7. b 8. d 9. d 10. b

Chapter 51

Study Break 51.1

1. Gross primary productivity is a measure of the total amount of solar energy converted into chemical energy by the producers in an ecosystem. Net primary productivity is the amount of chemical energy that remains after deducting the producers' maintenance costs from the gross primary productivity.

2. In terrestrial ecosystems, primary productivity may be influenced by the availability of light, water, and nutrients and by the environmental temperature. In aquatic ecosystems, primary productivity is often limited by the joint availability of light and nutrients in the same place.

3. Energy is lost from an ecosystem at every transfer between trophic levels because some of the energy is not assimilated; because organisms use some of the energy they assimilate for maintenance costs; and because biological processes are never 100 efficient.

4. The presence of a top predator can influence the interactions of organisms at lower trophic levels and an ecosystem's productivity by changing the relative abundances of organisms at lower trophic levels. These effects can reverberate through an ecosystem in a trophic cascade.

Study Break 51.2

1. In the generalized compartment model of biogeochemical cycling, nutrient pools are classified as either available or unavailable and as either organic or inorganic.

2. The global hydrologic cycle maintains its balance because the amount of water returned to the atmosphere by evaporation and transpiration is equal to the amount that falls as precipitation. Runoff from the land maintains the balance between terrestrial and marine components of the cycle.

3. Respiration, excretion, leaching, and the burning of fossil fuels move large quantities of carbon from an organic compartment to an inorganic compartment of an ecosystem.

4. Bacteria, cyanobacteria, and fungi drive the global nitrogen cycle through their activities in nitrogen fixation, ammonification, nitrification, and denitrification.

5. Marine sediments are Earth's main reservoir for phosphorus, which is recycled slowly after geological uplifting and erosion make it available to producers.

Study Break 51.3

1. The advantage of using conceptual models of ecosystem function is that they are a simplification of the processes that determine ecosystem function in nature. The disadvantage of these models is that they do not include processes that carry nutrients and energy out of one ecosystem and into another; neither do they include the nuts-and-bolts details of exactly how specific ecosystems function. Thus, conceptual models do not provide precise predictions about potential changes in ecosystem function.

2. Before constructing a simulation model of an ecosystem, ecologists must collect data about the population sizes of important species, the average energy and nutrient content of each, the food webs in which they participate, the quantity of food each species consumes, and the productivity of each population; the ecosystem's energy and nutrient gains and losses caused by erosion, weathering, precipitation, and runoff; and seasonal and annual variations in these factors.

Self-Test Questions
1. b 2. d 3. c 4. e 5. e 6. c 7. a 8. d 9. c 10. a

Chapter 52

Study Break 52.1

1. Because of Earth's spherical shape, sunlight striking the planet's surface is more concentrated near the equator than at the poles. As a result, temperatures are higher at low latitudes. The concentrated sunlight near the equator heats the atmosphere, causing air masses near the equator to rise, establishing three circulation cells in the Northern Hemisphere and three in the Southern Hemisphere.

2. Earth's fixed tilt on its axis causes seasonal variation in the amount of sunlight striking the temperate zones as the planet orbits the sun.

3. Dry conditions prevail at 30° north and south latitudes because sinking air masses warm as they descend, causing them to absorb water from the land.

4. Mountains affect local precipitation because rising air masses on the windward side of a mountain cool adiabatically and release moisture. When the air masses descend on the leeward side of a mountain, they warm and absorb moisture, causing a rain shadow.

Study Break 52.2

1. *Anolis* lizards in the Dominican Republic bask more frequently at high elevation than they do at low elevation.

2. Global warming will likely cause the geographical distributions of species to shift

or expand to higher latitudes and to higher elevations.

Study Break 52.3

1. Tropical rain forest and temperate rainforest are the terrestrial biomes that occur in habitats that receive the most rainfall.
2. Savannas, chaparral, and temperate grasslands are renewed by periodic fires.
3. Tropical rain forests and temperate rainforests have the tallest vegetation. Arctic and alpine tundra have the shortest vegetation.
4. Trees are usually evergreen in tropical rain forest, temperate rain forest, and taiga.

Study Break 52.4

1. Dissolved oxygen concentration is usually high in the headwaters of a stream, gradually diminishing as water flows into a river.
2. The factors that cause seasonal overturns in lakes include seasonal changes in environmental temperatures, variations in wind velocity, and the fact that water is densest at 4°C.
3. Oligotrophic lakes are better than eutrophic lakes for recreational purposes because the water in oligotrophic lakes is clear, whereas the water in eutrophic lakes is often clogged with strands of algae and cyanobacteria.

Study Break 52.5

1. The benthic province of the ocean includes all of the bottom sediments. The pelagic province includes all of the water.
2. Of all the marine biomes, estuaries experience the largest fluctuations in salinity over time.
3. Estuaries, the intertidal zone, and the upper layer of the oceanic pelagic zone receive substantial energy inputs from sunlight.
4. The benthos of the oceanic pelagic zone receives nutrients and energy from the detritus sinking from the upper layers of water.
5. Chemoautotrophic bacteria are the primary producers of hydrothermal vent communities and cold seep communities. Unlike photosynthetic organisms of the photic zone, they use hydrogen sulfide and other molecules, instead of sunlight, as an energy source for their chemosynthetic activity.

Self-Test Questions

1. c 2. e 3. b 4. a 5. d 6. a 7. d 8. a
9. a 10. d

Chapter 53

Study Break 53.1

1. Living systems are a storehouse of potentially useful genetic information because

naturally occurring compounds may prove to be useful in the treatment of disease, in the manufacture of new products, or in agriculture

2. Naturally occurring organisms provide many ecosystem services, such as the sequestration of carbon dioxide, fixation of nitrogen into forms that plants can absorb, recycling of nutrients with ecosystems, and the retention of water in ecosystems.

Study Break 53.2

1. Habitat fragmentation in eastern North America has affected breeding songbirds by reducing the variety of habitats available to them, increasing the frequency of brood parasitism of their nests, and increasing the rate of predation on their eggs and young.
2. The growing human population in south Florida has increased the likelihood of desertification there by withdrawing groundwater for agricultural, recreational, and residential uses faster than it is replenished.
3. Overexploitation of fish populations typically causes fishes to reach reproductive maturity at a smaller size and younger age; decreases population sizes; and sometimes leads to the extinction of populations.
4. By one estimate, extinction rates today may be 1000 times greater than the background extinction rate evident in the fossil record.

Study Break 53.3

1. Ecologists generally identify biodiversity hotspots as areas that include many endemic species that face the threat of extinction.
2. Conservation biologists are especially concerned about the rapid rate of deforestation in the tropics because tropical forests, although not very extensive in area, harbor many terrestrial species.

Study Break 53.4

1. Population bottlenecks—large, temporary reductions in a population's size—inevitably foster genetic drift, thereby reducing a population's genetic variability, which increases its likelihood of becoming extinct.
2. A population viability analysis allows a conservation biologist to identify the minimum population size that is likely to survive both predictable and unpredictable environmental change. It therefore specifies how many individuals must be conserved for the continued survival of the population and species.
3. Several small preserves would collectively experience more edge effects than one large preserve of the same total size.

Study Break 53.5

1. The Pine Bush habitat in New York state is an example of conservation through preservation.

2. The Royal Chitwan National Park has been judged a success because local residents benefit from the park's existence, and therefore support it, and populations of many animals, including tigers and rhinoceroses, have increased within its borders.
3. Economists can determine the economic values of specific ecosystem services and convince local governments that it is economically beneficial to preserve ecosystems and the services they provide.

Self-Test Questions

1. e 2. c 3. b 4. b 5. a 6. d 7. d 8. a
9. c 10. b

Chapter 54

Study Break 54.1

1. An instinctive behavior is a genetically or developmentally programmed response that appears in complete and functional form the first time it is used. A learned behavior is one that is dependent on having a particular kind of experience during development.
2. Marler demonstrated that singing the correct species song is a learned behavior by isolating some young male sparrows in soundproof cages, thereby preventing them from hearing their species song. Because the isolated males never learned to sing the correct song, Marler concluded that the experience of hearing the species song was necessary for song learning.

Study Break 54.2

1. The chicks of brood parasites stimulate their foster parents to feed them by engaging in exaggerated behaviors—opening their mouths and begging and peeping vigorously—that serve as sign stimuli, triggering feeding behavior in the parents.
2. Arnold demonstrated that the receptiveness of garter snakes to a meal of banana slugs had a genetic basis by breeding snakes that almost always eat banana slugs with snakes that rarely eat them. The behavior of the hybrid offspring was intermediate between the behaviors of the two parent populations.

Study Break 54.3

1. Tail wagging by a dog when it sees its owner pick up a leash is an example of classical conditioning.
2. Sleeping through an alarm clock is an example of habituation.

Study Break 54.4

1. The conclusion that certain neurons in a young male bird's brain are influenced only by acoustical signals from its own species and only during a critical period is supported by two observations: (1) young birds that did not hear

taped songs during the critical period never produced the full song of their species; and (2) young birds that heard recordings of *other* bird species' songs during the critical period never generated replicas of those songs as they matured.

2. After a zebra finch hears the song of a neighbor many times, cells in a nucleus in its forebrain habituate to that stimulus and stop responding to it.

3. The role of the ZENK enzyme in song learning is to program the nerve cells of the bird's brain to anticipate key acoustical events of potential biological importance.

Study Break 54.5

1. A high estrogen concentration in the brains of young male zebra finches stimulates the production of more neurons in the higher vocal center.

2. Juvenile hormone stimulates the production of octopamine, a protein that may help a foraging bee home in on the odors of flowers where it can collect nectar and pollen.

3. The loss of a territory by a male African cichlid fish causes its brain to produce less GnRH.

Study Break 54.6

1. The anatomy of the cricket's nervous system helps it avoid an approaching bat because sensory neurons on one side of the body send messages to motor neurons on the other side. When the cricket hears a bat approaching from the right, it automatically lifts a leg on the left side, blocking the left wing and causing the cricket to veer away from the oncoming bat's path.

2. When a fiddler crab detects movement above the midline of its visual field, it dashes into its burrow.

3. A star-nosed mole locates earthworms in its tunnels with touch receptors on the tentacles that sprout from its nose.

Self-Test Questions
1. b 2. a 3. d 4. b 5. c 6. e 7. d 8. c
9. b 10. b

Chapter 55

Study Break 55.1

1. When piloting, animals use familiar landmarks to guide their journey. When using compass orientation, animals use external environmental cues such as the position of the sun or stars as a compass to move in a particular direction, often over a specific distance or for a proscribed length of time. When navigating, animals use a compass as well as a mental map of their position in relation to their destination.

2. Seasonal changes in temperature and food availability are the most likely selection pressures responsible for the evolution of migratory behavior in birds.

Study Break 55.2

1. Wood lice accumulate in moist habitats because they move much less in moist habitats than they do in dry habitats.

2. The costs of maintaining a territory include the time and energy needed to defend territory borders, the possibility of being injured or killed during a territorial encounter, and the increased likelihood of being noticed and captured by a predator. The benefits of holding a territory include access to all of the resources found within the territory.

Study Break 55.3

1. Humans consciously use acoustical, visual, and tactile channels to communicate with each other.

2. A honeybee uses the waggle dance to communicate the location of a distant food source to its hive-mates.

3. A young wandering raven that discovers a food source in the territory of other ravens will call vigorously to attract other ravens from outside the territory. Collectively, the non-territory holders can overwhelm the defenses of the resident birds and then consume the food.

Study Break 55.4

1. For monogamous species, the males that are most attractive to females are those that can demonstrate their good genes with large showy morphological characteristics and elaborate behavioral displays and those that hold territories rich in resources.

2. Male sage grouse perform displays at a lek, trying to attract the attention of females. Female sage grouse go to a lek to evaluate the qualities of the males that are displaying there.

3. A female red-winged blackbird might settle in the territory of a male even in the presence of other resident females if the male's territory is very rich in resources.

Study Break 55.5

1. The social behavior of musk oxen and sawfly larvae includes cooperative defense of the group against predators.

2. Among the animals in a dominance hierarchy, the most dominant individuals are the most likely to reproduce.

3. The genetic system in eusocial insects promotes altruistic behavior because most individuals in a colony are more closely related to each other than are siblings in most other animal species.

Study Break 55.6

1. People may exhibit altruistic behavior to nonrelatives because their acts of charity may induce others to be charitable toward them.

2. Stepparents might provide fewer resources to their children than birthparents do because stepparents do not share as many genes with their stepchildren as birthparents share with their biological children.

Self-Test Questions
1. d 2. d 3. b 4. d 5. a 6. e 7. c 8. d
9. a 10. c

Appendix B
Classification System

The classification system presented here is based on a combination of organismal and molecular characters and is a composite of several systems developed by microbiologists, botanists, and zoologists. This classification reflects current trends toward a phylogenetic approach to taxonomy, one that incorporates the ever more detailed information about the relationships of monophyletic lineages provided by new molecular sequence data. In keeping with these trends, we have omitted reference to the traditional taxonomic categories, such as "class" and "order." Instead, we present the major monophyletic lineages in each of the three domains, and we indicate their relationships within a nested hierarchy that parallels that of traditional Linnaean classification.

Although researchers generally agree on the identity of the major monophyletic lineages, the biologists who study different groups have not established universal criteria for identifying the somewhat arbitrary taxonomic categories included in the traditional Linnaean hierarchy. As a result, a "class" or "order" of flowering plants may not be the equivalent of a "class" or "order" of animals. In fact, as described in *Unanswered Questions* at the end of Chapter 23, systematic biologists are shifting toward a more phylogenetic approach to taxonomy and classification, such as the one represented here.

Bear in mind that we include this appendix to introduce the diversity of life and illustrate many of the evolutionary relationships that link monophyletic groups. Like all phylogenetic hypotheses, this classification is open to revision as new information becomes available. Moreover, the classification is incomplete because it includes only those lineages that are described in Unit Four.

Prokaryotes and Eukaryotes

Organisms fall into two groups, prokaryotes and eukaryotes, based on the organization of their cells. Prokaryotes consist of the Domains Bacteria and Archaea and are characterized by a central region, the nucleoid, which has no boundary membrane separating it from the cytoplasm, and by membranes typically limited to the plasma membrane. Most prokaryotes are single-celled, although some are found in simple associations. All other organisms are eukaryotes, which make up the Domain Eukarya. Eukaryotes are characterized by cells with a central, membrane-bound nucleus, and an extensive membrane system. Some eukaryotes are single-celled, while others are multicellular.

Domain Bacteria

The largest and most diverse group of prokaryotes. Includes photoautotrophs, chemoautotrophs, and heterotrophs.

PROTEOBACTERIA Purple sulfur bacteria, purple nonsulfur bacteria, and some chemoheterotrophs

GREEN BACTERIA Green sulfur bacteria and green nonsulfur bacteria

CYANOBACTERIA Photoautotrophic Gram-negative bacteria that use the same chlorophyll as in plants

GRAM-POSITIVE BACTERIA Chemoheterotrophic bacteria with thick cell walls

SPIROCHETES Helically spiraled bacteria that move by twisting in a corkscrew pattern

CHLAMYDIAS Gram-negative intracellular parasites of animals, with cell walls that lack peptidoglycans

Domain Archaea

Prokaryotes that are evolutionarily between eukaryotic cells and the bacteria. Most are chemoautotrophs. None is photosynthetic. Originally discovered in extreme habitats, they are now known to be widely dispersed. Compared with bacteria, the Archaea have a distinctive cell wall structure and unique membrane lipids, ribosomes, and RNA sequences. Some are symbiotic with animals, but none is known to be pathogenic.

EURYARCHAEOTA Includes methanogens, extreme halophiles, and some extreme thermophiles

CRENARCHAEOTA Includes most of the extreme thermophiles, as well as psychrophiles; mesophilic species comprise a large part of plankton in cool, marine waters

KORARCHAEOTA Known only from DNA isolated from hydrothermal pools. As of this writing, none has been cultured and no species have been named.

Domain Eukarya

PROTOCTISTA A collection of single-celled and multicelled lineages, which are almost certainly not a monophyletic group. Some biologists consider the groups listed below to be kingdoms in their own right.

   Excavates Single-celled animal parasites that lack mitochondria and move using flagella; most have a hollow, ventral feeding groove

      Diplomonadida (diplomonads)—two nuclei; move by multiple free flagella

      Parabasala (parabasalids)—move by an undulating membrane and free flagella

   Discicristates Mostly single-celled, highly motile cells that swim using flagella, have disc-shaped mitochondrial inner membranes

      Euglenoids—free-living photosynthetic autotrophs

      Kinetoplastids—nonphotosynthetic, heterotrophs that live as animal parasites

Alveolates Characterized by small membrane-bound vesicles called alveoli in a layer under the plasma membrane

    Ciliophora (ciliates)—single-celled heterotrophs; swim by means of cilia

    Dinoflagellata (dinoflagellates)—single-celled marine heterotrophs or autotrophs; shell formed from cellulose plates

    Apicomplexa (apicomplexans)—nonmotile parasites of animals with apical complex for attachment and invasion of host cells

Heterokonts Characterized by two different flagella

    Oomycota (oomycetes)—water molds, white rusts, and mildews

    Bacillariophyta (diatoms)—single-celled; covered by a glassy silica shell

    Chrysophyta (golden algae)—colonial; each cell of the colony has a pair of flagella and is covered by a glassy shell consisting of plates or scales

    Phaeophyta (brown algae)—photoautotrophic protists

Cercozoa Amoebas with stiff, filamentous pseudopodia; some with outer shells

    Radiolaria (radiolarians)—heterotrophic; glassy internal skeleton with projecting raylike strands of cytoplasm

    Foraminifera (forams)—heterotrophic protists with shells consisting of organic matter reinforced by calcium carbonate

    Chlorarachniophyta (chlorarachniophytes)—green, photosynthetic amoebas that also engulf food

Amoebozoa Includes most of the amoebas and the slime molds

    Amoebas—single-celled; use non-stiffened pseudopods for locomotion and feeding

    Cellular slime molds—heterotrophs; primarily individual cells; move by amoeboid motion, or as a multicellular mass

    Plasmodial slime molds—heterotrophs; live as plasmodium, a large composite mass with nuclei in a common cytoplasm, that moves and feeds like a giant amoeba

Opisthokonts A single posterior flagellum at some stage in the life cycle

    Choanoflagellata (choanoflagellates)—motile protists with a single flagellum surrounded by collar of closely packed microvilli; likely ancestor of animals and fungi

Archaeplastida Red algae, green algae, and land plants, photosynthesizers with a common evolutionary origin

    Rhodophyta (red algae)—marine seaweeds, typically multicellular, reddish in color; with plantlike bodies

    Chlorophyta (green algae)—green photosynthetic single-celled, colonial, and multicellular protists that have the same photosynthetic pigments as plants; likely ancestor of land plants

Fungi Heterotrophic, mostly multicellular organisms with cell wall containing chitin and cell nuclei occurring in threadlike hyphae; life cycle typically includes both asexual and sexual phases, with sexual structures used as the basis for phylum-level classification. Single-celled species are known as yeasts.

    Zygomycota (zygomycetes)—terrestrial; asexual reproduction via nonmotile haploid spores formed in sporangia; sexual spores (zygospores) form in zygosporangia; aseptate hyphae

    Glomeromycota (glomeromycetes)—terrestrial; asexual reproduction via spores at the tips of hyphae; form mycorrhizal associations with plant roots

    Ascomycota (ascomycetes/sac fungi)—terrestrial and aquatic; sexual spores form in asci; asexual reproduction occurs via conidia (nonmotile spores); septate hyphae

    Basidiomycota (basidiomycetes)—terrestrial; reproduction usually via asexual basidiospores produced by basidia; septate hyphae

        Basidiomycetes: mushroom-forming fungi and relatives

        Teliomycetes: rusts

        Ustomycetes: smuts

    Chytridiomycota (chytrids)—mostly aquatic; asexual reproduction by way of motile zoospores; sexual reproduction via gametes produced in gametangia; hyphae mostly aseptate

    Conidial fungi—not a true phylum but a convenience grouping of species for which no sexual phase is known

    Microsporidia—single-celled sporelike parasites of animals, other groups; phylogeny uncertain

Plantae Multicellular autotrophs, mostly terrestrial, and most of which gain energy via photosynthesis; life cycle characterized by alternation of a gametophyte (gamete-producing) generation and sporophyte (spore-producing) generation

Nonvascular plants (bryophytes)—no vessels for transporting water and nutrients; swimming sperm require liquid water for sexual reproduction

    Hepatophyta (liverworts)—leafy or simple flattened thallus with rhizoids; no true leaves, stems, roots, or stomata (porelike openings for gas exchange); spores in capsules

    Anthocerophyta (hornworts)—simple flattened thallus, hornlike sporangia

    Bryophyta (mosses)—feathery or cushiony thallus; some with hydroids; spores in capsules

Seedless vascular plants—plants in which embryos are not housed inside seeds

    Lycophyta (club mosses)—simple leaves, cuticle, stomata, true roots; most species have sporangia on sporophylls; fertilization by swimming sperm

    Pterophyta (ferns, whisk ferns, horsetails)—*Ferns:* Finely divided leaves; sporangia in sori. *Whisk ferns:* Branching stem from rhizomes; sporangia on stem scales. *Horsetails:* hollow stem, scalelike leaves, sporangia in strobili.

Seed plants—vascular plants in which embryos develop within seeds

    Gymnosperms—seeds born on stems, on leaves, or under scales

    Cycadophyta (cycads)—shrubby or treelike with palmlike leaves; male and female strobili on separate plants

    Ginkgophyta (ginkgoes)—lineage with a single living species (*Ginkgo biloba*); tree with deciduous, fan-shaped leaves; male, female reproductive structures on separate plants

    Gnetophyta (gnetophytes)—shrubs or woody vinelike plants; male and female strobili on separate plants

    Coniferophyta (conifers)—predominant extant gymnosperm group; mostly evergreen trees and shrubs with needlelike or scalelike leaves; male and female cones usually on the same plant

    Anthophyta (angiosperms/flowering plants)—reproductive structures in flowers

        Monocotyledones (monocots)—grasses, palms, lilies, orchids and their relatives; a single cotyledon (seed leaf); pollen grains have one groove

Eudicotyledones (eudicots)—roses, melons, beans, potatoes, most fruit trees, others; two cotyledons; pollen grains have three grooves

Other major angiosperm lineages: magnoliids (magnolias and relatives); water lilies (Family Nymphaeaceae); *Amborella* (Family Amborellaceae)

ANIMALIA Multicellular heterotrophs; nearly all with tissues, organs, and organ systems; motile during at least part of the life cycle; sexual reproduction in most; embryos develop through a series of stages; many with larval and adult stages in life cycle

Parazoa Animals lacking tissues and body symmetry

Porifera (sponges)—multicellular; extract oxygen and particulate food from water drawn into a central cavity

Eumetazoa Animals possessing tissues and either radial or bilateral symmetry

Radiata—acoelomate animals possessing radial symmetry and two tissue layers

Cnidaria (cnidarians)—two tissue layers; single opening into gastrovascular cavity; nerve net; nematocysts for defense and predation; some sessile, some motile; most are predatory, some with photosynthetic endosymbionts; freshwater and marine

Hydrozoa: hydrozoans

Scyphozoa: jellyfishes

Cubozoa: box jellyfishes

Anthozoa: sea anemones, corals

Ctenophora (comb jellies)—two (possibly three) tissue layers; feeding tentacles capture particulate food; beating cilia provide weak locomotion; marine

Bilateria—animals possessing bilateral symmetry and three tissue layers

PROTOSTOMIA—acoelomate, pseudocoelomate, or schizocoelomate; many with spiral, indeterminate cleavage; blastopore forms mouth; nervous system on ventral side

Lophotrochozoa—many with either a lophophore for feeding and gas exchange or a trochophore larva

Ectoprocta (bryozoans)—coelomate; colonial; secrete hard covering over soft tissues; lophophore; sessile; particulate feeders; marine

Brachiopoda (lamp shells)—coelomate; dorsal and ventral shells; lophophore; sessile; particulate feeders; marine

Phoronida (phoronid worms)—coelomate; secrete tubes around soft tissues; sessile; particulate feeders; lophophore

Platyhelminthes (flatworms)—acoelomate; dorsoventrally flattened; complex reproductive, excretory, and nervous systems; gastrovascular cavity in many; free-living or parasitic, often with multiple hosts; terrestrial, freshwater, and marine

Turbellaria: free-living flatworms

Trematoda: flukes

Cestoda: tapeworms

Rotifera (wheel animals)—pseudocoelomate; microscopic; complete digestive system; well-developed reproductive, excretory, and nervous systems; particulate feeders; major components of marine and freshwater plankton

Nemertea (ribbon worms)—schizocoelomate; proboscis housed within rhynchocoel; complete digestive tract; circulatory system; predatory; mostly marine

Mollusca (mollusks)—schizocoelomate; many with trochophore larva; many with shell secreted by mantle; body divided into head–foot, visceral mass, and mantle; well-developed organ systems; variable locomotion; herbivorous or predatory; terrestrial, freshwater, and marine

Polyplacophora: chitons

Gastropoda: snails, sea slugs, land slugs

Bivalvia: clams, mussels, scallops, oysters

Cephalopoda: squids, octopuses, cuttlefish, nautiluses

Annelida (segmented worms)—schizocoelomate; many with trochophore larva; segmented body and organ systems; well-developed organ systems; many use hydrostatic skeleton for locomotion; some predatory, some particulate feeders, some detritivores; terrestrial, freshwater, and marine

Polychaeta: marine worms

Oligochaeta: freshwater and terrestrial worms

Hirudinea: leeches

Ecdysozoa—cuticle or exoskeleton is shed periodically

Nematoda (roundworms)—pseudocoelomate; body covered with tough cuticle that is shed periodically; well-developed organ systems; thrashing locomotion; many are parasitic on plants or animals; mostly terrestrial

Onychophora (velvet worms)—schizocoelomate; segmented body covered with cuticle; locomotion by many unjointed legs; complex organ systems; predatory; terrestrial

Arthropoda (arthropods)—schizocoelomate; jointed exoskeleton made of chitin; segmented body, some with fusion of segments in head, thorax, or abdomen; complex organ systems; variable modes of locomotion, including flight; specialization of numerous appendages; herbivorous, predatory, or parasitic; terrestrial, freshwater, and marine

Trilobita: trilobites (extinct)

Chelicerata: horseshoe crabs, spiders, scorpions, ticks, mites

Crustacea: shrimps, crayfishes, lobsters, crabs, barnacles, copepods, isopods

Myriapoda: centipedes, millipedes

Hexapoda: springtails and insects

DEUTEROSTOMIA—enterocoelomate; many with radial, determinate cleavage; blastopore forms anus; nervous system on dorsal side in many

Echinodermata (echinoderms)—secondary radial symmetry, often organized around five radii; hard internal skeleton; unique water vascular system with tube feet; complete digestive system; simple nervous system; no circulatory or respiratory system; generally slow locomotion using tube feet; predatory, herbivorous, particulate feeders, detritivores; exclusively marine

Asteroidea: sea stars

Ophiuroidea: brittle stars

Echinoidea: sea urchins, sand dollars

Holothuroidea: sea cucumbers

Crinoidea: feather stars, sea lilies

Concentricycloidea: sea daisies

Hemichordata (acorn worms)—pharynx perforated with gill slits; proboscis; complex organ systems; tube-dwelling in soft sediments; particulate or deposit feeders; exclusively marine

Chordata (chordates)—notochord; segmental body wall and tail muscles; dorsal hollow nerve chord; perforated pharynx; complex organ systems; variable modes of locomotion; extremely varied diets; terrestrial, freshwater, and marine

    Urochordata: tunicates, sea squirts

    Cephalochordata: lancelets

    Vertebrata: vertebrates

        Myxinoidea: hagfishes

        Petromyzontoidea: lampreys

        Placodermi: placoderms (extinct)

        Chondrichthyes: sharks, skates, and rays

Acanthodii: acanthodians

Actinopterygii: ray-finned fishes

Sarcopterygii: fleshy-finned fishes

Amphibia: salamanders, frogs, caecelians

Synapsida: mammals

Anapsida: turtles

Diapsida: sphenodontids, lizards, snakes, crocodilians, birds

This journal article reports on the movements of a female wolf during the summer of 2002 in northwestern Canada. It also reports on a scientific process of inquiry, observation and interpretation to learn where, how and why the wolf traveled as she did. In some ways, this article reflects the story of "how to do science" told in section 1.4 of this textbook. These notes are intended to help you read and understand how scientists work and how they report on their work.

1 Title of the journal, which reports on science taking place in Arctic regions.

2 Volume number, issue number and date of the journal, and page numbers of the article.

3 Title of the article: a concise but specific description of the subject of study—one episode of long-range travel by a wolf hunting for food on the Arctic tundra.

4 Authors of the article: scientists working at the institutions listed in the footnotes below. Note #2 indicates that P. F. Frame is the corresponding author—the person to contact with questions or comments. His email address is provided.

5 Date on which a draft of the article was received by the journal editor, followed by date on which a revised draft was accepted for publication. Between these dates, the article was reviewed and critiqued by other scientists, a process called peer review. The authors revised the article to make it clearer, according to those reviews.

6 ABSTRACT: A brief description of the study containing all basic elements of this report. First sentence summarizes the background material. Second sentence encapsulates the methods used. The rest of the paragraph sums up the results. Authors introduce the main subject of the study—a female wolf (#388) with pups in a den—and refer to later discussion of possible explanations for her behavior.

7 Key words are listed to help researchers using computer databases. Searching the databases using these key words will yield a list of studies related to this one.

8 RÉSUMÉ: The French translation of the abstract and key words. Many researchers in this field are French Canadian. Some journals provide such translations in French or in other languages.

9 INTRODUCTION: Gives the background for this wolf study. This paragraph tells of known or suspected wolf behavior that is important for this study. Note that (a) major species mentioned are always accompanied by scientific names, and (b) statements of fact or postulations (claims or assumptions about what is likely to be true) are followed by references to studies that established those facts or supported the postulations.

10 This paragraph focuses directly on the wolf behaviors that were studied here.

11 This paragraph starts with a statement of the hypothesis being tested, one that originated in other studies and is supported by this one. The hypothesis is restated more succinctly in the last sentence of this paragraph. This is the inquiry part of the scientific process—asking questions and suggesting possible answers.

---

1 ARCTIC

2 VOL. 57, NO. 2 (JUNE 2004) P. 196–203

3 # Long Foraging Movement of a Denning Tundra Wolf

4 Paul F. Frame,[1,2] David S. Hik,[1] H. Dean Cluff,[3] and Paul C. Paquet[4]

5 (Received 3 September 2003; accepted in revised form 16 January 2004)

---

6 ABSTRACT Wolves (*Canis lupus*) on the Canadian barrens are intimately linked to migrating herds of barren-ground caribou (*Rangifer tarandus*). We deployed a Global Positioning System (GPS) radio collar on an adult female wolf to record her movements in response to changing caribou densities near her den during summer. This wolf and two other females were observed nursing a group of 11 pups. She traveled a minimum of 341 km during a 14-day excursion. The straight-line distance from the den to the farthest location was 103 km, and the overall minimum rate of travel was 3.1 km/h. The distance between the wolf and the radio-collared caribou decreased from 242 km one week before the excursion to 8 km four days into the excursion. We discuss several possible explanations for the long foraging bout.

7 *Key words:* wolf, GPS tracking, movements, *Canis lupus*, foraging, caribou, Northwest Territories

8 RÉSUMÉ Les loups (*Canis lupus*) dans la toundra canadienne sont étroitement liés aux hardes de caribous des toundras (*Rangifer tarandus*). On a équipé une louve adulte d'un collier émetteur muni d'un système de positionnement mondial (GPS) afin d'enregistrer ses déplacements en réponse au changement de densité du caribou près de sa tanière durant l'été. On a observé cette louve ainsi que deux autres en train d'allaiter un groupe de 11 louveteaux. Elle a parcouru un minimum de 341 km durant une sortie de 14 jours. La distance en ligne droite de la tanière à l'endroit le plus éloigné était de 103 km, et la vitesse minimum durant tout le voyage était de 3,1 km/h. La distance entre la louve et le caribou muni du collier émetteur a diminué de 242 km une semaine avant la sortie à 8 km quatre jours après la sortie. On commente diverses explications possibles pour ce long épisode de recherche de nourriture.

Mots clés: loup, repérage GPS, déplacements, *Canis lupus*, recherche de nourriture, caribou, Territoires du Nord-Ouest

Traduit pour la revue *Arctic* par Nésida Loyer.

---

9 Introduction

Wolves (*Canis lupus*) that den on the central barrens of mainland Canada follow the seasonal movements of their main prey, migratory barren-ground caribou (*Rangifer tarandus*) (Kuyt, 1962; Kelsall, 1968; Walton et al., 2001). However, most wolves do not den near caribou calving grounds, but select sites farther south, closer to the tree line (Heard and Williams, 1992). Most caribou migrate beyond primary wolf denning areas by mid-June and do not return until mid-to-late July (Heard et al., 1996; Gunn et al., 2001). Consequently, caribou density near dens is low for part of the summer.

During this period of spatial separation from the main caribou herds, wolves must either search near **10** the homesite for scarce caribou or alternative prey (or both), travel to where prey are abundant, or use a combination of these strategies.

Walton et al. (2001) postulated that the travel of **11** tundra wolves outside their normal summer ranges is a response to low caribou availability rather than a pre-dispersal exploration like that observed in territorial wolves (Fritts and Mech, 1981; Messier, 1985). The authors postulated this because most such travel was directed toward caribou calving grounds. We report details of such a long-distance excursion by a breeding female tundra wolf wearing a GPS radio collar. We discuss the relationship of the excursion to movements of satellite-collared caribou (Gunn et al., 2001), supporting the hypothesis that tundra wolves make directional, rapid, long-distance movements in response to seasonal prey availability.

[1] Department of Biological Sciences, University of Alberta, Edmonton, Alberta T6G 2E9, Canada
[2] Corresponding author: pframe@ualberta.ca
[3] Department of Resources, Wildlife, and Economic Development, North Slave Region, Government of the Northwest Territories, P.O. Box 2668, 3803 Bretzlaff Dr., Yellowknife, Northwest Territories X1A 2P9, Canada; Dean_Cluff@gov.nt.ca
[4] Faculty of Environmental Design, University of Calgary, Calgary, Alberta T2N 1N4, Canada; current address: P.O. Box 150, Meacham, Saskatchewan S0K 2V0, Canada

12 This map shows the study area and depicts wolf and caribou locations and movements during one summer. Some of this information is explained below.

13 STUDY AREA: This section sets the stage for the study, locating it precisely with latitude and longitude coordinates and describing the area (illustrated by the map in Figure 1).

14 Here begins the story of how prey (caribou) and predators (wolves) interact on the tundra. Authors describe movements of these nomadic animals throughout the year.

15 We focus on the denning season (summer) and learn how wolves locate their dens and travel according to the movements of caribou herds.

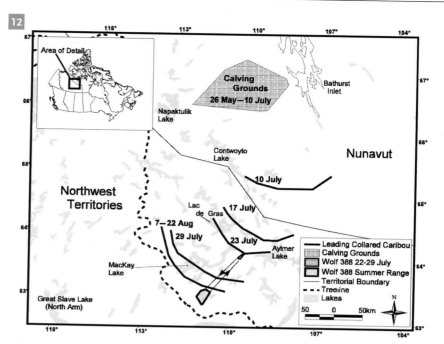

Figure 1. Map showing the movements of satellite radio-collared caribou with respect to female wolf 388's summer range and long foraging movement, in summer 2002.

13   Study Area

Our study took place in the northern boreal forest–low Arctic tundra transition zone (63° 30' N, 110° 00' W; Figure 1; Timoney et al., 1992). Permafrost in the area changes from discontinuous to continuous (Harris, 1986). Patches of spruce (*Picea mariana, P. glauca*) occur in the southern portion and give way to open tundra to the northeast. Eskers, kames, and other glacial deposits are scattered throughout the study area. Standing water and exposed bedrock are characteristic of the area.

14   *Details of the Caribou-Wolf System*

The Bathurst caribou herd uses this study area. Most caribou cows have begun migrating by late April, reaching calving grounds by June (Gunn et al., 2001;

Figure 1). Calving peaks by 15 June (Gunn et al., 2001), and calves begin to travel with the herd by one week of age (Kelsall, 1968). The movement patterns of bulls are less known, but bulls frequent areas near calving grounds by mid-June (Heard et al., 1996; Gunn et al., 2001). In summer, Bathurst caribou cows generally travel south from their calving grounds and then, parallel to the tree line, to the northwest. The rut usually takes place at the tree line in October (Gunn et al., 2001). The winter range of the Bathurst herd varies among years, ranging through the taiga and along the tree line from south of Great Bear Lake to southeast of Great Slave Lake. Some caribou spend the winter on the tundra (Gunn et al., 2001; Thorpe et al., 2001).

In winter, wolves that prey on Bathurst caribou do not behave territorially. Instead, they follow the herd throughout its winter range (Walton et al., 2001; Musiani, 2003). However, during denning (May– **15**

Foraging Movement of A Tundra Wolf   **197**

Table 1. Daily distances from wolf 388 and the den to the nearest radio-collared caribou during a long excursion in summer 2002.

Date (2002)	Mean distance from caribou to wolf (km)	Daily distance from closest caribou to den
12 July	242	241
13 July	210	209
14 July	200	199
15 July	186	180
16 July	163	162
17 July	151	148
18 July	144	137
19 July[1]	126	124
20 July	103	130
21 July	73	130
22 July	40	110
23 July[2]	9	104
29 July[3]	16	43
30 July	32	43
31 July	28	44
1 August	29	46
2 August[4]	54	52
3 August	53	53
4 August	74	74
5 August	75	75
6 August	74	75
7 August	72	75
8 August	76	75
9 August	79	79

[1] Excursion starts.
[2] Wolf closest to collared caribou.
[3] Previous five days' caribou locations not available.
[4] Excursion ends.

August, parturition late May to mid-June), wolf movements are limited by the need to return food to the den. To maximize access to migrating caribou, many wolves select den sites closer to the tree line than to caribou calving grounds (Heard and Williams, 1992). Because of caribou movement patterns, tundra denning wolves are separated from the main caribou herds by several hundred kilometers at some time during summer (Williams, 1990:19; Figure 1; Table 1).

Muskoxen do not occur in the study area (Fournier and Gunn, 1998), and there are few moose there (H.D. Cluff, pers. obs.). Therefore, alternative prey for wolves includes waterfowl, other ground-nesting birds, their eggs, rodents, and hares (Kuyt, 1972; Williams, 1990:16; H.D. Cluff and P.F. Frame, unpubl. data). During 56 hours of den observations, we saw no ground squirrels or hares, only birds. It appears that the abundance of alternative prey was relatively low in 2002.

17 Methods

Wolf Monitoring

We captured female wolf 388 near her den on 22 June 2002, using a helicopter net-gun (Walton et al., 2001). She was fitted with a releasable GPS radio collar (Merrill et al., 1998) programmed to acquire locations at 30-

minute intervals. The collar was electronically released (e.g., Mech and Gese, 1992) on 20 August 2002. From 27 June to 3 July 2002, we observed 388's den with a 78 mm spotting scope at a distance of 390 m.

Caribou Monitoring

In spring of 2002, ten female caribou were captured by helicopter net-gun and fitted with satellite radio collars, bringing the total number of collared Bathurst cows to 19. Eight of these spent the summer of 2002 south of Queen Maud Gulf, well east of normal Bathurst caribou range. Therefore, we used 11 caribou for this analysis. The collars provided one location per day during our study, except for five days from 24 to 28 July. Locations of satellite collars were obtained from Service Argos, Inc. (Landover, Maryland).

Data Analysis

Location data were analyzed by ArcView GIS software (Environmental Systems Research Institute Inc., Redlands, California). We calculated the average distance from the nearest collared caribou to the wolf and the den for each day of the study.

Wolf foraging bouts were calculated from the time 388 exited a buffer zone (500 m radius around the den) until she re-entered it. We considered her to be traveling when two consecutive locations were spatially separated by more than 100 m. Minimum distance traveled was the sum of distances between each location and the next during the excursion.

We compared pre- and post-excursion data using Analysis of Variance (ANOVA; Zar, 1999). We first tested for homogeneity of variances with Levene's test (Brown and Forsythe, 1974). No transformations of these data were required.

Results

Wolf Monitoring

Pre-Excursion Period: Wolf 388 was lactating when captured on 22 June. We observed her and two other females nursing a group of 11 pups between 27 June and 3 July. During our observations, the pack consisted of at least four adults (3 females and 1 male) and 11 pups. On 30 June, three pups were moved to a location 310 m from the other eight and cared for by an uncollared female. The male was not seen at the den after the evening of 30 June.

Before the excursion, telemetry indicated 18 foraging bouts. The mean distance traveled during these bouts was 25.29 km (± 4.5 SE, range 3.1–82.5 km). Mean greatest distance from the den on foraging

16 Other variables are considered—prey other than caribou and their relative abundance in 2002.

17 METHODS: There is no one scientific method. Procedures for each and every study must be explained carefully.

18 Authors explain when and how they tracked caribou and wolves, including tools used and the exact procedures followed.

19 This important subsection explains what data were calculated (average distance . . .) and how, including the software used and where it came from. (The calculations are listed in Table 1.) Note that the behavior measured (traveling) is carefully defined.

20 RESULTS: The heart of the report and the observation part of the scientific process. This section is organized parallel to the Methods section.

21 This subsection is broken down by periods of observation. Pre-excursion period covers the time between 388's capture and the start of her long-distance travel. The investigators used visual observations as well as telemetry (measurements taken using the global positioning system (GPS)) to gather data. They looked at how 388 cared for her pups, interacted with other adults, and moved about the den area.

22 The key in the lower right-hand corner of the map shows areas (shaded) within which the wolves and caribou moved, and the dotted trail of 388 during her excursion. From the results depicted on this map, the investigators tried to determine when and where 388 might have encountered caribou and how their locations affected her traveling behavior.

23 The wolf's excursion (her long trip away from the den area) is the focus of this study. These paragraphs present detailed measurements of daily movements during her two-week trip—how far she traveled, how far she was from collared caribou, her time spent traveling and resting, and her rate of speed. Authors use the phrase "minimum distance traveled" to acknowledge they couldn't track every step but were measuring samples of her movements. They knew that she went at least as far as they measured. This shows how scientists try to be exact when reporting results. Results of this study are depicted graphically in the map in Figure 2.

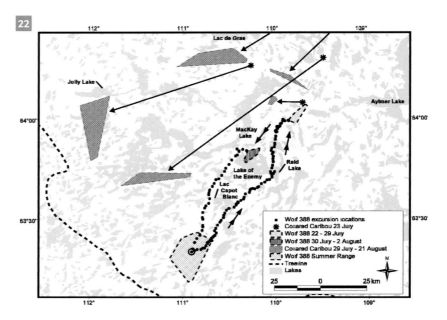

Figure 2. Details of a long foraging movement by female wolf 388 between 19 July and 2 August 2002. Also shown are locations and movements of three satellite radio-collared caribou from 23 July to 21 August 2002. On 23 July, the wolf was 8 km from a collared caribou. The farthest point from the den (103 km distant) was recorded on 27 July. Arrows indicate direction of travel.

bouts was 7.1 km (± 0.9 SE, range 1.7–17.0 km). The average duration of foraging bouts for the period was 20.9 h (± 4.5 SE, range 1–71 h).

The average daily distance between the wolf and the nearest collared caribou decreased from 242 km on 12 July, one week before the excursion period, to 126 km on 19 July, the day the excursion began (Table 1).

23 **Excursion Period:** On 19 July at 2203, after spending 14 h at the den, 388 began moving to the northeast and did not return for 336 h (14 d; Figure 2). Whether she traveled alone or with other wolves is unknown. During the excursion, 476 (71%) of 672 possible locations were recorded. The wolf crossed the southeast end of Lac Capot Blanc on a small land bridge, where she paused for 4.5 h after traveling for 19.5 h (37.5

km). Following this rest, she traveled for 9 h (26.3 km) onto a peninsula in Reid Lake, where she spent 2 h before backtracking and stopping for 8 h just off the peninsula. Her next period of travel lasted 16.5 h (32.7 km), terminating in a pause of 9.5 h just 3.8 km from a concentration of locations at the far end of her excursion, where we presume she encountered caribou. The mean duration of these three movement periods was 15.7 h (± 2.5 SE), and that of the pauses, 7.3 h (± 1.5). The wolf required 72.5 h (3.0 d) to travel a minimum of 95 km from her den to this area near caribou (Figure 2). She remained there (35.5 km2) for 151.5 h (6.3 d) and then moved south to Lake of the Enemy, where she stayed (31.9 km^2) for 74 h (3.1 d) before returning to her den. Her greatest distance from the den, 103 km, was recorded 174.5 h (7.3 d) after the excursion began, at 0433 on 27 July. She was 8 km from a collared caribou on 23 July, four days after the excursion began (Table 1).

The return trip began at 0403 on 2 August, 318 h (13.2 d) after leaving the den. She followed a relatively direct path for 18 h back to the den, a distance of 75 km.

The minimum distance traveled during the excursion was 339 km. The estimated overall minimum travel rate was 3.1 km/h, 2.6 km/h away from the den and 4.2 km/h on the return trip.

[24] Post-Excursion Period: We saw three pups when recovering the collar on 20 August, but others may have been hiding in vegetation.

Telemetry recorded 13 foraging bouts in the post-excursion period. The mean distance traveled during these bouts was 18.3 km (+ 2.7 SE, range 1.2–47.7 km), and mean greatest distance from the den was 7.1 km (+ 0.7 SE, range 1.1–11.0 km). The mean duration of these post-excursion foraging bouts was 10.9 h (+ 2.4 SE, range 1–33 h).

When 388 reached her den on 2 August, the distance to the nearest collared caribou was 54 km. On 9 August, one week after she returned, the distance was 79 km (Table 1).

Pre- and Post-Excursion Comparison

[25] We found no differences in the mean distance of foraging bouts before and after the excursion period (F = 1.5, df = 1, 29, p = 0.24). Likewise, the mean greatest distance from the den was similar pre- and post-excursion (F = 0.004, df = 1, 29, p = 0.95). However, the mean duration of 388's foraging bouts decreased by 10.0 h after her long excursion (F = 3.1, df = 1, 29, p = 0.09).

[26] *Caribou Monitoring*

Summer Movements: On 10 July, 5 of 11 collared caribou were dispersed over a distance of 10 km, 140 km south of their calving grounds (Figure 1). On the same day, three caribou were still on the calving grounds, two were between the calving grounds and the leaders, and one was missing. One week later (17 July), the leading radio-collared cows were 100 km farther south (Figure 1). Two were within 5 km of each other in front of the rest, who were more dispersed. All radio-collared cows had left the calving grounds by this time. On 23 July, the leading radio-collared caribou had moved 35 km farther south, and all of them were more widely dispersed. The two cows closest to the leader were 26 km and 33 km away, with 37 km between them. On the next location (29 July), the most southerly caribou were 60 km

farther south. All of the caribou were now in the areas where they remained for the duration of the study (Figure 2).

A Minimum Convex Polygon (Mohr and Stumpf, 1966) around all caribou locations acquired during the study encompassed 85 119 km².

Relative to the Wolf Den: [27] The distance from the nearest collared caribou to the den decreased from 241 km one week before the excursion to 124 km the day it began. The nearest a collared caribou came to the den was 43 km away, on 29 and 30 July. During the study, four collared caribou were located within 100 km of the den. Each of these four was closest to the wolf on at least one day during the period reported.

[28] Discussion

Prey Abundance

Caribou are the single most important prey of tundra [29] wolves (Clark, 1971; Kuyt, 1972; Stephenson and James, 1982; Williams, 1990). Caribou range over vast areas, and for part of the summer, they are scarce or absent in wolf home ranges (Heard et al., 1996). Both the long distance between radio-collared caribou and the den the week before the excursion and the increased time spent foraging by wolf 388 indicate that caribou availability near the den was low. Observations of the pups' being left alone for up to 18 h, presumably while adults were searching for food, provide additional support for low caribou availability locally. Mean foraging bout duration decreased by 10.0 h after the excursion, when collared caribou were closer to the den, suggesting an increase in caribou availability nearby.

Foraging Excursion

One aspect of central place foraging theory (CPFT) [30] deals with the optimality of returning different-sized food loads from varying distances to dependents at a central place (i.e., the den) (Orians and Pearson, 1979). Carlson (1985) tested CPFT and found that the predator usually consumed prey captured far from the central place, while feeding prey captured nearby to dependants. Wolf 388 spent 7.2 days in one area near caribou before moving to a location 23 km back towards the den, where she spent an additional 3.1 days, likely hunting caribou. She began her return trip from this closer location, traveling directly to the den. While away, she may have made one or more successful kills and spent time meeting her own energetic needs before returning to the den. Alternatively, it may have taken several attempts to make a kill,

24 Post-excursion measurements of 388's movements were made to compare with those of the pre-excursion period. In order to compare, scientists often use means, or averages, of a series of measurements—mean distances, mean duration, etc.

25 In the comparison, authors used statistical calculations (F and df) to determine that the differences between pre- and post-excursion measurements were statistically insignificant, or close enough to be considered essentially the same or similar.

26 As with wolf 388, the investigators measured the movements of caribou during the study period. The areas within which the caribou moved are shown in Figure 2 by shaded polygons mentioned in the second paragraph of this subsection.

27 This subsection summarizes how distances separating predators and prey varied during the study period.

28 DISCUSSION: This section is the interpretation part of the scientific process.

29 This subsection reviews observations from other studies and suggests that this study fits with patterns of those observations.

30 Authors discuss a prevailing theory (CBFT) which might explain why a wolf would travel far to meet her own energy needs while taking food caught closer to the den back to her pups. The results of this study seem to fit that pattern.

31 Here our authors note other possible explanations for wolves' excursions presented by other investigators, but this study does not seem to support those ideas.

32 Authors discuss possible reasons for why 388 traveled directly to where caribou were located. They take what they learned from earlier studies and apply it to this case, suggesting that the lay of the land played a role. Note that their description paints a clear picture of the landscape.

33 Authors suggest that 388 may have learned in traveling during previous summers where the caribou were. The last two sentences suggest ideas for future studies.

34 Or maybe 388 followed the scent of the caribou. Authors acknowledge difficulties of proving this, but they suggest another area where future studies might be done.

35 Authors suggest that results of this study support previous studies about how fast wolves travel to and from the den. In the last sentence, they speculate on how these observed patterns would fit into the theory of evolution.

36 Authors also speculate on the fate of 388's pups while she was traveling. This leads to . . .

---

which she then fed on before beginning her return trip. We do not know if she returned food to the pups, but such behavior would be supported by CPFT.

31 Other workers have reported wolves' making long round trips and referred to them as "extraterritorial" or "pre-dispersal" forays (Fritts and Mech, 1981; Messier, 1985; Ballard et al., 1997; Merrill and Mech, 2000). These movements are most often made by young wolves (1–3 years old), in areas where annual territories are maintained and prey are relatively sedentary (Fritts and Mech, 1981; Messier, 1985). The long excursion of 388 differs in that tundra wolves do not maintain annual territories (Walton et al., 2001), and the main prey migrate over vast areas (Gunn et al., 2001).

Another difference between 388's excursion and those reported earlier is that she is a mature, breeding female. No study of territorial wolves has reported reproductive adults making extraterritorial movements in summer (Fritts and Mech, 1981; Messier, 1985; Ballard et al., 1997; Merrill and Mech, 2001). However, Walton et al. (2001) also report that breeding female tundra wolves made excursions.

Direction of Movement

32 Possible explanations for the relatively direct route 388 took to the caribou include landscape influence and experience. Considering the timing of 388's trip and the locations of caribou, had the wolf moved northwest, she might have missed the caribou entirely, or the encounter might have been delayed.

A reasonable possibility is that the land directed 388's route. The barrens are crisscrossed with trails worn into the tundra over centuries by hundreds of thousands of caribou and other animals (Kelsall, 1968; Thorpe et al., 2001). At river crossings, lakes, or narrow peninsulas, trails converge and funnel towards and away from caribou calving grounds and summer range. Wolves use trails for travel (Paquet et al., 1996; Mech and Boitani, 2003; P. Frame, pers. observation). Thus, the landscape may direct an animal's movements and lead it to where cues, such as the odor of caribou on the wind or scent marks of other wolves, may lead it to caribou.

33 Another possibility is that 388 knew where to find caribou in summer. Sexually immature tundra wolves sometimes follow caribou to calving grounds (D. Heard, unpubl. data). Possibly, 388 had made such journeys in previous years and killed caribou. If this were the case, then in times of local prey scarcity she might travel to areas where she had hunted successfully before. Continued monitoring of tundra wolves may answer questions about how their food needs are met in times of low caribou abundance near dens.

34 Caribou often form large groups while moving south to the tree line (Kelsall, 1968). After a large aggregation of caribou moves through an area, its scent can linger for weeks (Thorpe et al., 2001:104). It is conceivable that 388 detected caribou scent on the wind, which was blowing from the northeast on 19–21 July (Environment Canada, 2003), at the same time her excursion began. Many factors, such as odor strength and wind direction and strength, make systematic study of scent detection in wolves difficult under field conditions (Harrington and Asa, 2003). However, humans are able to smell odors such as forest fires or oil refineries more than 100 km away. The olfactory capabilities of dogs, which are similar to wolves, are thought to be 100 to 1 million times that of humans (Harrington and Asa, 2003). Therefore, it is reasonable to think that under the right wind conditions, the scent of many caribou traveling together could be detected by wolves from great distances, thus triggering a long foraging bout.

Rate of Travel

35 Mech (1994) reported the rate of travel of Arctic wolves on barren ground was 8.7 km/h during regular travel and 10.0 km/h when returning to the den, a difference of 1.3 km/h. These rates are based on direct observation and exclude periods when wolves moved slowly or not at all. Our calculated travel rates are assumed to include periods of slow movement or no movement. However, the pattern we report is similar to that reported by Mech (1994), in that homeward travel was faster than regular travel by 1.6 km/h. The faster rate on return may be explained by the need to return food to the den. Pup survival can increase with the number of adults in a pack available to deliver food to pups (Harrington et al., 1983). Therefore, an increased rate of travel on homeward trips could improve a wolf's reproductive fitness by getting food to pups more quickly.

Fate of 388's Pups

36 Wolf 388 was caring for pups during den observations. The pups were estimated to be six weeks old, and were seen ranging as far as 800 m from the den. They received some regurgitated food from two of the females, but were unattended for long periods. The excursion started 16 days after our observations, and it is improbable that the pups could have traveled the distance that 388 moved. If the pups died, this would have removed parental responsibility, allowing the long movement.

Our observations and the locations of radiocollared caribou indicate that prey became scarce in the area of the den as summer progressed. Wolf 388 may have abandoned her pups to seek food for herself. However, she returned to the den after the excursion, where she was seen near pups. In fact, she foraged in a similar pattern before and after the excursion, suggesting that she again was providing for pups after her return to the den.

 A more likely possibility is that one or both of the other lactating females cared for the pups during 388's absence. The three females at this den were not seen with the pups at the same time. However, two weeks earlier, at a different den, we observed three females cooperatively caring for a group of six pups. At that den, the three lactating females were observed providing food for each other and trading places while nursing pups. Such a situation at the den of 388 could have created conditions that allowed one or more of the lactating females to range far from the den for a period, returning to her parental duties afterwards. However, the pups would have been weaned by eight weeks of age (Packard et al., 1992), so nonlactating adults could also have cared for them, as often happens in wolf packs (Packard et al., 1992; Mech et al., 1999).

Cooperative rearing of multiple litters by a pack could create opportunities for long-distance foraging movements by some reproductive wolves during summer periods of local food scarcity. We have recorded multiple lactating females at one or more tundra wolf dens per year since 1997. This reproductive strategy may be an adaptation to temporally and spatially unpredictable food resources. All of these possibilities require further study, but emphasize both the adaptability of wolves living on the barrens and their dependence on caribou.

Long-range wolf movement in response to caribou availability has been suggested by other researchers (Kuyt, 1972; Walton et al., 2001) and traditional ecological knowledge (Thorpe et al., 2001). Our report demonstrates the rapid and extreme response of wolves to caribou distribution and movements in summer. Increased human activity on the tundra (mining, road building, pipelines, ecotourism) may influence caribou movement patterns and change the interactions between wolves and caribou in the region. Continued monitoring of both species will help us to assess whether the association is being affected adversely by anthropogenic change.

40 Acknowledgements

This research was supported by the Department of Resources, Wildlife, and Economic Development, Government of the Northwest Territories; the Department of Biological Sciences at the University of Alberta; the Natural Sciences and Engineering Research Council of Canada; the Department of Indian and Northern Affairs Canada; the Canadian Circumpolar Institute; and DeBeers Canada, Ltd. Lorna Ruechel assisted with den observations. A. Gunn provided caribou location data. We thank Dave Mech for the use of GPS collars. M. Nelson, A. Gunn, and three anonymous reviewers made helpful comments on earlier drafts of the manuscript. This work was done under Wildlife Research Permit – WL002948 issued by the Government of the Northwest Territories, Department of Resources, Wildlife, and Economic Development.

41 References

BALLARD, W.B., AYRES, L.A., KRAUSMAN, P.R., REED, D.J., and FANCY, S.G. 1997. Ecology of wolves in relation to a migratory caribou herd in northwest Alaska. Wildlife Monographs 135. 47 p.

BROWN, M.B., and FORSYTHE, A.B. 1974. Robust tests for the equality of variances. Journal of the American Statistical Association 69:364–367.

CARLSON, A. 1985. Central place foraging in the red-backed shrike (*Lanius collurio* L.): Allocation of prey between forager and sedentary consumer. Animal Behaviour 33:664–666.

CLARK, K.R.F. 1971. Food habits and behavior of the tundra wolf on central Baffin Island. Ph.D. Thesis, University of Toronto, Ontario, Canada.

ENVIRONMENT CANADA. 2003. National climate data information archive. Available online: http://www.climate.weatheroffice.ec.gc.ca/Welcome_e.html

FOURNIER, B., and GUNN, A. 1998. Musk ox numbers and distribution in the NWT, 1997. File Report No. 121. Yellowknife: Department of Resources, Wildlife, and Economic Development, Government of the Northwest Territories. 55 p.

FRITTS, S.H., and MECH, L.D. 1981. Dynamics, movements, and feeding ecology of a newly protected wolf population in northwestern Minnesota. Wildlife Monographs 80. 79 p.

GUNN, A., DRAGON, J., and BOULANGER, J. 2001. Seasonal movements of satellite-collared caribou from the Bathurst herd. Final Report to the West Kitikmeot Slave Study Society, Yellowknife, NWT. 80 p. Available online: http://www.wkss.nt.ca/HTML/08_ProjectsReports/PDF/Seasonal MovementsFinal.pdf

HARRINGTON, F.H., and ASA, C.S. 2003. Wolf communication. In: Mech, L.D., and Boitani, L., eds. Wolves: Behavior, ecology, and conservation. Chicago: University of Chicago Press. 66–103.

HARRINGTON, F.H., MECH, L.D., and FRITTS, S.H. 1983. Pack size and wolf pup survival: Their relationship under varying ecological conditions. Behavioral Ecology and Sociobiology 13:19–26.

HARRIS, S.A. 1986. Permafrost distribution, zonation and stability along the eastern ranges of the cordillera of North America. Arctic 39(1):29–38.

HEARD, D.C., and WILLIAMS, T.M. 1992. Distribution of wolf dens on migratory caribou ranges in the Northwest

---

37 Discussion of cooperative rearing of pups and, in turn, to speculation on how this study and what is known about cooperative rearing might fit into the animal's strategies for survival of the species. Again, the authors approach the broader theory of evolution and how it might explain some of their results.

38 And again, they suggest that this study points to several areas where further study will shed some light.

39 In conclusion, the authors suggest that their study supports the hypothesis being tested here. And they touch on the implications of increased human activity on the tundra predicted by their results.

40 ACKNOWLEDGEMENTS: Authors note the support of institutions, companies and individuals. They thank their reviewers and list permits under which their research was carried out.

41 REFERENCES: List of all studies cited in the report. This may seem tedious, but is a vitally important part of scientific reporting. It is a record of the sources of information on which this study is based. It provides readers with a wealth of resources for further reading on this topic. Much of it will form the foundation of future scientific studies like this one.

Territories, Canada. Canadian Journal of Zoology 70:1504–1510.

HEARD, D.C., WILLIAMS, T.M., and MELTON, D.A. 1996. The relationship between food intake and predation risk in migratory caribou and implication to caribou and wolf population dynamics. Rangifer Special Issue No. 2:37–44.

KELSALL, J.P. 1968. The migratory barren-ground caribou of Canada. Canadian Wildlife Service Monograph Series 3. Ottawa: Queen's Printer. 340 p.

KUYT, E. 1962. Movements of young wolves in the Northwest Territories of Canada. Journal of Mammalogy 43:270–271.

———. 1972. Food habits and ecology of wolves on barren-ground caribou range in the Northwest Territories. Canadian Wildlife Service Report Series 21. Ottawa: Information Canada. 36 p.

MECH, L.D. 1994. Regular and homeward travel speeds of Arctic wolves. Journal of Mammalogy 75:741–742.

MECH, L.D., and BOITANI, L. 2003. Wolf social ecology. In: Mech, L.D., and Boitani, L., eds. Wolves: Behavior, ecology, and conservation. Chicago: University of Chicago Press. 1–34.

MECH, L.D., and GESE, E.M. 1992. Field testing the Wildlink capture collar on wolves. Wildlife Society Bulletin 20:249–256.

MECH, L.D., WOLFE, P., and PACKARD, J.M. 1999. Regurgitative food transfer among wild wolves. Canadian Journal of Zoology 77:1192–1195.

MERRILL, S.B., and MECH, L.D. 2000. Details of extensive movements by Minnesota wolves (Canis lupus). American Midland Naturalist 144:428–433.

MERRILL, S.B., ADAMS, L.G., NELSON, M.E., and MECH, L.D. 1998. Testing releasable GPS radiocollars on wolves and white-tailed deer. Wildlife Society Bulletin 26:830–835.

MESSIER, F. 1985. Solitary living and extraterritorial movements of wolves in relation to social status and prey abundance. Canadian Journal of Zoology 63:239–245.

MOHR, C.O., and STUMPF, W.A. 1966. Comparison of methods for calculating areas of animal activity. Journal of Wildlife Management 30:293–304.

MUSIANI, M. 2003. Conservation biology and management of wolves and wolf-human conflicts in western North America. Ph.D. Thesis, University of Calgary, Calgary, Alberta, Canada.

ORIANS, G.H., and PEARSON, N.E. 1979. On the theory of central place foraging. In: Mitchell, R.D., and Stairs, G.F., eds. Analysis of ecological systems. Columbus: Ohio State University Press. 154–177.

PACKARD, J.M., MECH, L.D., and REAM, R.R. 1992. Weaning in an arctic wolf pack: Behavioral mechanisms. Canadian Journal of Zoology 70:1269–1275.

PAQUET, P.C., WIERZCHOWSKI, J., and CALLAGHAN, C. 1996. Summary report on the effects of human activity on gray wolves in the Bow River Valley, Banff National Park, Alberta. In: Green, J., Pacas, C., Bayley, S., and Cornwell, L., eds. A cumulative effects assessment and futures outlook for the Banff Bow Valley. Prepared for the Banff Bow Valley Study. Ottawa: Department of Canadian Heritage.

STEPHENSON, R.O., and JAMES, D. 1982. Wolf movements and food habits in northwest Alaska. In: Harrington, F.H., and Paquet, P.C., eds. Wolves of the world. New Jersey: Noyes Publications. 223–237.

THORPE, N., EYEGETOK, S., HAKONGAK, N., and QITIRMIUT ELDERS. 2001. The Tuktu and Nogak Project: A caribou chronicle. Final Report to the West Kitikmeot/Slave Study Society, Ikaluktuuttiak, NWT. 160 p.

TIMONEY, K.P., LA ROI, G.H., ZOLTAI, S.C., and ROBINSON, A.L. 1992. The high subarctic forest-tundra of northwestern Canada: Position, width, and vegetation gradients in relation to climate. Arctic 45(1):1–9.

WALTON, L.R., CLUFF, H.D., PAQUET, P.C., and RAMSAY, M.A. 2001. Movement patterns of barren-ground wolves in the central Canadian Arctic. Journal of Mammalogy 82:867–876.

WILLIAMS, T.M. 1990. Summer diet and behavior of wolves denning on barren-ground caribou range in the Northwest Territories, Canada. M.Sc. Thesis, University of Alberta, Edmonton, Alberta, Canada.

ZAR, J.H. 1999. Biostatistical analysis. 4th ed. New Jersey: Prentice Hall. 663 p.

Glossary

abdomen The region of the body that contains much of the digestive tract and sometimes part of the reproductive system; in insects, the region behind the thorax.

abiotic Nonbiological, often in reference to physical factors in the environment.

abyssal zone The bottom sediments that lie permanently below deep ocean water.

acid precipitation Rainfall with low pH, primarily created when gaseous sulfur dioxide (SO_2) dissolves in water vapor in the atmosphere, forming sulfuric acid.

acoelomate A body plan of bilaterally symmetrical animals that lack a body cavity between the gut and the body wall.

acoustical signaling A means of animal communication in which a signaler produces a sound that is heard by a signal receiver.

active parental care Parents' investment of time and energy in caring for offspring after they are born or hatched.

adaptation, evolutionary The accumulation of adaptive traits over time.

adaptive radiation A cluster of closely related species that are each adaptively specialized to a specific habitat or food source.

adaptive trait A genetically based characteristic, preserved by natural selection, that increases an organism's likelihood of survival or its reproductive output.

adaptive zone A part of a habitat that may be occupied by a group of species exploiting the same resources in a similar manner.

adductor muscle A muscle that pulls inward toward the median line of the body; in bivalve mollusks, it pulls the shell closed.

adiabatic cooling A decrease in temperature without the actual loss of heat energy, occurring in air masses that expand as they rise in the atmosphere.

aerobe An organism that requires oxygen for cellular respiration.

African emergence hypothesis An hypothesis proposing that modern humans first evolved in Africa and then dispersed to other continents.

agar A gelatinous product extracted from certain red algae or seaweed used as a culture medium in the laboratory and as a gelling or stabilizing agent in foods.

age structure A statistical description or graph of the relative numbers of individuals in each age class in a population.

age-specific fecundity The average number of offspring produced by surviving females of a particular age.

age-specific mortality The proportion of individuals alive at the start of an age interval that died during that age interval.

age-specific survivorship The proportion of individuals alive at the start of an age interval that survived until the start of the next age interval.

albumin The portion of an egg that serves as the main source of nutrients and water for the embryo.

algin Alginic acid, found in the cell walls of brown algae.

allele frequency The abundance of one allele relative to others at the same gene locus in individuals of a population.

allometric growth A pattern of postembryonic development in which parts of the same organism grow at different rates.

allopatric speciation The evolution of reproductive isolating mechanisms between two populations that are geographically separated.

allopolyploidy The genetic condition of having two or more complete sets of chromosomes from different parent species.

alpine tundra A biome that occurs on high mountaintops throughout the world, in which dominant plants form cushions and mats.

alternation of generations The regular alternation of mode of reproduction in the life cycle of an organism, such as the alternation between diploid (sporophyte) and haploid (gametophyte) phases in plants.

altruism A behavioral phenomenon in which individuals appear to sacrifice their own reproductive success to help other individuals.

ammonification A metabolic process in which bacteria and fungi convert organic nitrogen compounds into ammonia and ammonium ions; part of the nitrogen cycle.

Amniota The monophyletic group of vertebrates that have an amnion during embryonic development.

amniote egg A shelled egg that can survive and develop on land.

anaerobe An organism that does not require oxygen to live.

anagenesis The slow accumulation of evolutionary changes in a lineage over time.

anapsid A member of the group of amniote vertebrates having no temporal arches and no spaces on the sides of the skull.

ancestral character A trait that was present in a distant common ancestor.

angiosperm A flowering plant. Its egg-containing ovules mature into seeds within protected chambers called ovaries.

animal behavior The responses of animals to specific internal and external stimuli.

Animalia The taxonomic kingdom that includes all living and extinct animals.

annulus In ferns, a ring of thick-walled cells that nearly encircles the sporangium and functions in spore release.

antenna A chemosensory appendage attached to the head of some adult arthropods.

anterior Indicating the head end of an animal.

antheridium (plural, antheridia) In plants, a structure in which sperm are produced.

Anthocerophyta The phylum comprising hornworts.

Anthophyta The phylum comprising flowering plants.

Anthropoidea The monophyletic lineage of primates comprising monkeys, apes, and humans.

aphotic zone Deeper water of a lake or ocean where sunlight does not penetrate.

apical meristem A region of unspecialized dividing cells at shoot tips and root tips of a plant.

aposematic coloration Bright, contrasting patterns that advertise the unpalatability of poisonous or repellant species.

appendicular skeleton The bones comprising the pectoral (shoulder) and pelvic (hip) girdles and limbs of a vertebrate.

aquatic succession A process in which debris from rivers and runoff accumulates in a body of fresh water, causing it to fill in at the margins.

Archaea One of two domains of prokaryotes; archaeans have some unique molecular and biochemical traits, but they also share some traits with Bacteria and other traits with Eukarya.

archegonium The flask-shaped structure in which bryophyte eggs form.

archenteron The central endoderm-lined cavity of an embryo at the gastrula stage, which forms the primitive gut.

Archosauromorpha A diverse group of diapsids that comprises crocodilians, pterosaurs, and dinosaurs (including birds).

arctic tundra A treeless biome that stretches from the boreal forests to the polar ice cap in Europe, Asia, and North America.

artificial selection Selective breeding of animals or plants to ensure that certain desirable traits appear at higher frequency in successive generations.

ascocarp A reproductive body that bears or contains asci.

ascus (plural, asci) A saclike cell in ascomycetes (sac fungi) in which meiosis gives rise to haploid sexual spores (meiospores).

asymmetrical Characterized by lack of proportion in the spatial arrangement or placement of parts.

atmosphere The component of the biosphere that includes the gases and airborne particles enveloping the planet.

atrial siphon A tube through which invertebrate chordates expel digestive and metabolic wastes.

atriopore The hole in the body wall of a cephalochordate through which water is expelled from the body.

atrium A body cavity or chamber surrounding the perforated pharynx of invertebrate chordates; also, one of the chambers that receive blood returning to the heart.

autopolyploidy The genetic condition of having more than two sets of chromosomes from the same parent species.

autotroph An organism that produces its own food using simple inorganic compounds from its environment and energy from the sun or from oxidation of inorganic substances.

autumn overturn A process in which winds mix the water in a lake vertically, equalizing the concentrations of dissolved gases and nutrients at all depths.

axial skeleton The bones comprising the head and trunk of a vertebrate: the cranium, vertebral column, ribs, and sternum (breastbone).

bacillus (plural, bacilli) A cylindrical or rod-shaped prokaryote.

background extinction rate The average rate of extinction of taxa through time.

bacteria One of the two domains of prokaryotes; collectively, bacteria are the most metabolically diverse organisms.

bacteriophage A virus that attacks bacteria; also known as a phage.

balanced polymorphism The maintenance of two or more phenotypes in fairly stable proportions over many generations.

basal angiosperm Any of the earliest branches of the flowering plant lineage; includes the star anise group and water lilies.

basidiocarp A fruiting body of a basidiomycete; mushrooms are examples.

basidiospore A haploid sexual spore produced by basidiomycete fungi.

basidium (plural, basidia) A small, club-shaped structure in which sexual spores of basidiomycetes arise.

Batesian mimicry The form of defense in which a palatable or harmless species resembles an unpalatable or poisonous one.

behavioral isolation A prezygotic reproductive isolating mechanism in which two species do not mate because of differences in courtship behavior; also known as ethological isolation.

behavioral repertoire The set of actions that an animal can perform in response to stimuli in its environment.

benthic province The bottom sediments in the ocean.

benthos Species living in and on the bottom sediments of the ocean.

bilateral symmetry The body plan of animals in which the body can be divided into mirror image right and left halves by a plane passing through the midline of the body.

binomial Relating to or consisting of two names or terms.

binomial nomenclature The naming of species with a two part scientific name, the first indicating the genus and the second indicating the species.

biodiversity The richness of living systems as reflected in genetic variability within and among species, the number of species living on Earth, and the variety of communities and ecosystems.

biodiversity hotspot An area where biodiversity is both highly concentrated and endangered.

biofilm A microbial community consisting of a complex aggregation of microorganisms attached to a surface.

biogeochemical cycle Any of several global processes in which a nutrient circulates between the abiotic environment and living organisms.

biogeographical realm A major region of Earth that is occupied by distinct evolutionary lineages of plants and animals.

biogeography The study of the geographical distributions of plants and animals.

biological evolution A process that causes the genomes of organisms to differ from those of their ancestors.

biological lineage An evolutionary sequence of ancestral organisms and their descendants.

biological magnification The increasing concentration of nondegradable poisons in the tissues of animals at higher trophic levels.

biological species concept The definition of species based upon the ability of populations to interbreed and produce fertile offspring.

bioluminescent An organism that glows or releases a flash of light, particularly when disturbed.

biomass The dry weight of biological material per unit area or volume of habitat.

biome A large scale vegetation type and its associated microorganisms, fungi, and animals.

biosphere The total of all ecosystems on Earth, including the lithosphere, hydrosphere, and atmosphere.

biota The total collection of organisms in a geographic region.

biotic Biological, often in reference to living components of the environment.

bipedalism The habit in animals of walking upright on two legs.

blastopore The opening at one end of the archenteron in the gastrula that gives rise to the mouth in protostomes and the anus in deuterostomes.

bone The densest form of connective tissue, in which living cells secrete the mineralized matrix of collagen and calcium salts that surrounds them; forms the skeleton.

book lungs Pocketlike respiratory organs found in some arachnids consisting of several parallel membrane folds arranged like the pages of a book.

boreal forest A biome that is a circumpolar expanse of evergreen coniferous trees in Europe, Asia, and North America.

brachiation A pattern of locomotion among primates in which an individual swings below branches from one handhold to another.

Bryophyta The phylum of nonvascular plants, including mosses and their relatives.

bryophyte A general term for plants (such as mosses) that lack internal transport vessels.

canines Pointed, conical teeth of a mammal, located between the incisors and the first premolars, that are specialized for biting and piercing.

capsid *See* coat.

capsule A slime coat typically composed of polysaccharide that is attached to many bacterial cells.

carapace A protective outer covering that extends backward behind the head on the dorsal side of an animal, such as the shell of a turtle or lobster.

carbon cycle The global circulation of carbon atoms, especially via the processes of photosynthesis and respiration.

carrageenan A chemical extracted from the red alga *Eucheuma* that is used to thicken and stabilize paints, dairy products such as pudding and ice cream, and many other creams and emulsions.

carrying capacity The maximum size of a population that an environment can support indefinitely.

catastrophism The theory that Earth has been affected by sudden, violent events that were sometimes worldwide in scope.

cellular respiration The process by which energy-rich molecules are broken down to produce energy in the form of ATP.

cellular slime mold Any of a variety of primitive organisms of the phylum Acrasiomycota, especially of the genus *Dictyostelium*; the life cycle is characterized by a slimelike amoeboid stage and a multicellular reproductive stage.

cephalization The development of an anterior head where sensory organs and nervous system tissue are concentrated.

cephalothorax The anterior section of an arachnid, consisting of a fused head and thorax.

chaparral A biome comprising a scrubby mix of short trees and shrubs that dominates coastal land between 30° and 40° latitude, where winters are cool and wet and summers hot and dry.

character displacement The phenomenon in which allopatric populations are morphologically similar and use similar resources, but sympatric populations are morphologically different and use different resources; may also apply to characters influencing mate choice.

charophyte A member of the group of green algae most similar to the algal ancestors of land plants.

chelicerae The first pair of fanglike appendages near the mouth of an arachnid, used for biting prey and often modified for grasping and piercing.

chemical signal Any secretion from one cell type that can alter the behavior of a different cell that bears a receptor for it; a means of cell communication."

chemoautotroph An organism that obtains energy by oxidizing inorganic substances such as hydrogen, iron, sulfur, ammonia, nitrites, and nitrates and uses carbon dioxide as a carbon source.

chemoheterotroph An organism that oxidizes organic molecules as an energy source and obtains carbon in organic form.

chemotroph An organism that obtains energy by oxidizing inorganic or organic substances.

chitin A polysaccharide that contains nitrogen and is present in the cell walls of fungi and the exoskeletons of arthropods.

choanocyte One of the inner layer of flagellated cells lining the body cavity of a sponge.

Choanoflagellata A group of minute, single-celled protists found in water; the flask-shaped body has a collar of closely packed microvilli that surrounds the single flagellum by which it moves and take in food.

circulatory vessel An element of the circulatory system through which fluid flows and carries nutrients and oxygen to tissues and remove wastes.

clade A monophyletic group of organisms that share homologous features derived from a common ancestor.

cladistics An approach to systematics that uses shared derived characters to infer the phylogenetic relationships and evolutionary history of groups of organisms.

cladogenesis The evolution of two or more descendant species from a common ancestor.

cladogram A branching diagram in which the endpoints of the branches represent different species of organisms, used to illustrate phylogenetic relationships.

claspers A pair of organs on the pelvic fins of male crustaceans and sharks, which help transfer sperm into the reproductive tract of the female.

class A Linnaean taxonomic category that ranks below a phylum and above an order.

classical conditioning A type of learning in which an animal develops a mental association between two phenomena that are usually unrelated.

classification An arrangement of organisms into hierarchical groups that reflect their relatedness.

cleavage Mitotic cell divisions of the zygote that produce a blastula from a fertilized ovum.

climate The weather conditions prevailing over an extended period of time.

climax community A relatively stable, late successional stage in which the dominant vegetation replaces itself and persists until an environmental disturbance eliminates it, allowing other species to invade.

climograph A graph that portrays the particular combination of temperature and rainfall conditions where each terrestrial biome occurs.

cline A pattern of smooth variation in a characteristic along a geographical gradient.

closed circulatory system A circulatory system in which the fluid, blood, is confined in blood vessels and is distinct from the interstitial fluid.

clumped dispersion A pattern of distribution in which individuals in a population are grouped together.

cnidocyte A prey-capturing and defensive cell in the epidermis of cnidarians.

coat A layer of protein that surrounds a virus particle; also known as a capsid.

coccus (plural, cocci) A spherical prokaryote.

coelom A fluid-filled body cavity in bilaterally symmetrical animals that is completely lined with derivatives of mesoderm.

coelomate A body plan of bilaterally symmetrical animals that have a coelom.

coenocytic Condition in which a single cell has many nuclei.

coevolution The evolution of genetically based, reciprocal adaptations in two or more species that interact closely in the same ecological setting.

cohort A group of individuals of similar age.

colony Multiple individual organisms of the same species living in a group.

commensalism A symbiotic interaction in which one species benefits and the other is unaffected.

community ecology The ecological discipline that examines groups of populations occurring together in one area.

comparative morphology Analysis of the structure of living and extinct organisms.

compass orientation A wayfinding mechanism that allows animals to move in a particular direction, often over a specific distance or for a prescribed length of time.

competitive exclusion principle The ecological principle stating that populations of two or more species cannot coexist indefinitely if they rely on the same limiting resources and exploit them in the same way.

complete digestive system A digestive system having a mouth at one end, through which food enters, and an anus at the other end, through which undigested waste is voided.

complete metamorphosis The form of metamorphosis in which an insect passes through four separate stages of growth: egg, larva, pupa, and adult.

complex virus A bacteriophage with a DNA genome that has a tail attached at one side of a polyhedral head.

compound eye The eye of most insects and some crustaceans, composed of many faceted, light-sensitive units called ommatidia fitted closely together, each having its own refractive system and each forming a portion of an image.

cone In cone-bearing plants, a cluster of sporophylls.

conidiophore A fungal hypha that gives rise to conidia.

conidium (plural, conidia) An asexually produced fungal spore.

Coniferophyta The major phylum of cone-bearing gymnosperms, most of which are substantial trees; includes pines, firs, and other conifers.

conjugation In bacteria, the process by which a copy of part of the DNA of a donor cell moves through the cytoplasmic bridge into the recipient cell where genetic recombination can occur. In ciliate protozoans, a process of sexual reproduction in which individuals of the same species temporarily couple and exchange genetic material.

conodont An abundant, bonelike fossil dating from the early Paleozoic era through the early Mesozoic era, now described as a feeding structure of some of the earliest vertebrates.

conservation biology An interdisciplinary science that focuses on the maintenance and preservation of biodiversity.

consumer An organism that consumes other organisms in a community or ecosystem.

continental climate Climate not moderated by the distant ocean.

continental drift The long-term movement of continents as a result of plate tectonics.

continuous distribution A geographical distribution in which a species lives in suitable habitats throughout a geographical area.

contractile vacuole A specialized cytoplasmic organelle that pumps fluid in a cyclical manner from within the cell to the outside by alternately filling and then contracting to release its contents at various points on the surface of the cell.

convergent evolution The evolution of similar adaptations in distantly related organisms that occupy similar environments.

coral reef A structure made from the hard skeletons of coral animals or polyps; found largely in tropical and subtropical marine environments.

cortex Generally, an outer, rindlike layer. In mammals, the outer layer of the brain, the kidneys, or the adrenal glands. In plants, the outer region of tissue in a root or stem lying between the epidermis and the vascular tissue, composed mainly of parenchyma.

courtship display A behavior performed by males to attract potential mates or to reinforce the bond between a male and a female.

cranium The part of the skull that encloses the brain.

Crenarchaeota A major group of the domain Archaea, separated from the other archaeans based mainly on rRNA sequences.

critical period A restricted stage of development early in life during which an animal has the capacity to respond to specific environmental stimuli.

cryptic coloration Coloration that allows an organism to match its background and hence become less vulnerable to predation or recognition by prey.

cuticle The outer layer of plants and some animals, which helps prevent desiccation by slowing water loss.

Cycadophyta A phylum of palmlike gymnosperms known as cycads; the pollen-bearing and seed-bearing cones (strobili) occur on separate plants.

cytoplasmic streaming Intracellular movement of cytoplasm.

decomposer A small organism, such as a bacterium or fungus, that reduces dead or dying organic material to its molecular components.

demographic transition model A graphical depiction of the historical relationship between a country's economic development and its birth and death rates.

demography The statistical study of the processes that change a population's size and density through time.

denitrification A metabolic process in which certain bacteria convert nitrites or nitrates into nitrous oxide and then into molecular nitrogen, which enters the atmosphere.

density-dependent Description of environmental factors for which the strength of their effect on a population varies with the population's density.

density-independent Description of environmental factors for which the strength of their effect on a population does not vary with the population's density.

derived character A new version of a trait found in the most recent common ancestor of a group.

descent with modification Biological evolution.

desert A sparsely vegetated biome that forms where rainfall averages less than 25 cm per year.

desertification A process in which large tracts of subtropical forest are cleared and overused, the groundwater table recedes to deeper levels, less surface water is available for plants, soil accumulates high concentrations of salts, and topsoil is eroded by wind and water.

determinate cleavage A type of cleavage in protosomes in which each cell's developmental path is determined as the cell is produced.

detritivore An organism that extracts energy from the organic detritus (refuse) produced at other trophic levels.

diapsid A member of a group within the amniote vertebrates having a skull with two temporal arches. Their living descendants include lizards and snakes, crocodilians, and birds.

dikaryon The life stage in certain fungi in which a cell contains two genetically distinct haploid nuclei.

diploblastic An animal body plan in which adult structures arise from only two cell layers, the ectoderm and the endoderm.

directional selection A type of selection in which individuals near one end of the phenotypic spectrum have the highest relative fitness.

Discicristates A protist group of single-celled, highly motile organisms that swim by means of flagella, and that have characteristic disc-shaped mitochondrial cristae (inner mitochondrial membranes).

disclimax community An ecological community in which regular disturbance inhibits successional change.

disjunct distribution A geographical distribution in which populations of the same species or closely related species live in widely separated locations.

dispersal The movement of organisms away from their place of origin.

dispersion The spatial distribution of individuals within a population's geographical range.

disruptive selection A type of natural selection in which extreme phenotypes have higher relative fitness than intermediate phenotypes.

domain The highest taxonomic category.

dominance hierarchy A social system in which the behavior of each individual is constrained by that individual's status in a highly structured social ranking.

dorsal Indicating the back side of an animal.

ecdysis Shedding of the cuticle, exoskeleton, or skin; molting.

ecological community An assemblage of species living in the same place.

ecological efficiency The ratio of net productivity at one trophic level to net productivity at the trophic level below it.

ecological isolation A prezygotic reproductive isolating mechanism in which species that live in the same geographical region occupy different habitats.

ecological niche The resources a population uses and the environmental conditions it requires over its lifetime.

ecological pyramid A diagram illustrating the effects of energy transfer from one trophic level to the next.

ecological succession A somewhat predictable series of changes in the species composition of a community over time.

ecology The study of the interactions between organisms and their environments.

ecosystem A biological community and the physical environment with which it interacts.

ecosystem ecology A ecological discipline that explores the cycling of nutrients and the flow of energy between the biotic components of an ecological community and the abiotic environment.

ecosystem services The ecological processes on which all life depends, which include decomposition of wastes, nutrient recycling, oxygen production, maintenance of fertile topsoil, and air and water purification.

ecosystem valuation A process in which ecosystem services are assigned an economic value.

ecotone A wide transition zone between adjacent communities.

ecotourism An activity in which visitors, often from wealthy countries, pay a fee to visit a nature preserve.

ectoderm The outermost of the three primary germ layers of an embryo, which develops into epidermis and nervous tissue.

ectomycorrhiza A mycorrhiza that grows between and around the young roots of trees and shrubs but does not enter root cells.

ectoparasite A parasite that lives on the exterior of its host organism.

edge effect A phenomenon in which the removal of natural vegetation disrupts the local physical environment, exposing the borders of the remaining habitat to additional sunlight, wind, and rainfall.

electrical signaling A means of animal communication in which a signaler emits an electric discharge that can be received by another individual.

electroreceptor A specialized sensory receptor that detects electrical fields.

embryo An organism in its early stage of reproductive development, beginning in the first moments after fertilization.

embryophyte Any plant in which the embryo is retained within maternal tissue.

emigration The movement of individuals out of a population.

endangered species A species in danger of extinction throughout all or a significant portion of its range.

endemic species A species that occurs in only one place on Earth.

endoderm The innermost of the three primary germ layers of an embryo, which develops into the gastrointestinal tract and, in some animals, the respiratory organs.

endomycorrhiza A mycorrhiza in which the fungal hyphae penetrate into cells of the root.

endoparasite A parasite that lives in the internal organs of its host organism.

endospore A small, metabolically inactive, asexual spore that develops within some bacterial cells when environmental conditions become unfavorable.

endosymbiont hypothesis The proposal that the membranous organelles of eukaryotic cells (mitochondria and chloroplasts) may have originated from symbiotic relationships between two prokaryotic cells.

endotoxin A lipopolysaccharide released from the outer membrane of the cell wall when a bacterium dies and lyses.

energy budget The total amount of energy that an organism can accumulate and use to fuel its activities.

enterocoelom In deuterostomes, the body cavity pinched off by outpocketings of the archenteron.

enveloped virus A virus that has a surface membrane derived from its host cell.

epilimnion The top layer of the limnetic zone in a lake.

epiphyte A plant that grows independently on other plants and obtains nutrients and water from the air.

equilibrium theory of island biogeography An hypothesis suggesting that the number of species on an island is governed by a give and take between the immigration of new species to the island and the extinction of species already there.

estuary A coastal habitat where tidal seawater mixes with fresh water from rivers, streams, and runoff.

ethology A discipline that focuses on how animals behave in their natural environments.

eudicot A plant belonging to the Eudicotyledones, one of the two major classes of angiosperms; their embryos generally have two seed leaves (cotyledons), and their pollen grains have three grooves.

Eukarya The domain that includes all eukaryotes, organisms that contain a membrane-bound nucleus within each of their cells; all protists, plants, fungi, and animals.

Euryarchaeota A major group of the domain Archaea, members of which are found in different extreme environments. They include methanogens, extreme halophiles, and some extreme thermophiles.

eusocial A form of social organization, observed in some insect species, in which numerous related individuals—a large percentage of them sterile female workers—live and work together in a colony for the reproductive benefit of a single queen and her mate(s).

eutrophic lake A lake that is rich in nutrients and organic matter.

evolutionary developmental biology A field of biology that compares the genes controlling the developmental processes of different animals to determine the evolutionary origin of morphological novelties and developmental processes.

evolutionary divergence A process whereby natural selection or genetic drift causes populations to become more different over time.

exoenzyme An enzymatic protein released by a bacterium that digests plasma membranes and causes cells of the infected host to rupture and die.

exoskeleton A hard external covering of an animal's body that blocks the passage of water and provides support and protection.

exotic species A nonnative organism.

exotoxin A toxic protein that leaks from or is secreted from a bacterium and interferes with the biochemical processes of body cells in various ways.

exploitative competition Form of competition in which two or more individuals or populations use the same limiting resources.

exponential model of population growth Model that describes unlimited population growth.

extinction The death of the last individual in a species or the last species in a lineage.

facilitation hypothesis An hypothesis that explains ecological succession, suggesting that species modify the local environment in ways that make it less suitable for themselves but more suitable for colonization by species typical of the next successional stage.

facultative anaerobe An organism that can live in the presence or absence of oxygen, using oxygen when it is present and living by fermentation under anaerobic conditions.

family A Linnaean taxonomic category that ranks below an order and above a genus.

family planning program A program that educates people about ways to produce an optimal family size on an economically feasible schedule.

feather A sturdy, lightweight structure of birds, derived from scales in the skin of their ancestors.

fixed action pattern A highly stereotyped instinctive behavior; when triggered by a specific cue, it is performed over and over in almost exactly the same way.

flagellum (plural, flagella) A long, thread-like, cellular appendage responsible for movement; found in both prokaryotes and eukaryotes, but with different structures and modes of locomotion.

flower The reproductive structure of angio-sperms, consisting of floral parts grouped on a stem; the structure in which seeds develop.

food chain A depiction of the trophic struc-ture of a community, a portrait of who eats whom.

food vacuole A membrane-bound sac used for digestion.

food web A set of interconnected food chains with multiple links.

fossil The remains or traces of an organism of a past geologic age embedded and preserved in Earth's crust.

founder effect An evolutionary phenomenon in which a population that was established by just a few colonizing individuals has only a fraction of the genetic diversity seen in the population from which it was derived.

frequency-dependent selection A form of natural selection in which rare phenotypes have a selective advantage simply because they are rare.

fruit A mature ovary, often with accessory parts, from a flower.

fruiting body In some fungi, a stalked, spore-producing structure such as a mushroom.

fundamental niche The range of conditions and resources that a population can possibly tolerate and use.

gametangium A cell or organ in which gametes are produced.

gamete A haploid cell, and egg or sperm. Haploid cells fuse during sexual reproduc-tion to form a diploid zygote.

gametic isolation A prezygotic reproductive isolating mechanism caused by incompatibil-ity between the sperm of one species and the eggs of another; may prevent fertilization.

gametophyte The phase of the plant life cycle that gives rise to haploid gametes (eggs and sperm).

ganglion A functional concentration of nervous system tissue composed principally of nerve-cell bodies, usually lying outside the central nervous system.

gastrovascular cavity A saclike body cavity with a single opening, a mouth, which serves both digestive and circulatory functions.

gemma (plural, gemmae) Small cell mass that forms in cuplike growths on a thallus.

gene flow The transfer of genes from one population to another through the move-ment of individuals or their gametes.

gene pool The sum of all alleles at all gene loci in all individuals in a population.

generalized compartment model A model used to describe nutrient cycling in which two criteria—organic *versus* inorganic nutrients and available *versus* unavailable nutrients—define four compartments where nutrients accumulate.

generation time The average time between the birth of an organism and the birth of its offspring.

genetic drift Random fluctuations in allele frequencies as a result of chance events; usually reduces genetic variation in a population.

genetic equilibrium The point at which neither the allele frequencies nor the geno-type frequencies in a population change in succeeding generations.

genotype frequency The percentage of indi-viduals in a population possessing a particu-lar genotype.

genus A Linnaean taxonomic category ranking below a family and above a species.

geographical range The overall spatial boundaries within which a population lives.

gill arch One of the series of curved support-ing structures between the slits in the pharynx of a chordate.

gill slit One of the openings in the pharynx of a chordate through which water passes out of the pharynx.

Ginkgophyta A plant phylum with a single living species, the ginkgo (or maiden-hair) tree.

Gnathostomata The group of vertebrates with moveable jaws.

gradualism The view that Earth and its living systems changed slowly over its history.

gradualist hypothesis The hypothesis that large changes in either geological features or biological lineages result from the slow, continuous accumulation of small changes over time.

Gram stain technique A technique of stain-ing bacteria to distinguish between types of bacteria with different cell wall compositions.

Gram-negative Describing bacteria that do not retain the stain used in the Gram stain technique.

Gram-positive Describing bacteria that appear purple when stained using the Gram stain technique.

greenhouse effect A phenomenon in which certain gases foster the accumulation of heat in the lower atmosphere, maintaining warm temperatures on Earth.

gross primary productivity The rate at which producers convert solar energy into chemical energy.

gymnosperm A seed plant that produces "naked" seeds not enclosed in an ovary.

habitat The specific environment in which a population lives, as characterized by its biotic and abiotic features.

habitat fragmentation A process in which remaining areas of intact habitat are reduced to small, isolated patches.

habituation The learned loss of responsive-ness to stimuli.

half-life The time it takes for half of a given amount of a radioisotope to decay.

haplodiploidy A pattern of sex determina-tion in insects in which females are diploid and males are haploid.

Hardy-Weinberg principle An evolutionary rule of thumb that specifies the conditions under which a population of diploid organ-isms achieves genetic equilibrium.

haustorium (plural, haustoria) The hyphal tip of a parasitic fungus that penetrates a host plant and absorbs nutrients from it; likewise in parasitic flowering plants, a root that can penetrate a host's tissues and absorb nutrients.

head The anteriormost part of the body, containing the brain, sensory structures, and feeding apparatus.

head–foot In mollusks, the region of the body that provides the major means of loco-motion and contains concentrations of nervous system tissues and sense organs.

helical virus A virus in which the protein subunits of the coat assemble in a rodlike spiral around the genome.

hemolymph The circulatory fluid of inverte-brates with open circulatory systems, includ-ing mollusks and arthropods.

Hepatophyta The phylum that includes liverworts and their bryophyte relatives.

herbivory The process in which herbivores consume plants.

heterochrony Changes in the relative rate of development of morphological characters.

heterosporous Producing two types of spores, "male" microspores and "female" megaspores.

heterotroph An organism that acquires energy and nutrients by eating other organ-isms or their remains.

heterozygote advantage An evolutionary circumstance in which individuals that are heterozygous at a particular locus have higher relative fitness than either homozygote.

historical biogeography The study of the geographical distributions of plants and animals in relation to their evolutionary history.

homeobox The region of a homeotic gene that corresponds to an amino acid section of the homeodomain.

homeodomain An encoded transcription factor of each protein that binds to a region in the promoters of the genes whose transcription it regulates.

homeotic gene Any of the family of genes that determines the structure of body parts during embryonic development.

hominid A member of a monophyletic group of primates, characterized by an erect bipedal stance, that includes modern humans and their recent ancestors.

Hominoidea The monophyletic group of primates that includes apes and humans.

homologies Characteristics shared by a set of species because they inherited them from their common ancestor.

homologous traits Characteristics that are similar in two species because they inherited the genetic basis of the trait from their common ancestor.

homoplasies Characteristics shared by a set of species, often because they live in similar environments, but not present in their common ancestor; often the product of convergent evolution.

homosporous Producing only one type of spore.

host A species that is fed upon by a parasite.

host race A population of insects that may be reproductively isolated from other populations of the same species as a consequence of their adaptation to feed on a specific host plant species.

hybrid breakdown A postzygotic reproductive isolating mechanism in which hybrids are capable of reproducing, but their offspring have either reduced fertility or reduced viability.

hybrid inviability A postzygotic reproductive isolating mechanism in which a hybrid individual has a low probability of survival to reproductive age.

hybrid sterility A postzygotic reproductive isolating mechanism in which hybrid offspring cannot form functional gametes.

hybrid zone A geographical area where the hybrid offspring of two divergent populations or species are common.

hydrologic cycle The global cycling of water between the ocean, the atmosphere, land, freshwater ecosystems, and living organisms.

hydrosphere The component of the biosphere that encompasses all the waters on Earth, including oceans, rivers, and polar ice caps.

hydrostatic skeleton A structure consisting of muscles and fluid that, by themselves, provide support for the animal or part of the animal; no rigid support, like a bone, is involved.

hypha (plural, hyphae) Any of the threadlike filaments that form the mycelium of a fungus.

hypolimnion The deep water of the profundal zone of a lake.

immigration Movement of organisms into a population.

imprinting The process of learning the identity of a caretaker and potential future mate during a critical period.

inbreeding A special form of nonrandom mating in which genetically related individuals mate with each other.

incisors Flattened, chisel-shaped teeth of mammals, located at the front of the mouth, that are used to nip or cut food.

incomplete metamorphosis In certain insects, a life cycle characterized by the absence of a pupal stage between the immature and adult stages.

incurrent siphon A muscular tube that brings water containing oxygen and food into the body of an invertebrate.

indeterminate cleavage A type of cleavage, observed in many deuterostomes, in which the developmental fates of the first few cells produced by mitosis are not determined as soon as cells are produced.

inhibition hypothesis An hypothesis suggesting that new species are prevented from occupying a community by whatever species are already present.

insight learning A phenomenon in which animals can solve problems without apparent trial-and-error attempts at the solution.

instinctive behavior A genetically "programmed" response that appears in complete and functional form the first time it is used.

interference competition Form of competition in which individuals fight over resources or otherwise harm each other directly.

intermediate disturbance hypothesis Hypothesis proposing that species richness is greatest in communities that experience fairly frequent disturbances of moderate intensity.

interspecific competition The competition for resources between species.

intertidal zone The shoreline that is alternately submerged and exposed by tides.

intraspecific competition The dependence of two or more individuals in a population on the same limiting resource.

intrinsic rate of increase The maximum possible per capita population growth rate in a population living under ideal conditions.

invertebrate An animal without a vertebral column.

karyogamy In plants, the fusion of two sexually compatible haploid nuclei after cell fusion (plasmogamy).

keeled sternum The ventrally extended breastbone of a bird to which the flight muscles attach.

keystone species A species that has a greater effect on community structure than its numbers might suggest.

kin selection Altruistic behavior to close relatives, allowing them to produce proportionately more surviving copies of the altruist's genes than the altruist might otherwise have produced on its own.

kinesis A change in the rate of movement or the frequency of turning movements in response to environmental stimuli.

kingdom Fungi The taxonomic kingdom that includes all living or extinct fungi, including chytrids, zygomycetes, glomeromycetes, ascomycetes, basidiomycetes, conidial fungi, and microsporidia. Fungi are thought to be more closely related to animals than to plants.

kingdom Plantae The taxonomic kingdom encompassing all living or extinct plants.

kingdom A Linnaean taxonomic category that ranks below a domain and above a phylum.

Korarchaeota A group of Archaea recognized solely on the basis of rRNA coding sequences in DNA taken from environmental samples.

***K*-selected species** Long-lived species that thrive in more stable environments.

landscape ecology The field that examines how large-scale ecological factors—such as the distribution of plants, topography, and human activity—influence local populations and communities.

larva A sexually immature stage in the life cycle of many animals that is morphologically distinct from the adult.

latent phase The time during which a virus remains in the cell in an inactive form.

lateral-line system The complex of organs and sensory receptors along the sides of many fishes and amphibians that detects vibrations in water.

learned behavior A response of an animal that is dependent upon having a particular kind of experience during development.

learning A process in which experiences stored in memory change the behavioral responses of an animal.

lek A display ground where males each possess a small territory from which they court attentive females.

Lepidosauromorpha A monophyletic lineage of diapsids that includes both marine and terrestrial animals, represented today by sphenodontids, lizards, and snakes.

lichen A single vegetative body that is the result of an association between a fungus and a photosynthetic partner, often an alga.

life cycle A series of developmental changes through which an organism passes from fertilization to death.

life history The lifetime pattern of growth, maturation, and reproduction that is characteristic of a population or species.

life table A chart that summarizes the demographic characteristics of a population.

lignin A tough, rather inert polymer that strengthens the secondary walls of various plant cells and thus helps vascular plants to grow taller and stay erect on land.

limiting nutrient An element in short supply within an ecosystem, the shortage of which limits productivity.

limnetic zone The sunlit, open water in a lake, beyond the zone where plants rooted in the bottom can grow.

lipopolysaccharide A large molecule that consists of a lipid and a carbohydrate joined by a covalent bond.

lithosphere The component of the biosphere that includes the rocks, sediments, and soils of the crust.

logistic model of population growth Model of population growth that assumes that a population's per capita growth rate decreases as the population gets larger.

lophophore The circular or U-shaped fold with one or two rows of hollow, ciliated tentacles that surrounds the mouth of brachiopods, bryozoans, and phoronids and is used to gather food.

Lycophyta The plant phylum that includes club mosses and their close relatives.

macroevolution Large-scale evolutionary patterns in the history of life, producing major changes in species and higher taxonomic groups.

macronucleus In ciliophorans, a single large nucleus that develops from a micronucleus but loses all genes except those required for basic "housekeeping" functions of the cell and for ribosomal RNAs.

magnoliids An angiosperm group that includes magnolias, laurels, and avocados; they are more closely related to monocots than to eudicots.

Malpighian tubules The main organs of excretion and osmoregulation in insects, helping them to maintain water and electrolyte balance.

mammary glands Specialized organs of female mammals that produce energy-rich milk, a watery mixture of fats, sugars, proteins, vitamins, and minerals.

mantle One or two folds of the body wall that lines the shell and secretes the substance that forms the shell in mollusks.

maritime climate Climate tempered by ocean winds.

marsupium An external pouch on the abdomen of many female marsupials, containing the mammary glands, and within which the young continue to develop after birth.

mass extinctions The disappearance of a large number of species in a relatively short period of geological time.

mating systems The social systems describing how males and females pair up.

mating type A genetically defined strain of an organism (such as a fungus) that can only mate with an organism of the opposite mating type; mating types are often designated + and −.

mechanical isolation A prezygotic reproductive isolating mechanism caused by differences in the structure of reproductive organs or other body parts.

medusa The tentacled, usually bell-shaped, free-swimming sexual stage in the life cycle of a coelenterate.

megaspore A plant spore that develops into a female gametophyte; usually larger than a microspore.

mesenteries Sheets of loose connective tissue, covered on both surfaces with epithelial cells, which suspend the abdominal organs in the coelom and provide lubricated, smooth surfaces that prevent chafing or abrasion between adjacent structures as the body moves.

mesoderm The middle layer of the three primary germ layers of an animal embryo, from which the muscular, skeletal, vascular, and connective tissues develop.

mesohyl The gelatinous middle layer of cells lining the body cavity of a sponge.

metamorphosis A reorganization of the form of certain animals during postembryonic development.

metanephridium The excretory tubule of most annelids and mollusks.

metapopulation A group of neighboring populations that exchange individuals.

microclimate The abiotic conditions immediately surrounding an organism.

microevolution Small-scale genetic changes within populations, often in response to shifting environmental circumstances or chance events.

micronucleus In ciliophorans, one or more diploid nuclei that contains a complete complement of genes, functioning primarily in cellular reproduction.

microspore A plant spore from which a male gametophyte develops; usually smaller than a megaspore.

microsporidium (plural, microsporidia) A fungal parasite of animals; many mycologists believe they make up a possible sixth phylum within the Kingdom Fungi.

migration The predictable seasonal movement of animals from the area where they are born to a distant and initially unfamiliar destination, returning to their birth site later.

mimic The species in Batesian mimicry that resembles the model.

mimicry A form of defense in which one species evolves an appearance resembling that of another.

minimum viable population size The smallest population size that is likely to survive both predictable and unpredictable environmental variation.

model The species in Batesian mimicry that is resembled by the mimic.

modern synthesis A unified theory of evolution developed in the middle of the twentieth century.

molars Posteriormost teeth of mammals, with a broad chewing surface for grinding food.

mold Asexual, spore-producing stage of many multicellular fungi.

molecular clock A technique for dating the time of divergence of two species or lineages, based upon the number of molecular sequence differences between them.

monocot A plant belonging to the Monoctoyledones, one of the two major classes of angiosperms; monocot embryos have a single seed leaf (cotyledon) and pollen grains with a single groove.

monogamy A mating system in which one male and one female form a long-term association.

monophyletic taxon A group of organisms that includes a single ancestral species and all of its descendants.

monsoon cycle A wind pattern that brings seasonally heavy rain to a region by blowing moisture-laden air from the sea to the land.

morphological species concept The concept that all individuals of a species share measurable traits that distinguish them from individuals of other species.

morphology The form or shape of an organism, or of a part of an organism.

mosaic evolution The tendency of characteristics to undergo different rates of evolutionary change within the same lineage.

motile Capable of self-propelled movement.

Müllerian mimicry A form of defense in which two or more unpalatable species share a similar appearance.

multiregional hypothesis An hypothesis proposing that after archaic humans migrated from Africa to many regions on Earth, their different populations evolved into modern humans simultaneously.

mutation A spontaneous and heritable change in DNA.

mutualism A symbiotic interaction between species in which both partners benefit.

mycelium A network of branching hyphae that constitutes the body of a multicellular fungus.

mycobiont The fungal component of a lichen.

mycorrhiza A mutualistic symbiosis in which fungal hyphae associate intimately with plant roots.

Nanoarchaeota A group of Archaea that was proposed based on rRNA sequence analysis of a thermophilic archaean found in a symbiotic relationship with an other thermophilic archaean; most probably a subgroup of the Euryarchaeota.

natural history The branch of biology that examines the form and variety of organisms in their natural environments.

natural selection A process by which alleles that increase the likelihood of survival and the reproductive output of the individuals that carry them become more common in subsequent generations.

natural theology A belief that knowledge of God may be acquired through the study of natural phenomena.

navigation A wayfinding mechanism in which an animal moves toward a specific destination, using both a compass and a "mental map" of where it is in relation to the destination.

nekton Animals that can actively swim against water currents.

nematocyst A coiled thread, encapsulated in a cnidocyte, that cnidarians fire at prey or predators, sometimes releasing a toxin through its tip.

neritic zone The shallow water of the oceans above the continental shelves.

nerve net A simple nervous system that coordinates responses to stimuli but has no central control organ or brain.

net primary productivity The chemical energy remaining in an ecosystem after a producer's cellular respiration is deducted.

neural crest A band of cells that arises early in the embryonic development of vertebrates near the region where the neural tube pinches off from the ectoderm; later, the cells migrate and develop into unique structures.

neuroscience The integrated study of the structure, function, and development of the nervous system.

neutral variation hypothesis An evolutionary hypothesis that some variation at gene loci coding for enzymes and other soluble proteins is neither favored nor eliminated by natural selection.

nitrification A metabolic process in which certain soil bacteria convert ammonia or ammonium ions into nitrites that are then converted by other bacteria to nitrates, a form usable by plants.

nitrogen cycle A biogeochemical cycle that moves nitrogen between the huge atmospheric pool of gaseous molecular nitrogen and several much smaller pools of nitrogen-containing compounds in soils, marine and freshwater ecosystems, and living organisms.

nitrogen fixation A metabolic process in which certain bacteria and cyanobacteria convert molecular nitrogen into ammonia and ammonium ions, forms usable by plants.

node The point on a stem where one or more leaves are attached.

nonvascular plant *See* bryophyte.

notochord A flexible rod-like structure, constructed of fluid-filled cells surrounded by tough connective tissue, which supports a chordate embryo from head to tail.

nucleoid The area within a prokaryotic cell containing the DNA.

null model A conceptual model that predicts what one would see if a particular factor had no effect.

obligate aerobe A microorganism that uses oxygen for cellular respiration and requires oxygen in its surroundings to support growth.

obligate anaerobe A microorganism that cannot use oxygen and can grow only in the absence of oxygen.

oceanic zone The deep ocean water beyond the continental shelves.

ocellus (plural, ocelli) The simplest eye, which detects light but does not form an image.

oligotrophic lake A lake that is poor in nutrients and organic matter, but rich in oxygen.

omnivore An animal that feeds at several trophic levels, consuming plants, animals, and other sources of organic matter.

open circulatory system An arrangement of internal transport in some invertebrates in which the vascular fluid, hemolymph, is released into sinuses, bathing organs directly, and is not always retained within vessels.

operant conditioning A form of associative learning in which animals learn to link a voluntary activity, an operant, with its favorable consequences, the reinforcement.

operculum A lid or flap of the bone serving as the gill cover in some fishes.

opposable big toe A big toe that can flex to contact the sole of the foot, allowing the foot to grasp objects in the environment.

optimal foraging theory A set of mathematical models that predict the diet choices of animals as they encounter a range of potential food items.

oral hood Soft fleshy structure at the anterior end of a cephalochordate that frames the opening of the mouth.

order A Linnaean taxonomic category of organisms that ranks above a family and below a class.

organismal ecology An ecological discipline in which researchers study the genetic, biochemical, physiological, morphological, and behavioral adaptations of organisms to their abiotic environments.

orthogenesis An obsolete theory that evolution is goal oriented, striving to perfect organisms.

oscula One or more openings in a sponge through which water is expelled.

ostracoderm One of an assortment of extinct, jawless fishes that were covered with bony armor.

outer membrane In Gram-negative bacteria, an additional boundary membrane that covers the peptidoglycan layer of the cell wall.

outgroup comparison A technique used to identify ancestral and derived characters by comparing the group under study to more distantly related species that are not otherwise included in the analysis.

ovary In plants, the ovule-bearing portion of a carpel that ripens into a fruit.

overexploitation The excessive harvesting of an animal or plant species, potentially leading to its extinction.

ovule In plants, the structure in a carpel in which a female gametophyte develops and fertilization takes place.

paedomorphosis A common form of heterochrony in which juvenile characteristics are retained in a reproductive adult.

paleobiology The study of ancient organisms.

parapatric speciation Speciation between populations with adjacent geographical distributions.

paraphyletic taxon A group of organisms that includes an ancestral species and some, but not all, of its descendents.

parapodia Fleshy lateral extensions of the body wall of aquatic annelids, used for locomotion and gas exchange.

parasite An organism that feeds on the tissues of or otherwise exploits its host.

parasitism A symbiotic interaction in which one species, the parasite, uses another, the host, in a way that is harmful to the host.

parasitoid An insect species in which a female lays eggs in the larva or pupa of another insect species, and her young consume the tissues of the living host.

parental investment The time and energy devoted to the production and rearing of offspring.

parthenogenesis A mode of asexual reproduction in which animals produce offspring by the growth and development of an egg without fertilization.

passive parental care The amount of energy invested in offspring—in the form of the energy stored in eggs or seeds or energy transferred to developing young through a placenta—before they are born.

pectoral girdle A bony or cartilaginous structure in vertebrates that supports and is attached to the forelimbs.

pedicellariae Small pincers at the base of short spines in starfishes and sea urchins.

pelagic province The water in a marine biome.

pellicle A layer of supportive protein fibers located inside the cell, just under the plasma membrane, providing strength and flexibility instead of a cell wall.

pelvic girdle A bony or cartilaginous structure in vertebrates that supports and is attached to the hind limbs.

peptidoglycan A polymeric substance formed from a polysaccharide backbone tied together by short polypeptides, which is the primary structural molecule of bacterial cell walls.

per capita growth rate The difference between the per capita birth rate and the per capita death rate of a population.

peritoneum The thin tissue derived from mesoderm that lines the abdominal wall and covers most of the organs in the abdomen.

permafrost Perpetually frozen ground below the topsoil.

phage *See* bacteriophage.

pharynx The throat. In some invertebrates, a protrusible tube used to bring food into the mouth for passage to the gastrovascular cavity; in mammals, the common pathway for air entering the larynx and food entering the esophagus.

phenotypic variation Differences in appearance or function between individual organisms.

pheromone A distinctive volatile chemical released in minute amounts to influence the behavior of members of the same species.

phloem The food-conducting tissue of a vascular plant.

phosphorus cycle A biogeochemcial cycle in which weathering and erosion carry phosphate ions from rocks to soil and into streams and rivers, which eventually transport them to the ocean, where they are slowly incorporated into rocks.

photic zone Surface water of a lake or ocean that sunlight penetrates.

photoautotroph A photosynthetic organism that uses light as its energy source and carbon dioxide as its carbon source.

photobiont The photosynthetic component of a lichen.

photoheterotroph An organism that uses light as the ultimate energy source but obtains carbon in organic form rather than as carbon dioxide.

phototroph An organism that obtains energy from light.

phylum (plural, phyla) A major Linnaean division of a kingdom, ranking above a class.

PhyloCode A formal set of rules governing phylogenetic nomenclature.

phylogenetic species concept A concept that seeks to delineate species as the smallest aggregate population that can be united by shared derived characters.

phylogenetic tree A branching diagram depicting the evolutionary relationships of groups of organisms.

phylogeny The evolutionary history of a group of organisms.

phytoplankton Microscopic, free-flowing aquatic plants and protists.

pilus (plural, pili) A hair or hairlike appendage on the surface of a prokaryote.

piloting A wayfinding mechanism in which animals use familiar landmarks to guide their journey.

pinacoderm In sponges, an unstratified outer layer of cells.

placenta A specialized temporary organ that connects the embryo and fetus with the uterus in mammals, mediating the delivery of oxygen and nutrients.

plasmid A small circle of DNA in the cytoplasm of certain prokaryotes, which often contains genes with functions that supplement those in the nucleoid and which can replicate independently of the nucleoid DNA and be passed along during cell division.

plasmodial slime mold A slime mold of the class Myxomycetes.

plasmodium The composite mass of plasmodial slime molds consisting of individual nuclei suspended in a common cytoplasm surrounded by a single plasma membrane.

plasmogamy The sexual stage of fungi during which the cytoplasms of two genetically different partners fuse.

plastron The ventral part of the shell of a turtle.

plate tectonics The geological theory describing how Earth's crust is broken into irregularly shaped plates of rock that float on its semisolid mantle.

pollen grain The male gametophyte of a seed plant.

pollination The transfer of pollen to a flower's reproductive parts by air currents or on the bodies of animal pollinators.

pollutant Materials or energy in a form or quantity that organisms do not usually encounter.

polyandry A polygamous mating system in which one female mates with multiple males.

polygamy A mating system in which either males or females may have many mating partners.

polygyny A polygamous mating system in which one male mates with many females.

polyhedral virus A virus in which the coat proteins form triangular units that fit together like the parts of a geodesic sphere.

polymorphism The existence of discrete variants of a character among individuals in a population.

polyphyletic taxa A group of organisms that belong to different evolutionary lineages and do not share a recent common ancestor.

polyploidy The condition of having one or more extra copies of the entire haploid complement of chromosomes.

polyp The tentacled, usually sessile stage in the life cycle of a coelenterate.

population All the individuals of a single species that live together in the same place and time.

population bottleneck An evolutionary event that occurs when a stressful factor reduces population size greatly and eliminates some alleles from a population.

population density The number of individuals per unit area or per unit volume of habitat.

population ecology The ecological discipline that focuses on how a population's size and other characteristics change in space and time.

population genetics The branch of science that studies the prevalence and variation in genes among populations of individuals.

population size The number of individuals in a population at a specified time.

population viability analysis A mathematical analysis used by conservation biologists to determine the minimum viable population size for threatened or endangered species.

posterior Indicating the tail end of an animal.

postzygotic isolating mechanism A reproductive isolating mechanism that acts after zygote formation.

preadaptation A characteristic evolved by an ancestral species that serves an adaptive but different function in a descendant species or population.

predation The interaction between animals and the animal prey they consume.

premolars Teeth located in pairs on each side of the upper and lower jaws of mammals, positioned behind the canines and in front of the molars.

prezygotic isolating mechanism A reproductive isolating mechanism that acts prior to the production of a zygote, or fertilized egg.

primary cell layers The ectoderm, mesoderm, and endoderm layers that form the embryonic tissues.

primary consumer An herbivore, a member of the second trophic level.

primary endosymbiosis In the model for the origin of plastids in eukaryotes, the first event in which a eukaryotic cell engulfed a photosynthetic cyanobacterium.

primary producer An autotroph, usually a photosynthetic organism, a member of the first trophic level.

primary succession Predictable change in species composition of an ecological community that develops on bare ground.

principle of monophyly A guiding principle of systematic biology that defines monophyletic taxa, each of which contains a single ancestral species and all of its descendants.

principle of parsimony A principle of systematic biology that states that a particular trait is unlikely to evolve independently in separate evolutionary lineages.

prion An infectious agent that contains only protein and does not include a nucleic acid molecule.

profundal zone The perpetually dark layer below the limnetic zone in a lake.

promiscuity A mating system in which individuals do not form close pair bonds, and both males and females mate with multiple partners.

protocell A primitive cell-like structure that has some of the properties of life and that might have been the precursor of cells.

Protoctista The kingdom that includes all the eukaryotes that are not fungi, plants, or animals.

protonema The structure that arises when a liverwort or moss spore germinates and eventually gives rise to a mature gametophyte.

pseudocoelom A fluid- or organ-filled body cavity between the gut (a derivative of endoderm) and the muscles of the body wall (a derivative of mesoderm).

pseudocoelomate A body plan of bilaterally symmetrical animals with a body cavity that lacks a complete lining derived from mesoderm.

pseudopod (plural, pseudopodia) A temporary cytoplasmic extension of a cell.

psychrophile An archaean or bacterium that grows optimally at temperatures in the range -10 to $-20°C$.

Pterophyta The plant phylum of ferns and their close relatives.

punctuated equilibrium hypothesis The evolutionary hypothesis that most morphological variation arises during speciation events in isolated populations at the edge of a species' geographical distribution.

pupa The nonfeeding stage between the larva and adult in the complete metamorphosis of some insects, during which the larval tissues are completely reorganized within a protective cocoon or hardened case.

pyramid of biomass A diagram that illustrates differences in standing crop biomass in a series of trophic levels.

pyramid of energy A diagram that illustrates the amount of energy that flows through a series of trophic levels.

pyramid of numbers A diagram that illustrates the number of individual organisms present in a series of trophic levels.

qualitative variation Variation that exists in two or more discrete states, with intermediate forms often being absent.

quantitative variation Variation that is measured on a continuum (such as height in human beings) rather than in discrete units or categories.

radial cleavage A cleavage pattern in deuterostomes in which newly formed cells lie directly above and below other cells of the embryo.

radial symmetry A body plan of organisms in which structures are arranged regularly around a central axis, like spokes radiating out from the center of a wheel.

radiometric dating A dating method that uses measurements of certain radioactive isotopes to calculate the absolute ages in years of rocks and minerals.

radula The tooth-lined "tongue" of mollusks that scrapes food into small particles or drills through the shells of prey.

rain shadow An area of reduced precipitation on the leeward side of a mountain.

random dispersion A pattern of distribution is which the individuals in a population are distributed unpredictably in their habitat.

realized niche The range of conditions and resources that a population actually uses in nature.

reciprocal altruism Form of altruistic behavior in which individuals help nonrelatives if they are likely to return the favor in the future.

red tide A growth in dinoflagellate populations that causes red, orange, or brown discoloration of coastal ocean waters.

reinforcement The enhancement of reproductive isolation that had begun to develop while populations were geographically separated.

relative abundance The relative commonness of populations within a community.

relative fitness The number of surviving offspring that an individual produces compared with the number left by others in the population.

reproductive isolating mechanism A biological characteristic that prevents the gene pools of two species from mixing.

reproductive strategy One of a set of behaviors that lead to reproductive success.

resource partitioning The use of different resources or the use of resources in different ways by species living in the same place.

rhizoid A modified hypha that anchors a fungus to its substrate and absorb moisture.

rhizome A horizontal, modified stem that can penetrate a substrate and anchor the plant.

ring species A species with a geographic distribution that forms a ring around uninhabitable terrain.

root An anchoring structure in land plants that also absorbs water and nutrients and (in some plant species) stores food.

root system An underground (or submerged) network of roots with a large surface area that favors the rapid uptake of soil water and dissolved mineral ions.

r-selected species A short-lived species adapted to function well in a rapidly changing environment.

salt marsh A tidal wetland dominated by emergent grasses and reeds.

saprobe An organism nourished by dead or decaying organic matter.

savanna A biome comprising grasslands with few trees, which grows in areas adjacent to tropical deciduous forests.

schizocoelom In protostomes, the body cavity that develops as inner and outer layers of mesoderm separate.

secondary consumer A carnivore that feeds on herbivores, a member of the third trophic level.

secondary endosymbiosis In the model for the origin of plastids in eukaryotes, the second event, in which a nonphotosynthetic eukaryote engulfed a photosynthetic eukaryote.

secondary productivity Energy stored in new consumer biomass as energy is transferred from producers to consumers.

secondary succession Predictable changes in species composition in an ecological community that develops after existing vegetation is destroyed or disrupted by an environmental disturbance.

seed The structure that forms when an ovule matures after a pollen grain reaches it and a sperm fertilizes the egg.

segmentation The production of body parts and some organ systems in repeating units.

selectively neutral *See* neutral variation hypothesis.

septum (plural, septa) A thin partition or cross wall that separates body segments.

sessile Unable to move from one place to another.

seta (plural, setae) A chitin-reinforced bristle that protrudes outward from the body wall in some annelid worms.

sex ratio The relative proportions of males and females in a population.

sexual dimorphism Differences in the size or appearance of males and females.

sexual selection A form of natural selection established by male competition for access to females and by the females' choice of mates.

shoot system Stems and leaves of a plant.

sign stimulus A simple cue that triggers a fixed action pattern.

simulation modeling An analytical method in which researchers gather detailed information about a system and then create a series of mathematical equations that predict how the components of the system interact and respond to change.

sink population In metapopulation analysis, a population that routinely declines in size after being replenished by immigrants from a source population.

slime layer A coat typically composed of polysaccharides that is loosely associated with bacterial cells.

SLOSS (*Single Large Or Several Small*) The debate among conservation biologists about the relative merits of establishing fewer large preserves or more numerous small ones.

social behavior The interactions that animals have with other members of their species.

soredium (plural, soredia) A specialized cell cluster produced by lichens, consisting of a mass of algal cells surrounded by fungal hyphae; soredia function like reproductive spores and can give rise to a new lichen.

sorus (plural, sori) A cluster of sporangia on the underside of a fern frond; reproductive spores arise by meiosis inside each sporangium.

source population In metapopulation analyses, a population that is either stable or increasing in size.

speciation The process of species formation.

species cluster A group of closely related species recently descended from a common ancestor.

species composition The particular combination of species that occupy a site.

species diversity A community characteristic defined by species richness and the relative abundance of species.

species richness The number of species that live within an ecological community.

species selection A type of natural selection that acts upon species rather than upon populations.

specific epithet The species name in a binomial.

spinneret A modified abdominal appendage from which spiders secrete silk threads.

spiral cleavage The cleavage pattern in many protostomes in which newly produced cells

lie in the space between the two cells immediately below them.

spiral valve A corkscrew-shaped fold of mucous membrane in the digestive system of elasmobranchs, which slows the passage of material and increases the surface area available for digestion and absorption.

spirillum (plural, spirilla) Any flagellated aerobic bacterium twisted helically like a corkscrew.

spongocoel The central cavity in a sponge.

sporangium (plural, sporangia) A single-celled or multicellular structure in fungi and plants in which spores are produced.

spore A reproductive structure, usually a single cell, that can develop into a new individual without fusing with another cell; found in plants, fungi, and certain protists.

sporophyll A specialized leaf that bears sporangia (spore-producing structures).

sporophyte The diploid generation in the plant life cycle; it produces haploid spores.

sporopollenin A tough polymer in the walls of spores and pollen grains, the presence of which helps such structures resist decay.

spring overturn The mixing of surface water with deep water in a lake or pond, causing oxygen at the surface to move to the bottom, and nutrients from the bottom to move to the surface.

squalene A liver oil found in sharks that is lighter than water, which increases their buoyancy.

stability The ability of a community to maintain its species composition and relative abundances when environmental disturbances eliminate some species from the community.

stabilizing selection A type of natural selection in which individuals expressing intermediate phenotypes have the highest relative fitness.

standing crop biomass The total dry weight of plants present in an ecosystem at a given time.

stapes The smallest of three sound-conducting bones in the middle ear of tetrapod vertebrates.

statocyst A mechanoreceptor in invertebrates that senses gravity and motion using statoliths.

stoma (plural, stomata) The opening between a pair of guard cells in the epidermis of a plant leaf or stem, through which gases and water vapor pass.

stratification Horizontal layering of sedimentary rocks beneath the soil surface.

strobilus *See* cone.

Credits

CHAPTER 19 **Page 401** Christopher Ralling. **19.1** (left) Courtesy George P. Darwin, Darwin Museum, Down House. (right) Down House and The Royal College of Surgeons of England. **19.2** (left) Wolfgang Kaehler/Corbis. (center) Kenneth W. Fink/Photo Researchers, Inc. (right) Dave Watts/A. N. T. Photo Library. **19.4** Rich Kirchner/Foto Natura/Photo Researchers, Inc. **19.6** (top) Charles R. Knight painting (negative CK21T), Field Museum of Natural History, Chicago. (bottom) Calvin Larsen/Photo Researchers, Inc. **19.7** (left) Hugo Willcox/Foto Natura/Minden Pictures. (right) Fred Hazelhoff/Foto Natura/Minden Pictures. **19.8** (b) D. Kaleth/Image Bank/Getty Images. (c) William Paton/Foto Natura/Photo Researchers, Inc. (d) Heather Angel/Natural Visions. **19.9** (a) Dr. P. Evans/Bruce Coleman. (b) Kevin Schafer/Corbis. (c) Mark Moffatt/Minden Pictures. (d) Alan Root/Bruce Coleman Ltd. **Page 408** William Perlman/Star Ledger/Corbis. **19.12** P. Morris/Ardea, London.

CHAPTER 20 **Page 419** © Mark Moffett/Foto Natura/Minden Pictures. **20.1** The Advertising Archives, London. **20.2** (a) George Bernard/Foto Natura/Photo Researchers, Inc. (b) Timothy A. Pearce, Ph.D./Section of Mollusks/Carnegie Museum of Natural History. Photograph by Mindy McNaugher. **20.4** Arthur Morris/VIREO. **20.5** (left) Eric Crichton/Bruce Coleman, Inc. (right) William E. Ferguson. **20.7** (top) David Neal Parks. (bottom) W. Carter Johnson. **20.8** Frans Lanting/Minden Pictures. **20.11** (a) Forrest W. Buchanan/Visuals Unlimited. (b) Gregory K. Scott/Photo Researchers, Inc. **20.12** (left) Heather Angel/Natural Visions. **20.13** (left) © 2008 Josef Hlasak.

CHAPTER 21 **Page 443** © Mickey Gibson/Animals, Animals—Earth Scenes. **21.1** Bruce Beehler. **21.6** Courtesy of James E. Lloyd. Miscellaneous Publications of the Museum of Zoology of the University of Michigan, 130:1–195, 1966. **21.7** (both) Reny Parker. **21.8** Jen and Des Bartlett/Bruce Coleman USA. **21.10** (left) Patrice Geisel/Visuals Unlimited. (center) Tom Van Sant/The Geosphere Project, Santa Monica, CA. (right) Fred Mc Connaughey/Photo Researchers, Inc. **Page 452** (both) Kenneth Y. Kaneshiro, University of Hawaii. **21.12** (left) © H. Clarke, VIREO/Academy of Natural Sciences. (right) Robert C. Simpson/Nature Stock. **21.14** Dr. Jim Smith, Michigan State University. **21.15** (left) Andrew Parkinson/Frank Lane Picture Agency. (right) Eric Soder/Foto Natura/Photo Researchers, Inc.

CHAPTER 22 **Page 463** Courtesy Lowcountry Geologic. **22.1** Rudwick, 1985. The Meaning of Fossils. University of Chicago Pres, Figure 3.1 p. 106 (partial). **22.2** (a) George H. H. Huey/Corbis. (b) Neville Pledge/South Australian Museum. (c) Jack Koivula/Photo Researchers, Inc. (d) Novosti/Photo Researchers, Inc. **22.3** David Noble/FPG/Getty Images. **22.4** (all) © 2001 PhotoDisc. **22.8** (both) Edward S. Ross. **Page 473** (figure e) Jack Dermid. **22.9** (a) Douglas P. Wilson/Eric and David Hosking. (b) Superstock, Inc. (c) E. R. Degginger. **22.12** (all) Alan Cheetham et al, Department of Paleobiology, National Museum of Natural History, Smithsonian Institution, Washington, D.C. **22.15** David Scott/SREL. **22.16** (both) Gary Head. **22.22** © Michael D. Shapiro and David Kingsley.

CHAPTER 23 **Page 491** Courtesy of U.S. Forest Service, Boise National Forest (Kathryn M. Beall photo). **23.1** (bottom) From L. W. Hackett, *Malaria in Europe*, Oxford University Press, 1937. © London School of Tropical Medicine & Hygiene. **23.3** (top) S. L. Collins and J. T. Collins. (bottom) The Amphibians and Reptiles of Missouri, by T. R. Johnson © 1987 by the Conservation Commission of the State of Missouri. Reprinted by permission. **23.6** (a) Nature's Images/Photo Researchers, Inc. (b) Neil Bowman/Frank Lane Picture Agency. (c) Millard Sharp/Photo Researchers, Inc. **23.12** (top) Thomas J. Lemieux, University of Colorado. (bottom) Sandra Floyd, University of Colorado.

CHAPTER 24 **Page 511** Dr. Ken Macdonald/SPL/Photo Researchers, Inc. **24.1** (bottom) Jeff Hester and Paul Scowen, Arizona State University, and NASA. (inset left) Stanley M. Awramik. **24.3** Photo by Chesley Bonestell. **24.5** (left) Dr. W. Hargreaves and D. Deamer. **24.6** Bill Bachmann, Photo Researchers, Inc. **24.9** Robert Trench, Professor Emeritus, University of British Columbia.

CHAPTER 25 **Page 525** © Phototake, Inc. **25.1** (all) Tony Brian, David Parker/SPL/Photo Researchers, Inc. **25.2** (all) David M. Phillips/Visuals Unlimited. **25.4** (both) T. J. Beveridge/Visuals Unlimited. **25.5** © Frank Dazzo, Michigan State University. **25.7** CNRI/SPL/Photo Researchers, Inc. **25.9** Dr. Terry J. Beveridge, Department of Microbiology, University of Guelph, Ontario, Canada/Biological Photo Service. **25.12** Hans Reichenbach, Gesellschaft for Biotechnologische Forschung, Braunsweig, Germany. **25.13** (a) Dr. Jeremy Burgess/SPL/Photo Researchers, Inc. (b) Tony Brian/SPL/Photo Researchers, Inc. (c) P. W. Johnson and J. McN. Sieburth, University of Rhode Island/Biological Photo Service. **25.14** David M. Phillips/Visuals Unlimited. **25.15** David M. Phillips/Visuals Unlimited. **25.16** (a) Barry Rokeach. (b) © Alan L. Detrick/Science Source/Photo Researchers, Inc. **25.17** R. Robinson/Visuals Unlimited. **25.19** (right) K. G. Murti/Visuals Unlimited. **25.20** © APHIS photo by Dr. Al Jenny.

CHAPTER 26 **Page 549** Steve Gschmeissner/SPL/Photo Researchers, Inc. **26.1** (a) Edward S. Ross. (b) Gary W. Grimes and Steven L'Hernault. (c) Steven C. Wilson/Entheos. (d) Wim van Egmond. **26.3** (top) Frieder Sauer/Bruce Coleman Ltd. (bottom) Redrawn from V. & J. Pearse and M. & R. Buchsbaum, *Living Invertebrates*, The Boxwood Press, 1987. **26.4** (a) © Dennis Kunkel Microscopy, Inc. (b) Dr. Dennis Kunkel/Visuals Unlimited. **26.5** (top) P. L. Walne nd J. H. Arnott, Planta, 77: 325–354, 1967. **26.6** (top) Oliver Meckes/Photo Researchers, Inc. **26.7** (bottom right) John Walsh/SPL/Photo Researchers, Inc. **26.8** Dr. David Phillips/Visuals Unlimited. **26.9** (a) Claude Taylor and the University of Wisconsin Dept. of Botany. (b) Heather Angel. (c) W. Merrill. **Page 559** (micrograph) Steven L'Hernault. (left) Sinclair Stammers/Photo Researchers, Inc. **26.10** Dr. John Cunningham/Visuals Unlimited. **26.11** Jan Hinsch/SPL/Photo Researchers, Inc. **26.12** (a) Ron Hoham, Dept. of Biology, Colgate University. (b) Lewis Trusty/Animals, Animals. (c) Jeffrey Levinton, State University of New York, Stony Brook. **26.14** (a) Wim van Egmond. (b) Courtesy of Allen W. H. Be and David A. Caron. (c) John Clegg/Ardea, London. (d) Redrawn from V. & J. Pearse and M. & R. Buchsbaum, *Living Invertebrates*, The Boxwood Press, 1987. **26.15** M. Abbey/Visuals Unlimited. **26.16** (a, b) Carolina Biological Supply. (c) Courtesy Robert R. Kay from R. R Kay, et al., Development, 1989 Supplement, pp. 81–90. © The Company of Biologists Ltd., 1989. **26.17** (a) Wim van Egmond. (b) Douglas Faulkner/Sally Faulkner Collection. **26.18** (a) Linda Sims/Visuals Unlimited. (b) Brian Parker/Tom Stack and Associates. (c) Manfrage Kage/Peter Arnold, Inc. **26.20** Dr. John Clayton, National Institute of Water and Atmospheric Research, New Zealand.

CHAPTER 27 **Page 575** Animals, Animals—Earth Scenes. **27.1** (a) Craig Wood/Visuals Unlimited. (b) Robert Potts, California Academy of Sciences. (c) © Craig Allikas/www.orchidworks.com. **27.2** ©Courtesy Microbial Culture Collection, National Institute for Environmental Studies, Japan. **27.3** (a) George S. Ellmore. (b) Jeremy Burgess/SPL/Photo Researchers, Inc. **27.4** Reprinted with permission from Elsevier. **27.10** (a) Martin Hutten/National Park Service. (b) Paul Stehr-green/National Park Service. (c) Wayne P. Armstrong, Professor of Biology and Botany, Palomar College, San Francisco, CA. **27.11** © clive@hiddenforest.co.nz. **27.12** (top left) Jane Burton/Bruce Coleman USA. **27.13** Dr. Judith Jernstedt, University California, Davis. **27.14** (b) Field Museum of Natural History, Chicago. **27.15** (a) © Ed Reschke/Peter Arnold, Inc. (b) Kathleen B. Pigg, Arizona State University. **27.16** (top) A. & E. Bomford/Ardea, London. (bottom) © Hubert Klein/Peter Arnold, Inc. **27.17** Kingsley R. Stern. **27.18** (a) William Ferguson. (b) W. H. Hodges. (c) Kratz/Zefa. **27.21** Carlton Ray/Science Photo Library/Photo Researchers, Inc.

Index

The letter i designates illustration; t designates table; **bold** *designates defined or introduced term.*

Abdomen, 655, 661, 661i
Abiotic, **1126**
Abrahamson, Warren G., 431, 432i
Absorption, by fungi, 608
Abyssal zone, **1221**, 1223i, 1224
Acacia *(Acacia cornigera)*, 1158, 1161i
Acanthamoeba, 563
Acetabularia, 567i
Acid precipitation, **1234**, 1235i
Acoelomate animal, **630**, 631i
Acorn worm, 671, 671i
Acoustical signal, **1276**
Acquired characteristics, inheritance of, 404
Actinopterygii, 681, 681i, 682i
Active parental care, **1155**
Adaptation
    angiosperm, 597–598
    constraints on adaptive evolution,
        438–440
    convergent evolution, 471, 471i, 473i
    defined, **438**
    by natural selection, 413, 416
Adaptive radiation, 452, **480**, 482i, 483
Adaptive trait
    defined, **410**, **438**
    evolution of, 438–440
    natural selection favoring of, 410
Adaptive zone, **480**
Adductor muscle, bivalve, **649**, 649i
Adiabatic cooling, **1206**
Aerobe, 531
Aflatoxin, 615
African cichlid fish *(Haplochromis
    burtoni)*, 1261–1263, 1262i
African Emergence Hypothesis, **705**–707
African Nile crocodile *(Crocodylus niloti-
    cus)*, 690i
African pancake tortoises *(Malacochersus
    tornieri)*, 1154
African sleeping sickness, 554–555, 555i
Agar, **567**
Age structure, **1128**–1129, 1146i, 1147i
Age-specific fecundity, **1131**, 1131t
Age-specific mortality, **1130**–1131, 1131t
Age-specific survivorship, **1130**–1131, 1131t
Agnathan, 675, 677–678, 677i, 678i
    conodonts and ostracoderms, 677–678,
        678i
    hagfishes and lampreys, 677, 677i
Agrawal, Anurag, 1178
Agriculture, benefits of biodiversity to,
    1231
Air circulation, 1206, 1206i
Air pollution, lichens as monitors of, 621
Albatross, 1276
Albumin, **686**
Algae
    brown, 561–562, 561i
    golden, 561, 561i
    green, 567–569, 567i, 568i
    red, 565–567, 566i
Algin, **561**–562
Aligning molecular sequences, 503, 505i
Allele frequency, **424**, 424t, 426–427
Allometric growth, **479**–480, 479i, 485
Allopatric speciation, **450**–451, 451i, 453i,
    459
Allopolyploidy, **456**–458, 457i

Alpine tundra, 1218i, **1219**
Alternation of generations, **579**, 579i
Altruism
    defined, **1282**
    genetic basis of, 1282–1283
    human, 1285
    kin selection, 1283, 1283i
    reciprocal, 1284, 1285
Alveolate, 555–558, 557i, 558i
Amanita muscaria, 617–618, 618i
Amazon basin, deforestation in, 1233i
Amber, 464, 465i
Amborella trichopoda, 506, 506i, 597, 597i,
    601
Ambystoma talpoideum, 480i
American alligator *(Alligator mississippien-
    sis)*, 690, 692
American porcupine *(Erethizon)*, 1154
Amino acid
    formation in Miller-Urey apparatus, 514
    formation near hydrothermal vents, 515
    in meteorites, 515
Amino acid sequence
    evolutionary relationships and, 415i
    molecular phylogenetic use of, 501, 503,
        504i
Ammonification, 1195, 1197t
Amniota, **675**
Amniote
    origin of, 686
    phylogeny, 686, 687i, 688
Amniote egg, **686**, 686i
Amoeba
    amoebozoan, 563–564, 564i
    cercozoan, 562
    defined, **562**
Amoeba proteus, 563, 564i
Amoebic dysentery, 564
Amoebozoa, 563–565, 564i, 565i
Amphibian, 684–685, 685i
Ampulla, echinoderm, 669, 670i
Anabaena, 1195
Anaerobe, 531
Anagenesis, **475**, 475i
Analogies (analogous characters), 496
Anapsid, **688**
Ancestral character, **496**–497
Andersson, Malte, 433, 434i
Angiosperm, 595–602
    adaptive success of, 596–597, 596i–598i,
        597–598
    basal, 596–597, 597i
    classification, 596–597, 596i–598i
    coevolution with pollinators, 598, 600,
        600i
    defined, **595**
    homeosis and origin of, 487
    origin of, 595–596, 596i, 601
    overview, 595
    phylogenetic tree, 596i
Anglerfish *(Himantolophus)*, 1225i
Angraecum sesquipedale, 600, 600i
Anguillicola crassus, 653i
Animal(s)
    ancestral forms, 629, 629i
    characteristics of, 628–629
    diversification of, 628
    diversity, history of, 484i
    eumetazoa with bilateral symmetry,
        641–663
    eumetazoa with radial symmetry,
        636–640

evolution, innovations in, 629–633
    body cavities, 630, 631i
    developmental patterns, 630,
        632–633, 632i
    segmentation, 633
    symmetry in body plan, 630, 631i
    tissues and tissue layers, 629–630
parazoa, 635–636, 635i
phylogeny and classification, 633, 634i,
    635
viral infection, 542–543, 543i
Animal behavior
    choice of behavior performed, 1265
    communication
        acoustical signals, 1276
        chemical signals, 1276, 1277i
        dance, 1277–1278, 1278i
        electrical signals, 1277
        evolutionary analysis, 1278
        tactile signals, 1277, 1277i
        visual signals, 1276
    defined, **1253**
    experience, influence of, 1265
    fixed action patterns, 1255
    genetic basis, 1256–1257, 1265
    hormones and, 1260–1263, 1261i, 1262i
    instinctive, 1254–1257, 1256i, 1257i
    learned, 1254–1255, 1257–1259, 1258i
    migration, 1269–1274, 1271i–1274i
    nervous system anatomy and,
        1263–1264, 1265i, 1266i, 1267–1268
    neurophysiological control, 1259–1260,
        1260i
    proximate causes, 1254
    reproduction
        mating systems, 1280
        parental investment, 1183
        sexual selection, 1279–1280
    social
        altruistic behavior, 1282–1284, 1283i
        benefits and costs of group living,
            1281–1282, 1281i, 1282i
        defined, 1281
        dominance hierarchies, 1282
        human, 1285–1287
        territoriality, 1275–1276
        ultimate causes, 1254
Animalia (kingdom), **628**
Ankylosaurus, 688
Annelid, 650–653
    anatomy, 650–651, 651i
    lineages
        Hirudinea, 652–653, 652i
        Oligochaeta, 651–652, 652i
        Polychaeta, 651, 652i
Annulus, **589**
Anolis, 691, 1209, 1211i
Anopheles mosquito, 491–492, 492i, 558,
    559, 662
Ant, 660i, 1158, 1161i, 1276–1277, 1277i
Anteater, 696
Antennae
    defined, **657**
    trilobite, 657
Anterior end, **630**
Antheridium
    ascomycete, 616
    description, **581**, 582i
    fern, 588i, 589
    liverwort, 582
    moss, 583, 584i
Anthocerophyta, **583**

Anthoceros, 583i
Anthophyta, **596**
Anthozoa, 638, 640i
Anthropoidea, **699**, 701
Antibiotic resistance, 419–420, 537
Antifreeze proteins, 1210
Antillean manatee *(Trichechus manatus)*, 697i
Antithamnion plumula, 566i
Antonovics, Janis, 451, 454i
Anura, 685, 685i
Apatosaurus, 688
Apes, 701, 701i
Aphotic zone, **1220**, 1220i
Apical meristem, **577**
Apicomplexa, 557–558, 559
Aposematic coloration, **1154**–1155, 1154i
Appendicular skeleton, **673**
Apple maggot *(Rhagoletis pomonella)*, 453, 454i
Aquadro, Chip, 439
Aquatic succession, **1171**
Arachnid, 657–658, 657i
Arbuscule, 614
Archaea (domain), 537–540
   characteristics of, 538–539, 538t
   Crenarchaeota, 540
   defined, **526**
   Euryarchaeota, 539–540, 539i
   Korarchaeota, 540
   origin of, 521–522
   virus infections of, 545
Archaefructus, 596, 596i, 601
Archaeopteris, 590, 591i
Archaeopteryx, 413i, 479, 692
Archaeplastida, 565–569, 566i–569i
Archegonium
   description, **581**, 582i
   fern, 588i, 589
   moss, 583, 584i
Archenteron, **632**, 632i
Archosauromorpha, **688**
Arctic tern *(Sterna paradisaea)*, 1270, 1271i
Arctic tundra, **1218**–1219, 1218i
Ardipithecus, 703
Aristotle, 402
Armadillo, 405, 405i
Armillaria ostoyae, 619
Arnold, Stevan, 1256–1257
Arthrobotrys dactyloides, 616i
Arthropod, 655–663
   anatomy, 655–656
   classification, 656, 657–663
   migrating, 1270, 1271i
   phylogenetics, 656
   subphyla
      Chelicerata, 657–658, 657i, 658i
      Crustacea, 658–659, 659i
      Hexapoda, 660i–663i, 661–663
      Myriapoda, 659–661, 660i
      Trilobita, 657, 657i
Artificial selection, **408**–409, 423i
Asci, 614, **615**, 615i
Ascocarp, 615i, **616**
Ascomycota (ascomycetes), 612t, 614–616, 615i–617i
Asexual reproduction
   fungi, 609, 609i, 613, 614, 614i, 616, 617i, 618
   hornworts, 583
   liverworts, 582–583, 583i
   mosses, 583, 584i
   parthenogenesis, 646
Aspergillus, 615
Assimilation efficiency, 1185–1186
Asteroid impact, 481
Asteroidea, 669, 670i
Asymmetrical body plan, **630**
Atlantic cod *(Gadus morhua)*, 1238i, 1239

Atmosphere
   defined, **1204**
   primordial Earth, 513–514
   reducing, 513–514, 523
Atmospheric cycles, 1191
ATP, in protocells, 516–517
Atrial siphon, **673**
Atriopore, **673**, 673i
Atrium, urochordate, **673**
Australian lungfish *(Neoceratodus forsteri)*, 683i
Australopithecus, 703–704, 704i
Autopolyploidy, **456**, 456i
Autotroph, **531**, 552, **1164**
Autumn overturn, **1220**, 1221i
Axial skeleton, **673**
Azotobacter, 532, 1195

Baboon, 700i, 1278, 1278i, 1282
Bacillariophyta, 560–561, 560i
Bacilli (prokaryote cell shape), **527**, 527i
Bacillus anthracis, 536
Bacillus subtilis, 535
Background extinction rate, **481**
Background selection, 439
Bacon, Sir Francis, 402
Bacteria (domain), 534–537
   antibiotic resistance, 537
   characteristics of, 538t
   chlamydias, 536, 536i
   cyanobacteria, 535, 535i
   defined, **526**
   Gram-positive bacteria, 535–536, 536i
   green bacteria, 535
   pathogenicity, mechanism of, 536–537
   Proteobacteria, 534–535, 535i
   size, 525, 526i
   spirochetes, 536
   virus infection of, 541, 542
Bacterial population growth, 1134–1135, 1136i
Bacteriophage, **541**, 542
Bahama woodstar hummingbird *(Calliphlox evelynae)*, 694i
Balanced polymorphism
   defined, **435**
   maintenance of, 435–436
      frequency-dependent selection, 436, 438i
      heterozygote advantage, 435, 436i
      selection in varying environments, 435–436, 437i
Balantidium coli, 556
Bald eagle *(Haliaeetus leucocephalus)*, 694i
Baltimore oriole *(Cterus galbula)*, 453i
Banana slug *(Ariolimax columbianus)*, 1256, 1257i
Banding pattern, chromosome, 458i, 459
Barcode system, DNA, 1242
Barnacle, 408, 659, 659i, 1157–1158, 1159i
Barton, Nick H., 459–460
Basal angiosperm, 596–597, 597i
Basidia, **617**
Basidiocarp, **618**
Basidiomycota (basidiomycetes), 612t, 617–619, 618i, 619i
Basidiospore, *618*, 619
Basket star, 669
Basking, 1209, 1211i
Bat, 600, 600i, 696, 697i, 1263–1264
Bates, Henry W., 1155
Batesian mimicry, **1155**, 1155i
Bath sponge *(Spongia)*, 635
Batrachochytrium dendrobatis, 612, 612i
Baum, David, 487
Bay checkerspot butterfly *(Euphydryas editha bayensis)*, 428, 1245, 1245i
Bees, pollination by, 600, 600i

Begging behavior, in herring gulls, 1256
Behavioral isolation, **448**, 448i
Bent grass *(Agrostis tenuis)*, 451, 453, 454i
Benthic province, **1221**, 1223i, 1224–1225, 1225i
Benthos, **1225**
Big Bang Theory, 511
Bilateral symmetry
   description, **630**, 630i
   ecdysozoan protosomes, 653–663, 653i–663i
   lophotrochozoan protosomes, 641–653, 641i–652i
Binary fission, 532
Binomial, **493**
Binomial nomenclature, **493**
Biodiversity
   benefits
      ecosystem services, 1231
      to humans, 1230–1231
   crisis, causes of
      deforestation, 1233, 1233i, 1234i
      desertification, 1233–1234, 1234i
      exotic species, 1235, 1237–1238, 1237i, 1238i
      extinction, 1239
      habitat fragmentation, 1232–1233, 1232i
      overexploitation, 1238–1239, 1238i
      pollution, 1234–1235, 1235i, 1236
   defined, **480**, **1230**
   historical, 482–483, 483i, 484i
   hotspots, 1239–1240, 1240i
   intrinsic worth, 1231
   macroevolutionary trends in, 480–483
   networks of interacting species, 1250
Biodiversity hotspots, **1239**–1240, 1241i
Biofilm, **532**–533, 533i
Biogeochemical cycle
   atmospheric, 1191
   carbon, 1192, 1194, 1194i–1195i, 1200
   defined, **1191**
   generalized compartment model, 1191–1192, 1191i
   hydrologic, 1192, 1192i
   nitrogen, 1194–1195, 1197–1198, 1197i, 1197t
   phosphorus, 1197–1198, 1198i
   sedimentary, 1191
Biogeographical realms, **471**, 471i
Biogeography
   continuous distribution, 470
   defined, **403**
   in development of evolutionary thought, 403
   disjunct distribution, 470–471
   dispersal, 471, 472–473
   equilibrium theory of island biogeography, 1175–1177, 1175i, 1176i
   historical, 413, 469–473
   realms, 471, 471i
   vicariance, 471, 472
Biological clock
   wayfinding and, 1271
Biological lineage, **413**
Biological magnification, **1236**
Biological species concept, **445**
Bioluminescent, **556**
Biomass
   description, **1183**–1184
   by ecosystem, 1186i
   of prokaryotes, 525
   pyramids of, 1188, 1188i
   standing crop, 1183, 1185, 1185t
Biome
   defined, **1212**
   freshwater, 1219–1221
      lakes, 1220–1221, 1220i, 1221i
      streams and rivers, 1219–1220, 1219i

marine, 1221–1225
  abyssal zone, 1224
  benthic province, 1224–1225, 1225i
  estuaries, 1222, 1223i
  intertidal zone, 1222–1223, 1224i
  neritic zone, 1223, 1224i
  oceanic zone, 1224
  zonation, ocean, 1221, 1223i
terrestrial, 1211–1219
  boreal forest/taiga, 1217–1218, 1218i
  chaparral, 1216, 1216i
  desert, 1215–1216, 1215i
  savanna, 1214–1215, 1215i
  temperate deciduous forest, 1217, 1217i
  temperate grassland, 1216–1217, 1216i
  temperate rain forest, 1218
  tropical forest, 1213–1214, 1213i
  tundra, 1218–1219, 1218i
Biosphere, 1203–1226
  defined, **1126, 1204**
  environmental variations
    air circulation, 1206, 1206i
    ocean currents, 1206, 1207i, 1208
    organismal responses to, 1209–1211
    precipitation, 1206, 1207i
    regional and local effects, 1208, 1208i, 1209i
    seasonality, 1205–1206, 1205i
    solar radiation, 1205, 1205i
    terrestrial biome distribution, 1212–1213, 1212i
  freshwater biomes, 1219–1221
    lakes, 1220–1221, 1220i, 1221i
    streams and rivers, 1219–1220, 1219i
  marine biomes, 1221–1225
    abyssal zone, 1224
    benthic province, 1224–1225, 1225i
    estuaries, 1222, 1223i
    intertidal zone, 1222–1223, 1224i
    neritic zone, 1223, 1224i
    oceanic zone, 1224
    zonation, ocean, 1221, 1223i
  terrestrial biomes, 1211–1219
    boreal forest/taiga, 1217–1218, 1218i
    chaparral, 1216, 1216i
    desert, 1215–1216, 1215i
    savanna, 1214–1215, 1215i
    temperate deciduous forest, 1217, 1217i
    temperate grassland, 1216–1217, 1216i
    temperate rain forest, 1218
    tropical forest, 1213–1214, 1213i
    tundra, 1218–1219, 1218i
Biosphere 2, 526
Biota, 471, 471i
Biotic, **1126**
Bipedal locomotion, **701**, 702, 702i
Bird
  colonial living, 1281–1282, 1281i
  diversity, 693–694, 694i
  DNA barcoding analysis, 1242
  evolution, 413, 413i
  key adaptations, 692, 693i
  mating systems, 1280
  migration, 1269–1272, 1271i, 1273i
  origin of, 692–693
  pollination by, 600, 600i
  sexual selection, 1279–1280, 1280i
  songbird decline, 1232–1233, 1232i
  songs, 1253–1255, 1254i, 1259–1261, 1260i, 1269, 1276
  territoriality, 1276
Bird flu, 541, 544
Bird of paradise (*Paradisaea raggiana*), 444i
Birth rate, per capita, 1135–1137
Bivalve, 647i, 648–649, 649i

Black field cricket (*Teleogryllus oceanicus*), 1263–1264, 1264i
Black rhinoceros (*Diceros bicornis*), 697i
Black widow (*Latrodectus mactans*), 658
Black-handed gibbon (*Hylobates agilis*), 701i
Blastopore, **632**, 632i
Blue and yellow macaw (*Ara arauna*), 443i
Blue jay (*Cyabicutta cristata*), 427i
Blue tit (*Parus caeruleus*), 1275, 1275i
Blue-footed booby (*Sula nebouxii*), 406i
Bluegill sunfish (*Lepomis macrochirus*), 1153, 1153i
Bodmer, Sir Walter, 431i
Body cavities, 630, 631i
Body plan
  asymmetrical, 630
  bilateral symmetry, 630, 631i
  diploblastic, 630, 637
  insect, 661, 661i
  radial symmetry, 630, 631i
  sac-within-a-sac, 629, 629i
  segmented, 633, 635
  simple, 635, 636i
  sponge, 635, 636i
  triploblastic, 630
  tube within a tube, 630
Body size, 478–479
Body temperature, mammalian, 695
Bone, **673**
Bonobo (*Pan paniscus*), 701
Bony fishes, 681, 681i–683i, 683
Book lung, 657i, **658**
Boore, Jeffry L., 656
Boreal forest, **1217**–1218, 1218i
Bormann, Herbert, 1193
Bottleneck, population, **428**, 428i, 429
Bovine spongiform encephalopathy (BSE), 545–546
Bowen, Brian W., 1190
Box jellyfish, 638
Brachiation, **701**
Brachiopoda, 641–642, 641i
Bradshaw, H. D., 455
Brain
  cephalopod, 650
  higher vocal center, 1260, 1261i
  mammalian, 695
Branchiostoma, 673, 675
Breeding grounds, migration to, 1269, 1271, 1272–1274, 1274i
Brenner, Sidney, 654
Brittle star, 669, 670i
Brood parasite, 1151–1152, 1152i, 1160, 1233
Brown algae, 561–562, 562i
Brown recluse spider (*Loxosceles reclusa*), 658
Brown-headed cowbird (*Molothrus ater*), 1160, 1233
Brugia, 645
Bryophyta, **583**
Bryophyte, 581–585
  defined, **577, 581**
  hornworts, 583, 583i
  liverworts, 582–583, 583i
  moss, 583–584, 584i
  overview, 581–582, 582i
BSE (bovine spongiform encephalopathy), 545–546
Buffon, le Comte de, 403
Bullock's oriole (*Cterus bullockii*), 453i
Bull's horn acacia (*Acacia cornigera*), 1158, 1161i
Burgess Shale formation, 627–628, 628i
Butterfly
  DNA barcoding analysis, 1242
  metapopulation, 1245
  mouthparts, 662i

Cactus finch (*Geospiza conirostris*), 433, 433i
Caddis fly, 497i
Caecelian, 685, 685i
Caenorhabditis elegans, 653, 654, 654i
Calamites, 585, 587i
Calcium, loss in watershed, 1193
Calliphlox evelynae, 600i
Camel (*Camelus dromedarius*), 697i
CAMP (cyclic adenosine monophosphate)
  role in development and differentiation, 566
Canada lynx (*Lynx candensis*), 1144–1145, 1144i
Candida albicans, 615, 616i
Canine, **695**
Cann, Rebecca, 706–707
Capsid, **541**
Capsule, **529**, 529i
Capybara (*Hydrochoerus hydrochaeris*), 697i
Carapace, **658**, **688**, 689i
Carbon, recycling of, 605, 608
Carbon cycle, 1192, 1194, **1194,** 1194i–1195i, 1200
Carbon dioxide, as greenhouse gas, 1196, 1235
Carbon sequestration, 1231
Carboniferous period, 585
Carnivora, 696
Carpel, 595
Carrageenan, **567**
Carroll, Sean, 485, 486
Carrying capacity, **1137**–1139, 1232
Carson, Hampton, 452
Cascade frog (*Rana cascadae*), 1275
Catastrophism, theory of, **404**
Caterpillar, 1154, 1154i
Cattle egret (*Bubulcus ibis*), 1158, 1159i
Caughey, Bryon, 545
Cavalli-Sforza, Luigi, 431i
Cell wall, prokaryotic, 528–529, 529i
Cells
  origin of
    archaean, 521–522
    eukaryotic, 519–523, 520i, 521i
    prokaryotic, 517–519, 519i
    protocells, 512, 515–517, 516i, 522
  protocells, 512
Cellular respiration, **1183**
Cellular slime mold, **564**, 565i, 566
Centipede, 659–660, 660i
Central nervous system, chordate, 672
Centruroides sculpuratus, 657i
Cepaea nemoralis, 420i, 421, 444
Cephalization, **630**
Cephalochordata, 673, 673i
Cephalopod, 647i, 650, 650i
Cephalothorax, **657**, 657i
Cercozoa, 562–563, 563i
Cerion snails, 420–421, 420i
Cestoda, 643, 643i, 644, 645i, 646
Cetacea, 696
Cetaceans, evolution of, 502
Chagas disease, 555
Chain fern (*Woodwardia*), 588i
Chang, Sherwood, 516
Chaparral, **1216**, 1216i
Chara, 569, 569i, 576i
Character displacement, **1157**–1158, 1158i, 1167
Charlesworth, Brian, 439
Charophyte, **569**, 569i, 575, 576
Cheetah (*Acinonyx jubatus*), 429
Cheetham, Alan, 476i, 477
Chelicerae, **657**, 658
Chelicerta, 657–658, 657i, 658i
Cheliped, 659, 659i
Chemical signal, **1276**, 1277i

Chemoautotroph, **531**, 531i
Chemoheterotroph, **531**, 531i
Chemotroph, **531**
Chen, Liaohai, 522
Cherry *(Prunus)*, 598i
Chestnut-backed chickadee *(Parus rufe-scens)*, 694i
Chestnut-headed oropendola *(Zarhynchus wagleri)*, 1151–1152, 1152i
Chicxulub crater, 481
Child abuse, 1285, 1287i
Chimpanzee *(Pan troglodytes)*, 479, 701, 701i, 1259
Chinese liver fluke *(Opisthorchis sinensis)*, 643i
Chitin, **606**, 655
Chiton, 647i, 648, 648i
Chlamydia trachomatis, 536, 536i
Chlamydias, 536, 536i
Chlorarachniophyta, 563
Chlorophyll, of bacteria, 535
Chlorophyta (green algae), 567–569, 567i, 568i
Chloroplast DNA (cpDNA), evolution rate of, 503
Chloroplasts, endosymbiont hypothesis and, 519, 520–521, 520i, 569–572, 570i
Choanocyte, sponge, **635**, 636i
Choanoflagellate, 569, 569i, 629
Chondrichthyes, 680–681, 680i
Chondromyces crocatus, 535i
Chondrus, 1167, 1169i
Chordate. *See also* Vertebrate(s)
  invertebrate, 672i, 673, 673i
  morphological innovations, 671–672, 672i
    dorsal hollow nerve chord, 672
    notochord, 672
    perforated pharynx, 672
    postanal tail, 672
    segmented body wall and tail, 672
  phylogeny, 676i
Chromosome
  banding patterns, 458i, 459
  human and great apes, differences among, 458–459, 458i
  prokaryotic, 527
Chrysophyta, 561, 561i
Chytridiomycota (chytrids), 612–613, 612i, 612t
Chytriomyces hyalinus, 612i
Cicada *(Graptopsalatsia nigrofusca)*, 655
Cilia
  of ctenophores, 639
  of protists, 553
Ciliophora, 555–556, 557i
Circulatory system
  annelid, 651, 651i
  cephalopod, 650
  closed, **650**
  mollusk, 648
  nemertean, 646
  open, **648**
Circulatory vessels, of nemerteans, **646**
Clade, **499**
Cladistics, 498–501, **499**, 499i–501i
Cladogenesis, **475**, 475i
Cladogram, **499**, 499i, 500i–501i
Cladonia rangiferina, 622i
Clam, 648–649, 649i
Claret cup cactus *(Echinocereus triglochi-dratus)*, 598i
Claspers, **680**
Classical conditioning, **1258**–1259
Classification
  by Aristotle, 402
  cladistics, 498–501, 499i–501i
  defined, **492**, 493
  phylogenetic trees, 497–498, 498i, 499i

Claviceps purpurea, 615
Clays, molecular aggregation on, 515–516
Cleaner wrasse *(Labroides dimidiatus)*, 1160i
Cleavage
  defined, **630**
  determinate, 631
  indeterminate, 631
  radial, 631, 632i
  spiral, 630, 632i
Clements, Frederic, 1161, 1162, 1212
Cleome droserifolia, 1131i, 1132
Cliff swallows *(Petrochelidon pyrrhonota)*, 1282
Climate
  continental, 1208
  defined, **1205**
  El Niño, 1203–1204, 1204i
  global warming, 1210–1211, 1226
  maritime, 1208
  microclimate, 1208
  ocean proximity, 1208, 1208i
  topography, 1208, 1208i
Climate change, 1226
Climax community, **1171**
Climograph, 1212–1213, 1212i
Cline, **446**, 447i
Closed circulatory system, **650**
Clostridium botulinum, 536
Clostridium tetani, 532i
Cloud forest, 1213–1214
Club moss, 586–587, 588i
Clumped dispersion, **1128**
Cnidarian, 637–638, 637i–640i
Cnidocyte, **637**, 637i, 638i
Coal Age, 585
Coal tit *(Parus ater)*, 1275, 1275i
Coat, virus, **541**
Cocci (prokaryote cell shape), **527**, 527i
Cochliobolus carbonum, 624
Cocoanut Grove nightclub, 419
Coconut palm *(Cocus nucifera)*, 1143i
Coelacanth *(Latimeria chalumnae)*, 683, 683i
Coelom
  description, **630**, 631i
  echinoderm, 669, 670i
  enterocoelom, 632, 632i
  schizocoelom, 632, 632i
Coelomate animal, **630**, 631i
Coencytic fungi, **613**
Coevolution, **600**, **1152**–1153, 1158, 1161i
Coho salmon *(Oncorhynchus kisutch)*, 1132
Cohort, **1130**
COI (cytochrome oxidase 1) gene, 1242
Coleochaete, 569
Colonial living, 1281–1282, 1281i
Colony, protist, 552
Coloration
  aposematic, 1154–1155, 1154i
  cryptic, 1154
Columbine *(Aquilegia)*, 487
Comb jellies, 638–639, 641i
Commensalism, **1158**, 1159i
Communication
  acoustical signals, 1276
  chemical signals, 1276, 1277i
  dance, 1277–1278, 1278i
  eavesdroppers, 1287
  electrical signals, 1277
  evolution of, 1276–1278, 1277i, 1278i
  evolutionary analysis, 1278
  tactile signals, 1277, 1277i
  visual signals, 1276
Community
  climax, 1171
  disclimax, 1173

Community ecology
  defined, **1126**
  population interactions, 1152–1160, 1152t
Comparative morphology, **413**–414
Compass orientation, **1271**, 1273i
Competitive exclusion principle, **1156**
Complete digestive tract, **646**
Complete metamorphosis, **663**, 663i
Complex virus, **542**, 542i
Compound eye, 656
Condensation reaction, 515, 516
Conditioned stimulus, 1259
Conditioning
  classical, 1258–1259
  operant, 1259
Cone
  club moss, **586**
  conifer, 593, 594i
Conidia, **616**
Conidial fungi, 619–620
Conidiophore, **616**, 617i
Coniferophyta, **593**
Conifers, 593–594, 594i
Conjugation
  in bacteria, 532
  in *Paramecium*, **556**, 557i
Connell, Joseph, 1157–1158, 1159i, 1166, 1167–1168
Conodont, **677**–678, 678i
Conservation biology
  conservation approaches, 1247–1249
    mixed-use conservation, 1248, 1248i
    preservation, 1248, 1248i
    restoration, 1248–1249
  defined, **1241**
  economic factors
    ecosystem valuation, 1250–1251
    ecotourism, 1249–1250
    local involvement, 1249
  genetic drift, implications of, 429
  landscape ecology, 1246–1247
  metapopulation dynamics, 1243, 1245, 1245i
  population ecology and behavior, 1242–1245
  population genetics, 1241
  population viability analysis (PVA), 1243, 1244
  species/area relationships, 1245–1246, 1246i
  systematics, 1241
Consumer, **1164**, 1165i, 1189–1190, 1189i
Continental climate, **1208**
Continental drift, **469**, 470i
Continental shelf, 1223
Continuous distribution, **470**
Contractile vacuole, **552**, 552i
Convergent evolution, 496
Cooksonia, 578, 578i, 585
Cooling, adiabatic, 1206
Cope, Edward Drinker, 478
Copepod, 659, 659i
Copernicus, Nicolaus, 402
Cope's Rule, 478
Coral fungus *(Ramaria)*, 618i
Coral reef
  biome, **1223**, 1224i
  disturbance of, 1168–1169, 1170i
Corals, 638, 640i
Coriolis effect, 1206
Corn *(Zea mays)*
  fungal pathogen, 624
  perennial, 1231, 1231i
  teosinte compared, 1231, 1231i
Corona, rotifer, 646
Cortex, brain, **695**
Cotyledon, 596
Courtship displays, 448, **1279**–1280, 1280i

Cowbird, 1151–1152, 1152i, 1160, 1233
Coyne, Jerry, 459
Crab, 655, 658, 659i
Crane fly, 660i
Craniate, 675
Cranium, **673**
Crayfish, 658
Crenarchaeota, **539**, 540
Cretaceous mass extinction, 481–482
Creutzfeldt-Jakob disease (CJD), 545–546
Cricket, 1263–1264, 1264i, 1287
Critical period, **1258**
Crocodilian, 688, 690, 690i, 692
Crowding, 1139–1141, 1140i
Cryptic coloration, **1154**
Cryptococcus neoformans, 618
Ctenophore, 638–639, 641i
Cubozoa, 638, 640i
Cuckoo, 1160
Currents, ocean, 1203–1204, 1204i, 1206,
    1207i, 1208
Cuticle
    plant, **576**, 577i
    roundworm, **653**
Cuvier, Georges, 404, 463–464, 464i
Cyanobacteria
    in eutrophic lake, 1221, 1222i
    heterocyst, 535i
    in lichens, 620
    metabolic activity, 535
    in *Paulinella*, 571
    photosynthesis, 535
    plastid evolution from, 570
Cyanophora paradoxa, 521
Cycad, 591, 592i
Cycadophyta, **591**
Cyst, 558
Cytochrome *c*, 415i
Cytoplasm, of prokaryotes, 528
Cytoplasmic streaming, 564, **607**

Dall mountain sheep *(Ovis dalli)*, 1131i,
    1132
Daly, Martin, 1285, 1287
Damschen, Ellen I., 1246, 1247i
Dance communication, 1277, 1278i
Darwin, Charles
    on artificial selection, 408–409
    barnacle study, 408
    Beagle voyage, 405–407, 405i
    on descent with modification, 410–411
    on earthworms, 652
    Galápagos finches and, 407, 407i, 410
    illustration of, 402i
    life as a scientist, 408
    Linnaean Society presentation, 401
    on natural selection, 410, 411
    observations and inferences, 409–410,
        410t
    *On the Origin of Species by Means of
        Natural Selection*, 402, 408,
        410–411, 444
    Transmutation of Species, 407
Dating
    radiometric, 468–469, 468i
    relative, 465
DDT, 1236
Deamer, David W., 516
Death cap mushroom *(Amanita phalloi-
    des)*, 618
Death rate, per capita, 1135–1137
Decapoda, 658–659, 659i
Deciduous forest
    temperate, 1217, 1217i
    tropical, 1213
Decomposer, **1164**, 1182
Defense behavior, social, 1281, 1281i
Deforestation, 1233, 1233i, 1234
Dehydration synthesis, 515

Delphinium decorum, 480, 481i
Delphinium nudicaule, 480, 481i
Demographic transition model,
    1146–1147, 1148i
Demography, 1129–1132
    defined, **1129**
    life tables, 1130–1131, 1131t
    survivorship curve, 1131–1132, 1131i
Denitrification, 1197, 1197t
Density-dependent factors, **1139**–1141,
    1140i–1142i
Density-independent factors, **1141**–1142,
    1141i
D'Erchia, Anna Maria, 698
Derived character, **496**–497
Dermatophagoides pteronyssinus, 657i
Descartes, René, 402
Descent with modification, **410**–411
Desert, **1215**–1216, 1215i
Desertification, **1233**–1234, 1234i
Determinate cleavage, **631**
Detrital food web, 1182, 1183i
Detritivore, **1164**, 1182
Deuteromycetes, 619–620
Deuterostome, 667–707
    developmental patterns, 630–633, 632i
    invertebrate, 668–671, 670i
    nervous system, 633
Development
    embryonic development pathways,
        evolution of, 416
    evolutionary developmental biology,
        483–488
    patterns in animals, 630–633, 632i
    patterns in insects, 662–663, 663i
Developmental biology, evolutionary,
    483–488
Developmental switch, 485–486
Diagnostic characters, 444, 444i
Diamond, Jared, 1246
Diapsid
    crocodilians, 690, 690i, 692
    defined, **688**
    diversification during mesozoic era,
        688
    sphenodontids, 689, 690i
    squamates, 689–690, 690i
Diatom, 560–561, 560i
Diatomaceous earth, 560–561
Dickerson, Richard E., 519
Diclonfenac, 1235
Didinium, 550i
Digestive system
    annelid, 651, 651i
    echinoderm, 669, 670i
    rotifers, 646
Digestive tract, complete, **646**
Digger wasp *(Philanthus triangulum)*,
    1270–1271, 1272i
Dikaryon, 609i, **610**, 613, 614i, 616, 617i,
    618, 619i
Dinoflagellate, 556–557, 558i
Dinosaur, 688
Diploblastic body plan, **630**, 637
Diploid organism, masking of recessive
    alleles in, 435, 435t
Diplomonadida, 553, 554i
Directional selection, 430i, **431**, 432i
Discicristates, **554**–555, 555i
Disclimax community, **1173**
Disheveled (dsh) gene, 1258
Disjunct distribution, **470**–471
Dispersal, **471**, 472
Dispersion, **1128**
Disruptive selection, **433**, 433i
Distal-less gene (*Dll*), 661
Disturbance
    ecological succession, 1170–1174,
        1172i, 1173i

effect on community characteristics,
    1167–1170, 1170i
    intermediate disturbance hypothesis,
        1169–1170, 1170i
Disturbance climax, 1173
Diversity, evolution of, 706
DNA
    development from RNA, 517
    sequence, molecular phylogenetic use
        of, 501, 503, 505
DNA barcoding, 1242
Dobzhansky, Theodosius, 1178
Dog
    classical conditioning, 1258–1259
    rabies in, 1235
Dolphin, 696, 697
Domain, **493**
Dominance
    hierarchies, **1282**
    incomplete, 424
    Donoghue, Michael, 506
    Dorsal surface, **630**
    Dosidicus gigas, 650i
    Double fertilization, 598, 599i
    Downy mildew, 549, 560
    Downy woodpecker (*Dendrocopus pubes-
        cens*), 432i
    Dragonfly, 660i, 663
    Drone fly (*Eristalis tenax*), 1155i
    Drosophila
        gametic isolation, 449
        hybrids, 449
        speciation in Hawaiian, 452
        temporal isolation, 447
        wingless/Wnt pathway, 1258
    Duboule, Denis, 486
    Duck-billed platypus (*Ornithorhynchus
        anatinus*), 667–668, 668i, 695,
        696i
    Dugong, 696
    Dunkleosteus, 679, 679i
    Dust mite, 657i, 658
    Dutch elm disease, 614

Eagle nebula, 512i
Earth
    age of, 511
    geological history, 469, 470i
    plate tectonics, 469, 470i
    primordial, 513, 513i
Earthworm, 652, 652i
Eastern gray kangaroo (*Macropus gigan-
    teus*), 696i
Eastern hemlock (*Tsuga canadensis*),
    1237–1238, 1238i
Eastern painted turtle (*Chrysemys picta*),
    689i
Eastern prairie fringed orchid (*Platan-
    thera leucophaea*), 598i
Ebbole, Dan, 624
Ecdysis in insects, 655
Ecdysozoa
    Arthropoda, 655–663, 655i, 657i–663i
    description, 633
    Nematoda, 653–654, 653i
    Onychophora, 654–655, 655i
Echidna, 695, 696i
Echinocereus, 471i
Echinoderm
    anatomy, 669, 670i
    groups
        Asteroidea, 669, 670i
        Concentricycloidea, 669
        Crinoidea, 670i, 671
        Echinoidea, 669, 670i, 671i
        Holothuroidea, 670i, 671
        Ophiuroidea, 669, 670i
    overview, 669
Echinoidea, 669, 670i, 671

Echolocation, 1263–1264
Ecological community
  boundaries, 1162–1163, 1162i
  characteristics
    change during succession, 1171, 1173
    disturbance, effects of, 1167–1170, 1170i
    plant growth forms, 1163, 1163i
    population interactions, effects on, 1166–1167, 1168i
    relative abundance, 1163–1164
    species diversity, 1164, 1164i, 1166
    species richness, 1163, 1164, 1166–1167, 1168i, 1169–1170, 1170i, 1174–1176, 1174i–1176i
    trophic structure, 1164–1166, 1165i
  defined, 1152
  ecological success in, 1170–1174, 1172i, 1173i
  ecotones, 1163
  individualistic view of, 1161–1163, 1162i
  interactive view of, 1161–1162, 1162i
  positive *versus* negative interactions, 1178
  species composition, 1161
  stability, 1166
Ecological efficiency, 1185–1186
  assimilation efficiency, 1185–1186
  harvesting efficiency, 1185
  production efficiency, 1186
Ecological isolation, 447
Ecological niche
  defined, 1156
  fundamental niche, 1157, 1157i, 1159i
  realized niche, 1157, 1157i, 1159i
Ecological pyramid, 1188, 1188i, 1189i
Ecological succession
  in animals, 1173, 1173i
  aquatic, 1171
  community characteristics, changes in, 1171, 1173
  defined, 1170
  disclimax community, 1173
  facilitation hypothesis, 1173
  inhibition hypothesis, 1173
  local, 1174
  primary, 1171, 1172i
  secondary, 1171
  tolerance hypothesis, 1173
Ecologist, 1126–1127
Ecology
  basic and applied, 1126
  community, 1126, 1160–1178
  defined, 1126
  ecosystem, 1126
  interaction networks, 1250
  organismal, 1126
  population, 1126, 1127–1149
  scale, importance of, 1148
Ecosystem, 1181–1200
  defined, 1182
  energy flow
    consumers, 1189–1190, 1189i
    ecological efficiency, 1185–1186
    ecological pyramids, 1188, 1188i, 1189i
    food webs, 1182, 1183i
    primary productivity, 1183–1186, 1184i, 1186i
    Silver Springs ecosystem, 1186, 1187i
    sunlight, 1182–1184
  Lake Eric, 1181–1182, 1182i
  modeling, 1199–1200, 1199i
  nutrient cycling, 1191–1199, 1200
    carbon cycle, 1192, 1194, 1194i–1195i, 1200
    generalized compartment model, 1191–1192, 1191i

hydrologic cycle, 1192, 1192i
nitrogen cycle, 1194–1195, 1197–1198, 1197i, 1197t
phosphorus cycle, 1197–1198, 1198i
Ecosystem ecology
  biodiversity, 1230
  defined, 1126
Ecosystem services, 1231
Ecosystem valuation, 1250–1251
Ecotherm, 1209
Ecotone, 1163
Ecotourism, 1249–1250
Ectoderm, 629–630, 642
Ectomycorrhizae, 621–622, 623i
Ectoparasite, 1160
Ectoprocta, 641–642, 641i
Edge effects, 1232, 1246, 1246i
Egg
  amniote, 686, 686i
  conifer, 593–594, 594i
Ehlers-Danlos syndrome, 425
Ehrlich, Paul R., 1245
Eimer's organ, 1265, 1266, 1266i
El Niño, 1203–1204, 1204i
El Niño Southern Oscillation (ENSO), 1204
Elasmobranchii, 680
Eldredge, Niles, 477
Electrical signal, 1277
Electroreceptor, 680
Eleodes longicollis, 1155, 1155i
Elephantiasis, 645, 645i
Ellis-van Crevald syndrome, 428–429, 434
Embryo, 629
Embryology, comparative, 414, 414i
Embryonic development pathways, evolution of, 416
Embryophyte, 577
Emigration, 1129
Emlen, Stephen, 1271, 1273i
Emu (*Dromaius novaehollandiae*), 403i
Endangered species, 1239
Endemic species, 471, 1240, 1240i
Endler, John, 1134–1135
Endocytosis, 520, 521i
Endoderm, 629, 642
Endomycorrhizae, 621, 622–623, 623i
Endoparasite, 1159–1160
Endoplasmic reticulum, origin of, 520, 521i
Endosperm, 598
Endospore, 532, 532i
Endosymbiont hypothesis, 519–521, 520i, 569–572, 570i
Endosymbiosis
  plastid evolution by, 519–521, 520i, 569–572, 570i
  primary, 570, 570i, 571
  secondary, 570i, 571
Endotherm, 1209–1210
Endotoxins, 536–537
Energy budget, 1132
Energy flow, ecosystem, 1182–1190
Energy-harnessing reaction pathways, development of, 516–517
Ensatina eschscholtzii, 446, 446i
Entamoeba histolytica, 564
Enterocoelom, 632, 632i
Enteromorpha, 1167, 1169i
Enveloped virus, 542, 542i, 543i
Environment, phenotypic variation due to, 421–422, 422i
Enzymes, 537
Ephedra, 592, 593i, 601
Epilimnion, 1220
Epiphyte, 581
Equilibrium theory of island biogeography, 1175–1177, 1175i, 1176i
Equus, 473, 474i, 475

Ergotism, 615
Erwin, Terry, 1241
Escape behavior, 1263–1264, 1264i, 1265i
Eschenmoser, Albert, 522
Escherichia coli
  food contamination by, 525
  pili, 530i
Essay on the Principles of Population (Malthus), 409
Estrogen, 1260, 1261i
Estuary, 1222, 1223i
Ethology, 1254
Eucheuma, 567
Eudicot, 596, 597, 598i
Euglena gracilis, 554, 555i
Eukarya (domain), 526, 538t
Eukaryote
  first, 571
  multicellularity, origin of, 522–523
Eukaryotic cells, origin of, 519–523, 520i, 521i
Eumetazoan
  with bilateral symmetry, 641–663
    ecdysozoan protosomes, 653–663, 653i–663i
    lophotrochozoan protosomes, 641–653, 641i–652i
  with radial symmetry, 636–640
    cnidarians, 637–638, 637i, 638i, 639i, 640i
    ctenophores, 638–639, 641i
Eupenicillium, 616i
Euphorbia, 471i
European beaver (*Castor fiber*), 406i, 407
European cuckoo (*Cuculus canorus*), 1257i
European mouflon sheep (*Ovis musimon*), 1139i
European nightjar (*Caprimulgus europaeus*), 694i
European rabbit (*Oryctolagus cuniculus*), 1125–1126, 1126i
European starling (*Sturnus vulgaris*), 1125, 1237, 1237i
Euryarchaeota, 539, 539–540, 539i
Eusocial insect, 1284–1285, 1284i
Eutrophic lake, 1221, 1222i
Evaporation, macromolecule formation by, 515
Everglades, 1234
Evergreen, 593
Evergreen coniferous forest, 1217–1218, 1218i
Evernia lichens, 621
Evolution
  of adaptive traits, 438–440
  angiosperms coevolution with animal pollinators, 598, 600, 600i
  animal, key innovations in, 629–633, 631i, 632i
  of antibiotic resistance, 419–420
  of birds, 413, 413i
  coevolution, 600, 1152–1153, 1158, 1161i
  of common themes, 706
  of communication, 1276–1278
  convergent, 471, 471i, 473i, 496
  defined, 402
  development of evolutionary thought, 401–418
  of diversity, 706
  ecological interactions and, 1148
  of embryonic development pathways, 416
  of fins, 679
  of flowering plants, 595–596
  of flowers, 601
  fossil record, 464–469
  geological time scale, 466t–467t
  of horses, 473, 474i, 475

of human social behavior, 1285–1287
of humans, 702–707, 702i–704i
of insecticide resistance, 412i
of jaws, 679, 679i
Lamarck's theory, 404
of leaves, 578, 578i
life history traits in guppies, 1134–1135
macroevolution, 463–490
microevolution, 419–442
misinterpretation of theory, 414–415
modern synthesis, 411–412
molecular, 416
mosaic, 496, 503
mutationism theory, 411
orthogenesis, 415
patterns of, 475, 475i
phylogenetic trees and, 633, 634i, 635
of plastids, 569–572, 570i
punctuated equilibrium hypothesis, 477
of reproductive behavior and mating
systems, 1279–1280
of social behavior, 1281–1287, 1281–1288
vertebrate, 674–675, 674i–676i
viral, 544
Evolutionary change, evidence for, 413–414
adaptation by natural selection, 413
comparative embryology, 414, 414i
comparative molecular biology, 414
comparative morphology, 413–414
fossil record, 413
historical biogeography, 413
Evolutionary developmental biology,
483–488
Evolutionary divergence, 410
Excavates, 553–554, 554i, 571
Excretory system, annelid, 651, 651i
Exoenzyme, 537
Exoskeleton
arthropod, 655
barnacle, 659
carapace, 658, 659i
crustacean, 658
defined, 655
ecdysis, 655i
Exotic species, 1235, 1237, 1237–1238,
1237i, 1238i
Exotoxin, 536
Exploitative competition, 1155
Exponential model of population growth,
1134–1137, 1136, 1136i
Extemophile, 538, 538i, 539
Extinction
background rate, 481, 1239
current rate, 1239
Cuvier and, 463–464
defined, 480
determinants of, 416
mass, 481–482, 483i, 1239
Miss Waldron's red colobus monkey,
1229–1230, 1230i
natural theology and, 402, 404
Extracellular digestion, by fungi, 608
Extracellular polymer substances (EPS),
533
Extreme thermophile, 540
Eye, compound, 656
Eyespot, Euglena, 554, 555i

Facilitation hypothesis, 1173
Facultative anaerobes, 531
Family, 493
Family planning programs, 1148
Feather, 692, 693i
Feather duster worm, 652i
Feather star, 670i, 671
Fecundity
age-specific, 1131, 1131t
versus parental care, 1133

Feeding
adaptations, 1153
defenses against herbivory and predation, 1153–1155, 1154i, 1155i
food choice, 1153, 1153i
optimal foraging theory, 1153
Fern, 588i, 589
Fernald, Russell, 1262–1263, 1262i
Fertilization, double, 598, 599i
Fiddler crab (Uca species), 1264, 1265i
Filarial worm, 645, 645i
Fin, 679, 679i
Fire ant, 660i
Firefly, behavioral isolation in, 448, 448i
Fish
anatomy, 682i
antifreeze proteins, 1210
evolution, 675
jawed, 678–683
actinopterygii, 681, 681i, 682i
chondrichthyes, 680–681, 680i
origin of, 679, 679i
sarcopterygii, 681, 683, 683i
osteolepiforms, 683–684, 684i
overexploitation, 1238–1239, 1238i
Fitzroy, Robert, 407
Five-lined skink (Eumeces fasiatus), 1131i,
1132
Fixed action pattern, 1255
Flabellina iodinea, 649i
Flagella
prokaryotic, 530, 530i
protist, 553
Flame cell system, 642, 642i
Flea, 660i
Fleming, Alexander, 419
Flight
adaptations for, 692, 692i
wing movements during, 692i
Floral organ identity program, 487
Florida panther (Puma concolor coryi), 1247
Flounder (Platichthys flesus), 682i
Flower
defined, 595
evolution of, 601
Flowering plants, 595–602
DNA barcoding analysis, 1242
molecular phylogenetic analysis,
505–506, 506i
Fluorescence, in dinoflagellates, 556
Food chain, 1164–1165
Food vacuole, 552, 552i
Food web
DDT concentrations in, 1236
description, 1165–1166, 1165i
detrital, 1182, 1183i
grazing, 1182, 1183i
interaction networks, 1250
Foraminifera (forams), 562–563, 563i
Forest
biomes
boreal forest/taiga, 1217–1218, 1218i
temperate deciduous, 1217, 1217i
temperate rain forest, 1218
thorn forest, 1215
tropical, 1213–1214, 1213i
carbon sequestration, 1231
deforestation, 1233, 1233i, 1234
habitat fragmentation, 1232–1233,
1232i
Fossil fuel, 585, 1196, 1231
Fossil record
ancestral and derived characters, determination of, 497
biological lineages, 413
completeness of, 465
as evidence of evolutionary change, 413
flowering plants, 595–596, 596i

geological time scale, 466t–467t
gradualist hypothesis, 477, 478i
for horses, 475
information provided by, 469
punctuated equilibrium hypothesis,
476i, 477
seedless vascular plants, 585
Fossils
Burgess Shale formation, 627–628, 628i
Cuvier and, 463–464, 464i
dating, 465, 468–469
defined, 403
examples, 465i
formation of, 464–465
organic matter in, 468–469
plants, 576, 578i, 581, 588i, 592i, 596,
596i, 601
role in development of evolutionary
thought, 403–404
trilobites, 657, 657i
Founder effect, 428–429, 450
Fragmentation, habitat, 1232–1233, 1232i,
1246, 1247i
Frequency-dependent selection, 436, 438i
Freshwater biomes, 1219–1221
lakes, 1220–1221, 1220i, 1221i
streams and rivers, 1219–1220, 1219i
Frog
calls, 1287
description, 685, 685i
habitat preferences, 1275
Fruit, 595
Fruit fly. See also Drosophila
Hox genes, 485i
temporal isolation, 447
Fruiting body, 534–535, 535i, 564
Fundamental niche, 1157, 1157i, 1159i
Fungal spore, 608–610, 609i
Fungi, 605–626
characteristics of, 606–610
as decomposers, 605–606, 608
defined, 605
diversity, 605, 606i, 611–612
groups, 610–620, 612t
Ascomycota (ascomycetes), 612t,
614–616, 615i–617i
Basidiomycota (basidiomycetes),
612t, 617–619, 618i, 619i
Chytridiomycota (chytrids), 610–620,
612i, 612t
conidial fungi (imperfect fungi/
deuteromycetes), 619–620
Glomeromycota (glomeromycetes),
612t, 614
microsporidia, 611, 620, 620i
Zygomycota (zygomycetes), 612t,
613–614, 614i
life cycle
basidiomycete, 618–619, 619i
generalized, 608, 609i
nutrient acquisition, 607–608
phylogeny of, 610–612, 610i
plant pathogens, 624
reproduction, 608–610, 609i
structure, 606–607, 607i
symbiotic associations
lichen, 620–621, 621i, 622i
mycorrhizae, 621–623, 623i
systems biology studies, 624
Futuyma, Douglas J., 416

Galagos, 699
Galápagos finches
adaptive radiation, 480, 482i
bill shape and food habits, 407, 407i, 410
character displacement, 1157, 1158i
disruptive selection in cactus finches,
433, 433i

Galápagos Islands
  Darwin and, 407
  finches, 1157, 1158i
  map of, 406i
Galápagos shark *(Carcharhinus galapagensis)*, 680i
Galilei, Galileo, 402
Gallmaking fly *(Eurosta solidaginis)*, 431–432, 432i
Gambusia, 492
Gametangium, **581**, 582i, 583, **613**, 614i
Gamete
  defined, **629**
  unreduced, **456**
Gametic isolation, 449
Gametophyte
  alternation of generations, 579, 579i
  angiosperm, 597, 598, 599i
  bryophyte, 581, 582i
  club moss, 586–587
  conifer, 593, 594, 594i
  defined, **579**
  fern, 588i, 589
  gymnosperm, 590, 591
  from heterosporous plants, 579
  from homosporous plants, 579
  hornworts, 583
  kelp, **561**
  liverwort, 582
  moss, 583, 584i
  seedless vascular plants, 585
Ganglion
  defined, **642, 672**
  in flatworms, 642, 642i
Gars, 681, 681i
Garter snake *(Thamnophis elegans)*, 1256, 1257i
Gastropod, 647i, 648, 649i
Gastrovascular cavity, **637**
Gause, G. F., 1156
Gemmae, **582**, 583i
Gemmule, 636
Gene flow, **426**, 426–428, 427i, 1244
Gene pool, 424
Generation time, **1129**, 1129i, 1137i
Genetic distance, 505
Genetic divergence, 454–456
Genetic drift
  as agent of microevolution, 428–429
  conservation implications, 429
  defined, **428**
  founder effect, 428–429
  population bottlenecks, 428, 428i
Genetic equilibrium, **425**
Genetic tool kit, 484
Genetic variation
  agents of microevolutionary change, 425–434, 425t
  balanced polymorphisms, 435–436
    frequency-dependent selection, 436, 438i
    heterozygote advantage, 435, 436i
    selection in varying environments, 435–436, 437i
  biodiversity, 1230
  gene flow effects, 426–428, 427i
  generation of, 422
  level in population, 422–423
  maintenance of, 435–437
  mutations as source of, 425–426
  natural selection effects, 430–433, 430i–433i
  neutral, 437
  recombination rate and, 439
  reduction with genetic drift, 428–429, 428i
Genetically modified crop, 1231

Genetics, population, 423–425, 424t
Genotype frequencies, **424**, 424t, 426–427, 433–434
Genus, **493**
Geoduck *(Panope generosa)*, 649i
Geographical range, **1127**
Geographical variation, in species, 445–447, 445i, 446i, 447i
Geography of speciation, 449–454, 451i, 453i, 454i
Geological strata, 465, 465i, 468
Geological time scale, 466t–467t
Geology
  catastrophism, 404
  gradualism, 404
  uniformitarianism, 404
German cockroach *(Blattella germanica)*, 494i
Giant saguaro *(Carnegia gigantea)*, 600i
Giant swallowtail butterfly *(Papilio cresphontes)*, 1154i
Giardia lamblia, 553, 554i, 571
Gibbon, 701, 701i
Gill arches, 672, 679, 679i
Gill slits, **671**, 671i
Gills
  bivalves, 649, 649i
  crustacean, 658
  mollusk, 647i, 648
Gingerich, Philip G., 502
Ginkgo biloba, 592, 592i
Ginkgophyta, **592**
Glacier retreat, 1171, 1172i
Gleason, Henry A., 1161, 1162
Global warming, 1196, 1210–1211, 1226, 1233, 1234
Glomeromycota (glomeromycetes), 612t, 614
Glomus versiforme, 623i
Glor, Rich, 507
Glyptodont, 405, 405i
Gnathostomata, **675**, 678–683
Gnetophyta, 592–593, 593i, 601
Gnetum, 592, 593i
Goldacre, R. J., 516
Golden algae, 561, 561i
Golden plover *(Pluvialis dominica)*, 694
Goldenrod plant *(Solidago altissima)*, 431
Goldingay, Russ, 1244
Goldschmidt, Richard, 487
Golgi complex, origin of, 520, 521i
Gonadotropin releasing hormone (GnRH), 1262–1263
Gondwana, 469, 470i, 472
Gorilla *(Gorilla gorilla)*, 701
Gould, Stephen Jay, 477, 487
Gradualism, **404**
Gradualist hypothesis, **477**, 478i
Grain borer beetle *(Rhizopertha dominica)*, 1139i
Gram stain technique, **528**
Gram-negative bacteria, **528**
Gram-positive bacteria, **528**, 535–536, 536i
Grand Banks, 1238
Grand Canyon, 465i
Grant, Peter, 433
Grasshopper, 661i, 662i, 663
Grasshopper mouse *(Onychiomys)*, 1155, 1155i
Grassland
  savanna, 1214–1215, 1215i
  temperate, 1216–1217, 1216i
Gray whale *(Eschrichtius robustus)*, 1270
Grazing food web, 1182, 1183i
Great American Interchange, 472
Great blue heron *(Ardea herodias)*, 404i
Great tit *(Parus major)*, 1140i, 1281

Great white shark *(Carcharodon carcharias)*, 680
Green algae, 567–569, 567i, 568i, 575, 576, 620, 621
Green bacteria, 535
Greenhouse effect, **1196**, 1233, 1234
Greylag geese *(Anser anser)*, 1258, 1258i
Grip, power *versus* precision, 702
Gross primary productivity, **1183**, 1200
Guano, 1199
Guerrant, Edward O., 481
Guinea pig, 698, 698i
Guppy *(Poecilia reticulata)*, 1134–1135
Gymnophilus, 606i
Gymnophiona, 685, 685i
Gymnosperm, 590–595
  ancestral forms, 590, 591i
  conifers, 593–594, 594i
  cycads, 591, 592i
  defined, **590**
  ginkgo, 592, 592i
  gnetophytes, 592–593, 593i
  overview, 590
  reproductive adaptations, 590–591

Habitat
  defined, **1127**
  patch, 1246, 1246i, 1247i
  of protists, 553
Habitat fragmentation, **1232**–1233, 1232i, 1246, 1247i
Habitat selection, 1274–1275, 1275i
Habituation, **1259**
Haddock *(Melanogrammus aegelfinus)*, 1238
Hadrosaurus, 688
Haeckel, Ernst, 629, 1126
Hagfish, 675, 677, 677i
Haikouichthys, 675
Hainsworth, F. Reed, 1209
Haldane, J. B. S., 513–514
Half-life, **468**
Hallucigenia, 627, 628i
Halophile, 540
Hamadryas baboon *(Papio hamadryas)*, 700i
Hamilton, William D., 1282
Haplodiploidy, **1284**–1285, 1284i
Haplorhini, 699–701, 700i, 701i
Hardy, G. H., 424
Hardy-Weinberg principle, **425**, 426–427
Harlequin frog/toad *(Atelopus)*, 612, 612i, 1154i
Harrison, Susan, 1245
Harvesting efficiency, 1185
Hatena, 571
Haustoria, **607**
Hawaiian Islands, fruit fly speciation in, 452
Hawkmoth, 600, 600i
Hawthorn *(Crataegus species)*, 453, 454i
Head
  arthropod segment, 655
  insect, 661, 661i
  viral, **542**
Head-foot, mollusk, **647**–648, 647i
Hebert, Paul, 1242
Heinrich, Bernd, 1278
Helical virus, **542**, 542i
Heliconius erato, 1155i
Heliconius melpone, 1155i
Helix pomatia, 649i
Hemichordata, 671, 671i
Hemlock woolly adelgid *(Adelges tsugae)*, 1237–1238
Hemocoel, 655
Hemoglobin, heterozygote advantage and, 435, 436i

Hemolymph
  of arthropods, 655
  defined, **648**
  in mollusks, 648
Hennig, Willi, 499
Henslow, John, 405
Hepatophyta, **582**
Herbivory
  defenses against, 1153–1155
  defined, **1153**
  interspecific competition among plants, effect on, 1167, 1169i
  by zooplankton, 1189
Hermaphroditic
  annelids, 651
  barnacle, 659
  ctenophores, 639
  defined, **636**
  snails, 648
  sponges, 636
  trematodes, 643i
  turbellaria, 643
Herring gull (*Larus argentatus*), 1255, 1256i
Hertz, Paul E., 1209, 1211i
Heterochrony, **480**, 485
Heterocyst, 535i
Heterokont, 558, 558i, 560–562, 560i–562i, 571
Heterosporous, **579**, 587
Heterotroph, **531**, 552, **628**, **1164**, 1192
Heterozygote advantage, **435**, 436i
Hexapoda, 660i–663i, 661–663
Hibiscus, 600i
Higher vocal center, 1260
Hileman, Lena, 487
Hinkle, Gregory, 611
Hippocrates, 491
Hirudinea, 652–653, 652i
Hirudo medicinalis, 652–653, 652i
Historical biogeography, **413**
HIV (human immunodeficiency virus), **542**, 542i
Holothuroidea, 670i, 671
Homeobox, **484**
Homeodomain, **484**
Homeosis, 487
Homeostasis
  environmental variation, responses to, 1209–1211, 1211i
Homeotic genes, **484**, 486, 487, 674–675, 674i
Homing pigeon (*Columba livia*), 1271
Hominids
  Ardipithecus, 703
  Australopithecus, 703–704
  bipedal locomotion, 702, 702i
  defined, **702**
  grip, power and precision, 702, 702i
  Homo erectus, 704–705, 704i
  Homo habilis, 704
  Homo neaderthalensis, 705, 706
  Homo sapiens, 705–707
  Orrorin tugenensis, 702–703
  Paranthropus, 703–704
  time line, 703i
Hominoidea, **701**
Homo erectus, 704–705, 704i
Homo habilis, 704
Homo neanderthalensis, 705, 706
Homologies (homologous characters), **495**–496, 496i
Homologous structures, 403i
Homologous traits, **413**–414
Homoplasies (homoplasious characters), 496
Homosporous, **579**, 587

Honeybee (*Apis mellifera*), 1155i, 1261, 1277, 1278i, 1284–1285, 1284i
Hooker, Joseph, 401
Hormones
  behavior and, 1260–1263, 1261i, 1262i
  seasonal migration, 1271–1272
Hornwort, 583, 583i
Horses, evolutionary lineage of, 473, 474i, 475
Horseshoe crab (Merostomata), 658, 658i
Horsetail, 589–590, 590i
Host, **1159**
Host race, **453**
House sparrow (*Passer domesticus*), 447i, 1125
Housefly, 662i
Hox genes, **484**, 485i, 486, 487, 633, 661, 674–675, 674i
Hubbard Brook watershed, 1193
Huey, Raymond B., 1209, 1211i
Human(s)
  biodiversity benefits to, 1230–1231
  chromosomal differences among great apes, 458–459, 458i
  evidence for stabilizing selection in, 431i
  evolution, 702–707, 702i–704i
  instinctive behaviors, 1255, 1255i
  origin, 705–707
  population growth, 1145–1149, 1145i–1148i
  skull, changes in, 479
  social behavior, 1285–1287
Hummingbird, 600i, 1209–1210
Humpback whale (*Megaptera novaeangliae*), 429, 1225i
Hutchinson, G. Evelyn, 1148
Hutton, James, 404
Huxley, Thomas, 411
Hybrid, polyploid, 456–458, 457i
Hybrid breakdown, **449**
Hybrid inviability, **449**
Hybrid sterility, **449**, 455
Hybrid zone, **451**, 453i
Hydra, 638, 638i
Hydrangea macrophylla, 422i
Hydroid, 584
Hydrologic cycle, 1192, 1192i
Hydrolytic enzyme, 608
Hydrosphere, **1204**
Hydrostatic skeleton
  annelid, 651
  arthropod, 655
  cnidarian, 637–638
  defined, **630**
Hydrothermal vent community, 515, 1225, 1225i
Hydrozoa, 638, 639i
Hyla species, 494–495, 495i
Hyomandibula, 495, 496i
Hyphae
  fungus, 606–607, 607i
  in Oomycota, 558, 560i
Hypolimnion, **1220**
Hyracotherium, 473, 474i, 475, 479

Ichthyosaurs, 688
Ichthyostega, 684, 684i
Iguanas, marine, 406i
Immigrant, 1243, 1245
Immigration, **1129**
Imperfect fungi, 619–620
Imprinting, **1258**, 1258i
Inbreeding, **434**
Incomplete dominance, 424
Incomplete metamorphosis, **663**, 663i
Incurrent siphon, **673**

Indeterminate cleavage, **631**
Indian pipe (*Monotropa uniflora*), 595i
Indigo bunting (*Passerina cyanea*), 1271, 1273i
Influenza virus, 544
Inheritance of acquired characteristics, 404
Inhibition hypothesis, **1173**
Insect
  anatomy, 661–662, 661i, 662i
  development patterns, 662–663, 663i
  diversity, 660i
  ecdysis, 655, 662
  eusocial, 1284–1285, 1284i
  exotic species, 1237–1238, 1238i
  mouthparts, 661–662, 662i
  parasitoids, 1160
  pollination by, 600, 600i
Insecticide resistance, evolution of, 412i
Insectivora, 696
Insight learning, **1259**
Instar, 662
Instinctive behavior, **1254**–1257, 1256i, 1257i
Interference competition, **1155**
Intermediate disturbance hypothesis, **1169**–1170, 1170i
Intermediate host, 644, 644i
Intersexual selection, 433, 434i
Interspecific competition, 1155–1158, 1156i, 1159i
  character displacement, 1157–1158, 1158i, 1167
  competitive exclusion, 1156
  defined, **1155**
  exploitative competition, 1155
  interference competition, 1155
  niche concept, 1156–1157, 1157i, 1159i
  resource partitioning, 1157–1158, 1157i, 1167
  species richness, reduction of, 1166–1167
Intertidal zone, **1221**, 1222–1223, 1223i, 1224i
Intrasexual selection, 433
Intraspecific competition, **1138**
Intrinsic rate of increase, **1137**, 1137i
Introduced species, 1125–1126, 1126i, 1235, 1237–1238, 1237i, 1238i
Introns, in archaeans, 521–522
Invertebrate(s)
  chordates, 672i, 673, 673i
  defined, **628**
  deuterostomes, 668–671, 670i
  diversity, genetic basis of, 662
  evolutionary relationships, 662
Irby, William S., 662
Islands
  allopatric speciation on, 450–451, 451i
  as biodiversity hotspots, 1240
  equilibrium theory of island biogeography, 1175–1177, 1175i, 1176i

Japanese blood fluke (*Schistosoma japonicum*), 644, 644i
Jarrow's spiny lizard (*Sceloporus jarrovi*), 1276
Jaws
  evolution of, 679, 679i
  mammalian specializations, 695
Jefferson, Thomas, 404
Jellyfish, 638, 640i
Johal, Guri, 624
Johnston, Wendy K., 518
Juvenile hormone, 1261

Karenia brevis, 558i
Karner Blue butterfly *(Lycaeides melissa samuelis)*, 1248i
Karyogamy, 609i, **610**, 613, 614i, 616, 617i, 619, 619i
Katydid *(Mimetica)*, 1154i
Keeled sternum, **692**, 693i
Kelp, 561, 1223, 1224i
Keystone species, **1167**
Killifish *(Rivulus hartii)*, 1134–1135
Kin selection, **1283**, 1283i
Kinesis, **1275**
Kinetoplast, 554, 555i
Kinetoplastid, 554–555, 555i
Kingdom, **493**
Kirtland's warbler *(Dendroica kirtlandii)*, 11601160i
Kissing Gourami *(Helostoma temmincki)*, 682i
"Knockout" mice, 1258
Komodo dragon *(Varanus komodoensis)*, 689
Korarchaeota, **539**, 540
Kramer, Elena M., 487
Kring, Matthias, 705
K-selected species, 1142–1143, 1143i, 1143t, 1167, 1171, 1173
Kudzu *(Pueraria lobata)*, 1237, 1237i
Kuru, 546
Kuspa, Adam, 566

La Niña, 1204
Labium, 661, 662i
Labrum, 661, 662i
Lactobacillus, 536
Lagomorpha, 696
Lahav, Noam, 516
Lake
    seasonal changes, 1220–1221, 1221i
    trophic nature of, 1221, 1222i
    zonation, 1220, 1220i
Lake Erie, 1181–1182, 1182i
Lamarck, Jean Baptiste de, 404
Laminaria, 562i
Lamprey, 677, 677i
Lancelet, 673, 673i
Landscape corridor, 1246–1247, 1247i
Landscape ecology, **1246**–1247
Larva
    defined, **629**
    insect, 663, 663i
Latent phase, of viral infection, 543
Lateral-line system, **680**
Laurasia, 469, 470i
Layne, John, 1264
Laysan albatross *(Phoebastria immutabilis)*, 694i
Leaf, evolution of, 578, 578i
LEAFY protein, 601
Leakey, Louis, 704
Learned behavior, **1254**–1255, 1257–1259
Learning
    classical conditioning, 1258–1259
    defined, **1257**
    habituation, 1259
    imprinting, 1258, 1258i
    insight, 1259
    operant-conditioning, 1259
Lecanora conizaeoides, 621
Leclerc, George-Louis, 403
Leech, 652–653, 652i
Leguminous plant, 1158, 1161
Leishmaniasis, 555
Leks, **1279**–1280, 1280i
Lemaître, George, 511
Lemming *(Lemmus lemmus)*, 1144
Lemur, 699, 700i
Lentic biome, 1219

Leopard frog *(Rana pipiens)*, 685i
Lepidodendron, 585, 587i
Lepidosauromorpha, **688**
Lepidosaurs, 688
LFY gene, 601
Lichen, **620**–621, 621i, 622i, 1171
Life
    inevitability of, 523
    origin of, 511–524
        energy-harnessing reactions, 516–517
        extraterrestrial, 512–513
        hydrothermal vents and, 515
        inevitability of life, 523
        nonliving to living transition, 512
        Oparin-Haldane hypothesis, 513–514
        organic molecule formation, 513–515, 514i
        protein-first hypothesis, 517, 522
        protocells, 512, 515–517, 516i, 522
        RNA world, 517, 518
        RNA-first hypothesis, 517
        timeline, 512i
    tree of, 410–411, 410i, 506–507, 506i
Life cycle
    algae
        brown, 561, 562i
        green, 568, 568i
    animal
        beef tapeworm, 645i
        hydrozoan, 639i
        Japanese blood fluke *(Schistosoma japonicum)*, 644, 644i
    bryophyte, 581
    fungi
        ascomycete, 616, 617i
        basidiomycete, 618–619, 619i
        generalized, 608, 609i
        zygomycete, 614i
    plant
        angiosperm monocot, 597, 599i
        conifer, 593–594, 594i
        fern, 588i
        general, 578–579, 579i
        moss, 584i
    Plasmodium, 559i
    protist, 552
Life history
    defined, **1132**
    early *versus* late reproduction, 1133
    energy budget, 1132
    evolution of traits in guppies, 1134
    fecundity *versus* parental care, 1133
    population size, effect on, 1142–1143, 1143i, 1143t
    reproductive episodes, number of, 1133
    variation among species, 1132–1133
Life table, **1130**–1131, 1131t
Light pollution, 1235, 1235i
Lignin, **577**, 578
Lijam, Nardos, 1258
Likens, Gene, 1193
Lilium, 599i
Limiting nutrient, **1184**
Limnetic zone, **1220**, 1220i
Limulus polyphemus, 658i
Linnaeus, Carolus, 402, 493
Lipid bilayer assembly, protocell, 516, 516i
Lipopolysaccharide, **528**, 536
Lithosphere, **1204**
Litt, Amy, 601
Littoral zone, **1220**, 1220i
Liverwort, 582–583, 583i
Lizard
    description, 689–690, 690i
    as model research organism, 691

predation on spiders, 1142, 1142i
    thermoregulation, 1209, 1211i
Lobaria verrucosa, 622i
Lobe-finned fish, 681, 683, 683i
Lobster, 658, 659i
Locust *(Locusta migratoria)*, 1140, 1141i
Loggerhead sea turtle *(Caretta caretta)*, 1188, 1190
Logistic model of population growth, **1137**–1139, 1138i, 1138t, 1139i
Longleaf pine *(Pinus palustris)*, 1247i
Long-tailed widowbird *(Euplectes progne)*, 434i
Lophophore, **641**–642
Lophotrochozoa, 633, 641–653
    feeding structure, 641–642, 641i
    phyla
        Annelida, 650–653, 651i, 652i
        Brachiopoda, 641–642, 641i
        Ectoprocta, 641–642, 641i
        Mollusca, 647–650, 647i–650i
        Nemertea, 646, 647i
        Phoronida, 641–642, 641i
        Platyhelminthes, 642–646, 642i–645i
        Rotifera, 646, 646i
Lorenz, Konrad, 1254, 1258
Lorises, 699
Losos, Jon, 691
Lotic biome, 1219
Lotka, Alfred J., 1155
Lubchenco, Jane, 1167
Luciferase, 556
Luciferin, 556
Lucy (hominid fossil), 703, 704i
Lumbricus, 652i
Luna moth, 660i
Lungfish, 683, 683i
Lycophyta, 585, **585**, 586–587, 588i
Lycopodium, 586–587, 588i
Lyell, Charles, 401, 404
Lyme disease *(Borrelia burgdorferi)*, 527
Lynx, 1144–1145, 1144i
Lysogenic cycle, 542
Lysozyme, 542
Lytic cycle, 542

MacArthur, Robert, 1148, 1166, 1175–1177
Macroevolution
    biodiversity trends, 480–483
        adaptive radiations, 480, 482i, 483
        extinction, 480–482, 483i
    Cuvier and, 464
    defined, **412**
    interpretation of lineages, 473–477, 474i–476i
    in modern synthesis view, 412
    modes of evolutionary change, 475, 475i
    morphological trends, 477–480
        allometric growth, 479–480, 479i
        complexity, 479
        heterochrony, 480
        novelties, 479–480
        paedomorphosis, 480, 480i, 481i
        preadaptation, 479
        size increase, 478–479
    tempo of morphological change, 475, 475i, 477, 478i
Macromolecules, formation on early Earth, 515–516
Macronucleus, 552i, **556**
Mad cow disease, 545–546
Madreporite, 669, 670i
Magnaporthe grisea, 624
Magnolia *(Magnolia grandiflora)*, 470–471, 727
Magnoliids, **596**, 597i
Maizel, Alexis, 601

Malaria, 435, 436i, 491–492, 558, 559, 1236
Malpighian tubule, **661**, 661i
Malthus, Thomas, 409
Mammal
  adaptive radiation, 480
  evolutionary history, 472–473
  phylogeny, 698, 698i
Mammary gland, **695**
Mammoth, 465i
Manatee, 696, 697i
Mandible, crustacean, 658
Mangrove, 1222, 1223i
Manta ray *(Manta birostris)*, 680, 680i
Mantle, 647i, **648**
Mantle cavity, 647i, 648
Malpighian tubule, 656
Marchantia, 582, 583i
Margulis, Lynn, 519
Marine biomes, 1221–1225
  abyssal zone, 1224
  benthic province, 1224–1225, 1225i
  estuaries, 1222, 1223i
  intertidal zone, 1222–1223, 1224i
  neritic zone, 1223, 1224i
  oceanic zone, 1224
  zonation, ocean, 1221, 1223i
Marine vertebrates, convergent evolution in, 473i
Maritime climate, **1208**
Mark-release-recapture technique, 1128i
Marler, Catherine, 1276
Marler, Peter, 1255
Marsh, Othniel C., 473, 475
Marsh marigold, 600i
Marshall, Larry G., 472
Marsupials (Metatheria), 472, 695–696, 696i
Marsupium, **696**
Mass extinction, **481**, 483i, 1239
Mastax, rotifer, 646
Mathews, Sarah, 506
Mating, nonrandom, 433–434
Mating system, 452, **1280**
Mating type, 609i, **610**
Matsuno, Koichiro, 515
Maxillae, crustacean, 658
Maxilliped, 659, 659i
Maximum likelihood methods, 505
Mayfly *(Hexagenia)*, 1182
Mayr, Ernst, 443–444, 445
McFadden, Dr. Geoff, 571
McGraw, William S., 1230
McNeilly, Thomas, 451, 454i
Mean, 421, 421i
Mechanical isolation, 448–449, 448i
Medullosa, 585, 587i
Medusae, **637**, 637i, 638, 639i
Megaspore, **593**, 594i
Membrane
  origin of eukaryotic, 520, 521i
  outer of Gram-negative bacteria, 528
Mesentery, **630**, 631i
Mesoderm, **630**, 632, 642
Mesophile, 538
Mesophyll, **635**
Metabolic rate, mammalian, 695
Metabolism, in protists, 552
Metamorphosis
  amphibian, 685
  complete, 663, 663i
  defined, **636**
  incomplete, 663, 663i
  sponge, 636
Metanephridia, **651**, 651i
Metapopulation, **1243**, 1245, 1245i
Meteorite, 512, 515
Methanococcus jannaschii, 539

Methanogen, 539–540, 539i
Methanosarcina, 539i
Metrarabdotos, 476i, 477
Micrasterias, 550i
Microclimate, **1208**
Microevolution, 419–442
  adaptation, 437–440
  agents of, 425–434, 425t
    gene flow, 426–428, 427i
    genetic drift, 428–429, 428i
    mutation, 425–426
    natural selection, 430–433, 430i–433i
    nonrandom mating, 433–434
  antibiotic resistance, 419–420
  defined, **412**, **420**
  maintenance of variation, 435–437
    balanced polymorphisms, 435–436, 436i, 437i, 438i
    diploidy, 435, 435t
    neutral variation hypothesis, 437
  population genetics, 423–425, 424t
  variation in natural populations, 420–423, 421i, 422i
    genetic, 422–423
    phenotypic, 420–422, 420i, 421i, 422i
    qualitative, 421, 421i
    quantitative, 421, 421i
Micronucleus, 552i, **556**
Microspore, **593**, 594i
Microsporidia, **611**, **620**, 620i
Migration, 1269–1274
  bird, 694
  defined, **1270**
  food supply variation and, 1272–1274, 1274i
  hormonal triggers, 1271–1272
  wayfinding mechanisms, 1270–1271
    compass orientation, 1271, 1273i
    navigation, 1271
    piloting, 1270–1271, 1272i
Mildew, downy, 549, 560
Mildew, powdery, 616
Milkweed, 1154
Miller, Stanley L., 514
Miller-Urey apparatus, 514, 514i
Millipede, 659–661, 660i
Mimic, **1155**
Mimicry, **1155**, 1155i
Minimum viable population size, **1243**, 1244
Miss Waldron's red colobus monkey *(Procolobus badius waldroni)*, 1229–1230, 1230i
Mite, 657, 658
Mitochondria, endosymbiont hypothesis and, 519–520, 520i
Mitochondrial DNA (mtDNA) studies
  arthropods, 656
  evolution rate of, 503
  human evolution, 705–707
  humpback whales, 429
  loggerhead sea turtles, 1190
  mammalian phylogeny and, 698
  Neanderthal, 705
Mixed-use conservation, 1248, 1248i
Mnium, 582i
Model, **1155**
Modeling, ecosystem, 1199–1200, 1199i
Modern synthesis, 411–412
Molar, **695**
Mold, **609**
Mole, 1264, 1265, 1266–1267, 1266i
Molecular biology, comparative, 414
Molecular clock
  defined, **503**
  multicellular eukaryotes, emergence of, 522
Mollusk, 647–648, 647–650, 647i

Monarch butterfly *(Danaus plexippus)*, 497i, 1272–1274, 1274i
Monkey, 699, 700i, 701
Monkey-flower *(Mimulus)*, 448–449, 448i, 455
Monocot, **596**, 597, 598i
Monogamous species, **1280**
Monogenoidea, 643
Monophyletic taxa, **497**, 498i
Monotremes (Prototheria), 695, 696i
Monsoon cycles, **1208**
Montane forest, tropical, 1213–1214
Moore, Michael, 1276
Moray eel *(Gymnothorax moringa)*, 682i
Morel *(Morchella esculenta)*, 615, 615i
Morphological species concept, **444–445**
Morphology
  comparative, 403, **413–414**
  defined, **403**
  macroevolutionary trends, 477–480
Mortality, age-specific, **1130–1131**, 1131t
Mosaic evolution, **496**, 503
Mosquitoes, 491–492, 492i, 558, 559, 662i
Moss, 576i, 583–584, 584i
Motile, **628**
Mouse
  activity level, genetic basis of, 423, 423i
  Hox genes, 485i
  knockout, 1258
Mule, 449
Müller, Fritz, 1155
Müllerian mimicry, **1155**, 1155i
Multicellularity, origin of, 522–523
Multiregional Hypothesis, **705–707**
Murphy, Dennis D., 1245
Mushroom, 606, 606i, 607, 607i, 617–619, 618i, 619i
Musk oxen *(Ovibos moschatus)*, 1281
Mussels, 648–649, 1167, 1168i
Mutation
  advantageous, 426
  as agent of microevolution, 425–426, 425t
  defined, **425**
  detrimental, 425, 439
  lethal, 425–426
  neutral, 426
  prezygotic reproductive isolation, 455
Mutationism, 411
Mutualism, **1158**, 1160i, 1161, 1161i
Mycelium, 558, 560i, 607, 607i
Mycobiont, **620–621**
Mycology, 606
Mycoplasma, 536
Mycoplasma genitalium, 527
Mycorrhizae, 621–623, 623i
Myers, Norman, 1239, 1240
Myers, Ransom A., 1239
Myllokunmingia, 675, 675i
Myriapoda, 659–661, 660i
Myrtle warbler *(Dendroica coronata)*, 444i
Myxinoidea, 675, 677
Myxobacteria, 534–535, 535i
Myxoma virus, 1126

Naked mole-rat *(Heterocephalus glaber)*, 1285, 1286
Nanoarchaeota, **539**
Nanoplankton, 561
Natural history, **402**
Natural selection
  adaptation by, 413, 416
  balanced polymorphisms, maintenance of, 435–436, 436i, 437i
  Darwin and, 410, 411
  defined, **410**, **430**
  human behavior, 1287
  kin selection, 1283, 1283i

Natural selection (Continued)
  modes
    directional, 430i, 431, 432i
    disruptive, 430i, 433, 433i
    stabilizing, 430i, 431–433, 431i, 432i
  recessive alleles masked by diploidy, 435, 435t
  relative fitness and, 430–431
Natural theology, **402**, 403, 404
Nautilus macromphalus, 650i
Nautiluses, 650, 650i
Navarro, Arcadi, 459–460
Navigation, 1271
Necrotizing fasciitis, 537
Nectar, 600
Neisseria gonorrhoeae, 530
Nekton, **1224**
Nematocyst, **637**, 637i, 638i
Nematode, 653–654, 653i
Nematode-trapping fungi, 616
Nemertea, 646, 647i
Nereis, 652i
Neritic zone, **1221**, 1223, 1223i, 1224i
Nerve cord, dorsal hollow, 672, 672i
Nerve net, **638**
Nervous system
  anatomy and behavior, 1263–1267
  annelid, 651, 651i
  bivalves, 649
  crustacean, 658
  development, hormone effects on, 1260–1261, 1261i
  nematode, 654
Nest predation, 1232i, 1233
Net primary productivity, **1183**, 1184, 1200
Neural crest, **673**
Neurophysiology, of behavior, 1259–1260, 1260i
Neuroscience, **1254**
Neurospora crassa
  hyphal branching in, 607
  life cycle, 616, 617i
  as research organism, 615
Neutral variation hypothesis, **437**
Newton, Sir Isaac, 402
Ngorongoro Conservation Area, 1248
Nielsen, Peter, 522
Nine-banded armadillo (*Dasypus novem-cinctus*), 405i, 1130
Nitella, 569
Nitrification, **532**, 1195, 1196t, 1197t
Nitrifying bacteria, 532
Nitrobacter, 532
Nitrogen cycle, 1194–1195, 1197–1198, 1197i, 1197t
  ammonification, 1195, 1197t
  defined, **1195**
  denitrification, 1197, 1197t
  human disruption, 1198
  nitrification, 1195, 1197t
  nitrogen fixation, 1195, 1197t
  in terrestrial ecosystem, 1197i
Nitrogen fixation, **531**–532, 535, 1195, 1197t
Nitrosomonas, 532
Node, **589**
Noise pollution, 1235
Nonhuman primates. *See* Primates
Nonrandom mating, 433–434
Nonvascular plant, **577**, 581–585
Noor, Mohamed, 439
Northern elephant seal (*Mirounga angustirostris*), 428, 428i, 1129, 1279
Northern spotted owl (*Strix occidentalis caurina*), 1130, 1249
Nostoc, 620, 1195
Notochord, **672**, 672i
Nucleoid, 527
Nucleus, origin of, 520, 521i
Nudibranch (sea slug), 648, 649i

Null model, **424**
Nutria (*Myocastor coypus*), 406i, 407
Nutrient
  cycling, 1191–1199, 1200
    carbon cycle, 1192, 1194, 1194i–1195i, 1200
    generalized compartment model, 1191–1192, 1191i
    hydrologic cycle, 1192, 1192i
    nitrogen cycle, 1194–1195, 1197–1198, 1197i, 1197t
    phosphorus cycle, 1197–1198, 1198i
  limiting, 1184
Nutritional modes, 531, 531i

Oak (*Quercus*), 427i
Oates, John, 1229
Obelia, 639i
Obligate aerobe, 531
Obligate anaerobe, 531
Oceanic zonation, 1221, 1223i
Oceanic zone, **1221**, 1223i, 1224
Ocelli (eye spots), **638**, 642, 642i
Octopamine, 1261
Octopus, 650, 650i
Odum, Howard T., 1187i
Oil spill, 1234–1235
Old man's beard (*Usnea trichodea*), 621, 621i
Oligochaeta, 651–652, 652i
Oligotrophic lake, **1221**
Olive baboon (*Papio anubis*), 1282i
Omnivore, **1164**
On the Origin of Species by Means of Natural Selection (Darwin), 402, 408, 410–411, 444
Onychophora, 627, 654–655, 655i
Oomycota, 558, 558i, 560, 560i
Opabinia, 627, 628i
Oparin, Aleksandr I., 513–514
Oparin-Haldane hypothesis, 513–514
Open circulatory system, **648**
Operant conditioning, **1259**
Operculum, **681**
Ophelia, 652i
Ophiostoma ulmi, 614
Ophisthokont, 569, 569i
Ophiuroidea, 669, 670i
Opossum (*Didelphis virginiana*), 473, 473i, 696
Optimal foraging theory, **1153**
Oral hood, **673**, 673i
Orange palm dart butterfly (*Cephrenes auglades*), 497i
Orangutan (*Pongo pygmaeus*), 701
Orca (*Orcinus orca*), 1165, 1165i
Orchid, 576i, 600i, 622
Order, **493**
Ordovician extinction, 481
Organic molecules, formation of early, 513–515, 514i
Organismal ecology, **1126**
Organismal traits, as systematic characters, 494–495, 495i
Orrorin tugenensis, 702
Orthogenesis, **414**
Oscula, **635**, 636, 636i
Ossicles, 669, 670i
Osteolepiform, 683–684, 684i
Ostracoderm, **678**, 678i
Ostrich (*Struthio camelus*), 403i
Ostrom, John, 479
Outer membrane, **528**
Outgroup comparison, 497, 497i
Ovary, 598, 599i
Overexploitation, 1238–1239, 1238i
Ovule
  angiosperm, 598, 599i
  conifer, 593, 594i

diploid, 456
gymnosperm, **590**
Oxidation-reduction reactions, of proto-cells, 516
Oxygen, prokaryotic metabolism of, 531
Oyster, 648–649
Oyster mushrooms (*Pleurotus ostreatus*), 617

Pääbo, Svante, 705
Pachycephalosaurus, 688
Pacific salmon (*Oncorhynchus*), 1270
Pacific yew tree (*Taxus brevifolia*), 1230–1231, 1230i
Paddlefish, 681
Paedomorphosis, **480**, 481i
Paine, Robert, 1167, 1168i
Paleobiology
  Cuvier and, 404, 463–464, 464i
  defined, **404**
  fossil record, 464–469
  historical biogeography, 469–473
Palp, bivalve, 649, 649i
Pampas, 1216
Pangaea, 469, 470i, 472
Panther, 1247
Parabasala, 553–554, 554i
Paramecium, 550i, 552i, 556, 557i, 1156, 1156i
Paranthropus, 703–704
Parapatric speciation, **451**, 453, 454i, 459
Paraphyletic taxa, **497**–498, 498i
Parapodia, **651**, 652i
Parasite, **1159**
Parasitism, 1141, **1159**–1160
Parasitoid, **1160**
Parazoa, 635–636, 635i
Parental care, 695, 1133
Parental investment, **1279**
Parmesan, Camille, 1210–1211, 1226
Parsimony, principle of, 498, 499, 505
Parthenogenesis, **646**
Passive parental care, **1155**
Patch, habitat, 1246, 1246i, 1247i
Paulinella, 571
Pavlov, Ivan, 1258
Pax-6 gene, 484–485
Peafowl (*Pavo cristatus*), 1279–1280, 1280i
Peat moss, 584
Pectoral girdle, **673**
Pedicellariae, **669**, 670i
Pedipalp, 657, 657i, 658
Pelagic province, **1221**, 1223i
Pellicle, **553**, 556
Pelvic girdle, **673**
Penguin, 473i
Penicillin, 419–420, 420i, 529
Penicillium, 615
Peptidoglycan, **528**, 529, 529i
Per capita growth rate, **1136**–1137, 1136i
Perdeck, A. C., 1256i
Peritoneum, **630**, 631i
Periwinkle snail (*Littorina littorea*), 1167, 1169i
Permian extinction, 481, 657
Petrified wood, 465i
Petromyzontoidea, 675, 677
Phaeophyta, 561–562, 562i
Phages, **541**, 542
Pharynx, chordate, 672
Phenotypic variation
  causes, 421–422, 422i
  defined, **421**
  maintenance of, 435–437
  qualitative, 421, 421i
  quantitative, 421, 421i
  in snails, 420–421, 420i
Pheromone, **1276**–1277, 1277i
Philodina roseola, 646i

Phloem
  angiosperm, 598
  defined, **578**
  in seedless vascular plants, 585
  whisk fern, 589
Phoronida, 641–642, 641i
Phoronopsis californica, 641i
Phospholipid, protocell, 516, 516i
Phosphorus, lake eutrophication and, 1222i
Phosphorus cycle, **1197**–1198, 1198i
Photic zone, **1220**, 1220i
Photoautotroph, **531**, 531i
Photobiont, **620**–621
Photoheterotroph, **531**, 531i
Photosynthesis
  in bacteria, 534, 535
  in brown algae, 561
  in dinoflagellates, 556
  in euglenoids, 554
  evolution of, 519
  freshwater biome, 1220–1221
  in golden algae, 561
  in green algae, 568
  open ocean, 1224
  primary productivity, 1183–1186, 1184i, 1186i, 1200, 1220–1221
  in prokaryotes, 528
  in red algae, 567
Phototroph, **531**
Phycobilin, 567
PhyloCode, **500**–501, 507
Phylogenetic species concept, **445**
Phylogenetic tree
  animal, 633, 634i, 635
  for arthropods, 656i
  cladograms, 499, 499i, 500i–501i
  comparative method, 493
  converting into a classification, 497–498, 498i, 499i
  defined, **492**
  for flowering plants, 596i
  land plants, 580i
  maximum likelihood approach, 505
  molecular phylogenetics, 503, 504i, 505
Phylogenetics
  classification and, 497–501
  molecular, 501–508
  monophyly, 497
  parsimony, 498, 505
  Protoctista, 551, 551i
Phylogeny
  animals, 633, 634i, 635
  defined, **492**
  molecular, 633
Phylum, **493**, **628**
Physarum, 550i, 565
Phytophthora infestans, 549, 560
Phytophthora ramorum, 549
Phytoplankton
  carbon sequestration, 1231
  defined, **553**
  diatoms, 560
  dinoflagellates, 556
  estuary, 1222
  golden algae, 561
  heterokonts, 571
  Lake Erie, 1182
  open ocean, 1224
  as primary producers, 1188
Pike-cichlid (*Crenicichla alta*), 1134–1135
Pili, **530**, 530i
Pillbug, 658
Pilobolus, 613, 615i
Piloting, **1270**–1271
Pinacoderm, **635**, 636, 636i
Pine (*Pinus*), 448, 593–594, 594i
Pine bush, 1248, 1248i
Pinworm (*Enterobius*), 653

Pitx1 gene, 488
Placenta, **695**
Placental mammals (Eutheria), 472, 696–697, 697i
Placoderm, 679
Planktonic crustacean, 659
Plant(s), 575–604
  adaptations to life on land, 576–581
  angiosperms, 595–602
    adaptive success of, 596–597, 596i–598i, 597–598
    classification, 596–597, 596i–598i
    coevolution with pollinators, 598, 600, 600i
    origin of, 595–596, 596i, 601
    overview, 595
    phylogenetic tree, 596i
  bryophytes, 581–585
    hornworts, 583, 583i
    liverworts, 582–583, 583i
    moss, 583–584, 584i
    overview, 581–582, 582i
  convergent evolution in, 471i
  diploid dominance, 578–579, 579i
  diversity, history of, 483i
  evolution of, 575–581, 577t
  gymnosperms, 590–595
    ancestral forms, 590, 591i
    conifers, 593–594, 594i
    cycads, 591, 592i
    ginkgo, 592, 592i
    gnetophytes, 592–593, 593i
    overview, 590
    reproductive adaptations, 590–591
  herbivory, 1153–1155
  introduced, 1237, 1237i
  life cycle, 578–579, 579i
    angiosperm monocot, 597, 599i
    conifer, 593–594, 594i
    fern, 588i
    general, 578–579, 579i
    moss, 584i
  mycorrhiza, 621–623, 623i
  phyla and major characteristics, 586t
  phylogenetic trees, 580i
  polyploidy, 456, 456i
  seedless vascular plants, 585–590
    ancestral forms, 585, 587i
    ferns, 588i, 589
    horsetails, 589–590, 590i
    lycophytes, 585, 586–587, 588i
    overview, 585
    whisk ferns, 589, 589i
  virus infection of, 543
Plantae (kingdom), **575**, 576i
Plasmid, 527
Plasmodial slime mold, **564**–565
Plasmodium, **564**, 565
Plasmodium species, 436i, 491, 558, 559, 1236
Plasmogamy, 609i, **610**, 613, 614i, 616, 617i, 618, 619i
Plastid
  endosymbiont hypothesis, 520–521, 520i
  evolution by endosymbiosis, 569–572, 570i
Plastron, **688**, 689i
Plate tectonics, **469**, 470i
Platy (*Xiphophorus maculatus*), 455
Platyhelminthes
  lineages
    Cestoda, 643, 643i, 644, 646
    Monogenoidea, 643
    Trematoda, 643, 643i, 644, 644i
    Turbellaria, 643, 643i
  organ systems, 642
  parasitic, 644–645, 644i, 645i
Platypus, 667–668, 668i, 695, 696i

Plesiosaurs, 688
Pleurobrachia, 641i
Plumatella repens, 641i
Pohorille, Andrew, 522
Pollen
  diploid, 456
  gymnosperm, 591
Pollen grain
  conifer, 593–594
  defined, **590**
  eudicot, 596, 596i
  as male gametophyte, 579
Pollen tube
  angiosperm, 599i
  conifer, 594, 594i
Pollination
  angiosperms coevolution with animal pollinators, 598, 600, 600i
  conifer, 593–594, 594i
  defined, **591**
  as gymnosperm adaptation, 591
Pollutants, **1234**
Pollution, 1234–1235, 1235i, 1236
Polyandry, **1280**
Polychaeta, 651, 652i
Polychlorinated biphenyl (PCB), 1236
Polygamous species, **1280**
Polygyny, **1280**
Polyhedral virus, **542**, 542i
Polymers, first, 522
Polymorphism
  balanced, 435–436, 436i, 437i
  defined, **421**
Polyp, **637**, 637i, 638, 639i
Polyphyletic taxa, **497**, 498i
Polyplacophora, 648, 648i
Polyploidy
  allopolyploidy, 456–458, 457i
  autopolyploidy, 456, 456i
  defined, **454**
Polyporus, 606i, 618i
Polytrichum, 584i
Ponderosa pine (*Pinus ponderosa*), 576i, 594i
Pools, 1219
Population
  age structure, 1128–1129, 1146i, 1147i
  carrying capacity, 1137–1139
  defined, **420**, **1126**
  density, 1127, 1127i
  dispersion, 1128, 1129i
  generation time, 1129, 1129i, 1137i
  geographical range, 1127
  habitat, 1127
  metapopulation, 1243, 1245, 1245i
  recruitment, 1168
  sex ratio, 1129
  sink, 1243, 1245, 1245i
  size, 1127, 1127i, 1128i
  source, 1243, 1245i
  variation, 420–423, 421i, 422i
    genetic, 422–423
    phenotypic, 420–422, 420i, 421i, 422i
    qualitative, 421, 421i
    quantitative, 421, 421i
Population bottleneck, **428**, 428i, 429
Population density, **1127**, 1127i
Population ecology
  age structure, 1128–1129, 1146i, 1147i
  defined, **1126**
  demography, 1129–1132
    life tables, 1130–1131, 1131t
    survivorship curve, 1131–1132, 1131i
  dispersal patterns, 1128, 1129i
  generation time, 1129, 1129i, 1137i
  human population growth, 1145–1149, 1145i–1148i
  life history, 1132–1133, 1134

Population ecology *(Continued)*
  population growth models, 1133–1139
    exponential growth, 1134–1137, 1136i
    logistic model, 1137–1139, 1138i, 1138t, 1139i
    population regulation
      cycles in population size, 1143–1145, 1144i
      density-dependent factors, 1139–1141, 1140i–1142i
      density-independent factors, 1141–1142, 1141i
      interacting environmental factors, 1142
      life history characteristics, 1142–1143, 1143i, 1143t
    population size and density, 1127, 1127i, 1128i
    sex ratio, 1129
Population genetics, **411**, 423–425, 424t
Population growth
  carrying capacity, 1137–1139
  generation time, 1129
  human, 1145–1149, 1145i–1148i
    age structure, 1146, 1147i
    economic development and, 1146–1147
    family planning programs, 1148
    zero population growth, 1147
  intraspecific competition and, 1138
  intrinsic rate of increase, 1137, 1137i
  models, 1133–1139
    exponential growth, 1134–1137, 1136i
    logistic model, 1137–1139, 1138i, 1138t, 1139i
  per capita rates, 1136–1137, 1136i
Population interactions, 1152–1160, 1152t
  coevolution, 1152–1153, 1158, 1161i
  community characteristics, effects on, 1166–1167, 1168, 1169i
  herbivory, 1153–1155
  interspecific competition, 1155–1158, 1156i, 1159i
    character displacement, 1157–1158, 1158i, 1167
    competitive exclusion, 1156
    exploitative competition, 1155
    interference competition, 1155
    niche concept, 1156–1157, 1157i, 1159i
    resource partitioning, 1157–1158, 1157i, 1167
  predation, 1153–1155
  symbiosis, 1158–1160
    commensalism, 1158, 1159i
    mutualism, 1158, 1160i, 1161, 1161i
    parasitism, 1159–1160
Population size
  carrying capacity, 1232
  cycles in, 1143–1145, 1144i
  description, **1127**, 1127i, 1128i
  extrinsic control, 1144
  intrinsic control, 1144
  K-selected species, 1142–1143, 1143i, 1143t
  minimum viable, 1243, 1244
  predator-prey interactions, 1143–1145, 1144i
  regulation
    crowding, 1139–1141, 1140i
    density-dependent factors, 1139–1141, 1140i–1142i
    density-independent factors, 1141–1142, 1141i
    interacting environmental factors, 1142

r-selected species, 1142–1143, 1143i, 1143t
Population viability analysis (PVA), **1243**, 1244
Porcupine *(Erethizon dorsatum)*, 473
Porifera, 635
Porocyte, 635, 636i
Porphyra, 567
Porpoise, 473i
Possingham, Hugh, 1244
Postelsia palmaeformis, 550i
Posterior end, **630**
Postzygotic isolating mechanisms, **447**, 447t, 449, 454–455
Potato cod *(Epinephelus tukula)*, 1160i
Potato famines, 549, 560
Powdery mildew, 616
Prado, Marco, 545
Prairie, 1216–1217, 1216i
Prakash, Om, 458, 458i
Praying mantid, 660i
Preadaptation, **479**
Precipitation
  monsoon cycles, 1208
  rain shadow, 1208, 1209i
  variations in, 1206, 1207i
Predation
  by cnidarians, 638i
  community stability and, 1166
  defenses against, 1153–1155, 1154i, 1155i
  defined, **1153**
  density-dependent population regulation and, 1141
  keystone species, 1167
  on migratory songbirds, 1232i, 1233
  species richness, effect on, 1166, 1167, 1168i
  trophic cascade, 1189–1190
Predator-prey model, 1144–1145, 1144i
Premolars, **695**
Preservation, 1248, 1248i
Prezygotic isolating mechanisms, 455, 456i
  behavioral isolation, 448, 448i
  defined, **447**
  ecological isolation, 447
  gametic isolation, 449
  mechanical isolation, 448–449, 448i
  temporal isolation, 447–448
Primary cell layers, **629–630**
Primary consumer, **1164**, 1165i, 1188
Primary endosymbiosis, **570**, 570i, 571
Primary host, 644, 644i
Primary producer, **1164**, 1165i, 1188
Primary productivity
  consumer regulation of, 1188–1189, 1189i
  by ecosystem, 1186i
  gross, 1183, 1200
  limnetic zone, 1220–1221
  net, 1183, 1184, 1184i, 1185–1186, 1186i, 1200
  variations in, 1184–1185, 1184i, 1186i
Primary succession, **1171**, 1172i
Primates
  Haplorhini, 699–701, 700i, 701i
  key derived traits, 697, 699
  phylogeny, 699i
  Strepsirhini, 699, 700i
Principle of monophyly, **497**
Principle of parsimony, **498**, 499, 505
Principles of Geology (Lyell), 404, 405
Prion, **545–546**
Proboscis worm, 646, 647i
Prodöhl, Paulo A., 1130
Production efficiency, 1186
Profundal zone, **1220**

Proglottid, 643i, 644, 645i, 646
Progymnosperm, 590
Prokaryote, 525–540
  biofilms, 532–533, 533i
  diversity, 526–527
  Domain Archaea, 537–540
    characteristics of, 538–539, 538t
    Crenarchaeota, 540
    Euryatchaeota, 539–540, 539i
    Korarchaeota, 540
    virus infections of, 545
  Domain Bacteria, 534–537
    antibiotic resistance, 537
    characteristics of, 538t
    chlamydias, 536, 536i
    cyanobacteria, 535, 535i
    Gram-positive bacteria, 535–536, 536i
    green bacteria, 535
    pathogenicity, mechanism of, 536–537
    Proteobacteria, 534–535, 535i
    spirochetes, 536
  fossil, 511, 512i
  metabolic activity, 526
    nitrogen metabolism, 531–532
    nutrition modes, 531, 531i
    oxygen use, 531
  phylogenetic tree, 534i
  reproduction, 532
  size, 525, 526i
  structure, 526–530, 528i
    capsule, 529, 529i
    cell walls, 528–529, 529i
    endospore, 532, 532i
    flagella, 530, 530i
    internal, 527–528
    nucleoid, 527
    pili, 530, 530i
    ribosomes, 527–528
    shape of cells, 527, 527i
    slime layer, 529
Prokaryotic cells, origin of, 517–519, 519i
Prokaryotic chromosome, 527
Promiscuous species, **1280**
Prophage, 542
Protein, use in molecular phylogenetics, 501, 503, 504i
Protein kinase, 566
Protein-first hypothesis, 517, 522
Proteobacteria, 534–535
Protist, 549–572
  classification, 551–552
  diversity of, 550, 550i, 552–553
  groups
    Alveolates, 555–558, 557i, 558i
      Apicomplexa, 557–558, 559
      Ciliophora, 555–556
      dinoflagellates, 556–557, 558i
    Amoebozoa, 563–565, 564i, 565i
      amoebas, 563–564, 564i
      cellular slime molds, 564, 565i
      plasmodial slime molds, 564–565
    Archaeplastida, 565–569, 566i–569i
      Chlorophyta (green algae), 567–569, 567i, 568i
      Rhodophyta (red algae), 565–567, 566i
    Cercozoa, 562–563, 563i
      Chlorarachinophyta, 563
      Foraminifera (forams), 562–563, 563i
      Radiolaria, 562, 563i
    Discicristates, 554–555, 555i
      euglenoids, 554, 555i
      kinetoplastids, 554–555, 555i
    Excavates, 553–554, 554i
      Diplomonadida, 553, 554i
      Parabasala, 553–554, 554i

Heterokonts, 558, 558i, 560–562, 560i–562i
  Bacillariophyta (diatoms), 560–561, 560i
  Chrysophyta (golden algae), 561, 561i
  Oomycota, 558, 558i, 560, 560i
  Phaeophyta (brown algae), 561–562, 562i
  Opisthokonts, 569, 569i
  habitat, 553
  metabolism, 552
  pathogenic, 549–550
  plastid evolution from endosymbionts, 569–572, 570i
  reproduction, 552
  structure, 552–553, 552i
Protocell, **512**, 515–517, 516i, 522
Protoctista, **550**
Protonema, **583**, 584i
Protostome
  developmental patterns, 630–633, 632i
  Ecdysozoa, 653–663, 653i–663i
  Lophotrochozoan, 641–653, 641i–652i
  nervous system, 632
Protostomia, 633
Protozoa, 554
Prusiner, Stanley, 545–546
Pseudocoelom, **630**, 631i, 646
Pseudocoelomate animal, **630**, 631i
Pseudomyrmex ferruginea, 1158, 1161i
Pseudonitzschia, 560
Pseudopod, **553**, 563, 564i
Psychrophile, **540**
Pteraspis, 678i
Pterophyta, **585**, 589
Pterosaurs, 688
Punctuated equilibrium hypothesis, 476i, **477**
Pupa, **663**, 663i
Purple nonsulfur bacteria, 534
Purple sulfur bacteria, 534
PVA (population viability analysis), **1243**, 1244
Pyramid of biomass, **1188**, 1188i
Pyramid of energy, **1186**, 1188i
Pyramid of numbers, **1188**, 1189i
Pyrenoid, 583
Pyrobolus, 540

Qualitative variation, **421**, 421i
Quantitative variation, **421**, 421i
Quillwort, 586
Quinoa (*Chenopodium quinoa*), 1143i

Rabbit, 1125–1126, 1126i
Rabies, 1235
Radial cleavage, **631**, 632i
Radial symmetry
  description, **630**, 630i
  echinoderms, 669
  eumetazoans with, 636–640
  secondary, 669
Radiolarian, 562, 563i
Radiometric dating, **468**–469, 468i
Radula, 647i, **648**
Rain forest
  temperate, **1218**
  tropical, **1213**, 1213i, 1214
Rain shadow, **1208**, 1209i
Ramalina, 621
Random dispersion, **1128**
Raphanobrassica, 458
Rasmussen, Steen, 522
Rat snake (*Elaphe obsoleta*), 445i
Rattlesnake (*Crotalus*), 690i, 1153
Raup, David, 481
Raven (*Corvus corax*), 1278
Rays, 680, 680i

Realized niche, **1157**, 1157i, 1159i
Recessive alleles, masking from evolution by diploidy, 435, 435t
Reciprocal altruism, **1284**, 1285
Recombination rate, 439
Recruitment, population, 1168
Red algae, 565–567, 566i
Red deer (*Cervus elaphus*), 1132
Red tide, **556**–557
Red-legged frog (*Rana aurora*), 1275
Red-spotted newt (*Notophthalmus viridescens*), 685i
Red-winged blackbird (*Agelaius phoenecius*), 1280
Reeve, H. Kern, 1286
Reinforcement, **451**, 1259
Relatedness, calculating degrees of, 1283i
Relative abundance, **1163**–1164, 1196
Relative fitness, **430**–431
Replication, RNA world and, 517, 518
Reproduction
  behavior, hormone effects on, 1261–1263, 1262i
  early *versus* late, 1133
  fungi, 608–610, 609i
    asexual, 609, 609i
    sexual, 609–610, 609i
  number of episodes, 1133
  prokaryotes, 532
  protists, 552
Reproductive behavior
  mating systems, 1280
  parental investment, 1183
  sexual selection, 1279–1280
Reproductive isolation
  genetic divergence, 454–456
  isolating mechanisms
    defined, **447**
    postzygotic, **447**, 449, 449i
    prezygotic, **447**–449, 448i
    table of, 447t
Reproductive strategies, **1279**
Resin, 593
Resource partitioning, **1157**–1158, 1157i, 1167
Respiration, cellular, 1183
Respiratory tree, sea cucumbers, 671
Restoration, 1248–1249
Reznick, David, 1134–1135, 1148
Rhea (*Rhea americana*), 403i
Rhinoceros, 697i
Rhinoceros beetle, 660i
Rhizobium, 529i, 1158, 1161, 1195
Rhizoid
  bryophyte, 581, 582i
  defined, **607**
  liverwort, **582**, 583
  moss, 583
  whisk fern, 589
Rhizome
  defined, **578**
  fern, 588i, 589
  horsetail, 590
  whisk fern, 589
Rhizopus stolonifer, 613, 614i, 615i
Rhodophyta (red algae), 565–567, 566i
Rhynchocoel, 646, 647i
Rhynia, 585, 587i
Ribbon worm, 646, 647i
Ribosome, prokaryotic, 527–528
Ribozyme, 517, 518
Rice blast, 624
Rickettsia rickettsii, 1236
Riffles, 1219
Ring canal, 669, 670i
Ring species, **446**
Ring-tailed lemur (*Lemur catta*), 700i
River biome, 1219–1220, 1219i

RNA
  RNA world, 517, 518
  RNA-first hypothesis, 517
  sequence, molecular phylogenetics use of, 501
Robinson, Gene E., 1265
Robinson, Scott K., 1232
Rocky shore habitat, 1222–1223, 1224i
Rodent (Rodentia), 696, 697i, 698
Romer, Alfred Sherwood, 679
Root, **578**
Root system, **578**
Rose (*Rosa*), 598i
Roseate spoonbill (*Ajaia ajaja*), 694i
Rotifera, 646, 646i
Roundworm
  filarial worms, 645, 645i
  Trichinella spiralis, 644–645, 645i
Royal Chitwan National Park, 1249, 1249i
Royal penguin (*Eudyptes schlegeli*), 1281–1282, 1281i
RRNA sequences, use in phylogenetics, 506i, 507
R-selected species, 1142–1143, 1143i, 1143t, 1167, 1171, 1173, 1237
Runs, 1219
Rust, white, 558
Ryan, Michael J., 1287
Rye (*Secale cereale*), 595i, 727

Saccharomyces cerevisiae, 539, 606i, 615–616, 624
Sacred lotus (*Nelumbo nucifera*), 597i
Sage grouse (*Centrocercus urophasianus*), 1279, 1280i
Sahara Desert, 1234, 1234i
Sahel, 1233–1234, 1234i
Salamander, 446i, 480, 480i, 685
Salinization, 1233, 1234
Salmonella typhi, 537
Salt marsh, **1222**, 1223i
Sand dollar, 669
Sandy shore habitat, 1222–1223
Saprobe, **607**
Sarcopterygii, 681, 683, 683i
Sargassum, 561
Savanna, **1214**–1215, 1215i
Sawfly (*Perga dorsalis*), 1281, 1281i
Scala Naturae, 402–403, 414
Scallop, 648–649
Scarlet hood (*Hygrophorus*), 618i
Scarlet macaw (*Ara chloroptera*), 443i
Scent marking, 1276
Schindler, David, 1221, 1222i
Schistosomiasis, 644
Schizocoelom, **632**, 632i
Schoener, Amy, 1176
Schoener, Thomas, 1142, 1142i, 1166, 1167
Scientific names, 493
Scolex, 643i, 646
Scorpion, 657i
Scrapie, 545
Scyphozoa, 638, 640i
Sea anemone, 638, 640i
Sea cucumber, 670i, 671
Sea daisies (Concentricycloidea), 669
Sea hare (*Aplysia*), 1259
Sea horse (*Hippocampus hudsonius*), 682i
Sea lilies, 671
Sea nettle (*Chrysaora* species), 640i
Sea otter (*Enhydra lutris*), 1242–1243, 1243i
Sea star, 669, 670i, 1167, 1168i
Sea urchin, 669, 670i, 671
Sea wasp (*Chironex fleckeri*), 640i
Seal
  reproduction, 696
  sex ratio, 1129

Seasonality, 1205–1206, 1205i
Seaweed, 552, 565, 567
Secondary consumer, **1164**, 1165i
Secondary contact, 451
Secondary endosymbiosis, 570i, 571
Secondary productivity, **1185**
Secondary succession, **1171**
Sedimentary cycles, 1191
Sedimentary rock, 403, 464
Sedimentation, 1191
Seed
    conifer, 594, 594i
    defined, **591**
    gymnosperm, 591, 591i
Seed coat, 594
Seed fern, 585, 587i
Seedless vascular plants, 585–590
    ancestral forms, 585, 587i
    ferns, 588i, 589
    horsetails, 589–590, 590i
    lycophytes, 585, 586–587, 588i
    overview, 585
    whisk ferns, 589, 589i
Segmentation
in annelids, 650, 651i
in chordates, 672
description, **633**
phylogeny and, 635
Selaginella, 586
Selection. *See also* Natural selection
    artificial, 408–409
    background, 439
    frequency-dependent, 436, 438i
    sexual, **433**, 434i, 455–456, 456i,
        1279–1280, 1280i
    species, 479
Selective breeding, 407–408
Selectively neutral, 437
Sensory drive, 1287
Sensory system, 1264, 1265i
Septa, 607, 650, **650**, 651i
Serengeti National Park, 1248
Sessile, **629**
Setae, **651**
Sex pili, 530
Sex ratio, **1129**
Sexual dimorphism, **433**
Sexual reproduction
    angiosperm, 599i
    conifers, 593–594, 594i
    ferns, 588i, 589
    fungi, 609–610, 609i, 613, 614i, 616,
        617i, 618–619, 619i
    gnetophytes, 592
    hornworts, 583
    liverworts, 582
    mosses, 583, 584i
    plants, 578–579
Sexual selection, **433**, 434i, 455–456, 456i,
    1279–1280, 1280i
Shark, 473i, 680, 680i
Shark, spiny, 679, 679i
Sheldon, Peter R., 478i
Shelf fungus, 618, 618i
Shelford, Victor, 1212
Shell, turtle, 688–689, 689i
Shepherd's purse (*Capsella bursa-pastoris*),
    1140i
Shimamura, Mitsuru, 502
Shoot system, **578**
Short tandem repeat (STR) loci, 1130
Short-eared owl (*Asio flammeus*), 1154i
Short-nosed echidna (*Tachyglossus aculea-
    tus*), 696i
Shrimp, 658
Siamang, 701
Sicheri, Frank, 1210
Sickle-cell disease, heterozygote advantage
    and, 435, 436i

Sign stimuli, **1255**
Silkworm moth (*Bombyx mori*), 1277
Sillén-Tullberg, Birgitta, 1281
Silverfish, 660i
Simberloff, Daniel, 1176, 1177
Simulation modeling, **1199**
SINEs (for Short/Interspersed Elements),
    502
Sink population, **1243**, 1245, 1245i
Siphon, 649, 649i, 650, 650i
Skate, 680
Skeleton
    bird, 692, 693i
    hydrostatic, 630
Slime layer, **529**, 529i
Slime mold
    cellular, 564, 565i, 566
    plasmodial, 564–565
    use in science, 565
SLOSS (Single Large Or Several Small),
    1246
Sloth, 464, 464i
Smith, Neal, 1151
Snail
    balanced polymorphisms in, 436, 437i
    Elysia, 521
    herbivory by, 1167, 1169i
    variability in populations of *Cepaea
        nemoralis*, 437i
Snake, 690, 690i
Snapdragon (*Antirrhinum*), 424, 424t,
    426–427
Snow geese (*Chen caerulescens*), 421i
Snowshoe hare (*Lepus americanus*),
    1144–1145, 1144i
Social behavior
    altruistic behavior, 1282–1284, 1283i
    benefits and costs of group living,
        1281–1282, 1281i, 1282i
    defined, 1281
    dominance hierarchies, 1282
    human, 1285–1287
Sogin, Mitchell L., 611
Soil, salinization of, 1233, 1234
Solar energy, 1182–1183
Solar radiation, 1205, 1205i
Song sparrow (*Melospiza melodia*), 1253,
    1254i
Songbirds, population declines in,
    1232–1233, 1232i
Sordino, Paolo, 486
Soredia, 621, 622i
Sorus, **589**
Sound spectrogram, 1254i
Source population, **1243**, 1245i
Southern magnolia (*Magnolia grandi-
    flora*), 597i
Sowbug, 658
Speciation, 443–462
    allopatric, 450–451, 450i, 451i, 453i, 459
    defined, **444**
    genetic mechanisms of, 454–460
        chromosome alterations, 458–460,
            458i
        genetic divergence, 454–456
        polyploidy, 456–458, 456i
    in Hawaiian fruit flies, 454
    parapatric, 451, 453, 454i, 459
    reproductive isolation, 447–449, 447t,
        448i, 449i
    sympatric, 453–454, 454i, 459
Speciation genes, 459, 507
Species
    asexual, 459
    biological species concept, 445
    geographical variation, 445–447,
        445i–447i
    morphological species concept,
        444–445

    phylogenetic species concept, 445
    subspecies, 445–446, 446i
Species cluster, **451**
Species composition, **1161**
Species diversity, **1164**, 1164i, 1166
Species pool, 1175
Species richness
    biodiversity, 1230
    defined, **1163**
    equilibrium theory of island biogeogra-
        phy, 1175–1177, 1175i, 1176i
    intermediate disturbance hypothesis,
        1169–1170, 1170i
    interspecific competition, effect of,
        1166–1167
    landscape corridors, effect of, 1247i
    latitudinal gradients, 1174–1175, 1174i
    predation, effect of, 1166, 1167, 1168i
    species diversity, 1164
Species selection, **479**
Species/area relationships, 1245–1246,
    1246i
Specific epithet, **493**
Spencer, Chris, 439
Sperm
    bryophyte, 581
    conifer, 594, 594i
    fern, 588i, 589
    gymnosperm, 590–591
    liverwort, 582
Sphagnum, 584
Sphenodontid, 689, 690i
Spicule, sponge, 635, 636i
Spider
    description, 657–658, 657i
    lizard predation on, 1142, 1142i
Spider monkey (*Ateles geoffroyi*), 700i
Spike moss, 586
Spiller, David, 1142, 1142i
Spinneret, 657i, 658
Spiny anteater, 696
Spiny lobster (*Panulirus*), 1270, 1271i
Spiny shark, 679, 679i
Spiral cleavage, **630**, 632i
Spiral valve, **680**
Spirilla (prokaryote cell shape), **527**, 527i
Spirochetes, 536
Spirogyra, 569
Sponge, 635–636, 635i
Spongocoel, **635**, 636, 636i
Sporangia
    club moss, 586
    conifer, 593
    defined, **579, 613**
    hornworts, 583
    horsetail, 590, 590i
    liverwort, 582
    moss, 583, 584i
    whisk fern, 589
Spore
    club moss, 586, 588i
    fungal, **608**–610, 609i
    horsetail, 590, 590i
    plant, **579**
    whisk fern, 589
Sporophyll, **586**
Sporophyte
    alternation of generations, 579, 579i
    angiosperms, 598
    bryophyte, 581, 582i
    club moss, 586–587, 588i
    defined, **579**
    fern, 588i, 589
    gymnosperm, 590, 591
    hornworts, 583
    horsetail, 590, 590i
    kelp, **561**
    liverwort, 582
    moss, 583, 584i

seedless vascular plants, 585
  whisk fern, 589
Sporopollenin, **576**
Sporozoite, 558
Spring overturn, **1220**, 1221i
Squalene, **680**
Squamate, 689–690, 690i
Squid, 650, 650i
Stability, community, **1166**
Stabilizing selection, 431–433, 431i, 432i
Standing crop biomass, **1183**, 1185, 1185t
Stanley, Steven, 478–479
Stapes, **684**
Staphylococcus, 419, 536
Star anise (*Illicium floridanum*), 597i
Star-nosed mole (*Condylura cristata*), 1264, 1265, 1266–1267, 1266i
Statocyst, **638**
Stegosaurus, 688
Stem, 578
Steppe, 1216
Stickel, Shawn K., 611
Stickleback (*Gasterosteus aculeatus*), 487–488, 488i
Stinker vase sponge (*Ircinia campana*), 635i
Stomata, 576–577, 577i
Stone, Anne, 705
Stoneking, Mark, 705, 706
Stonewort, 576i
Stork, Nigel, 1241
Stratification, **403**
Stream biome, 1219–1220, 1219i
Strepsirhini, 699, 700i
Streptococcus, 529, 536, 536i, 537
Strobilus
  conifer, 593
  cycad, 591
  description, **586**, 588i
  horsetail, 590, 590i
Stromatolites, **519**, 519i
Sturgeon, 681, 681i
Styracosaurus, 688
Subspecies
  defined, **445**
  rat snake, 445i
  ring species, 446, 446i
Succession. *See* Ecological succession
Sudden oak death, 549–550
Surgeonfish (*Acanthurus lineatus*), 1276
Survivorship, age-specific, **1130**–1131, 1131t
Survivorship curve, **1131**–1132, 1131i
Swamp sparrow (*Melospiza georgiana*), 1253, 1254i
Swell shark (*Cephaloscylium ventricosum*), 680i
Swim bladder, **681**, 682i
Swimmeret, 659, 659i
Switch, developmental, 485–486
Swordtail (*Xiphophorus helleri*), 455
Symbiosis, 1158–1160
  ciliates, 556
  commensalism, 1158, 1159i
  defined, **556**, **620**, **1158**
  diatoms, 560
  dinoflagellates, 556
  lichen, 620–621, 622i
  mutualism, 1158, 1160i, 1161, 1161i
  mycorrhizae, 621–623, 623i
  parasitism, 1159–1160
Symmetrical body plan, **630**
Sympatric speciation, 453–454, 454i, 459
Synapsid, **688**, 695
Syphilis, 536
Systematics, 491–510
  cladistics, 498–501, 499i–501i
  conservation biology and, 1241
  defined, **492**

goals, 492
information provided by, 492–493
Linnaean system of taxonomy, 493–494, 494i, 507
molecular phylogenetics, 501–507
phylogenetic trees, 493, 497–498, 498i, 499i, 503, 504i, 505
systematic characters
  behavioral traits, 494–495
  derived and ancestral characters, 496–497
  evaluating, 495–497, 496i, 497i
  homologous characters, 495–496, 496i
  molecular characters, 501–505
  morphological traits, 494
  organismal traits as, 494–495, 495i
  outgroup comparison, 497, 497i
  traditional evolutionary, 498
Systematics and the Origin of Species (Mayr), 444
Szostak, Jack, 522

Tactile signal, **1277**, 1277i
Taiga, **1217**–1218, 1218i
Tail, viral, **542**
Tapeworm, 643, 643i, 644, 645i, 646
Tarsiers (*Tarsius bancanus*), 699, 700i
Taubenberger, Jeffrey, 544
Taxis, **1275**
Taxol, 1230–1231, 1230i
Taxon, **493**
Taxonomic hierarchy, **493**
Taxonomy
  defined, **402**, **492**
  Linnaean system, 493–494, 494i, 507
Teeth, mammalian specializations in, 695
Teleosts, 681, 682i
Telson, 659, 659i
Temperate deciduous forest, **1217**, 1217i
Temperate grassland, **1216**–1217, 1216i
Temperate rain forest, **1218**
Temporal isolation, 447–448
Teosinte (*Zea diploperennis*), 1231, 1231i
Terebraulina septentrionalis, 641i
Terrestrial biomes, 1211–1219
  boreal forest/taiga, 1217–1218, 1218i
  chaparral, 1216, 1216i
  desert, 1215–1216, 1215i
  savanna, 1214–1215, 1215i
  temperate deciduous forest, 1217, 1217i
  temperate grassland, 1216–1217, 1216i
  temperate rain forest, 1218
  tropical forest, 1213–1214, 1213i
  tundra, 1218–1219, 1218i
Territory, **1260**, 1261–1263, 1262i, 1275–1276
Tertiary consumer, **1164**, 1165i
Test, 669
Tetrapoda, **675**
Tetrapods, evolution of, 683–684
Thallus
  description, **621**, 622i
  hornworts, 583
  liverwort, **582**, 583
Theissen, Günter, 487
Therapsid, 695
Thermocline, **1220**
Thermoregulation
  behavioral and physiological responses, 1209–1211, 1211i
Thorax
  arthropod segment, 655
  insect, 661, 661i
Thorn forest, **1215**
Thrips, 1141
Thrush (*Turdus ericetorum*), 436
Tick, 657, 658
Tilman, David, 1170

Tinbergen, Niko, 1254, 1256i, 1270–1271, 1272i
Tissue
  defined, **629**
  evolution of, 629–630
Tmesipteris, 589
Toad, 455, 685
Tobacco mosaic virus, 542i, 543
Tolerance hypothesis, **1173**
Tomato bushy stunt virus, 543
Tool-kit gene, 484, 485, 661
Topography, effect on climate, 1208, 1208i
Torpor, 1209–1210
Torsion, **648**
Tortoises, Galápagos, 406i, 407
Townsend, Colin R., 1170
Toxin
  of centipedes, 660
  of cnidarians, 637
  endotoxin, 536–537
  exotoxin, 536
  of spiders, 658
Toxoplasma, 558
Tracheal system, insect, **661**
Tracheid, 598
Tracheophyte, **577**
Traditional evolutionary systematics, 498
Transcription factors, flower development and, 601
Transmutation of Species (Darwin), 407
Transposable elements (TEs), 502
Trebouxia, 620
Tree fern, 588i, 589
Tree of life, 410–411, 410i, 506–507, 506i
Treefrog, 494–495, 495i
Trematoda, 643, 643i, 644, 644i
Treponema denticola, 536
Treponema pallidum, 536
Treponema vincentii, 536
Trichinella spiralis, 644–645, 645i
Trichocyst, **556**
Trichomonas vaginalis, 553–554, 554i, 571
Triploblastic body plan, **630**
Trivers, Robert, 1283–1284, 1284
Trochophore larvae, 648, 648i, 651
Trophic level, **1164**, 1165i
Tropic cascade, **1189**–1190
Tropical deciduous forest, **1213**
Tropical forest, **1213**–1214, 1213i
Tropical montane forest, **1213**–1214
Tropical rain forest, **1213**, 1213i, 1214
Tropics, **1205**
Truffle (*Tuber melanosporum*), 615, 622
Trypanosoma brucei, 554–555, 555i
Tuatara (*Sphenodon punctatus*), 689, 690i
Tube feet, 669, 670i
Tube worm, 1225, 1225i
Tulip (*Tulipa*), 598i
Tundra, **1218**–1219, 1218i
Túngara frogs, 1287
Turbellaria, 643, 643i
Turnover rates, **1188**
Turtle
  description, 688–689, 689i
  light pollution, 1235, 1235i
Tympanum, **684**
Typhus, 1236
Tyrannosaurus, 688

Ulothrix, 568i
Ultrabithorax (*Ubx*) gene, 661
Ulva, 567, 567i
Unconditioned stimulus, 1258
Undulating membrane, **553**, 554, 555i
Ungulate, 696
Uniform dispersion, **1128**
Uniformitarianism, **404**
Universe, origin of, 511

Unreduced gamete, **456**
Urey, Harold, 514
Uric acid, 686
Urochordata, 672i, 673
Urodela, 685, 685i
Uropods, 659, 659i
Use and disuse, principle of, 404

Vampire bat *(Desmodus rotundus)*,
    1283–1284
Van der Hoeven, Frank, 486
Van Leeuwenhoek, Anton, 555
Variation
    maintenance of, 435–437
        balanced polymorphisms, 435–436,
            436i, 437i, 438i
        diploidy, 435, 435t
        neutral variations, 437
    population, 420–423, 421i, 422i
        genetic, 422–423
        phenotypic, 420–422, 420i, 421i,
            422i
        qualitative, 421, 421i
        quantitative, 421, 421i
Vascular plants, **577**
Vascular tissue, plant, 577–578
Vázquez, Diego, 1250
Veldt, 1216
Veliger, 648
Velvet worm, 654–655
Venter, Craig J., 522, 539
Ventral surface, **630**
Venturia inaequalis, 614
Venus' flower basket *(Euplectella* species),
    636i
Venus' fly trap *(Dionaea muscipula)*, 1197
Vertebrae, **673**
Vertebral column, **673**
Vertebrate(s)
    agnathans, 677–678, 677i, 678i
        conodonts and ostracoderms,
            677–678, 678i
        hagfishes and lampreys, 677, 677i
    amniote origin and radiation, 686, 687i,
        688
    amphibians, 684–685, 685i
    birds, 692–694, 693i, 694i
    crocodilians, 690, 690i, 692
    defined, **628**
    jawed fishes, 678–683
        actinopterygii, 681, 681i, 682i
        chondrichthyes, 680–681, 680i
        origin of, 679, 679i
        sarcopterygii, 681, 683, 683i
    mammals, 695–707
        hominids, 702–707
        key adaptations, 695
        marsupials, 695–696, 696i
        monotremes, 695, 696i
        nonhuman primates, 697–701
        placentals, 696–697, 697i
    origin and diversification, 674–675,
        674i–676i
    phylogeny, 676i
    sphenodontids, 689, 690i
    squamates, 689–690, 690i
    tetrapods, early, 683–684, 684i
    turtles, 688–689, 689i
    unique tissues in, 673–674
Vessel elements, in angiosperms, 598
Vestigial structure, **403**
Vibrios (prokaryote cell shape), **527**, 527i

Vicariance, **471**, 472
Virion, **541**
Viroid, **544**–545
Virus, 540–544
    defined, **540**
    evolution of, 544
    infection
        of animal cells, 542–543, 543i
        of archaeans, 545
        of bacterial cells, 542
        of plant cells, 543
        treatment of, 543–544
    structure, 541–542, 542i
    table of major animal, 541t
Visceral mass, **647**, 647i
Visual signal, **1276**
Volterra, Vito, 1155
Volvox, 567i
Von Frisch, Karl, 1254, 1277
Vulpia fasciculata, 1140i
Vulture *(Gyps),* 1235

Wainwright, Patricia O., 611
Wake, Marvalee H., 706
Wald, George, 523
Wallace, Alfred Russel, 401, 402i, 408,
    471
Walrus *(Obodenus rosmarus),* 696, 697i
Wang, Bin, 566
Wasp *(Eurytoma gigantea),* 431–432, 432i
Water
    biomes
        freshwater, 1219–1221
        marine, 1221–1225
    evaporation of, 515
    hydrologic cycle, 1192, 1192i
    net primary productivity and, 1184,
        1184i
    of primordial Earth, 513
    water-splitting reaction in photosynthe-
        sis, 519
Water boatmen *(Sigara distincta),* 438i
Water flea *(Daphnia),* 1139i, 1140i
Water lily, 597i
Water mold, 549, 558, 558i, 560
Water pollution, 1234–1235
Water vascular system, 669, 670i
Watershed, **1193**
Wayfinding, 1270–1271
Webb, S. David, 472
Weis, Arthur E., 431
Welwitschia, 592, 593i, 601
West Nile virus, 545
Wetlands, **1219**
Whale, 502, 696, 697
Whale shark *(Rhincodon typus),* 680
Wheat *(Triticum),* 457, 457i, 595i, 598i
Whisk fern *(Psilotum),* 589, 589i
White pelican *(Pelecanus erythrorhynchos),*
    1281
White-crowned sparrow *(Zonotrichia
    leucophrys),* 1253, 1254, 1254i, 1255,
    1259, 1269, 1270i, 1278
White-egg bird's next fungus *(Crucibulum
    laeve),* 618i
Whitfield, L. Simon, 707
Whittaker, Robert, 1161
Whooping crane *(Grus americana),* 1241,
    1241i
Widowbird, 433, 434i
Wienberg, Wilhelm, 424
Wilcove, David S., 1232–1233

Williams, Ernest E., 691
Wilson, Allan, 706
Wilson, Edward O., 1175–1177, 1239
Wilson, Margo, 1285, 1287
Wind
    air circulation, 1206, 1206i
    ocean proximity and, 1208, 1208i
    topography and, 1208
Wingless/Wnt pathway, 1258
Wings
    convergent evolution and, 496
    homologous structures, 496, 496i
Winter flounder *(Pseudopleuronectes ameri-
    canus),* 1210
Woese, Carl R., 506i, 507, 537, 539
Wolcott, Charles, 627
Wolf, Larry, 1209
Wolf spider, 657i
Worm, Boris, 1239
Wrasse *(Thalassoma),* 450i
Wuchereria, 645, 645i

Xanthopan morgani praedicta, 600i
Xylem
    angiosperm, 598
    club moss, 586
    defined, **578**
    seedless vascular plants, 585
    whisk fern, 589

Yang, D. S. C., 1210
Yeast, **606**, 606i, 615–616, 616i
Yeast Systems Biology Network (YSBN),
    624
Yellow bush lupine *(Lupinus arboreus),*
    598i
Yellow-bellied gliding marsupial *(Petaurus
    australis),* 1244
Yellowfin tuna *(Thunnus albacares),* 682i
Yellowtail flounder *(Limanda ferruginea),*
    1238
Yellow-throated warbler *(Dendroica domi-
    nica),* 444i
Yohe, Gary, 1210–1211
Yolk, **686**
Yosemite National Park, 1249–1250
Yucca *(Yucca),* 1160i
Yucca moth *(Tegeticula),* 1160i
Yuma Myotis, 697i
Yunis, Jorge J., 458, 458i

Zamia, 592i
Zebra finch *(Taeniopygia guttata),*
    1259–1260, 1260i, 1261i
Zebra mussel *(Dreissina polymorpha),*
    1182
Zebrafish *(Danio rerio),* 486
Zebroid, 449, 449i
Zenk gene, 1260
Zero population growth, **1136–1137**, 1147
Zonation, ocean, 1221, 1223i
Zooplankton
    defined, **553**
    herbivores, 1189
    open ocean, 1224
    as primary consumers, 1188
Zygomycota (zygomycetes), 612t,
    613–614, 614i
Zygospore, **613**
Zygote
    defined, **629**
    fungal, 609–610, 609i